EXPERIMENTAL EVOLUTION

# EXPERIMENTAL EVOLUTION

*Concepts, Methods, and Applications
of Selection Experiments*

EDITED BY

Theodore Garland, Jr.
Michael R. Rose

UNIVERSITY OF CALIFORNIA PRESS
*Berkeley   Los Angeles   London*

University of California Press, one of the most distinguished university presses in the United States, enriches lives around the world by advancing scholarship in the humanities, social sciences, and natural sciences. Its activities are supported by the UC Press Foundation and by philanthropic contributions from individuals and institutions. For more information, visit *www.ucpress.edu*.

University of California Press
Berkeley and Los Angeles, California

University of California Press, Ltd.
London, England

Library of Congress Cataloging-in-Publication Data

Experimental evolution : concepts, methods, and applications of
selection experiments / edited by Theodore Garland, Jr. and Michael R. Rose.
      p.      cm.
    Includes bibliographical references and index.
    ISBN 978-0-520-24766-6 (cloth : alk. paper)—ISBN 978-0-520-26180-8 (pbk. : alk. paper)
    1. Evolution (Biology)—Research.    2. Natural selection—Research.
    3. Evolution (Biology)—Experiments.    4. Animal breeding—Experiments.
    I. Garland, Theodore, 1956–    II. Rose, Michael R. (Michael Robertson), 1955–

QH362.E97    2010
576.8'2072—dc22                                                          2009015362


16   15   14   13   12   11   10   09
10   9   8   7   6   5   4   3   2   1

The paper used in this publication meets the minimum requirements of ANSI/NISO Z39.48-1992 (R 1997) (*Permanence of Paper*). ∞

Cover illustration: Scanning electron microscope composite of a *Drosophila melanogaster* (commonly known as the fruit fly). © Nicholas Monu.

# CONTENTS

# LIST OF CONTRIBUTORS

**ALBERT F. BENNETT**
Department of Ecology & Evolutionary Biology
University of California
Irvine, California

**ANTHONY M. DEAN**
Department of Ecology, Evolution, & Behavior
BioTechnology Institute
University of Minnesota
St. Paul, Minnesota

**DANIEL E. DYKHUIZEN**
Department of Ecology & Evolution
Stony Brook University
Stony Brook, New York

**DOUGLAS J. EMLEN**
Division of Biological Sciences
University of Montana
Missoula, Montana

**SUZANNE ESTES**
Department of Biology
Portland Sate Univeristy
Portland, Oregon

**DAPHNE J. FAIRBAIRN**
Department of Biology
University of California
Riverside, California

**SAMANTHA E. FORDE**
Ecology & Evolution Department
University of California
Santa Cruz, California

**W. ANTHONY FRANKINO**
Department of Biology & Biochemistry
University of Houston
Houston, Texas

**JAMES D. FRY**
Department of Biology
University of Rochester
Rochester, New York

**DOUGLAS J. FUTUYMA**
Department of Ecology & Evolution
Stony Brook University
Stony Brook, New York

**THEODORE GARLAND, JR.**
Department of Biology
University of California
Riverside, California

**ERAN GEFEN**
Department of Biology
Haifa University—Oranim
Haifa, Israel

**ALLEN G. GIBBS**
School of Life Sciences
University of Nevada
Las Vegas, Nevada

**LAWRENCE G. HARSHMAN**
School of Biological Sciences
University of Nebraska
Lincoln, Nebraska

**JACK P. HAYES**
Department of Biology
University of Nevada
Reno, Nevada

**RAYMOND B. HUEY**
Department of Biology
University of Washington
Seattle, Washington

**DUNCAN J. IRSCHICK**
Department of Biology
Organismic & Evolutionary Biology Program
University of Massachusetts
Amherst, Massacusetts

**CHRISTINE M. JESSUP**
Center for Ecology and Evolutionary
University of Oregon
Eugene, Oregon

**TADEUSZ J. KAWECKI**
Department of Ecology & Evolution
University of Lausanne
Lausanne, Switzerland

**BENJAMIN KERR**
Department of Biology
University of Washington
Seattle, Washington

**PAWEL KOTEJA**
Institute of Environmental Sciences
Jagiellonian University, Gronostajowa
Krakow, Poland

**MARGARIDA MATOS**
Centro de Biologia Ambiental
Departamento de Biologia de Biologia Animal, Universidad de Lisboa
Lisbon, Portugal

**ROBERT C. McBRIDE**
Department of Ecology & Evolutionary Biology
Yale University
New Haven, Connecticut

**LAURENCE D. MUELLER**
School of Biological Sciences
University of California
Irvine, California

**TODD H. OAKLEY**
Department of Ecology, Evolution & Marine Biology
University of California Santa Barbara, California

**CASANDRA L. RAUSER**
Department of Ecology & Evolutionary Biology
University of California
Irvine, California

**DAVID REZNICK**
Department of Biology
University of California
Riverside, California

**JUSTIN S. RHODES**
Department of Psychology
University of Illinois
Beckman Institute
Urbana, Illinois

**DEREK A. ROFF**
Department of Biology
University of California
Riverside, California

**MICHAEL R. ROSE**
Department of Ecology & Evolutionary Biology
University of California
Irvine, California

**FRANK ROSENZWEIG**
Division of Biological Sciences
University of Montana
Missoula, Montana

**JOSIANE SANTOS**
Centro de Biologia Ambiental
Departamento de Biologia de Biologia Animal
Universidad de Lisboa Lisbon, Portugal

**GAVIN SHERLOCK**
Department of Genetics
Stanford University School of Medicine
Stanford, California

**ALEXANDER SHINGLETON**
Department of Zoology
Michigan State University
East Lansing, Michigan

**PEDRO SIMÕES**
Centro de Biologia Ambiental
Departamento de Biologia de Biologia Animal
Universidad de Lisboa Lisbon, Portugal

**JOHN G. SWALLOW**
Department of Biology
University of South Dakota
Vermillion, South Dakota

**HENRIQUE TEOTÓNIO**
Instituto Gulbenkian de Ciencia
Oeiras, Portugal

**MICHAEL TRAVISANO**
Department of Ecology, Evolution, & Behavior
University of Minnesota
St. Paul, Minnesota

**PAUL E. TURNER**
Department of Ecology & Evolutionary Biology
Yale University
New Haven, Connecticut

**ANTHONY J. ZERA**
School of Biological Sciences
University of Nebraska
Lincoln, Nebraska

**CLIFFORD W. ZEYL**
Department of Biology
Wake Forest University
Winston-Salem, North Carolina

# PART I

# INTRODUCTION TO EXPERIMENTAL EVOLUTION

# 1

# DARWIN'S OTHER MISTAKE

Michael R. Rose and Theodore Garland, Jr.

We are taught early in our education as evolutionists that Charles Darwin got the mechanism of heredity wrong. He supposed that there are an arbitrary number of ductile transmissible gemmules that migrate from the organs to the gonads, allowing the possibility of a kind of blending inheritance along with the inheritance of acquired characters. Of course, Gregor Mendel's discrete "hard" model for inheritance, which we now call genetics, turned out to be the correct mechanism for inheritance in eukaryotes. Furthermore, Darwin's mistake about inheritance probably cost the field of evolutionary biology some decades of delay. Genetics wasn't properly incorporated into evolutionary biology until the work of Fisher, Haldane, Wright, and Dobzhansky, in the period from 1910 to 1940 (Provine 1971; Mayr and Provine 1980). Regrettably, the person who may have seen that genetics supplied the mechanism of heredity that Darwinian evolution needed was Mendel, a humble monk who died unappreciated in 1884, some twenty years after he had worked out the basic principles of inheritance in plants. If Darwin had read Mendel with understanding in the 1860s, it is conceivable that much of modern evolutionary biology would have developed some fifty years earlier than it did, although such counterfactual speculation is of course essentially an idle exercise.

One of the common themes in the classroom presentation of Darwin's erroneous reasoning concerning heredity is the influence of his gradualist prejudices. It is well known that Darwin was in many respects a disciple of Charles Lyell, the leading gradualist geologist of nineteenth-century England. Lyell essentially founded modern scientific geology. Darwin was Secretary of the Geological Society early in his career, a scientific society dominated by Lyell's thinking, particularly his methodological strictures. The cardinal axiom in Lyell's geology was the idea that change in nature proceeds by gradual, observable, concrete mechanisms. In geology, such mechanisms are illustrated by erosion, subsidence, deposition, and the like. Darwin imported this style of thinking into biology. This led him to disparage the importance of discrete heritable variants, which he called "sports." That, in turn, prevented Darwin from giving appropriate attention to the hypothesis of discrete inheritance, leading evolutionary biology up a blind alley of blending inheritance. This was the famous mistake that is a key motif in the education of beginning evolutionary biologists.

Darwin's other mistake also came from his gradualist preconceptions. He repeatedly emphasized that natural selection acts only by slow accretion (Zimmer 2006). Darwin expected the action of selection within each generation to be almost imperceptible, even if thousands of generations of selection could evidently produce large differences between species: "natural selection will always act very slowly, often only at long intervals of time, and generally on only a very few of the inhabitants of the same region at the same time. I further believe, that this very slow, intermittent action of natural selection accords perfectly well with what geology tells us of the rate and manner at which the inhabitants of this world have changed" (Darwin, *Origin of Species*, first ed., chap. 4). Notably, the word *slowly* appears dozens of times in the *Origin*.

For modern scientists, at least, the problem with this assumption is that it implies that the action of natural selection will normally be very difficult to observe. Indeed, Darwin himself made no significant attempt to study natural selection in the wild. Instead, he studied the systematics of barnacles, bred pigeons, and crossed plants. He was certainly interested in both the long-term effects of evolution and the short-term effects of crosses, but he did not apparently seek out opportunities to study the process of natural selection itself. The closest he came to this was collecting an abundance of information on artificial selection from breeders, both agricultural and hobbyist, and discussions of their various results figure prominently in the *Origin*.

This dereliction did not persist, fortunately. The Illinois Corn Experiment began bidirectional selection on oil content in 1896 (Hill and Caballero 1992). W. F. R. Weldon (1901) published a pioneering study of selection in the wild on the morphology of estuarine crabs. Botanists, such as H. de Vries, began various selection experiments (Falconer 1992). In 1915, W. E. Castle published reasonably quantitative data on the response to "mass selection" on coat coloration in rats. In the 1930s, animal breeders such as Jay L. Lush took up the quantitative genetics theory developed initially by R. A. Fisher to implement well-designed breeding programs. Theodosius Dobzhansky started the "Genetics of Natural Populations" series of articles in the 1930s, studying selection on the chromosomal inversions of *Drosophila* in both wild and laboratory populations, often enlisting the aid of Sewall Wright. Ecological geneticists such as E. B. Ford and H. B. D. Kettlewell studied industrial melanism, one of our best examples of natural selection in the wild (Clarke 2003). Starting from this wide range of groundbreaking work, evolutionary biology has developed into a substantial body of empirically founded knowledge.

But there remains a tendency to adopt unthinkingly Charles Darwin's bias that natural selection is typically slow and difficult to observe. Very old patterns of research have persisted: studies of phylogenetics, genetic variation within and among populations, and occasional dramatic instances of natural selection in the wild have featured prominently in evolutionary research (Endler 1986; Hoekstra et al. 2001; Kingsolver et al. 2001). As these research paradigms have persisted, and indeed dominated within evolutionary biology, experimental evolution has been slow to develop as a research strategy. In 1976, M.R.R. did not consider trying selection for slowed aging in *Drosophila* because of the expectation that it would take too long to yield observable results. Reading about the results of an inadvertent and misinterpreted selection experiment by a neo-Lamarckian (Wattiaux 1968) in 1977 was the trigger that enabled M.R.R. to overcome his typical Darwinian inhibitions, leading to a deliberate test of Hamilton's (1966) analysis of the evolution of aging using laboratory evolution (Rose et al. 2004; Rose 2005). Now, of course, such laboratory evolution experiments on life-history characters are common in evolutionary biology. It was the pioneering work of Carol B. Lynch (1979; review in Lynch 1994) that convinced T.G. that selection experiments were actually practical for addressing classic questions in physiological ecology. Now, selection experiments of various types are common in evolutionary physiology (Bennett 2003; Garland 2003; Swallow and Garland

2005; Swallow et al. this volume). Yet as recently as 2005, T.G. observed colleagues discouraging graduate students from undertaking selection experiments in this area. Old biases die lingering deaths.

It is our conviction that the Darwinian inhibition about experimental research on evolution should now be resolutely discarded. This volume is effectively a brief in support of this view and naturally enough part of our campaign to foster selection experiments and experimental evolution as a central component of evolutionary biology. The Network for Experimental Research on Evolution (NERE), a University of California Multicampus Research Program, is our institutional effort to further the same end.

Indeed, experimental evolution is key to the ongoing effort to foster biology's reincarnation as a fully scientific field. It is only when evolutionary histories are known, controlled, and replicated that we can fairly claim to be performing rigorous experimental work. The biology of character X in inbred or mutant strain Y is like a beautiful painting: unique, intriguing, but of uncertain provenance or meaning. Any result with arbitrary strain Y may not be true of other strains or outbred populations of that species. And it will often be unclear how to sort out this situation. Strains M, Q, X, and Z might or might not have the same features. Individual outbred populations are marginally better, because they should have a broader set of genotypes, but they are still unique biological examples, of less reliability than postage stamps that are mass-produced to well-defined standards. If Ernest Rutherford could declare that science can be divided into physics and stamp collecting, then much of biology doesn't even rise to the level of stamp collecting. In its emphasis on hypothesis testing, quantitative trajectories, replication, and reproducibility, experimental evolution resembles physics more than it resembles most research in biology. We can only hope that both Darwin and Rutherford would have approved.

## DEFINITIONS AND CONCEPTS

What is *experimental evolution?* We use the term to mean research in which populations are studied across multiple generations under defined and reproducible conditions, whether in the laboratory or in nature (for recent overviews, see Bennett 2003; Garland 2003; Swallow and Garland 2005; Chippindale 2006; Garland and Kelly 2006). This intentionally general definition subsumes various types of experiments that involve evolutionary (cross-generational, genetically based) changes. At one end of the continuum, the study of evolutionary responses to naturally occurring events (e.g., droughts, fires, invasions, epidemics) may constitute a kind of adventitious experimental evolution, especially if these events occur repeatedly and predictably enough that the study can be replicated, either simultaneously or in subsequent years. One might also include "adaptations to the humanized landscape" (Bell 2008b), such as industrial melanism in moths (Clarke 2003). Next, we have "invasive species," which often invade repeatedly, thus allowing study of replicated events (Huey et al. 2005; Gilchrist and Lee 2007; Lee et al. 2007). Intentional "field introductions" involve populations placed in a new habitat in the wild or cases in which a population's habitat is altered by adding a predator, a pesticide, a food source, fertilizer, and so forth. The experimental population is then monitored across generations and compared with an unmanipulated control population (see Irschick and Reznick this volume).

"Laboratory natural selection" denotes experiments in which the environment of a laboratory-maintained population is altered (e.g., change of temperature, culture medium, food) as compared with an unaltered control population. "Laboratory culling" involves exposing an experimental population to a stress that is lethal (or sublethal) and then allowing the survivors (or the hardiest) to become the parents of the next generation. In all of the foregoing types of experiments, the investigator does not specifically measure and select individuals based on a particular phenotypic trait or combination of traits. Rather, selection is imposed in a general way, and the population has relatively great freedom to respond across multiple levels of biological organization (e.g., via behavior, morphology, physiology). "Multiple solutions" (different adaptive responses among replicate lines) are possible and even probable, depending on the kind of organism and experimental design.

In classical "artificial selection" or "selective breeding" experiments, individuals within a population are scored for one or more specific traits, and then breeders are chosen based on their score (e.g., highest or lowest). Depending on the level of biological organization at which selection is imposed—and the precision with which the phenotype is defined in practice—multiple solutions may again be common (Garland 2003; Swallow et al. this volume).

Domestication is an interesting (and ancient) type of experimental evolution that generally involves some amount of intentional selective breeding. In some cases, the process has been replicated enough times that general principles might be discerned (e.g., several species of rodents have been domesticated). Of course, whenever organisms are brought from the wild to the laboratory or agricultural setting, some amount of adaptation to the new conditions will occur, and this may be studied. Once domesticated, organisms may be the subject of additional selective breeding programs, with varying degrees of control and replication, leading to multiple breeds or lines. Simões et al. (this volume) discuss experimental evolutionary domestication of *Drosophila*.

More recently, the unintentional effects of various actions by human beings have been studied from the perspective that they constitute selective factors whose consequences may be predictable. Examples include changes in commercial fisheries (Hard et al. 2008), sport fishes (Cooke et al. 2007), and various ungulates that are hunted (e.g., Coltman et al. 2003).

What we are terming "experimental evolution" clearly covers a broad range of possible experiments. Historically and at present, different methodologies for experimental evolution have been and are being applied unequally across levels of biological organization (e.g., behavior, life history, physiology, morphology) and across kinds of organisms (e.g., bacteria, *Drosophila*, rodents). For example, experiments in the style of artificial selection often focus on organismal properties (e.g., physiological performance; see Swallow et al. this volume), whereas studies involving laboratory natural selection tend to focus more on testing genetic or evolutionary principles, with the organism serving perhaps mainly as a convenient conduit to such tests, as has been the case with evolutionary experiments on aging (Rauser et al. this volume) and sex (Turner et al. this volume). Indeed, as noted by one of our reviewers, many of the former types of experiments have their roots in the classic quantitative genetics literature. Of course, many artificial selection experiments have also been motivated by a desire to test aspects of quantitative-genetic theory (Falconer 1992; Hill and Caballero 1992; Bell 2008a, 2008b).

In any case, to qualify as experimental evolution, we require most if not all of the following fundamental design elements: maintenance of control populations, simultaneous replication, observation over multiple generations, and the prospect of detailed genetic analysis. In short, experimental evolution is evolutionary biology in its most empirical guise.

## MACROEVOLUTION, MICROEVOLUTION, AND THE ROLE OF SELECTION EXPERIMENTS

The experiments covered in this volume deal with evolving populations. Evolution within populations is traditionally referred to as "microevolution." The process of speciation (Fry this volume) forms a fuzzy boundary between microevolution and what is termed "macroevolution" (e.g., Charlesworth et al. 1982). Although definitions vary, *macroevolution* is generally used to refer to change at or above the level of the species, including long-term trends and biases that are observed in the fossil record (e.g., see www.talkorigins.org/faqs/macroevolution.html). Macroevolutionary phenomena are difficult to study experimentally because of the long time scales involved. One consequence of this is that many creationists accept microevolution as fact—how could they not if they drink cow's milk from a modern dairy or eat sweet corn?—but reject the fact of macroevolution.

Experimentally oriented biologists often use microevolutionary analyses to address hypotheses about macroevolution, and this volume includes a number of examples. For example, Swallow et al. (this volume) discuss selection experiments with rodents that are, in part, designed to test hypotheses about the evolution of mammalian endothermic homeothermy. On a seemingly unapproachable level, experimental evolution is now helping us to sort out alternative views about the evolutionary foundations of sex, as shown by Turner et al. (this volume), a process whose origin and evolutionary refinement have no doubt taken hundreds of millions of years.

Attempts to infer something about macroevolution from selection experiments face at least two important challenges. First, the organisms alive today are not the same as those living millions of years ago in which the phenomenon of interest occurred. For example, extant house mice are not the therapsid ancestors of mammals. If mice are somehow fundamentally different in "construction" from therapsids, then anything learned from experiments with the former may be misleading with respect to the latter. However, interspecific comparative studies (Garland et al. 2005) can be used to establish the generality of various features of organismal "design" and hence to make inferences about what is likely to have been similar versus different between mice (or any other putative model organism) and therapsids.

A second, and perhaps less tractable, limitation of selection experiments for addressing macroevolutionary hypotheses is that they simply may not last long enough to bear witness to all of the things that may have occurred over millions of years, such as the evolutionary consequences of very rare mutational events. For instance, the fixation of novel chromosomal translocations in a eukaryotic population is unlikely to be observed in an experimental evolution study. And is any study in experimental evolution, no matter how cleverly designed, likely to witness an event like the "capture" of a prokaryotic cell by another type of cell that led eventually to eukaryotic mitochondria? Nonetheless, long-term selection experiments (Travisano this volume) will only get longer in the future, and experiments focused on speciation (Fry this volume) and on adaptive radiation (Travisano this volume) have already achieved some notable successes, our caveats notwithstanding. Still, we must accept that mathematical or computer simulation models will need to substitute for, or at least supplement, experimental study of some (macro)evolutionary phenomena (e.g., Gavrilets and Vose 2005; see also Oakley this volume).

## OVERVIEW OF THIS VOLUME: EXCLUSIONS AND INCLUSIONS

We have intentionally excluded studies of plants from this volume, in part because fewer such studies exist and in part to keep the size of the volume within reasonable limits. We have also avoided the vast literature on selection experiments in the agricultural world, many of which involve plants (see *Plant Breeding* and *Plant Breeding Reviews*). Comparisons of inbred strains of mice and rats are innumerable (e.g., see *Behavior Genetics*), but they are only mentioned in this volume by Rhodes et al. and Swallow et al.; the latter also make some mention of comparisons of horse and dog breeds.

This volume is divided into five sections. The first includes this introductory chapter, a piece by Futuyma and Bennett that considers the place of experimental studies in evolutionary biology, and an overview of ways to model experimental evolution by Roff and Fairbairn. The second section considers distinguishable types of experimental evolution, ranging from bacteria (Dykhuizen and Dean), domestication (Simões et al.), through long-term experimental evolution, including adaptive radiation (Travisano), experimental studies of reverse evolution (Estes and Teotónio), and field experiments and introductions (Irschick and Reznick).

Part Three covers levels of observation in experimental evolution and includes chapters on fitness, demography, and population dynamics (Mueller), the life-history physiology of insects (Zera and Harshman), behavior and neurobiology (Rhodes and Kawecki), whole-organism performance and physiology (Swallow et al.), and, finally, genome evolution (Rosenzweig and Sherlock). Our intent in this part of the book is to show how experimental evolution can be used as a broad exploratory spatula with which to separate different horizontal layers of the scientific cake.

Part Four covers some exciting applications of selection experiments and experimental evolution, with reviews of studies that have used phages (Forde and Jessup), a range of organisms to study "allometry" (in the broad sense of the word: Frankino et al.), the evolution of sex (Turner et al.), physiological adaptation in laboratory environments (Gibbs and Gefen), the evolution of aging and late life (Rauser et al.), altruism and the levels of selection (Kerr), speciation (Fry), and experimental phylogenetics (Oakley). Here the field of experimental evolution is sliced up "vertically," with a diversity of experimental methods being used to address significant problems in evolutionary biology. Although many other useful applications of the techniques of experimental evolution could be included (e.g., fisheries management: Conover and Munch 2002; the evolution of antibiotic resistance: Krist and Showsh 2007), we feel that this miscellany of examples is at least instructive with respect to the scope of this burgeoning literature.

Part Five discusses the difficulties of balancing simplicity and realism in laboratory studies of natural selection, especially when they are intended to simulate selection in the wild (Oakley; Huey and Rosenzweig).

## SUMMARY

Experimental evolution is becoming a mainstream part of the biological sciences, beyond the confines of evolutionary biology, narrowly construed. For example, the journal *Integrative and Comparative Biology* recently published a symposium on "Selection Experiments as a Tool in Evolutionary and Comparative Physiology: Insights into Complex Traits" (Swallow and Garland 2005). In 2007, the journal *Physiological and Biochemical Zoology* published a "Focused Issue" on "Experimental Evolution and Artificial Selection." The response to the call for papers was so great that they ended up publishing papers in parts of three successive issues, including several by contributors to this volume. In 2008, the journal *Heredity* published a collection of six short reviews on microbial studies using experimental evolution (Bell 2008a). And experimental evolution is making it into the curriculum (e.g., box 13.3 in Moyes and Schulte 2006; Krist and Showsh 2007).

This volume provides further evidence that selection experiments have arrived. We hope that it serves to stimulate experimental evolutionary studies broadly. We also hope that it helps to improve the nature of such studies by careful attention to experimental design. In closing, we would appeal to the words of our colleagues regarding the importance of selection experiments and experimental evolution:

Ultimately, laboratory systems provide the best opportunity for the study of natural selection, genetic variation, and evolutionary response in the same population. . . . We suggest that the study of natural selection in a laboratory setting is the best method of making the link between natural selection and evolution and may thus permit predictive and rigorous study of adaptation. Houle and Rowe (2003, 50–51).

Selection experiments are irreplaceable tools for answering questions about adaptation and the genetic basis of adaptive trait clusters (i.e., repeated evolution of suites of traits in particular environments). (Fuller et al. 2005, 391)

## ACKNOWLEDGMENTS

T.G. thanks the National Science Foundation for continued support of his selection experiment (most recently IOB-0543429). T.G. also thanks his graduate students and other collaborators for suffering a lack of attention while I worked on this volume. M.R.R. thanks the British Commonwealth, NATO, the Natural Sciences and Engineering Research Council of Canada, the National Institutes of Health, the National Science Foundation, the University of California, and several private donors, including the Tyler family, for their support of his research on experimental evolution over more than thirty years. The numerous colleagues and students who have worked with M.R.R. on experimental evolution are listed in the authors' lists of Rose et al. (2004) and in the acknowledgments of Rose (2005), though neither listing is complete.

We also thank Chuck Crumly of the University of California Press for helping to make this volume happen. He is one of our favorite editors. We thank all of the book and chapter reviewers (some of whom remain anonymous to us), including E. Abouheif, A. F. Bennett, J. J. Bull, M. K. Burke, B. Charlesworth, A. K. Chippindale, D. J. Fairbairn, S. E. Forde, S. A. Frank, M. A. Frye, B. Gaut, C. K. Ghalambor, G. W. Gilchrist, S. A. Kelly, M. Konarzewski, R. C. McBride, L. M. Meffert, L. D. Mueller, L. Nunney, J. P. Phelan, D. E. L. Promislow, C. A. Reynolds, J. Rhodes, H. Rundle, E. Sober, M. Travisano, P. E. Turner, M. J. Wade, and J. B. Wolf. Over the years (that) we have been editing this book, we may have lost track of a reviewer, a supporter, or a commentator who has contributed to this effort; and if so, we hereby apologize.

## REFERENCES

Bell, G. 2008a. Experimental evolution. *Heredity* 100:441–442.

———. 2008b. *Selection: The Mechanism of Evolution*. 2nd ed. Oxford: Oxford University Press.

Bennett, A. F. 2003. Experimental evolution and the Krogh Principle: Generating biological novelty for functional and genetic analyses. *Physiological and Biochemical Zoology* 76:1–11.

Charlesworth, B., R. Lande, and M. Slatkin. 1982. A neo-Darwinian commentary on macroevolution. *Evolution* 36:474–498.

Chippindale, A. K. 2006. Experimental evolution. *In* C. Fox and J. Wolf, eds. *Evolutionary Genetics*. Oxford: Oxford University Press.

Clarke, B. 2003. The art of innuendo. *Heredity* 90:279–280.

Coltman, D. W., P. O'Donoghue, J. T. Jorgenson, J. T. Hogg, C. Strobeck, and M. Festa-Bianchet. 2003. Undesirable evolutionary consequences of trophy hunting. *Nature* 426:655–658.

Conover, D. O., and S. B. Munch. 2002. Sustaining fisheries yields over evolutionary time scales. *Science* 297:94–96.

Cooke, S. J., C. D. Suski, K. G. Ostrand, D. H. Wahl, and D. P. Philipp. 2007. Physiological and behavioral consequences of long-term artificial selection for vulnerability to recreational angling in a teleost fish. *Physiological and Biochemical Zoology* 80:480–490.

Endler, J. A. 1986. *Natural Selection in the Wild.* Princeton, NJ: Princeton University Press.

Falconer, D. S. 1992. Early selection experiments. *Annual Review of Genetics* 26:1–14.

Fuller, R. C., C. F. Baer, and J. Travis. 2005. How and when selection experiments might actually be useful. *Integrative and Comparative Biology* 45:391–404.

Garland, T., Jr. 2003. Selection experiments: An under-utilized tool in biomechanics and organismal biology. Pages 23–56 *in* V. L. Bels, J.-P. Gasc, A. Casinos, eds. *Vertebrate Biomechanics and Evolution.* Oxford: BIOS Scientific.

Garland, T., Jr., A. F. Bennett, and E. L. Rezende. 2005. Phylogenetic approaches in comparative physiology. *Journal of Experimental Biology* 208:3015–3035.

Garland, T., Jr., and S. A. Kelly. 2006. Phenotypic plasticity and experimental evolution. *Journal of Experimental Biology* 209:2234–2261.

Gavrilets, S., and A. Vose. 2005. Dynamic patterns of adaptive radiation. *Proceedings of the National Academy of Sciences of the USA* 102:18040–18045.

Gilchrist, G. W., and C. E. Lee. 2007. All stressed out and nowhere to go: Does evolvability limit adaptation in invasive species? *Genetica* 129:127–132.

Hamilton, W. D. 1966. The moulding of senescence by natural selection. *Journal of Theoretical Biology* 12:12–45.

Hard, J. J., M. R. Gross, M. Heino, R. Hilborn, R. G. Kope, R. Law, and J. D. Reynolds. 2008. Evolutionary consequences of fishing and their implications for salmon. *Evolutionary Applications* 1:388–408.

Hill, W. G., and A. Caballero. 1992. Artificial selection experiments. *Annual Review of Ecology and Systematics* 23:287–310.

Hoekstra, H. E., J. M. Hoekstra, D. Berrigan, S. N. Vignieri, A. Hoang, C. E. Hill, P. Beerlii P, and J. G. Kingsolver. 2001. Strength and tempo of directional selection in the wild. *Proceedings of the National Academy of Sciences of the USA* 98:9157–9160.

Houle, D., and L. Rowe. 2003. Natural selection in a bottle. *American Naturalist* 161:50–67.

Huey, R. B., G. W. Gilchrist, and A. P. Hendry. 2005. Using invasive species to study evolution. Pages 139–164 *in* D. F. Sax, S. D. Gaines, and J. J. Stachowicz, eds. *Species Invasions: Insights to Ecology, Evolution and Biogeography.* Sunderland, MA: Sinauer.

Kingsolver, J. G., H. E. Hoekstra, J. M. Hoekstra, D. Berrigan, S. N. Vignieri, C. E. Hill, A. Hoang, P. Gibert, and P. Beerli. 2001. The strength of phenotypic selection in natural populations. *American Naturalist* 157:245–261.

Krist, A. C., and S. A. Showsh. 2007. Experimental evolution of antibiotic resistance in bacteria. *American Biology Teacher* 69:94–97.

Lee, C. E., J. L. Remfert, and Y. M. Chang. 2007. Response to selection and evolvability of invasive populations. *Genetica* 129:179–192.

Lynch, C. B. 1980. Response to divergent selection for nesting behavior in *Mus musculus*. *Genetics* 96:757–765.

———. 1994. Evolutionary inferences from genetic analyses of cold adaptation in laboratory and wild populations of the house mouse. Pages 278–301 *in* C. R. B. Boake, ed. *Quantitative Genetic Studies of Behavioral Evolution*. Chicago: University of Chicago Press.

Mayr, E., and W. B. Provine, eds. 1980. *The Evolutionary Synthesis: Perspectives on the Unification of Biology*. Cambridge, MA: Harvard University Press.

Moyes, C. D., and P. M. Schulte. 2006. *Principles of Animal Physiology*. San Francisco: Pearson Benjamin Cummings.

Provine, W. B. 1971. *The Origins of Theoretical Population Genetics*. Chicago: University of Chicago Press.

Rose, M. R. 2005. *The Long Tomorrow: How Evolution Can Help Us Postpone Aging*. New York: Oxford University Press.

Rose, M. R., H. B. Passananti, and M. Matos, eds. 2004. *Methuselah Flies: A Case Study in the Evolution of Aging*. Singapore: World Scientific.

Swallow, J. G., and T. Garland, Jr. 2005. Selection experiments as a tool in evolutionary and comparative physiology: Insights into complex traits—An introduction to the symposium. *Integrative and Comparative Biology* 45:387–390.

Wattiaux, J. M. 1968. Cumulative parental age effects in *Drosophila subobscura*. *Evolution* 22:406–421.

Weldon, W. F. R. 1901. A first study of natural selection in *Clausilia laminata* (Montagu). *Biometrika* 1:109–124.

Zimmer, C. 2006. Evolution in a petri dish. *Yale Alumni Magazine*, May. www.yalealumnimagazine.com/issues/2006_05/evolution.html.

2

# THE IMPORTANCE OF EXPERIMENTAL STUDIES IN EVOLUTIONARY BIOLOGY

Douglas J. Futuyma and Albert F. Bennett

## METHODS OF STUDYING EVOLUTION

Experimental studies complement the several other major approaches to analyzing evolutionary processes. Each approach has both advantages and limitations. We first describe the advantages of experimental evolution, and then we contrast it with other approaches.

### EXPERIMENTAL EVOLUTION

The essence of experimental evolution is conceptually quite simple. For many generations, a series of replicated populations is exposed to a novel environment, while a parallel series of populations is maintained within the ancestral environment, thereby serving as experimental controls. The environmental novelty may involve alteration of any aspects of the abiotic, biotic, or demographic condition of the ancestral population. Usually only a single environmental variable is altered to keep the experiment as simple as possible. It is, however, feasible to change multiple environmental factors simultaneously. The novel experimental environment provides new selective conditions and hence promotes evolution. New genetic variants (produced through recombination, mutation, or other processes) may be differentially advantaged or disabled in the altered conditions, producing differential reproduction and increase of the favored genotypes.

Several different methods of environmental alteration are used in experimental evolutionary studies, including artificial truncation, culling, and laboratory natural selection (Rose et al. 1990; Garland 2003; Rose and Garland this volume). The distinctions among and advantages and limitations of each of these methods are discussed more fully elsewhere in this volume (especially Huey and Rosenzweig). They are all similar, however, in creating conditions favoring evolutionary alteration of the condition of the ancestral population. After a sufficient number of generations, the novel experimental populations may be compared directly with the controls (or in some cases directly to their own ancestors), and any variety of a priori hypotheses concerning evolution may thereby be tested. The longer the antecedents of the ancestral population have been maintained in the ancestral (control) condition, the more likely any differences observed within the novel experimental populations will be specific to the novel environmental alteration (see also Travisano this volume).

What is particularly advantageous about this approach to evolutionary studies? Its special strengths lie in the essence of any experiment: replication and control. By replicating the number of populations exposed to the novel environment, an investigator can, in effect, repeat the opportunity for evolutionary change and determine if the outcome has consistency. In the metaphor of S. J. Gould (1989), the tape of life and evolution can be replayed as often as desired to determine the similarity of its resulting products. Evolutionary lineages of natural populations are a series of unique events, and it is never possible to determine if things could have turned out differently or what factors were responsible for the observed differentiation. In experimental evolution, however, the diversity of responses can be examined directly. Through replicated measurements among experimental populations, this diversity can be analyzed statistically. Further, the novel experimental populations can as a group be statistically compared with the group of control populations (or, in some situations, with the ancestral population itself), with the number of degrees of freedom determined by the number of independent replicates of each. The statistical significance of any differences between the experimental and control groups can be evaluated to test evolutionary hypotheses.

Experimental evolution has certain additional attractive features. Unlike studies of historical evolution in the natural world, there is never any uncertainty about ancestral condition: the experimenters know the condition and composition of the ancestral population because they chose and observed it. Also in contrast to historical evolution, experimental evolution can define and control the degree of environmental change, limiting it to a single factor or any desired combination of factors. So many environmental aspects change simultaneously in the natural world that it is difficult to know to which aspect of the environment the population is adapting. Experimental evolution can isolate and analyze the adaptive response to specific environmental factors.

Experimental studies can also produce entirely new types of organisms for biological study, literally building a better mouse (Bennett 2003; Garland 2003; Swallow et al. this volume). By its nature, experimental evolution produces populations with traits that enhance function and overall fitness in the new selective environment. The resulting "improved" organisms are then subsequently available for functional and genetic analyses. The different experimental replicate populations may contain a diversity of adaptive solutions to their common environmental challenge, sometimes solutions not previously observed or expected (Bennett 2003; Garland 2003; Swallow et al. this volume). Given that biological systems are complex and their evolution is unpredictable, the emergent variability among selected populations can be especially informative about the possible range of adaptive mechanisms.

As with any methodology, experimental evolution has its limitations and drawbacks (see also Huey and Rosenzweig this volume; Oakley this volume). This approach, because of its emphases on control and replication, is more suited to laboratory rather than to natural situations. Although experimental evolutionary studies have been undertaken successfully in the field on natural populations of organisms (see, e.g., Reznick and Endler 1982; Reznick et al. 1990; Irschick and Reznick this volume), the large majority have been laboratory based. Consequently, they lack the multidimensionality of natural environments and sacrifice the ecological realism provided by studies carried out in nature. Further, because of their requirements for large population size and rapid reproduction, they have to be restricted to certain kinds of organisms that can be easily maintained and bred in quantity in the laboratory. Although the range of such organisms is phylogenetically broad, from viruses to mice, it is nonetheless limited, and many interesting taxa are largely excluded from consideration as experimental evolutionary subjects. Experimental evolution is not meant to reproduce, mimic, or predict evolution in nature precisely. Its principal utility is its ability to test evolutionary theory and hypotheses, and in that it is unmatched by alternative approaches (Bennett and Lenski 1999).

Experimental evolution can be contrasted with studies of natural populations of individual species and with comparisons among species.

## STUDIES ON NATURAL POPULATIONS

Most studies of natural populations draw inferences either from "snapshot" descriptions of patterns of variation or from temporal data on actively evolving populations. The great majority of studies of natural populations document patterns of variation, such as allele and genotype frequencies, or statistical data on phenotypic characters, within and among populations of a species. In some cases, the genetic component of phenotypic variation is described from samples brought into the laboratory or garden for controlled breeding. It is often attempted to infer the processes or history responsible for the observed patterns by comparing data with expectations from theory. For example, if a deficiency of heterozygotes and an excess of homozygotes is observed at many loci, it is reasonable to infer that the population is subject to inbreeding, and such patterns are indeed often seen in plants that practice some self-fertilization. A deficiency of genetic variation in a population, relative to other populations of the species, may signal recent reduction in population size or recent colonization; more detailed histories of population size and movement can be inferred by applying coalescent analysis and phylogeographic study of "gene trees."

There are several ways, moreover, in which inferences can be made about the effects of natural selection in natural populations on characters of interest (summarized in Endler 1986). For example, features may have evolved very recently, in response to identifiable selective factors, in populations introduced into new environments, or in response to anthropogenic environmental changes. The study of geographic variation within species, and its correlation with environmental variables, has been one of the most widely used and successful approaches to demonstrating adaptation (reviewed in Garland and Adolph 1991). Several methods of selection analysis have been used to describe the mode and strength of selection on variable loci or phenotypic traits. In almost all studies of natural populations, the analysis of factors that impinge on natural variation is used to make inferences about the causal history by which a feature has evolved. For instance, is the coloration (or size or behavior or thermal tolerance, etc.) of this organism a consequence of genetic drift or of natural selection? If selection, then what have been the selective factors?

As successful as these approaches have been, they have certain limitations that experimental evolutionary studies can often overcome (see also Irschick and Reznick this volume). The most important are concerns about adequacy of replication and control over confounding variables. Recent or ongoing evolution is often described for a single population or ecological event (e.g., spread of an introduced species from a single colonization), or for multiple populations that may not represent independent evolutionary changes if they are connected by gene flow. A similar problem of lack of independence may likewise be a concern in some studies of geographic variation. In contrast, experimental evolution enables replication of populations and therefore greater statistical confidence. Moreover, population replication may provide insights into the role of random events, such as the impact of different mutations in different lines.

In studying natural populations, investigators must be concerned that correlated variables might be responsible for genetic changes, rather than the variable they have identified. The confounding variables may be external to the population of organisms—environmental variables that are correlated with the one of interest—or internal, such as selection on genes physically linked to those that affect the character studied, or an unknown demographic history in which genetic drift has loomed large. Correlated variables are especially important if only a single population or event is studied, for there is then no possibility of separating the postulated independent variable from those with which it might be confounded (see also Garland and Adolph 1994). In contrast, the effects of historical contingency can be minimized, or at least identified, in replicated experimental populations. By manipulating population size, it may be possible to identify the relative effects of selection and genetic drift, a problem that is exceedingly difficult in natural populations. By the very nature of an experiment, environmental variables can be controlled or randomized, so that the effects of individual variables of interest can be isolated and intentionally altered, and assessed with reference to control populations. Consequently, the inferences that may be drawn about the factors responsible for evolution in experimental populations can be far more direct, and often can be asserted with greater confidence, than we can usually draw from natural populations.

## COMPARATIVE METHODS

Character similarity among taxa has long been taken as evidence of common evolutionary history, and the divergence of characters in otherwise similar groups, as evidence of adaptation. Studies of this type are generally termed "the comparative method." It is only within the past several decades, however, that it has been recognized that the comparative method must take into account the statistical nonindependence of taxa because of their common evolutionary history (Felsenstein 1985). A variety of different types of analytical methodologies (e.g., independent contrasts, generalized least-squares models, Monte Carlo computer simulations) has been developed for this purpose (see Garland et al. 2005 for a recent summary of modern comparative methods). These methods differ in their analytical details, sometimes in ways that affect the inferences that can be drawn. In some approaches, characters of interest (discrete or quantitative) are "mapped" onto a preexisting, independently derived phylogeny. With certain assumptions, ancestral nodal conditions can be estimated, and instances of evolutionary character reversal can be identified. The total pattern of historical diversification can then be examined for concordance of character change with historical or environmental or biotic changes within the taxa. In other approaches, statistical models (e.g., multiple regression) are applied in a way that acknowledges the likely nonindependence of residuals and can model that nonindependence in a variety of ways (reviewed in Lavin et al. in press).

The modern comparative method is our best tool for trying to understand the historical evolution of characters in the natural world, and it is widely accepted and employed. It has, however, some shortcomings (Garland and Adolph 1994; Garland et al. 2005). First, it is explicitly correlational and cannot directly demonstrate causality. Second, it depends crucially on the assumption of parsimony in character diversification, and we do not know if evolution is in fact parsimonious. Research on reverse evolution (see Estes and Teotonio this volume) suggests that this assumption will in fact lead to systematic error when evolution reverses direction. Finally, many of these methods are dependent on the assumption of a particular phylogeny of the group of interest (but see, e.g., Huelsenbeck and Rannala 2003). This phylogeny is only a hypothesis about ancestral relationships that cannot be directly observed, and it is susceptible to extensive revisions in the light of new information. Such phylogenetic revisions may completely change previous conclusions about the evolution of characters.

## TWO EXAMPLES OF THE CONTRIBUTION OF EXPERIMENTAL STUDIES TO UNDERSTANDING EVOLUTION

Here, to illustrate the power of the experimental approach, we discuss two areas of evolutionary thought and theory to which experimental studies have made substantial contributions: the presence and generality of trade-offs and the role of genetic drift during adaptation.

### ADAPTIVE GAIN AND CORRELATED LOSS

A long tradition in evolutionary thought claims that adaptive gain can be achieved only at a price. That is, as fitness increases in one respect (reproduction, environment, age, etc.), it decreases in other respects. These decrements are termed *trade-offs* (see also Frankino et al. this volume; Gibbs and Gefen this volume), which can be more precisely defined as a decline in nonselected fitness characters that accompanies adaptation to new selective conditions. Darwin (1859) referred to trade-offs as "compensation" and said that "in order to spend on one side, nature is forced to economise on the other side." The concept of a necessary reallocation of resources resulting from adaptation is still current; for example, "improvements cannot occur indefinitely, because eventually organisms come up against limitations. . . . At that point, improvements in one trait may be achievable only at the expense of others—there is a trade-off between the traits" (Sibly 2002; see also Futuyma and Moreno 1988; Stearns 1992; Futuyma 1998). The underlying basis for some trade-offs is relatively straightforward as the differential allocation of some limited resource, such as energy, space, or time. If a kilocalorie of energy is expended in one activity, then it is not available to support another activity; this is the sense in which Darwin discussed trade-offs. However, trade-offs may take other forms, including functional constraints (e.g., multiple effects of changes in circulating hormone levels, aerobic vs. anaerobic muscle structure) (see also Swallow et al. this volume).

Trade-offs are the presumptive basis for niche shifts (rather than niche expansion) along an environmental range, such as temperature, salinity, or acidity. If a population is adapting to one extreme of its realized niche, then the niche is expected to shift toward that extreme, with a consequent loss of function and fitness at the other end of the niche. Consequently, the niche is expected to shift along that environmental continuum. Adaptation to cold environments would therefore be expected to be accompanied by a decline in ability to persist in hot environments. Levins's (1968) principle of allocation provided the theoretical basis for such niche shifts, and many mathematical models concerning adaptive shifts during environmental change incorporate this principle (e.g., Lynch and Gabriel 1987; Pease et al. 1989; but see Gilchrist 1995).

Although the concept that constructive change may also entail loss has an intuitive appeal, operationally investigating trade-offs can be difficult. For a multidimensional niche, the existence of a trade-off cannot be conclusively disproved: if decrements in all functions examined cannot be demonstrated, then it can always be asserted that the decline occurred in some function that is yet to be investigated. What can be investigated is whether performance has declined in any portion of the range of the specific environmental variable under selection. For instance, during specialization, does performance decline at both niche extremes during adaptation to a constant intermediate environment? Selection experiments are a particularly powerful tool for investigations of such trade-offs because of their unique advantages of experimental replication and control of the selective environment. Fitness increments at the selective environmental variable can be measured and compared with correlated changes in fitness in other portions of the same environmental axis.

An example of an investigation of trade-offs using a selection experiment is provided by research on evolutionary adaptation in the bacterium *Escherichia coli*. A bacterial clone was first adapted to 37°C and 7.2 pH (Lenski et al. 1991) and then used as the ancestor for replicated experimental groups at a variety of different culture temperatures (Bennett et al. 1992; Mongold et al. 1996) and acidities (Hughes et al. 2007a, 2007b). The niche of the common ancestor, defined as the range over which it could grow and maintain itself in serial dilution culture of 100-fold per day, extended from 19.5°C to 42.2°C and from 5.4 to 8.0 pH. Experimental groups of six replicated populations were cultured in different environments (including the niche extremes of 20°C, 41.5°C, pH 5.4, and pH 8.0) for 2,000 generations. The fitness of each of these experimental groups relative to their common ancestor was measured directly by competition experiments in their new selective environments. The fitness increments provide a quantitative measure of their adaptation. At 20°C, fitness increased 8.7 percent; at 41.5°C, 33.5 percent; at pH 5.4, 20.0 percent; and at pH 8.0, 7.6 percent (Bennett and Lenski 1996; Hughes et al. 2007a).

Was this adaptation to one extreme of the niche accompanied by a loss of fitness at the other end of the niche? To determine whether trade-offs occurred, relative fitness of each group was measured at the other niche extreme (40°C for the 20°C group; 20°C for the 41.5°C group; pH 8.0 for the 5.4 pH group; and pH 5.4 for the 8.0 pH group). In contrast to the uniform pattern of fitness increments associated with the adaptive response, the trade-off response varied widely in both presence and magnitude. For the 20°C adapted group, fitness at 40°C declined significantly by 17 percent ($p = 0.02$) (Mongold et al. 1996), but the 41.5°C group did not experience any fitness loss at 20°C (Bennett and Lenski 1993). The acid-adapted (5.4 pH) group had a significant fitness decline of 23 percent ($p < 0.001$) at pH 8.0, while the base-adapted (8.0 pH) group measured at 5.4 pH experienced a significant decline in fitness of three of its six lines, an increment in one, and no significant change in the other two, resulting in no significant group trade-off effect (Hughes et al. 2007a).

Another, more extensive examination of thermal trade-offs (Bennett and Lenski 2007) was done on four groups of six lines each that were secondarily adapted to 20°C for 2,000 generations after previously adapting to diverse warmer thermal environments (32°, 37°, and 41.5°C, and an environment alternating between 32° and 41.5°C) (Mongold et al. 1996). Fitness relative to the progenitor of each line was measured at 40°C. Because of the multiple levels of replication possible within this system, it was possible to determine whether the trade-off effect was significant across all experimental populations and whether it occurred within each experimental population. Fitness at 40°C declined significantly in fifteen of the twenty-four lines; it did not change significantly in eight of the lines, and it actually increased significantly in one. For the entire group of twenty-four lines, mean fitness at 40°C declined 9.4 percent ($p < 0.001$), so there was a significant pattern of trade-off, even though it did not occur in 38 percent of the lines examined.

What can we learn from these results about adaptation and trade-off? It is apparent that trade-offs sometimes do occur as a correlated response to adaptation, as they did in these experiments during evolution in cold and acidic environments. In some situations, however, such as in adaptation to a hot environment, no discernible trade-offs were detected. In other situations (e.g., adaptation to alkalinity), there was such extreme variation among populations that the overall trade-off effect, while readily apparent in some lines, was not statistically significant overall. Even when an overall trade-off effect is significant, it may not be general. That is, even when evolution generally operates to produce trade-offs, it may not do so universally, and a large number of populations may demonstrate adaptation without correlated loss of function. These experimental results engender a more stochastic and nuanced view of adaptive evolution, in which generalities can be specifically tested and exceptions documented (see also Huey and Rosenzweig this volume).

## STUDYING THE ADAPTIVE ROLE OF GENETIC DRIFT

One of Ernst Mayr's most influential ideas was his proposal (1954) of founder-effect speciation, which he developed further in 1963 and christened "peripatric speciation" in 1982. Mayr suggested that evolution proceeds slowly in large, stable populations because of strong fitness interactions (epistasis) among loci; specifically, the prevalent alleles will be those that confer high fitness on the many different genetic backgrounds that they encounter in a large, genetically variable population. However, an isolated population founded by a few colonists will be much less genetically variable (the "founder effect," an instance of genetic drift), so selection may instead favor alleles that are good "soloists," those that increase fitness in concert with fewer segregating alleles at interacting loci. Consequently, a founder event (or bottleneck in population size) may initiate new paths of adaptive evolution, and these genetic changes, initiated by drift but directed by selection thereafter, might incidentally result in pre- or postzygotic reproductive isolation—speciation. Some years later, Hampton Carson (1968, 1975; also Carson and Templeton 1984) proposed a rather similar idea, the "founder-flush-crash" scenario, in which exponential population growth followed by severe population reduction results in major genetic reorganization: the relaxation of selection during population growth enables many recombinant genotypes to proliferate, of which only a few, perhaps very different from previously prevalent genotypes, survive the population crash. The population may thereby be set on a new trajectory of molding by natural and sexual selection that may result in speciation.

Many subsequent authors discussed these ideas in the context of Wright's adaptive landscape. The effect of a bottleneck on a population's genetic composition could be viewed as drift to a different position on the landscape; if this point were on the slope of a different adaptive peak than that of the parent population, selection would lead the population to a different equilibrium constitution. Mathematical theoreticians used this framework to analyze models that they thought represented Mayr's or Carson's verbal model. Depending on their assumptions, they concluded that speciation is quite likely (e.g., Gavrilets and Hastings 1996) or not at all likely (e.g., Barton and Charlesworth 1984; Charlesworth and Rouhani 1988). The differences in these authors' conclusions depend partly on assumptions about the nature of epistasis and the number of loci or characters involved, which have been very difficult to examine empirically.

Experiments, therefore, may be the best way to judge the likelihood that stochastic genetic changes owing to bottlenecks may initiate adaptive divergence and/or reproductive isolation from the large "parent" population. Such experiments typically have involved initiating replicate experimental populations with different numbers of founders, and scoring characters of interest after the passage of a number of generations at the same, large population size. (That is, the treatments differ in founder numbers, not equilibrium population numbers.) In one of the first and demonstrably less memorable such experiments, Futuyma (1970) exposed replicate populations of *Drosophila melanogaster*, initiated with two, ten, or thirty pairs of flies from the same base population, to competition with *Drosophila simulans*, and after ten generations scored them for population-level indicators of competitive ability (e.g., numbers or biomass produced from a standard number of parents placed with the competing species). Compared with control populations that were not subjected to interspecific competition, a few experimental lines displayed differences in competitive ability, but no effect of founder number was discerned.

Starting in the 1970s, many investigators tested for the evolution of incipient reproductive isolation in bottleneck experiments; most imposed multiple bottlenecks, more or less mimicking Carson's scenario (see reviews in Rice and Hostert 1993, Coyne and Orr 2004, and Rundle 2003; see also Fry this volume). For example, Agustí Galiana, Andrés Moya, and Francisco Ayala (1993; also Moya et al. 1995) set up multiple populations of *Drosophila pseudoobscura*, derived from flies collected in either Utah or Mexico, at several bottleneck sizes ($n$ = one, three, five, or nine pairs), and subjected them to multiple flush/crash cycles: in each cycle, populations grew exponentially for an average of six generations and then were reduced to $n$ pairs. The "ancestral" stocks were maintained at large size throughout. In cycles 4, 5, and 7, various experimental populations were tested for sexual isolation, either from one another or from their ancestral stock, by placing twelve pairs of virgin flies from each of two populations in an arena and scoring the numbers of matings. (Flies from the two populations were made distinguishable by clipping a small piece out of the wing, a procedure that did not alter the results.) Galiana et al. found, overall, that twenty-three of the forty-five experimental populations, viewed individually, displayed positive assortative mating (i.e., incipient reproductive isolation) when tested against other populations, and ten remained statistically significant when subjected to the Bonferroni correction for multiple tests. Significantly more populations in the small-bottleneck ($n$ = one or three) treatments showed assortative mating than in the large-bottleneck ($n$ = five, seven, or nine) treatments. Only positive assortative mating (i.e., mating within the same population) was observed; no population mated preferentially with members of different populations. (Such an outcome might have been expected by chance; indeed, positive and negative assortative mating were equally frequent in a rather similar experiment with *D. melanogaster*, performed by Howard Rundle [2003].) However, most individual pairs of populations did not evince assortative mating, and the authors concluded that their results "do not support the claim that the founder-flush-crash model identifies conditions very likely to result in speciation events." Perhaps the support or lack of support for the hypothesis lies in the eye of the beholder, for Templeton (1996) argued that this study does support Carson's scenario, but later experiments in *Drosophila* have led most students of speciation toward a similar negative conclusion (Coyne and Orr 2004).

As theoreticians turned their attention to the effects of population bottlenecks, another interesting hypothesis bearing on adaptation and speciation surfaced. It has long been known that if genetic variation in a trait is based on additively acting alleles, the additive genetic variance ($V_A$), which measures the correlation between parents and offspring and is consequently the kind of genetic variation that underlies the response to selection, is diminished within a population that is reduced in size. However, Charles Goodnight (1988; also Whitlock et al. 1993; Cheverud and Routman 1996) showed that if the phenotype is based substantially on interactions between alleles at different loci (i.e., on epistasis), the epistatic variance, $V_I$ (the component of the population's genetic variance that is caused by such interaction), is commonly transformed into additive genetic variance if the allele frequencies are altered by genetic drift during a bottleneck. Consequently, directional selection on a characteristic may be more effective in a bottlenecked population than in a stable, large population! If such characteristics, or the genes underlying them, affect reproductive isolation, steps toward speciation might be enhanced in founder populations for a quite different reason than either Mayr or Carson had envisioned.

Whether traits have the epistatic variance postulated by these models can be determined only by empirical study. Increases in $V_A$ in bottlenecked populations have been found in some, but not all, traits that have been studied. For example, Michael Wade et al. (1996) imposed artificial selection for pupal weight in fifteen populations of the flour beetle *Tribolium castaneum* after zero, two, or five generations of single-pair sister-brother matings. After eight generations of selection, they estimated heritability of both pupal weight and total fitness (as the number of adult offspring per individual) by offspring/parent regression. They found that the heritability of pupal weight declined in accord with the additive model, but the heritability of fitness in the bottlenecked lines equaled or exceeded that of outbred control populations. Similarly, Edwin Bryant, Lisa Meffert, and their colleagues have found that for several morphological measurements on houseflies (*Musca domestica*), $V_A$ increases in populations that have passed through one or more bottlenecks (e.g., Bryant and Meffert 1993). An important question is whether such increases are caused by conversion of $V_I$ to $V_A$ (which could enhance responses to natural selection) or simply by an increase, due to inbreeding, of homozygotes for deleterious recessive alleles, which would not contribute to adaptation. In a painstaking analysis of the components of genetic variance, Bryant and Meffert (1996) found that two characteristics that showed increased $V_A$ after bottlenecks have a high level of epistatic variance in the outbred base population, whereas this is not the case for two other characteristics, for which the bottleneck effect was not observed.

Experimental studies, in which genetic variance may be altered by genetic drift and natural selection, have revived the possibility that founder effects might initiate new evolutionary trajectories. The next question is whether or not such changes might also initiate evolution of reproductive isolation. Meffert and collaborators explored this question in a series of experiments with houseflies. In one study (Meffert and Bryant 1991), male courtship behavior was analyzed in fifteen bottlenecked lines. Although only one pair displayed significant assortative mating, many of the lines diverged in the frequency with which various courtship components were displayed. From a statistical point of view, these results do not encourage confidence that founder effects initiate reproductive isolation, but they are nonetheless biologically intriguing. One could entertain the analogy that a statistically trivial fraction of mutations enhance fitness, but mutation is nonetheless a foundation of adaptive evolution.

## FITTING THE EXPERIMENT TO THE QUESTION

These two examples illustrate the necessity of tailoring the design of an experimental evolutionary study to the question posed. The experiments with bacteria reviewed here could address questions about adaptation based on newly arisen mutations, because of the large populations and numbers of generations. Mutation-accumulation experiments have been carried out with *Drosophila* in order to estimate the rate at which variation arises, but most experimental evolutionary studies of *Drosophila*, such as the bottleneck experiments we have described, have characterized the effects of standing genetic variation in the ancestral population. A fly experiment modeled on the *E. coli* experiments we have described would call for a very long-term commitment, if the same number of generations were used. On the other hand, the rate of phenotypic response of an outbred population undergoing laboratory evolution can be two orders of magnitude faster per generation than that achievable with microbes (see Rose et al. 2004 for examples).

Conversely, the ideas tested in the bottleneck experiments described here pertain primarily to sexually reproducing populations, and it is generally supposed that, if founder effects are to have important consequences for subsequent evolutionary changes, the consequences should be evident within a few generations. If one postulated, instead, that drift-induced changes in a founder population's genetic constitution determined the selective advantage or disadvantage of subsequent new mutations, then bottleneck experiments would require extension or redesign. Before launching into the great commitment that an experimental evolutionary study typically requires, the investigator would do well to consider the question, and the background theory, very carefully.

## CONCLUSION

Of all methods used in evolutionary studies, experimental evolution mostly closely approximates the "strong inference" approach of studying science (Platt 1964). It satisfies all of the elements of the classical scientific method and provides, therefore, the most unarguable and convincing empirical analysis of evolutionary processes. It is not suited to answering all questions about evolution, particularly evolution in the natural world. But because of its powerful use of population replication and control, it can provide rigorous testing of evolutionary hypotheses and theories that formerly were matters only of assumption or speculation. It will become an approach of increasing utility and importance in future synthetic studies of evolutionary science.

## SUMMARY

The goals of experimental evolutionary research are to enhance our understanding of the bases and dynamics of evolution, and also to elucidate limitations on evolutionary processes. Experiments can be designed to estimate the effects of such factors as mutation, genetic drift, gene flow, and selection on genetic variation and on heritable phenotypic traits. Experiments can also cast light on variation in responses to selection and on the factors that may bias or constrain the ability of populations to respond to selection. Many evolutionary experiments probe the process of adaptation to specific environmental challenges. An important aspect of such studies is that they can provide insight into the genetic, molecular, cellular, and developmental bases and components of adaptive change—that is, into the mechanistic bases of adaptation.

## ACKNOWLEDGMENTS

A.F.B.'s research was supported by the NASA Astrobiology Institute (NASA Grant 632731, Center for Genomic and Evolutionary Studies on Microbial Life at Low Temperatures) and currently by National Science Foundation grant IOS-0748903.

## REFERENCES

Barton, N. H., and B. Charlesworth. 1984. Genetic revolutions, founder events, and speciation. *Annual Review of Ecology and Systematics* 15:133–164.

Bennett, A. F. 2003. Experimental evolution and the Krogh Principle: Generating biological novelty for functional and genetic analyses. *Physiological and Biochemical Zoology* 76:1–11.

Bennett A. F., and R. E. Lenski. 1993. Evolutionary adaptation to temperature. II. Thermal niches of experimental lines of *Escherichia coli*. *Evolution* 47:1–12.

———. 1996. Evolutionary adaptation to temperature. V. Adaptive mechanisms and correlated responses in experimental lines of *Escherichia coli*. *Evolution* 50:493–503.

———. 1999. Experimental evolution and its role in evolutionary physiology. *American Zoologist* 39:346–362.

———. 2007. Evolutionary trade-offs: an experimental test during temperature adaptation. *Proceedings of the National Academy of Sciences of the USA* 104:8649–8654.

Bennett, A. F., R. E. Lenski, and J. E. Mittler. 1992. Evolutionary adaptation to temperature. I. Fitness responses of *Escherichia coli* to changes in its thermal environment. *Evolution* 46:16–30.

Bryant, E. H., and L. M. Meffert. 1993. The effect of serial founder-flush cycles on quantitative genetic variation in the housefly. *Heredity* 70:122–129.

———. 1996. Nonadditive genetic structuring of morphometric variation in relation to a population bottleneck. *Heredity* 77:168–176.

Carson, H. L. 1968. The population flush and its consequences. Pages 123–137 *in* R. C. Lewontin, ed. *Population Biology and Evolution*. Syracuse, NY: Syracuse University Press.

———. 1975. The genetics of speciation at the diploid level. *American Naturalist* 109:83–92.

Carson, H. L., and A. R. Templeton. 1984. Genetic revolutions in relation to speciation phenomena: The founding of new populations. *Annual Review of Ecology and Systematics* 15:97–131.

Charlesworth, B., and S. Rouhani. 1988. The probability of peak shifts in a founder population. II. An additive polygenic trait. *Evolution* 42:1129–1145.

Cheverud, J. M., and E. J. Routman. 1996. Epistasis as a source of increased additive genetic variance in population bottlenecks. *Evolution* 50:1042–1051.

Coyne, J. A., and H. A. Orr. 2004. *Speciation*. Sunderland, MA: Sinauer.

Darwin, C. 1859. *On the Origin of Species by Means of Natural Selection*. London: Murray.

Felsenstein, J. 1985. Phylogenies and the comparative method. *American Naturalist* 125:1–15.

Futuyma, D. J. 1970. Variation in genetic response to interspecific competition in laboratory populations of *Drosophila*. *American Naturalist* 104:239–252.

———. 1998. *Evolutionary Biology*. 3rd ed. Sunderland, MA: Sinauer.

Futuyma, D. J., and G. Moreno. 1988. The evolution of ecological specialization. *Annual Review of Ecology and Systematics* 19:207–233.

Galiana, A., A. Moya, and F. J. Ayala. 1993. Founder-flush speciation in *Drosophila pseudoobscura*: A large-scale experiment. *Evolution* 47:432–444.

Garland, T., Jr. 2003. Selection experiments: An under-utilized tool in biomechanics and organismal biology. Pages 23–56 *in* V. L. Bels, J.-P. Gasc, and A. Casinos, eds. *Vertebrate Biomechanics and Evolution*. Oxford: BIOS Scientific.

Garland, T., Jr., and S. C. Adolph. 1991. Physiological differentiation of vertebrate populations. *Annual Review of Ecology and Systematics* 22:193–228.

———. 1994. Why not to do two-species comparative studies: Limitations on inferring adaptation. *Physiological Zoology* 67:797–828.

Garland, T., Jr., A. F. Bennett, and E. L. Rezende. 2005. Phylogenetic approaches in comparative physiology. *Journal of Experimental Biology* 208:3015–3035.

Gavrilets, S., and A. Hastings. 1996. Founder effect speciation: A theoretical assessment. *American Naturalist* 147:466–491.

Gilchrist, G. W. 1995. Specialists and generalists in changing environments. 1. Fitness landscapes of thermal sensitivity. *American Naturalist* 146:252–270.

Goodnight, C. J. 1988. Epistasis and the effect of founder events on the additive genetic variance. *Evolution* 42:441–454.

Gould, S. J. 1989. *Wonderful Life: The Burgess Shale and the Nature of History*. New York: Norton.

Huelsenbeck, J. P., and B. Rannala. 2003. Detecting correlation between characters in a comparative analysis with uncertain phylogeny. *Evolution* 57:1237–1247.

Hughes, B. S., A. J. Cullum, and A. F. Bennett. 2007a. Evolutionary adaptation to environmental acidity in experimental lineages of *Escherichia coli*. *Evolution* 61:1725–1734.

———. 2007b. Evolutionary adaptation to temporally fluctuating pH environments in *Escherichia coli*: Natural selection of generalists and specialists in transitions and acclimation. *Physiological and Biochemical Zoology* 80:406–421.

Lavin, S. R., W. H. Karasov, A. R. Ives, K. M. Middleton, and T. Garland, Jr. In press. Morphometrics of the avian small intestine, compared with non-flying mammals: A phylogenetic perspective. *Physiological and Biochemical Zoology*.

Lenski, R. E., M. R. Rose, S. C. Simpson, and S. C. Tadler. 1991. Long-term experimental evolution in *Escherichia coli*. I. Adaptation and divergence during 2,000 generations. *American Naturalist* 138:1315–1341.

Levins, R. 1968. *Evolution in changing environments*. Princeton, NJ: Princeton University Press.

Lynch, M., and W. Gabriel. 1987. Environmental tolerance. *American Naturalist* 129:283–303.

Mayr, E. 1954. Change of genetic environment and evolution. Pages 157–180 *in* J. Huxley, A. C. Hardy, and E. B. Ford, eds. *Evolution as a Process*. London: Allen and Unwin.

———. 1963. *Animal Species and Evolution*. Cambridge, MA: Harvard University Press.

———. 1982. Processes of speciation in animals. Pages 1–19 *in* C. Barigozzi, ed. *Mechanisms of speciation*. New York: Liss.

Meffert, L. M., and E. H. Bryant. 1991. Mating propensity and courtship behavior in serially bottlenecked lines of the housefly. *Evolution* 45:293–306.

Mongold J. A., A. F. Bennett, and R. E. Lenski. 1996. Evolutionary adaptation to temperature. IV. Adaptation of *Escherichia coli* at a niche boundary. *Evolution* 50:35–43.

Moya, A., A. Galiana, and F. J. Ayala. 1995. Founder-effect speciation theory: Failure of experimental corroboration. *Proceedings of the National Academy of Sciences of the USA* 92:3983–3986.

Pease, C. M., R. Lande, and J. J. Bull. 1989. A model of population growth, dispersal, and evolution in a changing environment. *Ecology* 70:1657–1664.

Platt, J. R. 1964. Strong inference: Certain systematic methods of scientific thinking may produce much more rapid progress than others. *Science* 146:347–353.

Reznick, D. N., H. Bryga, and J. A. Endler. 1990. Experimentally induced life-history evolution in a natural population. *Nature* 346:357–359.

Reznick, D., and J. A. Endler. 1982. The impact of predation on life history evolution in Trinidadian guppies (*Poecilia reticulata*). *Evolution* 36:160–177.

Rice, W. R., and E. E. Hostert. 1993. Laboratory experiments on speciation: What have we learned in 40 years? *Evolution* 47:1637–1653.

Rose, M. R., J. L. Graves, and E. W. Hutchison. 1990. The use of selection to probe patterns of pleiotropy in fitness characters. Pages 29–42 *in* F. Gilbert, ed. *Insect Life Cycles: Genetics, Evolution and Co-ordination*. New York: Springer.

Rose, M. R., H. B. Passananti, and M. Matos, eds. 2004. *Methuselah Flies: A Case Study in the Evolution of Aging*. Singapore: World Scientific.

Rundle, H. D. 2003. Divergent environments and population bottlenecks fail to generate premating isolation in *Drosophila pseudoobscura*. *Evolution* 57:2557–2565.

Sibly, R. M. 2002. Life history theory. Pages 623–627 *in* M. Pagel, ed. *The Encyclopedia of Evolution*. New York: Oxford University Press.

Stearns, S. C. 1992. *The Evolution of Life Histories*. New York: Oxford University Press.

Templeton, A. R. 1996. Experimental evidence for the genetic-transilience model of speciation. *Evolution* 50:909–915.

Wade, M. J., S. M. Shuster, and L. Stevens. 1996. Inbreeding: Its effect on response to selection for pupal weight and the heritable variance in fitness in the flour beetle, *Tribolium castaneum*. *Evolution* 50:723–733.

Whitlock, M. J., P. C. Phillips, and M. J. Wade. 1993. Gene interaction affects the additive genetic variance in subdivided populations with migration and extinction. *Evolution* 47:1758–1769.

# 3

# MODELING EXPERIMENTAL EVOLUTION USING INDIVIDUAL-BASED, VARIANCE-COMPONENTS MODELS

Derek A. Roff and Daphne J. Fairbairn

*Experimental evolution* as used in this volume encompasses both artificial selection and "laboratory" evolution in which populations are introduced into a novel environment and allowed to breed without any overt selection by the experimenter. Any selection that occurs in laboratory evolution experiments is assumed to be imposed by aspects of the novel environment. The major advantage of this latter approach over artificial selection is that the organisms are allowed to evolve relatively naturally in response to diverse selection acting on the whole phenotype, and hence the observed evolutionary processes may more closely mimic those that occur in nature (see also Futuyma and Bennett this volume; Gibbs and Gefen this volume; Huey and Rosenweig this volume; Rose and Garland this volume). In particular, allowing organisms to evolve naturally in the novel environment enables researchers to explore the coherence and coordination of traits in terms of both the immediate physiological interactions within an organism and their evolutionary potential. Artificial selection has been applied very extensively, although surprisingly almost entirely on single traits (Roff 2007; Bell 2008), whereas laboratory evolution is a relatively recent approach. It has been used to investigate evolution in cultures of unicellular eukaryotes, bacteria, and viruses (e.g., Travisano et al. 1995; Reboud and Bell 1997; Bennett and Lenski 1999; Messenger et al. 1999; Wichman et al. 2000) and in diverse multicellular organisms, such as nematodes, *Caenorhabditis elegans* (Cutter 2005), and fruit flies, *Drosophila melanogaster* (e.g., Stearns et al. 2000; Mery and Kawecki 2002; Rose et al. 2005).

The simple act of bringing organisms into the laboratory and keeping them in culture is itself a case of experimental evolution because the laboratory environment invariably differs from the native environment. The cultured organisms experience a novel selective regime, and, assuming that the requisite genetic variance is present, they are expected to evolve in response to this. This is a process of passive (i.e., not overtly selected by the experimenter) adaptation to the laboratory environment, or "domestication." One common example, observed in many insects (Danilevsky 1965), is loss of obligatory dormancy when laboratory environments are continuously favorable for growth and reproduction. Domestication of *D. melanogaster* leads to changes in both life history and physiology, although the selective factors favoring such changes are not well understood (Chippindale et al. 1997; Bochdanovits and de Jong 2003; Rose et al. 2005; Simões et al. this volume). Indeed, a significant problem in interpreting and predicting the course of experimental evolutionary experiments is the definition of what exactly is being favored in the new environment (Rose et al. 1990).

In any experimental evolution program, the researcher is faced with the problem of deciding on the population size per generation and number of generations over which to run the experiment. These issues have received considerable attention for artificial selection on single traits (Roff 1997) but remain relatively unexplored for multiple traits or laboratory evolution. The evolution of multiple traits is a primary focus of laboratory evolution experiments and studies of evolutionary changes in natural populations. Given the labor and costs of undertaking such studies, it behooves the researcher to develop a priori predictions from which to determine the appropriate experimental requirements. Genetic models of evolutionary change allow us to make such predictions.

There are three approaches to modeling the evolution of quantitative traits: (1) population-based models, (2) Mendelian-based models, and (3) variance-components models. Whereas the first approach is based on the population as the basic unit, the second two develop predictions using the individual as the basic model unit. Any of these approaches can be used to predict laboratory evolution and hence to inform experimental design. However, in this chapter, we focus on variance-components models because they are more flexible than population-based models and are easier to formulate and program than Mendelian-based models. We begin with a brief review of all three methods.

## MODELING APPROACHES

### POPULATION-BASED MODELS

Population-based models derive from the two central equations of quantitative genetics, the breeder's equation and its multivariate equivalent. For a single trait, the breeder's equation predicts the response to selection from the well-known relationship

$$R = \frac{\sigma_A^2}{\sigma_P^2} S = h^2 S, \tag{1}$$

where $R$ is response to selection (= difference between the population mean and the offspring mean), $\sigma_A^2$ is the additive genetic variance, $\sigma_P^2$ is the phenotypic variance, and $S$ is the selection differential (= the difference between the mean of the entire population before selection and the mean of the subset of selected parents). The ratio of the additive genetic to phenotypic variance is the narrow-sense heritability, $h^2$. From a modeling perspective, this equation is better recast in the form

$$X_{t+1} = (1 - h^2)X_t + h^2 X_{S,t}, \tag{2}$$

where $X_t$ is the mean trait value in the population at generation $t$ and $X_{S,t}$ is the mean value of the individuals selected in generation $t$. This model assumes nonoverlapping generations and the same value in both sexes. Modification of the equation to accommodate differences between the sexes is simple, but dealing with overlapping generations is not and is best dealt with using an individual-based model.

If the trait under study is normally distributed, or can be made so by transformation, and if the population size is large, then the above equations can be used to model the evolution of the mean value of a single trait. The equations predict that if the selection differential remains constant, the selected trait will change linearly over time. Furthermore, if equal selection is applied in the either direction, the response should be symmetrical (i.e., the rate of response should be independent of direction). Realized evolutionary trajectories frequently depart from these simple predictions, in part because traits are treated in isolation. (For a discussion of this and other circumstances in which asymmetrical and nonlinear responses may occur, see Roff 1997.)

The breeder's equation can be extended to consider multiple traits by taking into account the fact that traits are correlated due to both genetic and environmental sources. Thus, the response to selection is made up of a component due to direct selection on the trait and a component due to the indirect selection resulting from the correlation of the trait with other traits. In matrix form, the mean vector of response to selection, $\Delta \overline{Z}$, is given by

$$\Delta \overline{Z} = GP^{-1}S, \tag{3}$$

where $G$ is the matrix of additive genetic variances and covariances, $P^{-1}$ is the inverse of the matrix of phenotypic variances and covariances, and $S$ is the vector of selection differentials. The "expanded" version of this multivariate breeder's equation for two traits is

$$\begin{bmatrix} \Delta \overline{Z}_1 \\ \Delta \overline{Z}_2 \end{bmatrix} = \begin{bmatrix} \sigma^2_{G11} & \sigma^2_{G12} \\ \sigma^2_{G21} & \sigma^2_{G22} \end{bmatrix} \begin{bmatrix} \sigma^2_{P11} & \sigma^2_{P12} \\ \sigma^2_{P21} & \sigma^2_{P22} \end{bmatrix}^{-1} \begin{bmatrix} S_1 \\ S_2 \end{bmatrix}. \tag{4}$$

An important assumption of this model is that the distribution of traits is multivariate normal. Because equation (4) consists simply of series of constant terms, it defines a response to selection that is "linear" in multivariate space. Examples of the use of the population-based approach are to be found in Via and Lande (1985), Charlesworth (1990), van Tienderen (1991), Roff (1994), and Roff and Fairbairn (1999).

## MENDELIAN-BASED MODELS

The basic assumption of quantitative genetics is that quantitative traits are influenced by a large number of additively acting genes. It is this assumption that generates the equations discussed in the previous section. An alternate approach to modeling the evolution of quantitative traits is thus to use an individual-based model in which genetic transmission is explicitly modeled as Mendelian. Such a model would assume that each trait is composed of many (dozens or hundreds) of loci with two or more alleles per locus. Trait values can then be generated from the sum of the allelic values, plus a normally distributed environmental value with a mean of zero and a variance necessary to generate the required heritability. (Shared environments may influence the environmental covariance among relatives and can be readily incorporated.) Genetic correlations caused by pleiotropy are created by some genes affecting more than one trait. This approach is very computer-intensive but has the advantage that nonadditive effects can be directly programmed and mutation readily introduced. However, the results can be very sensitive to model details, such as the distribution of mutational effects, and programming can be quite difficult. Examples are given by Mani et al. (1990), Reeve (2000), Reeve and Fairbairn (2001), and Jones et al. (2003, 2004).

As noted in the previous section, the basic assumption of quantitative genetics is that a trait, $Z$, is made up of two normally distributed components,

$$Z = G + E, \tag{5}$$

where $G$ is the genetic value and $E$ is the environmental value. The two components are uncorrelated, with $G$ being distributed with mean $\mu_G$ and variance $\sigma_G^2$ and $E$ distributed with zero mean and variance $\sigma_E^2$. The assumption of uncorrelated effects may be violated in empirical data either by $G$-$E$ correlations or $G \times E$ interaction effects. (Because of the assumption of additivity, in the analysis of variation among families, $G$-$E$ correlations will appear as part of the $G$ component, whereas the $G \times E$ will be subsumed into the $E$ component.) As with common environmental effects discussed earlier, these components could be readily incorporated into the models if necessary. For simplicity, they have not been included in the models we use as illustrations in this chapter.

The phenotypic variance $\sigma_P^2$ is the sum of the genetic and environmental variances. The phenotypic value of an individual can be created by generating random normal values from normal generating functions with the appropriate means and variances. Such routines are available in most, if not all, statistical packages such as SAS or SPLUS, and in programming languages such as Matlab or Maple. A large number of individuals can typically be created by a single call: for example, in S-Plus the command rnorm(200, mean=5, sd=10) will generate 200 random normal values from a normal distribution with a mean of 5 and a standard deviation of 10. Thus, it is easy to generate individuals that satisfy the basic quantitative genetic assumptions. Genetic dominance can be introduced by using the theoretical contribution of the additive and dominance components given a known pedigree, but for simplicity, we will consider only additive effects in this chapter. We shall also assume that population sizes are large and that the genetic variances do not change as evolution proceeds. This is not a necessary requirement, but for many cases it will be a reasonable assumption and makes explanation of the approach more easily understood. The extension of equation (5) to multiple traits simply requires a move from the normal distribution to the multivariate normal distribution. Multivariate normal generating routines are also typically available in modern software packages such as R, S-Plus, and Matlab.

The advantages of the variance-components approach over the population-based approach are that changes in both means and the phenotypic distributions can be assessed and functional constraints (e.g., thresholds, see below) are readily accommodated. Application of the approach is straightforward for many phenotypic traits. However, there are three types of traits that require specialized treatment. For these traits, the phenotypic value as defined by the sum of the normally distributed additive genetic and environmental values is not the value of the phenotype that is actually expressed (hereafter, the "realized phenotype"). The first is a class of traits known as *threshold traits*, in which the realized phenotype consists of two or more discrete forms or states; examples include wing dimorphism in insects, horn dimorphism in some species of beetles (see also

Frankino et al. this volume), and susceptibility to disease (reviewed in Roff 1996). The threshold model resolves the apparent paradox of polygenic determination of discrete morphs by assuming an underlying normally distributed trait called the *liability*. Individuals with liability values exceeding a critical threshold develop into one morph, and those below the threshold develop into the alternate morph. The heritability of the liability and genetic correlations with other traits can be estimated from the frequency of the two discrete morphs and the phenotypic correlations with other traits (Roff 1997).

The second circumstance in which the realized phenotype is not the sum of the genetic and environmental values is when the trait has a limiting boundary; for example, fecundity cannot be less than zero. The threshold concept resolves this problem by assuming that values less than the limiting boundary have values equal to the limiting boundary, which in the case of fecundity would be zero. The realized phenotypic value is thus continuously distributed but may appear as a bimodal distribution as the mean trait value approaches the limiting boundary.

Finally, trait expression may be sex- or morph-specific (see also Frankino et al. this volume). For example, in most wing-dimorphic insects, the development of the flight muscles is suppressed in short-winged individuals. In a simulation model, flight muscle weight would thus be set to zero in this morph. These three categories of traits are easily incorporated into an individual-based model.

The variance-components approach has been used extensively to study the performance of statistical methods for estimating such genetic parameters as heritability and genetic correlations (e.g. Ronningen 1974; Olausson and Ronningen 1975; Roff and Preziosi 1994; Lynch 1999; Roff and Reale 2004; Roff 2001, 2006, 2007). We propose that this approach can also be extremely useful in simulating evolutionary responses in general (e.g., Roff 1975; Gilchrist 2000) and, specifically, in predicting the trajectory of experimental evolution. By taking an individual-based modeling approach, one is able to directly incorporate functional constraints, complexities of genetic architecture (e.g., the inclusion of threshold traits, continuously manifested traits, and simple Mendel single-locus models), and trait distributions that may be far from normal at the phenotypic level. We first present this approach for artificial selection on single traits and then expand it to include multivariate traits. Finally, we present an extensive example of how the approach can be applied in the design of a laboratory evolution study.

## SINGLE-TRAIT MODELS
### ARTIFICIAL SELECTION IN THE SINGLE-TRAIT MODEL

Although it is unlikely that single traits ever evolve without correlated changes in other traits, there is heuristic value in beginning with the simplest case of selection on a single trait with no correlated responses. The usual approach is to use the breeder's equation (equation [1] or [2]). Directional selection, as generally practiced in artificial selection experiments or expected in laboratory evolution, will reduce genetic variance, with mutation restoring at least some variance. Small populations will tend to lose variance through genetic drift and inbreeding. However, for populations larger than about one hundred individuals and evolution over thirty or less generations, which are desirable criteria for experimental evolution studies, changes in genetic variance and mutation may be ignored, though in principle they could be incorporated. In all the examples discussed in this chapter, we shall use large population sizes and assume a constant variance.

The breeder's equation is appropriate if one is interested primarily in changes in the mean and the assumption of normality can be made. However, if one is interested in changes in the trait distribution or if the phenotypic distribution is not normal, the breeder's equation can be cumbersome to use. In contrast, it is a relatively simple matter to simulate the response to selection using an individual-based model and follow the changing trait distribution, particularly when the manifested phenotype is decidedly not normal. As an example of such a trait, we will use flight muscle mass in long-winged sand crickets, *Gryllus firmus*. These crickets occur in both short-winged and long-winged forms. All short-winged forms lack flight muscles, and in long-winged crickets, the mass of the main flight muscles (the dorso-longitudinal muscles, hereafter DLM), is bimodally distributed (figure 3.1). In the population shown, about 44 percent of the long-winged females have no measurable DLM, while the others have larger, measurable DLM with variable mass. An excellent fit to this clearly nonnormal distribution is obtained using a truncated normal distribution with a power transformation; thus, $DLM^{3.18}$ is normally distributed with a mean of $1.5 \times 10^{-8}$, a standard deviation of $9.0 \times 10^{-8}$, and a truncation at zero (figure 3.1).

Now suppose we wish to model a population of $2N$ individuals ($N$ of each sex) with truncation selection of proportion $p$. For simplicity, we shall assume no differences between the sexes and that the trait is normally distributed on the measured scale (we address the issue raised by the DLM example after this discussion). We proceed as follows:

1. Set $\sigma_G^2$, $\sigma_E^2$, and the initial mean genetic value, $\mu_G$. In our example, these are set and 1, 1, and 0, respectively. The mean environmental value is set to zero. Operationally, it is frequently most convenient to input the heritability and phenotypic variance and compute the required components from $\sigma_G^2 = h^2 \sigma_P^2$ and $\sigma_E^2 = \sigma_P^2 - \sigma_G^2$.

2. Create four vectors, each of size $N$, two holding the additive genetic values of the males and females (say, **GM** and **GF**) and two holding the environmental values of the males and females (say, **EM** and **EF**). Genetic values are generated by drawing $2N$ random normal variates with mean $\mu_G$ and variance $\sigma_G^2$, and environmental values are generated by drawing $2N$ random normal variates with mean 0 and variance $\sigma_E^2$. Arbitrarily, the first $N$ are assigned to one sex and second to the other.

3. The phenotypic vectors—say, **PM** and **PF**—are created by adding the entries for the genetic and environmental values, so **PM**(i) = **GM**(i) + **EM**(i) and **PF**(i) = **GF**(i) + **EF**(i), where i runs from 1 to $N$.

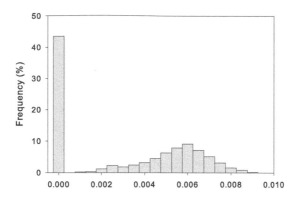

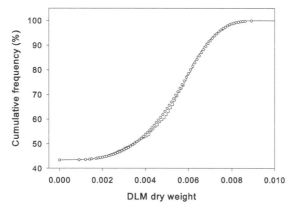

FIGURE 3.1
Distribution of dorso-longitudinal muscle weight in a sample ($N = 538$) of macropterous sand crickets. The lower panel shows the observed cumulative frequency distribution (o) and the fitted curve (—) obtained by assuming that DLM[3.18] is normally distributed with negative values set to zero.

4. Merge the phenotypic and genetic vectors into two matrices, one for males—say, **Male**—and one for females—say, **Female**—where the first column contains the phenotypic value and the second column the additive genetic values (this is arbitrary).

5. Sort each matrix into descending order according to the phenotypic values. The first 1 to $pN$ entries of each matrix are the selected parents for the next generation. Move these into separate matrices, say, **S.Male** and **S.Female**.

6. Assuming random mating of the selected males and females, there are several methods of generating the next generation. The simplest is to update the population mean genetic value of each sex using the mean of the genetic values of the selected males and females. A new generation is then created by cycling back to step 2.

An example assuming a normally distributed trait with a heritability of 0.5 and the top 30 percent of one hundred males and one hundred females being selected is shown in figure 3.2A. Because an individual-based model uses a finite population size, its results are stochastic. The mean phenotypic and genetic values of the total population and the selected parents thus vary among replicates, though all of the replicates show the directional change in trait value expected.

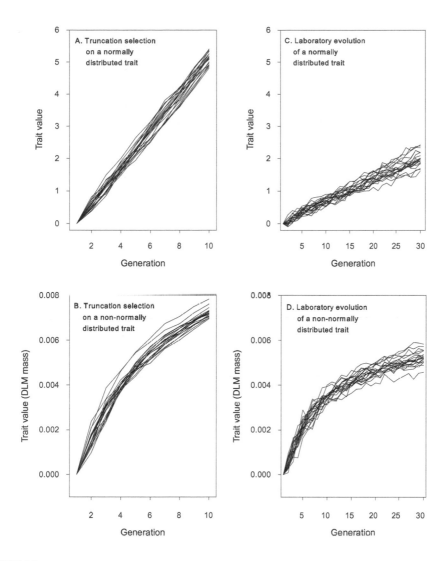

FIGURE 3.2

Replicate simulation runs predicting response to artificial truncation selection (A, B) or to natural selection in the laboratory environment (laboratory evolution, C, D). A, Truncation selection on a normally distributed trait with a heritability of 0.5 and the top 30 percent of one hundred males and one hundred females being selected. Variation in the initial mean trait values has been removed by subtracting the initial value from each generation of each replicate. B, Truncation selection as in (A) but on a trait that follows the distribution shown in figure 3.1. C, Laboratory evolution of a normally distributed trait with heritability 0.5 and fitness directly proportional to trait value. To avoid negative values, the initial mean is set at 6, which is three standard deviations from zero. Initial values have been subtracted as in (A). D, Laboratory evolution as in (C) but for a trait that follows the distribution shown in figure 3.1.

There is an extensive theory on the response of a single trait to truncation selection and the requirements in terms of sample size and generations required to obtain a statistically significant change (Roff 1997). Thus, there is little need for construction of individual-based models for simple cases such as that just illustrated here. However, this example serves to illustrate the general approach; all further examples are extensions of this approach.

If the trait, such as DLM mass, is nonnormally distributed on the measured scale but can be normalized by transformation, then the model must be adjusted in step 3 as follows: After creation of the phenotypic values on the underlying scale (i.e., the transformed scale), the phenotypic values on the realized scale (i.e., the scale of measurement), which we shall refer to as the *realized phenotypes*, are calculated. In the case of DLM mass, we apply the following rules:

$$\text{If } \mathbf{PM}(i) \leq 0, \text{ then } \mathbf{PM}(i) = 0;$$

$$\text{if } \mathbf{PM}(i) > 0, \text{ then } \mathbf{PM}(i) = \mathbf{PM}(i)^{1/3.18},$$

with the same rules applied to the females. Truncation selection is then done on the realized phenotypes, but the mean genetic value of the next generation is calculated using the genetic values on the underlying scale (note that because the selection is based on ranks, the transformation does not affect which individuals are selected). Simulation results for twenty replicates assuming one hundred males and one hundred females and $p = 0.3$ (as in the previous example) show that the expected change in mean trait value is nonlinear, with a reduction in rate of response over time (figure 3.2B). A visualization of the trait distribution across generations shows clearly a shift from a bimodal distribution to a unimodal, more normal distribution (figure 3.3). A simulation analysis is very valuable in this case in predicting a priori the result of artificial selection on both the mean trait value and its changing distribution.

## LABORATORY EVOLUTION IN THE SINGLE-TRAIT MODEL

Artificial selection typically uses truncation selection (i.e., a predetermined proportion of individuals from the top or bottom of the distribution are selected to be parents of the next generation), but this is unlikely to be the mode of selection in laboratory evolution. A more plausible mode of selection is for all individuals to contribute some offspring to the next generation, with their relative contribution being a function of the trait under selection,

$$\mu_{G,Male} = \sum_{i=1}^{N} \frac{\omega_{Male}\{PM(i)\}}{\sum_{j=1}^{N} \omega_{Male}\{PM(j)\}} GM(i)$$

$$\mu_{G,Female} = \sum_{i=1}^{N} \frac{\omega_{Female}\{PF(i)\}}{\sum_{j=1}^{N} \omega_{Female}\{PF(j)\}} GF(i)$$

$$(6)$$

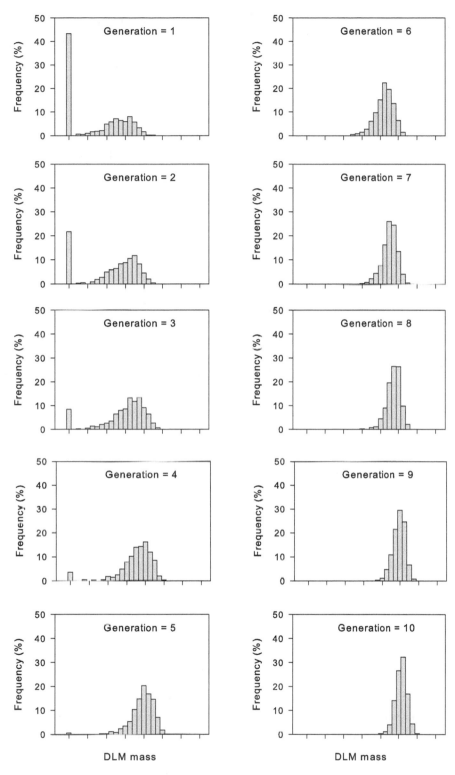

FIGURE 3.3

The change in the distribution of DLM weight as truncation selection for increased weight is applied. Parameter values are as in figure 3.2 except that population size is increased to five hundred males and five hundred females.

where $\omega_{Male}\{\ \}$, $\omega_{Female}\{\ \}$ are the fitness functions of males and females, respectively. In the long term, selection is likely to be stabilizing, although initially there will probably be a significant and possibly dominating linear component. Suppose, for example, the laboratory environment favors a change in body size (i.e., natural selection in the novel environment favors larger or smaller body size), and lifetime or net reproductive success is therefore proportional to body size. Assuming, for simplicity, that males and females have the same fitness function, we can incorporate this type of selection by eliminating step 5 (i.e., there is no need to sort the matrices) and calculating the mean genetic values of the two sexes in the next generation as

$$\mu_{G,Male} = \sum_{i=1}^{N} \frac{PM(i)}{\sum_{j=1}^{N} PM(j)} GM(i)$$

$$\mu_{G,Female} = \sum_{i=1}^{N} \frac{PF(i)}{\sum_{j=1}^{N} PF(j)} GF(i)$$

(7)

Assuming random mating and equal sex ratio, the mean genetic value of the population is

$$\mu_G = \frac{1}{2}\left(\mu_{G,Male} + \mu_{G,Female}\right). \tag{8}$$

Unlike the case of the artificial selection model, where selection intensity remains constant and hence response to selection is linear, in laboratory evolution the selection intensity will, in general, vary and the subsequent response will be nonlinear, particularly if selection is stabilizing. To test for nonlinear responses in the present simulation, we fitted linear and quadratic equations to the trait values over generations. In twelve of the twenty replicates, a significant quadratic term was found (figure 3.2C, though the curvature is clearly very slight).

An important question that can be addressed by this individual-based model is "How many generations are required to detect a statistical change in the mean phenotypic value?" For example, to ascertain the power to detect a change in five generations for the model presented above, we ran one thousand replications testing each run for a significant change using linear regression of trait value on generation. A generally accepted desirable level of power is 0.80 (Cohen 1988; Crawley 2002). In 75 percent of the replicates, a statistically significant change was detected, suggesting that five generations would not be an acceptable time span over which to follow the laboratory population. However, power jumped to 92 percent when six generations were monitored. Thus, the simulation model tells us that increasing the duration of the experiment by one generation would provide more than sufficient power to detect a significant response.

To model laboratory evolution for the nonnormal trait, we incremented each phenotypic value by 0.003 to avoid zero fitnesses. The general pattern of the simulated laboratory evolution was very similar to that for truncation selection (cf. figures 3.2B, 3.2D). In only four cases of the one thousand replications was a nonsignificant regression obtained when the population response was measured across five generations, indicating very high power (99.6 percent) to detect laboratory evolution in this circumstance. Laboratory evolution was also very evident in the shifting distribution of the realized phenotypic trait (not shown), as previously found in truncation selection (figure 3.3), adding another dimension to the analysis.

## MULTIPLE-TRAIT MODELS
### ARTIFICIAL SELECTION IN THE MULTIPLE-TRAIT MODEL

The response to selection of multiple traits is made up of a component due to direct selection on the trait and a component due to the indirect selection resulting from the correlation of the trait with other traits (equation [4]). To illustrate how the method previously given for modeling selection on a single trait can be expanded to include multiple traits, we shall first consider truncation selection on one or both of two traits, with a proportion $p$ of $N$ males and $N$ females selected based on their phenotypic rank.

1. As before, we require the additive genetic values and the environmental values. For simplicity, we again assume that these components are uncorrelated and each follows a multivariate normal distribution. The values of these distributions must be initially set. Most "canned" subroutines that calculate the multivariate normal distributions will require the input of a vector containing the means and a matrix containing the variances and covariances. To compute the variances and covariances, it is generally most convenient to input the heritabilities, correlations, and phenotypic variances. The required values are then estimated from

$$
\begin{aligned}
\sigma_{GII}^2 &= h_I^2 \sigma_{PII}^2 \\
\sigma_{EII}^2 &= \sigma_{PII}^2 - \sigma_{AII}^2 \\
\sigma_{PIJ}^2 &= r_{PIJ}\sqrt{\sigma_{PII}^2 \sigma_{PJJ}^2} \\
\sigma_{PJI}^2 &= \sigma_{PIJ}^2 \\
\sigma_{GIJ}^2 &= r_{GIJ}\sqrt{\sigma_{GII}^2 \sigma_{GJJ}^2} \\
\sigma_{GJI}^2 &= \sigma_{GIJ}^2 \\
\sigma_{EIJ}^2 &= \sigma_{PIJ}^2 - \sigma_{GIJ}^2 \\
\sigma_{EJI}^2 &= \sigma_{EIJ}^2
\end{aligned}
\tag{9}
$$

where the subscripts $I$ and $J$ denote the traits (thus, $II$ and $JJ$ denote the diagonals of matrix, which comprise the variances, while $IJ$ and $JI$ denote the off-diagonals, which comprise the covariances). The value of $\sigma_{EII}^2$ must be positive, which is not assured by all values of $\sigma_{PII}^2$ and $\sigma_{GII}^2$. It is therefore necessary to place a check in the program to ensure that the environmental variance is positive. Covariances can be negative, and hence no check is required for them.

2. Calculate the vectors of genetic and environmental values for the males and females by calling the multivariate normal generating subroutine (these are available in Matlab, SPLUS, R, and some Fortran packages). There is an important restriction that all the eigenvalues must be positive, or the subroutine will not work. This can be understood by a consideration of the geometry of a two-trait model. Variance-covariance matrices are symmetrical and can be converted, using principal components analysis, to a set of orthogonal axes designated by the eigenvectors. Each axis is made up of a linear combination of the individual traits (the principal component scores), and there are as many axes as there are traits. Thus, as illustrated in figure 3.4, for two traits, there are two axes, the major and minor axis. The variance in each principal component is given by the eigenvalue. If an eigenvalue is zero, there is no

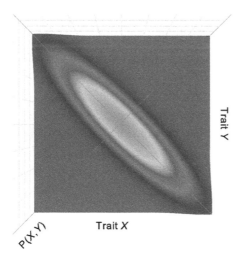

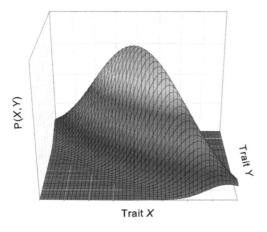

FIGURE 3.4
A quantitative genetic view of covariation between traits $X$ and $Y$ shown from two orientations. Each plot shows the bivariate normal probability distribution of values for each trait (where $P(X, Y)$ denotes the probability of the $XY$ combination), with the left plot showing it in "contour" perspective and the right plot showing it in "3D" mode. The solid lines in the left plot show the major and minor axes.

variance in the respective direction. Because the eigenvalues define variances, which by definition are positive, they cannot, when describing a multivariate normal distribution, be negative. Unfortunately, there appears to be no simple algorithm to find and fix the offending component when negative eigenvalues occur.

3. Merge the vectors of genetic and environmental values (the original values, not the principal components scores) into "male" and "female" matrices (e.g., column 1 for the genetic values and column 2 for the environmental), calculate the phenotypic value as the sum of the genetic and environmental values, and place these in the third column.

4. Apply selection. The method here will vary. In the simplest case, selection will be applied to only one trait. For this case, we sort the matrices using the phenotypic values of the target trait. For bivariate selection, we might use an index composed of some combination of the two variables: for example, one might use $X + Y$, where $X$ and $Y$ are the two standardized variables (Hazel 1943; Bulmer 1985). An alternative method is independent culling in which an independent selection threshold is set for each trait and individuals are selected only if they lie above all thresholds (Young and Weiler 1960; Bulmer 1985).

5. Calculate the mean genetic values for the selected parents, and cycle back to step 2 until the requisite number of generations have been completed.

As explained earlier, the predicted response in the means is a linear change with time, but we can also make a prediction concerning the covariation between the two traits. The phenotypic relationship between two traits can be described empirically by the simple linear regressions (OLS) between the two traits,

$$
Y = \left( \mu_Y - \frac{\sigma_{PXY}}{\sigma_{PX}^2}\mu_X \right) - \left( \frac{\sigma_{PXY}}{\sigma_{PX}^2} \right)X + \varepsilon
$$

$$
X = \left( \mu_X - \frac{\sigma_{PXY}}{\sigma_{PY}^2}\mu_Y \right) - \left( \frac{\sigma_{PXY}}{\sigma_{PY}^2} \right)Y + \varepsilon
$$

(10)

where $\mu_X, \mu_Y$ are the means of traits $X$ and $Y$, respectively; $\sigma_{PXY}$ is the phenotypic covariance between traits $X$ and $Y$, $\sigma_{PX}^2$ is the phenotypic variance of trait $X$, and $\varepsilon$ is a normally distributed error term (Roff et al. 2002). The first terms in parentheses define the intercept of the regression line, and the terms in the second set of parentheses define the slope. Because the designation of which trait is the dependent or independent variable is largely arbitrary, a better statistical model might be to define the covariation in terms of the structural relationship between the two variables. Two candidates to estimate this relationship are the major axis and reduced major axis regression, of which the second has better statistical properties (McArdle 1988). The formulae for the reduced major axis (RMA) equations for the present case (for positive relationships the signs would be reversed) are

$$Y = \left(\mu_Y + \frac{\sigma_{PY}}{\sigma_{PX}}\mu_X\right) - \left(\frac{\sigma_{PY}}{\sigma_{PX}}\right)X$$

$$X = \left(\mu_X + \frac{\sigma_{PX}}{\sigma_{PY}}\mu_Y\right) - \left(\frac{\sigma_{PX}}{\sigma_{PY}}\right)Y$$

(11)

Because the variances and covariances are assumed to remain constant, under either equation (10) or (11), the slope of the covariance function under short-term selection is expected to remain constant (Roff et al. 2002). However, as selection changes the mean trait values, $\mu_Y$ and $\mu_X$, it is expected to also change the intercepts of the covariance functions because the intercepts are functions of the means. (In principle, the two trait means could change such that the intercept value remained constant, but this is highly unlikely.) This analysis thus predicts that, under short-term or weak selection, the covariance function is expected to evolve by a shift in the intercept alone, defined as either the regression line or principal axis (Roff et al. 2002).

To illustrate this procedure, we simulated two traits, $X$ and $Y$, with the following arbitrary but realistic parameters: $h_X^2 = 0.4$, $h_Y^2 = 0.5$, $r_G = -0.7$, $r_P = -0.6$, $\mu_X = \mu_Y = 3$, $\sigma_{PX}^2 = 1$, $\sigma_{PY}^2 = 0.5$. Population size was set at five hundred males and five hundred females, with no sex-specific differences in trait expression. A sample distribution with the two least-squares regression lines and the RMA line is shown in figure 3.5A (top-left panel). Truncation selection of the top 30 percent was applied to $X$.

For comparison with the simulation results, we can predict the response to selection analytically using standard quantitative genetic equations (e.g., Roff 1997),

$$\mu_{X,t+1} = \mu_{X,t} + h_X^2 S$$

$$\mu_{Y,t+1} = \mu_{Y,t} + r_G h_X h_Y \frac{\sigma_{PY}}{\sigma_{PX}} S$$

(12)

where $S$ is the selection differential. Substituting in the parameter values gives predicted values for the means of traits $X$ and $Y$ of 8.22 and 0.69, respectively, after nine generations of selection. For OLS regressions, the intercept of $Y$ on $X$ is predicted to decline from 4.27 to 4.18 and the intercept of $X$ on $Y$ to increase from 5.55 to 8.80. The change in intercept for $Y$ on $X$ is very small and may not be statistically detectable, whereas that of $X$ on $Y$ is relatively large and more likely to be detected. For the RMA equations, the respective changes are 5.12 to 7.24 ($Y$ on $X$) and 6.50 to 9.19 ($X$ on $Y$). Neither change is as great as that for the simple linear regression of $X$ on $Y$.

The simulation results (figure 3.5B–H) show the expected linear response in trait means and reach the predicted values at generation 9 (figure 3.5B). The changes in the intercepts of the OLS regression also match the analytical predictions: the intercept of $Y$ on $X$ declines only slightly, while that of $X$ on $Y$ increases dramatically (figure 3.5C, E). As expected, the OLS slopes show no trend, although they do fluctuate because of

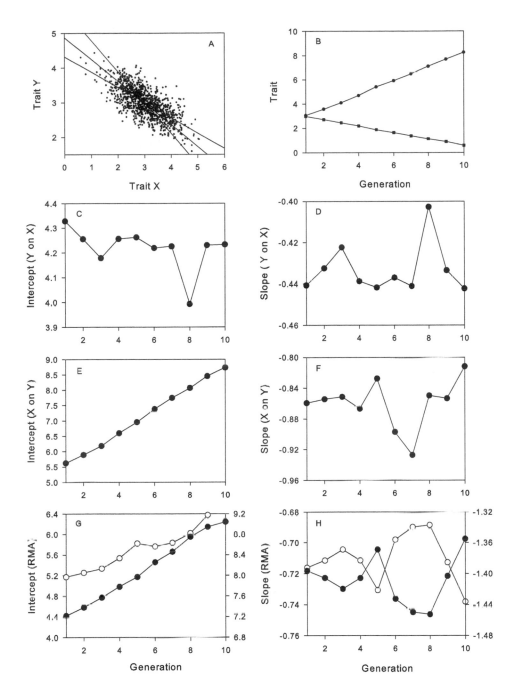

FIGURE 3.5

Simulation results of positive truncation selection on a single trait $X$ that is genetically correlated with another trait, $Y$. A, Scatter plot showing the phenotypic covariation between the two traits with the two least squares regression and reduced major axis lines. B, Mean values for traits $X$ (■) and $Y$ (•) as a function of generation of selection showing that selection for an increase in trait $X$ is accompanied by a correlated decrease in trait $Y$. B–F, The intercepts and slopes of the two least-square regression lines as a function of generation. G and H, The intercepts and slopes for the reduced major axis (RMA) regressions (• = $X$ on $Y$; ○ = $Y$ on $X$).

sampling error (figure 3.5D, F). The intercepts from the RMA analysis both showed the expected response, with the $Y$ on $X$ being somewhat more variable (figure 3.5G), and the slopes showing the expected lack of response (note that the slopes are simply the mirror images of each other: figure 3.5H). This simulation shows the advantage of using an individual-based, variance component model to investigate a priori the consequences of selection on trait values. With such a model, both the magnitude and the expected variance in the response can be predicted and used as a guide for experimental design and later statistical hypothesis testing. For example, based on the results of our example, one would be wise to use the regression of $X$ on $Y$ rather than $Y$ on $X$ to test the prediction that the intercept but not the slope will change under selection because the former is predicted to change more dramatically and with less variance. In the present example, the selection regime is simple enough that the appropriate regression can be determined theoretically; but in more complex models, simulation may be necessary. Even in the present example, simulation is useful to estimate the variability in the response, which is not easily done theoretically.

To examine the effect of simultaneous selection on both traits, we ran the simulation with truncation selection being applied to the index

$$\frac{(X - \mu_{PX})}{\sigma_{PX}} + \frac{(Y - \mu_{PY})}{\sigma_{PY}},$$

which is the standardized sum of the two trait values. As in the single-trait selection example, all estimates of the intercepts and slopes showed the predicted direction of responses (figure 3.6).

A very different pattern can emerge if one of the traits is not normally distributed. To illustrate this, we examine truncation selection on a trait that is distributed as shown in figure 3.1A and figure 3.3A for DLM weight, which we shall call trait $X$. We assume that trait $Y$ is negatively correlated to trait $X$ and is normally distributed. Because the distribution trait $X$ is bimodal, the bivariate plot also displays a bimodal pattern (figure 3.7A). Heritabilities and correlations were set as in the preceding example, except that the heritability of trait $X$ refers to the heritability on the underlying normally distributed scale (e.g., figure 3.1B). The means and phenotypic variances were set to approximately the values measured in our laboratory population for DLM weight ($X$) and ovary weight ($Y$), ($\mu_{PY} = 0.12$ g, $\mu_{PX} = 1.5 \times 10^{-8}$ g, $\sigma^2_{PY} = 0.002$, $\sigma^2_{PX} = 0.81 \times 10^{-14}$, where the $X$ values refer to the underlying scale). Truncation selection of 30 percent favoring larger $X$ produces a strong, nonlinear positive response in $X$ and an almost linear correlated negative response in $Y$ (figure 3.7B). As in the previous case, there are no consistent changes in the intercept and slope of the regressions of $Y$ on $X$ (figure 3.7C, D), but the regression of $X$ on $Y$ shows the predicted increase in intercept (figure 3.7E, F). In contrast to the previous simulation, both the intercepts and slopes of the RMA regressions changed, but the direction of change was reversed when the dependent and independent variables were reversed (figure 3.7G, H). This change in the slopes is not a consequence of the

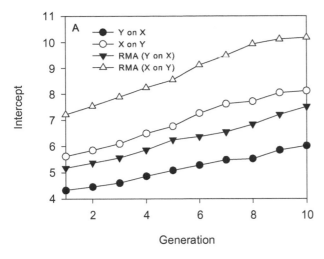

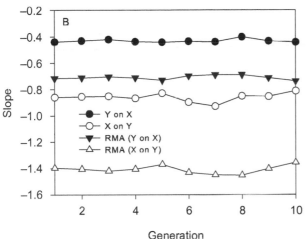

FIGURE 3.6
Simulation results for
changes in the intercept (A)
and slope (B) in response
to simultaneous truncation
selection on two normally
distributed traits, $X$ and $Y$,
using the standardized
sum of the two trait values
as the index of selection.
Circles: ordinary least-squares
regressions; triangles: reduced
major axis regressions.

bimodality in the distribution as it also occurs if animals with DLM equal to zero are excluded (results not shown). The reason for the change in the RMA slope is that although the phenotypic standard deviation of $X$ on the underlying scale remains constant, the expressed trait value has a decreasing phenotypic value (0.003 in generation 1, declining to 0.0006 by generation 10). This effect may also be evident in the OLS regressions, though in the present example any effect is slight. This example illustrates the profound influence that trait distribution can have on response to selection and the need to carefully model the proposed selection regime.

## LABORATORY EVOLUTION IN THE MULTIPLE-TRAIT MODEL

With multiple traits, fitness may be related to a single trait or to a composite of several traits; the particular form will depend on the circumstances of the study. However, correlations among traits will modulate the evolutionary responses, and thus all traits must

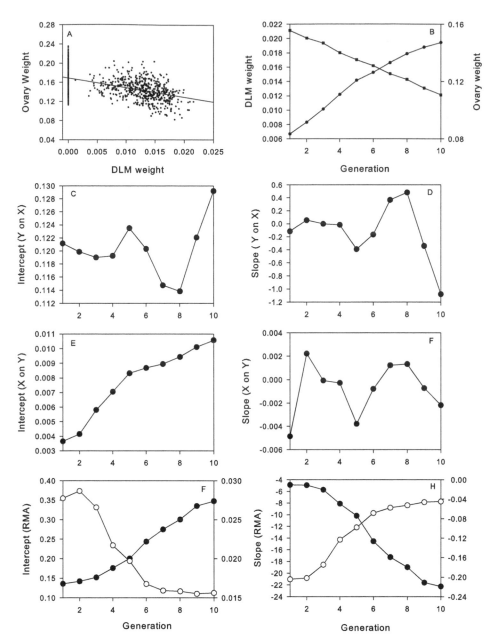

FIGURE 3.7

Results of an individual-based, variance-components model simulating truncation selection on a trait, X, with a nonnormal distribution as in figure 3.1, with correlated responses in trait Y. Proportion selected = 30 percent. A, Covariation of Y and X. B, Response of X (circles) and Y (squares) to selection. C and D, Changes in the slope and intercept of the OLS regression of Y on X (E) – (F) Changes in the slope and intercept of the OLS regression of X on Y, (G) – (H) Changes in the slope and intercept of the RMA regressions (● = X on Y; o = Y on X).

be considered. To illustrate how multiple traits can be dealt with in a laboratory evolution model, we present a case study of the laboratory evolution in the sand cricket.

## A CASE STUDY: PREDICTING LABORATORY EVOLUTION IN THE SAND CRICKET, *GRYLLUS FIRMUS*

The prediction of laboratory evolution of multiple traits is a central element of the study of experimental evolution. To this point, we have used simplified simulations to illustrate methodology and general trends, but these necessarily lack the complexity of a true experimental situation. In this section, we present a more complex, realistic model that shows how one can incorporate functional constraints into an individual-based, quantitative genetic simulation model and use the model to determine statistical requirements such as the appropriate number of generations over which to run the experiment. The system we use to illustrate the process is the evolution of reduced migratory capacity in a laboratory culture of the sand cricket, *Gryllus firmus*. This is a particularly apt example because it contains components that preclude the use of population-based or Mendelian-based models. It is also apt because the relevant traits and their interactions have been extensively studied and quantified, providing accurate estimates for inclusion in the model. This is an important consideration when there are multiple interactions.

### PROBLEM DESCRIPTION

*Gyllus firmus*, the sand or beach cricket, is a relatively large gryllid (adult weight ≈ 0.7 g) inhabiting sandy areas of the coastal southeastern United States (Alexander 1968; Kevan 1980; Harrison 1985). There are two distinct wing morphs, a flightless, short-winged (micropterous, hereafter SW) morph and a long-winged (macropterous, hereafter LW) morph that is typically capable of flight, at least during early adult life. Because the habitat of the sand cricket is ephemeral, only cricket lineages that produce at least some migrants (i.e., individuals that move to new patches) can persist in the long term. This need for migration among patches explains the existence of the LW morph. However, the co-occurrence of SW crickets is more problematic because LW individuals need not fly and hence can avoid the presumed costs of migration by simply remaining in their natal habitat if that habitat is still suitable (Fairbairn and Desranleau 1987). The presence of SW morphs in this and the numerous other insect species in which wing dimorphism is found provides key evidence that there is a cost not only of migratory flight itself but also of possession of the morphological and physiological capability of flight (Roff 1984, 1994; Roff and Fairbairn 1991).

In LW sand crickets, the production and maintenance of the DLM, as well as the production and storage of flight fuels, are energetically very expensive (Zera and Denno 1997; Marden 2000), as indicated by the increased metabolic rate of the LW morph (Zera and Harshman 2001; Crnokrak and Roff 2002). This high energetic cost of flight capability may compromise allocation of energy to reproductive traits such as fecundity in females and calling rate and testes size in males, reducing the reproductive success of LW relative to SW morphs. Trade-offs between flight capability and reproductive investment have been demonstrated in both male and female *G. firmus* (for females, see Roff 1984; Roff et al. 1997, 2002; Roff and Gelinas 2003; for males, see Crnokrak and Roff 1995, 1998a, 1998b, 2000; Roff et al. 2003). As noted previously (figure 3.7A), there is a highly significant negative relationship between fecundity (measured as ovary weight at day 7 posteclosion) and the weight of the DLM (Roff et al. 2002; Roff and Gelinas 2003). This trade-off is genetically determined and responsive to natural and artificial selection (reviewed in Roff and Fairbairn 2001, 2007a; Roff et al. 2002).

Flight capability in the sand cricket is not retained throughout adult life, the flight muscles being histolyzed in the first couple of weeks after the final molt. Histolysis of the flight muscles is accompanied by an increased rate of reproduction (Roff 1989), and the trade-off between fecundity and DLM weight reflects, at least to some extent, variation in the timing of histolysis of the flight muscles (Stirling et al. 2001). Thus, in the sand cricket, there are two trade-offs between flight capability and reproductive capacity. First, there is the trade-off between wing morph and reproductive capacity, expressed as the frequency of LW and SW adults. Second, within the LW morph, there is a trade-off between maintenance of fully functional DLM and reproductive capacity. These trade-offs involve a suite of morphological (e.g., LW vs. SW), physiological (e.g., state of the DLM, ovary development), and life-history (e.g., timing of the switch between migratory and reproductive modes) traits that coevolve in response to selection favoring increased or decreased migratory propensity (Fairbairn and Desranleau 1987; Fairbairn and Roff 1990; Roff and Fairbairn 1999; Roff et al. 1999, 2002).

In the natural habitat of *G. firmus*, selection favors a balanced polymorphism of wing morphs, with SW morphs favored in the short term within favorable habitats but LW morphs favored in the longer term. However, in the laboratory, there is no selection favoring flight because all reproduction occurs within the culture cages. Therefore, evolution in the laboratory environment would be expected to result in (1) a decrease in the frequency of the LW morph, (2) an increase in the proportion of LW individuals showing histolyzed or nonfunctional flight muscles, and (3) an increase in reproductive capacity manifest by an increase in ovary weight in females (Roff and Fairbairn 2007b).

Given estimates of the heritabilities and genetic correlations among the various traits, can we predict how long it would take for a statistically significant change in these parameters under laboratory culture? If we predict that the time span is fifty generations, then, because the number of generations per year in the lab is three to five, we are unlikely to detect changes within a time frame set by the usual granting agency. On the other hand, if changes are predicted to occur within five to fifteen generations, then we are in an excellent position to monitor significant evolutionary change and measure changes in other traits, such as changes in metabolic rate, within a reasonable time frame (two to three years).

## MODEL DESCRIPTION

Previously, we constructed a quantitative genetic simulation model using genetic parameter estimates from earlier data (Roff and Fairbairn 2007b). We extend this model here
to include changes in the proportion of SW and LW morphs and to statistically predict
how many generations are required before changes are statistically significant with a
power of at least 80 percent. We model the evolutionary change in four traits: wing
morph, ovary weight, and DLM weight and condition. DLM condition refers to whether
the muscles fully or partially histolyzed, and hence are not functional, or are fully intact
and functional. Wing morph is assessed at eclosion to the adult stage, while the other
three traits are measured on the seventh day following eclosion. The two DLM variables
are highly correlated, but we retained both in the model because stepwise multiple
regression, followed by cross-validation (i.e., splitting the data into multiple parts and
sequentially using one part to generate the regression model and the remaining data to
test the predictive power of the model; Roff 2006), indicated that both contribute independently to the trade-off between muscle mass and fecundity in LW females.

The suite of traits associated with wing dimorphism in the crickets includes the three
types of traits described earlier for which the sum of the normally distributed additive
genetic and environmental values does not represent the realized phenotype. Wing
dimorphism (LW/SW) and DLM condition (functional/nonfunctional) are examples of
threshold traits, the first class of traits.

The weight of the DLM and fecundity are examples of the second type of trait, those
with limiting boundaries. As described previously (figure 3.1), these traits have a limiting lower boundary because they cannot be less than zero. We resolved this problem by
assuming that values less than the limiting boundary have values equal to the limiting
boundary, here zero. The realized phenotypic value is thus continuously distributed but
may appear as a bimodal distribution as the mean trait value approaches the limiting
boundary.

Differences between SW and LW individuals represent the third category of traits,
morph-specific traits expressed in only one morph. In the present case, development of
the flight muscles is suppressed in SW individuals, and thus DLM weight and histolysis
status are both zero in these individuals. The underlying genetic and environmental
variances for these traits are masked in SW individuals but expressed in their LW relatives. Because the fecundity of SW females is genetically correlated with these two traits,
selection that, say, increased the proportion of LW females in the population (and hence
increased average DLM weight and reduced the frequency of muscle histolysis) would
result in a decrease in the fecundity of both LW and SW females (Roff et al. 1999). These
effects are incorporated into the individual-based, variance-components model through
the covariances between wing morph liability and the other traits. The expression of
pleiotropic genes may be conditional on the "environment" in which they occur (e.g., LW
or SW individuals), but they are still passed from parent to offspring.

In modeling the trajectory of adaptation to the laboratory environment, we assumed that the primary selective advantage of the SW morph in the laboratory environment is the increased fecundity of females resulting from the allocation of energy away from development and maintenance of flight capability. Our previous studies have uncovered no postmating paternity advantage for SW males (Roff and Fairbairn 1993), and the mating advantage accruing to them by virtue of higher calling rates is unlikely to be significant in the high densities of cage culture. Therefore, for the purposes of our predictive model, we assume that reproductive success among males in the laboratory stock cages does not differ between wing morphs. Given that males are unselected, the mean genetic value of offspring from the *i*th female is equal to one-half of her genetic value, $GF(i)$, plus one-half the population mean genetic value, $\mu_G$, which is the expected contribution of the male. The mean of the offspring is thus calculated by modifying equation (7) to

$$\mu_G = \frac{1}{2} \left\{ \mu_G + \sum_{i=1}^{N} \frac{PF(i)}{\sum\limits_{j=1}^{N} PF(j)} GF(i) \right\}. \tag{13}$$

The relative number of offspring that a female contributes to the next generation is equal to her fecundity ($PF_i$), as measured by her ovary weight (at seven days posteclosion), divided by the total fecundity of all $N$ females in the population.

## BACKGROUND ANALYSES AND DATA

The stock of *G. firmus* described in this section was collected from sandy areas in Gainesville, Florida. We measured proportion LW, DLM weight and condition and fecundity for females in the first two generations in the lab (sample 1) and at about the fifth generation (sample 2). Wing morph expression is temperature-sensitive in *G. firmus*, and we were able to suppress the expression of the LW morph by maintaining the laboratory stock at 25°C. This effectively eliminated selection on wing morph in our laboratory stock for more than ten generations (Roff and Fairbairn 2007b). We then raised the rearing temperature to 28°C, which resulted in expression of the LW phenotype and generated approximately an equal ratio of the two wing morphs. With the LW morph expressed, we predicted selection favoring the SW morph and hence laboratory evolution. To test this hypothesis, we measured our key variables after approximately ten generations at the higher temperature (sample 3), and we predicted no difference between samples 1 and 2, but significant evolutionary change in sample 3. The results confirmed our predictions: (1) the frequency of the LW morph significantly decreased (87.3 percent, 81.1 percent, and 46.5 percent, respectively); (2) within LW females, those from sample 3 had the lowest DLM weight and highest frequency of nonfunctional DLM; and (3) within LW females, those from sample 3 had the highest ovary weight. Because of the genetic correlation between the sexes, we also predicted and observed a smaller but statistically

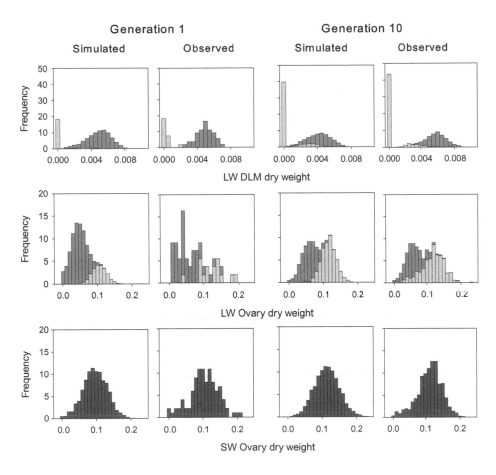

FIGURE 3.8

Predicted and observed distributions of trait values in female *G. firmus* during ten generations of laboratory evolution. In rows 1 and 2, grey bars indicate females with nonfunctional DLM; black bars indicate females with functional DLM.

significant increase in fecundity of the SW morph. The details of this analysis can be found in Roff and Fairbairn (2007b).

The laboratory evolution described here was largely serendipitous, the move to 28°C having been forced on us by changes in rearing facilities. There was no replication, and the exact number of generations over which the evolutionary changes took place is not known. Thus, whereas the results of the simulation model are consistent with the observed evolution, it would be desirable to test our predictions more rigorously by running a replicated experiment in which the changes are monitored every generation. This raises the question introduced at the beginning of this section: How many generations, given a realistic laboratory population size, should be required to observe a significant change? Is such an experiment feasible?

The heritabilities and genetic correlations used in the present model are given in table 3.1. The simulation model produces changes in wing morph frequency and trait distributions that show an excellent fit to the changes observed in our laboratory stock, assuming ten generations of evolution at 28°C (figure 3.8). This confirms that the model makes realistic predictions for changes in our target variables, but it does not permit estimates of power. To make these predictions, we used a realistic population size of three hundred per sex and ran multiple simulations to determine the average trajectories and the power of statistical tests to detect changes in morphological and physiological parameters. In a real-life situation, not all individuals can be measured each generation, so we assumed that a random sample of one hundred males and one hundred females would be measured each generation for proportion LW, ovary weight, DLM weight, and DLM status. The parameters of the regression of ovary weight on DLM weight were estimated from these data.

To determine the expected variation in trait values over fifteen generations, we ran one thousand replicates of the each simulation, commencing with the observed set of genetic parameters (table 3.1). Because each generation was created using a multivariate random normal distribution with means obtained from the selected populations, there was stochastic variation in the generated distributions, which mimics that expected in the real population. Using the data from the one thousand replicates, we calculated the expected response and the 95 percent confidence envelope of the response estimated for each generation as $\bar{x} \pm 2$ SD, where $\bar{x}$ is the mean value of the one thousand replicates at the given generation, and SD is the standard deviation. As can be seen in figure 3.9, with the exception of the intercept and slope of the trade-off function (central panels), there is no overlap between the upper confidence limit at the start of the simulation and the lower confidence limit at the fifteenth generation. The widening confidence region in the trade-off function as generations increase results from the declining number of LW females with measurable DLM. Both the mean intercept and slope of the trade-off

TABLE 3.1   Estimated Heritabilities (Diagonal Elements), Genetic Correlations (above Diagonal), and Phenotypic Correlations (below Diagonal) in Female *Gryllus firmus* for the Traits Incorporated into the Simulation Model

|  | Ovary Weight | Wing Morph Liability | DLM Weight | Histolysis Liability |
|---|---|---|---|---|
| Ovary weight | 0.41 | −0.67 | −0.72 | −0.71 |
| Wing morph liability | −0.25 | 0.71 | 0.96 | 0.95 |
| DLM weight | −0.63 | 0.60 | 0.47 | 0.99 |
| Histolysis liability | −0.66 | 0.53 | 0.79 | 0.70 |

NOTE: Sire estimates are for illustration only and from preliminary analysis of an unpublished half-sib experiment with 125 sires, 3 dams/sire, and an average of 5.3 female offspring per dam.

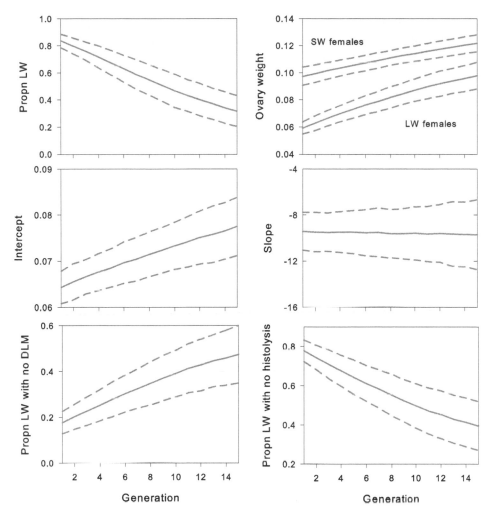

Predicted changes in trait values during fifteen generations of laboratory evolution in a simulated stock culture of the sand cricket, *Gryllus firmus*. Central panels show the intercept and slope of the regression of ovary weight on DLM weight. Dashed lines indicate ±2 standard deviations.

function change over the generations of laboratory evolution, but the change in the slope is minimal and not expected to be statistically detectable. The lack of response in the slope differs from the earlier results for laboratory evolution of the bivariate model; this difference is a consequence of changes in both parameter values and model components.

To estimate the statistical power of the design to detect changes in trait values, for each trait within each simulation, we subsampled one hundred females, as proposed in the experimental protocol described earlier, and calculated the regression of trait value on generation for generations 1 through 10 and for generations 1 through 15. We then estimated the power of a single replicate to detect change as the proportion of times a

TABLE 3.2 Estimated Power to Detect Evolutionary Change in Phenotypic Trait Values after 10 or 15 Generations of Laboratory Evolution in a Simulated Population of the Sand Cricket, *Gryllus firmus*

| | Power to Detect Change | |
| --- | --- | --- |
| Trait | *Generation 10* | *Generation 15* |
| Proportion LW | 1.00 | 1.00 |
| LW ovary weight | 1.00 | 1.00 |
| SW ovary weight | 0.69 | 1.00 |
| Proportion LW with no DLM | 0.91 | 0.99 |
| Proportion LW with fully functional DLM | 0.95 | 0.98 |
| Intercept of trade-off function | 0.78 | 0.98 |

significant regression was obtained. As previously noted, a generally accepted desirable level of power is 0.80 (Cohen 1988; Crawley 2002), which is satisfied for all but SW ovary weight and the trade-off intercept after ten generations, and for all traits after fifteen generations (table 3.2). These results clearly suggest that testing hypotheses about laboratory evolution in this system should be feasible within a three- to five-year time frame.

## CONCLUSION

Many studies have demonstrated that experimental evolution is a useful approach to understanding evolutionary change, but few have shown that such changes are quantitatively predictable. The testing of predictions for artificial selection on single traits has a long history, but the extension of the breeder's equation to selection on multiple traits has received relatively little testing (Roff 2007). Attempts to predict the time course of laboratory evolution are even fewer and have considered only the type of responses expected (selection for reduced diapause), not the evolutionary trajectories. The approach we have outlined in this chapter provides a framework for investigating the time course of evolutionary responses in experimental evolution studies.

As with any approach, some caveats must be borne in mind. For pedagogical simplicity, we have ignored some components, such as inbreeding depression, mutation, and genetic drift, but these can be added if necessary. We have assumed a constant G matrix, which is realistic for short-term evolutionary changes (about ten to twenty generations) but not for longer-term changes (see also Travisano this volume). The question of how the G matrix should change under selection is one that still requires investigation, both theoretically and empirically. Variability in response will depend not only on the variances and covariances but also on population size: for small population sizes, it will be necessary to incorporate genetic drift. An important assumption, present in all models, is that all relevant traits have been included in the model. Failure to include an important trait could lead to deviation between observed and predicted trajectories and should be one of the first assumptions checked. The advantages of using individual-based modeling using a variance-components approach over population-based or Mendelian models include computational simplicity and the ability to incorporate nonnormal distributions, functional constraints, and other deviations from the standard assumptions of quantitative genetic models. Furthermore, as illustrated by our example of laboratory evolution in *Gryllus firmus*, individual-based modeling permits estimates of the variance in the predicted trajectories and hence enables researchers to place confidence limits around the predicted responses. These tools have the potential to greatly enhance quantitative hypothesis testing in experimental evolution not only by facilitating accurate estimates of the power of a given experimental design, but also by permitting quantitative evaluation of specific models of the genetic architecture of evolutionary change.

## SUMMARY

Models for experimental evolution can be divided into three categories: population-based models, Mendelian-based models, and variance-components models. The first category focuses on changes in mean values, whereas the second two are individual-based models and permit analysis of changes in both the mean and the distribution of trait values. Because of their ease of implementation, we suggest that variance-components models are highly suitable for modeling experimental evolution and for determination of experimental design criteria, including sample size and duration of the experiment. This chapter discusses how to implement such models for both single- and multiple-trait cases and presents a detailed example using the evolution of reduced migratory capacity in a laboratory culture of the sand cricket, *Gryllus firmus*.

## ACKNOWLEDGMENTS

We thank the editors for inviting us to contribute to this volume, and we are grateful to Dr. George Gilchrist and two anonymous reviewers for their careful reviews of a preliminary draft of the chapter. This research was supported by National Science Foundation grant DEB-0445140.

## REFERENCES

Aggrey, S. E., C. Y. Lin, and K. M. Cheng. 1995. Size of breeding populations required for selection programs. *Theoretical and Applied Genetics* 91:553–556.

Alexander, R. D. 1968. Life cycles, specialization and related phenomena in crickets. *Quarterly Review of Biology* 43:1–41.

Bell, G. 2008. *Selection: The Mechanism of Evolution.* 2nd ed. Oxford: Oxford University Press.

Bennett, A. F., and R. E. Lenski. 1999. Experimental evolution and its role in evolutionary physiology. *American Zoologist* 39:346–362.

Bochdanovits, Z., and G. de Jong. 2003. Experimental evolution in *Drosophila melanogaster*: Interaction of temperature and food quality selection regimes. *Evolution* 57:1829–1836.

Bulmer, M. G. 1985. *The Mathematical Theory of Quantitative Genetics*. Oxford: Clarendon.

Charlesworth, B. 1990. Optimization models, quantitative genetics, and mutation. *Evolution* 44:520–538.

Chippindale, A. K., J. A. Alipaz, H. W. Chen, and M. R. Rose. 1997. Experimental evolution of accelerated development in *Drosophila*. 1. Development speed and larval survival. *Evolution* 51:1536–1551.

Cohen, J. 1988. *Statistical Power Analysis for the Behavioral Sciences*. Hillsdale, NJ: Erlbaum.

Crawley, M. J. 2002. *Statistical Computing: An Introduction to Data Analysis Using S-Plus*. Chichester, U.K.: Wiley.

Crnokrak, P., and D. A. Roff. 1995. Fitness differences associated with calling behaviour in the two wing morphs of male sand crickets, *Gryllus firmus*. *Animal Behaviour* 50:1475–1481.

———. 1998a. The contingency of fitness: An analysis of food restriction on the macroptery-reproduction trade-off in *Gryllus firmus*. *Animal Behaviour* 56:433–441.

———. 1998b. The genetic basis of the trade-off between calling and wing morph in males of the cricket, *Gryllus firmus*. *Evolution* 52:1111–1118.

———. 2000. The trade-off to macroptery in the cricket *Gryllus firmus*: A path analysis in males. *Journal of Evolutionary Biology* 13:396–408.

———. 2002. Trade-offs to flight capability in *Gryllus firmus*: The influence of whole-organism respiration rate on fitness. *Journal of Evolutionary Biology* 15:388–398.

Cutter, A. D. 2005. Mutation and the experimental evolution of outcrossing in *Caenorhabditis elegans*. *Journal of Evolutionary Biology* 18:27–34.

Danilevsky, A. S. 1965. *Photoperiodism and Seasonal Development of Insects*. London: Oliver and Boyd.

Fairbairn, D., and L. Desranleau. 1987. Flight threshold, wing muscle histolysis, and alary polymorphism: Correlated traits for dispersal tendency in the Gerridae. *Ecological Entomology* 12:13–24.

Fairbairn D. J., and D. A. Roff. 1990. Genetic correlations among traits determining migratory tendency in the sand cricket, *Gryllus firmus*. *Evolution* 44:1787–1795.

Frankham, R. 1990. Are responses to artificial selection for reproductive fitness characters consistently asymmetrical? *Genetical Research* 56:35–42.

Gilchrist, G. W. 2000. The evolution of thermal sensitivity in changing environments. Pages 55–70 *in* J. M. S. K. B. Storey, ed. *Cell and Molecular Responses to Stress. Vol. 1. Environmental Stressors and Gene Responses*. Amsterdam: Elsevier Science.

Harrison, R. G. 1985. Barriers to gene exchange between closely related cricket species. II. Life cycle variation and temporal isolation. *Evolution* 39:244–259.

Hazel, L. N. 1943. The genetic basis for constructing selection indices. *Genetics* 28:476–490.

Hill, W. G. 1985. Effects of population size on response to short and long term selection. *Zeitschrift für Tierzuchtung und Zuchtungsbiologie* 102:161–173.

Hill, W. G., and A. Caballero. 1992. Artificial selection experiments. *Annual Review of Ecology and Systematics* 23:287–310.

Jones, A. G., S. J. Arnold., and R. Borger. 2003. Stability of the G-matrix in a population experiencing pleiotropic mutation, stabilizing selection, and genetic drift. *Evolution* 57:1747–1760.

———. 2004. Evolution and stability of the G-matrix on a landscape with a moving optimum. *Evolution* 58:1639–1654.

Kevan, D. M. 1980. The taxonomic status of the Bermuda beach cricket (Orthopera: Gryllidae). *Systematic Entomologist* 5:83–95.

Lande, R. 1979. Quantitative genetic analysis of multivariate evolution applied to brain: body size allometry. *Evolution* 33:402–416.

———. 1982. A quantitative genetic theory of life history evolution. *Ecology* 63:607–615.

Lynch, M. 1999. Estimating genetic correlations in natural populations. *Genetical Research* 74:255–264.

Mani, G. S., B. C. Clarke, and P. R. Sheltom, P. R. 1990. A model of quantitative traits under frequency-dependent balancing selection. *Proceedings of the Royal Society of London B, Biological Sciences* 240:15–28.

Marden, J. H. 2000. Variability in the size, composition, and function of insect flight muscles. *Annual Review of Physiology* 62:157–178.

McArdle, B. H. 1988. The structural relationship: Regression in biology. *Canadian Journal of Zoology* 66:2329–2339.

Mery, F., and T. J. Kawecki. 2002. Experimental evolution of learning ability in fruit flies. *Proceedings of the National Academy of Sciences of the USA* 99:14274–14279.

Messenger, S. L., I. J. Molineux, and J. J. Bull. 1999. Virulence evolution in a virus obeys a trade-off. *Proceedings of the Royal Society of London B, Biological Sciences* 266:397–404.

Nicholas, F. W. 1980. Size of population required for artificial selection. *Genetical Research* 35:85–105.

Nomura, T. 1997. A simulation study on variation in response to selection and population size required for selection programmes. *Journal of Animal Breeding and Genetics—Zeitschrift für Tierzuchtung und Zuchtungsbiologie* 114:13–21.

Olausson, A., and K. Ronningen. 1975. Estimation of genetic parameters for threshold characters. *Acta Agricultural Scandinavica* 25:201–208.

Reboud, X., and G. Bell. 1997. Experimental evolution in *Chlamydomonas*. III. Evolution of specialist and generalist types in environments that vary in space and time. *Heredity* 78:507–514.

Reeve, J. P. 2000. Predicting long-term response to selection. *Genetical Research* 75:83–94.

Reeve, J. P., and D. J. Fairbairn. 2001. Predicting the evolution of sexual size dimorphism. *Journal of Evolutionary Biology* 14:244–254.

Roff, D. A. 1975. Population stability and the evolution of dispersal in a heterogeneous environment. *Oecologia* 19:217–237.

———. 1984. The cost of being able to fly: A study of wing polymorphism in two species of crickets. *Oecologia* 63:30–37.

———. 1986. The genetic basis of wing dimorphism in the sand cricket, *Gryllus firmus*, and its relevance to the evolution of wing dimorphisms in insects. *Heredity* 57:221–231.

———. 1989. Exaptation and the evolution of dealation in insects. *Journal of Evolutionary Biology* 2:109–123.

———. 1990. Selection for changes in the incidence of wing dimorphism in *Gryllus firmus*. *Heredity* 65:163–168.

———. 1994. Habitat persistence and the evolution of wing dimorphism in insects. *American Naturalist* 144:772–798.

———. 1996. The evolution of threshold traits in animals. *Quarterly Review of Biology* 71:3–35.

———. 1997. *Evolutionary Quantitative Genetics*. New York: Chapman and Hall.

———. 2001. The threshold model as a general purpose normalizing transformation. *Heredity* 86:404–411.

———. 2002. *Life History Evolution*. Sunderland, MA: Sinauer.

———. 2006. *Introduction to Computer-Intensive Methods of Data Analysis in Biology*. Cambridge: Cambridge University Press.

———. 2008. Comparing sire and dam estimates of heritability: Jackknife and likelihood approaches. *Heredity* 100:32–38.

Roff, D. A., P. Crnokrak, and D. J. Fairbairn. 2003. The evolution of trade-offs: geographic variation in call duration and flight ability in the sand cricket, *Gryllus firmus*. *Journal of Evolutionary Biology* 16:744–753.

Roff, D. A., and D. J. Fairbairn. 1991. Wing dimorphisms and the evolution of migratory polymorphisms among the Insecta. *American Zoologist* 31:243–251.

———. 1993. The evolution of alternative morphologies: Fitness and wing morphology in male sand crickets. *Evolution* 47:1572–1584.

———. 1999. Predicting correlated responses in natural populations: Changes in JHE activity in the Bermuda population of the sand cricket, *Gryllus firmus*. *Heredity* 83:440–450.

———. 2001. The genetic basis of dispersal and migration and its consequences for the evolution of correlated traits. Pages 191–202 *in* C. Clobert, J. Nichols, J. D. Danchin, and A. Dhondt, eds. *Causes, Consequences and Mechanisms of Dispersal at the Individual, Population and Community Level*. Oxford: Oxford University Press.

———. 2007a. The evolution and genetics of migration in insects. *BioScience* 57:155–164.

———. 2007b. Laboratory evolution of the migratory polymorphism in the sand cricket: Combining physiology with quantitative genetics. *Physiological and Biochemical Zoology* 80:358–369.

Roff, D. A., and M. B. Gelinas. 2003. Phenotypic plasticity and the evolution of trade-offs: the quantitative genetics of resource allocation in the wing dimorphic cricket, *Gryllus firmus*. *Journal of Evolutionary Biology* 16:55–63.

Roff, D. A., S. Mostowy, and D. J. Fairbairn. 2002. The evolution of trade-offs: Testing predictions on response to selection and environmental variation. *Evolution* 56:84–95.

Roff, D. A., and R. Preziosi. 1994. The estimation of the genetic correlation: The use of the jackknife. *Heredity* 73:544–548.

Roff, D. A., and D. Reale. 2004. The quantitative genetics of fluctuating asymmetry: A comparison of two models. *Evolution* 58:47–58.

Roff, D. A., G. Stirling, and D. J. Fairbairn. 1997. The evolution of threshold traits: A quantitative genetic analysis of the physiological and life history correlates of wing dimorphism in the sand cricket. *Evolution* 51:1910–1919.

Roff, D. A., J. Tucker, G. Stirling, and D. J. Fairbairn. 1999. The evolution of threshold traits: Effects of selection on fecundity and correlated response in wing dimorphism in the sand cricket. *Journal of Evolutionary Biology* 12:535–546.

Ronningen, K. 1974. Monte Carlo simulation of statistical-biological models which are of interest in animal breeding. *Acta Agricultural Scandanavica* 24:135–142.

Rose, M. R., J. L. Graves, and E. W. Hutchinson. 1990. The use of selection to probe patterns of pleiotropy in fitness characters. Pages 29–42 *in* F. Gilbert, ed. *Insect Life Cycles: Genetics, Evolution and Co-ordination.* London: Springer.

Rose, M. R., H. B. Passananti, A. K. Chippindale, J. P. Phelan, M. Matos, H. Teotonio, and L. D. Mueller. 2005. The effects of evolution are local: Evidence from experimental evolution in *Drosophila. Integrative and Comparative Biology* 45:486–491.

Stearns, S. C., M. Ackermann, M. Doebeli, and M. Kaiser. 2000. Experimental evolution of aging, growth, and reproduction in fruitflies. *Proceedings of the National Academy of Sciences of the USA* 97:3309–3313.

Stirling, G., D. J. Fairbairn, S. Jensen, and D. A. Roff. 2001. Does a negative genetic correlation between wing morph and early fecundity imply a functional constraint in *Gryllus firmus? Evolutionary Ecology Research* 3:157–177.

Travisano, M., V. F. Mongold, and R. E. Lenski. 1995. Long-term experimental evolution in *Escherichia coli.* III. Variation among replicate populations in correlated responses to novel environments. *Evolution* 49:189–200.

Van Tienderen, P. H. 1991. Evolution of generalists and specialists in spatially heterogeneous environments. *Evolution* 45:1317–1331.

Via, S., and R. Lande. 1985. Genotype-environment interaction and the evolution of phenotypic plasticity. *Evolution* 39:505–522.

Wichman, H. A., L. A. Scott, C. D. Yarber, and J. J. Bull. 2000. Experimental evolution recapitulates natural evolution. *Philosophical Transactions of the Royal Society of London B, Biological Sciences* 355:1677–1684.

Young, S. S. Y., and H. Weiler. 1960. Selection for two correlated traits by independent culling levels. *Journal of Genetics* 57:329 338.

Zera, A. J., and R. F. Denno. 1997. Physiology and ecology of dispersal polymorphism in insects. *Annual Review of Entomology* 42:207–230.

Zera, A. J., and L. G. Harshman. 2001. The physiology of life history trade-offs in animals. *Annual Review in Ecology and Systematics* 32:95–126.

# PART II

# TYPES OF EXPERIMENTAL EVOLUTION

# 4

# EXPERIMENTAL EVOLUTION FROM THE BOTTOM UP

Daniel E. Dykhuizen and Anthony M. Dean

A BOTTOM-UP ANALYSIS OF THE LACTOSE SYSTEM

OTHER APPROACHES

DISCUSSION

*Experimental Evolution: Concepts, Methods, and Applications of Selection Experiments,* edited by Theodore Garland, Jr., and Michael R. Rose. Copyright © by the Regents of the University of California. All rights of reproduction in any form reserved.

One of the most important uses of experimental evolution is the study of natural selection, its causes and its consequences. This involves mapping genotype onto phenotype (genetic architecture; Hanson 2006) and mapping phenotype onto fitness. Implicitly or explicitly, these mappings also include any impact the environment has on each.

Natural selection is studied using two very different approaches in experimental evolution, which we call "top down" and "bottom up." In the top-down approach, an experimenter sets up the environmental conditions, such as media, temperature, and growth regime, and allows the culture to evolve over thousands of generations (e.g., see Forde and Jessup this volume; Futuyma and Bennett this volume; Gibbs and Gefen this volume; Huey and Rosenzweig this volume). The usual output measured is change in fitness, but it also can be change in DNA sequence, change in protein expression, or change in any physiological process. The patterns that emerge are then compared to those expected from evolutionary theory.

In contrast, the bottom-up approach attempts to predict fitness from first principles. For example, two strains, genetically identical except for a specific genetic difference of interest, are placed in competition against each other, and fitness is measured over a short period, usually less than a hundred generations and before a selective sweep incorporates other genetic changes. Next, the genetic difference or the environment is changed, and fitness is measured again. The observed pattern of fitness differences is then compared against the pattern predicted from biochemistry and molecular biology.

In a classic example of the top-down approach, Bennett et al. (1992) studied temperature adaptation in *Escherichia coli*. A culture of *E. coli* was first evolved for two thousand generations in a serial transfer experiment—each day the culture was diluted 1 to 100 into fresh minimal medium plus glucose and shaken (not stirred) in a flask at 37°C (Lenski et al. 1991). A single bacterium from this culture was grown up as the common ancestor of multiple cultures that were then propagated for a further two thousand generations under identical conditions except for temperature: some were grown at 32°C, some at 37°C, some at 42°C, and some were shifted daily between 32° and 42°C. At the end of the experiment, fitness increases were measured in direct competition against the common ancestor (stored at –80°C).

Cultures grown at 37°C showed little increase in fitness under all temperature regimes. Those grown at 32°C showed large increases in fitness at this temperature and minor changes in fitness at the other temperatures. Likewise, those evolved at 42°C showed large increases in fitness at 42°C and minor changes in fitness at the other temperatures. Cultures that were subjected to daily shifts in temperature increased in fitness at all temperatures, though never as much as those grown at a single temperature, except for those grown at 37°C. The temperature-shift cultures were also fitter at 37°C than the culture evolved solely at this temperature (Bennett and Lenski 1999).

These results show that adaptation to temperature is specific to the culture temperature. Specialization arises simply because cultures adapt to their local environment—there is no evidence of fitness trade-offs at other temperatures. Additional experiments show that thermal niche breath does not change as cultures adapt to temperatures near the upper and lower thermal limits, and that evolution in a changing environment does not increase phenotypic plasticity (Bennett and Lenski 1999). Thus, well-conceived top-down experiments can be used to test and reject theories in evolutionary biology.

In an early example of a bottom-up experiment, Dykhuizen (1978) tested the energy conservation hypothesis, which states that organisms should benefit in direct proportion to the energy saved by not synthesizing compounds already present in the environment. *E. coli* strains, each carrying a gene disruption that abolished synthesis of tryptophan (an amino acid), were each placed in competition with a wild-type strain. Selection in the presence of excess tryptophan, or a tryptophan precursor, was measured during the intense competition for limiting glucose imposed by slow growth in chemostats.

As expected, selection favored the tryptophan auxotrophs. However, the response to selection was a hundred times that expected based solely on the cost of making all the tryptophan required by an *E. coli* cell. Providing indole instead of tryptophan was predicted to reduce selection response by 17 percent. Providing anthranilate, an earlier precursor of tryptophan, was predicted to reduce selection response by 33 percent. In fact, both precursors reduced the selective response by 50 percent. The energy conservation hypothesis was rejected decisively.

Were it not for the quantitative predictions provided by the bottom-up approach, the energy conservation hypothesis might well have been accepted. After all, selection always went in the direction expected. Similarly, the energy conservation hypothesis predicts that expression of useless protein should be deleterious, and, indeed, constitutive expression of the lactose operon during starvation on glucose is selected against (Novick and Weiner 1957; Andrews and Hageman 1976; Dykhuizen and Davies 1980). Yet recent experiments, designed to tease apart the various contributions to fitness, show that this selection is caused by the effects of transcription and/or translation and not by the energetic costs associated with sequestering amino acids in useless protein (Stoebel et al. 2008).

This short introduction suggests that our ideas of the causes of natural selection are prejudiced and rather naive. Furthermore, careful experiments that dissect the mechanistic basis of natural selection provide decisive tests of hypotheses. In this chapter, we will describe a system that we have used to understand the nature of natural selection (Dykhuizen and Dean 1990).

## A BOTTOM-UP ANALYSIS OF THE LACTOSE SYSTEM

We chose to study lactose metabolism by *E. coli* because it provides a tractable experimental system where mechanistic predictions about the direction and intensity of selection can be tested decisively. As will be seen, most of the characteristics expected of complex genetic architectures arise in this simple system.

*The Lactose Operon*   The *lac* operon (figure 4.1A) has long served as a model of prokaryotic gene expression (Jacob and Monod 1961; Miller and Reznikoff 1978). In the absence

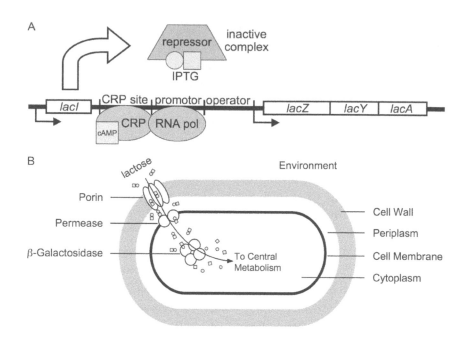

FIGURE 4.1

A, the lactose operon and its regulation; B, the lactose pathway (after Dykhuizen and Dean 1990).

of inducer, the *lacI* encoded repressor binds to the *lac* operator, physically blocking transcription. When inducer binds to the repressor, it produces an inactive complex with low affinity for the operator. With the blockage removed, transcription proceeds to synthesize a polycistronic *lacZYA* mRNA.

The natural inducer of the operon is allolactose, which is both synthesized from lactose by β-galactosidase and also hydrolyzed by it to glucose and galactose. Another natural substrate, galactosyl-glycerol, induces expression directly (Boos 1982). Isopropyl-thio-β-galactoside (IPTG) is an effective inducer of the lactose operon that cannot be metabolized by β-galactosidase and that we use to induce regulated operons.

Full induction involves a second regulatory system, the cAMP-CRP control system that regulates catabolite repression. In starving cells (e.g., cells growing in a chemostat), the cyclic AMP receptor protein (CRP) binds abundant cAMP. The resulting complex binds the promotor upstream of the RNA polymerase site to facilitate transcription initiation.

*The Lactose Pathway*   The lactose pathway (figure 4.1B) has three steps: passive diffusion of lactose through the porin pores of the outer cell wall into the periplasm, H$^+$ coupled import of lactose by the *lacY* encoded permease into the cytoplasm, and irreversible hydrolysis by the *lacZ*-encoded β-galactosidase. The products, glucose and galactose, enter central metabolism directly. The *lac* operon thus encodes only two of the three steps in the pathway.

*The Chemostat Environment*   Natural selection happens in an environment. In these experiments, the environment used is a phosphate buffered inorganic minimal medium with sugar for both carbon and energy as the sole limiting resource in a 30-milliliter Kubitschek-style chemostat (Dykhuizen 1993). Sterile hydrated air is pumped into the base of the chemostat to mix and aerate the culture. The generation time is set at two hours; the temperature at 37°C; and cell density, which is determined by the concentration of sugar in the fresh medium entering the chemostat, at $7.5 \times 10^7$ cells/milliliter. Thus, cells compete for a single limiting sugar (or sugars), and those that acquire the most have a higher nutrient flux, grow faster, and are selected for.

*From Fitness to Flux*   As Monod (1942) first demonstrated, growth rate ($\mu$) is proportional to the concentration of a growth-limiting nutrient. This implies

$$\mu = YJ, \tag{1}$$

where $Y$ is the yield coefficient (the number of "moles" of cells produced per mole of lactose consumed), and $J$ is the flux of lactose into the cell (the number of "moles" of lactose consumed per "mole" of cells per second). Defining relative fitness ($w_{wt}^i$) as the growth rate of strain $i$ relative to wild type, we find

$$w_{wt}^i = \mu_i/\mu_{wt} = Y_i J_i/Y_{wt} J_{wt} = J_i/J_{wt} = j_{wt}^i. \tag{2}$$

Mutations that affect flux do not impact yield, so the yield coefficients cancel. During starvation on lactose, we expect relative fitness to be determined by relative flux ($w_{wt}^i = j_{wt}^i$).

*From Flux to Enzyme Activities*   Following Kacser and Burns (1973, 1981), the flux of lactose through the pathway ($J$, the *in vivo* rate of lactose hydrolysis) is described by

$$J = \frac{L}{\dfrac{1}{D^{por}} + \left(\dfrac{K_m^{perm}}{V_{max}^{perm}}\right) \Big/ K_{eq}^{por} + \left(\dfrac{K_m^{\beta-gal}}{V_{max}^{\beta-gal}}\right) \Big/ (K_{eq}^{por} \cdot K_{eq}^{perm})} \tag{3}$$

(Dean 1989), where $L$ is the concentration of lactose in the environment, $D^{por}$ is the pseudo-first-order rate constant of lactose diffusion through the porin pores, the $K_m$'s are the Michaelis constants of the permease *(perm)* and $\beta$-galactosidase *($\beta$-gal)* enzymes, and the $V_{max}$'s are the maximum velocities (calculated on a per-cell basis) at each step. $K_{eq}^{por} = 1$ is the equilibrium constant associated with the passive diffusion lactose through the porins. $K_{eq}^{perm} - 444$ is the apparent equilibrium constant associated with H$^+$ symport of lactose by the permease. Lactose hydrolysis is essentially irreversible ($K_{eq}^{\beta-gal} \to \infty$), and downstream enzymes do not appear in this equation.

Equation (1) has the form

$$J = L/(1/E^{por} + 1/E^{perm} + 1/E^{\beta-gal}), \tag{4}$$

where the $E^j$'s are the "activities" at each step in the pathway. Equation (4) is essentially Ohm's law: $I = V/R = V/(1/c_1 + 1/c_2 + 1/c_3)$, where $I$ is current, $V$ is voltage, $R = \Sigma r_i$ is the total resistance of the circuit, and the $c_i = 1/r_i$ are the conductances of the components. In this analogy, flux is current, environmental lactose is voltage, and enzyme activities are conductances. Altering the activity of an enzyme by mutation is the equivalent of adjusting the volume on your stereo.

*Assumptions*   Equation (1) is a model of pure scramble competition for limiting lactose and is valid so long as there is no cross-feeding (where a substance excreted from one cell is used for growth by another) and no interference (where a substance excreted by one cell interferes with growth by another). In the presence of a strain metabolizing lactose, *lac⁻* strains are washed from the chemostat at the theoretical washout rate (Dean et al. 1988). Hence, neither cross-feeding nor interference occur in this experimental system.

Equation (1) also assumes that cells do not die, that mutations affect flux and not yield, and that no other mutations in the genomes affect fitness. Viable counts reveal that cells starving in chemostats have no detectable death rate (Dean et al. 1988). Mutations in *lac* do not affect yield, which remains constant regardless of sugar and genotype (Dykhuizen and Davies 1980). P1 transduction of *lac* operons into the *Δlac* strain DD320 ensures that genomic backgrounds are identical (Dykhuizen and Dean 1994).

Equation (3) is valid so long as *lac* expression is constitutive and the permease and the β-galactosidase operate far from saturation. Constitutive expression is ensured by using *lacI⁻* mutants or adding a small quantity of IPTG (a nonmetabolizable inducer) to the growth medium. Starvation growth in chemostats with lactose as the sole limiting nutrient ensures that both the permease and β-galactosidase operate far from saturation.

*The Biochemical Basis of Fitness*   Results from kinetic studies and competition experiments show that relative fitness is proportional to relative flux (figure 4.2)—namely,

$$w_{wt}^i = j_{wt}^i = \frac{\left( \frac{I}{D^{por}} + \left( \frac{K_m^{perm}}{V_{max}^{perm}} \right) \Big/ K_{eq}^{por} + \left( \frac{K_m^{\beta-gal}}{V_{max}^{\beta-gal}} \right) \Big/ (K_{eq}^{por} \cdot K_{eq}^{perm}) \right)_{wt}}{\left( \frac{I}{D^{por}} + \left( \frac{K_m^{perm}}{V_{max}^{perm}} \right) \Big/ K_{eq}^{por} + \left( \frac{K_m^{\beta-gal}}{V_{max}^{\beta-gal}} \right) \Big/ (K_{eq}^{por} \cdot K_{eq}^{perm}) \right)_{i}}. \tag{5}$$

Understand that there are no unknowns in equation (5)—*everything* has been measured. We did not fit the model—the straight line in figure 4.2 is the theoretical expectation. Understand that equation (5) is not an *ad hoc* guess at some plausible relationship (e.g., diversity begets stability). It is a statement of mechanism derived *ab initio*. In this experimental system, natural selection is reduced to biochemistry, and biochemistry can be used to predict the direction and intensity of natural selection precisely.

Fitness is a hyperbolic function of enzyme activity (figure 4.3). Wild-type β-galactosidase lies on a fitness plateau. Substantial changes in β-galactosidase activity have little effect on fitness even though the enzyme contributes to a flux under intense directional selection. The take-home message is clear: functional variation is necessary, but not sufficient, for selection to act. Neither the wild-type permease nor its porins lie near a fitness plateau (figure 4.3). The prediction that selection should target increases in permease activity and porin diffusibility (Dean 1989) has been experimentally confirmed

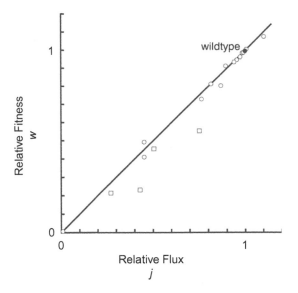

FIGURE 4.2
Relative fitness is directly proportional to relative flux and can be predicted solely from a knowledge of biochemistry. Wild-type K12 operon (dot), K12 mutants in β-galactosidase and permease (circles), and mutants of evolved β-galactosidase (*ebg*, squares). The data are from Dean et al. (1986), Dykhuizen et al. (1987), and Dean (1989).

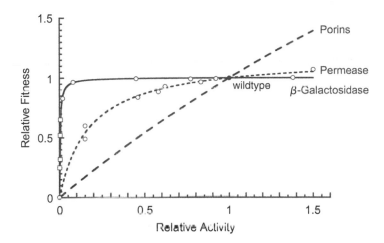

FIGURE 4.3
The same fitness data as in figure 4.2 but plotted as a function of relative enzyme activity (after Dykhuizen and Dean 1990).

(Tsen et al. 1996; Zhang and Ferenci 1999). Even in a pathway crucial to fitness, mutations having similar effects on activity at different steps can be subject to very different modes of evolution: both selection and neutrality.

*A "Just So" Story*   Why is β-galactosidase on a fitness plateau, whereas the permease and the porins are nowhere near one? Porins are protein pores embedded in the outer membrane of gram-negative bacteria that allow certain small molecules to diffuse from the environment into the periplasmic space while excluding other small molecules. In particular, they exclude gut bile salts that, were they to gain access to the bacteria's inner membrane, would emulsify it and kill the cell. The need to prevent bile salts from entering the periplasmic space evidently constrains the rate of lactose diffusion through the porins in natural environments.

Cells starving on lactose die when presented with a sudden excess of lactose (Dykhuizen and Hartl 1978; Wilson et al. 1981). During steady-state growth, the proton motive force is used to drive lactose into the cell up a concentration gradient. However, a sudden excess of lactose will just as easily drive protons into the cell. The rest of metabolism, still operating under starvation conditions, lacks the energy resources needed to regenerate the proton motive force which dissipates, killing the cells. Intermittent feeding by suckling mammalian infants may create the conditions needed for lactose killing—a sudden excess of lactose might select against starved gut *E. coli* expressing permeases of high activity.

The activities of the porins and permease have been constrained in the evolutionary history of naturally occurring bacteria. What constrains the activity of $\beta$-galactosidase? Maybe nothing. Perhaps directional selection has pushed $\beta$-galactosidase activity so far onto the fitness plateau that a limit of adaptation has been reached where further increases are selectively neutral. In this example, neutral evolution might be regarded as a consequence of the long-term action of strong directional natural selection (Hartl et al. 1985).

*The Control of Fitness*   Can laboratory selection drive all three steps to a limit of adaptation? Equation (5) can be rewritten as

$$w^i_{wt} = \frac{1}{\left( C^{por}_{wt}/e^{por}_i + C^{perm}_{wt}/e^{perm}_i + C^{\beta-gal}_{wt}/e^{\beta-gal}_i \right)}, \tag{6}$$

where $e^j_i$ is the activity at step $j$ in strain $i$ relative to that in the wild type, and $C^j_{wt} = dw/de^j_{wt}$ is the control coefficient at step $j$ in the wild type. As a tangent to a fitness curve at the point occupied by the wild type (figure 4.3), a control coefficient measures the degree to which a step limits flux (Kacser and Burns 1981; Dean 1989). For small changes in relative activity ($\delta e$), the selection coefficient is approximately $\delta w \approx C\delta e$. A small control coefficient implies that an enzyme lies in, or at least close to, its limit of adaptation. A large control coefficient implies that mutant enzymes will have functional consequences that might be subject to selection.

Each control coefficient has the form

$$C^j_{wt} = \frac{1/E^j_{wt}}{1/E^{por}_{wt} + 1/E^{perm}_{wt} + 1/E^{\beta-gal}_{wt}}. \tag{7}$$

Using the analogy with Ohm's law, we infer that a control coefficient is simply the proportion of the total resistance in the pathway that happens to be housed in a step. A sum of proportions is necessarily one,

$$\sum C^j_{wt} = 1. \tag{8}$$

This, the flux summation theorem (Kacser and Burns 1981), states that a finite amount of control is distributed among steps in the pathway regardless of the activities at each step. Selection cannot drive all three steps into the limit of adaptation so long as the lactose flux limits growth, as in a chemostat. Since selection is on the whole organism, if some other cellular function limited growth, the fitness control on the *lac* pathway could go to zero, and all steps would be at the limit of adaptation. In this case the lactose pathway would be saturated, which is impossible in a chemostat where the external concentration of lactose limits growth.

*Dominance*　Dominance is a logical consequence of enzyme kinetics (Keightley 1996). The enzyme activity of a heterozygote for a null mutant may be half that of the wild-type homozygote, but the flux in the heterozygote will be reduced by less than half (figure 4.4). Were *E. coli* diploid, heterozygotes would display complete dominance at the β-galactosidase, partial dominance at the permease, and codominance at the porins. These dominance relationships are a direct consequence of the distribution of control along the pathway.

*Pleiotropy*　A mutation that affects multiple characters is said to be *pleiotropic*. In principle, a change in the activity of any enzyme will affect both flux and the pools of metabolic intermediates. For example, doubling the activity of β-galactosidase has little effect on the flux because the intracellular pool of lactose is halved almost exactly ($J \approx (2E \cdot S)/2$). Halving porin activity halves fitness—and every flux in the cell must decrease in exact proportion because fitness is proportional to the lactose flux (figure 4.3). Pleiotropy is ubiquitous.

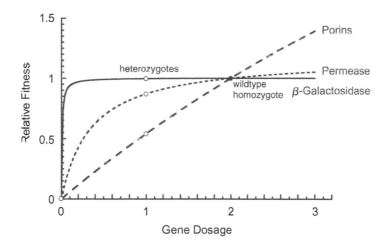

FIGURE 4.4
Were *E. coli* diploid, the three steps of the lactose pathway would show different dominance relationships. Heterozygotes of null mutants would show complete dominance at β-galactosidase, partial dominance at permease, and co-dominance at the porins.

TABLE 4.1    A Worked Example of Epistasis

| | Change in Activity $\delta e$ | Control Coefficient $C = dw/de$ | Selection Coefficient $\delta w \approx C\delta e$ |
|---|---|---|---|
| Background: wild type | | | |
| Porins | 0.1 | 0.819 | 0.0819 |
| Permease | 0.1 | 0.178 | 0.0178 |
| $\beta$-galactosidase | 0.1 | 0.003 | 0.0003 |
| Sum | | 1.000 | |
| Background: 10 $\times$ porin activity | | | |
| Porins | 0.1 | 0.312 | 0.0312 |
| Permease | 0.1 | 0.677 | 0.0677 |
| $\beta$-galactosidase | 0.1 | 0.011 | 0.0011 |
| Sum | | 1.000 | |

*Epistasis*    That the control coefficients are constrained to sum to one forces epistasis (Kacser and Burns 1981). For example, as ever more active porins are selected onto their fitness plateaus, their control coefficient diminishes; they become less "rate limiting" as they creep up their concave fitness function. Control now shifts to the permease and $\beta$-galactosidase, which passively become more "rate limiting." Selection at the permease and $\beta$-galactosidase intensifies (table 4.1). Previously neutral variation at $\beta$-galactosidase might now become exposed to selection. Hence, the selective neutrality of functionally distinct alleles at one step is conditional on the activities at other steps in the pathway. Conceivably, the evolution of many genes is characterized by alternating periods of selection and neutrality. Following a selective sweep at one gene, previously neutral alleles at a second gene become exposed to selection. As a succession of selective sweeps targets the second gene, a neutral polymorphism builds up at the first gene. The shifting and swaying of selection among steps in a pathway arise naturally from the kinetic effects of mutations and their impact on the distribution of control along the pathway.

*Genetic Architecture*    We have shown that the many characteristics of genetic architecture, pleiotropy, dominance, epistasis, constraints, selection, and neutrality are emergent properties of the simplest of biochemical systems. The traditional approach taken by evolutionary biologists discusses each of these terms as if they were separate, distinct phenomena. This is clearly misguided. They are epiphenomena generated by underlying metabolic architecture, and as such, they are inseparable from each other. To talk of one without consideration of the others is inappropriate because they are all simply facets of the same underlying processes.

*Genotype–by-Environment Interactions*    Genotype-by-environment interactions (G $\times$ E) arise whenever changes in fitness are produced by changes in environment. Growth on alternative pathway substrates provides a ready means to study G $\times$ E, precisely because the environmental variation is targeted directly at the genetic variation. Furthermore, selection on natural substrates (lactose and galactosyl-glycerol) can be contrasted with selection on synthetic substrates (lactulose, galactosyl-fructose, and methyl-galactoside) to determine if, as one might easily assume given Fisher's model of populations converging on a high fitness peak in a long-term environment, selection is more intense in novel environments.

Selection is indeed stronger on the synthetic sugars than the natural ones (figure 4.5A) (Silva and Dykhuizen 1993). This result lends credence to the idea that selection is more intense in novel environments. Application of the linear additive model of quantitative genetics to the natural logarithm of relative fitness allows the simple effects to be

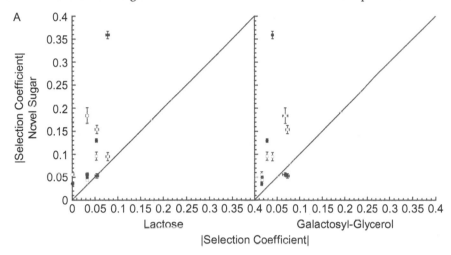

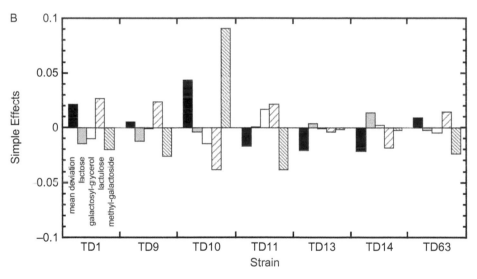

FIGURE 4.5

A, Selection is more intense in novel environments than typical environments. O Lactulose, Methyl-galactoside (after Silva and Dykhuizen 1994). B, The linear additive model of quantitative genetics can be used to partition the mean genetic effects (black bars) from the operon × sugar interactions (after Dean 1995).

partitioned (Dean 1995). Figure 4.5B shows that the G × E effects in this case are as large as, and commonly larger than, the mean genetic effects. Yet with few exceptions, the rank order of fitnesses remains unchanged despite the huge amount of G × E, because most of the G × E effects are created by redistribution of control along the pathway (discussed later).

Two mechanisms produce G × E (Dean 1995). In the first mechanism, changes in relative enzyme activity ($\delta e$) cause changes in selection ($\delta w$) because $\delta w \approx C\delta e$. The effect is most pronounced for strain TD10, where a change in the very direction of selection is evident (TD10 carries the *lac* operon transduced from ECOR16, a strain isolated from leopard scat; Ochman and Selander 1984). TD10 is favored over TD2 (which carries the wild-type K12 *lac* operon) on lactose, galactosyl-glycerol, and methyl-galactoside (figure 4.6) because its permease is very active toward these sugars ($\delta e > 0$). On lactulose, TD10 is selected against because its permease is less efficient than that of TD2 ($\delta e < 0$). But TD10 is exceptional; most changes in relative enzyme activity are small and cannot account for the huge amount of G × E.

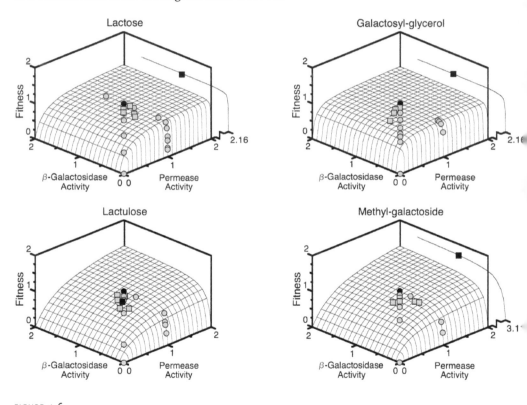

FIGURE 4.6

Adaptive landscapes of the lactose operon on four sugars, constructed using laboratory mutants (circles) and showing the positions of six operons from natural isolates (squares). These landscapes reveal the two causes of G × E in fitness. Only on lactulose does TD10 (black square) have lower activity than TD2 (dot), the result of strong G × E at the molecular level. A shift in control toward the β-galactosidase and permease causes the landscapes to become steeper on the novel sugars (lactulose and methyl-galactosidase) and is the major cause of G × E at *lac* (after Dean 1995).

TABLE 4.2    Fitness Control Coefficients on Various Sugars

| | Control Coefficient | | | |
| Sugar | Porin | Permease | β-Galactosidase | Sum |
|---|---|---|---|---|
| Lactose | 0.840 ± 0.028 | 0.156 ± 0.012 | 0.004 ± 0.0004 | 1 |
| Galactosyl-glycerol | 0.934 ± 0.009 | 0.060 ± 0.004 | 0.006 ± 0.0004 | 1 |
| Lactulose | 0.540 ± 0.072 | 0.408 ± 0.055 | 0.052 ± 0.0038 | 1 |
| Methyl-galactoside | 0.575 ± 0.043 | 0.383 ± 0.030 | 0.042 ± 0.0025 | 1 |

In the second mechanism, control is redistributed among steps in the pathway to cause changes in selection at all steps in the pathway. Both permease and β-galactosidase of the TD2 wild type have larger control coefficients ($Cs$) on the two synthetic sugars (lactulose and methyl-galactoside) than on the natural sugars (lactose and galactosyl-glycerol) (table 4.2). As a consequence, the adaptive landscape is steeper on the synthetic sugars (figure 4.6), and so the selection gradient is commensurately greater ($\delta w \approx C\delta e$). It is the redistribution of control among steps in the pathway that produces the huge amount of G × E. This mechanism also explains why there are so few changes in the rank order of fitnesses—on a monotonic concave fitness function, a change in a control coefficient affects only the magnitude of selection, never the sign.

Of course, the flux summation theorem dictates that as control shifts to the permease and the β-galactosidase, control shifts away from the porins (table 4.2). The porins become subject to less intense selection in the novel environments. Contrary to the prediction that selection is generally stronger in novel environments, any redistribution of control must act to increase the intensity of selection at some steps and reduce the intensity at others. The take-home message is clear: consistency between experimental results and theory provides an inadequate test of hypotheses when underlying assumptions are not first confirmed.

*Two Resources*    Bottom-up analysis can be expanded to include environmental complexity. As described earlier, strain TD10 is favored on methyl-galactoside and disfavored on lactulose when in competition with strain TD2. We applied resource-based competition theory (Stewart and Levin 1973; Tilman 1977, 1982) to determine whether the two strains might coexist while competing for the two sugars (Lunzer et al. 2002).

Assume growth rate depends linearly on the sum of the fluxes, $\mu = Y(J_{mg} + J_{lu})$. This seems reasonable as lactulose and methyl-galactoside are perfectly substitutable, and all products of the pathway enter central metabolism directly. With this assumption, it can be shown that the fitness of TD10 when *rare* is given by the following arithmetic mean:

$$w_{TD2.common}^{TD10.rare} = \frac{\mu_{TD10.rare}}{\mu_{TD2.common}} = w_{TD2.MG}^{TD10}\left[\frac{MG_o}{MG_o + LU_o}\right] + w_{TD2.LU}^{TD10}\left[\frac{LU_o}{MG_o + LU_o}\right], \tag{9}$$

where $MG_0$ and $LU_0$ are concentrations of the limiting resources *entering* the chemostat growth chamber, and $w_{TD2.MG}^{TD10} = \mu_{TD10.MG}/\mu_{TD2.MG}$ and $w_{TD2.LU}^{TD10} = \mu_{TD10.LU}/\mu_{TD2.LU}$ are the fitnesses on 100 percent methyl-galactoside and 100 percent lactulose, respectively (note: these fitnesses are independent of strain frequencies).

Inverting equation (9) reveals the fitness of TD2 when *common* to be a harmonic mean

$$
w_{TD10.rare}^{TD2.common} = \frac{1}{w_{TD2.common}^{TD10.rare}} = \frac{\mu_{TD10.rare}}{\mu_{TD2.common}}
$$

$$
= \frac{1}{\dfrac{1}{w_{TD10.MG}^{TD2}}\left[\dfrac{MG_0}{MG_0 + LU_0}\right] + \dfrac{1}{w_{TD10.LU}^{TD2}}\left[\dfrac{LU_0}{MG_0 + LU_0}\right]}. \qquad (10)
$$

A similar set of equations describes the fitness of TD2 when rare (an arithmetic mean) and TD10 when common (a harmonic mean). We conclude that the fitness of either strain is an arithmetic mean when rare and a harmonic mean when common.

Results from chemostat competition experiments between strains TD10 and TD2, conducted with the strains at high ($>0.9$) and low ($<0.1$) frequencies and at various ratios of lactulose to methyl-galactoside, closely match the theoretical predictions (figure 4.7A). A zone of coexistence is found where TD10 is favored when rare and disfavored when common. A least-squares fit to the model yields fitnesses on 100 percent methyl-galactoside of $w_{TD2.MG}^{TD10} = 1.316 \pm 0.003$ and on 100 percent lactulose of $w_{TD2.LU}^{TD10} = 0.904 \pm 0.002$. These fitnesses are independent of strain frequencies. At 28 percent methyl-galactoside, within the zone of coexistence, the equilibrium frequency of TD10 is given by

$$
p_{TD10} = \left[\frac{MG_0}{MG_0 + LU_0}\right]\frac{1}{1 - w_{TD2.LU}^{TD10}} + \left[\frac{LU_0}{MG_0 + LU_0}\right]\frac{1}{1 - w_{TD2.MG}^{TD10}}, \qquad (11)
$$

$p_{TD10} = 0.64$ or $Ln(p_{TD10}/p_{TD2}) = 0.56$. Selection trajectories above and below the observed equilibrium $Ln(p_{TD10}/p_{TD2}) = 0.50$ are of opposite sign, with the selection being more intense (i.e., the slopes steeper) the farther the frequencies are from equilibrium, as predicted (figure 4.7B). The direction, intensity, and frequency-dependent nature of selection across an environmental gradient, together with a zone of coexistence, are predictable solely from the fitnesses on the pure resources ($w_{TD2.LU}^{TD10}$ and $w_{TD2.MG}^{TD10}$) or, as in the case of the lactose operon, from knowledge of biochemistry ($\sum \frac{1}{E_{i.LU}}$ and $\sum \frac{1}{E_{i.MG}}$).

The reason why fitness is frequency-dependent has to do with the way strains interact with their environment. For example, the fluxes in TD2 when it dominates the population are

$$
J_{LU.wt} = \frac{lu}{\sum \frac{1}{E_{i.TD2.LU}}} = \frac{1}{YG}\left(\frac{LU_0}{LU_0 + MG_0}\right),
$$

$$
J_{MG.wt} = \frac{mg}{\sum \frac{1}{E_{i.TD2.MG}}} = \frac{1}{YG}\left(\frac{MG_0}{LU_0 + MG_0}\right), \qquad (12)
$$

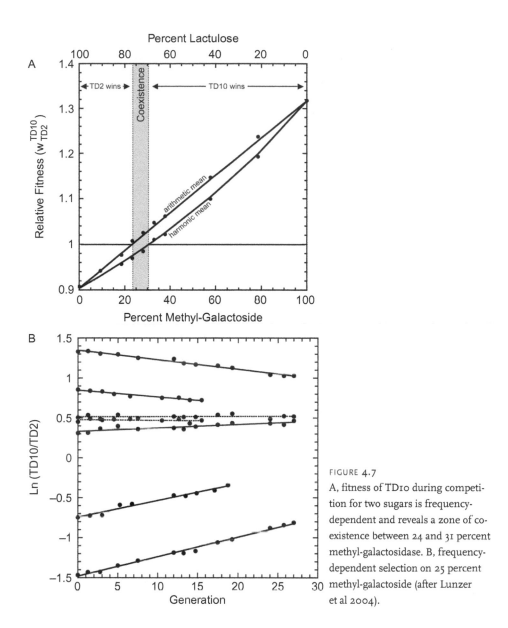

FIGURE 4.7

A, fitness of TD10 during competition for two sugars is frequency-dependent and reveals a zone of co-existence between 24 and 31 percent methyl-galactosidase. B, frequency-dependent selection on 25 percent methyl-galactoside (after Lunzer et al 2004).

where G is the mean population generation time (two hours). The terms on the right sides of equations (12) are constant; $LU_o$, $MG_o$, and $G$ are determined by the experimenter, and $Y$ is unaffected by mutations in the lactose operon. Hence, the fluxes of a pure population at steady-state in a chemostat are determined directly by the experimenter.

What does change is the resource concentrations in the growth chamber ($lu$ and $mg$). These are dependent on the activities of the enzymes in the strains ($\sum \frac{1}{E_{i,TD2.LU}}$ and $\sum \frac{1}{E_{i,TD2.MG}}$). A population dominated by an efficient lactulose consumer (TD2) will draw down the concentration of lactulose in the growth chamber far below that of methyl-galactoside, which TD2 consumes relatively inefficiently. This gives the rare methyl-galactoside specialist TD10 a competitive edge as the very resource on which it specializes is most abundant. Conversely, TD2 has a competitive edge when rare, because lactulose is abundant and methyl-galactoside scarce in a population dominated by TD10. Thus, genotype and environment cannot be treated as strictly separate entities. The two are inextricably linked, and it is the whole system that needs to be understood in context.

## OTHER APPROACHES

Bottom-up experiments in evolution can be done a number of different ways. We shall describe a few to illustrate the possibilities.

Walt Eanes and his coworkers (Eanes et al. 2006; Flowers et al. 2007) have studied glucose metabolism in *Drosophilia*. They wished to determine which enzymes had high control coefficients. To do this, they used low-activity and knock-out mutations in glycolytic enzymes and measured wing beat frequency of tethered male flies carrying these mutations as a proxy for flux and fitness. They showed that muscle-specific hexokinase (HEXA) and glycogen phosphorylase (GLYP) show evidence of having high control coefficients, while phosphoglucose isomerase, phosphoglucomutase (PGM), triosephosphate isomerase, pyruvate kinase, and trehalase did not (Eanes et al. 2006). Interestingly, sequence data suggest that PGM is under selection, while all the other enzymes look as if they are evolving neutrally (Flowers et al. 2007). Thus, even though the complexity of the glucose metabolism system makes it difficult to model at the present time, bottom-up experiments can be informative even in complex systems like this.

The genotype-phenotype map for single genes can be studied experimentally. Zhu et al. (2005) showed that the *E. coli* form of isocitrate dehydrogenase (IDH) that uses nicotinamide adenine dinucleotide phosphate (NADP) as a cofactor can be changed into the form that uses NAD by changing five amino acids. The engineered NAD-specific IDH was fitter when *E. coli* was grown in a glucose-limited chemostat, but much less fit when grown in an acetate-limited chemostat. These results support the idea that the evolution from an NAD form to an NADP form, which occurred three billion years ago, was an adaptation to permit growth on acetate.

Five point mutations in a particular β-lactamase increase the resistance to the β-lactam antibiotic, cefotaxime, 100,000-fold. Weinreich et al. (2006) constructed β-lactamases for each step of all trajectories that gave a positive selection at each step. Of the 120 possible trajectories, only 18 fit this criterion; and, if the probability of fixation of mutations depends on the magnitude of the fitness increase, only two trajectories are very likely.

The bottom-up approach can also be used from the perspective of interspecific comparisons (for a related overview, see Dean and Thornton 2007). The work of Shozo Yokoyama and his coworkers exemplify this approach (Yokoyama et al. 2006). They compare the wavelengths of maximal absorption across various species to understand similarities and differences in vision for organisms in different environments. For example, coelacanths, which live at depths of 200 meters, have two opsins with optimum light sensitivities of 478 and 485 nanometers. These span the narrow range of light, with a maximum intensity at about 480 nanometers, which penetrates to this depth (Yokoyama and Tada 2000). To do this work, the opsin genes are cloned, sequenced, and expressed in cell culture to obtain pure forms. Each amino acid change between the homologous opsins of different species is checked by in vitro mutagenesis and testing the resultant protein to distinguish amino acid changes important in adaptation from neutral changes.

## DISCUSSION

As our work on the lactose operon demonstrates, a bottom-up approach provides many insights into genetic architecture. Key among these is the realization that the replacement of one allele by another with different functional characteristics inevitably produces epistatic and dominance effects at other steps in the pathway, along with pleiotropic effects on dependent pathways and metabolite pools. The omnipresence of epistasis and pleiotropy in metabolism was first noted by Kacser and Burns (1981), who argued that Fisher's (1928, 1930) modifiers of dominance are not "a special class of genes acting only on the phenotype of the heterozygote" but rather "ordinary working enzymes with metabolic functions." Resurrecting an old model of Cornish-Bowden (1987), which showed that dominance is not inevitable when enzymes in a pathway become saturated with substrates, Bagheri and Wagner (2004) reached substantially the same conclusion: dominance can evolve, though not through the mechanism proposed by Fisher.

In contrast to biochemical models of metabolic flux, the structure of a quantitative genetics model of phenotype—with additive, dominance, $G \times E$, and pairwise epistatic effects—bears no relation to the etiology of any phenotype. Instead, a linear additive model is imposed on a highly interactive system, treating each epiphenomenon as if, like a blue sky, it were somehow real. Keightley (1989) concluded that variation in enzyme activities generates little nonadditive variance in flux despite the interactive nature of the underlying biochemical system. In our analysis of $G \times E$ (figure 4.5B), the linear additive model provided no insight into the underlying causes. Nor can the linear additive model of quantitative genetics cope with frequency-dependent selection produced by different genotypes modifying their environment in subtly different ways. Disconnected from reality, quantitative genetics cannot be used to dissect the architecture of real phenotypes.

Rather than focus on quantitative epiphenomena, we advocate detailed studies of the underlying systems. One can either take a bottom-up approach (from molecules to organismal phenotype), as we did at *lac*, or a top-down approach (from organismal phenotype to molecules). A recent study of beach mouse coat color used quantitative genetics to identify mutations of moderate effect and then moved toward a detailed molecular model of development (Hoekstra et al. 2006; Steiner et al. 2007)—tacit recognition that the genotype-phenotype map can only be understood in terms of causal mechanisms. Either way, top-down or bottom-up, the gulf from molecules to phenotypes needs to be bridged with rigorously tested mechanistic models of demonstrable predictive power (see also Dean and Thornton 2007).

Top-down and bottom-up approaches are not mutually exclusive and can act synergistically. We have found that a top-down experiment is very helpful in identifying the limits to a bottom-up approach. We predicted that a balanced polymorphism of specialists, lying as it does in a narrow zone of coexistence (figure 4.7A), should rapidly lose diversity once a fitter mutant of either strain arises and sweeps through the population (Lunzer et al. 2002). Long-term evolution experiments demonstrate that this prediction is wrong (Dykhuizen and Dean 2004). Even after 1,200 generations, chemostats fed by two galactosides were still inhabited by two specialists. Examination of the genetic changes (Zhong et al. 2004) shows that lactulose specialists amplify the *lac* operon, while methyl-galactose specialists constitutively express the *mgl* permease, regardless of genetic background (TD2 or TD10). Indeed, in several cultures of TD2 and TD10, the two specialists swapped resource preference: the lactulose specialist TD2 evolved into a methyl-galactoside specialist, while the methyl-galactoside specialists TD10 evolved into a lactulose specialist (Dykhuizen and Dean 2004).

The obsession with grand conceptualizations (e.g., the linear additive model, systems biology, abstract genetic architectures) should be abandoned in favor of studying specific systems that can be analyzed in detail. Though modest in scope, detailed studies of particular systems have led to remarkable advances in molecular biology. Where modern quantitative genetics differs little from Fisher's (1918) original ideas, the very concept of a gene has undergone several radical transformations at the hands of molecular geneticists (Beadle and Tatum 1941; Watson and Crick 1953; Jacob and Monod 1961; Chow et al. 1977; Nowacki et al. 2008). Detailed studies that proceed from molecules to organisms will allow genotype-phenotype maps to be realized. Once several are completed, general principles can be identified within each idiosyncratic system. In this way, evolutionary biology might progressively become a predictive science in the same way that physics and chemistry are predictive.

## SUMMARY

"Bottom-up" methods in experimental evolution are different from the "top-down" ones. In this chapter, we distinguish these methods and present the power of the bottom-up method through an example. Metabolic control theory is used to construct a biochemical model of fitness. The model is shown to predict precisely the direction and intensity of natural selection solely from knowledge of enzyme activity. Many characteristics of genetic architecture—pleiotropy, dominance, epistasis, constraints, selection, and neutrality—arise quite naturally as emergent properties of the underlying biochemistry. An extended model that describes competition for two resources accurately predicts the observed frequency-dependent selection and identifies a zone of coexistence where a protected polymorphism develops. Other examples of bottom-up experimental evolution studies are mentioned, noting that a precise mathematical model is not required for the method to be useful. We contrast this mechanistic approach with the linear additive model used by quantitative genetics.

## ACKNOWLEDGMENTS

This is contribution 1169 from the Graduate Studies in Ecology and Evolution, Stony Brook University. This study was supported by Public Health Service Grants to A.M.D. and D.E.D.

## REFERENCES

Andrews, K. J., and G. D. Hegeman. 1976. Selective disadvantage of nonfunctional protein synthesis in *Escherichia coli*. *Journal of Molecular Evolution* 8:317–328.

Bagheri, H. C., and G. P. Wagner. 2004. Evolution of dominance in metabolic pathways. *Genetics* 168:1713–1735.

Beadle, G. W., and E. L. Tatum. 1941. The genetic control of biochemical reactions in *Neurospora*. *Proceedings of the National Academy of Sciences of the USA* 27:499–506.

Bennett, A. F., and R. E. Lenski. 1999. Experimental evolution and its role in evolutionary physiology. *American Zoologist* 39:346–362.

Bennett, A. F., R. E. Lenski, and J. E. Mittler. 1992. Evolutionary adaptation to temperature. I. Fitness responses of *Escherichia coli* to changes in its thermal environment. *Evolution* 46:16–30.

Boos, W. 1982. Synthesis of (2R) glycerol-ortho-$\beta$-d-galactopyranoside by $\beta$-galactosidase. *Methods in Enzymology* 89:59–64.

Chow, L. T., R. E. Gelinas, T. R. Broker, and R. J. Roberts. 1977. An amazing sequence arrangement at the 5' ends of adenovirus 2 messenger RNA. *Cell* 12:1–8.

Cornish-Bowden, A. 1987. Dominance is not inevitable. *Journal of Theoretical Biology* 125:333–338.

Dean, A. M. 1989. Selection and neutrality in lactose operons of *Escherichia coli*. *Genetics* 123:441–454.

———. 1995. A molecular investigation of genotype by environment interaction. *Genetics* 139:19–33.

Dean, A. M., D. E. Dykhuizen, and D. L. Hartl. 1986. Fitness as a function of beta-galactosidase activity in *Escherichia coli*. *Genetical Research* 48:1–8.

———. 1988. Fitness effects of amino acid replacements in $\beta$-galactosidase of *Escherichia coli*. *Molecular Biology and Evolution* 5:469–485.

Dean, A. M., and J. W. Thornton. 2007. Mechanistic approaches to the study of evolution: the functional synthesis. *Nature Reviews Genetics* 8:675–688.

Dykhuizen, D. 1978. Selection for tryptophan auxotrophs of *Escherichia coli* in the glucose-limited chemostats as a test of the energy conservation hypothesis of evolution. *Evolution* 32:125–150.

———. 1993. Chemostats used for studying natural selection and adaptive evolution. *Methods in Enzymology* 224:613–631.

Dykhuizen, D., and M. Davies. 1980. An experimental model: bacterial specialists and generalists competing in chemostats. *Ecology* 61:1213–1227.

Dykhuizen, D. E., and A. M. Dean. 1990. Enzyme activity and fitness: Evolution in solution. *Trends in Ecology & Evolution* 5:257–262.

———. 1994. Predicted fitness changes along an environmental gradient. *Evolutionary Ecology* 8:524–541.

———. 2004. Evolution of specialists in an experimental microcosm. *Genetics* 167:2015–2026.

Dykhuizen, D. E., A. M. Dean, and D. L. Hartl. 1987. Metabolic flux and fitness. *Genetics* 115:25–31.

Dykhuizen, D., and D. Hartl. 1978. Transport by the lactose permease of *Escherichia coli* as the basis of lactose killing. *Journal of Bacteriology* 135:876–882.

Eanes, W. F., T. J. S. Merritt, J. M. Flowers, S. Kumagai, and C.-T. Zhu. 2006. Flux control and excess capacity in the enzymes of glycolysis and their relationship to flight metabolism in *Drosphilia melanogaster. Proceeding of the National Academy of Sciences USA* 103:19413–19418.

Fisher, R. A. 1918. The correlation between relatives on the supposition of Mendelian inheritance. *Transactions of the Royal Society of Edinburgh* 52:399–433.

———. 1928. The possible modification of the response of the wild type to recurrent mutations. *American Naturalist* 62:115–126.

———. 1930. The evolution of dominance in certain polymorphic species. *American Naturalist* 64:385–406.

Flowers, J. M., E. Sezgin, S. Kumagai, D. D. Duvernell, L. M. Matzkin, P. S. Schmidt, and W. F. Eanes. 2007. Adaptive evolution of metabolic pathways in *Drosophilia. Molecular Biology and Evolution* 24:1347–1354.

Hansen, T. F. 2006. The evolution of genetic architecture. *Annual Review of Ecology, Evolution, and Systematics* 37:123–157.

Hartl, D. L., D. E. Dykhuizen, and A. M. Dean. 1985. Limits of adaptation: The evolution of selective neutrality. *Genetics* 111:655–674.

Hoekstra, H. E., R. J. Hirschmann, R. A. Bundey, P. A. Insel, and J. P. Crossland. 2006. A single amino acid mutation contributes to adaptive beach mouse color pattern. *Science* 313:101–104.

Jacob, F., and J. Monod. 1961. Genetic regulatory mechanisms in the synthesis of proteins. *Journal of Molecular Biology* 3:318–356.

Kacser, H., and J. A. Burns. 1973. The control of flux. *Symposium for the Society of Experimental Biology* 27:65–104.

———. 1981. The molecular basis of dominance. *Genetics* 97:639–666.

Keightley, P. D. 1989. Models of quantitative variation of flux in metabolic pathways. *Genetics* 121:869–876.

———. 1996. A metabolic model for dominance and recessivity. *Genetics* 143:621–625.

Lenski, R. E., M. R. Rose, S. C. Simpson, and S. C. Tadler. 1991. Long-term experimental evolution in *Escherichia coli*. I. Adaptation and divergence during 2,000 generations. *American Naturalist* 138:1315–1341.

Lunzer, M., A. Natarajan, D. E. Dykhuizen, and A. M. Dean. 2002. Enzyme kinetics, substitutable resources and competition: From biochemistry to frequency dependent selection in *lac. Genetics* 162:485–499.

Miller, J. H., and W. S. Reznikoff. 1978. *The Operon*. Cold Spring Harbor, NY: Cold Spring Harbor Laboratory.

Monod, J. 1942. *Rescherches sur la croissance des cultures bactériennes*. Paris: Herman.

Novick, A., and M. Weiner. 1957. Enzyme induction as an all-or-none phenomenon. *Proceedings of the National Academy of Sciences of the USA* 43:553–566.

Nowacki M., V. Vijayan, Y. Zhou, K. Schotanus, T. G. Doak, and L. F. Landweber. 2008. RNA-mediated epigenetic programming of a genome-rearrangement pathway. *Nature* 451:153–158.

Ochman, H., and R. K. Selander. 1984. Standard reference strains of *Escherichia coli* from natural populations. *Journal of Bacteriology* 157:690–693.

Silva, P. J. N., and D. E. Dykhuizen. 1993. The increased potential for selection of the *lac* operon of *Escherichia coli*. *Evolution* 47:741–749.

Steiner, C. C., J. N. Weber, and H. E. Hoekstra. 2007. Adaptive variation in beach mice produced by two interacting pigmentation genes. *PLoS Biology* 5:1880–1889.

Steward, F. M., and B. R. Levin. 1973. Partitioning of resources and the outcome of interspecific competition: a model and some general conclusions. *American Naturalist* 107:171–198.

Stoebel, D. M., A. M. Dean, and D. E. Dykhuizen. 2008. The cost of expression of *Escherichia coli lac* operon proteins is in the process, not the products. *Genetics* 178:1653–1660.

Tilman, D. 1977. Resource competition between planktonic algae: An experimental and theoretical approach. *Ecology* 58:338–348.

————. 1982. *Resource Competition and Community Structure*. Princeton, NJ: Princeton University Press.

Tsen, S. D., S. C. Lai, C. P. Pang, J. I. Lee, and T. H. Wilson. 1996. Chemostat selection of an *Escherichia coli* mutant containing permease with enhanced lactose affinity. *Biochemistry and Biophysics Research Communications* 224:351–357.

Watson, J. D., and F. H. C. Crick. 1953. Molecular structure of nucleic acids: A structure for deoxyribose nucleic acid. *Nature* 171:737–738.

Weinreich, D. M., N. F. Delaney, M. A. DePristo, and D. L. Hartl. 2006. Darwinian evolution can follow only very few mutational paths to fitter proteins. *Science* 312:111–114.

Wilson, D. M., R. M. Putzrath, and T. H. Hastings. 1981. Inhibition of growth of *Escherichia coli* by lactose and other galactosides. *Biochimica et Biophysica Acta* 649:377–384.

Yokoyama, S., W. T. Starmer, Y. Takahashi, T. Tada. 2006. Tertiary structure and spectral tuning of UV and violet pigments in vertebrates. *Gene* 365:95–103.

Yokoyama, S., and T. Tada. 2000. Adaptive evolution of the African and Indonesian coelacanths to deep-sea environments. *Gene* 261:35–42.

Zhang, E., and T. Ferenci. 1999. OmpF changes and the complexity of *Escherichia coli* adaptation to prolonged lactose limitation. *FEMS Microbiology Letters* 176:395–401.

Zhong, S. B., A. Khodursky, D. E. Dykhuizen, and A. M. Dean. 2004. Evolutionary genomics of ecological specialization. *Proceedings of the National Academy of Sciences of the USA* 101:11719–11724.

Zhu, G. P., G. B. Golding, and A. M. Dean. 2005. The selective cause of an ancient adaptation. *Science* 307:1279–1282.

# 5

# EXPERIMENTAL EVOLUTIONARY DOMESTICATION

## Pedro Simões, Josiane Santos, and Margarida Matos

Many millennia before we understood the basic laws governing biological evolution, we bred our commensal species to our liking, whether for economic or leisure purposes. A range of species from plants to animals were thereby domesticated. In a sense, the longest-running experimental evolution projects are those of domestication, and they have produced an astonishing variety of animal breeds and plant varieties. Naturally enough, Darwin used cases of pigeon and dog domestication to illustrate the capacity of selection to produce evolutionary change in *The Origin of Species* (1859), and then he later expanded greatly on this theme in the volumes devoted specifically to domestication (Darwin 1883).

Domestication does not necessarily imply selection directed toward a single goal. In its broader sense, it means evolutionary change when wild populations are maintained in environments controlled, or at least strongly shaped, by human choices. Several features of such environments make the study of domestication interesting from the standpoint of evolutionary biology. Domesticated populations suffer more or less drastic changes in population structure, including size. The environment of domesticated populations is more stable than that of the wild, with a reduction in predation and interspecific competition. Relaxed selection may arise for a wide range of traits. But for other traits, selection may be greatly heightened, sometimes as a result of human intent, but sometimes not. Changes in age structure, a range of abiotic factors from temperature to nutrients, available space, and so on, may lead to significant changes in the components of fitness in domesticated populations. Domestication may thus be a cause of adaptive changes that are well worth analyzing.

Research in experimental evolution often entails the imposition of new selection regimes. These new selection regimes eventually lead to divergent evolution between the populations subject to them and control populations, the latter often being maintained under the antecedent selection regime imposed on the populations directly ancestral to the populations that are subjected to the new selection regime(s). The basic expectation is that sustained directional changes in the phenotypes of the populations subject to the new form of selection, when measured relative to the control populations, can be explained by the imposition of the new selection regime. Replicate populations can test whether such directional changes could be due to genetic drift, which is not expected to produce sustained directional changes on average (Rose et al. 1996). The fundamentals of such experimental evolution are more thoroughly addressed in other chapters of this volume and will not be reviewed further here (Futuyma and Bennett this volume; Huey and Rosenweig this volume; Rhodes and Kawecki this volume; Rose and Garland this volume; Swallow et al. this volume).

*Drosophila* is, of course, one of the commonly used organisms in experimental evolution. Studies of laboratory natural selection in *Drosophila* have characterized the evolution of populations subject to different densities (e.g., Mueller et al. 1993), demographic regimes (e.g., developmental rate—see Chippindale et al. 1997; age at reproduction—see Luckinbill et al. 1984; Rose 1984; Partridge and Fowler 1992; Roper et al. 1993;

Leroi et al. 1994), several stresses (e.g., starvation resistance—see Rose et al. 1992; Chippindale et al. 1996; Harshman et al. 1999; desiccation resistance—see Hoffmann and Parsons 1993; Gibbs et al. 1997; Folk and Bradley 2005; also see a brief review in Hoffmann and Harshman 1999), different temperatures (Kennington et al. 2003; Santos et al. 2004, 2005), and so on (for reviews, see Prasad and Joshi 2003; Chippindale 2006; Gibbs and Gefen this volume).

One of the goals of such studies is to characterize the potential of populations to respond directly to selection. By now, it is apparent that most *Drosophila* characters will respond significantly to direct selection. Of greater interest, therefore, is the pattern of indirect response to selection. Sometimes the aforementioned studies have revealed declines in functional characters that are not the target of selection, suggesting the presence of trade-offs or, less plausibly for large outbred populations, genetic correlations due to linkage disequilibrium. But such antagonistic indirect responses to selection are not the only possibility. At the start of adaptation to a novel environment, genotype-by-environment interactions are expected to entail significant positive genetic covariances among life-history traits (e.g., Service and Rose 1985; Stearns et al. 1991; de Jong 1993; Matos et al. 2000a; Chippindale et al. 2004).

Many organisms besides *Drosophila* have been studied with the same basic principles and goals: microorganisms (see Elena and Lenski 2003 for a review of studies of adaptation in microorganisms), vertebrates in the wild (e.g., Reznick and Ghalambor 2005; Irschick and Reznick this volume), other insects (e.g., *Tribolium*—see Wool 1987 for an example), among others. The long-term evolutionary studies of adaptation in *Escherichia coli* by Lenski and his collaborators are particularly noteworthy for the large number of generations that they commonly examine (e.g., Lenski 2004). Nevertheless, outbred *Drosophila*, like other sexual diploid organisms that have not been inbred, have the advantage of abundant standing genetic variation. This genetic variation is expected to be most abundant in large natural populations or in laboratory populations at the moment of their foundation from wild samples, since these samples will not have lost genetic variability due to either genetic drift with small population sizes or intense directional selection during initial domestication. This is one reason why studies of the evolutionary domestication of *Drosophila* are of interest, as they allow us to characterize the evolutionary dynamics of local adaptation in populations with considerable genetic variation at the start of selection.

Studies involving convergent evolution in *Drosophila* are much less abundant than studies of divergent evolution. Reverse evolution experiments have been done in lines previously derived from a common ancestor by divergent selection, where phenotypic reversion to the ancestral state is tested when the initial environmental conditions are resumed (e.g., Service et al. 1988; Graves et al. 1992; Teotónio and Rose 2000; Teotónio et al. 2002, 2004; Passananti et al. 2004). Such studies allow us to address the importance of the evolutionary history of populations as a determinant of their capacity to return to ancestral states. For further insight into the relevance of reverse evolution experiments, see Estes and Teotónio (this volume). In general, convergent evolution among populations subjected to a common selection regime is the intuitive expectation.

Adaptation to a novel, common environment is another way to test for convergent evolution. One such environment is the laboratory environment, where the evolution in captivity can be characterized as a particular case of evolutionary domestication. Detailed studies of domestication in *Drosophila* have appeared relatively recently in the scientific literature (e.g., Frankham and Loebel 1992; Latter and Mulley 1995; Hercus and Hoffmann 1999a, 1999b; Matos et al. 2000b, 2002, 2004; Sgrò and Partridge 2000; Hoffmann et al. 2001; Krebs et al. 2001; Woodworth et al. 2002; Gilligan and Frankham 2003; Reed et al. 2003; Griffiths et al. 2005; Simões et al. 2007, 2008). Most of these studies indicate that adaptation occurs during domestication, as revealed by improvement in one or several life history traits measured under the conditions of laboratory culture (Frankham and Loebel 1992; Latter and Mulley 1995; Hercus and Hoffmann 1999a, 1999b; Matos et al. 2000b, 2002, 2004; Sgrò and Partridge 2000; Woodworth et al. 2002; Gilligan and Frankham 2003). But there are some disagreements among the authors of such studies (e.g., Frankham and Loebel 1992, cf. Latter and Mulley 1995; Gilligan and Frankham 2003; Hoffmann et al. 2001, cf. Matos et al. 2000b, 2002, 2004; Griffiths et al. 2005; Simões et al. 2007, 2008). In particular, studies that employ a comparative approach (e.g., Frankham and Loebel 1992; Latter and Mulley 1995; Hoffmann et al. 2001; Woodworth et al. 2002; Gilligan and Frankham 2003; Griffiths et al. 2005) have often reached different conclusions from those obtained in studies of evolutionary trajectories (Matos et al. 2000b, 2002, 2004; Krebs et al. 2001). By a comparative approach, we mean the inference of the evolutionary dynamics of a population from comparisons among several contemporaneous populations at different stages of evolution, assuming that each one of them will represent the evolutionary state of a given population at a particular moment. Decoupling between such inferences and direct, real-time studies of experimental evolution may derive from several sources. In particular, evolutionary contingencies associated with founder effects may play an important role, among other sources of nonrepeatability of evolutionary patterns across populations. This will be a major theme of the present review. Another controversy concerns the use of long-established laboratory populations to test several evolutionary theories (see Promislow and Tatar 1998, Harshman and Hoffmann 2000; Sgrò and Partridge 2000; Hoffmann et al. 2001; Linnen et al. 2001). Finally, another common disagreement in the literature concerns the genetic mechanisms that cause the decline of some traits during laboratory domestication, specifically mutation accumulation, inbreeding depression, and genetic trade-offs (e.g., Latter and Mulley 1995; Shabalina et al. 1997; Bryant and Reed 1999; Sgrò and Partridge 2000; Hoffmann et al. 2001; Woodworth et al. 2002; Frankham 2005). We will discuss these issues in light of our own results.

Real-time studies of evolutionary trajectories during domestication test the assumption of convergence, as well as allowing the experimenter to tackle such important issues as the repeatability of the evolutionary dynamics of adaptation, the importance of founder effects in the process of laboratory adaptation, and the effects of long-term evolution in the laboratory. The study of evolutionary domestication is particularly helpful when it uses as starting populations different collections from the wild, samples that are expected to be highly variable sources of founders each time a study is conducted. By following the dynamic changes that occur within domesticating populations, it is possible to infer evolutionary rates and define evolutionary patterns directly. Though some short-term real-time studies of evolutionary trajectories have appeared in the *Drosophila* literature (e.g., Hercus and Hoffmann 1999b; Krebs et al. 2001), to our knowledge ours are the longest-term real-time studies of domestication in a sexual species that have been published to this point (Matos et al. 2000a, 2000b, 2002, 2004; Simões et al. 2007, 2008).

In this chapter, we start by reviewing our own results and then review other relevant studies, particularly focusing on the points of disagreement between laboratories already mentioned. We end with suggestions for future studies.

## EVOLUTIONARY DOMESTICATION: REAL-TIME STUDIES IN *DROSOPHILA SUBOBSCURA*

Since 1990 we have studied the evolutionary changes that occur during laboratory adaptation in the model organism *Drosophila subobscura*. However, we will focus here on experiments that were started in 1998, 2001, and 2005, using the population first domesticated in 1990 as a point of reference. In order to illustrate the type of results that we have obtained, we outline our results for just two adult traits, early fecundity and female starvation resistance, though we have studied a number of other functional characters.

### POPULATIONS AND EXPERIMENTAL DESIGNS

All our populations were founded from wild samples collected over one to several days using fermented fruit in traps. The first foundation was done in 1990, in an Adraga pinewood in Sintra, Portugal, from which we established our reference laboratory population for our subsequent studies of domestication (the "NB" populations). Later, in 1998, we collected flies from the same natural location, from which we established a second set of laboratory populations (labeled from here on "NW"). In 2001, we founded two new sets of populations, one from collections again in Adraga, Sintra (called "TW"), and another from a pinewood in Arrábida, some 50 kilometers from Adraga, on the other side of the Tagus River (called "AR") (see figure 5.1). The collections from Arrábida and Sintra were made synchronously, allowing us to follow the evolutionary dynamics of the two sets of populations in parallel simultaneous assays, with the same number of generations in the laboratory, an ideal situation for studying the effects of different wild-source populations on the process of domestication. In 2005, another foundation was made from a Sintra collection, establishing another set of populations (FWA) that will be used in an analysis presented in a later section.

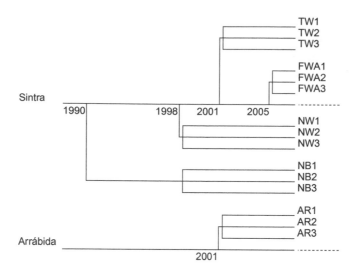

FIGURE 5.1

Phylogeny of the fifteen laboratory populations used in this chapter, indicating the original wild location. Our populations were obtained from the collection of flies of 1990 done in a pinewood of Adraga (Sintra), from which all foundations from Sintra were derived. By the time the last foundation was performed (FWA, 2005) the populations established in 1990, 1998, and 2001 were, respectively, at generations 181, 91, and 45. The foundations of Arrábida and Sintra, in 2001, were carried out synchronously. Three replicate populations were derived from each of the five foundations.

Each population was split into several replicate populations two generations after the collection of individuals from the wild, with the exception of the long-domesticated population founded in 1990, which was split up into replicates in 1998 at the same time as the then newly sampled flies were split. We label each replicate by a number. Thus, the NW1, NW2, and NW3 are the three populations derived from the "NW foundation" started with wild samples collected in Adraga in 1998.

From the moment our laboratory populations were founded, they were maintained in standard conditions, at discrete generations of twenty-eight days, close to the time of peak fecundity in *D. subobscura*, with control of medium, temperature, and population density. Our populations were maintained in numerous vials placed in racks within incubators, with care taken to avoid handling differences among populations during both maintenance and assays. Population sizes were typically about 1,200 (see details in Matos et al. 2002, 2004; Simões et al. 2007, 2008).

Adult assays were done periodically, both on the more recently introduced populations and on the longer-established (NB) populations. We will present here data for mean fecundity during the first week of life and female starvation resistance over generations 4 to 94 of the populations founded in 1998, and the corresponding generations 94 to 184 of the longer-established populations, as well as generations 3 to 48 of the populations founded in Sintra and Arrábida in 2001, when the longer-established populations were in their generations 139 to 184. Because the unit of evolutionary studies is the population and not the individual, our data analysis focuses on the averages of each replicate population using as source of error the heterogeneity among replicate populations.

## LONG-TERM EVOLUTIONARY DOMESTICATION

*Early Fecundity*　There was no significant phenotypic trend among the longer-established NB populations between generations 94 and 184, while the NW populations founded in 1998 showed a significant improvement in performance relative to the NB controls between their generations 4 and 94 (figure 5.2; average slope = 0.411, *t*-test, *p* = 0.02). A log-linear trend is even more significant (*p* = 0.007, data not shown). Altogether, these data indicate a clear, though not very quick, process of adaptation in early fecundity.

*Female Starvation Resistance*　In figure 5.3, we present the changes in female starvation resistance shown by the long-established populations between generations 94 and 184. Contrary to the data on fecundity, these populations show a significant decline in female starvation resistance (*t*-test, *p* = 0.03), with an average slope close to −0.05. This corresponds to a decline of about 0.11 percent per generation, and a decline of about 10 percent during the entire period.

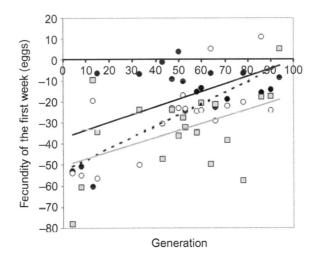

FIGURE 5.2

Fecundity during the first week of life in the populations founded in 1998 (NW) relative to the longer-established (NB) populations. Each data point is the difference between the average absolute values of each population and the same numbered longer established population. Replicate population 1: black circles, full black line; replicate population 2: open circles, broken line; replicate population 3: gray circles, gray line. All analyses of linear regressions used the individual slopes as data points in a *t*-test. NW populations show a steady increase in early fecundity throughout laboratory culture, corresponding to a significant pattern of convergence to longer-established reference populations.

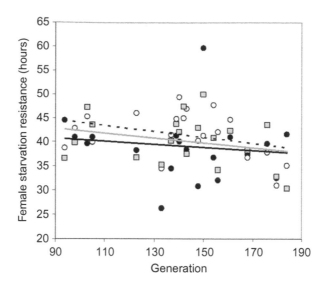

FIGURE 5.3

Female starvation resistance in the longer-established (NB) populations between generations 94 and 184. Replicate population 1: black circles, full black line; replicate population 2: open circles, broken line; replicate population 3: gray circles, gray line. The analysis used the individual slopes as data points in a *t*-test. NB populations show a significant decline in female starvation resistance with an average slope of –0.05, which corresponds to a decrease of about 0.11 percent per generation.

Ehiobu et al. (1989) found that viability in *D. melanogaster* decreased by about 0.96 percent for every 1 percent increase in the inbreeding statistic, *F*. In our case, assuming an effective population size of about five hundred individuals, we expect between generations 94 and 184 an increase in *F* value of about 8 percent, which corresponds to a decrease of 1.2 percent in female starvation resistance for every 1 percent increase in the inbreeding statistic *F*. These values can thus be explained by inbreeding depression alone.

There is no significant temporal change of female starvation resistance in the NW populations founded in 1998 relative to the long-established NB populations (average slope = −0.006, n.s.). Nevertheless, the NW populations also show a significant decline in starvation resistance when absolute values are analyzed (average slope = −0.05; *t*-test, $p < 0.01$, data not shown).

SHORTER-TERM EFFECTS OF FOUNDATION AND REPEATABILITY
OF EVOLUTION

*Early Fecundity*    Figure 5.4 presents the changes in fecundity over the first week of life for the populations founded in Sintra (TW) and Arrábida (AR) in 2001, relative to the

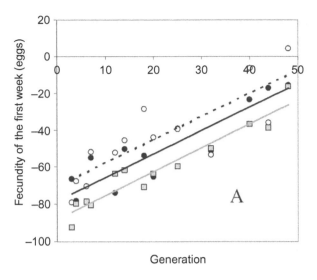

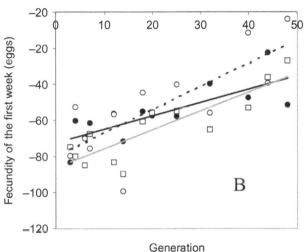

FIGURE 5.4
Fecundity of the first week of
life in the populations founded
in Sintra (A) and Arrábida (B)
in 2001, between 3 and 48,
relative to the longer-established
populations. The analysis of
each set of replicate populations
used the individual slopes as
data points in a *t*-test. Early
fecundity significantly increases
during domestication of both
sets of populations. There are
no significant differences in
evolutionary rate between the
two sets of populations.

long-established ND populations. Both regimes show significant linear improvement during the generations under study (average slope = 1.27, $p < 0.001$ for TW; average slope = 1.03, $p = 0.024$ for AR).

It is interesting to compare the slopes of these 2001 populations with the pattern of adaptation shown by the NW populations founded in 1998, over the equivalent generations 4 and 47. The NW populations had an average slope of 0.76 in that period relative to the reference NB populations, a result not significantly different from that obtained with the populations founded in 2001.

*Female Starvation Resistance*　There is no significant temporal linear increase in starvation resistance among the populations founded in Sintra and Arrábida in 2001, relative to the longer-established populations (TW, average slope = 0.114, n.s.; AR, average slope = 0.124, n.s.; data not shown). Interestingly, the analysis of the temporal changes during the first forty generations reveals a significant improvement in TW populations, though with a low rate (average slope = 0.06, $p$ = 0.04; see Simões et al. 2007), which suggests that heterogeneity between replicates with further analysis may have affected the statistical power. As for absolute values, both sets of populations showed a suggestion of a decline during the study, though it was not as clear as we had previously obtained for the NW populations founded in 1998 (TW, average slope = −0.09, n.s.; AR, average slope = −0.08, n.s.). Also, in contrast to the populations founded in 1998, there is no significant improvement of female starvation resistance when considering the first fourteen generations, relative to longer-established populations. This suggests that any improvement that may occur in the initial period of domestication differs among populations. In view of this result, it is apparent that generalizing from short-term studies of domestication can be misleading. This may explain some of the disparities in the conclusions of different studies of domestication (see the last section).

## BALANCE OF OUR STUDIES

In the three studies summarized here, there is clear adaptation to the laboratory, in the steady increase in early fecundity. The long-term study of the populations founded in 1998 also suggests that an adaptive plateau is being reached, indicated by a slowing down of the evolutionary rate. Our data indicate that inbreeding depression may play some role in the changes observed under domestication.

One of the odd features of our data is that starvation resistance initially increases and then decreases. This is the expected outcome if the additive genetic covariance between traits undergoing domestication changes through time, from positive (or less negative) to negative (or more negative) values. Initial positive covariances in a novel environment can derive from genotype-by-environment interactions, given the new selective scenario involved (see Service and Rose 1985; Chippindale 2006). As a consequence of the approach to an evolutionary equilibrium, most variation involving positive pleiotropy among traits will be likely exhausted, while antagonistic pleiotropy may allow the maintenance of additive variance, expressed as a negative genetic correlation in those traits. This change in genetic correlation may lead to a nonlinear, biphasic evolutionary pattern. In fact, the laboratory populations founded from Sintra in 1998 and 2001 show an initial phase with a significant increase in starvation resistance, while more generations show a shift to a negative slope. But it is possible that both inbreeding depression and selection act during the evolutionary changes of starvation resistance. The relative importance of these mechanisms of selection and inbreeding may in general change as a function of the initial composition of the population, selective pressures, and how long studies are conducted.

## COMPARATIVE STUDIES OF DOMESTICATION

Several studies have used the comparative method to study evolutionary patterns in laboratory adaptation as opposed to the analysis of evolutionary trajectories that we illustrated in the previous section. Here we discuss briefly such comparative studies.

### STATIC COMPARISONS OF LONG-ESTABLISHED VERSUS RECENTLY INTRODUCED POPULATIONS

Several studies of laboratory adaptation compare populations maintained in the laboratory for several generations with other populations, recently introduced from the wild. These studies do not present data on evolutionary dynamics. In some of these studies, the effects of population size, degree of inbreeding, and so forth, are also analyzed.

Hercus and Hoffmann (1999a) conducted a study involving interspecific hybrids between *Drosophila serrata* and *Drosophila birchii*. This study was short in duration and lacked adequate reference populations, but the results are suggestive. Populations that had been kept in the lab for seventeen to twenty generations were compared with populations derived from the same location that had spent just seven generations in the laboratory. Both fecundity and desiccation resistance were higher in the populations that had been in the lab longer, suggesting that desiccation resistance had increased without a trade-off with fecundity. It is a pity that these authors did not analyze the changes of these traits within each population over multiple generations, as they did for juvenile viability between generations 17 and 30, which showed a temporal increase in performance (Hercus and Hoffmann 1999b).

Woodworth et al. (2002) also analyzed the effects of both adaptation and inbreeding during evolutionary domestication. They founded laboratory populations of *D. melanogaster* at population sizes ranging from twenty-five to five hundred individuals and compared their performance after fifty generations in the laboratory, both in "benign" captive conditions and in "wild" competitive conditions. Several control populations were used in this study, some derived from the same location in later years. In benign conditions, populations of bigger size showed a higher performance, while those with the smallest population size performed poorly. In "wild" conditions, all laboratory populations had a lower performance than the recently derived populations. The authors concluded that both genetic adaptation and inbreeding depression were responsible for the poor performance of laboratory populations in the "wild" environment.

Another interesting study was conducted by Latter and Mulley (1995) using *D. melanogaster*. These authors analyzed the effects of both adaptation and inbreeding on reproductive ability in competitive and noncompetitive environments. They compared the performance of populations derived from the same wild-source population, but differing in the degree of inbreeding. Comparisons with recently introduced populations were also performed. Long-established populations were superior in competitive ability (assessed in competition experiments with a mutant, marked stock) in the laboratory relative to both recently introduced and inbred populations. Over about two hundred generations, there was a doubling of competitive fitness (estimated as the ratio between the competitive index of the longer-established populations and the more recent ones), even in populations with a population size of fifty during most generations. Comparing differences in performance as a function of the amount of inbreeding, the authors were able to disentangle effects of inbreeding from effects of selection. They concluded that both processes had acted in the inbred populations. Interestingly, fitness differences were minor in a noncompetitive environment, indicating the presence of genotype-by-environment interactions for these characters.

## EVOLUTIONARY DYNAMICS INFERRED FROM A COMPARATIVE APPROACH

In this experimental strategy, populations introduced into the laboratory environment at different times are compared synchronously at different stages of the adaptation process; with this data, the evolutionary trajectory of a single population adapting to the laboratory environment is inferred. The assumption is that the evolutionary pattern of the different populations used would be the same if they were compared directly over multiple generations, and so the performance of the most recently founded population will accurately reflect the early stages of adaptation of the previously founded populations. For example, Sgrò and Partridge (2000) compared life-history traits in populations of D. *melanogaster* founded three consecutive times from the same natural location, maintained in either bottles or population cages. The analyses revealed marked changes in some of the traits but few changes in most of them. Differences were found between populations maintained in cages and bottles as a function of time in the laboratory. Development time increased during laboratory culture. The authors advanced the hypothesis that this might be due to higher larval competition in laboratory culture, based on the fact that this increase was particularly seen in cages, with higher larval densities. Early fecundity increased with bottle culture, while late fecundity decreased. However, with cage culture, the fecundity patterns were less clear. This was assumed to be due to the truncation of the adult period in bottle culture, enhancing the relative focus of natural selection on the early adult period. The authors propose that this led to a decrease in late fecundity by either mutation accumulation or antagonistic pleiotropy.

Using the same three sets of populations and a new one from a recent foundation, Hoffmann et al. (2001) tested the hypothesis that stress resistance is lost during laboratory adaptation. The most recently founded populations showed higher starvation and desiccation resistance than the previously founded ones, a result that was interpreted as a marked evolutionary decline in resistance for both stresses during laboratory adaptation. According to the authors, the rapidity of the response ruled out mutation accumulation as a possible explanation for the pattern obtained. They propose that the most likely explanation is that resistance to starvation and desiccation was lost as a correlated response to selection on early fertility, as a result of a negative genetic correlation between stress resistance and fecundity traits.

To investigate the genetic dynamics of adaptation to captivity, Gilligan and Frankham (2003) also used the comparative approach, measuring the fitness of several independently founded populations of *D. melanogaster*, derived from the same natural site in consecutive years, relative to a genetically marked stock. The authors inferred a curvilinear pattern of adaptation, with an increase of captive fitness reaching 3.3 times the initial fitness after eighty-seven generations of laboratory adaptation.

Griffiths et al. (2005) studied the effects of laboratory adaptation in *D. birchii* using isofemale lines established from collections made in the same four natural locations over three consecutive years. They concluded that time in laboratory culture influenced evolutionary responses for some traits but not others. For example, there was an increase in starvation resistance and development time in the laboratory lines, while recovering time following a cold shock decreased. On the other hand, heat knockdown resistance and wing size were not affected. The authors argue that collections made in different locations and the use of isofemale lines can overcome the limitations of using a classic comparative approach (e.g., Sgrò and Partridge 2000; Hoffmann et al. 2001). Nevertheless, the data on development time presented in this study clearly illustrate some of the limitations of this approach, in that the data of one of the sets of lines were quite different from the others. The authors attributed this to changes in the genetic composition of the wild populations.

Although some traits appear to give consistent results across studies (e.g., increased fecundity and development time during laboratory adaptation), others, such as stress resistance, do not. This may not only be due to the different genetic composition of the populations analyzed but also to methodological issues (see later discussion).

## TESTING COMPARATIVE METHODS USING TRAJECTORY DATA

We will now test the validity of the comparative approach with our own data, as we now have several sets of populations founded at different times and know their actual evolutionary trajectories. The question is, can evolutionary dynamics be correctly inferred using comparative data only?

In a recent study, Matos et al. (2004) tested the consistency of results using both the comparative and temporal methods applied to the study of domestication. Although the comparative method proved to be quite accurate for the analysis of robust evolutionary patterns, such as those of fecundity traits, it can lead to problems with less predictable traits. This applies clearly to starvation resistance. Our own studies of real-time evolution suggest that starvation resistance is a trait that has complex evolutionary trajectories during domestication, rendering short-term and comparative studies problematic. It is also a trait that has given disparate results among laboratories in studies that infer evolutionary changes from comparisons among contemporaneous populations. For example, while Hoffmann et al. (2001) found a consistent decline of this trait over generations with laboratory culture, the study by Griffiths et al. (2005) finds an improvement during laboratory adaptation.

We can illustrate this problem using new data that we have collected from a new 2005 foundation from Sintra, the same location where the TW, NW, and NB populations were derived (see figure 5.1).

At generation 3 after foundation, we made our first assay with these more recently founded populations (which we call FWA), as well as TW (at their corresponding 48th generation), NW (in the lab for 94 generations), and NB (the longer-established populations, for 184 generations in the laboratory). The plots for both early fecundity and female starvation resistance are presented in figure 5.5. In that figure we also plot the data obtained in our previous study of TW populations, when these were in their 4th generation (the earliest assay conducted in that study, involving simultaneous assays of NW and NB populations, by that time in their 50th and 140th generations), using the same methodology.

Our comparative analysis of fecundity does give similar results to those of our real-time evolution studies, with clear-cut differences between populations as a function of how many generations they have been in the lab, even though they derive from different foundations. There is also robustness of results among the plots using our most recent data and those of the previous study (TW populations at generation 4). But contrary to these fecundity results, starvation resistance shows differences between the two studies: the assay at generation 4 of the TW populations suggests stability of this trait, while the most recent data present evidence for a decline with generations. These data illustrate one of our points about the limitations of a comparative approach: if the values of the TW populations in their generation 4 were close to the ones presented by our most recent populations (assuming the differences to be purely genetic, which is obviously simplifying), the inferred trend might even be positive. In fact, the data of an assay done at generation 6 of the TW populations present such a shift relative to NB values, with TW populations having lower values than these populations, though bigger than NW (see Matos et al. 2004). This does not correspond to any trend in the actual evolutionary trajectories. The problems of a comparative approach are thus clearly revealed by our data. The differences among comparative studies in the evolution of starvation resistance contrast with the more repeatable patterns obtained with evolutionary trajectories. This suggests that the comparative approach to experimental evolution can yield

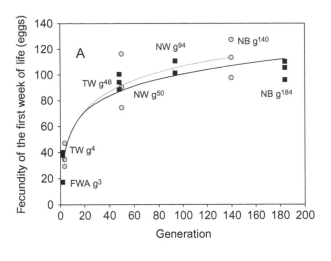

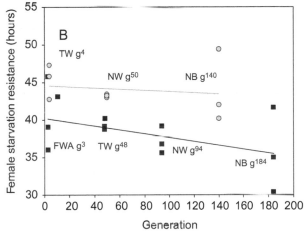

FIGURE 5.5

Comparative plots of the values of fecundity of the first week of life (A) and female starvation resistance (B) of independently founded populations as a function of number of generations in the laboratory. Gray circles and gray lines: data from assays done at generation 4 after foundation of TW (NW at generation 50; NB at generation 140); Black squares and black lines: data from assays done at generation 3 after foundation of FWA (TW at generation 48; NW at generation 94 and NB at generation 184). Fecundity comparative plots are remarkably similar to those obtained using real-time evolutionary trajectories. These results are also robust among comparative plots using data from different studies. Contrary to this, female starvation resistance shows differences between the two comparative studies: the one using data from TW at generation 4 suggests stability for this trait, while the one for FWA at generation 3 shows evidence for a decline with generations.

misleading results. Indeed, the use of contemporaneous populations as "surrogates" for the evaluation of the phenotypic state of a given population through time (e.g., Frankham and Loebel 1992; Hoffmann et al. 2001; Gilligan and Frankham 2003; Griffiths et al. 2005) rests on several untested a priori assumptions that may not always apply. For example, it is often assumed that founder effects and random genetic drift during adaptation are negligible (as proposed in Sgrò and Partridge 2002; but see Matos and Avelar 2001; Woodworth et al. 2002). Furthermore, comparative studies often lack appropriate reference populations, and this prevents the disentangling of the evolutionary mechanisms involved, particularly in traits exhibiting complex evolutionary trajectories, as our studies of starvation resistance illustrate.

We conclude that the comparative approach is not the appropriate tool with which to study the detailed dynamics of domestication. A proper approach is to follow the temporal, evolutionary changes in captivity of a population, since its foundation from the wild, using the methods of experimental evolution, of which experimental evolutionary domestication is a particular case. This is not to say, obviously, that comparisons of evolutionary dynamics presented by different populations are not a fundamental approach of such studies, as it is by such analyses that the repeatability of the evolutionary dynamics under study can be tested and the degree of predictability of evolution measured.

## GENERAL ISSUES
### THE PROBLEM OF COMPLEX EVOLUTIONARY TRAJECTORIES

The evolutionary trajectories that we have adduced indicate that starvation resistance is evolving through both selection and drift mechanisms during the domestication of *D. subobscura*. It seems likely to us that these mechanisms might generate nonlinear evolutionary trajectories for any particular functional character during longer-term laboratory evolution. How much each of these mechanisms affects the trajectory of a particular character may be rather unpredictable. Novel environments pose difficult evolutionary challenges for both organism and experimenter, challenges that may give rise to genotype-by-environment interactions that in turn generate novel additive genetic covariances among traits.

How repeatable is evolution? Our data across three different studies of detailed characterization of adaptation to the laboratory suggest general repeatability of evolutionary processes and patterns, though also disparity of results for particular traits. This contingency is apparently related to the relevance of these traits with fitness: early fecundity is clearly a very important fitness component, while this is not necessarily the case for starvation resistance. Also, short- and long-term studies can give different results. Our conclusions add to a body of data indicating that although evolution is a global process, its specific outcomes often cannot be generalized (see Rose et al. 2005).

### APPLICATION TO CONSERVATION

Recent interest in characterizing the evolutionary changes of populations from the moment they are brought to the laboratory arises from both their general significance for the study of biological evolution and the need to characterize the specific effects of captivity for the purpose of conservation (Gilligan and Frankham 2003). Not all agree as to what studies of adaptation during captivity can tell us about the impact of such evolution for conservation purposes. Genotype-by-environment interactions will limit considerably extrapolations from the laboratory even to zoo and enclosure environments.

We thus certainly cannot extrapolate the findings of evolutionary change in the laboratory to what will occur when populations are reintroduced in the wild (see Shabalina et al. 1997). The evolutionary genetic complexity of functional traits does not allow reliable inference (cf. Woodworth et al. 2002; Reed et al. 2003). As a safeguard, the best strategy may be to avoid prolonged captivity, minimizing concomitant evolutionary changes (Frankham 1995; Woodworth et al. 2002; Gilligan and Frankham 2003; Rodriguez-Ramilo et al. 2006, cf. with Shabalina et al. 1997).

## WHAT HAVE WE LEARNED ABOUT DOMESTICATION FROM EXPERIMENTAL EVOLUTION?

Most studies of evolutionary domestication indicate that adaptation occurs during domestication, as can be inferred from improvement in such traits as juvenile viability (Hercus and Hoffmann 1999b), early fecundity (e.g., Hercus and Hoffmann 1999a; Matos et al. 2000b, 2002, 2004; Sgrò and Partridge 2000), competitive ability (Frankham and Loebel 1992; Latter and Mulley 1995), and noncompetitive fitness (Woodworth et al. 2002). Some studies differ over the rate of adaptation during captivity (e.g., Frankham and Loebel 1992, cf. Latter and Mulley 1995), and short-term studies may be misleading, as we have shown here. Our studies suggest that domestication can involve complex evolutionary trajectories. We have shown that disparate results among studies of domestication may be due to different methodologies, specifically the limitations of a comparative approach (e.g., Latter and Mulley 1995; Hoffmann et al. 2001; Gilligan and Frankham 2003; Griffiths et al. 2005) versus studies of evolutionary trajectories (Matos et al. 2000b, 2002, 2004; Krebs et al. 2001; Simões et al. 2007). In our view, multiple evolutionary mechanisms can be involved in domestication, and their specific relevance will probably vary from case to case.

From an applied standpoint, the study of adaptation to captivity has received progressively more attention in the conservation literature. There is still a substantial need for basic research on the evolutionary and genetic mechanisms relevant to conservation programs, where these mechanisms range from direct and correlated adaptive responses to inbreeding and drift. The experimental study of domestication is a particularly useful vein for such basic research.

## ARE LAB FLIES DEGENERATE?

Some have argued that laboratory populations that have been established for many generations are of little use for evolutionary studies (Promislow and Tatar 1998; Harshman and Hoffmann 2000; Linnen et al. 2001). Such a view is based, at least in part, on the idea that experimental evolution studies necessarily try to extrapolate results from laboratory populations to evolution in the wild. This is not correct. Experimental evolution is about potential genetic changes in response to defined selection regimes. In particular, some have argued that the ability to select for delay of senescence suggests that alleles with different effects at late ages have accumulated in laboratory populations maintained using short generation times, to a much higher extent than would occur in overlapping generations (Promislow and Tatar 1998; Linnen et al. 2001). While this is indeed expected, we find this criticism ironic in that, to our view, this is one more reason why populations maintained with discrete generations may be the best material to test for the mechanism of accumulation of mutations (Rose and Matos 2004; see also Rauser et al. this volume). After all, this is one of the important tools of experimental evolution, allowing selection to generate differences between the average phenotypes of populations that permit us to infer underlying evolutionary mechanisms (see also Futuyma and Bennett this volume).

More generally, there is no reason to assume that the laboratory environment is not a particular kind of environment or that laboratory populations are not simply natural populations evolving in that environment (Matos et al. 2000a; but see Huey and Rosenzweig this volume, for a different view). As a final note, and as Darwin already understood, evolutionary domestication illustrates the power of natural selection as a process that leads both to the adaptation and the diversity of organisms, as a function of the peculiarities of each environment, whether controlled by humans or not.

**SUMMARY**

This chapter provides a general overview of the field of experimental evolutionary domestication, focusing on studies using *Drosophila* as a model organism. In its general evolutionary sense, domestication means evolutionary change when wild populations are maintained in environments controlled by human choices. It is thus an evolutionary process that is worth analyzing. Here we review the most relevant findings in the field of evolutionary domestication in *Drosophila*, analyzing the evolutionary changes that occur when populations are placed in a human-controlled environment, such as a laboratory. We present our own experiments on evolutionary domestication in the laboratory, going back almost two decades, and discuss our results in the context of relevant literature. We focus on the effects of natural selection, inbreeding, and genetic drift on evolving populations. We compare the results of different research groups, particularly given the common disparities among results involving less relevant fitness traits. We point out the limitations of a comparative approach that relies on inferences based solely on contemporaneous populations, the predominant method used in studies of laboratory adaptation. We argue that the best approach is experimental domestication, in which real-time trajectories are monitored, as is usually the case in experimental evolution studies. After considering applications to conservation, we conclude with a discussion of the debate over the use of long-established laboratory populations for evolutionary research in general.

## ACKNOWLEDGMENTS

This study was partially financed by Fundação para a Ciência e a Tecnologia (FCT) project number POCTI/BSE/33673/2000 and by FCT and POCI 2010 project number POCI-PPCDT/BIA-BDE/55853/2004 (both with the coparticipation of FEDER). J.S. had a BTI grant, and P.S. had a PhD grant (SFRH/BD/10604/2002) from FCT.

## REFERENCES

Bryant, E. H., and D. H. Reed. 1999. Fitness decline under relaxed selection in captive populations. *Conservation Biology* 13:665–669.

Chippindale, A. K. 2006. Experimental evolution. Pages 482–501 *in* C. Fox and J. Wolf, eds. *Evolutionary Genetics: Concepts and Case Studies*. Oxford: Oxford University Press.

Chippindale, A. K., J. A. Alipaz, H.-W. Chen, and M. R. Rose. 1997. Experimental evolution of accelerated development in *Drosophila*. 1. Development speed and larval survival. *Evolution* 51:1536–1551.

Chippindale, A. K., T. J. F. Chu, and M. R. Rose. 1996. Complex trade-offs and the evolution of starvation resistance in *Drosophila melanogaster*. *Evolution* 50:753–766.

Chippindale, A. K., A. L. Ngo, and M. R. Rose. 2004. The devil in the details of life-history evolution: Instability and reversal of genetic correlations during selection on *Drosophila* development. *Journal of Genetics* 82: 133–145.

Darwin, C. 1859. *On the Origin of Species by Means of Natural Selection, or the Preservation of Favoured Races in the Struggle for Life*. London: Murray.

———. 1883. *The Variation of Animals and Plants under Domestication*. 2nd ed. New York: Appleton.

de Jong, G. 1993. Covariance between traits deriving from repeated allocations of a resource. *Functional Ecology* 7:75–83.

Ehiobu, N. G., M. E. Goddard, and J. F. Taylor. 1989. Effect of rate of inbreeding on inbreeding depression in *Drosophila melanogaster*. *Theoretical Applied Genetics* 77:123–127.

Elena, S. F., and R. E. Lenski. 2003. Evolution experiments with microorganisms: The dynamics and genetic bases of adaptation. *Nature Reviews Genetics* 4:457–469.

Folk, D. G., and T. J. Bradley. 2005. Adaptive evolution in the lab: Unique phenotypes in fruit flies comprise a fertile field of study. *Integrative and Comparative Biology* 45:492–499.

Frankham, R. 2005. Genetics and extinction. *Biological Conservation* 126:131–140.

Frankham, R., and D. A. Loebel. 1992. Modeling problems in conservation genetics using captive *Drosophila* populations: Rapid genetic adaptation to captivity. *Zoo Biology* 11:333–342.

Gibbs, A. G., A. K. Chippindale, and M. R. Rose. 1997. Physiological mechanisms of evolved desiccation resistance in *Drosophila melanogaster*. *Journal of Experimental Biology* 200:1821–1832.

Gilligan, D. M., and R. Frankham. 2003. Dynamics of adaptation to captivity. *Conservation Genetics* 4:189–197.

Graves, J. L., E. C. Toolson, C. Jeong, L. N. Vu, and M. R. Rose. 1992. Desiccation, flight, glycogen, and postponed senescence in *Drosophila melanogaster*. *Physiological Zoology* 65:268–286.

Griffiths, J. A., M. Schiffer, and A. A. Hoffmann. 2005. Clinal variation and laboratory adaptation in the rainforest species *Drosophila birchii* for stress resistance, wing size, wing shape and development time. *Journal of Evolutionary Biology* 18:213–222.

Harshman, L. G., and A. A. Hoffmann. 2000. Laboratory selection experiments using *Drosophila*: What do they really tell us? *Trends in Ecology & Evolution* 15:32–36.

Harshman, L. G., A. A. Hoffmann, and A. G. Clark. 1999. Selection for starvation resistance in *Drosophila melanogaster*: Physiological correlates, enzyme activities and multiple stress responses. *Journal of Evolutionary Biology* 12:370–379.

Hercus, M. J., and A. A. Hoffmann. 1999a. Desiccation resistance in interspecific *Drosophila* crosses: Genetic interactions and trait correlations. *Genetics* 151:1493–1502.

———. 1999b. Does inter-specific hybridization influence evolutionary rates? An experimental study of laboratory adaptation in hybrids between *Drosophila serrata* and *Drosophila birchii*. *Proceedings of the Royal Society of London B, Biological Sciences* 266:2195–2200.

Hoffmann, A. A., R. Hallas, C. Sinclair, and L. Partridge. 2001. Rapid loss of stress resistance in *Drosophila melanogaster* under adaptation to laboratory culture. *Evolution* 55:436–438.

Hoffmann, A. A., and L. G. Harshman. 1999. Desiccation and starvation resistance in *Drosophila*: Patterns of variation at the species, population and intrapopulation levels. *Heredity* 83:637–643.

Hoffmann, A. A., and P. A. Parsons. 1993. Direct and correlated responses to selection for desiccation resistance: A comparison of *Drosophila melanogaster* and *D. simulans*. *Journal of Evolutionary Biology* 6:643–657.

Kennington, W. J., J. R. Killeen, D. B. Goldstein, and L. Partridge. 2003. Rapid laboratory evolution of adult body size in *Drosophila melanogaster* in response to humidity and temperature. *Evolution* 57:932–936.

Krebs, R. A., S. P. Roberts, B. R. Bettencourt, and M. E. Feder. 2001. Changes in thermotolerance and hsp70 expression with domestication in *Drosophila melanogaster*. *Journal of Evolutionary Biology* 14:75–82.

Latter, B. D. H., and J. C. Mulley. 1995. Genetic adaptation to captivity and inbreeding depression in small laboratory populations of *Drosophila melanogaster*. *Genetics* 139:255–266.

Lenski, R. 2004. Phenotypic and genomic evolution during a 20,000-generation experiment with the bacterium *Escherichia coli*. Pages 225–265 *in* J. Janike, ed. *Plant Breeding Systems*. New York: Wiley.

Leroi, A. M., W. R. Chen, and M. R. Rose. 1994. Long-term laboratory evolution of a genetic trade-off in *Drosophila melanogaster*. II. Stability of genetic correlations. *Evolution* 48:1258–1268.

Linnen, C., M. Tatar, and D. Promislow. 2001. Cultural artifacts: A comparison of senescence in natural, laboratory-adapted and artificially selected lines of *Drosophila melanogaster*. *Evolutionary Ecology Research* 3:877–888.

Luckinbill, L. S., R. Arking, M. J. Clare, W. C. Cirocco, and S. A. Buck. 1984. Selection for delayed senescence in *Drosophila melanogaster*. *Evolution* 38:996–1003.

Matos, M., and T. Avelar. 2001. Adaptation to the laboratory: comments on Sgrò and Partridge. *American Naturalist* 158:655–656.

Matos, M., T. Avelar, and M. R. Rose. 2002. Variation in the rate of convergent evolution: Adaptation to a laboratory environment in *Drosophila subobscura*. Journal of *Evolutionary Biology* 15:673–682.

Matos, M., C. Rego, A. Levy, H. Teotónio, and M. R. Rose. 2000a. An evolutionary no man's land. *Trends in Ecology & Evolution* 15:2067.

Matos, M., M. R. Rose, M. T. Rocha Pité, C. Rego, and T. Avelar. 2000b. Adaptation to the laboratory environment in *Drosophila subobscura*. *Journal of Evolutionary Biology* 13:9–19.

Matos, M., P. Simões, A. Duarte, C. Rego, T. Avelar, and M. R. Rose. 2004. Convergence to a novel environment: Comparative method versus experimental evolution. *Evolution* 58:1503–1510.

Mueller, L. D., J. L. Graves, and M. R. Rose. 1993. Interactions between density-dependent and age-specific selection in *Drosophila melanogaster*. *Functional Ecology* 7:469–479.

Partridge, L., and K. Fowler. 1992. Direct and correlated responses to selection on age at reproduction in *Drosophila melanogaster*. *Evolution* 46:76–91.

Passananti, H. B., D. J. Deckert-Cruz, A. K. Chippindale, B. H. Le, and M. R. Rose. 2004. Reverse evolution of aging in *Drosophila melanogaster*. Pages 296–322 *in* M. R. Rose, H. B. Passananti, and M. Matos, eds. *Methuselah Flies: A Case Study in the Evolution of Aging*. Singapore: World Scientific.

Prasad, N. G., and A. Joshi. 2003. What have two decades of laboratory life-history evolution studies on *Drosophila melanogaster* taught us? *Journal of Genetics* 82:45–76.

Promislow, D., and M. Tatar. 1998. Mutation and senescence: Where genetics and demography meet. *Genetica* 102/103:299–314.

Reed, D. H., E. H. Lowe, D. A. Briscoe, and R. Frankham. 2003. Fitness and adaptation in a novel environment: Effect of inbreeding, prior environment, and lineage. *Evolution* 57:1822–1828.

Reznick, D. N., and C. K. Ghalambor. 2005. Selection in nature: Experimental manipulations of natural populations. *Integrative and Comparative Biology* 45:456–462.

Rodriguez-Ramilo, S. T., P. Moran, and A. Caballero. 2006. Relaxation of selection with equalization of parental contributions in conservation programs: An experimental test with *Drosophila melanogaster*. *Genetics* 172:1043–1054.

Roper, C., P. Pignatelli, and L. Partridge. 1993. Evolutionary effects of selection on age at reproduction in larval and adult *Drosophila melanogaster*. *Evolution* 47:445–455.

Rose, M. R. 1984. Laboratory evolution of postponed senescence in *Drosophila melanogaster*. *Evolution* 38:1004–1010.

Rose, M. R., and M. Matos. 2004. The creation of Methuselah flies by laboratory evolution. Pages 3–9 *in* M.R. Rose, H. B. Passananti and M. Matos, eds. *Methuselah Flies: A Case Study in the Evolution of Aging*. Singapore: World Scientific.

Rose, M. R., T. J. Nusbaum, and A. K. Chippindale. 1996. Laboratory evolution: the experimental Wonderland and the Cheshire Cat syndrome. Pages 221–241 *in* M. R. Rose and G. V. Lauder, eds. *Adaptation*. San Diego, CA: Academic Press.

Rose, M. R., H. B. Passananti, A. K. Chippindale, J. P. Phelan, M. Matos, H. Teotónio, and L. D. Mueller. 2005. The effects of evolution are local: Evidence from experimental evolution in *Drosophila*. *Integrative and Comparative Biology* 45:486–491.

Rose, M. R., L. N. Vu, S. U. Park, and J. L. Graves. 1992. Selection on stress resistance increases longevity in *Drosophila melanogaster*. *Experimental Gerontology* 27:241–250.

Santos, M., W. Céspedes, J. Balanyà, V. Trotta, F. C. F. Calboli, A. Fontdevila, and L. Serra. 2005. Temperature-related genetic changes in laboratory populations of *Drosophila subobscura*: Evidence against simple climatic-based explanations for latitudinal clines. *American Naturalist* 165:258–273.

Santos, M., P. Fernández Iriarte, W. Céspedes, J. Balanyà, A. Fontdevila, and L. Serra. 2004. Swift laboratory thermal evolution of wing shape (but not size) in *Drosophila subobscura* and its relationship with chromosomal inversion polymorphism. *Journal of Evolutionary Biology* 17:841–855.

Service, P. M., E. W. Hutchinson, and M. R. Rose. 1988. Multiple genetic mechanisms for the evolution of senescence in *Drosophila melanogaster*. *Evolution* 42:708–716.

Service, P. M., and M. R. Rose. 1985. Genetic covariation among life-history components: The effect of novel environments. *Evolution* 39:943–945.

Sgrò, C. M., and L. Partridge. 2000. Evolutionary responses of the life history of wild-caught *Drosophila melanogaster* to two standard methods of laboratory culture. *American Naturalist* 156:341–353.

———. 2002. Laboratory adaptation of life history in *Drosophila*. *American Naturalist* 158:657–658.

Shabalina, S. A., L. Y. Yampolskya, and A. S. Kondrashov. 1997. Rapid decline of fitness in panmictic populations of *Drosophila melanogaster* maintained under relaxed natural selection. *Proceedings of the National Academy of Sciences of the USA* 94:13034–13039.

Simões, P., M. R. Rose, A. Duarte, R. Gonçalves, and M. Matos. 2007. Evolutionary domestication in *Drosophila subobscura*. *Journal of Evolutionary Biology* 20:758–766.

Simões, P., J. Santos, I. Fragata, L. D. Mueller, M. R. Rose, and M. Matos. 2008. How repeatable is adaptive evolution? The role of geographical origin and founder effects in laboratory adaptation. *Evolution* 62:1817–1829.

Stearns S. C., G. de Jong, and B. Newman. 1991. The effects of plasticity on genetic correlations. *Trends in Ecology & Evolution* 6:20–26.

Teotónio, T., and M. R. Rose. 2000. Variation in the reversibility of evolution. *Nature* 408:463–466.

Teotónio, T., M. Matos, and M. R. Rose. 2002. Reverse evolution of fitness in *Drosophila melanogaster*. *Journal of Evolutionary Biology* 15:608–617.

———. 2004. Quantitative genetics of functional characters in *Drosophila melanogaster* populations subjected to laboratory selection. *Journal of Genetics* 83:265–277.

Woodworth, L. M., M. E. Montgomery, D. A. Briscoe, and R. Frankham. 2002. Rapid genetic deterioration in captive populations: Causes and conservation implications. *Conservation Genetics* 3:277–288.

Wool, D. 1987. Differentiation of island populations: A laboratory model. *American Naturalist* 129:188–202.

# 6

# LONG-TERM EXPERIMENTAL EVOLUTION AND ADAPTIVE RADIATION

Michael Travisano

## WHAT IS LONG-TERM EVOLUTION?

Long-term evolution studies are selection experiments that explore evolutionary consequences. Rather than focusing on the potential for selection to act, which is typical of short-term selection studies, long-term studies emphasize the process and eventual outcome of evolution.

The major benefit of this type of study is that evolutionary outcomes can be directly studied in terms of the processes that gave rise to them, such as mutation and adaptation, leading to a better understanding of the mechanisms underlying evolutionary change. The focus on consequences, rather than potential, typically involves multiple sequential genetic differences, rather than only one or a few, an approach motivated by the inherent nonlinearity of biological systems. In scope, long-term studies bridge the gap between phylogenetic approaches and short-term experimental studies of evolution, allowing the identification of the broad patterns of evolutionary change as well as their underlying mechanisms.

## THE NEED FOR LONG-TERM EVOLUTION STUDIES

Since Darwin, evolutionary biologists have marveled at the diversity of life and the amazing "fit" of organisms to their environments. The best-known examples are adaptive radiations, in which a single lineage diversifies into many ecologically distinct species (Grant 1999). The niche-specific adaptations of derived lineages provide ideal examples of the potential for evolutionary fine-tuning to environmental conditions (Seehausen 2006). The commonalities among the derived species of an adaptive radiation provide a striking counterpoint to the adaptive diversity among these species, illuminating the specificity of adaptive evolution (Schluter 2000). Indeed, the apparent fit of organisms to their environments has frequently blinded biologists to the limits of adaptation and the importance of other evolutionary factors (Gould and Lewontin 1979). Perhaps one of the major sources of such overestimation of the impact of adaptation is the inability of biologists to directly observe the evolutionary mechanisms that underlie adaptations as they first evolve. Support for this interpretation is evident from the great value of studies of adaptive radiation, studies that have disproportionately contributed to evolutionary understanding (Travisano and Rainey 2000; Schluter 2000). Familiar examples include Darwin's finches (Petren et al. 1999), the three-spined stickleback (Colosimo et al. 2005), and African cichlids (Seehausen 2006). In these and other adaptive radiation studies, the impacts of selection and adaptation are assessed relative to the ancestral state and the relevant ecological conditions, both of which are often known with some degree of confidence.

Comparative approaches, such as most studies of adaptive radiation, provide insight into past evolutionary change (recent review in Garland et al. 2005). Paleontological and molecular phylogenetic studies illuminate the avenues that evolution has taken, providing a list of answers to prior selective questions. Such answers demonstrate the potential for adaptation to mold organismal structure and function. Unfortunately, the "selective questions" that produced the answers are often unknown and typically can only be indirectly inferred from these answers, even in studies of adaptive radiation. Hence, comparative studies provide limited insight as to why a particular evolutionary outcome occurred and what other evolutionary outcomes might have been possible. There are numerous examples of difficulties arising from uncertainty concerning prior selective conditions in understanding patterns in processes in adaptation, one of the best being the evolution of feathers. Although feathers are clearly adaptive for avian flight, that is unlikely to have been the case initially (Prum and Brush 2002), and uncertainty persists with respect to both the origin and the evolutionary history of feathers (Perrichot et al. 2008). This uncertainty, despite the significance of feathers for a highly successful family (Aves) and their occasional fossilization, results in part from an absence of extant "living fossils" (think monotremes and marsupials for mammals) as well as the complexity of feathers among extant species (Maderson and Homberger 2000). Traits that have undergone dramatic modification, such as feathers, and for which there are no longer living species with the ancestral trait state, can be particularly difficult to assess comparatively. Of course, experimental evolution is unlikely to address the specific circumstances of feather evolution, as long-term selection requires species that have short generation times. All study systems have limitations. Even so, the problem of feather evolution exemplifies the difficulties that arise in applying the comparative approach to the evolution of novel traits whose selective benefits may have changed.

Short-term selection studies are an attempt to overcome some of the difficulties of comparative approaches (Futuyma and Bennett this volume; Rose and Garland this volume). By maintaining replicate populations in experimentally defined conditions, both the causes of selection and the evolutionary responses can be studied directly. Moreover, selective cause and response can be studied in conjunction, allowing direct study of the causation of evolutionary change, rather than correlation or indirect assessments of causation. Even so, most selection studies involve relatively few generations, typically about ten to one hundred. Relatively rapid evolutionary changes can be investigated over such lengths of time, but even rapid evolutionary changes represent only the most immediate type of adaptation. As evolutionary change frequently involves epistasis, pleiotropy, and contingency, extrapolating from short-term studies to longer-term evolutionary patterns is problematic (see also Huey and Rosenzweig this volume regarding various other limitations of short-term studies in experimental evolution). Genetic correlations between traits change, so that basing long-term evolutionary predictions on the response to selection over relatively few generations can lead to erroneous inferences (Travisano 1997; Chippindale et al. 2003; Phelan et al. 2003). Moreover, the patterns of diversification that are studied comparatively involve time periods vastly longer than those of short-term evolution experiments, leading some to question the potential of the two different approaches to complement each other (Gould 2002).

Long-term evolution experiments can overcome the limited scope of traditional short-term selection experiments, simply by expanding the time over which the experiment is carried out. The association of organism and environment can be studied in an open-ended way, rather than focusing only on the anticipated evolutionary responses to particular conditions, which shorter-term experiments are forced to do. The broader scope of long-term studies greatly improves direct experimental investigation of a central topic of evolutionary biology: the causality of adaptation (Travisano 2001; Dean and Thornton 2007).

It is important to understand just how difficult it is to unravel the causality of adaptation. The series of evolutionary events and the causes for each step must be identified (Sokurenko et al. 2004). The complexity of biological systems ensures that understanding adaptation is inherently a multilevel pursuit. Selection acts on fitness, but a particular adaptation might involve protein structure, gene expression, or any number of other aspects of biological function. For example, microbial resistance to the antibiotic tetracycline can occur by at least thirty-six different genes that encode for at least three different physiological mechanisms (table 6.1): active efflux of the antibiotic from the cell, protection of the target, and enzymatic inactivation (Chopra and Roberts 2001). Although the multiplicity of genes and physiological modes illustrated by this case is ideal for testing the generality of trade-offs, it greatly complicates understanding the causality of adaptation. Even for such traits as antibiotic resistance, where the fitness benefits of resistance are evident, multiple avenues for selective responses complicates our understanding evolutionary causality because the mechanistic functions of resistance alleles are often uncertain, at least initially. The theory of evolution by natural selection provides a causal framework for adaptation and permits the development of powerful mathematical analyses of evolution. But this theory does not address the underlying functional mechanisms by which adaptation proceeds. However, the occurrence of particular selective outcomes critically depends on the mechanisms by which evolution occurs (see also Swallow et al. this volume).

## SUCCESSES AND CHALLENGES FOR LONG-TERM EVOLUTION

Long-term experimental evolution studies invariably use model organisms. There is typically a wealth of available biological information about model systems, information that is essential to successful interpretation of responses to the selection that is imposed experimentally. Not all model organisms are ideal for experimental evolution, as relatively easy propagation and short generation times are key for carrying out a successful long-term evolutionary experiment. Most studies use insects, typically *Drosophila*, or microbes. Greater biological complexity and a closer phylogenetic association with other organisms of general interest, such as *Homo sapiens*, are the key benefits for using insects or other metazoans, but these trade off with far fewer generations than can be obtained with microbes, with generation time:generation time ratios of 1:100+, insects relative to microbes, just to give one contrast. Indeed, most nonmicrobial long-term experiments are extensions of short-term selection experiments that have persisted over many years.

TABLE 6.1  Resistance Mechanism for Characterized
Tetracycline Resistance Genes

| Efflux[a] | Ribosomal Protection[b] | Enzymatic[c] | Unknown |
|---|---|---|---|
| otr(B) | otr(A) | tet(X) | tet(U) |
| otr(C) | tet | tet(34) | |
| tcr | tet(M) | tet(M37) | |
| tet(A) | tet(O) | | |
| tet(B) | tet(Q) | | |
| tet(C) | tet(S) | | |
| tet(D) | tet(T) | | |
| tet(E) | tet(W) | | |
| tet(G) | tet(32) | | |
| tet(H) | tet(36) | | |
| tet(J) | tetB(P) | | |
| tet(K) | | | |
| tet(L) | | | |
| tet(V) | | | |
| tet(Y) | | | |
| tet(Z) | | | |
| tet(30) | | | |
| tet(31) | | | |
| tet(33) | | | |
| tet(35) | | | |
| tet(38) | | | |
| tet(39) | | | |
| tet(41) | | | |

NOTE: From Chopra and Roberts 2001. Each gene product has less than ≤79% amino acid identity with any other tetracycline resistance gene.

[a] Tetracycline is expelled from the cell.
[b] The target of tetracycline activity is blocked.
[c] Tetracycline is inactivated.

## DIVERSIFICATION AND SPECIALIZATION

A large literature, both theoretical and experimental, addresses the evolution of specialization and diversification. Multiple contending hypotheses concern the mechanisms underlying these evolutionary patterns, and these hypotheses are difficult to fully evaluate via traditional means because specialization and diversification are typically studied long after they have occurred.

A striking finding from long-term microbial evolution is that diversity readily evolves. The rapidity, extent, and persistence of diversity in microbial evolution experiments was largely unanticipated by prior evolutionary research, including most microbial studies. The surprising amount of diversity in microbial studies arises from varied modes of adaptation to the same environment, the large extent to which even an apparently simple environment can be subdivided, and the effect of subtle impacts of the environment on evolutionary outcomes.

## CLONAL INTERFERENCE

One of the first observations of microbial experimental evolution studies was "periodic selection," the purging of diversity within a population as a result of selective sweeps to fixation by adaptive mutations (Atwood et al. 1951). In the absence of recombination, periodic selection was thought to preclude the persistence of diversity within microbial populations, and repeated periodic selection events can result in continued loss of diversity (Dykhuizen and Hartl 1983). This paradigm was consistent with the "classical" model of adaptation (Dobzhansky 1955), developed prior to neutral theory (Kimura 1968), a model that persisted even after the observation of large amounts of genetic variation in bacterial populations (Lewontin 1974). The potential for multiple avenues of adaptation was generally underappreciated (but see Levin 1972), and bacterial population structure was poorly known (as it generally remains), making the causes of genetic variation difficult to evaluate (Selander and Whittam 1983). Nevertheless, diversity persists in larger populations due to the appearance of multiple contemporaneous adaptive mutations in different lineages within a population, resulting in co-occurring selective sweeps that prevent purging of extant diversity (Hill and Robertson 1966; Gerrish and Lenski 1998). Diversity can thus persist because multiple beneficial mutations occur roughly contemporaneously (figure 6.1).

Substantial empirical and theoretical efforts have been made to assess the importance of competition associated with the different mutant lineages, since such "clonal interference" not only causes the persistence of variation but also reduces rates of fixation for adaptive mutations (e.g., Miralles et al. 1999; Imhof and Schlötterer 2001; Wilke 2004; de Visser and Rozen 2006; Park and Krug 2007).

## ADAPTIVE DIVERSIFICATION

The importance of chance during adaptation is typically unexamined. While all adaptations are the result of both deterministic selective factors and stochastic factors associated with genetic variation, such as mutation, studying the impact of chance on past adaptations is difficult. Indeed, the role of chance in evolution is primarily studied experimentally for traits that have approximate selective neutrality. Nevertheless, adaptations have the potential to be strongly impacted by chance since mutational effects are

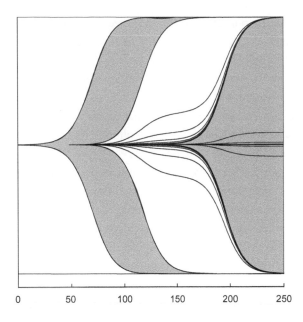

FIGURE 6.1

Clonal interference maintains diversity. A simulation of three rounds of adaptation, shaded then white then shaded, of roughly 10 percent fitness increases each time. Each interior curved line represents a mutant, and the area between lines corresponds to the total fraction of the population consisting of the mutant class associated with the line to the left. In the first round, a single mutant occurs and reaches about 90 percent before the next more fit mutant arises. In the second round, additional equally fit mutants arise and all increase in frequency until a third set of more fit mutations appears. The number of new adaptive mutants arising for both the second and third rounds is based on a strongly declining number of remaining adaptive mutations. Clonal interference in microbial populations is likely to lead to even greater amounts of sustained diversity than shown here, as there are likely many more adaptive mutations occurring contemporaneously, consistent with some models of ecological neutrality (Hubbell 2001).

context-dependent. Different mutations can potentially have disparate effects on different genetic backgrounds and in different environments. For example, if there are multiple potential adaptive mutations, as mentioned earlier, then the fixation of one versus another can potentially have long-term evolutionary consequences. This is because the selective effect of subsequent mutations will be impacted by the chance fixation of a particular prior adaptive mutation. Fortunately, long-term evolution with microbes encompasses enough evolutionary scope to include the impact of both adaptation and chance.

The first long-term evolution experiment to explicitly examine chance impacts on adaptation demonstrated that inclusion of chance is key to understanding evolutionary outcomes. Replicate populations of *Escherichia coli* were initiated from the same common ancestor and maintained under uniform selective conditions, serial batch culture with glucose as the sole usable carbon and energy source (Lenski et al. 1991; Lenski and

Travisano 1994). As the populations were initiated using a single *E. coli* genotype, all adaptation could only occur via the appearance of *de novo* mutations during the course of selection, and all differences among populations could also only arise via mutation. The replicate populations were large. Each was refreshed daily, by a hundred-fold dilution of the prior day's culture, with $5 \times 10^6$ the minimum census population size. Large population sizes increase the variety of mutations that appear and promote parallelism as the appearance of the same mutations among replicates increases with population size. Both parallelism and divergence were observed.

Adaptation was rapid, with an average of about 30 percent improvement in competitive ability in the selected environment after two thousand generations (Lenski et al. 1991). This estimate was obtained by performing head-to-head competitions of derived populations against the common ancestral genotype (figure 6.2). Derived and ancestral individuals were differentiated using a marker, arabinose utilization, that had no fitness impact in the selected environment. Arabinose utilizing (Ara⁺) *E. coli* form white colonies on tetrazolium arabinose (TA) solid agar, while *E. coli* incapable of using arabinose form red colonies (Ara⁻). The arabinose marker was embedded in the selection experiment (half were Ara⁺, half Ara⁻), the genotypes being isogenic besides the

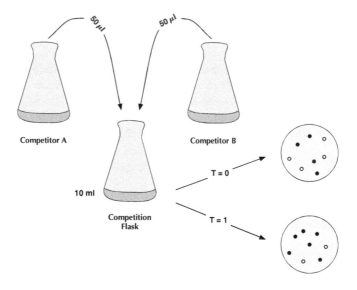

FIGURE 6.2

Competition assay schematic. Competitors are simultaneously introduced into a single competition environment, which is immediately assayed for density of both competitors ($T = 0$). Competitors are distinguished by a color marker: red (shown as black) versus white. After twenty-four hours, the competition environment is again sampled to assess competitor densities ($T = 1$). Fitness is the ratio of number of divisions of both competitors over the sampling period. The simplicity of the culture conditions allows for strict control of the competition environment, permitting the experimenter either to closely follow the selective conditions of the experimental evolution or to change these conditions in specific ways.

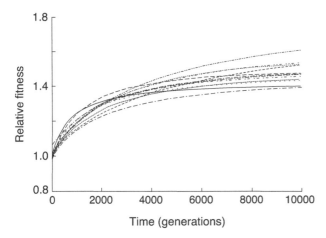

FIGURE 6.3
Trajectory of mean fitness for
each of the twelve replicates over
ten thousand generations. Each
curve is the best hyperbolic fit to
data obtained every five hundred
generations (Lenski and
Travisano 1994).

arabinose marker, and competitions between derived and ancestral individuals used opposing arabinose marker states.

The rate of adaptation declined over the course of selection, and by ten thousand generations, the total improvement in average fitness was about 50 percent relative to that of the common ancestor (figure 6.3). This decline in rate of fitness improvement is consistent with a decline in the number of available beneficial mutations and a decline in the fitness benefits conferred by each mutation. The dynamics of adaptation were largely parallel, with statistically significant fitness differences yielding an among-population standard deviation of ~0.05 after ten thousand generations. While chance was essential for the appearance of *de novo* selectively advantageous mutations, it had only a moderate impact on the improvement in fitness in the selected environment. The high degree of phenotypic parallelism found in these experiments reflects both the fixation of similar mutations affecting the same traits as well as constraints on the rate of fitness improvement.

Parallel adaptive evolution of genome sequences was particularly evident in the ribose operon (Cooper et al. 2001), which led to elimination of D-ribose catalytic function. The loss of catalytic function in all twelve replicate populations by two thousand generations involved the same mechanism, partial deletion of the *rbs* operon mediated by an insertion element (IS*150*) located immediately upstream in the unselected ancestral genotype (figure 6.4). The amount of operon lost varied among replicates, from almost complete operon deletion to loss of only the initial portion, but in every case, this resulted in loss of catalytic function and an approximate fitness improvement of 1.5 percent. Deletion presumably occurred via transposition of another IS*150* element into the *rbs* operon, at the downstream end of the deleted portion, followed by recombination between the two insertion elements. Although the fitness improvement observed was small, its rapid and repeated occurrence was promoted by an increased rate of mutational loss of ribose catalytic function due to the presence of the insertion element. The

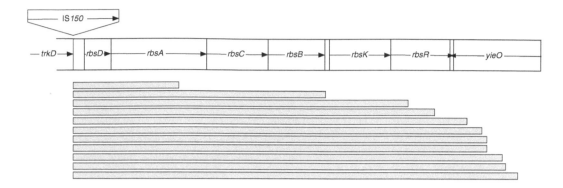

FIGURE 6.4

*E. coli* ribose (*rbs*) operon and deletion genotypes. The ancestral gene structure of the ribose catabolic operon is given, *rbsD* to *rbsR*, with flanking genes. Loss of ribose catabolism by operon deletion occurred in twelve of twelve replicate populations during two thousand generations of selection in glucose-limited medium. Isolates from eleven of the replicate populations were sequenced, and deletions were mapped (lower gray bars). In every case, the upstream deletion mapped to an insertion sequence (IS*150*), which facilitated deletion, even though the extent of downstream deletion varied considerably. Redrawn from Cooper et al. (2001).

combination of both localized increased mutation rate and consistent competitive benefit in the selected environment led to the rapid and parallel adaptation.

Parallelism was also evident in the general loss of catabolic function during the course of selection, as there were consistent and parallel losses of catabolic function by twenty thousand generations (Cooper and Lenski 2000). In principle, there are two potential causes for such parallelism, causes similar to those associated with aging: mutation accumulation or antagonistic pleiotropy (Rauser et al. this volume). Loss of function can result from the fixation of mutations that are neutral in the selected environment, glucose-supplemented medium, but that have negative impacts for catabolism of novel nutrients, a case of mutation accumulation. Alternatively, parallel catabolic loss could have occurred due to fixation of mutations across populations that are beneficial in the selected environment and simultaneously lead to a loss of catabolic function, a form of antagonistic pleiotropy. The two mechanisms can be discriminated by the dynamics of evolutionary change, as mutation accumulation is driven by rates of neutral mutation, while pleiotropy is associated with the evolutionary rates characteristic of adaptive substitutions. The overwhelming majority of catabolic losses occurred early in the evolution, by two thousand generations, consistent with the rapid rate characteristic of adaptation. If mutation accumulation had been largely responsible, catabolic losses would have been more evenly spaced throughout the selection experiment. The losses of catabolic function were also largely independent of differences in mutation rate. Three replicates evolved about fiftyfold higher mutation rates, at 2,500, 3,000, and 8,500 generations, but these three replicates experienced neither higher rates of catabolic loss nor higher rates of adaptation. Hence, antagonistic pleiotropy was probably the dominant factor causing parallel evolution and loss of catabolic function (Cooper and Lenski 2000).

Divergence was observed and pleiotropy was likely the dominant factor causing diversity across replicates. Prior short-term bacterial selection experiments had identified nutrient transport as a frequent target of selection during growth in single nutrient environment, the fitness of an individual bacterium being tightly associated with its ability to acquire resources (Dykhuizen et al. 1986). Because of the different physiological modes of nutrient uptake in *E. coli*, Travisano et al. (1995b) anticipated that the fitness of the glucose-selected populations would vary among novel nutrient environments. Adaptations improving glucose uptake could carry over to improved uptake of those carbohydrates brought into cells by the same physiological mechanisms (Travisano and Lenski 1996). Glucose transport through the inner membrane in *E. coli* primarily occurs by the phosphotransferase system (PTS), which concomitantly transports and phosphorylates transported carbohydrates, employing a multistep phosphorylation cascade from a central metabolic intermediate (phosphoenolpyruvate). Uptake of non-PTS carbohydrates does not involve the PTS phosphorylation cascade; potentially more importantly, non-PTS carbohydrate uptake is generally inhibited by glucose transport and catabolism (Saier 1989). These physiological interconnections underlying carbohydrate transport

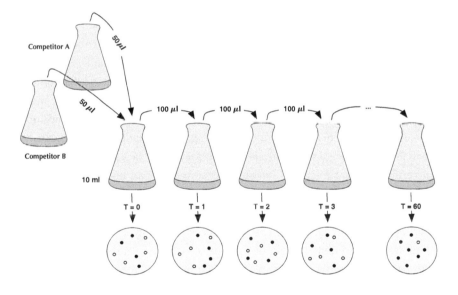

FIGURE 6.5
Collection of independent adaptive mutants. Replicate cultures were initiated with a mix of Ara[+] and Ara[−] variants, which were otherwise isogenic. The arabinose marker is selectively neutral (Lenski et al. 1991; Travisano et al. 1995). The mix was propagated daily and periodically sampled onto TA indicator agar, which distinguishes the variants by color. During propagation, a consistent deviation from the initial ratio ($T = 0$) indicated the appearance of a beneficial mutation in the increasing marker subpopulation (Rozen et al. 2002; Ostrowski et al. 2005). The cultures were propagated up to four hundred generations.

suggested that fitness assays in other, novel, single-carbohydrate media would provide clues as to modes of adaptation to the glucose-limited selective environment.

Abundant diversity was observed among the evolved replicate populations in some novel carbohydrate environments (Travisano et al. 1995b; Travisano and Lenski 1996). The extent of diversity was largely dependent on nutrient transport, with less diversity when glucose was replaced with another PTS carbohydrate that shares the same mechanisms of inner membrane nutrient uptake as glucose. In these novel carbohydrate environments, adaptations to the selected environment were also generally beneficial.

But there was greater diversity among replicates when transport mechanisms for the novel carbohydrate differed from glucose. In these environments, some adaptations to the selected environment were beneficial, but others were deleterious. Since each replicate had an independent evolutionary history, different mutations fixed in different replicates, and the different mutations had different pleiotropic effects. As with the parallel catabolic losses discussed earlier, the rapidity of diversification observed in selected lines is far faster than anticipated for mutation accumulation (Travisano and Lenski 1996). Moreover, even those carbohydrates sharing the PTS uptake mechanism with glucose have nutrient-specific uptake genes, and most PTS and non-PTS carbohydrates have roughly the same number of nutrient-specific coding and regulatory regions (about five thousand bases). The parallelism in novel PTS nutrients, but diversity among non-PTS nutrients, indicates that the primary cause of divergence is not the accumulation of mutations neutral in the selected environment and deleterious in novel environments.

Here we see that the same factor, pleiotropy, yields both parallelism and divergence depending on the trait in question. The difference could only be readily observed in an experimental system that includes both the potential for adaptation and the potential for truly independent evolutionary histories, thanks to replication. Simultaneous evaluation of both adaptation and chance impacts were essential for understanding the evolutionary outcome, which has led to a broader understanding of the importance of pleiotropy.

The major limitation of this analysis of pleiotropy is the lack of independent experimental verification, but microbial long-term experiments can be subsequently investigated and repeated to directly address specific issues. The potential for simultaneous parallel and divergent phenotypic responses via pleiotropy has been examined in a related set of lineages (Ostrowski et al. 2005). The same ancestral genotype was used as that of the prior long-term experiment (Lenski et al. 1991; Lenski and Travisano 1994), but each replicate was initiated with a polymorphic neutral marker (figure 6.5) so that genotypes with single beneficial mutations could be identified (see Rozen et al. 2002). Initiating a replicate with two genotypes, isogenic except for a neutral marker, allows easy observation of the rapid changes in frequency that are associated with beneficial mutants sweeping through the replicate. Within four hundred generations, twenty-seven independent higher fitness genotypes were obtained for the glucose-limited environment. The mutants were assayed in the novel carbohydrate environments previously identified as illustrating generally parallel or divergent correlated responses (Travisano and Lenski 1996).

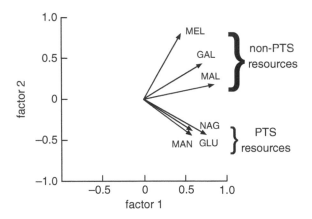

FIGURE 6.6

Principal components analysis plot for fitness in six nutrients: GAL, galactose; GLU, glucose; MAL, maltose; MAN, mannitol; MEL, melibiose; and NAG, N-acetylglucosamine. PTS nutrients share a resource uptake pathway, while non-PTS nutrients use a diverse set of uptake pathways. The first and second components respectively explain 46.0 percent and 24.3 percent of the fitness variation among the different nutrients for the mutants. The vectors for the three PTS nutrients are all similar to one another, while the three non-PTS nutrient vectors differ from the non-PTS nutrients and one another. Redrawn from Ostrowski et al. (2008).

Numerous pleiotropic effects were observed, the majority being positive, but antagonistic pleiotropy was also observed.

A principal components analysis associated pleiotropic effects with mechanisms of carbohydrate uptake (figure 6.6). As expected from the initial long-term study, fitness in PTS nutrients largely mirrored that in the selected glucose environment, while fitnesses in novel non-PTS nutrients differed between nutrients and from that with the PTS nutrients. The genetic mechanisms underlying these pleiotropic responses were resolved (Ostrowski et al. 2008), based on identification of the beneficial mutations found in the initial long-term experiment. Previously, beneficial mutations in the long-term selection study were identified by screening isolates for insertion events (Schneider et al. 2000), a fraction of which were beneficial and served as candidate loci for subsequent screening in the independently selected replicate long-term isolates (Woods et al. 2006). Having identified five loci frequently associated with adaptation in the long-term isolates, these gene regions were then sequenced from each of the twenty-seven independently derived genotypes. Twenty-one mutations were observed that had undergone selective substitution, and all but one of the genotypes fixed by selection contained a single mutational difference (table 6.2). While there was near parallelism, across and within loci having mutations, for fitness effects in the selected environment, there was divergence among replicates. Phenotypic divergence was caused both by mutations arising at different loci and different mutations within a locus. Different beneficial mutations had largely parallel effects in the selected environment, but divergent responses in novel environments.

TABLE 6.2　Twenty-one Mutations Discovered by Sequencing of Five Candidate Loci in 27 Independently Selected Isolates

| Gene | Position | Mutation | Amino Acid Change |
|------|----------|----------|-------------------|
| hokB-sokB | NA | ::IS150 | (insertion) |
| nadR | 30 | A → △[a] | (deletion) |
| nadR | 169 | ::IS150 | (insertion) |
| nadR | 186–189[b] | G → △ | (deletion) |
| nadR | 186–189[b] | G → △ | (deletion) |
| nadR | 931 | A → G | Lys → Glu |
| pbpA-rodA | −828[c] | C → A | (noncoding) |
| pykF[d] | 1153 | C → A | Arg → Ser |
| spoT | 316 | C → T | Leu → Phe |
| spoT | 990 | G → A | Met → Ile |
| spoT | 1226 | T → C | Phe → Ser |
| spoT | 1249 | A → C | Ile → Leu |
| spoT | 1249 | A → C | Ile → Leu |
| spoT | 1324 | A → C | Thr → Pro |
| spoT | 1324 | A → C | Thr → Pro |
| spoT | 1370 | G → T | Trp → Leu |
| spoT | 1715 | A → C | Lys → Thr |
| spoT | 1724 | G → T | Arg → Leu |
| spoT | 1769 | A → C | Lys → Thr |
| spoT | 1993 | C → T | Arg → Cys |
| spoT[d] | 1994 | G → A | Arg → His |

NOTE: From Ostrowski et al. 2008. Three identical mutations were found in two isolates each (boxed).

[a] △ indicates a deletion mutation.
[b] Deletion of one G in a string of 4 G's.
[c] Number of bps upstream from the start of the pbpA-rodA operon.
[d] Pair of mutations found in one isolate.

## ADAPTIVE RADIATION

Studies of diversity arising from adaptive radiation often focus on unique adaptations. In particular, studies of adaptive radiation frequently involve species that are endemic to specific environments. Repurposing the investigation of adaptive radiations to microbial selection experiments is surprisingly fruitful. The same general approach is taken, studying the modes of adaptation to particular features of the (laboratory) environment by studying the unique aspects of the endemic (laboratory) species. A major advantage of laboratory studies of adaptive radiation is the potential to make connections among biochemical mechanisms, genetic details, and phenotypic effects. Making these connections allows the mechanisms underlying adaptation, and adaptive radiation, to be directly studied.

Adaptive radiation occurs readily in long-term experimental evolution. In spatially homogenous minimal nutrient environments, resource subdivision can be observed that appears to involve resource energy content (e.g., Helling et al. 1987; Souza et al. 1997; Rozen and Lenski 2000; Friesen et al. 2004; Maharjan et al. 2007). Specialists on high-energy content resources (six-carbon carbohydrates) grow poorly or not at all on low-energy content resources (two to three carbons), while genotypes that are capable of growing on low-energy resources are inferior competitors on high-energy resources. The evolution of specialists and generalists is central to understanding biological diversity, and their evolution is readily understood in these experimental evolution studies: there is a trade-off between resource breadth and competitive ability on the preferred resource. Moreover, the underlying mechanisms for this trade-off have been partially elucidated (Rosenzweig et al. 1994; Spencer et al. 2007). In one instance in a temporally varying environment, derepression of growth on low-energy resources allows for faster nutrient switching once the preferred nutrient concentration declines (Spencer et al. 2007). Further molecular analysis of the underlying mechanisms associated with trade-offs in resource use will illuminate the factors affecting resource specialization.

The most dramatic phenotypic diversification has been observed in the case of the rapid evolution of *Pseudomonas fluorescens* (Rainey and Travisano 1998) in a nutrient-rich spatially structured environment. In as little as three days, a single genotype of *P. fluorescens* diversifies into at least three morph phenotypes that are easily differentiated when sampled onto nutrient-rich solid medium and allowed to form colonies. Each morph can be grown in isolation and occupies distinct portions of the microcosm: the air-broth interface at the top, the broth phase, or the bottom of the microcosm. The morph phenotypes are heritable and persist when transferred from plate to plate, without passage through the selective environment. Two key aspects of the environment lead to diversification: spatial structure and nutrient richness. The absence of either precludes initial diversification or leads to loss of diversity if removed after diversification (Rainey and Travisano 1998; Travisano and Rainey 2000). In a nutrient-rich spatially structured environment, diversity persists due to competition among the morph phenotypes via negative frequency dependence: rare genotypes are more fit than common genotypes. The negative frequency dependence is due to the niche preferences of the three morph types for the different microcosm regions.

Perhaps the most distinctive of the three morphs is the wrinkly spreader (WS) morph (figure 6.7). WS colonies grow from genotypes able to colonize the air-broth interface, via production of an extracellular polysaccharide (Goymer et al. 2006) and overproduction of an adhesin (Spiers et al. 2003). WS genotypes readily stick to one another and to the microcosm glass sides, thereby allowing mat formation at the air-broth interface. This stickiness provides WS genotypes access to both oxygen and nutrients, at a cost of reduced migration and resource use for production of the polysaccharide. Regulation of the extracellular polysaccharide occurs via the Wsp operon, a seven-gene single transcriptional unit. Of particular importance is that Wsp regulation of polysaccharide

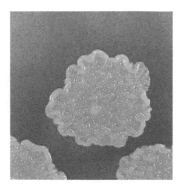

FIGURE 6.7
A wrinkly spreader (WS)
colony and the localization
of growth of WS genotypes
in static broth culture
(Rainey and Travisano
1998). Photos courtesy of
C. F. Landry.

appears to occur by modulation of a dynamic equilibrium, where Wsp proteins switch
between active and inactive states. Evaluation of this type of regulatory structure
indicates that it is particularly sensitive to bimodal evolution, resulting in either overexpression of polysaccharide (WS phenotype) or no expression (ancestral phenotype). Of
twenty-six independently derived WS genotypes, thirteen contained mutations in a single gene, *wspF*, that reduces or eliminates its inhibition of polysaccharide production
(Bantinaki et al. 2007).

These experimental studies of adaptive radiation have identified ecological circumstances that promote and maintain niche diversity. Confidence in the generality of the
conclusions is obtained via the replicated nature of the studies; specifically, it is clear that
the results are not idiosyncratic effects of an unknown and uncontrollable factor. Beyond
this, the underlying biochemistry and genetics have been uncovered for some aspects
of the niche specialization. This level of detail provides the type of insight into the causes
of trade-offs, adaptive radiation, and adaptive evolution in general that only long-term
replicated studies of experimental evolution can provide.

## CAUTION IS REQUIRED

The inability to tie evolutionary responses to underlying genetic and physiological mechanisms has repeatedly led to misunderstanding and confusion about evolutionary
processes (see also Rauser et al. this volume; Swallow et al. this volume). There have
been frequent attempts to invoke additional processes or mechanisms, in addition to
natural selection, to explain the fit of an organism to its environment. Some of these
attempts have involved faulty analysis of experimental evolution. The directed mutation
hypothesis (Cairns et al. 1988) is a relatively recent example, in which some biologists
hypothesized that the utility of mutations affects the joint probability of their occurrence
and expression, so that mutations are "directed" by their utility. The principal claim that

generated this particular controversy was that mutations encoding catabolic abilities occur more frequently when they are beneficial than when neutral or deleterious (Hall 1988). In two studies, bacterial populations were placed into single-resource environments, in which the bacteria were incapable of catabolizing the single available resource. As they starved, mutants arose in the bacterial populations that had acquired the necessary mutations for growth on the only available resource. The large number of beneficial mutants and their positive correlation with the time spent starving appeared to suggest that the mutations had occurred specifically because of their selective benefits. This interpretation directly contradicted a central premise of both molecular and evolutionary biology that the initial appearance of mutations is unaffected by their selective value (Luria and Delbrück 1943).

A lack of knowledge of the mechanistic basis for the ostensible cases of directed mutation left an opening for an alternative hypothesis that superficially appeared more parsimonious than natural selection (Stahl 1988). A fuller understanding was achieved by additional experiments and analysis that demonstrated that erroneous assumptions about the growth and rate of spontaneous mutation were largely responsible for the apparent neo-Lamarckian observations (e.g., Mittler and Lenski 1992). The population sizes used in the original directed mutation experiments of Cairns et al. (1988) were larger than assumed, and spontaneous mutation rates vary depending on environmental conditions— experimental problems illustrating the care necessary for interpreting selection studies. The actuarial errors in the original studies arose due to a lack of proper evolutionary controls and insufficient appreciation of the potential for subtle ecological effects to have profound evolutionary consequences. Unfortunately experimental evolution studies are prone to faulty analysis, largely because investigators assume that they have greater control of the experimental conditions than they actually achieve. Responses to selection are determined by actual selective conditions, not the selective conditions supposed by the experimenter. The differences can be subtle but are critical for interpreting outcomes, as shown by the work of Mittler and Lenski (1992) as well as Leroi et al. (1994).

## HOW LONG IS LONG ENOUGH?

Multiyear experimental evolution is not new, as there have been several long-term agricultural studies (e.g., Laurie et al. 2004), although such studies rarely top one hundred generations. Given the disparity in calendar time among long-term experiments, the calendar duration of an experimental study is a poor criterion by which to determine how long a experiment should be carried out. There is no fine distinction between short- and long-term experimental evolution. It is generally accepted that experiments running through multiple rounds of adaptation (e.g., several selective sweeps) are long-term.

But a more appropriate criterion may be evolutionary context. Experiments that only involve "ecological" dynamics, such as a competition experiment for estimating fitness differences among genotypes (Dean et al. 1988), are essentially short-term. Short-term selection studies typically focus on the potential for an individual phenotype to evolve, not on the evolutionary consequences of the phenotype for the population. For an experiment that addresses niche specialization, the evolutionary consequences of niche specialization is necessary to qualify as a long-term study and for making nontrivial evolutionary conclusions about the outcomes. Insufficient appreciation of the evolutionary time necessary to understand trait evolution is unfortunately easy in experimental evolution studies. Evolutionary responses are subject to preexisting genetic variation and linkage, both of which can yield short-term evolutionary responses that differ from long-term selection. Great caution is necessary while interpreting evolutionary responses that are dependent on preexisting characteristics of populations, such as established genetic correlations among traits. The evolutionary origin or persistence of such phenomena may involve environmental factors beyond those involved in the selection experiment, and any evolutionary responses need to be evaluated in terms of the genetic materials and environmental conditions involved.

## RECOMMENDATIONS

The goals of long-term selection studies are varied, so recommendations for carrying out such studies are necessarily general. A clearly stated hypothesis is extremely helpful, as it provides a good basis for any scientific undertaking. Without a good hypothesis, the number of potential problems is very large, many of which will not be discovered until after much time has already been invested. Among the many potential problems, a poor hypothesis could lead an investigator to initiate a selection experiment with inappropriate genotypes, insufficient replication, or an uninteresting selection regime. These problems will not only compromise the merits of the study, but they will often require substantial additional efforts be made to even partially overcome the initial limitations. A major strength of good long-term selection studies is the choice of an interesting research problem, a strength that is lost without a clear hypothesis.

Of course, it is essential to choose an appropriate experimental system. The system should be easily handled and easily controlled; otherwise, unintended environmental variation will be introduced, and substantial time will be spent simply propagating the experiment rather than examining the selective responses. The time necessary for replication also needs to be considered, as propagation and analysis of replicates is time-consuming and potentially costly, but limited replicated can limit generality and statistical power. In addition, the biological system itself should be well characterized, allowing better understanding of the selective responses. The greater the experimental control, replication, and knowledge of the system, the more power for observing and correctly interpreting interesting results.

It is also critical that the investigator evaluate results within the scope of the experiment. A potential problem with long-term studies is the tendency to overstate the results, particularly given that the use of an experimental approach tends to imbue confidence in the conclusions. However, the results observed for any one study pertain only to the genetic and environmental conditions under which the selection was performed. Experimental evolution studies have demonstrated that merely two thousand generations of relatively benign selection is sufficient for such prior selective history to impact evolutionary responses (Travisano et al. 1995a). Similarly, responses to selection can be strongly environment-dependent, often in unanticipated ways (Travisano 1997). Appropriate hypotheses at the outset of a selection experiment can help with drawing appropriate conclusions, by delimiting the scope of the experiment.

Long-term experimental evolution is a particularly useful approach for understanding the causes of evolutionary change. The ability to perform replicate experiments under tightly controlled conditions provides substantial explanatory power. For many topics, long-term studies are the only rigorous method available than can discriminate among alternative hypotheses. Long-term experiments allow direct observation of adaptation, investigation of the mechanistic basis of adaptations, and evaluation of the environmental and genetic factors that affect evolutionary responses. The limitations of long-term studies have thus far been the ability of investigators to appropriately design experiments and interpret results. A general observation from the many diverse long-term evolution experiments that have been performed is that evolutionary theory consistently underestimates the diversity of evolutionary responses of actual populations, illustrating the potential for long-term studies to expand the horizons of evolutionary biology.

## SUMMARY

Long-term experimental evolution is a direct approach for investigating the causality of evolutionary change. It is often the only experimental approach available to address many evolutionary questions. The ability to directly study adaptation greatly improves the clarity of the results and conclusions. Unlike traditional genetic studies that focus on single genes, experimental evolution allows broad-based investigation of biological phenomena. Long-term studies with microbes have demonstrated that the range of evolutionary possibilities are often far greater than anticipated from current evolutionary theory. A major strength of long-term selection studies is the flexibility in choice of research topic, as the lineages of interest are obtained by investigator motivated selection, rather than from extant diversity. Adaptive radiation is one of the major features of evolution that can be analyzed in detail using long-term experimental evolution.

## ACKNOWLEDGMENTS

The editors, Theodore Garland, Jr. and Michael R. Rose, were extremely patient during the writing of this chapter. M. R. Rose and two anonymous reviewers made many helpful comments on this and an earlier version. For the portions of the chapter relating my research, I am grateful to Rich Lenski, Paul Rainey, and Michael Doebeli for many lively discussions.

## REFERENCES

Atwood, K., L. Schneider, and F. Ryan. 1951. Periodic selection in *Escherichia coli*. *Proceedings of the National Academy of Sciences of the USA* 37:146–155.

Bantinaki, E., R. Kassen, C. Knight, Z. Robinson, A. Spiers, and P. Rainey. 2007. Adaptive divergence in experimental populations of *Pseudomonas fluorescens*. III. Mutational origins of wrinkly spreader diversity. *Genetics* 176:441–453.

Cairns, J., J. Overbaugh, and S. Miller. 1988. The origin of mutants. *Nature* 335:142–145.

Chippindale, A. K., J. A. Alipaz, and M. R. Rose. 2004. Experimental evolution of accelerated development in *Drosophila*. 2. Adult fitness and the fast development syndrome. Pages 165–192 in M. R. Rose, H. B. Passananti, and M. Matos, eds. *Methuselah Flies: A Case Study in the Evolution of Aging*. Singapore: World Scientific.

Chippindale, A., A. L. Ngo, and M. Rose. 2003. The devil in the details of life-history evolution: Instability and reversal of genetic correlations during selection on *Drosophila* development. *Journal of Genetics* 82:133–145.

Chopra, I., and M. Roberts. 2001. Tetracycline antibiotics: Mode of action, applications, molecular biology, and epidemiology of bacterial resistance. *Microbiology and Molecular Biology Reviews* 65:232–260.

Colosimo, P. F., K. E. Hosemann, S. Balabhadra, G. Villarreal Jr., M. Dickson, J. Grimwood, J. Schmutz, R. M. Myers, D. Schluter, and D. M. Kingsley. 2005. Widespread parallel evolution in sticklebacks by repeated fixation of ectodysplasin alleles. *Science* 307:1928–1933.

Cooper, V., and R. Lenski. 2000. The population genetics of ecological specialization in evolving *Escherichia coli* populations. *Nature* 407:736–739.

Cooper, V., D. Schneider, M. Blot, and R. Lenski. 2001. Mechanisms causing rapid and parallel losses of ribose catabolism in evolving populations of *Escherichia coli* B. *Journal of Bacteriology* 183:2834–2841.

de Visser, J. A., and D. E. Rozen. 2006. Clonal interference and the periodic selection of new beneficial mutations in *Escherichia coli*. *Genetics* 172:2093–2100.

Dean, A., D. Dykhuizen, and D. Hartl. 1988. Fitness effects of amino acid replacements in the beta-galactosidase of *Escherichia coli*. *Molecular Biology and Evolution* 5:469–485.

Dean, A., and J. Thornton. 2007. Mechanistic approaches to the study of evolution: The functional synthesis. *Nature Reviews Genetics* 8:675–688.

Dobzhansky, T. 1955. A review of some fundamental concepts and problems of population genetics. *Cold Spring Harbor Symposium Quantitative Biology* 20:1–15.

Dykhuizen, D., A. M. Dean, and D. Hartl. 1986. Fitness as a function of $\beta$-galactosidase activity in *Escherichia coli*. *Genetical Research* 48:1–8.

Dykhuizen, D., and D. Hartl. 1983. Selection in chemostats. *Microbiological Reviews* 47:150–168.

Friesen, M., G. Saxer, M. Travisano, and M. Doebeli. 2004. Experimental evidence for sympatric ecological diversification due to frequency-dependent competition in *Escherichia coli*. *Evolution* 58:245–260.

Garland, T., Jr., A. F. Bennett, and E. L. Rezende. 2005. Phylogenetic approaches in comparative physiology. *Journal of Experimental Biology* 208:3015–3035.

Gerrish, P. J., and R. E. Lenski. 1998. The fate of competing beneficial mutations in an asexual population. *Genetica* 102/103:127–144.

Gould, S. J. 2002. *The Structure of Evolutionary Theory*. Cambridge, MA: Harvard University Press.

Gould, S. J., and R. Lewontin. 1979. The spandrels of San Marco and the Panglossian paradigm: A critique of the adaptationist programme. *Proceedings of the Royal Society of London B, Biological Sciences* 205:581–598.

Goymer, P., S. Kahn, J. Malone, S. Gehrig, A. Spiers, and P. Rainey. 2006. Adaptive divergence in experimental populations of *Pseudomonas fluorescens*. II. The role of the GGDEF regulator WspR in evolution and development of the wrinkly spreader phenotype. *Genetics* 173:515–526.

Grant, P. R. 1999. *Ecology and Evolution of Darwin's Finches*. Princeton, NJ: Princeton University Press.

Hall, B. 1988. Adaptive evolution that requires multiple spontaneous mutation. I. Mutations involving an insertion sequence. *Genetics* 120:887–897.

Helling, R., C. Vargas, and J. Adams. 1987. Evolution of *Escherichia coli* during growth in a constant environment. *Genetics* 116:349–358.

Hill, W. G., and A. Robertson. 1966. The effect of linkage on limits to artificial selection. *Genetical Research* 8:269–294.

Hubbell, S. P. 2001. *The Unified Neutral Theory of Biodiversity and Biogeography*. Princeton, NJ: Princeton University Press.

Imhof, M., and C. Schlötterer. 2001. Fitness effects of advantageous mutations in evolving *Escherichia coli* populations. *Proceedings of the National Academy of Sciences of the USA* 98:1113–1117.

Laurie, C., S. Chasalow, J. Ledeaux, R. McCarroll, D. Bush, B. Hauge, C. Lai, D. Clark, T. Rocheford, and J. Dudley. 2004. The genetic architecture of response to long-term artificial selection for oil concentration in the maize kernel. *Genetics* 168:2141–2155.

Lenski, R., M. R. Rose, S. Simpson, and S. Tadler. 1991. Long-term experimental evolution in *Escherichia coli*. I. Adaptation and divergence during 2,000 generations. *American Naturalist* 138:1315–1341.

Lenski, R., and M. Travisano. 1994. Dynamics of adaptation and diversification: A 10,000-generation experiment with bacterial populations. *Proceedings of the National Academy of Sciences of the USA* 91:6808–6814.

Leroi, A. M., A. K. Chippindale, and M. R. Rose. 1994. Long-term laboratory evolution of a genetic trade-off in *Drosophila melanogaster*. I. The role of genotype × environment interaction. *Evolution* 48:1244–1257.

Levin, B. R. 1972. Coexistence of two asexual strains on a single resource. *Science* 175:1272–1274.

Lewontin, R. C. 1974. *The Genetic Basis of Evolutionary Change*. New York: Columbia University Press.

Luria, S., and M. Delbrück. 1943. Mutations of bacteria from virus sensitivity to virus resistance. *Genetics* 28:491–511.

Maderson, P. F. A., and D. G. Homberger. 2000. The evolutionary origin of feathers: A problem demanding interdisciplinary communication. *American Zoologist* 40:455–460.

Maharjan, R., S. Seeto, and T. Ferenci. 2007. Divergence and redundancy of transport and metabolic rate-yield strategies in a single *Escherichia coli* population. *Journal of Bacteriology* 189:2350–2358.

Medrano-Soto, A., G. Moreno-Hagelsieb, P. Vinuesa, J. Christen, and J. Collado-Vides. 2004. Successful lateral transfer requires codon usage compatibility between foreign genes and recipient genomes. *Molecular Biology and Evolution* 21:1884–1894.

Miralles, R., P. Gerrish, A. Moya, and S. Elena. 1999. Clonal interference and the evolution of RNA viruses. *Science* 285:1745–1747.

Mittler, J., and R. Lenski. 1992. Experimental evidence for an alternative to directed mutation in the *bgl* operon. *Nature* 356:446–448.

Ostrowski, E., D. Rozen, and R. Lenski. 2005. Pleiotropic effects of beneficial mutations in *Escherichia coli. Evolution* 59:2343–2352.

Ostrowski, E., R. Woods, and R. Lenski. 2008. The genetic basis of parallel and divergent phenotypic responses in evolving populations of *Escherichia coli. Proceedings Biological Sciences* 275:277–284.

Park, S.-C., and J. Krug. 2007. Clonal interference in large populations. *Proceedings of the National Academy of Sciences of the USA* 104:18135–18140.

Perrichot, V., L. Marion, D. Néraudeau, R. Vullo, and P. Tafforeau. 2008. The early evolution of feathers: Fossil evidence from Cretaceous amber of France. *Proceedings Biological Sciences* 275:1197–1202.

Petren, K., B. R. Grant, and P. R. Grant. 1999. A phylogeny of Darwin's finches based on microsatellite DNA length variation. *Proceedings of the Royal Society of London B, Biological Sciences* 266:321–329.

Phelan, J. P., M. A. Archer, K. A. Beckman, A. K. Chippindale, T. J. Nusbaum, and M. R. Rose. 2003. Breakdown in correlations during laboratory evolution. I. Comparative analyses of *Drosophilia* populations. *Evolution* 57:527–535.

Prum, R. O., and A. H. Brush. 2002. The evolutionary origin and diversification of feathers. *Quarterly Review of Biology* 77:261–295.

Rainey, P., and M. Travisano. 1998. Adaptive radiation in a heterogeneous environment. *Nature* 394:69–72.

Reed, D., and E. Bryant. 2000. The evolution of senescence under curtailed life span in laboratory populations of *Musca domestica* (the housefly). *Heredity* 85:115–121.

Rosenzweig, R., D. Treves, R. Sharp, and J. Adams. 1994. Microbial evolution in a simple unstructured environment: Genetic differentiation in *Escherichia coli. Genetics* 137:903–917.

Rozen, D., J. A. De Visser, and P. Gerrish. 2002. Fitness effects of fixed beneficial mutations in microbial populations. *Current Biology* 12:1040–1045.

Rozen, D., and R. Lenski. 2000. Long-term experimental evolution in *Escherichia coli*. VIII. Dynamics of a balanced polymorphism. *American Naturalist* 155:24–35.

Saier, M. H., Jr. 1989. Protein phosphorylation and allosteric control of inducer exclusion and catabolite repression by the bacterial phosphoenolpyruvate:sugar phosphotransferase system. *Microbiological Reviews* 53:109–120.

Schluter, D. 2000. *The Ecology of Adaptive Radiation*. Oxford: Oxford University Press.

Schneider, D., E. Duperchy, E. Coursange, R. E. Lenski, M. Blot. 2000. Long-term experimental evolution in *Escherchia coli*. IX. Characterization of IS-mediated mutations and rearrangements. *Genetics* 156:477–488.

Seehausen, O. 2006. African cichlid fish: A model system in adaptive radiation research. *Proceedings Biological Sciences* 273:1987–1998.

Selander, R. K., and T. S. Whittam. 1983. Protein polymorphism and the genetic structure of populations. Pages 89–114 *in* M. Nei and R. K Koehn, eds. *Evolution of Genes and Proteins*. Sunderland, MA: Sinauer.

Sokurenko, E., M. Feldgarden, E. Trintchina, S. Weissman, S. Avagyan, S. Chattopadhyay, J. Johnson, and D. Dykhuizen. 2004. Selection footprint in the FimH adhesin shows pathoadaptive niche differentiation in *Escherichia coli. Molecular Biology and Evolution* 21:1373–1383.

Souza, V., P. E. Turner, and R. E. Lenski. 1997. Long-term experimental evolution in *Escherichia coli*. V. Effects of recombination with immigrant genotypes on the rate of bacterial evolution. *Journal of Evolutionary Biology* 10:743–769.

Spencer, C., M. Bertrand, M. Travisano, and M. Doebeli. 2007. Adaptive diversification in genes that regulate resource use in *Escherichia coli. PLoS Genetics* 3:e15.

Spiers, A., J. Bohannon, S. Gehrig, and P. Rainey. 2003. Biofilm formation at the air-liquid interface by the *Pseudomonas fluorescens* SBW25 wrinkly spreader requires an acetylated form of cellulose. *Molecular Microbiology* 50:15–27.

Stahl, F. 1988. Bacterial genetics: A unicorn in the garden. *Nature* 335:112–113.

Travisano, M. 1997. Long-term experimental evolution in *Escherichia coli*. VI. Environmental constraints on adaptation and divergence. *Genetics* 146:471–479.

———. 2001. Evolution: Towards a genetical theory of adaptation. *Current Biology* 11:R440–442.

Travisano, M., and R. Lenski. 1996. Long-term experimental evolution in *Escherichia coli*. IV. Targets of selection and the specificity of adaptation. *Genetics* 143:15–26.

Travisano, M., J. A. Mongold, A. F. Bennett, and R. E. Lenski. 1995a. Experimental tests of the roles of adaptation, chance, and history in evolution. *Science* 267:87–90.

Travisano, M., and P. Rainey. 2000. Studies of adaptive radiation using model microbial systems. *American Naturalist* 156:S35–S44.

Travisano, M., F. Vasi, and R. Lenski. 1995b. Long-term experimental evolution in *Escherichia coli*. III. Variation among replicate populations in correlated responses to novel environments. *Evolution* 49:189–200.

Wilke, C. 2004. The speed of adaptation in large asexual populations. *Genetics* 167: 2045–2053.

Woods, R., D. Schneider, C. Winkworth, M. Riley, and R. Lenski. 2006. Tests of parallel molecular evolution in a long-term experiment with *Escherichia coli. Proceedings of the National Academy of Sciences of the USA* 103:9107–9112.

# 7

# THE EXPERIMENTAL STUDY
# OF REVERSE EVOLUTION

## Suzanne Estes and Henrique Teotónio

Darwin was the first to suggest that selection of small differences among individuals for characters related to survival and reproduction was sufficient to promote directional evolutionary change in those characters. Evolution, understood as common descent of organisms with modification over time, was quickly accepted by the scientific community. However, the role of natural selection in the evolution of biodiversity was controversial since the inheritance of phenotypic information both during development and transmission across generations was unexplained.

The major problem confronted by evolutionary theory derived from the observation that when individuals were selected, character values in their offspring showed partial *regression* to the population's average character state (figure 7.1; for terms in italic type, see the glossary at the end of this chapter). Hence, while Darwin believed that natural selection was the dominant mechanism in evolution, it was not envisioned by others as strong enough to change character states away from those of previous generations. Furthermore, existing knowledge about heredity failed to explain how individuals could exhibit character states absent in their parents but present in more distant relatives

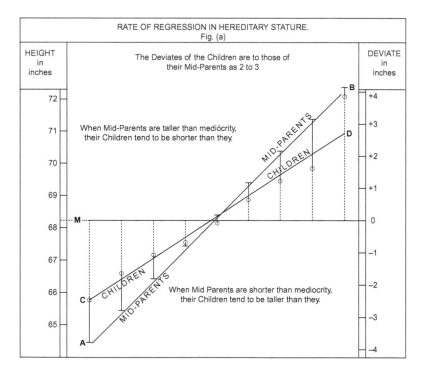

FIGURE 7.1

Galton's original regression of offspring character values (human height) onto parent character values, and the reciprocal regression of parental onto offspring character values (from Galton 1885).

(i.e., *atavisms*). The empirical observation of such reversions motivated the research agenda during the next fifty years. Gayon (1998) gives an excellent account of this period. We summarize only a few key events here.

While his motives and approach were occasionally dubious, Galton (1887) led the way in studying how descriptors of character variation could be related to resemblance between relatives and to the population to which they belonged. He established that the distribution of character states in a population frequently followed a Laplace-Gaussian shape, which could be regenerated by sampling sufficiently large numbers of individuals from the distributions of previous generations. Importantly, when different parents were selected from such a distribution, the average deviation of offspring from the mean phenotype was less than that of the parents (figure 7.1); furthermore, the ratio of the deviations of offspring values from their parents, to the deviations of different parental averages from the population grand mean were invariant. Galton inferred from these observations that heredity could be separated into two components—variability among parental inheritance types and reversions—with differences among individuals of the same population explained by a mixture of the two.

Reversions were initially noted to support the prevailing belief that individual species retained an intrinsic property that explained the inheritance of phenotypic information. Under Galton's conceptualization, the role of natural selection was limited to stabilizing phenotypic distributions around this typological ancestor. He dismissed Darwin's idea that natural selection could transform populations by working on small, individual differences because both parental and offspring distributions were scaled to the same (invariable) average population value. If a population happened to diverge from the ancestral type, reversions would accumulate until the average character value returned to that of the ancestor. The gains of selection could therefore only be maintained if an ancestral type mutated to a radically different type, which was believed possible based on periodic observations of especially abnormal individuals.

Contemporaneously, Weissman (1883) coined the term "panmixia" to describe the process whereby free mating among individuals of a population leads to the degeneration of organismal integrity, primarily manifested by atrophy or disappearance of anatomical characters (Dollo 1893; Fong et al. 1995; Porter and Crandall 2003). In Weissman's view, natural selection was required to prevent such anatomical reversion, but it could also be an effectual agent of change, capable of creating a sustained bias in intrapopulation variation across generations, and on a longer time scale, in differences accumulating among species. Weissman provides the first clear exposition of *reverse evolution*, albeit conceptually equivalent to *degenerative evolution* in his view. However, this selectionist approach continued to leave unexplained the origin of heritable variation among individuals, a necessary condition for the operation of natural selection in a population. Why atavisms and reversions existed remained poorly understood.

At the start of the twentieth century, an understanding of heredity according to Mendelian rules finally provided an explanation for the problem of reversions. Heredity was defined as the phenotypic information contained in pairs of indivisible factors that could be segregated and transmitted across generations independently of one another. Considerable experimental effort was expended to arrive at this definition (see Provine 1971; Gayon 1998; Sarkar 1998); we mention only two studies that epitomize the resolution of the problem of reversions.

First, the experiments of Johannsen (1903), who made use of *artificial selection* to define the hereditary basis of continuous characters, were key to this resolution. Studying seed size in the common bean, *Phaseolus vulgaris*, Johannsen propagated individual plants for two generations by self-fertilization. In the first generation, he observed that within each of nineteen self-fertilized lineages, there was complete reversion of offspring character values to the parental average—that is, regression was maximal. There was only partial regression of offspring values to the overall population average. In the second generation, Johannsen again observed a complete reversion of average offspring character values to those of the parents. However, the average offspring values did not regress at all to the average population value, as observed in the preceding generation. Hence, there was constancy and integrity of the heredity material across the generations. Johannsen inferred that within a particular selfing lineage, individual character differences must originate outside the organism. Together with other experiments from this era, this study helped pave the way for a clear conceptual distinction between genotype, defined as an individual's unique constitution of heredity factors, and phenotype, defined as the individual character states resulting from expression of a genotype in a given environmental context.

In 1908, the theoretical equilibria attained by hereditary factors across generations in a panmictic population were derived (Hardy 1908; Weinberg 1908; Provine 1971). Hardy and Weinberg demonstrated that, in the absence of natural selection, the transmission of pairs of hereditary factors across generations was stable. Because of this, the instantaneous, partial reversions of offspring character values to the population average observed by Johanssen and others must have an environmental origin. Furthermore, it became understood that repeated operation of natural selection could indeed cause a population's average phenotype to change by small steps and become differentiated from that of its ancestral population. There was no intrinsic hereditary property of the species or of populations, only of the individual.

The findings of Hardy, Weinberg, Johannsen, and others established that, without selection, heredity would not alter character states during transmission. In the presence of selection, average character values of offspring could remain differentiated from the population average. An understanding of how Mendelian inheritance operates in populations also shed light on atavisms: if different hereditary factors made different contributions to the phenotype, then each individual could have a unique combination of these factors, with rare individuals expressing a particular combination that resulted in the expression of an ancestral phenotype.

## HISTORICAL BACKGROUND

With the problem of reversions at the population and individual levels resolved, biologists turned their efforts to describing more carefully the nature of heredity and natural selection. The operation of selection in nature remained a contentious issue that gave rise to a research program dedicated to verifying its occurrence and quantifying its role relative to other mechanisms of evolution, such as genetic drift and migration (e.g., Weldon 1895; Fisher and Ford 1926; Wright and Dobzhansky 1946; Kettlewell 1973). For simple Mendelian characters with discrete phenotypes, most observations were made in the laboratory where experiments sought to demonstrate the action of selection among genotypes, in agreement with theoretical predictions being developed at the time (Teissier 1954; see Wright 1977, chapter 9). In the case of quantitative characters, experimenters typically selected individuals from a distribution of phenotypes irrespective of how well they survived and reproduced. Only experiments where the characters of interest were related to population growth rates can be considered to have studied natural selection, but this condition was rarely met (cf. Bell 1997).

A common experimental technique in studies of continuous characters was to apply *reverse selection* to characters having initially been selected in an opposite direction of phenotypic change for some number of generations. Castle's experiments with Piebald rats perhaps mark the beginning of modern experimental evolution studies of this kind (Castle and Phillips 1914). Applying thirteen generations of artificial selection for specific coat color patterns, Castle and colleagues observed as the phenotypic distributions of different lineages moved beyond the bounds of the ancestral range of phenotypes. When artificial selection was applied in the opposite direction, reversion to ancestral character states was only partially achieved and at a slower rate than forward artificial selection. Since the gains of artificial selection were stable, Castle concluded that selection slowly weeded out some or most of the large number of variants present in the original stock. Hence, there were limits on the evolutionary response that could be achieved due to limits on heredity and selection.

Following Castle's and others' experiments, evolutionary genetics research was able to determine the origin of genetic variability by quantifying mutation rates based on the appearance of (usually morphological) abnormalities, while the origin of variation due to recombination was quantified via gametic disequilibrium among several of the same phenotypes. The genetics of quantitative characters were identified by their additive, dominance, or epistatic effects when regressing phenotypic variability onto allelic content; and the number of genes underlying phenotypic variation were estimated by rates of selection response. This base of empirical knowledge motivated the development of theory to address the efficacy of natural selection in sorting genetic variation (Bohren et al. 1966; Hill and Robertson 1966; Maynard Smith and Haigh 1974; Barton and Partridge 2000).

It was not until the mid-1980s that appropriate theoretical models and experimental design strategies became available for studying the process of reverse evolution. The work of Lande and colleagues, among others, brought evolutionary theory to a new level of sophistication by incorporating explicit genetic architectures, summarized in matrices of additive genetic variances and covariances among quantitative characters, together with models of natural selection on these characters (Lande 1976, 1979, 1980; Barton and Turelli 1989, section 4). Meanwhile, formal theory for the evolution of life-history and related characters helped clarify the patterns of selection that can arise with several types of experimental treatments in the laboratory (e.g., Charlesworth 1980; Rose 1991).

The design of selection experiments was immediately improved by delineation of the necessary conditions to detect significant evolutionary responses. For example, starting experimental populations came to be clearly defined as outbred or inbred, sexual or asexual. In most microbial studies, the initial replicate populations were identical clones, with genetic variation originating solely from mutation over the course of thousands of generations of laboratory culture (e.g., Lenski et al. 1991). Most *Drosophila* studies of experimental evolution, by contrast, exploited standing genetic variation present in the base populations, which allowed rapid and extensive responses of selected characters (e.g., Rose et al. 2005).

Other important changes in experimental practice included the standard replication of populations undergoing identical treatment, which permitted researchers to distinguish the consequences of selection and genetic drift using both the mean and variance of evolutionary responses. This practice also made the individual population the unit of experimental observation, one of the more important accomplishments of experimental evolution. Second, enforcing specific life cycles within experimental populations permitted the study of characters that could serve as reasonable proxies for *fitness*, a difficult or impossible task in wild populations. Lastly, researchers became aware of a series of common laboratory-induced artifacts, in particular the nonlinearity of the genotype-environment relationship, which experimenters strived to control with varying degrees of success (Leroi et al. 1994; Rose et al. 1996; Prasad and Joshi 2003).

## THE USE OF CONTROLS IN REVERSE EVOLUTION EXPERIMENTS

Experimental tests of reverse evolution are often predicated on the assumption that derived populations will at least partially revert to the same phenotype or genotype as their ancestor population. This expectation makes experimental reverse evolution unique relative to other evolution experiments in which the expected evolutionary state of derived populations is unknown, derived from evolutionary theory, or anticipated from prior experiments of similar design. Thus, reverse evolution experiments have a well-defined expected outcome.

A difficulty with reverse evolution experiments arises from variability in the degree to which control populations accurately reflect the ancestral evolutionary state. In microbial reverse evolution experiments, the control populations are the ancestor populations them-selves, stored in a nonevolving state. Whenever such preservation of experimental organisms is possible—by cryopreservation or vernalization, for example—the results from comparisons between control and derived populations are not confounded with nonsystematic differentiation of actively cultured ancestor populations that may occur over the duration of a reverse evolution experiment (see discussion in Lynch et al. 1999). Storing populations in nonevolving conditions is, however, not a warranty against confounded results as selection during and after storage can bias control genotypes. This is of less concern when the ancestor control population is composed of a unique genotype, as in the case of microbial experiments that start from single clones (e.g., Lenski et al. 1991; Bull et al. 1997).

When storage is not possible and control populations must be maintained in culture, then reverse evolution experiments can only begin when control populations have attained mutation-selection equilibria. For neutral diploid genotypes, these equilibria are achieved after $4N_e$ generations. And while equilibriums are expected to be achieved in considerably less time for selected alleles, on the order of tens of generations (Crow and Kimura 1970), results from reverse evolution experiments will be difficult to interpret except when long-established control populations are available or when the control population is composed of a single genotype, in which case the concept of an equilibrium loses significance.

The difficulties surrounding appropriate controls are not unique to reverse experimental evolution. But because the reverse evolution design is such a powerful one when the ancestral state is accurately reflected by the control, the value of good controls is particularly high in this case.

## REVERSE EVOLUTION AND THE INFERENCE OF EVOLUTIONARY RELATIONSHIPS

A recent study reported the repeated evolution of wing structures in stick insects (order Phasmatodea) from wingless ancestors, which had themselves evolved from winged ancestors (Whiting et al. 2003). This group of insects has apparently retained the ontogenic and physiological capacity to repeatedly lose and gain a phenotype over millions of years. Convincing examples of such repeated character reversibility within clades are fairly uncommon (see examples in Tchernov et al. 2000; Wiens 2001; Ober 2003; Chippindale et al. 2004). Is this indicative of the rarity of reverse evolution or rather a reflection of the short-comings of methods for phylogenetic reconstruction? Standard phylogenetics uses comparative methods that rely on parsimony, which assumes that evolution involves a minimum number of character state changes and/or similar evolutionary rates across lineages. As with *parallel evolution*, reverse evolution will increase instances of *convergence*, and therefore *homoplasy*, in the clades being compared: identical character states will be uninformative; and when they occur, the number of character state changes will be underestimated while the number of *synapomorphies* will be overestimated (figure 7.2).

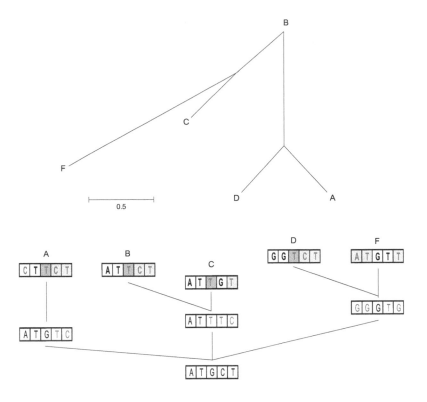

FIGURE 7.2

Top diagram shows the phylogenetic reconstruction obtained by neighbor-joining methods of the nucleotide sequences with the evolutionary history depicted in the bottom diagram. Capital letters A-F indicate derived (observed) sequences from the ancestral "ATGCT" sequence. Grey lettering indicates all nucleotide changes that occurred over the course of evolution; light grey boxes denote reversions, while dark grey boxes denote parallel nucleotide changes. The bar in the top diagram is scaled to 0.5 nucleotide changes. Note that with extensive homoplasy, evolutionary rates are underestimated and incorrect ancestries are inferred.

In principle, these problems can be alleviated by using molecular sequences due to the fact that each nucleotide or amino acid is considered an independent character (e.g., Felsenstein 2003; Thornton et al. 2003). However, little is known about the extent to which natural selection or genetic drift can facilitate parallel and reverse evolution at the molecular level. Identity in character state can arise by convergence due to selection or by common ancestry. Furthermore, we lack knowledge about the constraints imposed on molecular evolution by, for example, different mutation and recombination rates along the genomes of clades under study (see Baer et al. 2005). Ignorance about these issues will continue to make phylogenetic reconstruction a difficult undertaking, even with the large body of data generated from molecular sequencing (Felsenstein 2003). Because experimental evolution allows complete control over evolutionary history, this approach can characterize the occurrence of reversions and parallelisms, which in turn can facilitate the incorporation of explicit evolutionary models into phylogenetic reconstruction by use of maximum-likelihood methods (Felsenstein 1981; Hillis et al. 1992; Bull et al. 1993; Fleming et al. 1993; Thornton 2003).

At this point, it is useful to consider a few case studies of experimental reverse evolution in some detail. Bull and colleagues followed adaptation in the single-stranded DNA bacteriophage, Φ-X174, to high temperatures in either one of two bacterial hosts (*Escherichia coli* or *Salmonella enterica*) starting from the same ancestral clone (Bull et al. 1997). Since the genome of this phage is only 5.4 kilobases, complete nucleotide sequencing of derived and ancestral lineages, and therefore a comparison of imposed evolutionary history with inferred phylogenies, was possible. Convergence by parallel changes was extensive both among replicate populations adapting to the same bacterial host and between populations adapting to different hosts. Such changes accounted for approximately half of all nucleotide changes observed. Two replicates were further derived from one of the *S. enterica* populations and adapted in *E. coli*. These results indicated that, again, about 50 percent of the changes were convergent with the changes observed in the lineages selected only in *E. coli* during the first phase. In one of the replicate populations, however, convergent changes were mostly due to reversions (90 percent of them), while reversion and parallel changes were observed with near equal frequency (40:60, respectively) in the other population. Finally, when one of the replicate populations adapted to *E. coli* was further passaged in this bacterium, of the eight additional changes observed, only one was a reversion to the ancestral clone state. With these evolutionary patterns, phylogenetic reconstruction did not match the true evolutionary history of the populations: one population that had evolved in *Salmonella* was grouped with the clade that evolved in *E. coli*, indicating that evolutionary rates were incorrectly estimated. Furthermore, the true evolutionary history was rejected as a good description of relationships among populations using standard statistical models. In a follow-up study by Crill et al. (2000) specifically designed to reveal the extent of reverse evolution as a result of repeated bacterial host switching, convergence by reversion to both the ancestral sequence and to the intermediate sequence was substantial, with reversions accounting for 20 percent of the changes. However, with limited sequence data and low replication in both studies, phylogenetic reconstruction may have failed in part due to lack of statistical power.

The prevalence of reverse evolution during adaptation to novel environments is still inadequately explored at the molecular level. A study of the foot-and-mouth disease single-stranded RNA virus showed that, out of more than twenty nucleotide changes that occurred with selection for high growth rates from a poorly growing ancestral clone, only one was a reversion (Escarmis et al. 1999). Another study with HIV viruses found that adaptation to antiviral drugs was correlated with a single amino acid change in a protease protein, but that there was no reversion to the ancestral amino acid upon *relaxed selection* (Borman et al. 1996). Finally, MacLean et al. (2004) found that catabolism was largely reversible to ancestral states in yeast populations evolving in structured environments, presumably reflecting a reversal of DNA sequence as well.

Considerably more data is available on the frequency of parallel as opposed to reversion changes. Another study using Φ-X174 by Wichman and colleagues described the nucleotide changes occurring during thirteen thousand generations of adaptation to the *Salmonella* host described earlier. After conducting laboratory evolution with two experimental populations, the authors observed extensive convergence by parallelism with the aforementioned Φ-X174 studies (Wichman et al. 2005; cf. Wichman et al. 1999, 2000): thirty of seventy-one missense changes, fifteen of fifty-eight silent changes, and four of twelve intergenic changes along the genome were common among studies. Furthermore, preliminary work conducted in different labs showed that 50 percent of the observed molecular changes that occurred during about seventy generations of evolution in two replicate populations were parallelisms; however, the order in which these changes accumulated within the two populations was dissimilar. Interestingly, extensive parallelism was also observed during experimental adaptation to similar temperature conditions between Φ-X174 and a related virus, S3, isolated from *Salmonella* (Wichman et al. 2000). In agreement with these findings, a recent study with RNA viruses also reported considerable convergence during adaptation to several demographic environments (Cuevas et al. 2002). Conversely, studies using the double-stranded T7 DNA virus revealed that convergence in populations exposed to a mutagen was minimal, and as a consequence, accurate phylogenetic reconstruction using restriction-enzyme sites was possible (Hillis et al. 1992; Bull et al. 1993).

Comparable work outside viruses is scarce. The recent study by Woods et al. (2006) sought a characterization of convergence in *Escherichia coli* evolved for twenty thousand generations in a constant environment (see also Lenski et al. 1991; Travisano et al. 1995b; Pelosi et al. 2006). Approximately 0.5 percent of a 4.5-megabase genome was compared for nucleotide changes among twelve derived populations and their ancestral clone, with little parallelism observed: out of the thirty-eight mutations found, one was present in three replicates, while another mutation was found in two replicates. There was far more parallelism at the gene (ORF) level, however, with two sets of mutations observed at nonsynonymous positions of two genes in all twelve populations. Fitting stochastic null models to the data, the authors concluded that natural selection was responsible for some of these parallel changes. These findings substantiate earlier results by Treves et al. (1998) showing that *E. coli* maintained in glucose-limited medium evolved considerable polymorphism in acetate cross-feeding phenotypes by parallel changes at the gene level. Unlike the findings of Woods et al. (2006), however, these authors also reported a high degree of convergence at the nucleotide sequence level, although given the limited amount of data in Treves et al. (1998), the generality of this pattern in *E. coli* remains unknown. See Sègre et al. (2006) for a similar study in yeast.

These results suggest that parallelism at the nucleotide sequence level may be less frequent in *E. coli* than in Φ-X174. What does this difference tell us about the biology of these organisms? One possibility is that increased genomic complexity in *E. coli* with respect to the size and number of mutational targets with similar or redundant functional consequences acts to reduce the possible number of parallelisms at the nucleotide sequence level in *E. coli* as compared to viruses (Woods et al. 2006; cf. Sègre et al. 2006). If true, the distinction between genotype and phenotype may be almost nonexistent in Φ-X174 but more pronounced in *E. coli*. However, while it may be safe to assume that all phenotypically important features of mutational trajectories through DNA sequence space have been thoroughly explored in virus experiments (cf. Bull et al. 1997; Crill et al. 2000; Wichman et al. 2000, 2005), the same is probably not true for *E. coli*. This is because the large number of generations observed in some selection experiments (more than twenty thousand in one case) might still be two of orders of magnitude lower than required to ensure each unique nucleotide mutation in a genome of size 4.5 megabases has been sampled—a number that scales with population size (*E. coli* population sizes $\sim 10^{6-8}$; Crow and Kimura 1970; Lenski et al. 1991). If this reasoning is valid, then the two organisms have not experienced equivalent evolutionary times or effective mutation rates.

An important question that remains virtually unexplored is the role of recombination in shaping molecular sequence diversity. While recombination was completely absent in the *E. coli* studies, recombination among different genotypes is thought to have occurred in some of the viral evolution experiments when multiple phage infected the same host cell. As a result, some viral populations could have maintained considerable molecular diversity within replicates (Wichman et al. 1999, 2005; see also Escarmis et al. 1999). Nonmicrobial studies addressing the role of recombination and mutation are essentially nonexistent. Deckert-Cruz et al. (1997) documented parallel evolution of electrophoretic alleles in replicate populations of *Drosophila melanogaster*, while Passananti et al. (2004) demonstrated reverse evolution for electrophoretic alleles at one of the same loci. Because the same electromorph can be encoded and expressed by different DNA sequences, ability to determine the source of such convergence in recombining organisms is limited. Neither the level of recombination sufficient to create redundant genetic structures nor the minimal structural unit of inheritance during evolution (i.e., nucleotide, sequence motif, locus, or groups of loci) is known for any natural or experimental population. Whether heterogeneity of recombination rates within organisms is important for the evolution of phenotypes and how such heterogeneity determines the physical organization of the DNA units that have phenotypic effects are completely open questions.

In addition to genome size and local mutation and recombination rates, selective conditions will also determine the extent of convergence at the molecular sequence level. Most laboratory evolution experiments have been conducted with extremely large populations with little demographic structure, allowing the operation of mass natural selection (e.g., Lenski et al. 1991; Wichman et al. 2000). Reversions like those observed for Φ-X174 by Crill et al. (2000) are particularly instructive in this regard because they demonstrate selection at specific nucleotide sites: it may be that under mass selection conditions, convergence is favored over divergence (Hill et al. 1993). But, as Bull et al. (1997) observed, parallelisms can occur among populations selected in different environments, while reversions can occur among populations selected continuously in the same environment (see also Rainey and Rainey 2003). These findings indicate that the molecular sequence changes of particular populations are demonstrably contingent not only on their long-term evolutionary histories but also on the environment they currently occupy (cf. Cohan 1984; see also the next section).

Interpreting these results within the context of inferring common ancestry in the face of convergence, it appears that, depending on the organism under study, a measure of caution is warranted when applying current phylogenetic methods. If convergence is taken to be the hallmark of natural selection, perhaps phylogenetic inference should be restricted to molecular characters with well-understood stochastic properties with respect to the frequency of convergence in the absence of common ancestry and natural selection (e.g., see models in Kimura 1983). One could conclude from the results that these methods are destined for failure in experiments with organisms with small genomes and simple organization, such as many viruses. For organisms of greater complexity, however, convergence may not be sufficiently common to bias phylogenetic reconstruction: identity in nucleotide sequences should mainly reflect common ancestry. But these conclusions remain speculative.

## CONTINGENCY AND REVERSE EVOLUTION

When phenotypically diversified populations are returned to their common ancestral environments, several outcomes are possible (figure 7.3). How can the contingencies of evolutionary change be quantified and related to the genetic and environmental contexts of the evolving populations? *Adaptive landscape* models provide a useful framework in which to think about evolutionary change, including reverse evolution. Their use as conceptual tools has enjoyed a recent resurgence reflecting, in part, an appreciation of their ability to unify research both at different levels of biological organization (Lande 1976, 1979; Rice 2004; Colegrave and Buckling 2005) and at several temporal scales (Schluter 2000; Arnold et al. 2001; Estes and Arnold 2007).

In some versions, such as that of Simpson (1944), the adaptive landscape depicts the relationship between the average fitness and the average character state of a population (cf. Lande 1976, 1979; Arnold et al. 2001, figure 8). On one hand, patterns of genetic variance and covariance among characters will influence the average character state of a population and its location within the landscape since the distribution of phenotypes is a reflection of distribution of genotypes. On the other hand, the external environment will establish the specific relationships between phenotypic states and fitness. These two complex assemblages of factors, genetic and environmental, will together determine the topography of an adaptive landscape, which will in turn help dictate phenotypic evolution. Since adaptive landscapes are defined for individual populations, they embody the concept

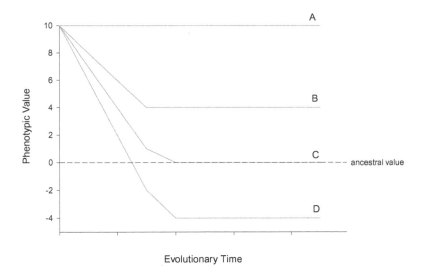

FIGURE 7.3

Possible evolutionary phenotypic trajectories of reverse evolution. The dashed line at zero represents the ancestral character value, and populations are assumed to have diverged to phenotypic values of ±10. Line A depicts no evolutionary response and implies that the character is neutral. Line B shows partial convergence to the ancestral value, with an initially rapid response followed by stasis. Line C shows linear convergence to the ancestral value. Finally, line D shows rapid convergence followed by a surpassing of the ancestral value and stasis at a new value.

of contingency in evolutionary theory (cf. Price 1972; Lande 1976, 1979; Barton and Turelli 1987; Arnold et al. 2001; Kirkpatrick et al. 2002; Rice 2004). A landscape's topography can be smooth or rugged, the importance of the distinction being that evolution can proceed quite differently on each type. For example, a smooth, single-peaked landscape across a given phenotypic range will promote deterministic convergence of all populations with similar phenotypic distributions to a similar fitness optimum. This situation can also allow divergent populations occupying different phenotypic spaces to converge on identical phenotypic states via different routes. Alternatively, if multiple fitness peaks are separated by intermediate phenotypic states associated with low fitness, different evolutionary trajectories may be taken by such populations, with some becoming trapped on suboptimal fitness peaks (e.g., Burch and Chao 2000).

Although in its infancy, experimental evolution is the most powerful empirical method for the characterization of adaptive landscapes. It can achieve this goal by manipulating the genetic architecture that gives rise to patterns of genetic variance and/or by manipulating the environment in which populations evolve (for reviews, see Elena and Lenski 2003; Colegrave and Buckling 2005). In this section, we review a sampling of important studies that reveal general features of adaptive landscapes in the context of reverse evolution.

Phillips and colleagues (2001; Whitlock et al. 2002) followed the effect of genetic drift on reverse evolution in morphological characters. If changes in certain character values are inconsequential for organismal fitness, genetic drift alone will be responsible for phenotypic change in the short term. The adaptive landscape will in these cases be flat and featureless, such that initially identical replicate populations are expected to wander away from one another in character space in a manner describable by Brownian motion (see figure 6 in Arnold et al. 2001). The final outcome of the evolutionary process depends on the time since initial divergence, the effective population size, and the genetic variance-covariance patterns in the ancestral population, the latter of which are summarized in the *G-matrix* (Lande 1979). The G-matrices of a suite of wing morphological characters were measured in a large number of lines of *D. melanogaster* generated by a single episode of sib mating. In accordance with theoretical predictions, the authors documented substantial divergence in G-matrices, with the average size and shape of the G-matrix for fifty-two descendant lineages being proportional to that of the outbred population from which these lines were derived—that is, the ancestral control (Phillips et al. 2001). Moreover, a subset of lineages exhibiting divergent G-matrices was found to have diverged further from the ancestral state after twenty generations of maintenance at moderate population size (Whitlock et al. 2002). Importantly, there was no tendency for the new populations to converge to the ancestral state for any genetic variance component. These data show that even mild genetic drift can alter the genetic associations among characters and that such chance events can persist long enough to influence future evolution, at least with moderate population sizes. An obvious corollary of this finding is that reverse evolution is highly unlikely under such conditions, even in the short term (cf. Gould 1989).

That changes in G-matrices can affect future evolution has been shown repeatedly (e.g., Shiotsugu et al. 1997; Phelan et al. 2003; Rose et al. 2005; cf. Bohren et al. 1966; Gromko 1990; Houle 1992). In the context of reverse evolution, a recent study by Vermuelen and Bijlsma (2006) reported the outcome of 220 generations of relaxed selection following just 6 generations of artificial selection for increased or decreased virgin life span in *D. melanogaster* (Zwann et al. 1995). A negative genetic correlation between virgin life span and early fecundity observed in the original selected populations disappeared during relaxed selection and was accompanied by the evolution of several new correlations among life-history characters. Significantly, when a new round of artificial selection was applied to virgin life span in populations derived from crosses among the relaxed-selection populations, the original negative correlation between virgin life span and egg production was not recovered. Thus, not only can historical chance events change the genetic constitution of populations and determine evolutionary changes as in Whitlock et al. (2002), but historical patterns of selection can do so as well (also see Buckling et al. 2003).

The long-term laboratory evolution study initiated by Lenski and colleagues in the late 1980's has demonstrated that parallel changes in phenotypes are probable but not guaranteed during adaptation to a constant environment (e.g., Lenski et al. 1991; Travisano et al. 1995b). Twelve populations of *E. coli*, derived from a single ancestral clone, underwent laboratory evolution during serial passage in a glucose-limited medium. Significant parallel changes in competitive growth rates and cell size were observed, with both characters exhibiting a rapid initial increase over one thousand generations followed by a dampened response (Lenski et al. 1991; Lenski and Travisano 1994). However, significant variation among replicate populations for both characters persisted even after more than two thousand generations of adaptation (Travisano et al. 1995b; Travisano and Lenski 1996; see also Maharjan et al. 2006 for a recent independent experiment in glucose-limited chemostat populations of *E. coli*). In a follow-up study, replicate populations were created by initiating three clones from each of the twelve populations after two thousand generations of evolution on glucose medium. These were then maintained in a novel maltose-limited environment for another thousand generations (Travisano et al. 1995a). This design allowed an assessment of the relative effects of current selection versus each population's unique history in the glucose-limited medium, while also accounting for chance effects. Growth rates in the new environment quickly increased, indicating an increasing contribution of current selection for this character. Cell size, however, did not change significantly. Since this character was found to be uncorrelated with growth rates in maltose-limited environments, its evolutionary dynamics were mainly explained by historical effects and, increasingly over the course of the experiment, by chance effects (Travisano et al. 1995a). The same patterns have been found in *Drosophila* experiments where characters more closely related to fitness in current environments show convergence coincident with diminishing effects of past selection, while those unrelated to fitness are prone to maintain historical changes and to be influenced by chance (Joshi et al. 2003). In terms of adaptive landscape models, ruggedness is prevalent, but similar fitnesses nonetheless appear to be attained over the course of experimental evolution under a common culture regime.

One of the few characterizations of adaptive landscapes during reverse evolution has been conducted using *D. melanogaster* as a model (Teotónio and Rose 2000, 2001; Teotónio et al. 2002, 2004). Briefly, four groups, comprising five replicate populations each, were adapted to several demographic and stress environments for more than one hundred generations. During this time, populations became differentiated for several life-history, physiological, and biochemical phenotypes (Rose 1984; Rose et al. 1992, Chippindale et al. 1997). These populations were then returned to their common ancestral environment for another fifty generations. With repeated assays during the course of reverse selection, essentially every evolutionary pattern depicted in figure 7.3 was observed for at least one character. Reverse evolution to the ancestral phenotypic state was exhibited by some characters in some groups of populations, but it was hardly universal.

As in other studies, genetic correlations between characters varied widely in magnitude and sometimes even sign among the experimental populations. Furthermore, all groups of populations appeared to have achieved similar fitness peaks after the fiftieth generation of the experiment, at least as inferred from male reproductive success (Teotónio et al. 2002). Characters presumed to be more closely associated with fitness tend to show more successful reversion to ancestral phenotypic states. However, variance decomposition to discern the contributions of historical versus current selection revealed a complicated pattern in which the evolutionary dynamics of most characters were significantly influenced by chance events, and in which selection in the ancestral environment erased the effects of their divergent selective histories in some cases but not in others (Teotónio et al. 2002). Thus, as in the microbial evolution experiments, adaptive landscapes in these experiments were found to be rugged: the reverse-evolved populations generally adapted to their ancestral environments, but they did so by a multitude of different phenotypic changes that resulted in similar levels of fitness. And again, the pattern of reverse evolution was found to be more contingent and less predictable for phenotypes of loose association with fitness.

## THE POPULATION GENETICS OF REVERSE EVOLUTION

What do cases of convergent versus nonconvergent reverse evolution imply about the distribution of genotypes within a population? How is the phenotypic information encoded in nucleotide sequences transformed into genetic variances and covariances that determine evolutionary convergence? Adaptive landscapes are obtained by averaging the mapping functions relating each character state to fitness over the phenotypic distribution in a particular population (Arnold et al. 2001). Greater genetic variation, if it translates into greater phenotypic variance, can lead to a smooth landscape and easy movement of a population within it (cf. Teotónio et al. 2006). As already noted, limitations on genetic variability can create a rugged adaptive landscape and thereby impede reverse evolution. But ample genetic variation is not by itself sufficient to generate a smooth landscape. Antagonistic pleiotropy (AP), accumulation of neutral alleles, nonlinear interactions among loci (epistasis), and negative frequency-dependent selection can all maintain genetic variability while constraining the evolution of populations in rugged adaptive landscapes.

### ANTAGONISTIC PLEIOTROPY AND ACCUMULATION
### OF NEUTRAL ALLELES

Antagonistic pleiotropy arises when alleles have opposite fitness effects in different environments or when alleles exhibit trade-offs among two or more fitness characters in a given environment. Populations harboring such alleles may maintain considerable genetic variance for particular characters while experiencing compromised evolutionary potential (Rose 1982, 1983, 1985). Evidence for fitness trade-offs between environments and between characters is sufficiently abundant that current theory for the evolution of demographic characters such as developmental time, fecundity, viability, and life span is often predicated on the existence of AP genotypes (Charlesworth 1980; Rose 1991; Prasad and Joshi 2003). Negative genetic covariances between fitness characters can, of course, also be generated by physical linkage among alleles with opposing effects at different loci and by the nonrandom transmission of gametic types (Lande 1980). The extent to which populations in evolution experiments are subject to negative genetic covariances due to linkage is not generally known, but the situation can be expected in cases of low recombination, such as asexual *microbes* and small populations of sexual species.

In the context of reverse evolution, a line of inquiry that can be traced to Weissman (discussed at the start of this chapter; Fong et al. 1995; Porter and Crandall 2003) questions the degree to which degenerative evolution results from the expression of AP alleles as opposed to the accumulation of novel alleles of neutral fitness effect. A few studies serve to illustrate this question. For example, Service et al. (1988) found that certain characters underwent reversion to ancestral states far more rapidly than others in *D. melanogaster*. This pattern is indicative of AP since the number of generations necessary for the accumulation of new neutral alleles is about an order of magnitude higher than that required for the fixation of AP genotypes segregating in the original population. More recently, Cooper and Lenski (2000) found that in *E. coli* populations derived from a single clone and adapted to glucose-limited media (discussed earlier), loss of certain metabolic activities was more pronounced in earlier evolutionary phases than in later ones, and the evolutionary responses of populations with high versus normal mutation rates were roughly equivalent. These results suggest that the loss of metabolic activities unnecessary for cell growth is mainly due to antagonistic pleiotropic mutations (see Duffy et al. 2007 for a virus example). However, a recent characterization of the same *E. coli* populations has shown that mutations that were beneficial in glucose-limited medium had an overwhelming tendency to be beneficial in other environments as well (Ostrowski et al. 2005; cf. Remold and Lenski 2004). It is therefore unclear whether AP alleles are the main cause of degeneration of fitness-related characters during reverse evolution.

The accumulation of degenerative alleles has been demonstrated when experimental evolution has been conducted by passaging single individuals or mated pairs from ancestral populations in the near absence of selection (e.g., Mukai et al. 1972; Mackay et al. 1994; Funchain et al. 2000; Vassilieva et al. 2000) and in natural populations of asexual species (e.g., Paland and Lynch 2006), humans (Eyre-Walker et al. 2006), and nonrecombining organellar genomes (Lynch and Blanchard 1998; Denver et al. 2000). Accumulation of such mutations causes steady declines in fitness-related characters over time (reviewed in Lynch et al. 1999). The extent to which long-term accumulation of deleterious alleles prevents reverse evolution has been incompletely studied, but the findings of Estes and Lynch (2003) suggest that reverse evolution of fitness is at least in some cases possible even after more than two hundred generations of spontaneous mutation accumulation (see later discussion). Furthermore, deleterious alleles accumulating in this type of experiment tend to exhibit positively correlated effects on fitness characters (Houle et al. 2004; Estes et al. 2005 and references therein). A tendency toward positive genetic correlations among characters has also been observed when outbred populations are subjected to novel laboratory environments (Service and Rose 1985). Alleles with positively correlated fitness effects may thus be abundant in populations that are not close to mutation-selection equilibrium.

EPISTASIS

Another population genetic mechanism that may affect reverse evolution is epistasis, defined as the nonlinear interaction of effects among alleles at different loci (see Wolf et al. 2000). In the large *D. melanogaster* reverse evolution experiments described previously, contingency was readily evident (Teotónio and Rose 2000; Teotónio et al. 2002). One population-genetic explanation for these results is the evolution of different favorable epistatic interactions, and therefore linkage disequilibrium, among loci in different populations over the course of diversifying selection. If reversibility in the ancestral environment involved the breakup of such beneficial epistatic interactions, evolution could be constrained and contingent on previous selection history.

To test this possibility, Teotónio and Rose (2000) conducted crosses between populations with different selection histories, and then cultured these hybrid populations in parallel with their parent populations for fifty generations in the ancestral environment. This experimental manipulation reduced linkage disequilibrium by creating new combinations of alleles, which might be expected to allow populations to more easily return to ancestral states. There was, however, no difference in rates of reverse evolution or in phenotypic outcomes between hybrid and nonhybrid populations (Teotónio and Rose 2000). Furthermore, explicit genetic models fitted to crosses among differentiated populations found little evidence for epistasis in the original phenotypic differentiation of the diversified populations (Teotónio et al. 2004). In this case, conditions of mass natural selection on large outbred populations were employed, and the average epistatic effects were zero. However, even when the average effect is additive, variance in epistatic effects can have profound impacts on evolutionary change in populations with different demographic structures than those employed (cf. Phillips et al. 2000; Meffert 2000).

Some of the most compelling evidence that epistasis is involved in cases of reverse evolution comes from studies of evolution following the acquisition of antibiotic or bacteriophage resistance in bacteria (for reviews, see Maisnier-Patin and Andersson 2004; Colegrave and Buckling 2005). Mutations conferring resistance are often found to be costly if their carriers are returned to ancestral drug- or phage-free environments (e.g., Lenski 1988). It has been shown repeatedly that reverse evolution is frequently achieved through changes at secondary sites—that is, through epistatic mutations that compensate for the fitness costs of the initial resistance mutations. These mutations are called compensatory (Phillips et al. 2000) because they can to some extent compensate for the deleterious effects of other mutations (Kimura 1985). Conversely, such evolution is accomplished much less frequently by back mutations that exactly restore ancestral sequences. In addition, a variety of studies have demonstrated that the beneficial effects of compensatory mutations are generated through their nonlinear interactions with deleterious mutations, and that the compensatory changes are deleterious or neutral in other genetic backgrounds (e.g., Lenski 1988; Levin et al. 2000; Moore et al. 2000; Rokyta et al. 2002; but see Schrag et al. 1997). These results are also consistent with the rapid evolution of suppressor mechanisms observed in a variety of laboratory mutant stocks (Prelich 1999; reviewed in Whitlock et al. 2003).

A particularly illustrative example of the ubiquity of epistasis is the recent study by Poon and Chao (2005), in which the rate of compensatory mutation was quantified for replicate populations of bacteriophage Φ-X174. They found that an average of nine compensatory mutations were available to neutralize the effects of a given deleterious missense substitution. As discussed by several authors (e.g., Prelich et al. 1999; Whitlock et al. 2003; Maisnier-Patin and Andersson 2004), the effects of many point mutations can be suppressed not only by secondary mutations in other genes and biochemical pathways, but also by a variety of intragenic changes. This suggests that within-locus epistasis is quite common. In agreement with this, Poon and Chao (2005) found that about half of all the compensatory mutations fixed in their populations arose in the same gene as the original deleterious mutation. This is perhaps not surprising given that the Φ-X174 genome contains only eleven genes; however, a likelihood analysis of suppressor mutation suggests that the number of such intragenic mutations available to mask the effects of a given mutation (about ten) is approximately the same in viruses, prokaryotes, and eukaryotes (Poon et al. 2005). The size and the complexity of structural organization of genomes may prove to have little influence on the preponderance of epistasis.

These results documenting extensive reverse evolution of fitness via epistatic mutations are contrary to those of Crill et al. (2000), who found that reverse evolution of Φ-X174 growth rates on *E. coli* after adapting to *Salmonella* was mostly due to back mutations rather than second site mutations (see also Whatmore et al. 1995). The exact reasons for this disparity are unknown, but several possibilities exist. The most likely explanation is that the evolving populations used by Crill et al. were large enough for back mutations—whose selection coefficients tend to be larger than those of compensatory mutations—to arise at sufficiently high frequencies to be utilized in the process of reverse evolution.

The 1999 study of Burch and Chao suggests that this may indeed be the case. Beginning with an isogenic population of Φ-6 bacteriophage displaced from its adaptive optimum by a single highly deleterious mutation, the authors found that ancestral growth rates were attained in one mutational step, presumably a back mutation, in evolving populations of large size, whereas an increasing number of mutations were responsible for the sometimes incomplete fitness restoration in smaller populations (i.e., the spectrum of mutations fixed during adaptation changed with bottleneck size). This study demonstrated that, even in small genomes, compensatory mutations arise more often than back mutations, implying a great deal of epistasis for fitness. Furthermore, these results demonstrate that adaptive landscapes become smoother and easier to traverse with large population sizes and greater genetic variation, in accord with evolutionary theory (Fisher 1946; Weber 1996; Weinreich and Chao 2005).

Recent studies in several microbes have confirmed the basic result of Burch and Chao (1999) and have provided insight into other properties of compensatory evolution. For example, it has been shown that the selective effects of fixed compensatory mutations increase with larger bottleneck sizes (Maisnier-Patin et al. 2002; Rozen et al. 2002), and that the likelihood and rate of compensation will depend on the initial distance of a population from its optimum, with less fit genotypes recovering fitness faster than more fit ones (Moore et al. 2000; Poon and Chao 2005). Finally, because the particular form of epistasis (e.g., positive, negative) will determine the combined fitness effect of mutant alleles (Phillips et al. 2000), it is expected that the likelihood and rate of compensation will depend on the number of deleterious mutations harbored by a population, and on the type and magnitude of epistasis that exists (1) between multiple deleterious mutations, (2) between the compensatory mutations that arise to moderate the effects of deleterious mutation(s), and (3) between the deleterious and compensatory mutation(s) (Moore et al. 2000; Sanjuán et al. 2005).

How relevant are the above microbial data to sexual organisms with larger genomes, more complex development and physiology, lower mutation rates, and smaller long-term effective population sizes (McKenzie 1993)? Estes and Lynch (2003) documented the reverse evolution of ancestral growth rates in populations of the nematode, *Caenorhabditis elegans*, that had undergone an average of 240 single-worm bottlenecks and thus suffered reduced fitness as a result of the random accumulation of deleterious mutations. These populations were returned to their ancestral laboratory selection environment (i.e., mass natural selection of large populations). Although there was considerable variation in evolutionary response among lines, recovery of average levels of ancestral fitness occurred in fewer than eighty generations of reverse selection (Estes and Lynch 2003). Their findings also agree with those from microbial studies outlined earlier, in which evolution of fitness was shown to proceed at a faster pace in populations beginning at lower fitness states. While it is likely that the reverse evolution of fitness in these populations resulted from the accumulation of compensatory rather than back mutations, this has now been verified. New experiments with a subset of the same mutationally degraded lines—this time evolved in replicate—indicate that reverse evolution of growth rates is not necessarily repeatable because populations that previously reverted to ancestral fitness levels did not always do so in the new set of experiments. However, when complete recovery of ancestral fitness has occurred, it is highly repeatable among the independent replicate populations within a line. What this pattern implies about evolution at levels of DNA sequence and transcription is the subject of ongoing research (S. Estes, C. Palmer, J. Anderson, P. Phillips, and D. Denver, unpublished data).

While AP and epistasis result from structural characteristics of alleles within genotypes that affect fitness, frequency-dependent selection can be viewed as resulting from non-linear interactions of effects among different genotypes within a population, particularly when the fitness of a genotype depends on the frequency other genotypes in the population. The reverse evolution study of Rainey and Travisano (1998) illustrates such a case. The authors found that when ten replicate populations of *Pseudomonas fluorescens*, all initiated from a single genotype, were cultured in spatially heterogeneous environments, each one rapidly diversified into an identical set of three dominant genotypes expressing different colony morphologies. Each morph was specialized to occupy a particular depth in the medium. The selective mechanism maintaining this diversity was apparently dependent on which genotypes were interacting, such that no single evolved genotype could outgrow the other two. When the growth medium was homogenized, complete reverse evolution of morphology was observed; therefore, natural selection imposed by environmental heterogeneity was apparently required to maintain diversity (see also Rainey and Rainey 2003).

Similar phenomena have been observed for other bacterial experimental systems (e.g., Rozen and Lenski 2000; Kerr et al. 2002). A recent study reported the evolution of high levels of phenotypic and genotypic diversity within populations of *E. coli* adapting to a seemingly homogeneous chemostat environment (Maharjan et al. 2006). Because negative frequency-dependent selection maintains population genetic diversity—diversity that may include ancestral genotypes—it can be viewed as a potential mechanism for fostering convergent reverse evolution.

## OUTLOOK

We have presented an overview of the experimental study of reverse evolution highlighting the historical development of the field and important recent results. Here, organized in terms of three major outstanding questions, we summarize the current state of knowledge with regard to the study of reverse evolution and its significance for other disciplines of evolutionary biology. We also discuss ways in which laboratory experimental evolution studies can help address some of these issues.

### IS THERE A GENERAL SET OF CONDITIONS UNDER WHICH PHENOTYPIC REVERSIBILITY IS GUARANTEED?

Although reverse evolution is in principle always possible, it will be promoted by smooth, single-peaked adaptive landscapes. The true number of dimensions of these landscapes will, of course, be much larger than the easily visualized two- and three-dimensional examples commonly depicted (see Phillips and Arnold 1989; Gavrilets 1997). But considering the simplest case of two characters, available data from natural populations suggest that saddle shapes that promote diversification of populations may not be uncommon (e.g., Blows and Brooks 2003). Perhaps it should not be surprising, then, that neither convergent reverse evolution nor repeatability at the phenotypic level is consistently observed in natural populations (but for examples on ecological time scales, see Kettlewell 1973; Grant and Grant 2002). While no adaptive landscape has yet or will likely ever be completely characterized, the data reviewed here imply that even general features of landscapes promoting or constraining convergence during reverse evolution may be difficult to identify.

Experimental and observational studies have thoroughly established that evolutionary reversals of phenotypes such as morphologies and life-history characters are indeed possible even over fairly long periods of time (Stanley and Yang 1987; Jackson and Cheetham 1994), but they are certainly not guaranteed even when populations are returned to common ancestral environments (Bull and Charnov 1985; Teotónio and Rose 2000; Whitlock et al. 2002; Rainey and Rainey 2003). A general message from the results reviewed here, then, is that the likelihood of returning to an ancestral character state will depend heavily on how closely the character in question is related to fitness— that is, on the strength and direction of selection on the character in the ancestral environment (Travisano et al. 1995a; Teotónio et al. 2002), and possibly on how far a population has diverged from the ancestral state.

If a particular phenotype is successfully resurrected during reverse evolution, the probability that it involved the same genotype as that of the ancestor, thereby constituting convergent reverse evolution at all levels of biological organization, will depend on whether the underlying genes have been preserved by natural selection for a related function. Reversion will be more likely under these conditions than it would be if the genes have been completely released from selection and have undergone extensive mutation accumulation. The results of Marshall et al. (1994) suggest that it may take upward of ten million years for mutational decay to render phenotypic evolution irreversible (see also Lande 1978). If the genes underlying a lost character have been conserved for other functions, heterochronic and spatial changes in the expression of these genes leading to the reappearance of an ancestral character state may be relatively easy to achieve. This is a possibility that could benefit from an experimental evolution approach, particularly one employing a metazoan model system in which the reverse evolution of populations that have experienced varying numbers of generations of divergence is followed.

Although evolution by natural selection is hardly infallible, as existing background extinction rates attest, it has nevertheless found solutions to many of earth's environmental challenges over the last few billion years. Thus, the repeated evolution of phenotypes, particularly those with close ties to fitness in the ancestral environment, may in some cases be best explained by the frequency of reversibility in the external environment. In fact, it may be that true irreversibility of most phenotypes requires major genomic revolutions such as polyploidization or genetic changes leading to radical breeding system differences (Bull and Charnov 1985). While many aspects of the environment (e.g., climate) may be reversible and even cyclical, it is unclear how often an exact set of ancestral ecologies ever appear for a second time outside clinical environments, given the complex interactions that would have to be re-created (but see Losos et al. 1998). Some of the studies reviewed here suggest further that reverse evolution can result in convergence even when populations have diverged in constant environments (Bull et al. 1997; Maharjan et al. 2006). This is certainly an issue deserving of further experimental treatment by, for example, determining the role that differences in phenotypic plasticity play in the likelihood of reverse evolution (Teotónio et al. 2006).

## WHAT ARE THE GENETIC MECHANISMS
## OF REVERSE EVOLUTION?

While ancestral phenotypes can be regained, many of the studies reviewed here indicate that these reversals are often realized through the evolution of genotypes different from those of the ancestor. According to Gavrilets and others (Gavrilets 1997; Gavrilets and Gravner 1997; Gavrilets 2004), genotypic landscapes, relating genotypes directly to fitness, can be accurately described by a hypercube—one resembling a block of Swiss cheese where the holes represent nearly neutral trajectories through sequence space that can connect distant precincts. On this metaphor, reverse evolution might be even less likely than on a traditional adaptive landscape beset with valleys of low fitness (see Wright 1932). This is because replicate populations could potentially move with ease to isolated regions of high fitness, while initially divergent populations will be highly unlikely to converge on the same genetic solution. The outcome will depend on the number of holes, a distinctly empirical problem. Apparently, landscapes in relatively simple laboratory environments can in some cases allow reverse evolution at the level of genotype but be sufficiently rugged or porous to prevent it in other cases. Furthermore, if the location of peaks or holes changes with environmental setting, as some studies indicate (e.g., Björkman et al. 2000; Teotónio et al. 2002; Remold and Lenski 2004), predicting evolutionary responses to an ancestral (or any other) environment will be challenging at best. On the other hand, such environmental effects may allow evolutionary trajectories to escape from domains of attraction in hypercubes that might otherwise prevent convergence during reverse evolution.

The integration of whole-genome and other molecular techniques into evolution experiments is allowing an improved characterization of the effects of individual alleles, and unprecedented insight into the molecular genetic underpinnings of adaptive evolution. Studies following the adaptation of populations to both new and ancestral selective environments have demonstrated that parallel and convergent phenotypic evolution can be realized through the evolution of different genotypes, but that the number of different adaptive solutions may still be quite limited (e.g., Bull et al. 1997; Wichman et al. 1999;

Cuevas et al. 2002; Novella and Ebendick-Corpus 2004; see Weinrich et al. 2006 for a simulation study). Furthermore, many different nucleotide changes may lead to similar gene expression patterns and adaptive phenotypes (Thacker et al. 1995; Cooper et al. 2003). Additional information on the genetic mechanisms of reverse evolution in replicate, divergent populations would better inform us about the nature and likelihood of integrated evolution across definable levels of organization. In addition, such studies would yield important information about the genetic complexity of adaptation (Orr and Coyne 1992), the general importance of epistasis in defining fitness landscapes (e.g., Gavrilets 1997; Weinreich et al. 2005, 2006).

Although pertinent data are scant, convergence may become more likely with increased genomic complexity and greater redundancy among different biochemical, ontogenic, and physiological pathways. This is simply because the statistical probability of retracing ancestral steps to a certain genotypic state by parallel changes diminishes with the number of loci affecting phenotypic expression. It thus seems improbable that patterns of reverse evolution will be highly integrated across definable levels of organization in organisms other than viruses. Furthermore, theory predicts that mutational effect distributions will be altered by organismal complexity: the probability that a mutation of given effect size will be favorable becomes exceedingly small as the number of interacting components of an organism's genotype or phenotype increases (Fisher 1930; Orr 2000; Poon and Otto 2000). This may imply that reverse evolution by the accumulation of new mutations will be more difficult to achieve in complex organisms.

Bacteria and viruses have thus far been the workhorses of experimental evolution with few exceptions. Parallel experiments in multicellular eukaryotes are needed to assess the effect of increased organismal and genetic complexity on the patterns and mechanisms of reverse evolution (e.g., Orr 2000; Welch and Waxman 2003). The scant empirical data available for compensatory evolution in animals (Estes and Lynch 2003) and a likelihood analysis of suppressor mutation literature on a variety of organisms (Poon et al. 2005) suggest that the size and complexity of structural organization of genomes may prove to have little influence on the preponderance of epistasis and therefore on compensatory mutation. But we need to know whether other types of reverse evolution results from viruses and other microbes are general, or whether organisms with larger genomes, greater developmental and physiological complexity, more diverse genetic functions, regular genetic recombination and repair, and smaller long-term effective population sizes exhibit different patterns of evolution. These questions are becoming more accessible with the large-scale genotypic and phenotypic characterization techniques mentioned earlier. Phylogenetic inference will also have much to gain from these studies.

Several generalities are emerging with respect to the reverse evolution of fitness, a topic that has been studied mainly from the perspective of discerning the genetics of evolving pathogen populations. For example, it is clear that the recovery of fitness proceeds mainly by the accumulation of compensatory epistatic mutations rather than by back mutations—at least in viruses and bacteria (Whitlock et al. 2003 and references therein). Such compensatory substitutions can be highly repeatable, as illustrated by the consistency with which certain nucleotide changes arise during the evolution of resistance. However, several issues remain to be addressed with respect to the genetics of reverse evolution by compensation. These include the molecular physiology of compensation, compensatory mutation rates in organisms other than viruses, and the properties of mutations that cannot be compensated. Further experimental treatment in a metazoan system where a broader array of suppression mechanisms is expected to be available for buffering the effects of deleterious mutations would be useful. Finally, given the recently established importance of epistasis in defining accessible mutational trajectories (Weinreich et al. 2005) and the fact that compensatory effects may be both environment- (Björkman et al. 2000; Remold and Lenski 2004) and sex-specific (Pischedda and Chippindale 2005), further characterization of the effects of random mutations is in order (e.g., Rokyta et al. 2005; Burch et al. 2007).

Having just argued the necessity of further study on compensatory reverse evolution, it is important to consider how often we should expect compensatory evolution to actually occur in natural settings. As discussed by Whitlock and colleagues (2003), if the dominant mode of selection on a quantitative character is stabilizing such that an intermediate value is favored, deleterious mutations that perturb the mean away from this optimum can potentially be ameliorated by changes at a large number of other sites or loci. Studies have confirmed that, indeed, a number of epistatic mutations are potentially available to mask the effects of a given deleterious mutation fixed in a population. They have also shown that the likelihood of compensation increases with a population's starting distance from the optimum. However, it is unclear how often natural populations are required to evolve from states of extremely low fitness (McKenzie 1993). Moreover, if a population contains genetic variation, it is uncertain whether evolution in this case will ever proceed by the substitution of mutations that specifically compensate for the effects of segregating deleterious alleles, or whether evolution instead proceeds by the appearance of novel genotypes by recombination. There are no data on this topic. It may be that compensatory evolution will only be of importance for populations that have fixed generally deleterious mutations, which is a scenario with relevance for an increasing number of species facing inbreeding depression and mutational meltdown (cf. Lynch et al. 1995).

## WHAT ARE THE EFFECTS OF RECOMBINATION ON THE LIKELIHOOD OF REVERSE EVOLUTION?

Apart from other issues that render the general applicability of microbial studies to other organisms questionable, such as their extremely large population sizes, the absence of recombination is one of the most glaring. Indeed, the role of recombination in both divergence and convergence remains a perennial problem in evolutionary theory. In strictly asexual organisms, genomes are selected *in toto*, and integration at several levels of organization might be more extensive than in sexual species where natural selection only operates on additive gene effects (Fisher 1946; Rice 2004).

As previously noted, viral recombination was apparently occurring in the study of Wichman et al. (2005), but it was uncontrolled and of unknown frequency. For a nonrecombining population, reaching a higher fitness peak some distance away in sequence space would require the population to do so one mutational step at a time. Between the necessity of waiting for new beneficial mutations to arise (quite possibly in a particular order) and the fact that many intermediate states may be associated with low fitness and thus never achieve high frequencies, such dramatic peak shifts seem implausible. With recombination, however, new assemblages of alleles can be brought together, making instantaneous peak shifts possible. On the other hand, recombination can once again break up the favorable combination of alleles that made the peak shift achievable. The effects of sex and the influence of recombination rates on the likelihood and rate of reverse evolution constitute a gaping hole in our knowledge. Sex is expected to facilitate reverse evolution just as it does general adaptation. However, if populations have evolved dramatically different genotype-by-environment interactions from those of their ancestors, sexual recombination may not be able to successfully recreate ancestral phenotypes (cf. Teotónio et al. 2006).

How important variation in epistatic effects is for reverse evolution of sexual populations also continues to be an open question. If hybridization occurs among independently evolved populations that have fixed different alleles, reverse evolution can be achieved through genetic compensation. How often this situation occurs in nature is unknown. Interestingly, recent computer simulation work suggests that synergistic negative epistasis can evolve as a by-product of sexual recombination (Azevedo et al. 2006). Because this form of epistasis among deleterious alleles is thought to be required for sex to be favored by selection, this is, as the authors say, "a case of evolution forging its own path." Deleterious mutations exhibiting this form of epistasis were shown to be more readily compensated than others (Sanjuán et al. 2005). It is therefore possible that epistasis could in some cases be important for the reverse evolution of sexual as well as asexual populations. Experimental evolution could shed light on these issues by the manipulation of recombination rates during adaptation.

## SUMMARY

Observations of phenotypic reversions (atavisms and the regression of offspring to parental phenotypes) motivated most of the evolutionary research that succeeded Darwin until their bases were demystified by the rules of Mendelian heredity and Hardy-Weinberg equilibria. In the period that followed, experiments employing relaxed and reverse artificial selection were conducted to characterize the genetic structure of quantitative characters and to determine the operation of natural selection in laboratory mutant stocks. Only within the last twenty years has reverse evolution again come to be studied in its own right. The resurgence in interest in this topic is due mainly to a sophistication of evolutionary theory and experimental design. We review recent evolution experiments that bear on the problem of inference of common ancestry in the face of reverse evolution (homoplasy), the twin problems of contingency and irreversibility of phenotypic states deriving from the historical nature of evolution, and on the characterization of population genetic mechanisms of reverse evolution. We finish with a discussion of how experimental reverse evolution can provide a fruitful approach to studying several outstanding problems in evolutionary biology.

## ACKNOWLEDGMENTS

We thank the editors of this volume for the invitation, and several reviewers for their comments on draft versions of the manuscript. This work was supported by a National Science Foundation grant (DEB-0625211) and a Faculty Enhancement Grant from Portland State University (S.E.) and Fundação para a Ciência e a Tecnologia, FEDER, and Fundação Calouste Gulbenkian (H.T.).

## GLOSSARY

ADAPTIVE LANDSCAPE A theoretical construct that results from integrating the relationships between average fitness and phenotypes in a given environment, with the distribution of those phenotypes in a given population.

ARTIFICIAL SELECTION An experimental technique used to after particular phenotypes and to determine the hereditary basis of quantitative characters, in which individuals expressing particular phenotypic values are selected to contribute to the next generation often irrespective of how well they survive and reproduce.

ATAVISM The rare and often incomplete reappearance of a long-absent ancestral phenotype in a member of a derived population.

CONVERGENCE Used here to refer to an identity in character state due to parallel evolution among two or more populations. Parallel changes can be reversions or diversions from the ancestral character state due to deterministic processes or to the stochastic nature of sampling genotypes across generations; in contrast, identity in character state can be due to common ancestry.

DEGENERATIVE EVOLUTION Evolution leading to the loss of phenotypes such as complex morphological structures or physiological processes; an alternate definition of reverse evolution in some texts, but it does not necessarily involve a return to an ancestral character state.

FITNESS An operational measure of evolutionary success of a genotype or phenotype. Fitness can be equated with expected population growth rates under specific demographic models or with expected extinction rates. See also Mueller (this volume).

G-MATRIX A square matrix of $n$ dimensions, in which the diagonal cells correspond to additive genetic variances for each of the $n$ phenotypes, while the off-diagonal cells correspond to additive genetic covariances for all pairwise interactions of the $n$ phenotypes; a statistical descriptor of inheritance in multivariate quantitative genetics.

HOMOPLASY Nonhomologous similarity in character states among organisms due to reverse and parallel substitutions.

MICROBE A generic term used here to distinguish viruses, bacteria, and other unicellular organisms from multicellular organisms.

PARALLEL EVOLUTION A process whereby two or more populations undergo the same genotypic or phenotypic changes. Parallel evolution can occur through both reversions and diversions from the common ancestor of the populations under study.

REGRESSION A statistical technique originating with Francis Galton whereby the mean of one variable may be predicted by the distribution of a second, related one (simple linear regression); or the biological phenomenon in which a population of offspring exhibits a complete or partial return to the mean character state of their parental generation.

RELAXED SELECTION The cessation of a particular artificial selection scheme when experimenters are no longer selecting the genetic contributors to the next generation based on a phenotype; the near or complete absence of any form of selection experienced when organisms are artificially evolved in benign environments and/or extremely small population sizes such that genetic drift becomes the dominant evolutionary force; or, less common, the result of a natural population experiencing a new selective environment in which a currently expressed character(s) is (are) neutral.

REVERSE EVOLUTION A process whereby one population acquires the phenotypes or genotypes of an ancestor population.

REVERSE SELECTION Selection on phenotypes in the opposite direction of selection imposed previously by artificial selection or by selection inferred to have occurred in the past.

SYNAPOMORPHY A shared character state derived from a common ancestor that unites the members of a clade.

## REFERENCES

Arnold, S. J., M. E. Pfrender, and A. G. Jones. 2001. The adaptive landscape as a conceptual bridge between micro- and macro-evolution. *Genetica* 112/113:9–32.

Azevedo, R. B. R., R. Lohaus, S. Srinivasan, K. K. Dang, and C. L. Burch. 2006. Sexual reproduction selects for robustness and negative epistasis in artificial gene networks. *Nature* 440:87–90.

Baer, C. F., F. Shaw, C. Steding, et al. 2005. Comparative evolutionary genetics of spontaneous mutations affecting fitness in rhabditid nematodes. *Proceedings of the National Academy of Sciences of the USA* 102:5785–5790.

Barton, N., and L. Partridge. 2000. Limits to natural selection. *BioEssays* 22:1075–1084.

Barton, N., and M. Turelli. 1989. Evolutionary quantitative genetics: How little do we know? *Annual Review of Genetics* 23:337–370.

Bell, G. 1997. *Selection: The Mechanism of Evolution*. New York: Chapman and Hall.

Björkman, J., I. Nagaev, O. G. Berg, D. Hughes, and D. I. Andersson. 2000. Effects of environment on compensatory mutations to ameliorate costs of antibiotic resistance. *Science* 287:1479–1482.

Blows, M. W., and R. Brooks. 2003. Measuring nonlinear selection. *American Naturalist* 162:815–820.

Bohren, B. B., W. G. Hill, and A. Robertson. 1966. Some observations on asymmetrical correlated responses to selection. *Genetical Research Cambridge* 7:44–57.

Borman, A. M., S. Paulous, and F. Clavel. 1996. Resistance of human immunodeficiency virus type 1 to protease inhibitors: Selection of resistance mutations in the presence and absence of the drug. *Journal of General Virology* 77:419–426.

Buckling, A., M. A. Wills, and N. Colegrave. 2003. Adaptation limits diversification of experimental bacterial populations. *Science* 302:2107–2109.

Bull, J. J., M. R. Badgett, H. A. Wichman, J. P. Huelsenbeck, D. M. Hillis, A. Gulati, C. Ho, and I. J. Molineaux. 1997. Exceptional convergent evolution in a virus. *Genetics* 147:1497–1507.

Bull, J. J., and E. L. Charnov. 1985. On irreversible evolution. *Evolution* 39:1149–1155.

Bull, J. J., C. W. Cunningham, I. J. Molineaux, et al. 1993. Experimental molecular evolution of bacteriophage T7. *Evolution* 47:993–1007.

Burch, C. L., and L. Chao. 1999. Evolution by small steps and rugged landscapes in the RNA virus φ-6. *Genetics* 151:921–927.

——— . 2000. Evolvability of an RNA virus is determined by its mutational neighborhood. *Nature* 406:625–628.

Burch, C. L., S. Guyader, D. Samarov, and H. Shen. 2007. Experimental estimate of the abundance and effects of nearly neutral mutations in the RNA virus phi-6. *Genetics* 176:467–476.

Castle, W. E., and J. C. Phillips. 1914. *Piebald Rats and Selection: An Experimental Test of Selection and the Theory of Gametic Purity in Mendelian Crosses*. Carnegie Institution of Washington, Publication No. 195. Washington, DC: Carnegie Institution.

Charlesworth, B. 1980. *Evolution in Age-Structured Populations*. Cambridge: Cambridge University Press.

Chippindale, A. K., J. A. Alipaz, H.-W. Chen, and M. R. Rose. 1997. Experimental evolution of accelerated development in *Drosophila*. I. Development speed and larval survival. *Evolution* 51:1536–1551.

Chippindale, A. K., J. A. Alipaz, and M. R. Rose. 2004. Experimental evolution of accelerated development in *Drosophila*. 2. Adult fitness and the fast development syndrome. *In* M. R. Rose, H. B. Passananti, and M. Matos, eds. *Methuselah Flies: A Case Study in the Evolution of Aging*. Singapore: World Scientific.

Cohan, F. M. 1984. Can uniform selection retard random genetic divergence between isolated conspecific populations? *Evolution* 38:495–504.

Colegrave, N., and A. Buckling. 2005. Microbial experiments on adaptive landscapes. *BioEssays* 27:1167–1173.

Cooper, V. S., and R. E. Lenski. 2000. The population genetics of ecological specialization in evolving *Escherichia coli* populations. *Nature* 407:736–739.

Cooper, T. F., D. E. Rozen, and R. E. Lenski. 2003. Parallel changes in gene expression after 20,000 generations of evolution in *Escherichia coli*. *Proceedings of the National Academy of Sciences of the USA* 100:1072–1077.

Crill, W. D., H. A. Wichman, and J. J. Bull. 2000. Evolutionary reversals during viral adaptation to alternating hosts. *Genetics* 154:27–37.

Crow, J. F., and M. Kimura. 1970. *An Introduction to Population Genetics Theory*. New York: Harper and Row.

Cuevas, J. M., S. F. Elena, and A. Moya. 2002. Molecular basis of adaptive convergence in experimental populations of RNA viruses. *Genetics* 162:533–542.

Deckert-Cruz, D. J., R. H. Tyler, J. E. Landmesser, and M. R. Rose. 1997. Allozymic differentiation in response to laboratory demographic selection of *Drosophila melanogaster*. *Evolution* 51:865–872.

Denver, D. R., K. Morris, M. Lynch, L. L. Vassilieva, and W. K. Thomas. 2000. High direct estimate of the mutation rate in the mitochondrial genome of *Caenorhabditis elegans*. *Science* 289:2342–2344.

Dollo, L. 1893. The laws of evolution. *Bulletin of the Society of Belgian. Geology and Paleontology* 7:164–166.

Duffy, S., C. L. Burch, and P. E. Turner. 2007. Evolution of host specificity drives reproductive isolation among RNA viruses. *Evolution* 61:2614–2622.

Elena, S. F., and R. E. Lenski. 2003. Evolution experiments with microorganisms: The dynamics and genetic bases of adaptation. *Nature Reviews Genetics* 4:457–469.

Escarmís, C., M. Dávila, and E. Domingo. 1999. Multiple molecular pathways for fitness recovery of an RNA virus debilitated by operation of Muller's Ratchet. *Journal of Molecular Biology* 285:495–505.

Estes, S., B. Ajie, M. Lynch, and P. C. Phillips. 2005. Spontaneous mutational correlations for life-history, morphological and behavioral characters in *Caenorhabditis elegans*. *Genetics* 170:645–653.

Estes, S., and S. J. Arnold. 2007. Resolving the paradox of stasis: Models of stabilizing selection account for divergence on all timescales. *American Naturalist* 169:227–244.

Estes, S., and M. Lynch. 2003. Rapid recovery in mutationally degraded lines of *Caenorhabditis elegans*. *Evolution* 57:1022–1030.

Eyre-Walker, A., M. Woolfit, and T. Phelps. 2006. The distribution of fitness effects of new deleterious amino acid mutations in humans. *Genetics* 173:891–900.

Felsenstein, J. 1981. Evolutionary trees from DNA sequences: A maximum likelihood approach. *Journal of Molecular Evolution* 17:368–376.

———. 2003. *Inferring Phylogenies*. Sunderland, MA: Sinauer.

Fisher, R. A. 1930. *The Genetical Theory of Natural Selection*. 2nd ed. New York: Dover.

———. 1946. *Statistical Methods for Research Workers*. 10th ed. Edinburgh: Oliver and Boyd.

Fisher, R. A., and E. B. Ford. 1926. Variability of species. *Nature* 118:515–516.

Fleming, J. E., G. S. Spicer, R. C. Garrison, and M. R. Rose. 1993. Two dimensional protein electrophoretic analysis of postponed aging in *Drosophila. Genetica* 91:183–198.

Fong, D. W., T. C. Kane, and D. C. Culver. 1995. Evolution of vestigial and nonfunctional characters. *Annual Review of Ecology and Systematics* 26:249–268.

Funchain, P., A. Yeung, J. Lee Stewart, R. Lin, M. M. Slupska, and J. H. Miller. 2000. The consequences of growth of a mutator strain of *Escherichia coli* as measured by loss of function among multiple gene targets and loss of fitness. *Genetics* 154:959–970.

Galton, F. 1887. *Natural Inheritance*. London: Macmillan.

Gavrilets, S. 1997. Evolution and speciation on holey landscapes. *Trends in Ecology & Evolution* 12:307–312.

———. 2004. *Fitness Landscapes and the Origin of Species*. Princeton, NJ: Princeton University Press.

Gavrilets, S., and J. Gravner. 1997. Percolation on the fitness hypercube and the evolution of reproductive isolation. *Journal of Theoretical Biology* 184:51–64.

Gayon, J. 1998. *Darwinism's Struggle for Survival: Heredity and the Hypothesis of Natural Selection.* Cambridge: Cambridge University Press.

Gould, S. J. 1989. *Wonderful Life: The Burgess Shale and the Nature of History.* New York: Norton.

Grant, P. R., and B. R. Grant. 2002. Unpredictable evolution in a 30-year study of Darwin's finches. *Science* 296:707–711.

Gromko, M. H. 1990. Unpredictability of correlated response to selection: Pleiotropy and sampling interact. *Evolution* 49:685–693.

Hardy, G. H. 1908. Mendelian proportions in a mixed population. *Science* 28:49–50.

Hill, W. G., and A. Robertson. 1966. The effect of linkage on limits to artificial selection. *Genetical Research Cambridge* 8:269–294.

Hillis, D. M., J. J. Bull, M. E. White, M. R. Badgett, and I. J. Molineux. 1992. Experimental phylogenetics: generation of a known phylogeny. *Science* 255:589–592.

Houle, D. 1992. Comparing evolvability and variability of quantitative traits. *Genetics* 130:195–204.

Houle, D., K. A. Hughes, D. K. Hoffmaster, J. Ihara, S. Assimacopoulos, D. Canada, and B. Charlesworth. 1994. The effects of spontaneous mutation on quantitative traits. I. Variances and covariances of life history traits. *Genetics* 138:773–785.

Jackson, J. B. C., and A. H. Cheetham. 1994. Phylogeny reconstruction and the tempo of speciation in the cheilostome Bryozoa. *Paleobiology* 20:407–723.

Johannsen, W. 1903. *Über Erblichkeit in Populationen und in reinen Linien.* Jena: Fischer.

Joshi, A., R. B. Castillo, and L. D. Mueller. 2003. The contribution of ancestry, chance, and past and ongoing selection to adaptive evolution. *Journal of Genetics* 82:147–162.

Kerr, B., M. A. Riley, M. W. Feldman, et al. 2002. Local dispersal promotes biodiversity in a real-life game of rock-paper-scissors. *Nature* 418:171–174.

Kettlewell, H. B. D. 1973. *The Evolution of Melanism.* Oxford: Clarendon Press.

Kimura, M. 1985. The role of compensatory neutral mutations in molecular evolution. *Journal of Genetics* 64:7–19.

———. 1983. *The Neutral Theory of Molecular Evolution.* Cambridge University Press.

Kirkpatrick, M., T. Johnson, and N. Barton. 2002. General models of multilocus evolution *Genetics* 161:1727–1750.

Lande, R. 1976. Natural selection and random genetic drift in phenotypic evolution. *Evolution* 30:314–334.

———. 1978. Evolutionary mechanisms of limb loss in Tetrapods. *Evolution* 32:73–92.

———. 1979. Quantitative genetic analysis of multivariate evolution, applied to brain:body allometry. *Evolution* 33:402–416.

———. 1980. The genetic covariance between characters maintained by pleiotropic mutations. *Genetics* 94:203–215.

Lenski, R. E. 1988. Experimental studies of pleiotropy and epistasis in *Escherichia coli.* II. Compensation for maladaptive effects associated with resistance to virus T4. *Evolution* 42:433–440.

Lenski, R. E., M. R. Rose, S. C. Simpson, and S. C. Tadler. 1991. Long-term experimental evolution in Escherichia coli. 1. Adaptation and divergence during 2000 generations. *American Naturalist* 138:1315–1341.

Lenski, R. E., and M. Travisano. 1994. Dynamics of adaptation and diversification: A 10,000 generation experiment with bacterial populations. *Proceedings of the National Academy of Sciences of the USA* 91:6808–6814.

Leroi, A. M., W. Royal Chen, and M. R. Rose. 1994. Long-term laboratory evolution of a genetic life-history trade-off in *Drosophila melanogaster*. 2. Stability of genetic correlations. *Evolution* 48:1258–1268.

Levin, B. R., B. Perrot, and N. Walker. 2000. Compensatory mutations, antibiotic resistance and the population genetics of adaptive evolution in bacteria. *Genetics* 154:985–997.

Losos, J. B., T. R. Jackman, A. Larson, K. de Queiroz, and L. Rodríguez-Schettino. 1998. Contingency and determinism in replicated adaptive radiations of island lizards. *Science* 279:2115–2118.

Lynch, M., and J. Blanchard. 1998. Deleterious mutation accumulation in organelle genomes. *Genetica* 102/103:29–39.

Lynch, M., J. Blanchard, D. Houle, T. Kibota, S. Schultz, L. Vassilieva, and J. Willis. 1999. Spontaneous deleterious mutation. *Evolution* 53:645–663.

Lynch, M., J. Conery, and R. Bürger. 1995. Mutation accumulation and the extinction of small populations. *American Naturalist* 146:489–518.

Mackay, T. F. C., J. D. Fry, R. F. Lyman, and S. F. Nuzhdin. 1994. Polygenic mutation in *Drosophila melanogaster*: Estimates from response to selection of inbred strains. *Genetics* 136:937–951.

MacLean, R. C., G. Bell, and P. B. Rainey. 2004. The evolution of a pleiotropic fitness trade-off in *Pseudomonas fluorescens*. *Proceedings of the National Academy of Sciences of the USA* 101:8072–8077.

Maharjan, R., S. Seeto, L. Notley-McRobb, and T. Ferenci. 2006. Clonal adaptive radiation in a constant environment. *Science* 313:514–517.

Maisnier-Patin, S., and D. I. Andersson. 2004. Adaptation to the deleterious effects of anti-microbial drug resistance mutations by compensatory evolution. *Research Microbiology* 155:360–369.

Maisnier-Patin, S., O. G. Berg, L. Liljas, and D. I. Andersson. 2002. Compensatory adaptation to the deleterious effect of antibiotic resistance in *Salmonella typhimurium*. *Molecular Microbiology* 46:355–366.

Marshall, C. R., E. C. Raff, and R. A. Raff. 1994. Dollo's law and the death and resurrection of genes. *Proceedings of the National Academy of Sciences of the USA* 91:12283–12287.

Maynard Smith, J., and J. Haigh. 1974. The hitch-hiking effect of a favourable gene. *Genetical Research* 23:23–35.

McKenzie, J. A. 1993. Measuring fitness and intergenic interactions: The evolution of resistance to diazinon in *Lucilia cuprina*. *Genetica* 90:227–237.

Meffert, L. M. 2000. The evolutionary potential of morphology and mating behavior: the role of epistasis in bottlenecked populations. Pages 177–196 *in* J. B. Wolf, E. D. Brodie III, and M. J. Wade, eds. *Epistasis and the Evolutionary Process*. Oxford: Oxford University Press.

Moore, F. B. G., D. E. Rozen, and R. E. Lenski. 2000. Pervasive compensatory adaptation in *Escherichia coli*. *Proceedings of the Royal Society of London B, Biological Sciences* 267:515–522.

Mukai, T., S. I. Chigusa, L. E. Mettler, and J. F. Crow. 1972. Mutation rate and dominance of genes affecting viability in *Drosophila melanogaster*. *Genetics* 72:335–355.

Novella, I. S., and B. E. Ebendick-Corpus. 2004. Molecular basis of fitness loss and fitness recovery in vesicular stomatitis virus. *Journal of Molecular Biology* 342:1423–1430.

Ober, K. A. 2003. Arboreality and morphological evolution in ground beetles (Carabidae: Harpalinae): Testing the taxon pulse model. *Evolution* 57:1343–1358.

Orr, H. A. 2000. Adaptation and the cost of complexity. *Evolution* 54:13–20.

Orr, H. A., and J. A. Coyne. 1992. The genetics of adaptation: A reassessment. *American Naturalist* 140:725–742.

Ostrowski, E. A., D. E. Rozen, and R. E. Lenski. 2005. Pleiotropic effects of beneficial mutations in *Escherichia coli*. *Evolution* 59:2343–2352.

Paland, S., and M. Lynch. 2006. Transitions to asexuality result in excess amino acid substitutions. *Science* 311:990–992.

Passananti et al. 2004. Reverse evolution of aging in *Drosophila melanogaster*. Pages 296–322 in M. R. Rose, M. Matos, and H. Passananti, eds. *Methuselah Flies: A Case Study in the Evolution of Aging. Singapore*: World Scientific.

Pelosi, L., L. Kuhn, D. Güetta, J. Garin, J. Geiselmann, R. E. Lenski, and D. Schneider. 2006. Parallel changes in global protein profiles during long-term experimental evolution in *Escherichia coli*. *Genetics* 173:1851–1869.

Phelan, J. P., M. A. Archer, K. A. Beckman, et al. 2003. Breakdown in correlations during laboratory evolution. I. Comparative analyses of *Drosophila* populations. *Evolution* 57:527–535.

Phillips, P. C., and S. J. Arnold. 1989. Visualizing multivariate selection. *Evolution* 43:1209–1222.

Phillips, P. C., S. P. Otto, and M. C. Whitlock. 2000. Beyond the average: The evolutionary importance of gene interactions and variability of epistatic effects. Pages 20–38 in J. B. Wolf, E. D. Brodie III, and M. J. Wade, eds. *Epistasis and the Evolutionary Process*. Oxford: Oxford University Press.

Phillips, P. C., M. C. Whitlock, and K. Fowler. 2001. Inbreeding changes the shape of the genetic covariance matrix in *Drosophila melanogaster*. *Genetics* 158:1137–1145.

Pischedda, A., and A. Chippindale. 2005. Sex, mutation and fitness: Asymmetric costs and routes to recovery through compensatory evolution. *Journal of Evolutionary Biology* 18:1115–1122.

Poon, A., and L. Chao. 2005. The rate of compensatory mutation in the DNA bacteriophage Φ-X174. *Genetics* 170:989–999.

Poon, A., B. H. Davis, and L. Chao. 2005. The coupon collector and the suppressor mutation: Estimating the number of compensatory mutations by maximum likelihood. *Genetics* 170:1323–1332.

Poon, A. F. Y., and S. P. Otto. 2000. Compensating for our load of mutations: Freezing the meltdown of small populations. *Evolution* 54:1467–1479.

Porter, M. L., and K. A. Crandall. 2003. Lost along the way: The significance of evolution in reverse. *Trends in Ecology & Evolution* 18:541–547.

Prasad, N. G., and A. Joshi. 2003. What have two decades of laboratory life-history studies on *Drosophila melanogaster* taught us? *Journal of Genetics* 82:45–76.

Prelich, G. 1999. Suppression mechanisms: Themes from variations. *Trends in Genetics* 15:261–266.

Price, G. R. 1972. Fisher's "fundamental theorem" made clear. *Annals of Human Genetics* 36:129–140.

Provine, W. B. 1971. *The Origins of Theoretical Population Genetics*. Chicago: University of Chicago Press.

Rainey, P. B., and K. Rainey. 2003. Evolution of cooperation and conflict in experimental bacterial populations. *Nature* 425:72–74.

Rainey, P. B., and M. Travisano. 1998. Adaptive radiation in a heterogeneous environment. *Nature* 394:69–72.

Remold, S. K., and R. E. Lenski. 2004. Pervasive joint influence of epistasis and plasticity on mutational effects in *Escherichia coli*. *Nature Genetics* 36:423–426.

Rice, S. H. 2004. *Evolutionary Theory*. Sunderland, MA: Sinauer.

Rice, W. 2002. Experimental tests of the adaptive significance of sexual recombination. *Nature Review Genetics* 3:241–251.

Riehle, M. M., A. F. Bennett, and A. D. Long. 2005. Changes in gene expression following high-temperature adaptation in experimentally evolved populations of *E. coli*. *Physiological and Biochemical Zoology* 78:299–315.

Rokyta, D. R., M. R. Badgett, I. J. Molineaux, and J. J. Bull. 2002. Experimental genomic evolution: Extensive compensation for a loss of DNA activity in a virus. *Molecular Biology Evolution* 19:230–238.

Rokyta, D. R., P. Joyce, and S. B. Caudle. 2005. An empirical test of the mutational landscape model of adaptation using a single-stranded DNA virus. *Nature Genetics* 37:441–444.

Rose, M. R. 1982. Antagonistic pleiotropy, dominance, and genetic variation. *Heredity* 48:63–78.

———. 1983. Further models of selection with antagonistic pleiotropy. Pages 47–53 *in* H. I. Freedman and C. Strobeck, eds. *Population Biology*. New York: Springer.

———. 1984. Laboratory evolution of postponed senescence in *Drosophila melanogaster*. *Evolution* 38:1004–1010.

———. 1985. Life-history evolution with antagonistic pleiotropy and overlapping generations. *Theoretical Population Biology* 28:342–358.

———. 1991. *Evolutionary Biology of Aging*. New York: Oxford University Press.

Rose, M. R., T. J. Nusbaum, and A. K. Chippindale. 1996. Laboratory evolution: The experimental wonderland and the Cheshire Cat syndrome. Pages 221–241 *in* M. R. Rose and G. V. Lauder, eds. *Adaptation*. San Diego, CA: Academic Press.

Rose, M. R., H. B. Passananti, A. K. Chippindale, J. P. Phelan, M. Matos, H. Teotónio, and L. D. Mueller. 2005. The effects of evolution are local: Evidence from experimental evolution in *Drosophila*. *Integrative and Comparative Biology* 45:486–491.

Rose, M. R., L. N. Vu, S. V. Park, and J. L. Graves Jr. 1992. Selection of stress resistance increases longevity in *Drosophila melanogaster*. *Experimental Gerontology* 27:241–250.

Rozen, D. E., J. A. G. M. de Visser, and P. J. Gerrish. 2002. Fitness effects of fixed beneficial mutations in microbial populations. *Current Biology* 12:1040–1045.

Rozen, D. E., and R. E. Lenski. 2000. Long-term experimental evolution in *Escherichia coli*. VIII. Dynamics of a balanced polymorphism. *American Naturalist* 155:24–35.

Sanjuán, R., J. M. Cuevas, A. Moya, and S. F. Elena. 2005. Epistasis and the adaptability of an RNA virus. *Genetics* 170:1001–1008.

Sarkar S. 1998. *Genetics and Reductionism*. Cambridge: Cambridge University Press.

Schluter, D. 2000. *The Ecology of Adaptive Radiations*. Oxford: Oxford University Press.

Schrag, S. J., V. Perrot, and B. R. Levin. 1997. Adaptation to the fitness costs of antibiotic resistance in *Escherichia coli*. *Proceedings of the Royal Society of London B, Biological Sciences* 264:1287–1291.

Segrè A.V., A.W. Murray, and J.-Y. Leu. 2006. High-resolution mutation mapping reveals parallel experimental evolution in Yeast. *PLoS Biology* 4:e256.

Service, P. M., E. W. Hutchinson, and M. R. Rose. 1988. Multiple genetic mechanisms for the evolution of senescence. *Evolution* 42:708–716.

Service, P. M., and M. R. Rose. 1985. Genetic covariation among life-history components: The effect of novel environments. *Evolution* 39:943–945.

Shiotsugu, J., A. M. Leroi, H. Yashiro, M. R. Rose, and L. D. Mueller. 1997. The symmetry of correlated responses in adaptive evolution: An experimental study using *Drosophila*. *Evolution* 51:163–172.

Simpson 1944. *Tempo and Mode in Evolution*. New York: Columbia University Press.

Stanley, S. M., and X. Yang. 1987. Approximate evolutionary stasis for bivalve morphology over millions of years: A multivariate, multilineage study. *Paleobiology* 13:113–139.

Tchernov, E., O. Rieppel, H. Zaher, et al. 2000. A fossil snake with limbs. *Science* 287:2010–2012.

Teissier, G. 1954. Conditions d'équilibre d'un couple d'alleles et supériorité des heterozygotes. *Cahiers Recherche Academie des Sciences Paris* 238:621–623.

Teotónio, H., M. Matos, and M. R. Rose. 2002. Reverse evolution of fitness in *Drosophila melanogaster*. *Journal of Evolutionary Biology* 15:608–617.

———. 2004. Quantitative genetics of functional characters in *Drosophila melanogaster* populations subjected to laboratory selection. *Journal of Genetics* 83:265–277.

Teotónio, H., and M. R. Rose. 2000. Variation in the reversibility of evolution. *Nature* 408:463–466.

———. 2001. Perspective: Reverse evolution. *Evolution* 55:653–660.

Teotónio, H., M. R. Rose, and S. R. Proulx. 2006. Phenotypic plasticity and evolvability: An empirical test with experimental evolution. *In* D. Whitman and T. N. Ananthakrishnan, eds, *Phenotypic Plasticity of Insects*. Plymouth, U.K.: Science.

Thacker, C., K. Peters, M. Srayko, and A. M. Rose. 1995. The bli-4 locus of *Caenorhabditis elegans* encodes structurally distinct kex2/subtilisin-like endoproteases essential for early development and adult morphology. *Genes Development* 9:956–971.

Thornton, J. W., E. Need, and D. Crews. 2003. Resurrecting the ancestral steroid receptor: Ancient origin of estrogen signaling. *Science* 301:1714–1717.

Travisano, M., and R. E. Lenski. 1996. Long-term experimental evolution in *Escherichia coli*. 4. Targets of selection and the specificity of adaptation. *Genetics* 143:15–26.

Travisano, M., J. A. Mongold, A. F. Bennett, and R. E. Lenski. 1995a. Experimental tests of the roles of adaptation, chance, and history in evolution. *Science* 267:87–90.

Travisano, M., F. Vasi, and R. E. Lenski. 1995b. Long-term experimental evolution in *Escherichia coli*. 3. Variation among replicate populations in correlated responses to novel environments. *Evolution* 49:189–200.

Treves, D. S., S. Manning, and J. Adams. 1998. Repeated evolution of an acetate-crossfeeding polymorphism in long-term populations of *Escherichia coli*. *Molecular Biology and Evolution* 15:789–797.

Vassilieva, L. L., A. M. Hook, and M. Lynch. 2000. The fitness effects of spontaneous mutations in *Caenorhabditis elegans*. *Evolution* 54:1234–1246.

Vermeulen, C. J., and R. Bijlsma. 2006. Changes in genetic architecture during relaxation in *Drosophila melanogaster* selected on divergent virgin lifespan. *Journal of Evolutionary Biology* 19:216–227.

Weber, K. E. 1996. Large genetic change at small fitness cost in large populations of *Drosophila melanogaster* selected for wind tunnel flight: Rethinking fitness surfaces. *Genetics* 144:205–213.

Weinberg, W. 1908. Über den Nachweis der Vererbung beim Menschen. *Jahresh Verein F vateri. Naturk Württem.* 64:369–382.

Weinreich, D. M., N. F. Delaney, and M. A. DePristo, and D. L. Hartl. 2006. Darwinian evolution can follow only very few mutational paths to fitter proteins. *Science* 312:111–114.

Weinreich, D. M., R. A. Watson, and L. Chao. 2005. Sign epistasis and genetic constraint on evolutionary trajectories. *Evolution* 59:1165–1174.

Weissman, F. L. A. 1883. *Über die Vererbung.* Jena: Fischer.

Welch, J., and D. Waxman. 2003. Modularity and the cost of complexity. *Evolution* 57:1723–1734.

Weldon, W. F. R. 1895. Attempt to measure the death rate due to selective destruction of *Carcinus moenas* with respect to a particular dimension. *Proceedings of the Royal Society of London* 58:360–379.

Whatmore, A. M., N. Cook, G. A. Hall, S. Sharpe, E. W. Rud, and M. P. Cranage. 1995. Repair and evolution of nef in vivo modulates simian immunodeficiency virus virulence. *Journal of Virology* 69:5117–5123.

Whiting M. F., S. Bradler, and T. Maxwell. 2003. Loss and recovery of wings in stick insects *Nature* 421:264–267.

Whitlock, M. C., C. K. Griswold, and A. D. Peters. 2003. Compensating for the meltdown: The critical effective size of a population with deleterious and compensatory mutations. *Annals Zoological Fennici* 40:169–183.

Whitlock, M. C., P. C. Phillips, and K. Fowler. 2002. Persistence of changes in the genetic covariance matrix after a bottleneck. *Evolution* 56:1968–1975.

Wichman, H. A., M. R. Badgett, L. A. Scott, C. M. Boulianne, and J. J. Bull. 1999. Different trajectories of parallel evolution during viral adaptation. *Science* 285:422–424.

Wichman, H. A., J. Wichman, and J. J. Bull. 2005. Adaptive molecular evolution for 13,000 phage generations: A possible arms race. *Genetics* 170:19–31.

Wichman, H. A., C. D. Yarber, L. A. Scott, and J. J. Bull. 2000. Experimental evolution recapitulates natural evolution. *Philosophical Transactions of the Royal Society of London B, Biological Sciences* 355:1677–1684.

Wiens, J. J. 2000. *Phylogenetic Analysis of Morphological Data.* Washington, DC: Smithsonian Institution Press.

Wolf, J. B., E. D. Brodie III, and M. J. Wade. 2000. *Epistasis and the Evolutionary Process.* Oxford: Oxford University Press.

Woods, R., D. Schneider, C. L. Winkworth, et al. 2006. Tests of parallel molecular evolution in a long-term experiment with *Escherichia coli*. *Proceedings of the National Academy of Sciences of the USA* 103:9107–9112.

Wright, S. 1932. The roles of mutation, inbreeding, crossbreeding and selection in evolution. Pages 356–366 *in* D. F. Jones, ed. *Proceedings of the Sixth International Congress of Genetics.* Menasha, WI: Brooklyn Botanic Garden.

———. 1977. *Evolution and the Genetics of Populations. Vol. 3. Experimental Results and Evolutionary Deductions.* Chicago: University of Chicago Press.

Wright, S., and T. Dobzhansky. 1946. Genetics of natural populations. XII. Experimental reproduction of some of the changes caused by natural selection in certain populations of *Drosophila pseudoobscura. Genetics* 31:125–156.

Zwaan, B. J., R. Bijlsma, and R. F. Hoekstra. 1995. Artificial selection for developmental time in *Drosophila melanogaster* in relation to the evolution of aging: Direct and correlated responses. *Evolution* 49:635–648.

# 8

# FIELD EXPERIMENTS, INTRODUCTIONS, AND EXPERIMENTAL EVOLUTION

*A Review and Practical Guide*

Duncan J. Irschick and David Reznick

The field of evolutionary biology has primarily adopted a descriptive approach throughout its history, in large part due to the difficulty of replicating evolutionary processes under controlled conditions. However, experimental approaches provide a powerful tool kit for researchers to disentangle cause and effect (Hurlbert 1984; Futuyma and Bennett this volume; Huey and Rosenzweig this volume). In the context of evolution, experimental approaches are especially attractive because of the potential for replicating rarely observed evolutionary forces, such as natural selection, drift, or the occurrence of movement into a novel environment (Garland and Kelly 2006). Laboratory experimental approaches to evolution have long been fashionable (e.g., this volume; Bradley et al. 1999; Teotónio et al. 2004), and manipulative studies have often been used for testing ideas about adaptation (Sinervo and Basolo 1996; Travis and Reznick 1998). In this vein, field experiments offer a complementary opportunity for addressing evolutionary questions in more natural circumstances (Reznick and Ghalambor 2005). Field experiments come in different forms, but the most widely discussed method in the context of evolution is field introduction of live animals, either intentional or unintentional (Reznick and Ghalambor 2005; Strauss et al. 2006). Moreover, field experiments examining evolution are likely to rise in popularity as researchers become increasingly aware that species can rapidly evolve to novel environments in an era of dramatic climate change (Carroll et al. 2007).

Within the broader context of experimental evolution, field introductions play an interesting role for several reasons. First and foremost, field introductions represent an ecological point of comparison for laboratory studies that artificially manipulate population dynamics. As the breadth of this volume shows, laboratory studies of experimental evolution are diverse and widespread. However, no laboratory experiment, no matter how complex, can replicate the structure of natural surroundings or the balance of different types of selection and trade-offs that characterize natural settings. Long-term studies of adaptation in nature invariably reveal that adaptation represents a compromise among diverse forms of selection (Reznick and Travis 1996). In laboratory settings, trade-offs between traits are often absent or weakened, and therefore potential constraints on evolution are often relaxed. In nature, strong trade-offs can act as a strong constraint on evolutionary change, and understanding the nature of this constraint is essential for predicting evolutionary responses to environmental perturbations. On the other hand, any field study presents its own set of issues that must be overcome; for example, the complexity of field settings can be problematic when attempting to discern cause and effect because of the large number of uncontrolled variables. A key point of this review is to discuss ways that researchers can cope with this seemingly overwhelming complexity.

Our goal is to provide a practical guide for researchers interested in experimental evolution and field introductions. We do not comprehensively review the topic of field introductions, as other essays have recently reviewed the topic of experimental field manipulations (Reznick and Ghalambor 2005) and the evolutionary consequences of introduced species (e.g., invasive species, Strauss et al. 2006). We begin by differentiating different kinds of field "experiments" in regard to the study of evolution, and we discuss specific considerations for these different kinds of studies. We also provide brief encapsulations of well-known examples of field introductions that provide an overview of the range of possible study methods that researchers can access.

We define two classes of experiments. We first discuss what we term "replacement" experimental methods, which employ fake (replica) organisms that are then assessed for environmental effects (e.g., predator bites). We believe that such methods are underappreciated as part of the experimental study of evolution, but we argue that they play an essential role for understanding of how selection operates. The second class of methods is the introduction of live organisms into novel habitats. Strictly speaking, experimental approaches entail purposeful manipulation. By this definition, field experiments that manipulate the environment and then examine the evolution of target organisms in response to the manipulation are rare (Reznick and Ghalambor 2005). However, if one relaxes this definition, one can also effectively use a wider range of "manipulations" as examples of experimental evolution, even if they are not the product of planned experiments.

Within this second class, we distinguish between "human-planned" and "unplanned natural" introductions. "Human-planned" introductions represent rare and intentionally designed field experiments, such as those conducted by Reznick et al. (1990) and Losos et al. (1997). "Unplanned natural" introductions represent the other extreme, such as in the case of invasive species or the intentional human introductions of species for biological control, and so forth. We note that these latter introductions are "intentional" but to a large extent unplanned, as there is typically little forethought as to the number of colonizing individuals or to the collateral impact of these individuals to different components of the ecosystem. Besides examining each of these categories and discussing their advantages and disadvantages, in the second part of the chapter we address the unique advantages of field introductions for addressing evolutionary issues.

## METHODS FOR FIELD MANIPULATIONS
### REPLACEMENT METHODS

#### BACKGROUND

These methods probe how selection operates in a natural setting and enable one to make inferences about how and why certain features of organisms have evolved. Replacement methods consist of placing synthetic replicas of real organisms in the natural environment, primarily to quantify the impact of predation. Replacement methods offer an interesting and underutilized method for assessing the strength of natural selection on natural populations. Because replica models are not living organisms, one cannot typically study how predation causes evolutionary change, and thus, these methods stand apart from the other field experiments noted here. Despite this shortfall, these methods are unparalleled for studying the impact of predation, which is perhaps the most powerful selective force in nature, on natural populations. Natural predation events are infrequently observed, and replacement methods enable researchers to study predation in a quantitative context, as opposed to the rare (and often biased) observation of single predation events. Moreover, in some cases, researchers have been able to discern different kinds of predators by examining teeth marks left on the replica models (Brodie 1993; McMillan and Irschick, in press). In short, these methods offer one more class of experimental field techniques allowing researchers to understand the causes of mortality in their study population. Given that the majority of selection studies (see Kingsolver et al. 2001 for a review) present little or no information on potential causes of mortality, one should not discount the ability of replacement methods to fill this void. As discussed later, replacement methods also offer severe limitations that should be weighed carefully against potential benefits.

*EXAMPLES*

Several studies have successfully used clay or plasticine replicas for assessing the relative risk of mortality in natural populations (Brodie 1993; Pfenning et al. 2001; Husak et al. 2006; McMillan and Irschick, in press). Replicas are typically left in the field for a set period of time (e.g., seventy-two hours), after which the numbers and types of bites are recorded. Replacement studies can be "controlled" in an experimental sense, such as by introducing nondescript items (i.e., a round piece of clay) that serve as a comparison to clay models that mimic actual prey.

Lincoln Brower and colleagues (Brower et al. 1967; Cook et al. 1967) were early pioneers in the use of this type of manipulation to characterize selection in nature. They were interested in the evolution of model-mimic complexes in Lepidoptera and in the role of bird predation in selecting for the evolution of mimicry. They performed a series of experiments in which they created artificial mimics and modified nonmimics, and then released them into the natural environment to assess their survivorship using mark-recapture methods. They also scored recaptured moths to determine if they had survived attacks by birds. In one series of experiments, the fore and hind wings of palatable mimics were painted with conspicuously colored stripes on the outer margin that made them resemble unpalatable models; controls were painted in the same area, but with colors that made them conspicuous without mimicking an unpalatable species. Their goal was to compare survival between the experimental and control moths to see if mimetic moths had a higher probability of recapture and hence survival. The overarching message of these experiments was that the mimics sometimes have a transient advantage, but the data also revealed how difficult it is to perform such manipulations in nature. Predators quickly learned that mimics were edible, so the advantage only lasted for a short while. It also proved difficult to generate sufficient sample sizes without increasing prey density to the point that might cause changes in predator behavior.

Edmund D. Brodie III (Brodie 1993), as well as David Pfenning and his colleagues (Pfenning et al. 2001), have used replicas of snakes to test ideas regarding the evolution of mimicry. Coral snakes are highly venomous snakes that display bright alternating colored bands as a warning signal to predators. Although there exist a variety of different coral snake species, each with unique color patterns, certain species are notable for coexisting alongside nonvenomous "mimics" (typically colubrid snakes) that have evolved similar banding and color patterns (Brodie 1993). Brodie (1993) used clay replicas to test the hypothesis that predators should avoid not only venomous coral snakes (as judged by banding and color patterns on the clay replica) but also replicas of their mimics, which co-occur geographically with the coral snakes and have color patterns that are similar to those of the coral snakes. As predicted, predators avoided not only replicas showing the actual coral snake pattern but also replicas showing the mimic pattern. Predators regularly attacked a "control" replica that mimicked a harmless and similarly sized snake. The inference that predators avoided the coral snake mimics is thus derived from their lower susceptibility to attack relative to the nonmimetic "control" replicas.

The coral snake–mimic story is fascinating because of its complex geographic patterns. Alongside the eastern seaboard of the United States, coral snakes occur in the southern region, but not the northern. However, potential mimics, such as the milk snake *(Lampropeltis triangulum)* and the scarlet snake *(Cemaphora)* occur both in southern and northern regions, and therefore, one might expect that the mimics would exhibit colors similar to coral snakes in the South, but not the North. Red milk snakes *(Lampropeltis triangulum)* found north of the range of the coral snake have a non-mimetic, saddle-back pattern of blotches; populations that co-occur with coral snakes are banded in a fashion that mimics coral snakes. Likewise, other species of apparently mimetic snakes, such as the scarlet snake, have ranges that broadly overlap with the coral snake. Pfenning et al. (2001) tested whether the distribution of mimetic versus nonmimetic king snakes was shaped by patterns of mimic avoidance by placing the same set of replicas along a North-South gradient along the eastern seaboard of the United States. Pfenning et al. (2001) showed that the relative "avoidance" (as judged by the number of predator bites) of the coral snake pattern, as well as the mimic pattern that has apparently evolved along side it, diminishes from South to North, which is consistent with the known distribution of coral snakes. This work supports the view that predators are adapted to local prey assemblages and that banded color patterns of southern populations of some harmless snakes that co-occur with coral snakes evolved to be mimetic of coral snakes.

The risk of injury to an individual in some species may be caused by conspecifics, such as through male-male combat for territories. Relatively few studies have simultaneously quantified the intensity of sexual and natural selection, and even fewer have examined how the relative intensities of both processes change both seasonally (i.e., breeding season vs. nonbreeding season) and across different populations (Svensson et al. 2006). For animals with distinct breeding seasons, one might predict that the relative intensities of sexual and natural selection should peak at the same time (i.e., during the breeding/growing season). From a spatial perspective, there is reason to predict trade-offs in the relative intensity of sexual versus natural selection. In populations in which predation is high, the intensity of sexual selection, particularly male competition, may be dampened because of the inherent risks assumed by males when competing (e.g., conspicuous displays).

Recent work in Duncan Irschick's laboratory tested these possibilities using replica models of green anole lizards *(Anolis carolinensis)*. Within polygynous lizards, such as green anoles, male competition is the predominant factor that determines reproductive opportunities (Tokarz 1995). Green anoles will often bite one another during agonistic fights, which take place primarily during the spring breeding season (Jenssen et al. 2001; Lailvaux et al. 2004). McMillan and Irschick (in press) placed clay models mimicking adult male green anoles in natural habitats, and then observed bites from both predators (e.g., birds) and competing male green anoles. This experiment was replicated across different seasons (spring, fall, and winter) and in two divergent populations. One population was an urban site in which male competition was predicted to be unusually intense, and predation was predicted to be relatively low (Bloch and Irschick 2006), whereas the other population was a more natural field site in which natural predators were more common, although the intensity of sexual selection was poorly understood. As predicted, the intensity of male competition (the number of bites from green anole males) was highest during the spring, intermediate in the fall, and lowest during the winter, and predator bites followed a similar pattern. These results support the view that sexual and natural selection "peak" during the warmer breeding season, and then together decline during the nonbreeding season. On the other hand, although there was some evidence that the intensity of male competition was extremely high in the urban population with few predators, the intensity of male competition was also very high in the more natural field site with more predators.

Finally, Husak et al. (2006) used clay models of collared lizards *(Crotaphytus collaris)* in three different populations that differ in their average degree of coloration. Male collared lizards exhibit bright colors that act as sexual signals, and females are far less colorful. The role of color in these collared lizards is not well understood, but in many male lizards, there is strong selection for bright colors that may signal male quality to females (Lailvaux and Irschick 2006). Therefore, any predation cost on bright colors implies a cost associated with a sexually selected trait. Husak et al. (2006) tested the hypothesis that more colorful individuals, particularly those that contrast with their environment, suffer an increased cost of predation by building colorful clay models and noting the number of predator marks within each of the three populations. They found that at each of the three populations, more colorful individuals that were in contrast to their environment incurred higher rates of predator bites compared to less colorful models of female lizards. This work characterizes one side of the potential interplay between sexual and natural selection, and it implies a definite cost from the perspective of natural selection on sexually selected traits. If male brightness evolved because it enhanced mating success, then this study shows that it did so in the face of increased susceptibility to predation.

The clay models offer one primary advantage over the use of live animals: because the replicas do not move, and because they are designed to show marks, one can obtain reasonable estimates of the degree of predation or male competition that might otherwise be difficult, or impossible, to acquire. The "modified" animals employed by Brower and colleagues offer the ability to manipulate the phenotypes of natural populations yet maintain some of the mobility and natural behavior of live animals, thus enhancing the ability to determine how selection operates on populations via predation.

The clay model method, in particular, presents shortcomings that need to be considered by researchers. First, the index of predation intensity gained is unlikely to represent a complete profile of all potential predators. For example, some predators focus on movement or smell, and therefore replicas may not be effective for examining how such predators capture prey. Second, the interpretation of the bites themselves can be challenging. While the approximate size and shape of the bites provide rough information on the kind of predator involved (e.g., bird vs. rodent), they provide limited information about the species of predator or the attributes of individual predators, such as their age or size, and hence limit one's ability to make inferences about their likely experience. Third, one cannot discount the possibility that the same predator or conspecific might bite a model multiple times. Therefore, analysis of bites should be conservative and perhaps consider just the proportion of replicas that are attacked rather than the number of attacks per replica. The motive of the bites is not known; one must consider the possibility that some bites may arise more out of curiosity than an actual intent to kill and eat.

## LIVE ANIMAL INTRODUCTIONS

*BACKGROUND*

Live animal introductions (i.e., transplant experiments) involve the transfer, either intentional or unintentional, of live organisms from their source population to another geographic location. Here we distinguish between "human-planned" introductions, usually done in the context of a field experiment, and "unplanned natural" experiments, which amount to the introduction (either planned or unplanned, but usually undertaken haphazardly) of species into novel environments. Intentional introductions of animals often occur for reasons such as biological control, and unintentional introductions often represent "invasive" species that have moved into new habitats by various means. We do not review plant studies here even though experimental transplant experiments have been widely used by plant biologists. However, plant studies have primarily been used as a way of evaluating the extent to which natural populations of plants differ genetically from one another and are adapted to their local environment, rather than to study how plants evolve in response to "novel" environments.

*"Planned" Experiments*    There are three successful examples of investigators transplanting populations to habitats from which the species was initially absent, then evaluating the evolution of the introduced population by using the source population as a control and evaluating the differences between the source and derived populations.

Losos et al. (1997) examined the evolutionary outcome from a planned introduction of a propagule of the naturally invasive anole *A. sagrei* onto replicate small Bahamian islands. During 1977 and 1981, Thomas Schoener and his colleagues introduced propagules of five or ten adult *A. sagrei* lizards onto fourteen islands from a larger "mainland" island that represented the "source" of the invasion. Losos et al. (1997) evaluated evolution by comparing the morphological characteristics of samples of the "mainland" (control) source versus the introduced island populations. During the intervening 10–14 years, the introduced *A. sagrei* population proliferated and also apparently evolved a significantly different phenotype relative to the mainland control. Losos et al. (1997) detected a slight but statistically significant decrease in hind limb length in the introduced population relative to the mainland source. This phenotypic change was potentially important because of the influence of hind limb length on both sprint speed (Irschick and Losos 1999) and average perch diameter usage. Losos et al. (1997) interpreted the smaller hind limbs as evidence of adaption to the use of relatively narrow surfaces, which were prevalent on the small island. Later work (Losos et al. 2000) suggested that this change in limb proportions may have arisen via plasticity because young *A. sagrei* that were raised on narrow twigs tended also to have relatively shorter limbs as adults. Regardless, this work represents a case of either rapid evolution in morphology or potentially a plastic adaptive response.

Reznick and colleagues (Reznick and Endler 1982; Reznick and Bryga 1987; Reznick et al. 1990) have used transplant studies with Trinidad guppies to evaluate the role of predation in guppy life-history evolution. Guppies are found in natural communities where they occur with different abundances and species of predators. Reznick and colleagues contrasted guppies from low predation environments and one type of high-predation community. Low-predation environments are either headwater streams or portions of streams above barrier waterfalls that exclude predators but not guppies. High-predation environments are generally found in the lower regions of drainages, below any such barriers. This contrast between low- and high-predation environments is replicated across drainages. Reznick and colleagues, and John Endler before them, transplanted guppies from a high-predation environment over barrier waterfalls that excluded both guppies and predators. Reznick and colleagues also attained the reciprocal treatment by transplanting predators over a waterfall that excluded predators, but not guppies. Their primary interest was in testing basic tenets of life-history theory—namely, whether the risk of mortality from predation (or any other extrinsic source of mortality) affected the evolution of reproductive effort and age at maturity. Specifically, increased risk of mortality is predicted to favor those individuals that mature at an earlier age and have higher levels of reproductive effort. Thus, over human life spans, these predictions were upheld in the transplanted populations, providing strong support for key aspects of life-history theory (Reznick et al. 1997). This work represents one of the most compelling examples for how rapid evolution by natural selection can occur (Hendry and Kinnison 1999).

A recent study reveals strong evidence for rapid evolution resulting from transplanted animals, and it also shows that the evolution of brand-new morphological traits can appear in human time scales (Herrel et al. 2008). In 1971, a team of biologists lead by Evitar Nevo performed a reciprocal transplant between two islands (Nevo et al. 1972). Prior to the transplant, each of these adjacent islands (Pod Mrcaru and Pod Kopiste) had their own population of lizards (*Podarcis melliselensis* and *P. sicula*, respectively). These two species overlap broadly across various parts of the Adriatic coast and are competitors, with *P. sicula* typically displacing *P. melliselensis* out of available habitats. Nevo et al. performed a reciprocal transplant; they placed five male-female pairs on each island, with the source being the other island. In 2004, these islands were resampled, and genetic analyses showed that only the introduction of the invasive *P. sicula* onto Pod Mrcaru (formerly *P. melliselensis* habitat) was successful, and the original ten descendants had resulted in more than five thousand individuals (A. Herrel, unpublished data). In comparisons between the source island of Pod Kopiste and the introduced site of Pod Mrcaru, individuals on the latter introduced site had relatively longer and wider jaws, harder bites, and enlarged gut structures (caecal valves) that were not present in the source population (Herrel et. al. 2008). Dietary analysis showed that the introduced *P. sicula* consumed more plant matter than lizards on their source island, and the enlarged jaws and enhanced bites were likely adaptive for allowing lizards to more effectively bite and chew tough fibrous plant matter. Moreover, the presence of novel caecal valves in the introduced *P. sicula* is consistent with a plant-based diet, as such structures are typically found in herbivorous lizard families (Herrel et al. 2008). The fact that the caecal values were found in extremely young juvenile *P. sicula* from Pod Mrcaru is evidence both for genetic adaptation and against the possibility of a plastic effect from lizards eating plants from birth. While the other transplant studies noted earlier (brown anoles and Trinidad guppies) have each documented rapid evolution, Herrel et al.'s (2008) study is unique in revealing that populations can evolve novel and complex physical traits within human life spans.

A final, somewhat divergent example merits mention. Malhotra and Thorpe (1991) performed a manipulative experiment with different geographic races of the anole lizard *Anolis oculatus* on the island of Dominica. They transplanted samples of different races (e.g., highland, lowland) into alternative habitats and monitored survival in large enclosures. They found that the degree to which the introduced race survived was a function of how different that habitat was from its original habitat. In other words, lizards placed in a very different habitat from the one they were from (i.e., highland lizards placed into a lowland environment enclosure) were less likely to survive compared to lizards placed in environment enclosures that were similar to their own. This study departs from the research described earlier in not clearly demonstrating evolution, although the study is valuable in demonstrating that natural selection seems to favor the fit between lizards and the habitat in which they originally evolved.

*"Unplanned Natural" Experiments*  There are numerous examples of species moving into novel habitats (either human induced or not), but relatively few well-understood examples of how such introductions have resulted in evolutionary change.

One of the most widespread forms of purposeful human introductions of animals consists of introduced fish stocks. Because there is good information on the dates of the introductions of fish stocks, one can examine both the extent of overall phenotypic and genetic differentiation, and also the potential for formation of new species, a process that could likely become more commonplace in the context of global pattern of climate change. Hendry et al. (2000) examined several populations of sockeye salmon (*Oncorhynchus nerka*) that were relatively recently derived (introduction dates: 1937–1945 for one population, ca. 1957 for another) from an "ancestral" wild stock, and showed strong evidence for evolution of reproductive isolation in one more "derived" stock. The introduced fish have apparently diverged into distinct populations that either migrate upstream to breed or breed on beaches in the lake that they were introduced to. These different breeding sites incorporate a difference in the balance between sexual selection for males with laterally compressed bodies versus natural selection for males with shallower bodies that can survive upstream migration; the population that breeds in the lake has males with far deeper bodies than the population that migrates upstream to breed. This change in body shape, in association with female preferences, is now associated with a reduced probability of interbreeding between the two ecotypes. This result suggests that human manipulation of populations may be causing the evolution of reproductive isolation. Such isolation is in turn a key step in the formation of a new species.

The soapberry bug represents one of the most well-cited examples of unintentional introductions that have resulted in rapid evolution. Soapberry bugs (*Jadera haematoloma*) suck juice from small fruits (soapberries) contained within inflated seed pods. In order for the bug to successfully feed, its beak needs to be long enough to access the seed inside of the seed pod. For example, the fruits of the soapberry plant are like an inflated paper lantern with a seed placed near the center. Natural populations of bugs that feed on different species of plants have beak lengths that are in proportion to how inflated the seed pod is. Those that feed on plants with relatively small seed containers have relatively short beaks in comparison to those that feed on plants with highly inflated seed containers. Scott Carroll and colleagues (Carroll and Boyd 1992; Carroll et al. 1998; Carroll et al. 2003) have studied the ecology and evolution of a dramatic shift from soapberry bugs using an ancestral large fruit to a new host. The new host in Florida is the introduced golden rain tree, which has relatively flat seed pods in comparison to those found on the native soapberry vine. The range of dates of the introduction of golden rain trees is known. The evolutionary response of the bugs was evaluated in two ways; with the measurement of periodic samples of bugs that were preserved in museum collections and through the comparison of live bugs collected from both hosts. The museum collections enabled Carroll and his colleagues to evaluate the time course of change in beak length and other aspects of morphology. The comparisons of live bugs from the alternative hosts has enabled them to show that not only has beak length changed, but the genetic architecture of this phenotypic difference has also evolved, and in a complex fashion. Furthermore, these different populations of soapberry bugs have each evolved to perform best (feed and grow) on their host plants. The native population has significantly higher fitness on the native host, while the "transplanted" population has significantly higher fitness on the introduced host. This example shows how adaptation to new ecological conditions can be both rapid and complex in terms of overall differences in the phenotype and the genotype.

Some species of copepods (small invertebrates, *Eurytemora affinis*) have invaded freshwater locations from natural saltwater habitats in the Gulf Coast and the Atlantic Ocean (Lee 1999). This repeated invasion of freshwater environments from saltwater represents a dramatic physiological change in the capacity of these invertebrates to adapt to the low salinity of fresh water. Lee (1999) studied the invasive pathways of these copepods in several ways. First, she reconstructed the phylogenetic pattern of invasion based on mitochondrial DNA sequences. Her tree included samples of copepods collected from throughout their range, as well as samples from several freshwater sites containing invasive populations of copepods. She found that the invasive populations were always genetically more similar to the native populations found within close proximity of the invaded locality than to those found elsewhere. From this pattern, she inferred that each freshwater invasion was derived from that local population. This phylogenetic tree thus clearly confirmed multiple invasions into freshwater habitats from saltwater, indicating the facility with which these animals are able to adapt to fresh water. Laboratory experiments focused on whether all of the invasive populations were equally capable of tolerating fresh water (i.e., no salinity). For the most part, invasive populations of copepods were superior in tolerating freshwater conditions compared to marine-derived populations, suggesting that local adaptation has occurred in a relatively brief period of time. This view of genetic divergence driven by natural selection was supported by evidence that, for the few cases evaluated, Lee detected reproductive isolation between the invaders and presumed ancestors. There was, however, some variation in the ability of invasive populations to tolerate freshwater environments, suggesting significant genetic variation for the capacity to invade new areas (Lee and Bell 1999).

Two final noteworthy examples are rapid evolution in wing morphology in *Drosophila* and the evolution of geographic races in house sparrows. In the old world, *Drosophila subobscura* exhibits a clinal increase in wing size with latitude. These flies were introduced into the New World (South America) around 1980, but no parallel cline was observed in these introduced populations one decade after their introduction; yet between 1990 and 2000, a rapid cline came into existence that is similar to that observed in the Old World (Huey et al. 2000). This study suggests that evolution is both predictable and rapid, at least in terms of geographic clines.

A classic example of evolution resulting from rapid range expansion is the European house sparrow (*Passer domesticus*). This species did not colonize North America until 1852 when an initial breeding population was found in New York. That population of house sparrows rapidly expanded their geographic range, reaching Vancouver by 1900, Mexico around 1933, and Central America around 1974. This species now occurs across most of the continent, and during this rapid range expansion, this species also rapidly evolved into geographic races that differ in color as well as wing and bill morphology (Johnston and Selander 1962). These rapid physical changes, which occurred over about fifty years (Johnston and Selander 1962) are convergent with native species and seem to follow predictive geographic rules (e.g., Bergmann's rule). Taken together, both bodies of work show that widespread geographic variation can arise quickly, suggesting local adaptation at multiple different scales.

### CONSIDERATIONS

"Human-planned" and "unplanned natural" introductions each present their own set of rewards and challenges for biologists to consider. The primary advantage of human-planned introductions is that they provide the opportunity to observe ecological interactions and their possible evolutionary responses under the watchful eye of the human experimenter, as opposed to operating on nature's own whim and time scale. Such manipulations offer the researcher at least the potential for controlling the nature of the introduced population (e.g., body size, sex ratio), although in many cases, decisions on initial conditions may be made more on practical grounds than on experimental ones. Most important, the researcher has the opportunity to quantify the intensity of natural selection and direction and tempo of evolutionary change.

The primary disadvantage to planned introductions is also an advantage of most unplanned natural introductions: time. Although rapid evolution is well documented over quite short time periods (less than fifteen years, Hendry and Kinnison 1999), the likelihood that significant evolution will occur within a practical time interval may be small. Such introductions thus benefit from prior knowledge of the likely intensity of selection and the biology of the study organism, particularly its population dynamics and generation time. Other potential disadvantages must also be considered. Some organisms may not transplant successfully because of aspects of their behavior or reproductive biology. For example, survival and successful reproduction may be too closely tied to some prior knowledge of the territory. Another concern is the nature of the actual founding populations. Care must be taken to ensure that the introduced animals are representative of the population that they are derived from and that will be used as a control for evaluating evolution after the introduction.

In all human-planned introductions and some natural unplanned introductions, researchers are afforded an opportunity to study local adaptation when one knows both the source of the introduced population and when the introduction was made. In cases like the salmon work by Hendry and colleagues (2000), researchers have been able to study the evolution of organisms over a time scale of decades that would not be accessible to an individual investigator. This means that researchers can study evolution in organisms with relatively long generation times, at least relative to a guppy or *Anolis* lizard, and hence expand our knowledge beyond what one could argue are special cases. However, key details on any introduction are often difficult to nail down, such as when it occurred; the nature of the introduced population; whether any subsequent, possibly inadvertent, introductions were made; and what the history of population growth of the introduced population was. Moreover, there is substantial difficulty in establishing a control for the evaluation of evolution in the introduced population. In the salmon example, we can evaluate the fish from the hatchery that they were derived from to provide a control, but we will have at best incomplete information on what that population was like at the time that the introduction was made or the extent to which that population may have changed in the interim.

The nature of invasive species is also inherently fascinating for evolutionary biologists because of their tantalizing similarity to natural patterns of colonization, which have characterized natural ecosystems for millions of years (Strauss et al. 2006). A significant amount of research has centered on traits of invasive species and whether good invaders exhibit unique physiological and ecological traits, for example. This also suggests caution against over interpretation, as evolutionary changes in invasive species may not be typical of most species.

A final challenge is to define the actual cause-and-effect relationships that underlie any observed evolution. Any manipulation of the environment potentially incurs a combination of direct and indirect effects. For example, guppies that are transplanted from a high-predation site into a previously guppy-free low-predation site above a barrier waterfall will experience a reduction in mortality rate, which is the presumed source of selection for life-history evolution. However, they will also predictably experience an increase in population density and a reduction in per capita resource availability (Reznick et al. 2001). Defining the relative importance of the direct effect of predators, versus the potential indirect effects, is an added hurdle that has yet to be cleared.

## WHAT CAN WE LEARN FROM FIELD EXPERIMENTS THAT WE CANNOT LEARN FROM LABORATORY STUDIES?

Prior authors (e.g., Hurlbert 1984) have emphasized the importance of rigorous experimental design during field experiments (e.g., controlled conditions, replication), which is often difficult to attain. This debate has also manifested itself in prior arguments about the efficacy of historical versus contemporary approaches (Leroi et al. 1994; Reeve and Sherman 1993; but see Larson and Losos 1996 for an opposing view). These divergent views have also revealed themselves in the debate over studies of "experimental evolution," which are typically conducted in controlled laboratory experiments with organisms that have short generation times and can be easily bred (e.g., Bennett et al. 1990; Travisano et al. 1995). Following others (e.g., Reznick and Ghalambor 2005; Carroll et al. 2007), we take the perspective that introductions—whether planned, unplanned, intentional, or unintentional—present a source of material for understanding rapid evolution. We also feel that the field studies outlined in this review are complementary to controlled laboratory experiments with microorganisms. As the breadth of the chapters in this volume shows, such experimental laboratory approaches, particularly in the context of evolution, have blossomed tremendously over the past ten years or so. In direct contrast to this rapid growth in laboratory experimental studies of evolution, we have not observed a similar growth in the number of carefully designed field introduction on animal species. In this context, we list several areas that field experimental studies of evolution have the unique capacity to address.

MECHANISMS OF POPULATION ESTABLISHMENT

While research has established some salient features of how small groups of individuals become established in novel areas, we know little about the manner in which these populations grow and adapt to new surroundings. Introduction experiments offer a window into processes that impact the earliest stages of population establishment, growth, and genetic differentiation, and they may also provide information about speciation (Otte and Endler 1989; Templeton 1989). Most natural or unnatural field experiments involve a clear "before and after" approach in which the populations are compared as point samples between time periods (Losos et al. 1997; Herrel et al. 2008), and therefore there is little information on early stages of population establishment. Furthermore, some authors (Garland and Kelly 2006) have noted the importance of plastic responses to the environment in studies of experimental evolution. Plastic effects are typically controlled for in laboratory studies of experimental evolution, but these effects are likely to be important in natural settings for enabling organisms to survive in difficult circumstances. For example, the rapid evolution of novel gut structures (caecal valves) in *Podarcis* lizards (Herrel et al. 2008) begs the question about when these structures first arose in the population, how many individuals first evolved them, and what the genetic architecture underlying this complex trait was composed of (this last point is important because Herrel et al. [2008] did not clearly demonstrate that caecal valves possessed a genetic basis). Transplants with repeated sampling (i.e., yearly) over long time periods offer the promise for understanding these early stages of growth, plastic responses, and genetic differentiation, and the delayed hope for an evolutionary response that is discernible to human eyes.

Selection during the early course of population establishment is likely to be dynamic, and there is potential for interplay between habitat choice and morphological evolution.

For instance, in manipulative studies with a predator (the curly-tailed lizard, *Leiocephalus carinatus*) and its prey (the brown anole, *A. sagrei*), selection initially favored longer limb length in *Anolis* when they are first exposed to a predator, but selection later favored shorter limbs as prey populations shift to a more arboreal lifestyle, possibly to elude the predator (Losos et al. 2006). Intensive sampling of transplanted populations can capture these subtle interrelationships among habitat use, morphology, and species interactions. Ideally, therefore, researchers should gather ecological data on transplanted populations over a period of time to determine if there is a dynamic interplay for selection on morphology, habitat use, and the interaction between the two.

RAPID EVOLUTION

In an era of increasing global changes in habitat and climate, the topic of "rapid" evolution has become prominent in the evolutionary community (Hendry and Kinnison 1999; Carroll et al. 2007). Past reviews have shown that selection in nature can often be surprisingly strong (Endler 1986; Hoekstra et al. 2001; Kingsolver et al. 2001; Hereford et al. 2004), leading to many cases of "rapid" evolution observed in the modern era (Hendry and Kinnison 1999). As a caveat, direct observations of evolutionary change yield faster estimates than those seen in the fossil record or inferred from historical introductions. The reason is that longer-term estimates average intervals when the rates vary, when direction may be reversed, and when there is no evolution. Furthermore, long-term field studies show that the pace of evolution can be remarkably high in one direction in one year and then show the opposite trend the other year (Grant 1999). In other words, researchers should interpret short-term evolutionary studies with caution.

Nonetheless, as noted in a recent special issue of *Functional Ecology* devoted to "Evolution on Ecological Time-Scales," researchers are increasingly turning from the view that evolution is a historical phenomenon that cannot be studied in human life spans (Carroll et al. 2007). Recent work also shows that species may be able to evolve novel structures in human life spans, although more work is needed to verify this exciting possibility (Herrel et al. 2008). Moreover, in an era of ever-increasing human contact, the rapid movement of animal and plant species ensures that invasive species will continue to challenge efforts to preserve "natural" habitats and suites of species. Added to this phenomenon of the movement of species is the potential for both introduced species, and the species they affect, to evolve, in essence providing a moving target for researchers interested in biological control (Carroll et al. 2007).

Quantifying the "rate" of evolution is challenging given the diversity of different traits and taxa that researchers have examined. Previous authors have primarily used two standardized metrics: the Darwin and the Haldane. Hendry and Kinnison (1999) presented data for such values across widely disparate taxa (fish, insects, lizards) from both the fossil record and present-day field introductions and microevolution. Based on their analysis, as well as simple calculations completed in Herrel et al. (2008), all three of the planned field experiments discussed here provide some of the most rapid estimates of evolution discussed to date (Hendry and Kinnison 1999; Herrel et al. 2008), suggesting that the "payoff" for such difficult studies is substantial. Moreover, some of the other commonly discussed examples of rapid evolution come from studies of introduced species (Hendry and Kinnison 1999), such as in salmon and birds. The reasons for such rapid evolution in introduced species seems obvious: in order for species to effectively live in a new environment, rapid change is often necessary, such as in salinity preferences in copeopods (Lee 1999), yet there is also the possibility that the capacity for such rapid evolution is a hallmark of highly invasive species and is not typical for species that have more specific habitat preferences (Cox 2004).

## NATURAL AND UNNATURAL HABITAT ALTERATIONS

Habitat alteration, in the form of either natural catastrophic events or human alteration, has long been recognized has having a significant impact on ecological dynamics and, more recently, evolutionary dynamics (Hendry et al. 2006). Catastrophic events, such as major storms, volcanic explosions, and earthquakes, among others, have played an important role in the "unraveling" of the history of life (Gould 1989); in an era of global warming, such events promise to become increasingly important for modern ecological communities. For example, one cited consequence of global warming is a greater frequency and strength of hurricanes, such as those that commonly occur on the Gulf Coast of North America. Recent work has shown that major hurricanes in the Bahamas have the potential to dramatically reshape lizard populations, particularly if the hurricane strikes during periods when populations are vulnerable, such as during the reproductive season (Schoener et al. 2004). How these disturbances affect the evolution of these populations is not yet known however and is a topic of current investigation (J. Losos, personal communication). In the context of field experiments and evolution, catastrophes have the potential to provide a nearly unlimited number of natural experiments, and they certainly have the potential to speed up rates of evolution. We believe that this avenue has barely been explored, but the possibilities are unlimited. Especially valuable would be studies that can show how catastrophic events alter the course by which selection operates, like a powerful wind that alters the course of a ship. This kind of data would likely only emerge from researchers that already have good long-term field data and data on which a catastrophe fortuitously (or not so fortuitously, for the study organisms) strikes.

While not spectacular in the same vein as a hurricane or earthquake, dramatic human disturbance can be catastrophic to local animal and plant populations, and recent work by Hendry et al. (2006) hints that such disturbance may actually be dampening the rate of evolutionary divergence. They examined two populations of Darwin's finches, one of which occurs in relatively natural settings apart from human disturbance, whereas the other has experienced significant human disturbance over the past thirty years. Hendry et al. (2006) found that human activity may be "dampening" morphological variation in the human-disturbed population by diminishing the presence of discrete bird morphs, whereas the more natural population experienced a higher degree of morphological variation. This finding highlights an unexpected consequence of catastrophic changes to habitat structure—namely, a slowing down of evolution.

## CONCLUSION

The increasing trend of global warming and climate change on earth means that future transplant experiments in nature offer the potential for detecting rapid evolution over human life spans (Carroll et al. 2007), and we would like to see this class of data grow both in number and taxonomic breadth (two of the three transplant studies mentioned here were on lizards, one on fish). In this vein, we point toward several areas of research that would expand our knowledge base on how small populations become established, grow, and evolve, in novel environments.

First, more research that examines processes of population establishment, growth, and genetic change would give a window into whether plastic responses (Garland and Kelly 2006) enable organisms to survive or, alternatively, whether evolution begins to occur almost immediately.

Second, research that places different genetic lines into different environments to test alternative hypotheses on how different genetic backgrounds respond to selection would be welcome. Because different genotypes might respond differently to the same selective pressure, intentional field introductions of a mix of genotypes, along with subsequent sampling, would establish not just selection on phenotypes but also selection on genotypes, and perhaps plastic responses as well (Goodnight 1988). This avenue offers the best way for researchers to test classic population genetic models of population divergence and speciation in a natural setting, and the rapidly expanding genetic tool kit available to researchers should make such manipulations possible in a wide array of organisms.

Third, evolutionary ecologists should also consider joining forces with natural resource managers who regularly monitor natural populations and their patterns of life and death. Given recent work showing very strong selection on horn morphology in bighorn rams (Coltman et al. 2002), there is ample opportunity for observing humans as a strong selective force on animal populations, with the potential for subsequent evolution.

Finally, we remind the reader of the importance of carefully considering the ethical implications of natural introductions. Because of rapid climate change on earth, introductions of certain species, even into seemingly remote locations, might be disastrous, especially for highly invasive species that are introduced near the edge of their ranges. We note that in all three planned transplant experiments (experiments with *Anolis* lizards, Trinidad guppies, and Italian wall lizards), each of these species was unlikely to disturb the larger ecological community following the transplant. Ensuring that field experiments are conducted with minimal long-term consequences for other plant or animal species should be of paramount importance.

## SUMMARY

Field experimental evolution studies have served as an important complement to more common descriptive evolutionary studies. Our goal is to provide a practical guide for researchers interested in experimental evolution and field introductions. We distinguish between "replacement" experimental methods, which employ fake (replica) organisms that are assessed for environmental effects (e.g., predator bites), and the introduction of live organisms into novel habitats. While relatively few studies have purposefully designed experiments for examining evolutionary change, there are many examples of "unplanned natural" experiments, such as human introductions for pest control. Replacement experiments provide a useful method for studying the effects of predation, and they have been used to test ideas about mimicry, the contributions of natural and sexual selection, and the costs of extravagant colors. Planned experiments show that evolution in introduced populations can be extremely rapid, suggesting that the initial investment in time toward such experiments is worthwhile. Recent studies also show that even new phenotypic traits can evolve over human life spans. Finally, different populations of invasive species are not equally successful for evolving effective specializations for new habitats.

## ACKNOWLEDGMENTS

This work was supported by a National Science Foundation (NSF) grant to D.J.I. (IOB 0421917) and NSF Grants DEB-0416085 and DEB-0623632EF to D.R.

## REFERENCES

Bennett, A. F., K. M. Dao, and R. E. Lenski. 1990. Rapid evolution in response to high-temperature selection. *Nature* 346:79–81.

Bloch, N., and D. J. Irschick. 2006. An analysis of inter-population divergence in visual display behaviour of the green anole lizard (*Anolis carolinensis*). *Ethology* 112:370–378.

Bradley, T. J., A. E. Williams, and M. R. Rose. 1999. Physiological responses to selection for desiccation resistance in *Drosophila melanogaster*. *American Zoologist* 39:337–345.

Brodie, E. D. I. 1993. Differential avoidance of coral snake banded patterns by free ranging avian predators in Costa Rica. *Evolution* 47:227–235.

Brower, L. P., L. M. Cook, and H. J. Croze. 1967. Predator responses to artificial Batesian mimics released in a neotropical environment. *Evolution* 21:11–23.

Carroll, S. P., and C. Boyd. 1992. Host race radiation in the soapberry bug: Natural history, with the history. *Evolution* 46:1052–1069.

Carroll, S. P., Dingle, H., and T. R. Famula. 2003. Rapid appearance of epistasis during adaptive divergence following colonization. *Proceedings of the Royal Society of London B, Biological Sciences* 270 (Suppl.):S80–S83.

Carroll, S. P., A. P. Hendry, D. N. Reznick, and C. W. Fox. 2007. Evolution on ecological time-scales. *Functional Ecology* 21:387–393.

Carroll, S. P., S. P. Klassen, and H. Dingle. 1998. Rapidly evolving adaptations to host ecology and nutrition in the soapberry bug. *Evolutionary Ecology* 12:955–968.

Coltman, D. W., M. Festa-Bianchet, J. T. Jorgenson, and C. Strobeck. 2002. Age-dependent sexual selection in bighorn rams. *Proceedings of the Royal Society of London B, Biological Sciences* 269:265–172.

Cook, L. M., L. P. Brower, and H. J. Croze. 1967. Accuracy of a population estimation from multiple recapture data. *Journal of Animal Ecology* 36:57–60.

Cox, G. 2004. *Alien Species and Evolution*. Washington, DC: Island.

Endler, J. A. 1986. *Natural Selection in the Wild*. Princeton, NJ: Princeton University Press.

Garland, T., Jr., and S. A. Kelly. 2006. Phenotypic plasticity and experimental evolution. *Journal of Experimental Biology* 209:2234–2261.

Goodnight, C. J. 1988. Epistasis and the effect of founder events on the additive genetic variance. *Evolution* 42:441–454.

Gould, S. J. 1989. *Wonderful Life: The Burgess Shale and the Nature of History*. New York: Norton.

Grant, P. R. 1999. *Ecology and Evolution of Darwin's Finches*. Princeton, NJ: Princeton University Press.

Hendry, A. P., and M. T. Kinnison. 1999. The pace of modern life: Measuring rates of contemporary microevolution. *Evolution* 53:1637–1653.

Hendry, A. P., J. K. Wenburg, P. Bentzen, E. C. Volk, and T. P. Quinn. 2000. Rapid evolution of reproductive isolation in the wild: Evidence from introduced salmon. *Science* 290:516–518.

Herrel, A., K. Huyghe, B. Vanhooydonck, T. Backeljau, K. Breugelmans, I. Grbac, R. Van Damme, and D. J. Irschick. 2008. Rapid large scale evolutionary divergence in morphology and performance associated with the exploitation of a novel dietary resource in the lizard *Podarcis sicula*. *Proceedings of the National Academy of Sciences of the USA* 105:4792–4795.

Hendry, A. P., Grant, P. R., Grant, B. R., Ford, H. A., Brewer, M. J., and Podos, J. 2006. Possible human impacts on adaptive radiation: Beak size bimodality in Darwin's finches. *Proceedings of the Royal Society of London B, Biological Sciences* 273:1887–1894.

Hereford, J., T. F. Hansen, and D. Houle. 2004. Comparing strengths of directional selection: How strong is strong? *Evolution* 58:2133–2143.

Hoekstra, H. E., J. M. Hoekstra, D. Berrigan, S. N. Vignieri, A. Hoang, C. E. Hill, P. Beerlii, and J. G. Kingsolver. 2001. Strength and tempo of directional selection in the wild. *Proceedings of the National Academy of Sciences of the USA* 98:9157–9160.

Huey, R. B., G. W. Gilchrist, and M. Carlsen. 2000. Rapid evolution of a latitudinal cline in body size in an introduced fly. *Science* 287:308–309.

Hurlbert, S. H. 1984. Pseudoreplication and the design of ecological field experiments. *Ecological Monographs* 54:187–211.

Husak, J. F., J. M. Macedonia, S. F. Fox, and R. C. Sauceda. 2006. Predation cost of conspicuous male coloration in collared lizards (*Crotaphytus collaris*): An experimental test using clay-covered model lizards. *Ethology* 112:572–580.

Irschick, D. J., and J. B. Losos. 1999. Do lizards avoid habitats in which their performance is submaximal? The relationship between sprinting capabilities and structural habitat use in Caribbean anoles. *American Naturalist* 154:293–305.

Jenssen, T. A., M. B. Lovern, and J. D. Congdon. 2001. Field testing the protandry-based mating system for a lizard: Does the model organism have the right model? *Behavioral Ecology and Sociobiology* 50:162–172.

Johnston, R. F., and R. K. Selander. 1962. House sparrows: Rapid evolution of races in North America. *Science* 144:548–550.

Kingsolver, J. G., H. E. Hoekstra, J. M. Hoekstra, D. Berrigan, S. N. Vignieri, C. E. Hill, A. Hoang, P. Gibert, and P. Beerli. 2001. The strength of phenotypic selection in natural populations. *American Naturalist* 157:245–261.

Lailvaux, S. P., A. Herrel, B. Vanhooydonck, J. J. Meyers, and D. J. Irschick. 2004. Performance capacity, fighting tactics, and the evolution of life-stage male morphs in the green anole lizard (*Anolis carolinensis*). *Proceedings of the Royal Society of London B, Biological Sciences* 271:2501–2508.

Lailvaux, S., and D. J. Irschick. 2006. A functional perspective on sexual selection: Insights and future prospects. *Animal Behaviour* 72:263–273.

Larson, A., and J. B. Losos. 1996. Phylogenetic systematics of adaptation. Pages 187–220 *in* M. R. Rose and G. V. Lauder, eds. *Adaptation*. San Diego, CA: Academic Press.

Lee, C. E. 1999. Rapid and repeated invasions of fresh water by the copepod *Eurytemora affinis*. *Evolution* 53:1423–1434.

Lee, C. E., and M. A. Bell. 1999. Causes and consequences of recent freshwater invasions by saltwater animals. *Trends in Ecology & Evolution* 14:284–288.

Leroi, A. M., M. R. Rose, and G. V. Lauder. 1994. What does the comparative method reveal about adaptation? *American Naturalist* 143:381–402.

Losos, J. B., D. A. Creer, D. Glossip, R. Goellner, A. Hampton, G. Roberts, N. Haskell, P. Taylor, and J. Etling. 2000. Evolutionary implications of phenotypic plasticity in the hindlimb of the lizard *Anolis sagrei*. *Evolution* 54:301–305.

Losos, J. B., T. W. Schoener, R. B. Langerhans, and D. A. Spiller. 2006. Rapid temporal reversal in predator-driven natural selection. *Science* 314:1111.

Losos, J. B., K. I. Warheit, and T. W. Schoener. 1997. Adaptive differentiation following experimental island colonization in *Anolis* lizards. *Nature* 387:70–73.

Malhotra, A., and R. S. Thorpe. 1991. Experimental detection of rapid evolutionary response in natural lizard populations. *Nature* 353:347–348.

McMillan D., and D. J. Irschick. In press. An experimental test of predation and sexual selection pressures on the green anole lizard (*Anolis carolinensis*). *Journal of Herpetology*.

Nevo, E., G. Gorman, M. Soulé, S. Y. Yang, R. Clover, and V. Jovanovic. 1972. Competitive exclusion between insular Lacerta species (Saurta, Lacertidae). *Oecologia* 10:183–190.

Otte, D., and J. A. Endler. 1989. *Speciation and Its Consequences*. Sunderland, MA: Sinauer.

Pfennig, D. W., W. R. Harcombe, and K. S. Pfennig. 2001. Frequency-dependent Batesian mimicry: Predators avoid look-alikes of venomous snakes only when the real thing is around. *Nature* 410:323–323.

Reeve, H. K., and P. W. Sherman. 1993. Adaptation and the goals of evolutionary research. *Quarterly Review of Biology* 68:1–32.

Reznick, D. N., and H. Bryga. 1987. Life-history evolution in guppies. 1. Phenotypic and genotypic changes in an introduction experiment. *Evolution* 41:1370–1385.

Reznick, D. N., H. Bryga, and J. A. Endler. 1990. Experimentally induced life-history evolution in a natural population. *Nature* 346:357–359.

Reznick, D., M. J. Butler, and H. Rodd. 2001. Life-history evolution in guppies. VII. The comparative ecology of high- and low-predation environments. *American Naturalist* 157:126–140.

Reznick, D. N., and J. A. Endler. 1982. The impact of predation on life history evolution in Trindadian guppies (*Poecilia reticulata*). *Evolution* 36:160–177.

Reznick, D. N., and C. K. Ghalambor. 2005. Selection in nature: Experimental manipulations of natural populations. *Integrative and Comparative Biology* 45:456–462.

Reznick, D. N., F. H. Shaw, F. H. Rodd, and R. G. Shaw. 1997. Evaluation of the rate of evolution in natural populations of guppies (*Poecilia reticulata*). *Science* 275:1934–1937.

Reznick, D., and J. Travis. 1996. The empirical study of adaptation in natural populations. Pages 243–289 *in* M. R. Rose and G. Lauder, eds. *Adaptation*. San Diego, CA: Academic Press.

Schoener, T. W., D. A. Spiller, and J. B. Losos. 2004. Variable ecological effects of hurricanes: The importance of timing for survival of lizards on Bahamian islands. *Proceedings of the National Academy of Sciences of the USA* 101:177–181.

Sinervo, B., and A. L. Basolo. 1996. Testing adaptation using phenotypic manipulations. Pages 148–185 *in* M. R. Rose and G. Lauder, eds. *Adaptation*. San Diego, CA: Academic Press.

Strauss, S. Y., J. A. Lau, and S. P. Carroll. 2006. Evolutionary responses of natives to introduced species: what do introductions tell us about natural communities? *Ecology Letters* 9:357–374.

Svensson, E. I., F. Eroukhmanoff, and M. Friberg. 2006. Effects of natural and sexual selection on adaptive population divergence and premating isolation in a damselfly. *Evolution* 60:1242–1253.

Templeton, A. R. 1989. Founder effects and the evolution of reproductive isolation. Pages 329–344 in L.V. Giddings, K. Kaneshiro, and W. Anderson, eds. *Genetics, Speciation and the Founder Principle*. New York: Oxford University Press.

Teotónio, H., M. Mateos, and M. R. Rose. 2004. Quantitative genetics of functional characters in *Drosophila melanogaster* populations subjected to laboratory selection. *Journal of Genetics* 83:265–277.

Tokarz, R. R. 1995. Mate choice in lizards: A review. *Herpetological Monographs* 9:17–40.

Travis, J., and D. N. Reznick. 1998. Experimental approaches to the study of evolution. Pages 437–459 *in* W. J. Resetarits and J. Bernardo, eds. *Experimental Ecology*. New York: Oxford University Press.

Travisano, M., J. A. Mongold, A. F. Bennett, and R. E. Lenski. 1995. Experimental tests of the roles of adaptation, chance, and history in evolution. *Science* 267:87–90.

# PART III

# LEVELS OF OBSERVATION IN EXPERIMENTAL EVOLUTION

# 9

# FITNESS, DEMOGRAPHY, AND POPULATION DYNAMICS IN LABORATORY EXPERIMENTS

Laurence D. Mueller

An important theme in this review will be the importance of a close balance between theory and experiments. Theory has as a goal the creation of simple and general principles. Understandably, therefore, theoreticians are often loath to spend too much time and energy worrying about the details of any specific organism's biology. However, experimental population biologists must determine if the assumptions of the specific theories that they wish to test are biologically unrealistic for the particular experimental organism that they wish to study. It is perilous for the experimental biologist to rely on the theoretician to determine if a particular organism or experimental design is appropriate for a test of the theory that they are both interested in.

Another important aspect of the work reviewed here is the interweaving of problems in ecology and evolution. Early in the history of these fields, topics like population dynamics were relegated to the realm of ecology, while the fields of evolution and population genetics considered problems like fitness components and methods for their estimation, ignoring ecological context. But over the course of the twentieth century, it became clear that population dynamics are affected by the organisms involved within what had been considered an "ecological" time frame. This idea was first treated in detail in the theoretical work of MacArthur (1962), but later the evolution of population growth rates became the focus of experimental studies (Mueller and Ayala 1981c). In a similar vein, the impact of life cycles and population regulation on fitness measurements and evolution was first clearly demonstrated by the work of Prout (1965, 1971a, 1971b, 1980). While this commingling of ecological and evolutionary research is still developing, there has already been significant progress over the last thirty years (see also Irschick and Reznick this volume).

Some of the earliest work in experimental population biology was carried out early in the twentieth century by Raymond Pearl using laboratory populations of *Drosophila*. His choice of fruit flies as a model system was in part an accident. A fire destroyed Pearl's laboratory and his mouse colony, forcing him to reconsider his plan to pursue questions in population biology using mice. On the advice of T. H. Morgan, he decided to study fruit flies. By 1919, Pearl had his experimental fly populations established, and experimental ecology was born.

The most enduring legacy of Pearl's work is not his experimental findings; in fact, his experimental methods are by today's standards unacceptably imprecise. Pearl's legacy was actually his focus on the interaction between experiment and theory. Pearl had interests in human population growth and so always viewed his fly experiments as models for other organisms. Pearl was also interested in using simple models to describe his experimental results. Because these theoretical models, like logistic population growth, were offered as general principles of population dynamics by Pearl, they caught the attention and interest of many people. Ultimately, they led others to consider more seriously the techniques used for determining population growth rates experimentally.

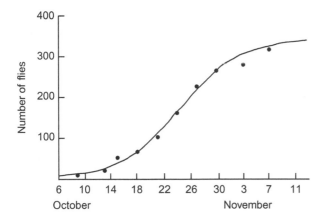

FIGURE 9.1
Population growth of a laboratory population of *D. melanogaster*. The circles are observed numbers and the line is the fitted logistic curve (taken from Pearl 1927).

Pearl was interested in human population growth, but he thought that there must be a universal law of population growth that would apply to all organisms, including humans and fruit flies. He began his experimental research with *D. melanogaster*. His experimental results apparently closely followed the simple logistic equation (figure 9.1). The techniques used to maintain Pearl's flies were somewhat haphazard, however. He would supply food to the flies at irregular intervals in varying amounts whenever it seemed as if food was needed. Pearl described his procedure this way: "The second type of experiment is one in which an attempt is made to add food as the supply is used up. The technical difficulties of doing this satisfactorily with a *Drosophila* population are considerable but by sufficient care they can be overcome in large degree" (Pearl 1927).

This procedure evidently lacked a proper protocol for the systematic renewal of resources, the replacement of bottle environments, or census taking. In fact, the experimental procedures are so vague that one can imagine that judgments about food addition could be unduly influenced by the numbers of flies produced in the most recent census. Clearly, the techniques cannot be replicated by another scientist. Many of these problems in experimental technique are discussed by Sang (1949). It would be another thirty-five years before *Drosophila* was again the subject of serious population dynamic experiments.

At the same time as Pearl was conducting his experimental research, many of the theoretical results for population genetics with and without overlapping generations were established. The connection between Mendelian genes and quantitative characters was developed by Fisher (1918). Selection in populations with discrete nonoverlapping generations was explored by Haldane (1927a). Two important papers (Norton 1928; Haldane 1927b) developed measures of fitness in age-structured populations. These papers demonstrated the importance of age-specific mortality and fecundity in determining fitness (see Rauser et al. this volume).

Pearl was not apparently directly motivated by the papers of Haldane and Norton, but he nevertheless started collecting experimental data on age-specific survival and fecundity (Pearl et al. 1927). Pearl's work established that age-specific survival depends not only on the current total population density but also on the past history of population densities that an individual has experienced. A similar relationship between the current dynamics of a population and its past environments has also been recently described by Benton et al. (2006). Almost all theoretical work has ignored this biological finding, most likely because it greatly complicates the mathematical analysis. Fortunately, Pearl et al. (1927) did show that the current population density has a much greater impact on present survival then does past density. However, Pearl's work had shown that even an empirical problem as basic as the estimation of fitness in populations without age structure presented subtle complications that took some time for most population biologists to appreciate.

## EMPIRICAL MEASURES OF FITNESS

Simple theory regarding the action of natural selection was well developed before experimental tests of this theory were even attempted. In keeping with the separation of ecological and evolutionary thought before 1960, little consideration was given then to the specifics of the life cycle of an organism when interpreting experimental data in evolutionary biology. Prout (1965) was the first to illustrate these complications and used a typical set of experimental data collected by Polivanov (1964). Polivanov used a simple model to interpret his experimental results. The model (figure 9.2A) assumes a single locus with two alleles and therefore three genotypes, $A_1A_1$, $A_1A_2$, and $A_2A_2$. Assuming random mating, zygote genotype frequencies will be in Hardy-Weinberg proportions, but after selection operates these frequencies are perturbed to $X_{11}$, $X_{12}$, and $X_{22}$ for the genotypes $A_1A_1$, $A_1A_2$, and $A_2A_2$, respectively. If the fitness of the three genotypes are

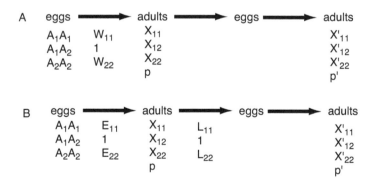

FIGURE 9.2
Two generations of a simple life cycle. A, Here the three genotypes at a single diallelic locus experience viability selection sometime between the egg and the adult census stage. B, In this example, selection at a single locus is affected by viability selection prior to the adult census stage and by fertility selection immediately after the adult census stage.

$W_{11}$, I, and $W_{22}$, respectively, then it is relatively simple to show that the homozygote fitness ($W_{11}$), under this model, should be

$$2\frac{X'_{11}(1 - p)}{X'_{12}p},$$

where $p$ is the frequency of the $A_1$ allele among the zygotes.

Prout (1965) pointed out that even simple laboratory populations of *Drosophila* can violate the implicit assumptions of these simple population genetic models, in turn leading to serious problems with the interpretation of fitness estimates. The major problem is that even under carefully controlled lab conditions, the life cycle of the fruit fly is still more complicated than these simple models assume (Prout 1965). Polivanov used two common third chromosome mutants stocks of *D. melanogaster* for his experimental estimates of fitness. *Stubble* is a dominant phenotypic mutant that causes the thorax bristles to be half their normal size. In addition, it acts as a recessive lethal. The recessive mutant *ebony* causes a dark coloration of the body. When dealing with these real genetic variants, there is no guarantee that the fitness effects of these deleterious alleles will be limited to egg-to-adult viability, which the simple population genetic model employed assumes. For instance, one possibility is that these mutants affect both egg-to-adult survival and male and female fertility (figure 9.2B). When this is true, the estimated fitness will typically be incorrect by a large amount, and there can be a spurious inference of frequency dependence (figure 9.3).

These observations led Prout (1969, 1971a, 1971b) to outline a detailed methodology for estimating fitness and its components. This methodology required that important

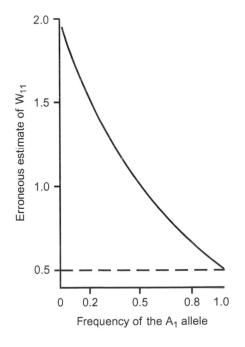

FIGURE 9.3
The erroneous value of fitness given by the equation, $2\dfrac{X'_{11}(1 - p)}{X'_{12}p}$, when there are actually neglected early and late components of fitness as shown in figure 9.2B. The correct value of fitness is 0.5. Adapted from Prout (1965).

biological details of the life cycle of study organisms be taken into account. These techniques were further refined by Christiansen (Christiansen and Frydenberg 1973; Christiansen 1980). These methods have been applied to polymorphisms on the fourth chromosome of *Drosophila melanogaster* (Bungaard and Christiansen 1972) and the esterase polymorphism in the blenny *Zoarces* (Christiansen et al. 1973, 1978).

Before the advent of molecular techniques like protein gel electrophoresis, naturally occurring genetic variation was studied in experimental lines derived from specific crosses with specially constructed mutants. One method devised to study variation for fitness-related traits was to examine a collection of individual fruit flies that were all homozygous for the same second or third chromosome sampled from nature (Sved 1971; Sved and Ayala 1970; Tracey and Ayala 1974). While this technique resulted in large fitness differences that could be easily measured, these homozygous genotypes were hardly likely to be found in nature or for that matter in outbred lab populations. In addition to producing homozygosity for 20 to 30 percent of all genes, the chromosome extraction process usually resulted in genetic variation from marker stocks being introduced into the study lines.

Not surprisingly, these studies showed that making whole chromosomes in *Drosophila* homozygous resulted in large declines in fecundity (Marinkovic 1967), viability (Dobzhansky et al. 1963), and male virility (Britnacher 1981) relative to heterozygous genotypes. Sved and Ayala (1970) devised experimental techniques for allowing populations to complete their entire life span in a population cage, so that after many generations, net fitness estimates could be made from the equilibrium frequency of marked chromosomes. These studies revealed that viability, although often equated to fitness, only accounted for a small part of the reduced fitness of these homozygous genotypes. Adult fitness components, especially virility, contributed substantially to net fitness (Britnacher 1981).

These laboratory genotypes were also used to study the interactions between separate chromosomes on net fitness. For instance, Seager et al. (1982) estimated the fitness of effects of homozygosity on the second chromosome and the third chromosome of *D. melanogaster*. They found that the fitness of genotypes homozygous for both the second and third chromosome was generally higher than models of independent gene action predicted.

Recently, significant improvements in the techniques of chromosome extraction and fitness estimation have been made (Fowler et al. 1997; Barton and Partridge 2000; Gardner et al. 2001, 2005). These improvements include (1) studying the fitness of chromosomal heterozygotes, (2) backcrossing the extracted chromosome lines to an outbred population that has already been adapted to the lab environment, and (3) making replicate estimates of fitness for each chromosome. It is still hard to determine if the fitness variation detected by the techniques used by Partridge and her collaborators is representative of natural fitness variation. This is because these new techniques still suffer from several shortcomings, which include (1) the fitness effects of a multiply inverted marker chromosome; (2) nonrandom sampling of chromosomes (most or all chromosomes had recessive lethal effects); and (3) populations maintained with overlapping generations, despite the use of a discrete-time model to estimate fitness, although Barton and Partridge outline conditions under which they argue this approximation might work.

Clark et al. (1981) studied fitness among a variety of two-locus mutant genotypes. Their analysis also attempted to estimate preadult and adult components of fitness, as shown in figure 9.2. Clark et al. noted that with two loci, ignoring the early or late components of fitness could lead to spurious estimates of epistasis, in addition to the potential for artifactual estimation of frequency dependence noted by Prout (1965). They found that estimates of epistasis for these laboratory mutant systems were often significantly different from zero, although the sign of these epistatic effects was not consistently positive or negative.

The primary goal of the chromosome extraction methods just described was to infer genetic variation for fitness or fitness components in natural populations. Population geneticists have also been interested in getting estimates of fitness for whole populations of wild-type individuals. These populations might be genetically variable but be adapted to different environments or have different geographic origins. Several surrogates of fitness, like biomass or productivity, have been proposed for whole population estimates (Carson 1961a, 1961b). *Productivity* usually refers to the total number of individuals produced by a genotype or population under controlled conditions, whereas *biomass* is simply the wet or dry weight of all the individuals. Productivity is still occasionally used as a measure of fitness (e.g., Houle et al. 1994). Productivity can provide a convenient index of fitness for a wide variety of genotypes or genetically differentiated populations. Haymer and Hartl (1983) tested the utility of such productivity measures of fitness by comparing them to more traditionally based fitness estimates. They measured fitness by measuring biomass, productivity, and the direct competition of genotypes for an array of extracted second chromosomes. The results show a very weak correlation between the more direct competitive estimates of fitness and biomass or productivity. In the next section, on population dynamics, I will discuss one possible explanation for this effect.

Perhaps the most sophisticated method for making fitness estimates of wild-type genomes is the cytogenetic cloning, or "hemiclone," method (Chippindale et al. 2001). This method requires the sophisticated array of attached-X and compound autosome stocks available in *D. melanogaster*. However, with these tools nearly the entire genome of a single fly (comprising X, second, and third chromosomes) can be placed into the genetic background of an appropriate stock population. This then permits fitness estimates of a single naturally occurring haploid genome against a wide array of natural genetic backgrounds.

Using the hemiclone technique for fitness estimation has permitted the estimate of fitness effects of genomes in males and female contexts. These studies have revealed disparate and often antagonistic fitness effects of genes in males and females, effects that may help explain the evolution of extreme secondary sexual characteristics in males (Chippindale et al. 2001; Pischedda and Chippindale 2006).

EXPERIMENTAL SYSTEMS

The commentary by Sang on Pearl's work emphasized the idea that any study of population dynamics in the laboratory will need a well-defined system for resource replacement or movement of individuals to new environments. Implicit in these experimental design issues are also decisions about whether the experimental population will have overlapping or discrete generations. One early experimental system for studying the population dynamics of *Drosophila* was the serial transfer system (Ayala 1965). This system consisted of a breeding adult population with overlapping generations, although it was first analyzed as if it were a discrete generation system. New recruits were collected at weekly intervals from cultures that had eggs laid in them one, two, three, and four weeks ago. Early attempts were made to use discrete-time models to study the adult population size (Hasting et al. 1981). However, the *Drosophila* serial transfer system cannot be modeled by first-order difference equations due to the complicated sampling structure used for collecting new recruits (Mueller and Ayala 1981a; Mueller and Joshi 2000, chapter 3). Nevertheless, it is not difficult to create discrete-generation populations of *Drosophila* (Mueller et al. 2000).

In contrast to the discrete-generation life cycles of *Drosophila*, most unicellular organisms used for ecological and evolutionary research are maintained on a continuous reproductive schedule. Bacteria (Lenski and Levin 1985; Bohannan and Lenski 2000) and communities consisting of unicellular algae and rotifers (Fussmann et al. 2000, 2003, 2005) are maintained in systems with a continuous flow of nutrients called *chemostats*. These systems permit continuous reproduction and thus can be modeled with continuous-time equations.

*Tribolium* is another well-studied experimental system that allows the maintenance of all life stages—eggs, larvae, pupae, and adults—in the same flour medium. At regular time intervals, often once a month, all life stages can be censused and the flour changed. Without additional experimental intervention, the adult population will have overlapping generations. The interesting aspect of ecological studies with *Tribolium* is the ease of obtaining simultaneous census counts for all life stages by using different size sieves to filter the flour. The duration and number of larval instars are affected by a number of genetic and environmental factors, especially temperature, humidity, and food (King and Dawson 1972).

One important model of *Tribolium* population dynamics is the larva-pupa-adult (LPA) model (Dennis et al. 1995). This model combines the egg and larval stages together and ignores the relatively weak density dependence of both larval and adult mortality as well as fecundity. Nevertheless, as discussed in the next section, this model has proven to be especially useful.

Nicholson (1954a, 1954b, 1957) pioneered the use of blowflies *(Lucilia cuprina)* as a model organism. Nicholson kept blowflies in large cages capable of supporting populations of ten thousand or more adults. The larval and adult food sources were separately controlled. Adults received both a sugar and protein food source and were allowed to live indefinitely. Thus, these blowfly populations consisted of overlapping generations that were counted at regular time intervals. Blowflies, *Tribolium*, and *Drosophila* suffer from the common liability that there is no simple way to assess the age of the adults. The next section considers models of population dynamics for these specific experimental systems.

## ECOLOGICAL MODELS

Experimental studies of *Drosophila* population dynamics inevitably involve both adult and larval life stages. The effects of crowding on both life stages will ultimately be important for population dynamics. Larval crowding affects both survival to the adult stage and ultimate adult size (Bakker 1961). Since female fecundity is highly correlated with adult size, the effects of larval crowding will carry over into the next generation. In addition, there are substantial effects of adult crowding on female fecundity.

Bakker's work motivated the development of several theoretical models of competition and population dynamics (de Jong 1976; Nunney 1983; Mueller 1988a). Mueller's population dynamic model can be used to determine the conditions for stability of the population at its carrying capacity. In general terms, this model revealed that *Drosophila* populations may become unstable and enter fixed point cycles, or even chaos, if adults were provided with abundant resources and larvae were provided with low levels of food (Mueller 1988a; Mueller and Huynh 1994). These predictions about the dependence of stability on food provisioning were experimentally verified (Mueller and Huynh 1994), and later the model was used to study the evolution of population stability (Mueller and Joshi 2000).

Laboratory populations of blowflies often display strong and regular cyclic fluctuations with respect to population density (Nicholson 1954b, 1957). Blowflies show strong density-dependent fecundity and larval mortality. However, there is about a twenty-day time delay between the current adult density and its effect on future adult density. Blowfly population size cycles are strongly affected by the relative levels of larval and adult food levels, so they are in some aspects similar to *Drosophila* laboratory populations (Mueller and Joshi 2000, chapter 4). However, the conditions that are stabilizing for blowflies in the laboratory are low levels of food for both adults and larvae. High levels of larval food and low levels of adult food do not stabilize blowfly population dynamics as they do for *Drosophila*.

The growth of *Tribolium* populations is regulated by strong density-dependent cannibalism of eggs and pupae. Experimental manipulation of these rates of cannibalism leads to fine control of population stability and the strength of the underlying LPA model of *Tribolium* population dynamics (Costantino et al. 1995, 1997), a very different population-dynamic system from that of blowflies or *Drosophila*. While the level of ecological modeling for *Drosophila*, *Tribolium*, and blowflies is equally high, most evolutionary studies, described in the next few sections, have employed *Drosophila*.

The modern union of ecological theories of density-dependent population dynamics and natural selection was started by MacArthur (1962) and then refined by Anderson (1971), Charlesworth (1971), Clarke (1972), and Roughgarden (1971). Under this theory, genotypic fitness is taken to be density-dependent. For a single locus with multiple alleles, the fitness of genotype $A_iA_j$ is given by

$$W_{ij} = 1 + r_{ij} - r_{ij}NK_{ij}^{-1},$$

where $r_{ij}$ is the genotypic intrinsic rate of increase, $K_{ij}$ is the genotypic specific value of the carrying capacity of the logistic equation, and $N$ is the total population size. The outcome of natural selection inherently depends on the environment with this specification of fitness. In uncrowded environments, genotypes with the highest values of $r$ are favored, while in crowded environments the genotypes with the highest values of $K$ are favored. This theory can be used to explain variation in life-history traits if one assumes that there are trade-offs such that phenotypes with high values of $r$ have low values of $K$ and vice versa.

## MEASUREMENTS OF THE DENSITY DEPENDENCE OF FITNESS

The relationship between density-dependent rates of population growth and fitness was studied with extraction lines like the ones reviewed in the previous section (Mueller and Ayala 1981b). Mueller and Ayala (1981b) measured density-dependent rates of population growth in twenty-five lines, each homozygous for a different second chromosome that had been sampled from nature. These growth rates were compared to the growth of an outbred population to determine relative fitness (Mueller and Ayala 1981b). At low density, the fitness-reducing effects of inbreeding are observed (figure 9.4). However, at high density, relative fitness of the average inbred line is no different from that of the outbred line using population growth rates as a measure of fitness. This is surprising, since the population cage studies that demonstrated the severe reduction in fitness caused by inbreeding were carried out at high population density (Seager et al. 1982).

These observations can be understood by looking at models that incorporate density-independent natural selection with density-dependent population growth (Prout 1980). This type of model doesn't require that there be no difference between isogenic lines in their density-dependent survival and fecundity, only that it be small relative to the density-independent fitness differences. Let's assume population dynamics are described by the discrete-time hyperbolic model,

$$N_{t+1} = \left[\frac{\widetilde{S}}{1 + sFN_t}\right]FN_t,$$

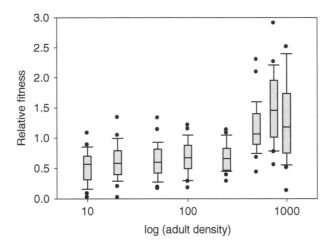

FIGURE 9.4

Box plots of the relative fitnesses of twenty-five inbred lines of *Drosophila melanogaster* based on density-dependent rates of population growth. From Mueller and Ayala 1981b. The growth rate of each line is divided by the growth rate of the genetically variable outbred line to get the relative fitness value. The box boundaries denote the 25th and 75th percentiles, the bars indicate the 10th and 90th percentiles, and the points show outliers that exceed the 10th and 90th percentiles. The median is indicated by a line through the box.

where $N_t$ is the adult population size at time $t$, $F$ is the per capita fecundity, and $\widetilde{S}$ and $s$ measure density-dependent survival from egg to adult. Thus, after random mating, there are a total of $FN_t$ eggs produced. A fraction

$$\frac{\widetilde{S}}{1 + sFN_t}$$

of these eggs survive to become adults in the next generation. Under this model, the equilibrium population size, or carrying capacity, is

$$\frac{\widetilde{S}F - 1}{sF}.$$

When fecundity is high, as it is for young *Drosophila* females, then the carrying capacity simplifies to approximately $\frac{\widetilde{S}}{s}$. In other words, the carrying capacity and, by extension, population growth rates near the carrying capacity are insensitive to changes in fecundity. Therefore, if differences in female fecundity are a large part of the fitness reduction of inbred *Drosophila* lines, it is not surprising that their growth rates at high density do not differ. However, large differences in fecundity will affect population growth rates at low densities, since nearly all eggs will survive to become adults. In *D. pseudoobscura* homozygous for whole second chromosomes, decreased fecundity accounts for about a 20 percent fitness reduction (Marinkovic 1967). This type of effect may also explain why productivity does not reliably reflect fitness differences among chromosomal homozygotes in *Drosophila*.

The major lesson from this work is that the fitness consequences of genetic variation may not be apparent in all population growth measurements. In the case of genetic variation affecting fecundity, it appears that this will at least affect population growth rates at low density. However, there can be other large differences in fitness components among genotypes that have no effect on population growth rates. In *D. melanogaster*, for instance, it is known that lines homozygous for second chromosomes suffer a substantial reduction in male virility relative to heterozygous males. Male virility is a frequency-dependent fitness component, and therefore in a population of equivalent males, as long as all females are fertilized—which is what we normally observe—we would not expect low male virility to have any effect on population growth rates at any density. In conclusion, we see that as a general surrogate for fitness, population growth rates have many limitations and will often incorrectly estimate the true fitness of a genotype.

The *Drosophila* population-dynamic model discussed here predicted that density-dependent natural selection would favor increases in competitive ability (Mueller 1988a). Crowding is also expected to increase the equilibrium adult numbers and under some circumstances could result in the evolution of smaller body size, contrary to verbal theories of *r*- and *K*-selection (Mueller 1988a). A simple analysis would suggest that larger body size should be favored under crowded conditions since this would buffer the organism against variations in food availability. However, a more detailed analysis of the relationship between size, fitness, and density in *Drosophila* reveals that in food-limited environments, the ability to pupate at a smaller size may be advantageous.

The relationship between fitness and density may be complicated and depend in nontrivial ways on the details of the organism's life history. These observations suggest that experimental work must be founded on theories that have taken the experimental organisms peculiar life-history traits into account. The use of general theories lacking biological specificity, like *r*- and *K*-selection, to predict the outcomes of evolution for specific experimental systems is unlikely to be successful.

## EVOLUTION OF GROWTH RATES

Even if growth rates are not good measures of fitness, this in itself does not show that growth rates will not evolve as suggested by the theory of Roughgarden (1971) under the appropriate conditions. This idea has been tested experimentally with *Drosophila* (Mueller and Ayala 1981c; Mueller et al. 1991). These tests involved experimentally manipulating adult and larval densities and creating one set of replicate populations, called *r*'s, that were kept at low adult and larval density and another set, called *K*'s, that were kept at high adult and larval density. Each of these environmental treatments was replicated threefold, and natural selection was allowed to change the genetic composition of these populations.

After just eight generations, these populations showed trade-offs in their density-dependent growth rates at high and low densities (Mueller and Ayala 1981c), (figure 9.5). This experiment was repeated after 198 generations of *r*- and *K*-selection by rederiving two new sets

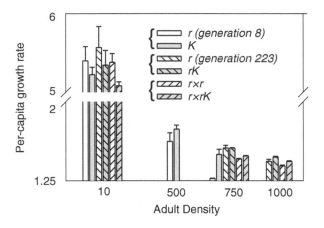

FIGURE 9.5

The per capita growth rates at four adult densities for populations cultured at low density ($r$, $r \times r$) and populations cultured at high densities ($K$, $rK$, and $r \times rK$; from Mueller and Ayala 1981c and Mueller et al. 1991). The bars are standard errors. The derivation of the various lines is described in the text. The measurements for the $r$- and $K$-populations shown as solid histograms were made after eight generations of selection. The measurements for the other populations were made after 223 generations of selection in the $r$-environment.

of populations called $rK$ and $r \times rK$ (see Mueller et al. 1991 for details of their derivation). Both sets of new populations also showed the evolution of trade-offs in population growth rates after twenty-five generations of further evolution (Mueller et al. 1991; figure 9.5).

The changes in population growth rates due to density-dependent natural selection can involve the evolution of several phenotypes. In accordance with the predictions of the *Drosophila* model (Mueller 1988a), larval competitive ability increases in populations kept under crowded larval conditions (Mueller 1988b). Egg-to-adult viability was also affected by the evolution of increased pupation height in the crowded cultures (Mueller and Sweet 1986; Joshi and Mueller 1993).

Crowding *Drosophila* larvae is expected to cause competition for food and space. It has also been observed that, over the span of one generation, crowded larval environments show a temporal decline in quality (Borash et al. 1998). Ammonia levels increase over time, while food and ethanol levels decrease. This complexity appears to be responsible for a genetic polymorphism in crowded populations. Very early-developing genotypes have high feeding rates but low tolerance to ammonia, while late-developing genotypes feed more slowly and can tolerate higher ammonia levels. There may be many natural environments that exhibit similar patterns of temporal decay (Borash et al. 1998).

In *Drosophila*, larval competitive ability is determined primarily by the larval feeding rate (Burnet et al. 1977; Joshi and Mueller 1988; Fellowes et al. 1998). Better competitors feed at a faster rate. However, it appears that feeding fast decreases the efficiency of food utilization (Mueller 1990; Joshi and Mueller 1996). Feeding rate also responds to selection for several other larval stressors, including ammonia in the larval environment (Borash et al. 2000), urea in the larval food (Borash et al. 2000), parasitoids (Fellowes et al. 1999), and reduced time for larval development (Borash et al. 2000; Prasad et al. 2001). These seemingly disparate results indicate that the larval energy budget is sensitive to changes in larval feeding rates. Lowering feeding rates, which occurs in larvae adapted to urea, ammonia, and parasitoids, may increase efficiency and provide larvae with the energy needed to meet the demands of a stressful environment (Mueller et al. 2005).

Luckinbill (1978) studied density-dependent natural selection in *Escherichia coli* by creating cultures that underwent exponential growth to simulate low-density, or *r*-selection, conditions. High-density, or *K*-selection, conditions were created by letting populations grow exponentially followed by periods of maintenance at saturation density. Luckinbill observed that *K*-selected bacteria grew faster than *r*-selected bacteria at all test densities. Vasi et al. (1994) studied the evolution of *E. coli* in seasonal environments that were similar to Luckinbill's *K*-selected environment. Vasi et al. used their data to estimate the parameters for a model of bacterial population dynamics, and then showed that these populations had evolved traits that would be most important during the exponential growth phase of the environment, while parameters that would be most important during the periods of saturation density had not changed. Thus, Luckinbill's results may simply reflect differences in the intensity of selection during the exponential growth phase rather than differences in selection at high and low density.

The evolution of population dynamics is of great practical interest for conservation biology (see Saccheri and Hanski 2006). Genetic changes that may affect either the equilibrium population size or the ability of a population to grow at low densities may in turn have an impact on the persistence of a population over time. Examples of genetic variation in natural populations that affects their dynamics are hard to come by, but they do in fact exist. For instance, a genetic polymorphism for horn shape in Soay sheep appears to affect density-dependent rates of population growth (Moorcroft et al. 1996). More recently, *Pgi* polymorphisms in the butterfly *Melitaea cinxia* have been implicated in their population growth (Hanski and Saccheri 2006).

## EVOLUTION OF POPULATION STABILITY

If density-dependent rates of population growth evolve, then it makes sense that population stability might in turn evolve, since both ultimately depend on nonlinear responses to density. The first test of this idea came from Mueller et al. (Mueller and Joshi 2000; Mueller et al. 2000). These tests placed populations with different selection histories in environments that caused population cycling. Although there was clear evidence of evolution in both population carrying capacity and larval feeding rates, there was no discernible change in the stability properties of any populations. Thus, if stability does evolve, it does so much more slowly than other phenotypes that are affected by population crowding.

In designing their experiments, Mueller et al. (2000) chose techniques that would minimize the effects of inbreeding, because it is known that this will cause a decline in female fecundity, which in theory could secondarily increase population stability. A result of this kind was in fact obtained by Prasad et al. (2001). They selected populations for rapid development and early reproduction. These populations showed a reduction in female fecundity relative to controls. Prasad et al. observed increased stability in the rapidly developing populations compared to their controls, as predicted by the simple theories discussed earlier (Mueller and Huynh 1994; Mueller et al. 2000). This same phenomenon may have been responsible for the evolution of increased stability in Nicholson's blowfly experiments (Stokes et al. 1988).

More complicated predator-prey systems also demonstrate the impact of evolution on population dynamics. Fussman et al. (2003, 2005) studied a rotifer-algal system in chemostats. They found that the rotifers evolved lower rates of sexual reproduction, and the algae evolved higher rates of herbivore resistance. However, herbivore resistance was accompanied by reduced growth rates.

## DISCUSSION

A major goal of experimental evolution is to simplify the conditions under which evolution occurs in order to effectively study how evolution operates. Much of the research outlined here has shown that, even in apparently simple laboratory settings, understanding the ecological and evolutionary forces at work can be quite tricky. Without doubt, the real world is usually even more complicated. The work of Fussman et al. reveals the complication of interacting species that are both capable of evolving. The effects on population dynamics of such coevolution could be substantial.

An area of research that is still understudied is the dynamics of populations with age structure. Although the experimental systems of *Tribolium* and blowflies have age-structured populations, the extant models of these systems have assumed that all adults are equivalent. While that assumption might be adequate for these particular experimental systems, it does not show that age structure is generally unimportant. Of course, severe practical problems make the study of age structure with model systems technically difficult, although not impossible (see Mueller and Joshi 2000, chapter 6).

The study of adaptation in different laboratory environments and studies of demography could benefit from fitness estimates using the hemiclone technology developed for *Drosophila*. One difficult problem with demography is that measurements of age-specific survival cannot be made on single individuals. However, hemiclones could be used to estimate demographic parameters for individual genotypes. This could permit direct estimates of genetic and environmental variation in demographic parameters that are crucial components for demographic theories of late life (Mueller et al. 2003).

Overall, few of the simple models and empirical predictions that were first developed concerning relationships among fitness, age structure, and density dependence have survived the scrutiny that laboratory experiments have afforded. The study of the joint action of population dynamics and natural selection has thus made salutary progress, at least in this respect.

A recurring theme in experimental population biology has been the wealth of unanticipated effects that are detected. This collective history argues strongly for the pursuit of research with experimental systems since the prospect of revealing, defining, and understanding these complications in uncontrolled natural conditions is low.

## SUMMARY

Experimental laboratory systems that combine elements of population biology have contributed to our understanding of many basic problems in ecological and evolutionary biology, particularly with respect to their interface. Some of these problems include the partitioning of fitness into components, epistatic interactions affecting fitness, ecological factors that determine population stability, phenotypic evolution due to density-dependent natural selection, and the role of evolution in molding population stability. Laboratory systems are designed to be simple, but the factors that affect the evolution and ecology of these systems can still be quite complicated. Consequently, experimental research in ecology and evolutionary biology will continue to make important contributions to our understanding of the basic principles of these fields.

## ACKNOWLEDGMENTS

I thank two anonymous referees for unusually helpful comments on the manuscript and Michael Rose for reminding me that commas have not outlived their usefulness.

## REFERENCES

Anderson, W. W. 1971. Genetic equilibrium and population growth under density-regulation selection. *American Naturalist* 105:489–498.

Ayala, F. J. 1965. Relative fitness of populations of *Drosophila serrata* and *Drosophila birchii*. *Genetics* 51:527–544.

Bakker, K. 1961. An analysis of factors which determine success in competition for food among larvae of *Drosophila melanogaster*. *Archives Néerlandaises de Zoologie*. 14:200–281.

Barton, N. H., and L. Partridge. 2000. Measuring fitness by means of balancer chromosomes. *Genetical Research, Cambridge* 75:297–313.

Benton, T. G., S. J. Plaistow, and T. N. Coulson. 2006. Complex population dynamics and complex causation: Devils, details and demography. *Proceedings of the Royal Society of London B, Biological Sciences* 273:1173–1181.

Bohannan, B. J., and R. E. Lenski. 2000. Linking genetic change to community evolution: Insights from studies of bacteria and bacteriophage. *Ecology Letters* 3:362–377.

Borash, D. J., A. G. Gibbs, A. Joshi, and L. D. Mueller 1998. A genetic polymorphism maintained by natural selection in a temporally varying environment. *American Naturalist* 151:148–156.

Borash, D. J., Teotónio, H., M. R. Rose, and L. D. Mueller. 2000. Density-dependent natural selection in *Drosophila*: Correlations between feeding rate, development time, and viability. *Journal of Evolutionary Biology* 13:181–187.

Britnacher, J. G. 1981. Genetic variation and genetic load due to the male reproductive component of fitness in *Drosophila*. *Genetics* 97:719–730.

Bundgaard, J., and F. B. Christiansen. 1980. Dynamics of polymorphisms. I. Selection components in an experimental population of *Drosophila melanogaster*. *Genetics* 71:439–460.

Burnet, B., D. Sewell, and M. Bos. 1977. Genetic analysis of larval feeding behavior in *Drosophila melanogaster*. II. Growth relations and competition between selected lines. *Genetical Research* 30:149–161.

Carson, H. L. 1961a. Heterosis and fitness in experimental populations of *Drosophila melanogaster*. *Evolution* 15:496–509.

Carson, H. L. 1961b. Relative fitness of genetically open and closed populations of *Drosophila robusta*. *Genetics* 46:553–567.

Charlesworth, B. 1971. Selection in density-regulated populations. *Ecology* 52:469–474.

Chippindale, A. K., J. R. Gibson, and W. R. Rice. 2001. Negative genetic correlation for adult fitness between sexes reveals ontogenetic conflict in *Drosophila*. *Proceedings of the National Academy of Sciences of the USA* 98:1671–1675.

Christiansen, F. B. 1980. Studies on selection components in natural populations using population samples of mother-offspring combinations. *Hereditas* 92:199–203.

Christiansen, F. B., and O. Frydenberg. 1973. Selection component analysis of natural polymorphisms using population samples including mother-offspring combinations. *Theoretical Population Biology* 4:425–445.

Christiansen, F. B., O. Frydenberg, and V. Simonsen. 1973. Genetics of *Zoarces* populations. IV. Selection component analysis of an esterase polymorphism using samples including mother-offspring combinations. *Hereditas* 73:291–304.

———. 1978. Genetics of *Zoarces* populations. X. Selection component analysis of the *Est*III polymorphism using samples of successive cohorts. *Hereditas* 87:129–150.

Clark, A. G., M. W. Feldman, and F. B. Christiansen. 1981. The estimation of epistasis in components of fitness in experimental populations of *Drosophila melanogaster*. I. A two-stage maximum likelihood model. *Heredity* 46:321–346.

Clarke, B. 1972. Density-dependent selection. *American Naturalist* 106:1–13.

Costantino, R. F., J. M. Cushing, B. Dennis, and R. A. Desharnais. 1995. Experimentally induced transitions in the dynamic behaviour of insect populations. *Nature* 375:227–230.

Costantino, R. F., R. A. Desharnais, J. M. Cushing, and B. Dennis. 1997. Chaotic dynamics in an insect population. *Science* 275:389–391.

de Jong, G. 1976. A model of competition for food. I Frequency dependent viabilities. *American Naturalist* 110:1013–1027.

Dennis, B., R. A. Desharnais, J. M. Cushing, and R. F. Costantino. 1995. Nonlinear demographic dynamics: Mathematical models, statistical methods, and biological experiments. *Ecological Monographs* 65:261–281.

Dobzhansky, T., B. Spassky, and T. Tidwell. 1963. Genetics of natural populations. XXXII. Inbreeding and mutational loads in natural populations of *Drosophila pseudoobscura*. *Genetics* 48:361–373.

Fellowes, M. D. E., A. R. Kraaijeveld, and H. C. J. Godfray. 1998. Trade-off associated with selection for increased ability to resist parasitoid attack in *Drosophila melanogaster*. *Proceedings of the Royal Society of London B, Biological Sciences* 265:1553–1558.

———. 1999. Association between feeding rate and parasitoid resistance in *Drosophila melanogaster*. *Evolution* 53:1302–1305.

Fisher, R. A. 1918. The correlation between relatives on the supposition of Mendelian inheritance. *Transactions of the Royal Society of Edinburgh* 52:399–433.

Fowler, K., C. Semple, N. H. Barton, and L. Partridge. 1997. Genetic variation for total fitness in *Drosophila melanogaster*. *Proceedings of the Royal Society of London B, Biological Sciences* 264:191–199.

Fussmann, G. F., S. P. Ellner, and N. G. Hairston. 2003. Evolution as a critical component of plankton dynamics. *Proceedings of the Royal Society of London B, Biological Sciences* 270:1015–1022.

Fussmann, G. F., S. P. Ellner, N. G. Hairston, L. E. Jones, K. W. Shertzer, and T. Yoshida. 2005. Ecological and evolutionary dynamics of experimental plankton communities. Pages 221–243 *in* R. A. Desharnais, ed. *Advances in Ecological Research*, vol. 37. Amsterdam: Elsevier.

Fussmann, G. F., S. P. Ellner, K. W. Shertzer, and N. G. Hairston. 2000. Crossing the Hopf bifurcation in a live predator-prey system. *Science* 290:1358–1360.

Gardner, M., K. Fowler, L. Partridge, and N. H. Barton. 2001. Genetic variation for preadult viability in *Drosophila melanogaster*. *Evolution* 55:1609–1620.

———. 2005. Genetic variation for total fitness in *Drosophila melanogaster*: Complex yet replicable patterns. *Genetics* 169:1553–1571.

Haldane, J. B. S. 1927a. A mathematical theory of natural and artificial selection. Part IV. *Proceedings of the Cambridge Philosophical Society* 23:607–615.

———. 1927b. A mathematical theory of natural and artificial selection. Part V. Selection and mutation. *Proceedings of the Cambridge Philosophical Society* 23:838–844.

Hanski, I., and I. J. Saccheri. 2006. Molecular-level variation affects population growth in a butterfly metapopulation. *PLoS Biology* 4:e129.

Hastings, A., Serradilla, J. M., and Ayala, F. J. 1981. Boundary-layer model for the population dynamics of single species. *Proceedings of the National Academy of Sciences of the USA* 78:1972–1975.

Haymer, D. S., and D. L. Hartl. 1983. The experimental assessment of fitness in *Drosophila*. II. Comparison of competitive and noncompetitive measures. *Genetics* 104:343–352.

Houle, D., K. A. Hughes, D. K. Hoffmaster, J. Ihara, S. Assimacopolous, D. Canada, and B. Charlesworth. 1994. The effects of spontaneous mutation on quantitative traits. I. Variance and covariance of life history traits. *Genetics* 138:773–785.

Joshi, A., and L. D. Mueller, 1988. Evolution of higher feeding rate in *Drosophila* due to density-dependent natural selection. *Evolution* 42:1090–1093.

———. 1993. Directional and stabilizing density-dependent natural selection for pupation height in *Drosophila melanogaster*. *Evolution* 47:176–184.

———. 1996. Density-dependent natural selection in *Drosophila*: Trade-offs between larval food acquisition and utilization. *Evolutionary Ecology* 10:463–474.

King, C. E., and P. S. Dawson. 1972. Population biology and the *Tribolium* model. *Evolutionary Biology* 5:1789–1805.

Lenski, R. E., and B. R. Levin. 1985. Constraints on the coevolution of bacteria and virulent phage: A model, some experiments, and predictions for natural communities. *American Naturalist* 125:585–602.

Luckinbill, L. S. 1978. *r*- and *K*-selection in experimental populations of *Escherichia coli*. *Science* 202:1201–1203.

MacArthur, R. H. 1962. Some generalized theorems of natural selection. *Proceedings of the National Academy of Sciences of the USA* 48:1893–1897.

Marinkovic, D. 1967. Genetic loads affecting fecundity in natural populations of *Drosophila pseudoobscura*. *Genetics* 56:61–71.

Moorcroft, P. R., S. D. Albon, J. M. Pemberton, I. R. Stevenson, and T. H. Clutton-Brock. 1996. Density-dependent selection in a fluctuating ungulate population. *Proceedings of the Royal Society of London B, Biological Sciences* 263:31–38.

Mueller, L. D. 1988a. Density-dependent population growth and natural selection in food limited environments: The *Drosophila* model. *American Naturalist* 132:786–809.

———. 1988b. Evolution of competitive ability in *Drosophila* due to density-dependent natural selection. *Proceedings of the National Academy of Sciences of the USA* 85:4383–4386.

———. 1990. Density-dependent natural selection does not increase efficiency. *Evolutionary Ecology* 4:290–297.

Mueller, L. D., and F. J. Ayala. 1981a. Dynamics of single species population growth: Experimental and statistical analysis. *Theoretical Population Biology* 20:101–117.

———. 1981b. Fitness and density dependent population growth in *Drosophila melanogaster*. *Genetics* 97:667–677.

———. 1981c. Trade-off between *r*-selection and *K*-selection in *Drosophila* populations. *Proceedings of the National Academy of Sciences of the USA* 78:1303–1305.

Mueller, L. D., M. D. Drapeau, C. S. Adams, C. W. Hammerle, K. M. Doyal, A. J. Jazayeri, T. Ly, S. A. Beguwala, A. R. Mamidi, and M. R. Rose. 2003. Statistical tests of demographic heterogeneity theories. *Experimental Gerontology* 38:373–386.

Mueller, L. D., D. G. Folk, N. Nguyen, P. Nguyen, P. Lam, M. R. Rose, and T. Bradley. 2005. Evolution of larval foraging behaviour in *Drosophila* and its effects on growth and metabolic rate. *Physiological Entomology* 30:262–269.

Mueller, L. D., P. Z. Guo, and F. J. Ayala. 1991. Density-dependent natural selection and trade-offs in life history traits. *Science* 253:433–435.

Mueller, L. D., and P. T. Huynh. 1994. Ecological determinants of stability in model populations. *Ecology* 75:430–437.

Mueller, L. D., and A. Joshi. 2000. *Stability in Model Populations*. Princeton, NJ: Princeton University Press.

Mueller, L. D., A. Joshi, and D. J. Borash. 2000. Does population stability evolve? *Ecology* 81:1273–1285.

Mueller, L. D., and V. F. Sweet. 1986. Density-dependent natural selection in *Drosophila*: Evolution of pupation height. *Evolution* 40:1354–1356.

Nicholson, A. J. 1954a. Compensatory reactions of populations to stresses, and their evolutionary significance. *Australian Journal of Zoology* 2:1–8.

———. 1954b. An outline of the dynamics of animal populations. *Australian Journal of Zoology* 2:9–65.

———. 1957. The self adjustment of populations to change. *Cold Spring Harbor Symposium on Quantitative Biology* 22:153–173.

Norton, H. T. J. 1928. Natural selection and Mendelian variation. *Proceedings of the London Mathematical Society* 28:1–45.

Nunney L. 1983. Sex differences in larval competition in *Drosophila melanogaster*: The testing of a competition model and its relevance to frequency dependent selection. *American Naturalist* 121:67–93.

Pearl, R. 1927. The growth of populations. *Quarterly Review of Biology* 2:532–548.

Pearl, R., J. R. Miner, and S. L. Parker. 1927. Experimental studies of life. XI. Density of population and life duration in *Drosophila*. *American Naturalist* 61:289–317.

Pischedda, A., and A. K. Chippindale. 2006. Intralocus sexual conflict diminishes the benefits of sexual selection. *PLoS Biology* 4:e356.

Polivanov, S. 1964. Selection in experimental populations of *Drosophila melanogaster* with different genetic backgrounds. *Genetics* 50:81–100.

Prasad, N. G., M. Shakarad, D. Anitha, M. Rajamani, and A. Joshi. 2001. Correlated responses to selection for faster development and early reproduction in *Drosophila*: The evolution of larval traits. *Evolution* 55:1363–1372.

Prout, T. 1965. The estimation of fitness from genotypic frequencies. *Evolution* 19:546–551.

———. 1971a. The relation between fitness components and population prediction in *Drosophila*. I. The estimation of fitness components. *Genetics* 68:127–149.

———. 1971b. The relation between fitness components and population prediction in *Drosophila*. II. Population prediction. *Genetics* 68:151–167.

———. 1980. Some relationships between density-independent selection and density-dependent population growth. *Evolutionary Biology* 13:1–68.

Roughgarden, J. 1971. Density-dependent natural selection. *Ecology* 52:453–468.

Saccheri, I. J., and I. Hanski. 2006. Natural selection and population dynamics. *Trends in Ecology & Evolution* 21:341–347.

Sang, J. H. 1949. Population growth in *Drosophila* cultures. *Biological Review* 25:188–219.

Seager, R. D., F. J. Ayala, and R. W. Marks. 1982. Chromosome interactions in *Drosophila melanogaster*. II. Total fitness. *Genetics* 102:485–502.

Stokes, T. K., W. S. C. Gurney, R. M. Nisbet, and S. P. Blythe. 1988. Parameter evolution in a laboratory insect population. *Theoretical Population Biology* 34:248–265.

Sved, J. A. 1971. An estimate of heterosis in *Drosophila melanogaster*. *Genetical Research* 18:97–105.

Sved, J. A., and F. J. Ayala. 1970. A population cage test for heterosis in *Drosophila pseudoobscura*. *Genetics* 66:97–113.

Tracey, M. L., and F. J. Ayala. 1974. Genetic load in natural populations: Is it compatible with the hypothesis that many polymorphisms are maintained by natural selection? *Genetics* 77:569–589.

Vasi F., M. Travisano, and R. E. Lenski. 1994. Long term experimental evolution in *Escherichia coli*. II. Changes in life-history traits during adaptation to a seasonal environment. *American Naturalist* 144:432–456.

# 10

# LABORATORY SELECTION STUDIES OF LIFE-HISTORY PHYSIOLOGY IN INSECTS

Anthony J. Zera and Lawrence G. Harshman

The physiological basis of life-history traits and trade-offs has been a long-standing issue in the study of life-history evolution (Fisher 1930; Tinkle and Hadley 1975; Townsend and Calow 1981; Zera and Harshman 2001; Harshman and Zera 2007). Research on this topic has attempted to identify components of physiology (function at various levels of biological organization: Swallow et al. this volume) that contribute significantly to variation in individual life-history traits, such as early-age fecundity, and trade-offs between traits, such as the commonly observed negative correlation between early-age fecundity and longevity. The main goal of this research has been to illuminate the mechanisms of life-history evolution by identifying, for example, functional causes of trait interactions that explain why life-history traits often evolve in concert (e.g., trade-off).

Early studies of life-history physiology mainly focused on energetic correlates of life-history traits, such as reproductive effort in lizards and birds at the level of variation among species or populations (Tinkle and Hadley 1975; Drent and Daan 1980). Starting in the 1980s, the introduction of quantitative genetics and laboratory selection revolutionized the study of life-history evolution by shifting the focus to within-species genetic variation, genetic covariation, and microevolution (Rose et al. 1996; Zera and Harshman 2001). Because of their short generation time and amenability to genetic analysis, insects have figured prominently in these studies. Physiological analysis of laboratory-selected lines now comprises a major focus of research in life-history evolution and evolutionary physiology (Service et al. 1985; Service 1987; Garland and Carter 1994; Rose and Bradley 1998; Gibbs 1999; Feder et al. 2000; Zera and Harshman 2001; Harshman and Zera 2007; Zera et al. 2007a; Gibbs and Gefen this volume; Huey and Rosenzweig this volume; Swallow et al. this volume).

Because of the central role played by resource allocation in hypotheses concerning the physiological causes of life-history variation and trade-offs (figure 10.1), a primary focus of early functional characterizations of laboratory-selected lines was whole-organism energetics such as metabolic rate and energy reserves (Service 1987; Rose and Bradley 1998). However, the scope of physiological analysis has dramatically expanded in the past decade with studies investigating additional topics such as endocrine regulation, parasitism and infection, and resistance to oxidative damage (all discussed in this chapter). Furthermore, physiological studies of laboratory-selected lines have expanded beyond analysis of whole-organism physiology to include investigations of resource allocation and other topics at the level of flux through pathways of intermediary metabolism, enzyme activities, and gene expression. These physiological-genetic studies have contributed significantly to our understanding of the mechanisms underlying the generation of life-history variation, the functional causes of life-history trade-offs, and the evolution of life histories.

We summarize the current status of major topics of research on the physiological causes of life-history traits and trade-offs conducted during the past twenty-five years on

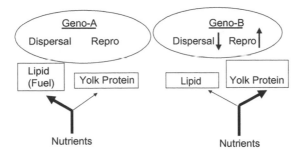

FIGURE 10.1

Illustration of the classic life-history allocation trade-off model. According to this model, allocation of limited internal resources to somatic functions (aspects of dispersal such as flight fuel in this example) occurs at the expense of resources allocated to reproduction. A genotype (genotype B) that exhibits increased allocation of resources to one organismal function (reproduction in this case) must do so by decreasing allocation to other organismal functions or body compartments (aspects of dispersal in this case).

laboratory-selected lines of insects. Most detailed studies have been undertaken using two experimental models: the fruit fly *Drosophila melanogaster* and crickets of the genus *Gryllus*. Hallmarks of the *Drosophila* work are the wide diversity of selection experiments and physiological traits investigated. The hallmark of the *Gryllus* work is the unusual depth of physiological analysis of two integrated laboratory selection studies. The third part of this chapter focuses on several notable physiological studies of laboratory selection on life-history traits in the Lepidoptera.

## LABORATORY SELECTION: STRENGTHS, LIMITATIONS, CAVEATS

Detailed descriptions of the various types of laboratory selection can be found in other chapters of this volume (Gibbs and Gefen this volume; Huey and Rosenzweig this volume) and earlier publications (e.g., Gibbs 1999; Harshman and Hoffmann 2000; Garland 2003; Swallow and Garland 2005). Briefly, there are two main types of laboratory selection: artificial selection and laboratory natural selection. In artificial selection, a specific phenotype is measured on each individual in a population sample, and a subset of those individuals, usually the upper or lower end of the phenotypic distribution, is chosen as breeders for the next generation. Prominent examples include selection on the activity of the endocrine regulator juvenile hormone esterase in *Gryllus* (Zera and Zhang 1995), selection on feeding rate in *Drosophila* (Foley and Luckinbill 2001), and selection on life history and morphology in *Bicyclus* (Zilstra et al. 2004). In laboratory natural selection, including "culling" (see Garland 2003), a specific phenotype is not measured on individuals each generation, nor are specific breeders selected by the investigator based on a specific, measured phenotype. Rather, freely breeding populations are raised under different environments, and the genetic contribution to the next generation is determined by natural selection as the populations adapt to the different environments. Examples include selection studies of life history in *Drosophila* reviewed in Rose et al. (1996; see also Mueller this volume; Rauser et al. this volume; Simões et al. this volume). Artificial selection is best viewed as a genetic manipulation that allows the estimation of specific quantitative-genetic parameters such as heritability, direct and correlated responses to selection, and genetic correlation, in addition to producing lines selected for a specific phenotype that can be subjected to subsequent genetic and functional analyses. Laboratory natural selection more closely approximates the action of natural selection in the field.

One strength of laboratory selection is the use of replicated lines raised under controlled laboratory conditions, which provides important information on the dynamics of selection under specific, known environments (Rose et al. 1996; Harshman and Hoffmann 2000; Garland 2003). Such information is much more difficult to obtain under field conditions where environmental variables are much less controlled. Another great strength of laboratory selection is the production of lines that differ to a much greater degree in individual or suites of life-history phenotypes than do individuals of an unselected population. This magnification of phenotypic differences increases the resolving power of functional analyses investigating the physiological causes of life-history trait variation and covariation. The existence of multiple lines allows the consistency of the response to selection to be ascertained and the discovery of "multiple solutions" (Garland 2003). Moreover, laboratory selection provides material for investigations of the physiological-genetic correlates of life-history traits and trade-offs, which is especially relevant for understanding the functional basis of life-history microevolution (Zera and Harshman 2001; Zera et al. 2007a).

Physiological investigations of life-history traits have typically consisted of a "top-down" approach in which the physiological correlates of selection on life-history traits have been studied (Rose and Bradley 1998; Zhao and Zera 2002; Zera and Zhao 2003; Zilstra et al. 2004). A less common but equally important approach has been a "bottom-up" approach in which aspects of physiology have been directly selected and effects on life-history or other whole-organism traits ascertained by identifying correlated responses to selection (Zera and Zhang 1995; Zera 2006; Foley and Luckinbill 2001; see also Dykhuizen and Dean this volume). A particularly powerful approach has been the use of multiple, complementary selection studies. In some cases, life-history traits and physiological traits thought to regulate these life histories have been directly selected in independent experiments (e.g., wing polymorphism and juvenile hormone esterase activity in *Gryllus*; reviewed in Zera 2006). In other cases, artificial selection has been applied independently to each of several correlated traits to determine the extent to which trait correlations can be broken and the physiological reasons why the correlations can or cannot be broken (Zilstra et al. 2004; Frankino et al. this volume). Each of these examples will be discussed in detail here.

There are also a number of important limitations and caveats to the use of laboratory selection, which have been reviewed elsewhere (Gibbs 1999; Harshman and Hoffman 2000; Gibbs and Gefen this volume; Huey and Rosenzweig this volume). Some of these caveats are relevant to assessing the physiological basis of life-history traits studied in selected lines.

One issue is that laboratory environments might inadvertently predispose laboratory selection to a particular outcome or give rise to various unanticipated artifacts. For example, water and food are typically continuously abundant in laboratory selection studies, and it is not clear that this is the case in the field. This might explain why laboratory selection for stress resistance often results in lines that have evolved adaptations for increased nutrient accumulation rather than energy conservation (Harshman and Hoffmann 2000), in contrast to adaptations that have evolved in the field (Hoffmann and Parsons 1991). Physiological adaptations produced under laboratory selection are sometimes different from those seen in the field. A classic example is resistance to desiccation in *Drosophila* in the laboratory that has evolved by increasing bulk water, in contrast to adaptations in the field that involve other mechanisms such as reduced rates of water loss (Gibbs 1999; Harshman and Hoffmann 2000; Gibbs and Gefen this volume).

The response to selection can be strongly influenced by the particular genetic variants that are present in the base population from which selected lines are generated. For example, genetic differences among base populations used in laboratory selection appear to be a major factor responsible for differences in the response to selection observed in different laboratories (see tables 1 and 2 of Harshman and Hoffmann 2000).

Finally, selected lines typically differ by many genes; these differences make identification of the response to selection difficult unless complementary methods of analysis are used (discussed later; see also Rhodes and Kawecki this volume). In short, although laboratory selection is a powerful approach in experimental evolution, it is not a panacea; it is but one of several important experimental approaches in evolutionary biology that has its own particular set of strengths and weakness. Thus, where possible, laboratory selection should be augmented by complementary approaches such as field studies, phylogenetic analyses (e.g., see Garland et al. 2005, de Magalhães et al. 2007), and the multitude of other experimental manipulations available (e.g., hormone manipulation, pharmacological agents to inhibit specific enzymes, transgenics, gene knockouts using RNAi, microarrays).

## MODEL SPECIES AND RESEARCH TOPICS

### DROSOPHILA

*Drosophila melanogaster* has been used extensively for physiological studies of laboratory-selected lines, most notably for physiological investigations of extended longevity and environmental stress resistance (Gibbs and Gefen this volume; Huey and Rosenzweig this volume; Mueller this volume; Rauser et al. this volume). Part of the reason it has been so commonly used is that it is an outbreeding species with a short generation time that is straightforward to culture and manipulate in the laboratory. The fly is known for the genetic tools available for researchers including a vast number of mutations and extensive genomic resources.

*SELECTION FOR EXTENDED LONGEVITY*

Increased life span in selected lines relative to control lines has repeatedly been produced in laboratory evolution experiments, but the underlying mechanisms of longevity extension are not well understood. At least six *D. melanogaster* selection experiments have generated lines of flies with extended life span. The most common type of experiment has been to impose a late age of reproduction. For example, Rose (1984) used females that were four days old as the source of progeny for the next generation in control lines. In the selected lines, the age of breeding for the first generation of selection was at twenty-eight days of adult life, eventually increasing to seventy days. The survival of flies from these lines selected for extended longevity and control lines is shown in figure 18.2 of Rauser et al. (this volume). A contemporaneous experiment was conducted using late age of reproduction to extend life span (Luckinbill et al. 1984). In the present chapter, the Rose (1984) lines will be referred to as the Rose lines, and the Luckinbill et al. (1984) lines will be referred to as the Luckinbill lines.

Additional experiments have been conducted to extend life span by selection on age of reproduction (Partridge and Fowler 1992; Partridge et al. 1999; Bubliy and Loeschcke 2005). Extended longevity was directly selected in one experiment using families with a high average fecundity as breeders for each generation of selection (Zwann et al. 1995). Relatively long life span was also generated in a laboratory natural selection experiment by varying the level of extrinsic adult mortality (Gasser et al. 2000). Lines that experience low levels of mortality evolved to become long-lived.

Reproduction

Decreased early-age female reproduction has almost always been observed in lines selected for increased life span compared to unselected lines. Reduced early-age fecundity is found in *Drosophila melanogaster* studies (Rose 1984; Luckinbill et al. 1984; Zwann et al. 1995; Partridge et al. 1999; Gasser et al. 2000) and in laboratory selection that extended the life span of a Tephitid fly *(Bactrocera cucurbitae)* (Miyatake 1997). The importance of reproduction for extended longevity in lines selected for extended life span has been experimentally demonstrated; Sgrò and Partridge (1999) blocked egg production in selected and unselected line females using irradiation in some experiments and a mutation that blocks oocyte maturation in other experiments. By using two methods to ablate egg production, the investigators circumvented the issue of potential idiosyncratic results associated with any one perturbation. When selected and control line females

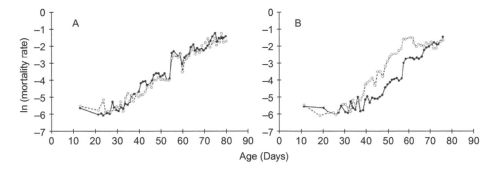

FIGURE 10.2

There were no differences in survival of sterile female *D. melanogaster* from lines selected for extended longevity (panel A, open circles) and sterile control-line females (panel A, solid circles). Differences in survival of fertile females from lines selected for extended longevity (panel B, open circles) and fertile females from control lines (panel B, solid circles). Figures from Sgrò and Partridge (1999).

were made sterile, the female survival differences between selected and control lines disappeared (figure 10.2). The differences in acceleration of mortality in midlife in the selected versus control lines were gone. In addition, the deceleration of mortality at late ages was not observed in the sterile females, indicating that removal of the negative effects of reproduction causes the plateau in survival at older ages (but see also the contrasting results reviewed by Rauser et al. this volume). This work has general implications for understanding the demography of aging in other species, including human beings.

In spite of the importance of reduced early-age reproduction in life span selection experiments, and the existence of other differences in reproduction between the selected and control lines (e.g., Service 1993), few studies have been directed toward understanding the underlying mechanisms in insects. As an exception, studies have investigated aspects of the female reproductive system—namely, ovarioles and oocytes. *Ovarioles* are the subdivisions of the ovary down which oocytes mature. Given the age-specific differences in egg production, for example, very high egg production in the control lines early in life and low in the lines selected for extended longevity, one might expect a difference in the number of ovarioles if they are correlated with fecundity. In the Rose lines, selected-line females have relatively small ovaries when females are young (Rose et al. 1984; see also Chippindale et al. 1997). In a study of ovariole number during the first four days of life, Carlson et al. (1998) observed that the selected females have more ovarioles. In a study conducted at various ages across the life span of longevity-selected and control females, Carlson and Harshman (1999b) found that that early in adult life, selected-line females have more ovarioles than unselected females, suggesting that ovariole number is associated with lifetime egg production. Using the set of lines selected for extended longevity by low extrinsic mortality, Gasser et al. (2000) also observed that females from long-lived strains had more ovarioles, but this difference disappeared when weight was used as a covariate in the analysis. In general, the number of ovarioles in lines selected for extended longevity was correlated with lifetime egg production, which is considerably higher in the selected lines. An intriguing result of the lifetime study of oocyte maturation in Carlson and Harshman (1999b) was the persistence of the previtellogenic gradient of oocyte maturation. In *D. melanogaster*, there are fourteen stages of oocyte maturation defined by changes in morphology and size. Oocytes are produced by stem cells near the tip of the ovariole; the first seven stages are previtellogenic, and during this phase, no yolk protein has been taken up. Typically, each ovariole has all seven stages in a gradient of maturation. In aging females, the gradient of maturation begins to recede, and the terminal stage might be stage 6 or 5 or an even earlier stage. Selected-line females undergo this type of aging considerably more slowly than the unselected females (Carlson and Harshman 1999b; figure 10.3). The egg-cell progenitor stem cells in these females may be more numerous, more persistent, or more active in terms of generating eggs. A compelling area of future research is to study stem cells in these lines as they might provide important insight into stem cell biology and aging that could have implications for basic science and applied human health implications.

Another study of reproduction investigated yolk protein mRNA abundance in the Rose selected and unselected lines. The motivation for this study was the disposable soma theory of aging (Kirkwood 1977; Kirkwood and Rose 1991; see also Rauser et al. this volume). This theory argues that the germ line is immortal in the sense that it is passed on from generation to generation, whereas the soma is ephemeral. The disposable soma theory argues that somatic maintenance has lower priority than reproduction.

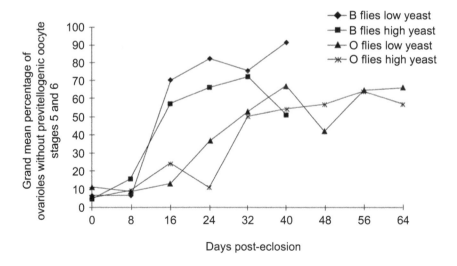

FIGURE 10.3

In female *D. melanogaster* selected for extended longevity, the gradient of oocyte maturation is preserved until much later in life in long-lived females (O flies) than in control line B females (B flies). Figure from Carlson and Harshman (1999a).

Hence, aging occurs because resources that could be used for somatic maintenance are instead used for reproduction. Yolk protein, used to provision the egg for embryogenesis, is the most abundant soluble protein in the body of reproductively active females; presumably it is costly for females to make this protein. Female insects make yolk protein in the fat body, a major insect organ that plays a role analogous to the liver in vertebrates. Higher Diptera, such as *Drosophila*, also make yolk protein in the reproductive tract. Using the Rose lines, Carlson and Harshman (1999a) investigated yolk protein mRNA abundance in fat body versus ovaries as a measure of the amount of yolk protein produced during day 1 and day 4 of adult life. Under the food conditions used to raise larvae and maintain adults in the Rose selection experiment ("low yeast" in Carlson and Harshman 1999a), selected-line, four-day-old females produced relatively less yolk protein in the fat body than in the reproductive tract compared to control females. The presumed reduction of fat body synthesis of a costly protein might reflect reduced relative energy use in the body of selected-line females, perhaps due to increased somatic maintenance underlying extended longevity.

A study was conducted on aspects of the endocrine system of the Rose lines selected for extended life span. Levels of ecdysteroids in the body and levels of the activity of enzymes that metabolize juvenile hormone were measured. The ovary is the source of steroids in adult females, which in insects are known as *ecdysteroids*. Juvenile hormone is secreted by a neurosecretory organ near the brain in adult *Drosophila*. These secreted hormones are thought of as "global organizing hormones" because they play broad roles in controlling development, reproduction, and other adult body functions. Juvenile hormone levels in insects are controlled, in part, by hormone catabolism, and there are two major classes of hormone-degrading enzymes: juvenile hormone esterases (JHEs) and juvenile hormone epoxide hydrolases (JHEHs). The activities of juvenile hormone catabolic enzymes and ecdysteroid titers were investigated in the Rose set of lines (Harshman 1999). There was no difference in JHE or JHEH activity between selected and control lines on either day 1 or day 4 posteclosion. With respect to ecdysteroid titers, there were relatively low levels in selected females on the first day of adult life, but not by day 4. Slowed maturation of the ovary in long-lived females (Carlson et al. 1998; Carlson and Harshman 1999b) might be the reason for the lower ecdysteroid level on the first day of adult female life. Harshman (1999) is the only study of endocrine regulators in a *Drosophila* selection experiment. Given the potential for juvenile hormones, ecdysteroids, insulin, and other hormones to mediate life-history evolution (Finch and Rose 1995; Zera 2006; Zera et al. 2007a), studies on the endocrine system in lines selected for altered longevity are surprisingly scarce.

Another area of limited data is the physiology of reproduction in males in selected and control lines. Males live longer in selected lines, but it is not clear if reproductive system products (sperm, accessory gland proteins) are diminished in selected lines compared to control lines as fecundity is in females. Accessory gland proteins are passed from males to females at the time of mating, and they have many physiological effects on females (Kubli 1996; Wolfner 2002). Several of the male accessory gland proteins cause a reduction in female longevity (Wigby and Chapman 2005; Mueller et al. 2007). It would be interesting to know if the negative effect of male accessory gland proteins on female survival is moderated, or absent, in lines selected for extended longevity, or if females have evolved resistance to these proteins.

### Energy Storage Compounds

Lipid levels were measured in four sets of lines selected for extended longevity and glycogen was measured in one set (Rose et al. 1984; Luckinbill et al. 1984; Zwann et al. 1995; Gasser et al. 2000). Lipid and glycogen were elevated in the Rose lines (Service 1987). Later work confirmed that flies from the long-lived lines have more total lipid than the control lines (Djawdan et al. 1996; Simmons and Bradley 1997). The selected lines have greater lifetime egg production, but the control females reproduce many more eggs early in life under poor conditions (Leroi et al. 1994) and thus might deplete energy stores needed for long-term somatic maintenance. When females from the selected lines were allowed to feed on different levels of live yeast, their fecundity increased appreciably more than control females (Simmons and Bradley 1997). Metabolic studies and energetic analyses indicated that little of the stores of energy compounds in females was allocated to egg production; the females obtained the requisite nutrients for egg production from the consumption of live yeast. Food resources are not limiting for *Drosophila* in typical laboratory environments, and flies apparently can consume as much food as needed for a given level of reproduction (Rose and Bradley 1998). In only the Rose line selection experiment was lipid found to be elevated, and thus it is not clear that lipid abundance is a generally important contributor to extended longevity in *Drosophila* selection experiments. However, in lines selected in the same laboratory for starvation resistance, extended longevity was an indirect response to selection (Rose et al. 1992), and the starvation resistance lines had high lipid levels (Djawdan et al. 1998).

### Respiration Rate

The rate of living theory of aging is the idea that each individual has a limited amount of some feature—for example, heartbeats or breaths—and that death is the result of using up this allotment (for a phylogenetic comparative study, see de Magalhães et al. 2007). A version of this theory is based on the recognition that the process of metabolism generates oxygen radicals, and on the generally accepted idea that aging is caused by oxidative damage. Thus, one might expect metabolism to be reduced in individuals from selected lines compared to control lines. A number of studies have investigated respiration in adult flies using the Rose lines selected for extended longevity and control lines. Respiration was found to be lower in the selected lines than control lines at young ages, but not in older flies (Service 1987). These experiments were conducted in vials that were

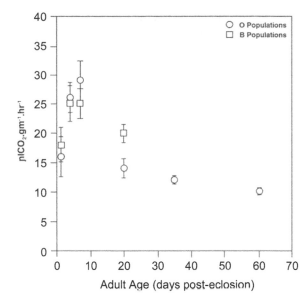

FIGURE 10.4
Female *D. melanogaster* selected
for extended longevity (O populations) exhibited little difference in
respiration rate compared to control line females (B populations).
Figure from Djawdan et al. (1996).

much smaller than the containers used in the process of selection; in subsequent studies, respiration was measured in containers of the same size as those used for selection, and no differences in respiration were found between selected and control lines (Djawdan et al. 1996). When flies were provided with a yeast supplement, control females exhibited a somewhat higher rate of respiration than selected females (Simmons and Bradley 1997). Djawdan et al. (1997) found little difference in respiration rate among selected and control lines when their mass was adjusted by removing water, lipid, and carbohy-drates (figure 10.4). In this set of selected lines, there is little evidence that a decrease in respiration contributes to relatively long life span.

Respiration rate has also been studied in other lines selected for extended longevity. A study on the Luckinbill lines found no difference in respiration rate between selected and control lines (Arking et al. 1988). Given the longer life span of the selected flies, it was noted that the total respiration potential was expanded by response to selection (Arking and Dudas 1989), which is opposite to the prediction of the rate of living theory. Investi-gations of longevity and respiration were conducted on recombinant inbred lines derived from a Luckinbill line selected for extended life span. Khazaeli et al. (2005) found that midlife (adult days 16 and 29) respiration and longevity mapped to the same quantitative trait loci (QTLs) regions on the X chromosome and the two major auto-somes. At these ages, elevated respiration rate was positively correlated with life span. Early-age (day 5) and late-life (day 47) metabolic rates were not correlated with life span. In a second QTL mapping experiment using the Luckinbill lines, no correlation was found between metabolic rate and life span (Van Voorhies et al. 2004). In both QTL studies, there was no evidence that supports the rate of living theory. Selection for increased life span, due to lowering the intrinsic mortality rate, resulted in decreased

(7 percent) mass specific metabolic rate in young flies, but not in late life (Gasser et al. 2000). In general, there is little support for the rate of living theory from *Drosophila* selection experiments and only modest support for the notion that the selected flies have evolved longer life by reducing metabolic rate.

### Molecular Function

Increased survival under oxidative stress has been documented in the Luckinbill selected lines (Dudas and Arking 1995) and the Rose lines (Harshman and Haberer 2000). Reciprocally, selection for oxidative stress resistance results in extended longevity (Vettraino et al. 2001). Given the connection between oxidative stress resistance and longevity, there was strong motivation to investigate the molecular basis of the relationship between these traits in lines selected for extended longevity.

The Luckinbill extended longevity lines provide consistent evidence for a longevity-promoting role for antioxidant enzymes. mRNA levels corresponding to antioxidant enzymes (superoxide dismutase [SOD], catalase [CAT], and xanthine dehydrogenase) were elevated in the Luckinbill lines, as were SOD and CAT enzyme activities (Dudas and Arking 1995; Force et al. 1995). This was some of the earliest experimental evidence that supported the free radical theory of aging (Harman 1956), which hypothesizes that aging occurs as a result of damage to macromolecules. There was elevated levels of CuZnSOD throughout life in a line selected for extended longevity (figure 10.5), and there was reduced damage to lipids (lipid peroxides) and apparently to proteins (protein carbonyls) as a function of age in a Luckinbill selected line (Arking et al. 2000). A reversal of selection

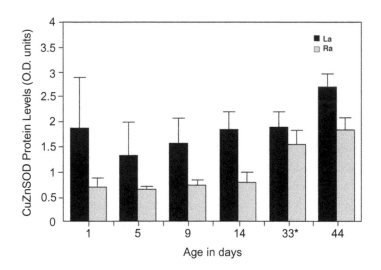

FIGURE 10.5
*Drosophila melanogaster* from a line selected for extended life span (La) differed in superoxide dismutase protein abundance (CuZnSOD) at midlife relative to control line flies (Ra). Figure from Arking et al. (2000).

resulted in a decrease in antioxidant enzyme activities to unselected line levels, but there was no decrease in the activity of eleven general metabolic enzymes (Arking et al. 2000). In the Rose lines, selection for extended longevity increased the frequency of the SOD allele with the highest activity, but there were no differences in longevity among SOD genotypes (Tyler et al. 1993). In addition to antioxidant enzymes, activities of a variety of enzymes of intermediary metabolism were elevated in the Luckinbill lines (Luckinbill et al. 1999; Buck and Arking 2002). One possible consequence of elevated activities of enzymes of intermediary metabolism is increased reducing potential in the selected lines, which could play a role in providing NADPH for antioxidant enzyme activity (Arking et al. 2002).

Mitochondria are the major source of reactive oxygen species (ROS) that cause oxidative damage, and mitochondrial function differs between long-lived and unselected Luckinbill lines. There was less age-specific hydrogen peroxide production in a long-lived line (Ross 2000). It is interesting that the selected lines have up-regulated antioxidant enzyme activities even though they might produce relatively low levels of ROS. There is additional data indicating a role for mitochondria in extended life span (Soh et al. 2007). Specifically, crosses were conducted to make hybrid strains between a long-lived strain and its progenitor line. One hybrid strain consisted of mitochondria from the long-lived line and isogenic (identical) nuclear chromosomes from the unselected line; the other hybrid strain consisted of the reciprocal combination of chromosomes and mitochondria. Four lines (long-lived, unselected, and two hybrid strains) were tested on dilute diets that can confer extended longevity by dietary restriction. Mitochondria from the long-lived strain enhanced the longevity of the unselected progenitor strain, indicating that chromosomes and mitochondria both contribute to longevity.

There is evidence that heat shock proteins are affected by selection for extended life span. In the Rose lines, *hsp23* and especially *hsp22* mRNA levels were elevated across a range of adult ages in all of the lines selected for extended life span compared to unselected lines (Kurapati et al. 2001). Also in the Rose lines, hsp26 expression is elevated, and an allele with a P element that disrupts transcription is more abundant in control lines (Chen et al. 2007). However, these lines did not differ in resistance to heat stress (Service et al. 1985). A different selection experiment resulted in a decline in heat-induced production of heat shock protein 70 (Hsp70) in the lines selected for increased longevity (Norry and Loeschcke 2003). There could be a negative effect of elevated Hsp70 on some fitness component other than survival; for example, diminished egg hatch can be a deleterious consequence of elevated Hsp70 expression (Silbermann and Tatar 2000).

## Learning

An intriguing area of emerging research is the relationship between learning (see also Rhodes and Kawecki this volume) and life span. When lines of D. *melanogaster* are selected for improved learning ability, their life span decreases by 15 percent (Burger et al. 2008).

Reciprocally, when the Luckinbill lines were tested, flies from the long-lived lines were found to have a 40 percent reduction in learning early in life (Burger et al. 2008). Longevity and intelligence are often regarded as highly desirable traits; the potentially negative consequences of these traits on each other are not often considered. This work is a reminder that there is no free lunch in the world of life-history evolution. The underlying mechanism is not known. Perhaps learning requires a considerable amount of energy, which is then not available for somatic maintenance. Another possibility is that forming neuronal connections has a deleterious effect.

SELECTION FOR RESISTANCE TO ENVIRONMENTAL STRESS

Laboratory selection experiments have been conducted for resistance to environmental stress, and life-history traits were often affected. For example, Chippindale et al. (1996) found that flies from lines selected for starvation resistance were highly resistant to starvation at the time of adult eclosion. This was due to increased lipid accumulation during the juvenile stage, which, in turn, was correlated with an extension of development time. Many of these experiments are summarized in the chapter by Gibbs and Gefen (this volume; see also Huey and Rosenzweig this volume). This section of the present chapter will focus on the correlated response to selection in terms of enzyme activities or changes in abundance of nonenzymatic proteins.

Several selection experiments on *Drosophila* have been conducted in which temperature was the agent of selection (see also Gibbs and Gefen this volume; Huey and Rosenzweig this volume). In one such experiment, *D. melanogaster* was cultured for more than twenty years at three temperatures—18°, 25°, and 28°C (Cavicchi et al. 1985)—and Hsp70 levels were analyzed in the lines (Bettencourt et al. 1999). Even though Hsp70 protein level and regulation of expression of the gene that encodes it are remarkably evolutionarily conserved, the abundance of the protein did change in the selected lines. Interestingly, the 28°C lines exhibited considerably lower levels of Hsp70 even though this protein is well known to protect cellular function at higher temperatures. One possible reason for this counterintuitive response to temperature adaptation is that development is very fast at this temperature, and anything that slows it down would be maladaptive; Hsp70 has a negative effect on growth and reduces survival to the adult stage (Krebs and Feder 1997). This result indicates an intimate connection between the mechanism underlying the response to selection for temperature adaptation and expression of a life-history trait.

Two experiments have examined the effect of selection for heat knock-down (temporary incapacitation by high temperature) resistance and effects on heat shock proteins. Folk et al. (2006) selected for knock down performance in *D. melanogaster* and investigated the abundance of two heat shock proteins (Hsp70 and Hsc70). The abundance of heat shock proteins differed between lines selected at high or low temperatures compared to control lines. They found a pattern of abundance of the two proteins that suggests control of induction of Hsp70 possibly for the purpose of protecting female reproduction. There is evidence that egg hatch is diminished when heat shock protein level is elevated (Silbermann and Tatar 2000). Pappas et al. (2006) also selected for knock-down resistance and found that Hsp70 level was reduced early in life, whereas knock-down resistance decreased throughout life in both selected and control lines. Taken together, the evidence from laboratory selection experiments indicates that an important protein that protects organisms from thermal extremes (Hsp70) does not necessarily respond to selection as expected based on its role in cellular physiology. One possible reason is that there are negative interactions between this protein and life-history traits.

## SELECTION FOR RESISTANCE TO PARASITOIDS, PARASITES, AND PATHOGENS

Selection experiments are useful for investigation of environmentally relevant trade-offs (Fry 2003). Selection experiments for resistance to parasitoids, parasites, and microbial pathogens have been conducted using a variety of taxa. *Drosophila melanogaster* has been used to investigate selection for resistance to a fungal pathogen *(Beauveria bassiana)* (Kraaijveld and Godfray 2008), as well as for resistance to the parasitoids *Leptopilina boulardi* (Fellowes et al. 1998) and *Asobara tabida* (Kraaijveld and Godfray 1997). A snail *(Biomphalaria glabrata)* was selected for resistance to a schistosome *(Schistosoma mansoni)* (Webster and Woodhouse 1998). A mosquito *(Aedes aegypti)* was selected for resistance to a malaria parasite *(Plasmodium gallinaceum)* (Yan et al. 1997). The Indian meal moth *(Plodia interpunctella)* was selected for resistance to a granulosis virus (Boots and Begon 1993). There was a significant direct response to selection in all but one (Kraaijveld and Godfrey 2008) of these experiments. Genetically correlated responses to selection included the following: increased development time and a reduction in egg viability in the Indian meal moth; decreased development time, smaller body size, a smaller blood meal, and fewer eggs in mosquitoes; fertility reduction in resistant snails; and lower *Drosophila* larval survival under competitive conditions. Although the outcomes are diverse, there is a consistency in that fitness was usually reduced in environments without parasites.

Limited physiological analyses have been conducted on these selected lines. Fellowes et al. (1999) determined that feeding rate was reduced in *D. melanogaster* larvae selected for resistance to *Leptopilina*. Kraaijveld et al. (2001) determined that hematocyte density was higher in selected populations resistant to Asobara. Perhaps the two observations are due to a trade-off, because food resources are needed for increased hematocyte density in larvae for defense, but these larvae eat more slowly, making them poorer competitors (Kraaijveld et al. 2001). Increased hematocyte density is correlated with resistance of different *Drosophila* species to parasitoids, and thus the response to selection for resistance in laboratory populations is apparently similar to the response to selection in natural populations.

Territoriality is a behavioral life-history trait that affects reproductive success (for a general review of selection experiments on behavioral traits, see Rhodes and Kawecki this volume). *Drosophila melanogaster* laboratory selection experiments have allowed investigators to determine the impact of genes that play roles in various functional categories. The initial laboratory selection study on territoriality set the stage for genetic studies on aggression in this species (Hoffmann 1989). Subsequently, two selection experiments were conducted on *D. melanogaster* territoriality, and the selected lines were analyzed by microarrays followed by tests of mutations in candidate genes. A transcriptome analysis found that 1,539 genes differed between one set of selected and control lines (Edward et al. 2006). Twelve genes were identified that had opposite effects in male and females, and that varied in a consistent pattern (e.g., up in the high-aggression line and down in the low-aggression line). Some examples of candidate genes related to aggression included the following categories of biological function: circadian rhythm, learning and memory, courtship behavior, response to stress, and nervous system development. P-element mutations in nineteen candidate genes were tested for their effect on territoriality. Surprisingly, fifteen of these genes were found to have significant effects, and all of them were novel functional categories in the area of research on aggression. Dierick and Greenspan (2006) selected for aggressive behavior; and after twenty-one generations of selection, microarray analysis was performed on the heads of selected and control line males. At a stringent level of statistical analysis, eighty genes showed a consistent difference between selected and control lines. None of these genes was associated with serotonin metabolism, which was unexpected given the strong connection between serotonin and aggression (Dierick and Greenspan 2006). Similarly, there was no difference in total serotonin in the heads of selected and control flies. P-element mutations in candidate genes from the microarray studies were introgressed into a standard laboratory stock and tested for altered territoriality. One of these genes, a cytochrome P450, had a consistent impact on territoriality. Taken together, these studies indicate a new functional approach for analysis of life-history traits that relies on laboratory selection experiments. First, a selection experiment is conducted—in this case, on a trait that this generally important in behavior and life history. Second, a microarray study is conducted to identify candidate genes. Third, mutations in candidate genes are tested for their effect on the trait of interest.

One of the goals of studies on reproductive physiology is to understand the reproductive interactions of males and females, as such interactions could be responsible for a variety of evolutionary process, including sexually antagonistic coevolution and speciation (Fry this volume). A selection experiment has made an important advance on this front. Miller and Pitnick (2002) selected on male sperm length and female seminal receptacle length. Sperm length and seminal receptacle size both responded up and down to selection. Longer sperm were better competitors, and increasing seminal receptacle size drove the evolution of even longer sperm. This was quite likely due to postcopulatory cryptic selection of sperm by females. This form of sexual selection (see also Swallow et al. this volume) could underlie the evolution of widely varied sperm length in the genus *Drosophila* (Miller and Pitnick 2002).

## SELECTION FOR RESISTANCE TO A HORMONE ANALOGUE

Flatt and Kawecki (2007) subjected lines of *D. melanogaster* to laboratory natural selection by including or excluding methoprene, a JH agonist, in the larval medium. Selected lines lived longer than unselected controls, but there was no detectable decrease in early-age fecundity, which would be expected if JH regulates the trade-off between early-age fecundity and life span. In the absence of physiological studies of differences in JH regulatory mechanisms between control and selected lines, it is difficult to assess the significance of these results with respect to the endocrine mechanisms underlying life-history variation and evolution. For example, the response to selection on the JH analogue in the larval diet might have been accomplished by increased activity of detoxification enzymes, leading to increased metabolism of this analogue.

Although not investigated in selected lines, the role of JH in regulating life-history trade-offs has been extensively studied using laboratory-generated mutants of *Drosophila* (reviewed in detail in Harshman and Zera 2007; Zera et al. 2007a). However, the role of JH is currently unclear, complex, and controversial (Harshman and Zera 2007). A major reason for the rudimentary state of knowledge of the role of JH in life-history variation and trade-offs in *Drosophila* is the complexity of JH endocrinology in this species, coupled with its small size. Higher *Diptera* contain a unique JH (JH bis-epoxide), not found in other insect groups, which can be the major JH in the blood, in addition to a variety of other JHs and JH-like molecules (Zera et al. 2007a). There is a paucity of fundamental endocrine information on blood levels, or whole organism levels, of various JHs or JH-like compounds during various life cycle stages. An earlier study of the JH titer in whole-body extracts of *Drosophila* (Bownes and Rembold, 1987) measured JH-III, but this molecule is now known not to be the major contributor to the JH titer in *Drosophila*. The small size of *Drosophila* also is a major impediment to obtaining key endocrine information (e.g., blood hormone levels), required to test hypotheses regarding variation in systemic endocrine regulation and life-history trade-offs. Only very rarely has a hormone titer been measured in the blood of *Drosophila* (Richard et al. 2005).

## SELECTION AFFECTING FEEDING RATE

Foley and Luckinbill (2001) directly selected on feeding rate in laboratory lines of *D. melanogaster*, which in turn modified caloric intake, energy reserves, and various life-history traits as indirect responses. Replicate lines with reduced rate of feeding exhibited reduced lipid content at adult eclosion, lower early-age fecundity, but greater lifetime fecundity and greater longevity. These are the same trade-offs seen when life span was directly selected (see earlier discussion), suggesting that a significant amount of the response to selection observed in classic life-history laboratory selection studies in *D. melanogaster* (e.g., Rose et al. 1996) may have been due to genetic changes in nutrient acquisition (but see Chippindale et al. 1993). Comparison of feeding rates between various laboratory lines obtained in these studies would be very informative in assessing this hypothesis.

*Drosophila melanogaster* has been subjected to selection at high density, which results in a trade-off between food acquisition and food use (Mueller 1990, this volume; Joshi and Mueller 1996). An underlying mechanism is not faster movement of food through the gut.

Nutrient acquisition, the combination of consumption and assimilation, is a key component of energy budgets (Withers 1992; Karasov and Martinez del Rio 2007), and variation in nutrient acquisition can strongly influence the expression of allocation trade-offs (de Jong and van Noordwijk, 1992; Zera and Harshman, 2001). Early classic studies of life-history physiology involved careful analyses of energy budgets, including nutrient input (e.g., Congdon et al. 1982). However, the importance of assessing nutrient input, by either controlling or quantifying consumption, in energetic analyses of life-history laboratory selection studies of *Drosophila* has been surprisingly underappreciated (Rose and Bradley 1998; Zera and Harshman 2001). A number of *Drosophila* studies have emphasized the importance of variation in nutrient input as an important contributor to life-history differences between laboratory-selected lines (Rose and Bradley 1998; Simmons and Bradley 1997). However, because consumption and assimilation have not been quantified or controlled in *Drosophila* laboratory selection studies, assessing the relative contribution of genetic differences in nutrient allocation versus acquisition to life-history differences between laboratory-selected lines is currently problematic (Zera and Harshman 2001; Lee et al. 2008). More rigorous investigations of nutrient consumption and assimilation, such as the stable isotopic analyses of Min et al. (2006) and studies of Lee et al. (2008) in which overall consumption and the intake of total calories, as well as specific nutrients, were carefully monitored, should be models for future investigations of the role of nutrient input and life-history selection in *Drosophila*.

## GRYLLUS

### SELECTION ON THE TRADE-OFF BETWEEN REPRODUCTION AND DISPERSAL CAPABILITY

For over four decades, wing polymorphism has been viewed as a classic example of a life-history trade-off involving the differential allocation of resources to the soma versus ovaries in morphs adapted for flight at the expense of reproduction, and vice versa (Zera 1984, 2004; Roff 1996; Zera and Denno 1997; Zera and Harshman 2001). In *Gryllus*

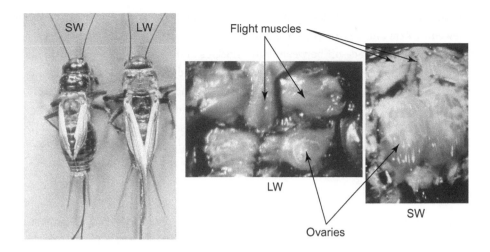

FIGURE 10.6
Characteristics of flight-capable (LW) and flightless (SW) morphs of the cricket *Gryllus firmus*. Left panel: Morph-specific variation in the length of hind wings used for flight (forewings removed). Middle and right panels: Dissections of same-aged morphs (day 5 of adulthood) illustrating large, functional (pink) flight muscles but small ovaries in the LW morph, and underdeveloped, nonfunctional muscles, but large ovaries in the SW morph. Note that "LW" in this figure is equivalent to "LW(f)" (long-winged, functional flight muscles) used in some other figures. Figure is from Zera (2005).

(see also Roff and Fairbairn this volume), the polymorphism basically consists of a flight-capable morph that has long wings and functional flight muscles (LW, also designated LW(f)) and a flightless, morph with short wings and underdeveloped flight muscles (SW) (figure 10.6). Some LW(f) individuals histolyze (degenerate) their flight muscles producing another flightless morph, termed LW(h) (Zera et al. (1996a), which is not dealt with in this chapter. Importantly, the SW morph has substantially elevated early-age reproduction compared with the LW morph and thus trades off early-age fecundity with dispersal capability. The great power of wing polymorphism as an experimental model is that morphs within the same species exhibited differences in life-history traits, and their physiological correlates, that are as great as those between species. This exceptional magnitude of phenotypic variation coupled with the large size of *Gryllus*, extensive background on cricket physiology, and amenability to experimental manipulation make this group a powerful model for dissecting the physiological-genetic underpinnings of life-history variation and trade-offs.

Wing polymorphism in *Gryllus* has been especially well studied with respect to three physiological aspects of life-history variation and trade-offs: (1) modifications of intermediary metabolism that underlie specializations for dispersal versus reproduction in adults; (2) the endocrine regulation of morph development, and the trade-off between reproduction and flight capability in adults; and (3) whole-organism nutrient acquisition (combined effects of consumption and assimilation) and allocation, especially to ovarian and flight muscle growth and maintenance. Here we mainly focus on newer metabolic and endocrine aspects. Whole-organism nutrient acquisition and allocation have been reviewed in Zera and Denno (1997) and in Zera and Harshman (2001). Earlier physiological work on the trade-off between flight muscles and ovaries (Zera et al 1996a) are reviewed in Zera and Harshman (2001) and in Roff and Fairbairn (2007). In addition, recent laboratory evolution studies of flight muscles and ovaries are discussed in Roff and Fairbairn (2007). The kinds of metabolic and endocrine studies undertaken in *Gryllus* have not been intensively studied in *Drosophila* selection experiments, and thus the *Gryllus* work strongly complements the *Drosophila* studies.

## Intermediary Metabolism

Evolutionary modification of intermediary metabolism is expected to be an important factor in life-history evolution (Zera 2005; Zera and Zhao 2006). Key constituents of life histories such as egg yolk protein or the energy reserves required for enhanced somatic maintenance are the products of pathways of metabolism. Thus, evolutionary changes in life histories are expected to be strongly contingent on evolutionary changes in the flow of nutrients through pathways whose end products contribute to reproduction, somatic maintenance, and so forth. In addition, pathways of intermediary metabolism are strongly interconnected; thus, increased diversion of nutrients through one arm of metabolism (e.g., oxidation of fatty acids or amino acids for energy) is expected to influence nutrient flow through other sections of metabolism (e.g., conversion of fatty acids into triglyceride energy reserves, or conversion of amino acids into protein), thus giving rise to trade-offs (figure 10.1). Which pathways, enzymes, and regulatory controls of various metabolic pathways have been modified during the course of life-history evolution? This has not been investigated in detail, except for some studies (discussed earlier) of activities or transcript abundance of enzymes of intermediary metabolism in lines of *D. melanogaster* divergently selected for various life histories (Harshman and Schmidt 1998; Dudas and Arking 1995; Arking et al. 2000). During the last decade, detailed investigations of lipid and amino acid metabolism in artificially selected lines of the wing-dimorphic cricket *Gryllus firmus* have provided in-depth information on alterations of intermediary metabolism that underlie life-history evolution.

Artificial selection on life-history traits in *G. firmus* differs from most other life-history selection experiments (e.g., in *Drosophila*) in that the character selected (long-winged vs. short-winged morph) was a discontinuous (polymorphic) rather than continuous trait. Extensive studies by Roff and colleagues have investigated various genetic aspects of wing polymorphism in *Gryllus* using laboratory selection (Roff 1986, 1994; also see references in Roff and Fairbairn 2007, this volume). In the Zera lab, replicate lines, produced in three separate blocks, or independent selection trails were obtained that were nearly pure-breeding (frequency of the selected morph >90 percent) for either the LW morph or the SW morph (Zera 2005; figure 10.7). Ovarian growth and egg production is 100 to 400 percent higher in females of SW- vs. LW-selected lines during the first one to two weeks of

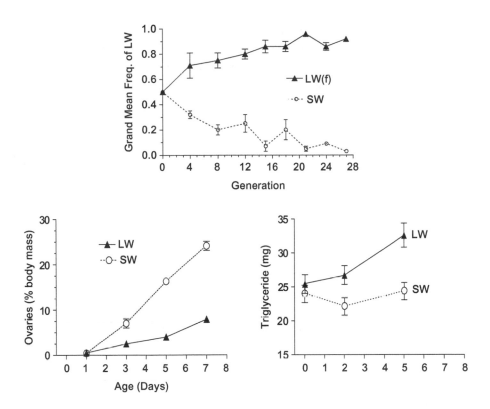

FIGURE 10.7

Upper panel: Response to selection on wing morph in *G. firmus*. Symbols are grand mean morph frequencies (means of 3 replicate selected lines), and bars are standard errors of the three line means. See text for additional information on the selection experiment. Lower panels: Left: Ovarian mass of LW (same as LW(f) in some other figures) and SW-selected females; Right panel: Whole-body triglyceride content (adjusted for whole-body dry mass by ANCOVA) of LW- and SW-selected females. Top panel is from Zera (2005), and bottom panels are from Zera and Larsen (2001).

adulthood. Differences of a similar magnitude in ovarian mass have been reported in other laboratory selection studies of *Gryllus* (see references in Roff and Fairbairn 2007).

Biochemical studies of life-history evolution in artificially selected lines of *G. firmus* have focused primarily on lipid metabolism and to a lesser degree on amino acid metabolism (discussed later). The focus on lipid metabolism was motivated by the importance of lipids in many organismal functions, including flight, reproduction, and somatic maintenance (Beenakkers et al. 1985; Zera 2005). Thus, changes in lipid metabolism are expected to be important components of life-history evolution (Zera 2005).

*Lipids*   Most previous studies of life-history physiology (e.g., in *Drosophila*) considered lipid primarily as a somatic energy reserve (Zera and Harshman 2001). However, lipid, in fact, is a heterogeneous class of molecules with many functions. Triglyceride, the most abundant component of whole-organism lipid (Beenakkers et al. 1985), is an energy reserve, while phospholipid, the second-most abundant lipid class (Beenakkers et al. 1985) plays a structural role, being the major component of cellular membranes. Importantly, triglyceride and phospholipid are both in high concentration in eggs as well as in the soma and thus have reproductive as well as somatic functions. An important aspect of the *Gryllus* studies is that they have investigated individual lipid classes in both reproductive and somatic body compartments.

*Gryllus firmus* females from LW-selected lines have considerably higher (30–40 percent) triglyceride content than females from SW-selected lines (figure 10.7; Zera and Larsen 2001). Importantly, elevated triglyceride content accumulates in the LW-selected lines during the first week of adulthood, precisely when the ovaries of the LW morph are growing much more slowly than those in the SW morph. Thus, lipid accumulation may be an important cost of flight capability that directly impacts fecundity in a negative manner. By contrast, the SW morph has about a 13 percent greater phospholipid content, which is consistent with the greater egg production in SW females. Eggs are especially rich in phospholipid (Beenakkers et al. 1985), which is required for membrane biosynthesis in developing embryos.

Extensive feeding studies have demonstrated that LW and SW morphs of selected lines of *G. firmus* and LW and SW morphs of other *Gryllus* species only differ to a slight degree (<10 percent) in nutrient consumption or assimilation (Zera and Denno 1997; Zera et al. 1998; Zera and Brink 2000; Zera and Harshman 2001; A. J. Zera, unpublished data). Furthermore, morphs do not differentially assimilate various macronutrients, such as carbohydrate, lipid, or protein, from the diet. These feeding studies are important because they demonstrate that differences between LW and SW morphs in various macronutrients (e.g., lipid) must arise primarily from morph differences in metabolic conversions of the same internal resource pool. Such information on nutrient input is not available from studies of *Drosophila* in which energy reserves have been compared between selected lines (see the *Drosophila* section).

*Pathway Flux and Activities of Enzymes of Lipid Metabolism*    Radiotracer studies using lipid precursors ($^{14}$C-palmitic acid, and $^{14}$C-acetate) have demonstrated that flux through the pathway of fatty acid biosynthesis is greater in females of LW- versus SW-selected lines of *G. firmus*, resulting in a greater overall production of lipid (mainly composed of fatty acids) in the LW morph (Zhao and Zera 2002; Zera 2005). In addition, LW- and SW-selected females differ in the types of lipid produced and the organs to which newly biosynthesized lipid classes are allocated.

The main pathway of glyceride biosynthesis bifurcates after the production of fatty acids, and fatty acids can be converted into either triglyceride or phospholipid (figures 10.8 and 10.9). Fatty acids are converted to a much greater degree into triglyceride versus phospholipid in the LW morph, while the opposite occurs in the SW morph. In this respect, it is notable that triglyceride is the main flight fuel in *Gryllus* and phospholipid is an important component in eggs. Even though SW females produce less total fatty

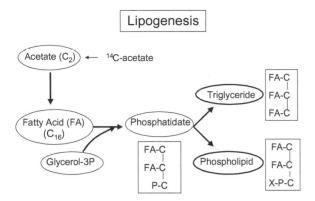

FIGURE 10.8

Simplified diagram of lipid biosynthetic pathways. New fatty acids, such as the sixteen-carbon palmitic acid, are first made from the two-carbon compound, acetate (*de novo* pathway). The pathway of lipid biosynthesis then diverges: three fatty acids can be linked to glycerol-3-phosphate (phosphate group subsequently removed) to produce triglyceride from the intermediate phosphatidic acid, or two fatty acids can be linked to glycerol-3-phosphate to produce phospholipid. "X" indicates that the phosphate group is modified in phospholipid. $^{14}$C-acetate indicates the entry point of injected radiolabeled acetate into the pathway of lipid biosynthesis in studies of Zhao and Zera (2002).

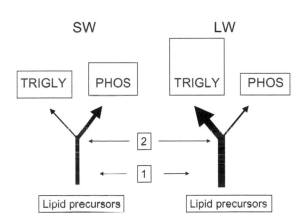

FIGURE 10.9

Schematic representation of two major differences in lipid biosynthesis between LW- and SW-selected lines of the cricket *G. firmus*: (1) LW lines produce more fatty acid (denoted by the thicker base of the "Y"), leading to greater production of total lipid (triglyceride and phospholipid; and (2) LW lines divert a greater amount of fatty acid to the production of triglyceride versus phospholipid (trade-off).

acid (and hence less total lipid) than LW females, they produce more phospholipid, because of the greater proportional diversion of fatty acids through the phospholipid arm of glyceride biosynthesis. This is a robust result that occurs in each of three pairs (blocks) of LW- and SW-selected lines. To our knowledge, this is the first demonstration of a genetic trade-off at the level of flux through a pathway of intermediary metabolism that underlies a genetic trade-off at the level of whole-organism life history (Zhao and Zera 2002; Zera 2005). Radiotracer studies also indicate that newly biosynthesized triglyceride is preferentially allocated to somatic tissues, while newly biosynthesized phospholipid is preferentially allocated to the ovaries (Zhao and Zera 2002). These studies indicate that trade-offs involving lipid are more complex than indicated by previous studies of *Drosophila* that have not distinguished between the types of lipid produced or the body compartments in which the lipids are found. Taking these factors into account also provides a more accurate assessment of energetic costs involved in various life-history adaptations (see the discussion in Zhao and Zera 2002).

Enzymological studies have demonstrated that the elevated production of fatty acids in LW- versus SW-selected lines is due to substantial increases in the specific activities of numerous enzymes of various pathways contributing to lipid biosynthesis (figure 10.10). These include enzymes of the *de novo* pathway of fatty acid biosynthesis (e.g., fatty acid synthase, ATP-citrate lyase), the pentose shunt (e.g., 6-phosphogluconate dehydrogenase, which produces reduced NADPH, a key component in lipid biosynthesis), and other enzymes producing NADPH (e.g., NADP-dependent isocitrate dehydrogenase) (Zera and Zhao 2003; Zera 2005). Thus, evolutionary modulation of lipid metabolism in morphs of *G. firmus* has come about by global changes in numerous enzymes of lipid metabolism. This result is consistent with predictions of metabolic control analysis (see also Dykhuizen and Dean this volume), which proposes that substantial changes in flux through pathways of metabolism most often require changes in multiple enzymes of the pathway (Fell 2003; see discussions in Zera and Zhao 2003; Zera 2005).

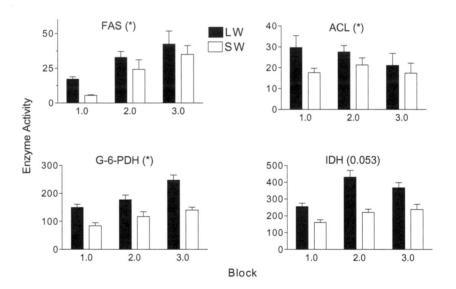

FIGURE 10.10

Activities of four enzymes involved in lipid biosynthesis in LW- and SW-selected lines of *G. firmus*. Note the consistently higher activity of each enzyme in LW-selected lines of each of the three blocks (independent selection trials). Asterisks in parentheses denote results of paired t-tests demonstrating significant genetic differences in enzyme activities (i.e., consistently higher activities in LW versus SW selected lines across the three blocks (2 df in each test). FAS = fatty-acid synthase; ACL = ATP-citrate lyase; G-6-PDH = glucose-6-phosphate dehydrogenase; IDH = NADP-isocitrate dehydrogenase. Figure is from Zera (2005).

Ongoing research on lipid metabolism and life-history trade-offs in *Gryllus* is focusing on the biochemical and molecular causes of increased activities of enzymes involved in lipogenesis. The goal of these studies is to determine the mechanistic link between variation at the molecular and biochemical levels for organ-specific activities of various enzymes. A variety of factors may be responsible for the differences in activities of the various lipogenic enzymes between LW and SW lines (figure 10.10). These include morph-specific differences in gene expression (e.g., transcript abundance), which can lead to differences in enzyme concentration between selected lines. Alternatively, differences in primary amino acid sequence or posttranslational modification of the mature enzyme could produce morph differences in catalytic properties. Using real-time PCR, increased expression of two lipogenic genes (ATP-citrase lyase and NADP-dependent isocitrate dehydrogenase) have been documented in LW- versus SW-selected lines (R. Schilder and A. J. Zera, unpublished data).

*Coordination of Metabolic Traits*   Although hormones regulate many aspects of metabolism, the extent to which change in endocrine regulation is responsible for the alterations in intermediary metabolism that underlie variation in life-history traits or trade-offs remains understudied. Juvenile hormone regulates many aspects of metabolism, reproduction, and development in insects (Gilbert et al. 2005). Methoprene, a juvenile hormone analogue, produced a striking SW phenocopy when applied to LW females (Zera and Zhao 2004). The extent to which the hormonally manipulated phenotype approximated the unmanipulated SW morph with respect to numerous aspects of lipid metabolism (triglyceride level, lipogenic enzyme activities, relative rates of triglyceride and phospholipid biosynthesis, rate of fatty acid oxidation), as well as ovarian growth and size of flight muscles, was remarkable. These results suggest that the strong covariation among activities of many enzymes of intermediary metabolism, aspects of reproduction, and aspects of morphology related to dispersal in *G. firmus* result from pleiotropic effects of genetically variable endocrine regulators (Zera and Zhao 2004). Because hormone manipulation is a gross-level technique (Zera 2007), the specific endocrine mechanisms, or even the identity of the hormone(s) responsible for the results observed in Zera and Zhao (2004), must await more detailed endocrine studies. These results have important implications for the ongoing debate regarding the relative importance of hormonal regulation versus nutrient allocation as physiological drivers of life-history evolution. Rather than being alternate explanations, as proposed by some workers, both are likely to be important causal components that act in concert (see discussion in Harshman and Zera 2007).

*Amino Acid Metabolism and Life-history Trade-offs*   Subsequent studies in *Gryllus* have investigated the role of amino acid metabolism in the trade-off between early age reproduction and flight capability in LW and SW-selected lines (Zera and Zhao 2006). Amino acids can be oxidized for energy, used to produce protein for either somatic or reproductive (e.g., yolk protein) functions, or converted into other macromolecules such as various lipids or carbohydrates for storage. Because of these many important functions, pathways of amino acid metabolism are expected to be important targets of divergent selection on life histories. Except for recent experiments in *Gryllus* (discussed later), variation in amino acid metabolism in the context of life-history microevolution has not been studied in detail. By contrast, amino acid metabolism and life-history adaptation has been a prominent focus of physiological-ecological studies of life-history variation in insects in the field (Boggs 1997; O'Brien et al. 2002).

Using the radiolabeled amino acid glycine, consistent differences were observed in amino acid metabolism between pairs of LW- and SW-selected lines of *G. firmus* (Zera and Zhao 2006). Metabolic trade-offs (differential diversion of amino acid through bifurcating pathways of metabolism), as well as allocation trade-offs (differential distribution of end products—e.g., proteins—to somatic vs. reproductive organs), have been documented. Females of LW-selected lines preferentially oxidized amino acids for energy, or they preferentially converted them into lipid or somatic protein as opposed to using amino acids for biosynthesis of ovarian protein. The opposite situation occurred in SW females.

One of the most interesting findings of biochemical studies of lipid and amino acid metabolism in *Gryllus* is the existence of trade-offs at the level of whole blocks of intermediary metabolism (Zera and Zhao 2006). That is, LW lines selected for reduced early-age reproduction and diversion of nutrients to the soma are characterized by reduced oxidation of fatty acids (spared for enhanced conversion into triglyceride flight fuel) but increased oxidation of amino acids. There may be less demand for conversion of amino acids into protein in LW females because of substantially reduced egg production in that morph, relative to SW females (figures 10.6 and 10.7), thus allowing enhanced oxidation of amino acids for energy production. Alternatively, increased oxidation of amino acids may simply be a mechanism for dissipating an excess nutrient, as described in Warbrick-Smith et al. (2006) for the diamondback moth, *Plutella xylostella*. SW-selected lines of *G. firmus*, on the other hand, are characterized by reduced oxidation of amino acids (spared for enhanced conversion into ovarian proteins) but increased oxidation of fatty acids, possibly to obtain energy to drive the greater biosynthesis of ovarian protein or to remove an unneeded nutrient (Warbrick-Smith et al. 2006). These findings indicate a remarkable remodeling of metabolism in morphs of *G. firmus*, resulting in each morph conserving the nutrient required for its specialized function (e.g., fatty acid and lipid for flight in the LW morph).

Endocrine Regulation

For decades, evolutionary modification of endocrine regulation has been considered an important physiological aspect of life-history evolution (Gould 1977; Matsuda 1987; Nijhout and Wheeler 1982; Ketterson and Nolan 1992; Finch and Rose 1995; Zera and Huang 1999; Zera and Harshman 2001; West-Eberhard 2003; Harshman and Zera 2007). Only recently, however, have detailed evolutionary-genetic studies of endocrine variation been undertaken in the context of life-history microevolution. The most extensive studies have involved artificial laboratory selection of insects. Because laboratory selection of endocrine traits has been reviewed in detail recently (Zera 2006; Zera et al. 2007a), we will only summarize the main points relevant to this volume on experimental evolution, and will mainly discuss aspects not dealt with in detail in earlier reviews.

*Endocrine Regulation of the Development of Life-history Morphs* The evolutionary genetics of endocrine variation and trade-offs in insects has been studied in detail in species of *Gryllus* over the past two decades (Zera and Tiebel 1989; Zera and Huang 1999; Zera 2004, 2006; Zera et al. 2007a). A unique aspect of the *Gryllus* studies is the use of multiple, complementary artificial selection experiments focused on juvenile hormone (JH), juvenile hormone esterase (JHE, an enzyme that degrades and regulates the blood JH titer), and ecdysteroids (a class of hormones structurally related to ecdysone). These hormones and hormonal regulators (JHE) are thought to be involved in regulating the expression of morph-specific aspects of life history and morphology, in wing-polymorphic insects (Zera 2004) and in other ecologically important polymorphisms (e.g., phase, or caste, polymorphism; Nijhout 1994; Hartfelder and Emlen 2005). These hormones also regulate many general aspects of development, physiology, behavior, and reproduction in insects (Nijhout 1994; Gilbert et al. 2005). In this section, we will focus on the endocrine control of morph-specific traits in lines artificially selected for the LW or SW morph of *G. firmus* and its congener, *G. rubens*. In the next section, we will focus on complementary studies in which a regulator of JH, JHE, has been directly selected in a wing monomorphic congener, *Gryllus assimilis*.

The most widely held endocrine-evolutionary model of wing polymorphism postulates that an elevated hemolymph (blood) level of juvenile hormone (JH) and/or a decreased level of ecdysteroids during the latter part of nymphal (juvenile) development specifies the emergence of the short-winged, as opposed to the long-winged, morph (see Zera 2004, 2006; Zera et al. 2007a for extensive discussions of this model). Moreover, in adults, an elevated level of JH or ecdysteroids in SW females is thought to cause the greater ovarian growth and depression of dispersal-related traits (e.g., reduced biosynthesis of triglyceride flight fuel) in that morph. The first detailed test of this hypothesis involved endocrine characterization of lines artificially selected for the LW or SW morphs in each of two closely related species of *Gryllus* (for a more detailed description of the selection experiment in *G. firmus*, see this chapter's earlier discussion and also Zera 2006). This analysis resulted in the first endocrine analysis of a complex (multitrait) genetic polymorphism in insects and thus is particularly relevant to the evolution of complex, adaptive phenotypes. Other endocrine studies of complex polymorphism, such as castes in social insects or phases in locusts, have focused on polyphenism (environmentally induced polymorphism). Relative to LW-selected lines, SW-selected lines exhibited a slightly elevated hemolymph JH titer, a dramatically reduced ecdysteroid titer, and, most prominently, substantially reduced activity of the regulatory enzyme JHE

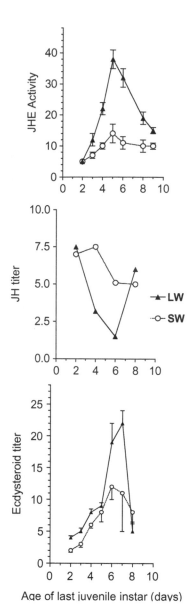

FIGURE 10.11

Endocrine differences between LW- and SW-selected lines of *G. rubens* during the last juvenile instar. Top panel: blood activities of juvenile hormone esterase (JHE; nmol/min/ml); middle panel: blood titer (concentration) of juvenile hormone (JH; pg/ul); lower panel: blood titer of ecdysteroids (nmol, 20-OH ecdysone equivalents). Note the higher concentration of JH, lower activity of JHE (enzyme that degrades JH), and lower titer of ecdysteroids in SW-selected females. See text for additional explanation. Figure is from Zera (2006).

(figure 10.11; for details see Zera and Tiebel 1989; Zera 2004, 2006; Zera et al. 2007a). These results are, in general, consistent with the hypothesis that evolution of the SW phenotype results, at least in part, from evolutionary changes in JH and/or ecdysteroid regulation (Zera 2004, 2006; Zera et al. 2007a). Subsequent studies primarily focused on JHE activity because it was the JH regulator that most dramatically differed between LW- and SW-selected lines (Zera and Tiebel 1989; Fairbairn and Yadlowski 1997; Zera and Huang, 1999; Zera 2006; Zera et al. 2007a), in addition to strongly co-varying with wing morph in crosses between LW and SW lines (Zera and Tiebel 1989).

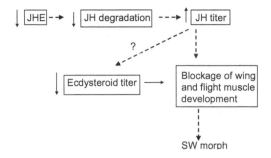

FIGURE 10.12

Standard endocrine model illustrating the factors responsible for endocrine variation in LW- and SW-selected lines of *Gryllus* during the last instar (see figure 10.11) and their effects on morph development. Up and down solid arrows refer to endocrine characteristics in SW relative to LW females. Relative to LW females, reduced JHE activity lowers *in vivo* rate of JH degradation giving rise to a JH titer that is elevated for a longer period of time during the last stadium. This elevated titer is thought to block full development of flight muscles and wings giving rise to a morph with shortened wings and underdeveloped flight muscles. Reduced ecdysteroid titer, which may be due to elevated JH titer (denoted by "?") or to other factors, also may cause reduced size of wings and flight muscles. Note, as discussed in the text and in Zera (2006), more recent experiments suggest that an elevated JH titer may alter development of flight muscles but not wings. Figure is from Zera (2006).

Endocrine characterizations identified causal physiological components of differences in JHE activity between replicate LW- and SW-selected lines of *G. firmus* (figure 10.12).[2] For example, the six- to eightfold reduced hemolymph JHE activity in SW versus LW(f)-selected lines was due to both a decrease in whole-organism JHE activity (50 percent of the indirect response) and decreased amount of JHE activity in the hemolymph (50 percent of the response). Furthermore, *in vivo* radioisotopic studies demonstrated that the reduced JHE activity was associated with reduced *in vivo* degradation of juvenile hormone in SW versus LW lines. This result demonstrates the causal link between modulation of JHE activity and modulation of *in vivo* JH metabolism, a key functional link not often made in evolutionary-genetic studies of enzyme activity variation.

*Endocrine Regulation of the Trade-off between Reproduction and Dispersal in Adult Morphs* A long-standing and widely held idea in the field of life-history evolution is that evolutionary modification of the endocrine control of reproduction is an important physiological cause of the evolution of many adult life-history traits (Gould 1977; Matsuda 1987; Finch and Rose 1995; Flatt et al. 2005; West-Eberhard 2003; Zera et al. 2007a). For example, morphs specialized for increased egg production, such as the flightless morph of wing-polymorphic insects, or queens of social insects, are thought to have evolved by chronic elevation of the titer of reproductive hormones, such as juvenile hormone, the major gonadotropin in insects (Gould 1977; Nijhout and Wheeler 1982; Zera 2004).

The first detailed test of the aforementioned endocrine hypothesis of wing polymorphism demonstrated a much more complex situation: LW- and SW-selected morphs of *G. firmus*, differed in the presence/absence of a circadian rhythm for the hemolymph JH titer, rather than simply differing consistently (noncircadian) in the JH blood titer (Zera and Cisper 2001; Zhao and Zera 2004; Zera 2006; figure 10.13). This discovery underscores the likelihood that endocrine mechanisms regulating life-history trade-offs are much more complex than presently suspected. The circadian rhythm for the JH titer in the LW dispersing morph possibly regulates physiological or behavioral aspects of flight, which occurs only during the night in *Gryllus*. Interestingly, the titer of ecdysteroids (another important group of gonadotropins in insects; Nijhout 1994; Gilbert et al. 2005) is consistently elevated (noncircadian) in SW versus LW adult females in selected lines (Zera and Bottsford 2001; Zhao and Zera 2004). This group of hormones may have taken on a more important role in reproduction, freeing JH to regulate morph-specific aspects of behavior, similar to the situation recently proposed for bees (Hartfelder et al. 2002). Recent studies have also shown that the morph-specific circadian rhythm occurs in field populations of *G. firmus* and other cricket species sampled in the field (Zera et al. 2007b). The phenomenon of a morph-associated, genetically polymorphic, circadian JH titer represents an exciting new

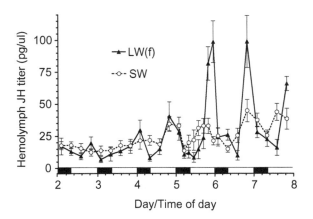

FIGURE 10.13

Morph-specific hemolymph (blood) JH titer circadian rhythm in *G. firmus*. Note the strong daily rhythm of the JH titer in the LW(f) morph (= LW morph in other figures and in the text), and the relatively temporally invariant pattern in the SW morph. The low-amplitude cycle in the SW morph is due to a daily cycle in the total blood volume. Light and dark bars on the *x*-axis denote light and dark portion of photoregime. Day refers to day since molt to adult. Figure is from Zhao and Zera (2004).

experimental trait that can be used to investigate the interface between endocrine circadian rhythms and life-history microevolution.

## SELECTION ON THE ENDOCRINE REGULATOR, JHE

Concurrent with endocrine characterizations of lines of *Gryllus* selected for LW or SW morphs, complementary artificial selection studies were undertaken on a wing-monomorphic congener, *Gryllus assimilis*, a species containing only LW individuals (figure 10.14). In these studies, JHE was directly selected for higher or lower blood activity in either juveniles (early last stadium, a time in development when JHE activities differ dramatically between LW and SW *G. firmus*) or in adults. Correlated responses to selection were measured for a wide variety of endocrine, enzymological, morphological, and life-history traits in adults and juveniles (Zera and Zhang 1995; Zera et al. 1996b, 1998; Zera 2006). A *Jhe* gene has now been sequenced in *G. assimilis* (Crone et al., 2007) and its differential expression is correlated with JHE activity differences between the selected lines (Anand et al. 2008). Because of space constraints, we provide only three examples of how complementary, artificial selection studies comprise a powerful experimental tool to investigate the functional relationship between genetic variation of JHE activity and life-history variation. Artificial selection studies of the butterfly, *Bicyclus* (discussed later; see also Koch et al. 1996; Zilstra et al. 2004; Brakefield and Frankino, 2009; Frankino et al. this volume), provide another set of examples of the power of complementary artificial selection studies in the analysis of life-history physiology.

### To What Extent Does Variation in JHE Activity Account for Variation in the Expression of Differences in Life History and Morphology between LW and SW Morphs?

A major problem in assessing the endocrine causes of the various life-history and morphological differences between LW- and SW-selected lines of *G. firmus* is that the lines differ in multiple endocrine regulators (JH titer, ecdysteroid titer, JHE activity). Which endocrine factors are causal, and which are correlative? Lines produced by directly selecting on JHE activity help to answer this question because they differ in some (JHE activity) but not other endocrine factors (ecdysteroid titers are equivalent; A. Anand and A. J. Zera, unpublished data) that differ significantly between LW- and SW-selected lines. Lines directly selected for reduced JHE activity exhibited significantly faster rate of juvenile development (a trait often influenced by JH) and significantly smaller flight muscles (as occurs in the SW morph), but no differences in the length of the wings (Zera 2006). Thus, changes in JH regulation are most closely correlated genetically with differences in flight muscles but not in the length of the wings. This result is consistent with some experimental manipulation studies in which applied JH at specific times in development alters development of flight muscles but not wings (Zera and Tanaka

1996). These parallel selection studies also demonstrate that differences in JHE activity alone cannot fully account for the differences in flight muscles between LW and SW morphs: although the magnitude of difference in JHE activity is the same between LW and SW selected lines as between lines selected for high or low JHE activity, differences in the mass of flight muscles is much greater between the morphs. Despite the utility of artificial selection, additional experimental techniques, which more precisely target and disrupt the action of specific genes, such as RNAi, will be needed to fully dissect the influences of changes in specific aspects of endocrine regulation on life-history and morphological variation in *Gryllus*.

### To What Degree Must Enzyme Activities Be Altered by Natural Selection to Affect Physiological Processes and Expression of Life-history Traits?

This is a fundamental question in life-history physiology, as well as a long-standing question in enzyme microevolution (Zera et al. 1985; Dykhuizen and Dean 1990; Zera, 2006; see also Dykhuizen and Dean this volume). In order for evolutionary changes in JHE activity to cause the evolution of morphological and life-history traits, JHE activity must be changed sufficiently to change the *in vivo* degradation of JH. Only by doing so can changes in JHE activity modulate the JH titer and expression of JH-regulated traits. The magnitude of required change in JHE activity is unclear and difficult to predict, given the often complex, nonlinear relationships between enzyme activity and higher-level *in vivo* physiological processes (Fell 2003; Dykhuizen and Dean 1990; Kacer and Burnes 1979). To empirically investigate this issue, *in vivo* JH degradation was measured periodically during direct selection on hemolymph JHE activity (Zera and Zhang 1995; Zera et al. 1996b). A measurable difference in *in vivo* JH degradation between selected lines was not observed until high- and low-activity lines diverged by greater than five-fold in JHE activity (figure 10.14). This experiment helps to explain why such a large-magnitude difference in JHE activity (often six- to eightfold) occurs between LW- and SW-selected lines.

### To What Extent Are Endocrine Regulators of Life Histories Correlated across Life Cycle Stages?

A key issue in evolutionary endocrinology, which is relevant to life-history physiology, is the extent to which aspects of endocrine regulation are correlated across life cycle stages. The existence of these correlations, or lack thereof, is expected to strongly influence the degree to which hormonally regulated life-history traits in different life cycle stages (e.g., rate of juvenile development and adult fecundity) are free to evolve independently versus constrained to evolve in concert. Very little information is available on endocrine correlations, or their impact on life history traits, and complementary selection studies on JHE activity in *G. assimilis* have provided information on this topic (figure 10.14). Selection

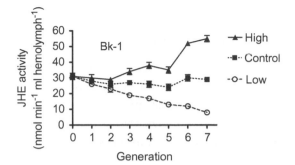

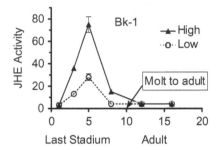

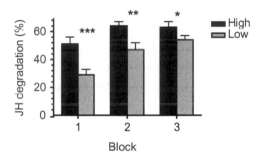

FIGURE 10.14
Direct and correlated responses to direct
selection on hemolymph (blood) juvenile hormone esterase (JHE) activity in
*Gryllus assimilis* in one of three blocks
(independent selection trails). Top
panel: Direct response to selection on
blood JHE activity on day 3 of the last
juvenile stadium. Middle panel: Correlated responses to selection on JHE on
other days of the last stadium not
directly selected, and in adults. Note the
strong correlated responses to selection
on other days of the last stadium, but no
correlated responses in adults. Bottom
panel: Correlated responses in *in vivo*
juvenile hormone degradation to direct
selection on JHE activity. Note the significantly elevated *in vivo* metabolism of
JH in lines selected for increased JHE
activity. Correlated responses in the two
lower panels were measured during
generation 6 (middle panel) or generation 7 (lower panel) of selection.
Figure is from Zera (2006).

on JHE activity on a particular day (day 3) of the last juvenile stage resulted in equivalent
changes in JHE activity on each of four days of that stage, but not during any day in
adulthood. Complementary selection on JHE activity in adults resulted in changes in
JHE activity during adulthood, but not during the last juvenile stage (Zera and Zhang
1995; Zera et al. 1998; Zera 2006). Thus, traits regulated by JHE in adults and juvenile
have the capability to evolve independently. On the other hand, strong, functionally
important genetic correlations between different components of JH regulation within
the same life cycle stage, such as blood JHE activity and JH binding, also have been identified (Zera et al. 1996; discussed in Zera 2006). These results on endocrine correlations
have important implications regarding the extent to which life cycle stages can evolve in
a modular manner (West-Eberhard 1993).

## SELECTION ON EYESPOT SIZE AND DEVELOPMENT RATE IN BICYCLUS

An important model in experimental studies of adaptive polyphenism—that is, discrete morphs produced by specific signals of environments to which they are adapted—is the butterfly *Bicyclus anynana*. This species exhibits two morphs that differ in morphology (eyespot pattern of the wings) and life history (development rate; longevity) and that are adapted to dry versus wet seasons of tropical Africa. Brakefield and coworkers (Koch et al. 1996; Zilstra et al. 2004; Brakefield and Frankino, 2009; Frankino et al. this volume) have investigated the microevolution of the polyphenism by conducting a variety of complementary artificial selection studies similar to the studies undertaken in *Gryllus*. In the first study, they artificially selected for larger or smaller eyespot size in a population in which eyespot size was intermediate between the seasonal phenotypes and in a thermal environment intermediate between the wet and dry seasons. Starting from the same base population, they also selected for fast or slow development. Lines selected for large eyespots exhibited faster rate of development and lines selected for faster rate of development exhibited larger eyespots. The ecdysteroid titer rose faster during development and was higher in lines selected for either large eyespot or faster rate of development, while injection of ecdysteroid into pupae from the line with small eyespots increased both development rate and the size of eyespots (Koch et al. 1996). From these results, the authors initially concluded that eyespot size and development rate were genetically correlated due to pleiotropic effects of the genetically variable ecdysteroid titer. This correlation was not surprising since ecdysteroids induce molting and thus influence development rate in many insects (Nijhout 1994), in addition to influencing wing pattern coloration in many insects (Nijhout 1994) However, a subsequent ingenious study indicated a more complex situation.

Zilstra et al (2004) were able to decouple eyespot size and development rate by divergently selecting on all four two-trait combinations, such as large eyespots and fast development rate, large eyespots and slow development rate, and so forth. Unexpectedly, the ecdysteroid titer covaried genetically with development rate but not with eyespot size, suggesting that variation in the titer of this hormone causes variation in rate of development but not in eyespot size. However, injection of ecdysteroids still increased eyespot size as well as rate of development. The authors concluded that the ecdysteroid titer primarily regulated rate of development, and that eyespot size could be modulated by a variety of ecdysone-dependent and ecdysone-independent mechanisms.

In a similar manner to studies of JHE and wing morph in *Gryllus*, multiple, complementary selection studies in *Bicyclus* have been a powerful tool to investigate the endocrine-genetic basis of life history and morphology. These studies also illustrate two important points that are often forgotten in selection studies. First, genetically correlated traits are not necessarily causally related (also see the earlier section on *Gryllus* studies). Second, experimental manipulation is a powerful complementary tool to artificial selection (also see discussion in Zera et al. 2007a).

Growth rate and body size are important life-history characteristics. The physiology of body size regulation in *Manduca sexta*, a major experimental model in insect endocrinology, has been a topic of research since the 1970s (Nijhout 1994). The extensive physiological and endocrine information for this species has provided the basis for quantitative-genetic and physiological analyses of body size and growth rate microevolution in *Manduca* employing laboratory selection. In a recent series of papers, Davidowitz, Nijhout, and colleagues proposed a physiological model that accounts for over 95 percent of the regulation and phenotypic plasticity of body size in *M. sexta* (Davidowitz et al. 2005). Three main physiological factors comprise the model, two of which are components of endocrine regulation (Davidowitz et al. 2005; for review, see Zera et al. 2007a). Ongoing artificial selection is investigating how these various components underlie laboratory selection on body size, growth, and development rate. An important issue being investigated is the degree to which the three physiological factors constrain or enable the response to simultaneous directional selection on body size and development rate (see discussion in Davidowitz et al. 2005).

## CONCLUSION

Physiological analysis of lines obtained by laboratory selection has contributed significantly to our understanding of the functional causes of life-history evolution. Extended longevity in *Drosophila* appears to have evolved in the laboratory by up-regulation of the antioxidant defense system and reduced production of reactive oxygen species. However, respiration rate does not appear to have been reduced as a mechanism to lower generation of reactive oxygen species. Low early-age reproduction is clearly a key aspect of extended longevity, but the mechanisms that link these two life-history traits are poorly understood. Negative effects of a heat-shock protein on life-history traits apparently constrain the heat-shock protein response to selection by elevated temperature. Thus, correlations with life-history traits can strongly influence the evolution of biochemical adaptations. Laboratory evolution studies of territoriality are emerging as a model system for the study of aggression.

With respect to the long-standing topic of energetics and life-history trade-offs, studies during the last decade have demonstrated strong correlations between energy reserves, most notably lipid, and specific life histories in selected lines of *Gryllus* and *Drosophila*. Investigations of lipid and amino acid metabolism in selected lines of *Gryllus* have identified activities of enzymes, flux through pathways of metabolism, and alterations in endocrine regulation that underlie the trade-off between early-age reproduction and dispersal. We still have a relatively rudimentary understanding of the mechanisms by which evolutionary changes in nutrient acquisition influence life-history microevolution.

The past decade has seen the first detailed information on the endocrine basis of life-history variation and trade-offs in laboratory selected lines. Top-down and bottom-up approaches have identified variation in specific endocrine regulators, most notably aspects of juvenile hormone and ecdysteroid signaling, that are associated with the divergent development of morphs of *Gryllus* that differ in life histories. Endocrine differences between adult dispersing and reproductive morphs have also been identified, but these are more complex than previously suspected. Endocrine mechanisms responsible for variation in individual life-history traits, and trade-offs between traits, are still poorly understood, especially in *Drosophila*.

An important positive aspect of current functional studies of life-history evolution in the laboratory is the increasing variety of techniques that are being brought to bear on this topic. These complementary approaches are significantly extending the scope and power of traditional functional analyses of life-history evolution. Some examples from this article include the use of microarrays and perturbations of candidate genes in the study of territoriality, and the use of laboratory mutants and irradiation to investigate the impact of reproduction on survival in lines selected for longevity. The increasing use of additional techniques, such as mutational analysis, transgenics, RNAi, endocrine manipulations, and pharmaceutical agents (e.g., enzyme inhibitors) will expand our ability to identify specific mechanism underlying the evolution of life histories in the laboratory.

## SUMMARY

Physiological analysis of laboratory-selected lines has been a powerful tool to identify the functional basis of life-history evolution. Elevated activities of antioxidant enzymes and altered mitochondrial function appear to reduce oxidative damage to proteins and lipids. These traits are important functional components of the evolution of extended longevity. Reduced metabolic rate does not appear to be an important factor in extending life span. Studies of an important heat shock protein (Hsp70) indicate how negative correlations of life-history traits constrain the degree of involvement of this protein in the adaptive response to elevated temperature. Studies on territoriality illustrate the use of genetics and genomics to investigate function in selection experiments. Enzymological, radiotracer, and endocrine studies of lipid metabolism in wing-polymorphic crickets have identified alterations in specific pathways of lipid and amino acid metabolism that underlie a classic life-history trade-off between reproduction and dispersal. Complementary direct selection on life-history traits or endocrine regulators of life-history traits in *Gryllus* crickets has identified aspects of endocrine regulation that underlie production of morphs differing in life histories. Complementary selection studies in *Bicyclus* butterflies have illuminated the functional nature of correlations between life-history and morphological traits. Despite the preeminence of *D. melanogaster* in selection studies of life histories, the endocrine control of life-history traits and their trade-offs are poorly understood in the species. The increasing use of complementary approaches to laboratory selection, including microarrays, perturbation of candidate genes, and the use of laboratory-generated mutants, are significantly extending the scope and power of traditional functional analyses of life-history evolution. These and other techniques will expand our ability to identify specific mechanisms that underlie the evolution of life histories in the laboratory.

## ACKNOWLEDGMENTS

A.J.Z. gratefully acknowledges the support of the National Science Foundation (most recently by grants IBN-0516973 and IBN-0212486), which has continuously funded research on *Gryllus* in the Zera laboratory during the past eighteen years. L.G.H. acknowledges support from the National Science Foundation (NSF DEB-ESP0346476), the National Institutes of Health (1 RO1 DK074136), and the Army Office of Research (W911NF-07-1-0370).

## REFERENCES

Anand, A., E. J. Crone, and A. J. Zera. 2008. Tissue and stage-specific juvenile hormone esterase (JHE) and epoxide hydrolase (JHEH) enzyme activities and *Jhe* transcript abundance in lines of the cricket *Gryllus assimilis* artificially selected for plasma JHE activity: Implications for JHE microevolution. *Journal of Insect Physiology* 54:1323–1331.

Arking, R., S. Buck, D.-S. Hwang-Bo, and M. Lane. 2002. Metabolic alterations and shifts in energy allocation are corequisites for expression of extended longevity genes in *Drosophila*. *Annals of the New York Academy of Sciences* 959:251–262.

Arking, R., S. Buck, R. A. Wells, and R. Pretzlapf. 1988. Metabolic rate in genetically based long lived strains of *Drosophila*. *Experimental Gerontology* 23:59–76.

Arking, R., V. Burde, K. Graves, R. Harl, E. Feldman, A. Zeevi, S. Soliman, A. Saraiya, S. Buck, J. Vettraino, K. Sathrasala, N. Wehr, and R. L. Levine. 2000. Forward and reverse selection for longevity in *Drosophila* is characterized by alteration of antioxidant gene expression and oxidative damage pattern. *Experimental Gerontology* 35:167–185.

Arking, R. S., and S. P. Dudas. 1989. Review of genetic investigations into the aging processes of *Drosophila*. *Geriatric Bioscience* 37:757–773.

Beenakkers, A., D. Van der Horst, and W. Van Marrewijk. 1985. Insect lipids and lipoproteins, their role in physiological processes. *Progress in Lipid Research* 24:19–67.

Bettencourt, B. R., M. E. Feder, and S. Cavicchi. 1999. Experimental evolution of Hsp70 expression and thermotolerance in *Drosophila melanogaster*. *Evolution* 53:484–492.

Boggs, C. L. 1997. Reproductive allocation from reserves and income in butterfly species with differing adult diets. *Ecology* 78:181–191.

Boots, M., and M. Begon. 1993. Trade-offs with resistance to a grandulosis virus in the Indian Meal Moth, examined by a laboratory evolution experiment. *Functional Ecology* 7:528–534.

Bownes, M., and H. Rembold. 1987. The titer of juvenile hormone during the pupal and adult stages of the life cycle of *Drosophila melanogaster*. *European Journal of Biochemistry* 164:709–712.

Brakefield, P., and W. Frankino. 2009. Polyphenisms in Lepidoptera: Multidisciplinary approaches to studies of evolution and development. *In* T. Ananthakrishnan and D. Whitman, eds. *Phenotypic Plasticity in Insects: Mechanisms and Consequences*. Plymouth, U.K.: Science. Pages 337–368.

Bubliy, O. A., and V. Loeschcke. 2005. Correlated responses for stress resistance and longevity in a laboratory population of *Drosophila melanogaster*. *Journal of Evolutionary Biology* 18:789–803.

Buck, S. A., and R. Arking. 2002. Metabolic alterations in genetically selected *Drosophila* strains with different longevities. *Journal of the American Aging Association* 24:151–162.

Burger, J. M. S., K. Munjong, J. Pont, and T. Kawecki. 2008. Learning ability and longevity: A symmetrical evolutionary trade-off. *Evolution* 62:1294–1304.

Carlson, K. A., and L. G. Harshman. 1999a. Extended longevity lines of *Drosophila melanogaster*: Abundance of yolk protein gene mRNA in fat body and ovary. *Experimental Gerontology* 34:173–184.

———. 1999b. Extended longevity lines of *Drosophila melanogaster*: Characterization of oocyte stages and ovariole numbers as a function of age and diet. *Journal of Gerontology, Biological Sciences* 54A:B432–B440.

Carlson, K. A., T. J. Nusbaum, M. R. Rose, and L. G. Harshman. 1998. Oocyte maturation and ovariole number in lines of *Drosophila melanogaster* selected for postponed senescence. *Functional Ecology* 12:514–520.

Cavicchi, S., D. Guerra, G. Giorgi, and C. Pezzoli. 1985. Temperature-related divergence in experimental populations of *Drosophila*. I. Genetic and developmental basis of wing size and shape variation. *Genetics* 109:665–689.

Chen, B., J.-C. Walser, T. H. Rodgers, R. S. Sobota, M. K. Burke, M. R. Rose, and M. E. Feder. 2007. Abundant, diverse, and consequential P elements segregate in promoters of small heat-shock genes in *Drosophila* populations. *Journal of Evolutionary Biology* 20:2056–2066.

Chippindale, A. K., T. J. F. Chu, and M. R. Rose. Complex trade-off and the evolution of starvation resistance in *Drosophila melanogaster*. *Evolution* 50:753–766.

Chippindale, A. K., A. M. Leroi, S. B. Kim, and M. R. Rose. 1993. Phenotypic plasticity and selection in *Drosophila* life history evolution. I. Nutrition and the cost of reproduction. *Journal of Evolutionary Biology* 6:171–193.

Chippindale, A. K., A. M. Leroi, H. Sang, D. J. Borash, and M. R. Rose. 1997. Phenotypic plasticity and selection in *Drosophila*. II. Diet, mates, and the cost of reproduction. *Journal of Evolutionary Biology* 10:269–293.

Congdon, J. D., A. E. Dunham, and D. W. Tinkle. 1982. Energy budgets and life histories of reptiles. Pages 233–271 in C. Gans, ed. *Biology of the Reptilia*. New York: Academic Press.

Crone, E., A. Zera, A. Anand, J. Oakeshott, T. Sutherland, R. Russell, L. Harshman, F. Hoffman, and C. Claudianos. 2007. JHE in *Gryllus assimilis*: Cloning, sequence-function associations and phylogeny. *Insect Biochemistry and Molecular Biology* 37:1359–1365.

Davidowitz, G., D. A. Roff, and H. F. Nijhout. 2005. A physiological perspective on the response of body size and development time to simultaneous directional selection. *Integrative and Comparative Biology* 45:525–531.

de Jong, G., and A. van Noordwijk. 1992. Acquisition and allocation of resources: Genetic (co)variances, selection, and life histories. *American Naturalist* 139:749–770.

de Magalhães, J. P., J. Costa, and G. M. Church. 2007. An analysis of the relationship between metabolism, developmental schedules, and longevity using phylogenetic independent contrasts. *Journal of Gerontology: Biological Sciences* 62A:149–160.

Dierick, H. A., and R. J. Greenspan. 2006. Molecular analysis of flies selected for aggressive behavior. *Nature Genetics* 38:1023–1031.

Djawdan, M., A. K. Chippindale, M. R. Rose, and T. J. Bradley. 1998. Metabolic reserves and evolved stress resistance in *Drosophila melanogaster*. *Physiological and Biochemical Zoology* 71:584–594.

Djawdan, M., M. R. Rose, and T. J. Bradley. 1997. Does selection for stress resistance lower metabolic rate? *Ecology* 78:828–837.

Djawdan, M., T. T. Sugiyama, L. K. Schlaeger, T. J. Bradley, and M. R. Rose. 1996. Metabolic aspects of the trade-off between fecundity and longevity in *Drosophila melanogaster*. *Physiological Zoology* 69:1176–1195.

Drent, R. H., and S. Daan. 1980. The prudent parent: energetic adjustments in avian breeding. *Ardea* 68:225–252.

Dudas, S. P., and R. A. Arking. 1995. A coordinate upregulation of antioxidant gene activities is associated with the delayed onset of senescence in a long-lived strain of *Drosophila*. *Journal of Gerontology Series A: Biological Sciences and Medical Sciences* 50A:B117–B127.

Dykhuizen, D. E., and A. M. Dean. 1990. Enzyme activity and fitness: Evolution in solution. *Trends in Ecology & Evolution* 5:257–262.

Edwards, A. C., S. M. Rollmann, T. J. Morgan, and T. F. C Mackay. 2006. Quantitative genomics of aggressive behavior in *Drosophila melanogaster*. *PLoS Genetics* 2:1386–1395.

Fairbairn, D. J., and D. E. Yadlowski. 1997. Coevolution of traits determining migratory tendency: correlated response of a critical enzyme, juvenile hormone esterase, to selection on wing morphology. *Journal of Evolutionary Biology* 10:495–513.

Feder, M. E., A. F. Bennett, and R. B. Huey. 2000. Evolutionary physiology. *Annual Review of Ecology and Systematics* 31:315–341.

Fell, D. 2003. *Understanding the Control of Metabolism*. Seattle, WA: Portland.

Fellowes, M. D. E., A. R. Kraaijveld, and H. C. J. Godfray. 1998. Trade-off associated with selection for increased ability to resist parasitoid attack in *Drosophila melanogaster*. *Proceedings of the Royal Society: Biological Sciences* 265:1553–1558.

Finch, C. E., and M. R. Rose. 1995. Hormones and the physiological architecture of life history evolution. *Quarterly Review of Biology* 70:1–51.

Fisher, R. A. 1930. *The Genetical Theory of Natural Selection*. New York: Dover.

Flatt, T., and T. J. Kawecki. 2007. Juvenile hormone as a regulator of the trade-off between reproduction and life span in *Drosophila melanogaster*. *Evolution* 61:1980–1991.

Flatt, T., M.-P. Tu, and M. Tatar. 2005. Hormonal pleiotropy and the juvenile hormone regulation of *Drosophila* development and life history. *BioEssays* 27:999–1010.

Foley, P. A., and L. S. Luckinbill. 2001. The effects of selection for larval behavior on adult life-history features in *Drosophila melanogaster*. *Evolution* 55:2493–2502.

Folk, D. G., P. Zwollo, D. R. Rand, and G. W. Gilchrist. 2006. Selection on knockdown performance in *Drosophila melanogaster* impacts thermotolerance and heat-shock response differently in females and males. *Journal of Experimental Biology* 209:3964–3973.

Force, A. G., T. Staples, T. Soliman, and R. Arking. 1995. A comparative biochemical and stress analysis of genetically selected *Drosophila* strains with different longevities. *Developmental Genetics* 17:340–351.

Fry, J. D. 2003. Detecting ecological trade-offs using selection experiments. *Ecology* 84:1672–1678.

Garland, T., Jr. 2003. Selection experiments: an under-utilized tool in biomechanics and organismal biology. Pages 23–65 *in* V. Bels, J.-P. Gasc, and A. Casinos, eds. *Vertebrate Biomechanics and Evolution.* Oxford: BIOS Scientific.

Garland, T., Jr., A. F. Bennett, and E. L. Rezende. 2005. Phylogenetic approaches in comparative physiology. *Journal of Experimental Biology* 208:3015–3035.

Garland, T., Jr., and P. A. Carter. 1994. Evolutionary physiology. *Annual Review of Physiology* 56:579–621.

Gasser, M., M. Kaiser, D. Berrigan, and S. C. Stearns. 2000. Life-history correlates of evolution under high and low mortality. *Evolution* 54:1260–1272.

Gibbs, A. G. 1999. Laboratory selection for the comparative physiologist. *Journal of Experimental Biology* 202:2709–2718.

Gilbert, L., K. Iatrou, and S. Gill. 2005. *Comprehensive Molecular Insect Science. Vol. 3 Endocrinology.* Amsterdam: Elsevier.

Gould, S. J. 1977. *Ontogeny and Phylogeny.* Cambridge, MA: Belknap.

Graves, J. L., E. C. Toolson, C. Joong, L. N. Vu, and M. R. Rose. 1992. Desiccation, flight, glycogen and postponed senescence in *Drosophila melanogaster. Physiological Zoology* 65:268–286.

Harman, D. 1956. Aging: A theory based on free radical and radiation chemistry. *Journal of Gerontology* 11:298–300.

Harshman, L. G. 1999. Investigation of the endocrine system in extended longevity lines of *Drosophila melanogaster. Experimental Gerontology* 34:997–1006.

Harshman, L. G., and B. A. Haberer. 2000. Oxidative stress resistance: A robust correlated response to selection in extended longevity lines of *Drosophila melanogaster? Journal of Gerontology Series A: Biological Sciences and Medical Sciences* 55A:B415–B417.

Harshman, L. G., and A. A. Hoffmann. 2000. Laboratory selection experiments using *Drosophila*: What do they really tell us? *Trends in Ecology & Evolution* 15:32–36.

Harshman, L. G., and A. J. Zera. 2007. The cost of reproduction: the devil in the details. *Trends in Ecology & Evolution* 22:80–86.

Hartfelder, K., M. M. G. Bitondi, W. C. Santana, and Z. I. P. Simones. 2002. Ecdysteroid titer and reproduction in queens and workers of the honey bee and a stingless bee: Loss of ecdysteroid function at increasing levels of sociality. *Insect Biochemistry and Molecular Biology* 32:211–216.

Hartfelder, K., and D. Emlen. 2005. Endocrine control of insect polyphenism. Pages 651–703 *in* L. Gilbert, K. Iatrou, and S. S. Gill, eds. *Comprehensive Molecular Insect Science.* Amsterdam: Elsevier.

Hoffmann, A. 1989. Selection for territoriality in *Drosophila melanogaster*: Correlated responses in mating success and other fitness components. *Animal Behaviour* 38:23–34.

Hoffmann, A. A., and P. A. Parsons. 1991. *Evolutionary Genetics and Environmental Stress.* Oxford: Oxford University Press.

Hoffmann, A. A., M. Scott, L. Partridge, and R. Hallas. 2003. Overwintering in *Drosophila melanogaster*: Outdoor field cage experiments on clinal and laboratory selected populations help to elucidate traits under selection. *Journal of Evolutionary Biology* 16:614–623.

Joshi, A., and L. D. Mueller. 1996. Density-dependent natural selection in *Drosophila*: Trade-offs between larval food acquisition and utilization. *Evolutionary Ecology* 10:463–474.

Karasov, W. H., and C. Martinez del Rio. 2007. *Physiological Ecology: How Animals Process Energy, Nutrients, and Toxins.* Princeton, NJ: Princeton University Press.

Kascer, H., and J. A. Burns. 1979. Molecular democracy: Who shares the controls? *Transactions of the Biochemical Society* 7:1149–1160.

Ketterson, E. D., and V. Nolan Jr. 1992. Hormones and life histories: An integrative approach. *American Naturalist* 140:S33–S62.

Khazaeli, A. A., W. V. Voorhies, and J. W. Curtsinger. 2005. Longevity and metabolism in *Drosophila melanogaster. Genetics* 169:231–242.

Kirkwood, T. B. L. 1977. Evolution of aging. *Nature* 270:301–304.

Kirkwood, T. B. L., and M. R. Rose. 1991. Evolution of senescence: Late survival sacrificed for reproduction. *Philosophical Transactions of the Royal Society of London B, Biological Sciences* 332:B15–B24.

Koch, P., P. Brakefield, and F. Kesbeke. 1996. Ecdysteroids control eyespot size and wing colour pattern in the polyphenic butterfly, *Bicyclus anynana* (Lepidoptera: Satyridae). *Journal of Insect Physiology* 42:223–230.

Kraaijveld, A. R., and H. C. J. Godfray. 1997. Trade-off between parasitoid resistance and larval competitive ability in *Drosophila melanogaster. Nature* 389:278–280.

———. 2008. Selection for resistance to a fungal pathogen in *Drosophila melanogaster. Heredity* 100:400–406.

Kraaijveld, A. R., E. C. Limentani, and H. C. J. Godfray. 2001. Basis of the trade-off between parasitoid resistance and larval competitive ability in *Drosophila melanogaster. Nature* 389:278–280.

Krebs, R. A., and M. E. Feder. 1997. Deleterious consequences of Hsp70 overexpression in *Drosophila melanogaster* larvae. *Cell Stress Chaperones* 2:60–71.

Kristensen, T. N., V. Loeschcke, and A. A. Hoffmann. 2007. Can artificially selected phenotypes influence a component of field fitness? Thermal selection and fly performance under thermal extremes. *Proceedings Biological Sciences* 274:771–778.

Kubli, E. 1996. The *Drosophila* sex-peptide: A peptide pheromone involved in reproduction. *Advances in Developmental Biochemistry* 4:99–128.

Kurapati, R., H. B. Passananti, M. R. Rose, and J. Tower. 2000. Increased *hsp22* RNA level in *Drosophila* lines genetically selected for increased longevity. *Journal of Gerontology Series A: Biological Sciences and Medical Sciences* 55A:B552–B559.

Lee, K. P., S. J. Simpson, F. J. Clissold, R. Brooks, J. W. O. Ballard, P. W. Taylor, N. Soran, and D. Raubenheimer. 2008. Lifespan and reproduction in *Drosophila*: New insights from nutritional geometry. *Proceedings of the National Academy of Sciences of the USA* 105:2498–2503.

Leroi, A. M., A. K. Chippindale, and M. R. Rose. 1994. Long-term evolution of a genetic life-history trade-off in *Drosophila. Evolution* 48:1880–1899.

Luckinbill, L. S., R. Arking, M. J. Clare, W. C. Cirocco, and S. A. Buck. 1984. Selection for delayed senescence in *Drosophila melanogaster*. *Evolution* 38:996–1004.

Luckinbill, L. S., T. Z. Grudzien, A. Rhine, and G. Weisman. 1989. The genetic basis of adaptation to selection for longevity in *Drosophila melanogaster*. *Evolutionary Ecology* 3:31–39.

Luckinbill, L. S., V. Riha, S. Rhine, and T. A. Grudzien. 1990. The role of glucose-6-phosphate dehydrogenase in the evolution of longevity in *Drosophila melanogaster*. *Heredity* 65:29–38.

Matsuda, R. 1987. *Animal Evolution in Changing Environments with Special Reference to Abnormal Metamorphosis*. New York: Wiley-Interscience.

Miller, G. T., and S. Pitnick. 2002. Sperm-female coevolution in *Drosophila*. *Science* 298:1230–1233.

Min, K.-J., M. F. Hogan, M. Tatar, and D. M. O'Brien. 2006. Resource allocation and soma in *Drosophila*: A stable isotope analysis of carbon from dietary sugar. *Journal of Insect Physiology* 52:763–770.

Miyatake, T. 1997. Genetic trade-off between early fecundity and longevity in *Bactrocera cucurbitae* (Diptera: Tephritidae). *Heredity* 78: 93–100.

Mueller, L. D. 1990. Density-dependent selection does not increase efficiency. *Evolutionary Ecology* 4:290–297.

Mueller, J. L., J. L. Page, and M. F. Wolfner. 2007. An ectopic expression screen reveals *Drosophila* seminal fluid protein's protective and toxic effects. *Genetics* 175:777–783.

Nijhout, H. F. 1994. *Insect Hormones*. Princeton, NJ: Princeton University Press.

Nijhout, H. F., and D. Wheeler. 1982. Juvenile hormone and the physiological basis of insect polymorphism. *Quarterly Review of Biology* 57:109–133.

Norry, F. M., and V. Loeschcke. 2003. Heat-induced expression of a molecular chaperone decreases by selecting for long-lived individuals. *Experimental Gerontology* 38:673–681.

O'Brien, D. M., M. L. Fogel, and C. L. Boggs. 2002. Renewable and nonrenewable resources: Amino acid turnover and allocation to reproduction. *Proceedings of the National Academy of Sciences of the USA* 99:4413–4418.

Pappas, C., D. Hyde, K. Bowler, V. Loeschcke, and J. G. Sorensen. 2006. Post-eclosion decline in "knock-down" thermal resistance and reduced effect of heat hardening in *Drosophila melanogaster*. *Comparative Biochemistry and Physiology A, Molecular and Integrative Physiology* 146:355–359.

Partridge, L., and K. Fowler. 1992. Direct and correlated responses to selection on age at reproduction in *Drosophila melanogaster*. *Evolution* 46:76–91.

Partridge, L., N. Prowse, and P. Pignatelli. 1999. Another set of responses and correlated responses to selection on age at reproduction in *Drosophila melanogaster*. *Proceedings of the Royal Society of London B, Biological Sciences* 266:255–261.

Richard, D. S., R. Rybczynski, T. G. Wilson, Y. Wang, M. L. Wayne, Y. Zhou, L. Partridge, and L. G. Harshman. 2005. Insulin signaling is necessary for vitellogenesis in *Drosophila melanogaster* independent of the roles of juvenile hormone and ecdysteroids: Female sterility of the *chico[1]* insulin signaling mutation is autonomous to the ovary. *Journal of Insect Physiology* 51:455–464.

Roff, D. A. 1986. The genetic basis of wing dimorphism in the sand cricket, *Gryllus firmus* and its relevance to the evolution of wing dimorphism in insects. *Heredity* 57:221–231.

———. 1996. The evolution of threshold traits in animals. *Quarterly Review of Biology* 71:3–35.

Roff, D. A., and D. Fairbairn. 2007. Laboratory evolution of the migratory polymorphism in the sand cricket: Combining physiology with quantitative genetics. *Physiological and Biochemical Zoology* 80:358–369.

Rose, M. R. 1984. Laboratory evolution of postponed senescence in *Drosophila melanogaster*. *Evolution* 38:1004–1010.

Rose, M. R., and T. J. Bradley. 1998. Evolutionary physiology of the cost of reproduction. *Oikos* 83:443–451.

Rose, M. R., M. L. Dorey, A. M. Coyle, and P. M. Service. 1984. The morphology of postponed senescence in *Drosophila melanogaster*. *Canadian Journal of Zoology* 62:1576–1580.

Rose, M. R., T. J. Nusbaum, and A. K. Chippindale. 1996. Laboratory evolution: The experimental wonderland and the Cheshire cat syndrome. Pages 221–241 *in* M. R. Rose and G. V. Lauder, eds. *Adaptation*. San Diego, CA: Academic Press.

Rose, M. R., L. N. Vu, S. U. Park, and J. L. Graves. 1992. Selection on stress resistance increases longevity in *Drosophila melanogaster*. *Experimental Gerontology* 27:241–250.

Ross, R. E. 2000. Age specific decreases in aerobic efficiency associated with an increase in oxygen free radical production in *Drosophila melanogaster*. *Journal of Insect Physiology* 46:1477–1480.

Scheirs, J., K. Jordaens, and L. D. Bruyn. 2005. Have genetic trade-offs in host use been overlooked in arthropods? *Evolutionary Ecology* 19:551–561.

Service, P. M. 1987. Physiological mechanisms of increased stress resistance in *Drosophila melanogaster* selected for postponed senescence. *Physiological Zoology* 60:321–326.

———. 1993. Laboratory evolution of longevity and reproductive fitness components in male fruit flies. *Evolution* 47:387–399.

Service, P. M., E. W. Hutchinson, M. D. MacKinley, and M. R. Rose. 1985. Resistance to environmental stress in *Drosophila melanogaster* selected for postponed senescence. *Physiological Zoology* 58:380–389.

Sgrò, C. M., and L. Partridge. 1999. A delayed wave of death from reproduction in *Drosophila*. *Science* 286:2521–2524.

Silbermann, R., and M. Tatar. 2000. Reproductive costs of heat shock proteins in transgenic *Drosophila melanogaster*. *Evolution* 54:2038–2045.

Simmons, F. H., and T. J. Bradley. 1997. An analysis of resource allocation in response to dietary yeast in *Drosophila melanogaster*. *Journal of Insect Physiology* 43:779–788.

Soh, J. W., S. Hotic, and R. Arking. 2007. Dietary restriction in *Drosophila* is dependent on mitochondrial efficiency and constrained by pre-existing extended longevity. *Mechanisms of Ageing and Development* 128:581–593.

Swallow, J. G., and T. Garland, Jr. 2005. Selection experiments as a tool in evolutionary and comparative physiology: Insights into complex traits: An introduction to the symposium. *Integrative and Comparative Biology* 45:387–390.

Tinkle, D. W., and N. F. Hadley. 1975. Lizard reproductive effort: Caloric estimates and comments on its evolution. *Ecology* 56:427–434.

Townsend, C. R., and P. Calow. 1981. *Physiological Ecology. An Evolutionary Approach to Resource Use*. Oxford: Blackwell Scientific.

Tyler, R. H., H. Brar, M. Singh, A. Latorre, J. L. Graves, L. D. Mueller, M. R. Rose, and F. J. Ayala. 1993. The effect of superoxide dismutase alleles on aging in *Drosophila*. *Genetica* 91:143–149.

Van Voorhies, W. A., A. Z. Khazaeli, and J. W. Curtsinger. 2004. Testing the "Rate of Living" model: Further evidence that longevity and metabolic rate are not inversely correlated in *Drosophila melanogaster. Journal of Applied Physiology* 97:1915–1922.

Vettraino, J., S. Buck, and R. Arking. 2001. Direct selection for paraquat resistance results in a different longevity phenotype. *Journal of Gerontology* 56A:B415–B425.

Warbrick-Smith, J., S. T. Behmer, K. P. Lee, D. Raubenheimer, and S. J. Simpson. 2006. Evolving resistance to obesity in an insect. *Proceedings of the National Academy of Sciences of the USA* 103:14045–14049.

Webster, J. P., and M. E. J. Woodhouse. 1999. Cost of resistance: relationship between reduced fertility and increased resistance in a snail-schistosome system. *Proceedings of the Royal Society of London B, Biological Sciences* 266:391–396.

West-Eberhard, M. 2003. *Developmental Plasticity and Evolution*. Oxford: Oxford University Press.

Wigby, S., and T. Chapman. 2005. Sex peptide causes mating costs in female *Drosophila melanogaster. Current Biology* 15:315–321.

Withers, P. C. 1992. *Comparative Animal Physiology*. Fort Worth, TX: Saunders College.

Wolfner, M. F. 2002. The gifts that keep on giving: Physiological functions and evolutionary dynamics of male seminal proteins in *Drosophila. Heredity* 88:85–93.

Yan, G., D. W. Severson, and B. M. Christensen. 1997. Costs and benefits of mosquito refractoriness to malaria parasites: Implications for genetic variability of mosquitoes and genetic control of malaria. *Evolution* 51:441–450.

Zera, A. J. 1984. Differences in survivorship, development rate and fertility between the long-winged and wingless morphs of the waterstrider, *Limnoporus canaliculatus. Evolution* 36:1023–1032.

———. 2004. The endocrine regulation of wing polymorphism: State of the art, recent surprises, and future directions. *Integrative and Comparative Biology* 43:607–616.

———. 2005. Intermediary metabolism and life history trade-offs: Lipid metabolism in lines of the wing-polymorphic cricket, *Gryllus firmus*, selected for flight capability vs. early age reproduction. *Integrative and Comparative Biology* 45:511–524.

———. 2006. Evolutionary genetics of juvenile hormone and ecdysteroid regulation in *Gryllus*: A case study in the microevolution of endocrine regulation. *Comparative Biochemistry and Physiology Part A* 144:365–379.

———. 2007. Endocrine analysis in evolutionary-developmental studies of insect polymorphism: Use and misuse of hormone manipulation. *Evolution and Development* 9:499–513.

Zera, A. J., and J. Bottsford. 2001. The endocrine-genetic basis of life-history variation: Relationship between the ecdysteroid titer and morph-specific reproduction in the wing-polymorphic cricket, *Gryllus firmus. Evolution* 55:538–549.

Zera, A. J., and T. Brink. 2000. Nutrient absorption and utilization by wing and flight muscle morphs of the cricket *Gryllus firmus*: Implications for the trade-off between flight capability and early reproduction. *Journal of Insect Physiology* 46:1207–1218.

Zera, A. J., and G. Cisper. 2001. Genetic and diurnal variation in the juvenile hormone titer in a wing-polymorphic cricket: Implications for the evolution of life histories and dispersal. *Physiological and Biochemical Zoology* 74:293–306.

Zera, A. J., and L. G. Harshman. 2001. Physiology of life history trade-offs in animals. *Annual Review of Ecology and Systematics* 32:95–126.

Zera, A. J., L. G. Harshman, and T. Williams. 2007a. Evolutionary endocrinology: The developing synthesis between endocrinology and evolutionary genetics. *Annual Review of Ecology, Evolution and Systematics* 38:793–817.

Zera, A. J., and Y. Huang. 1999. Evolutionary endocrinology of juvenile hormone esterase: Functional relationship with wing polymorphism in the cricket, *Gryllus firmus. Evolution* 53:837–847.

Zera, A. J., J. G. Koehn, and J. G. Hall. 1985. Allozymes and biochemical adaptation. Pages 633–674 *in* G. A. Kerkut and L. I. Gilbert, eds. *Comprehensive Insect Physiology. Vol. 10. Biochemistry and Pharmacology.* Oxford: Pergamon.

Zera, A. J., and A. Larsen. 2001. The metabolic basis of life history variation: Genetic and phenotypic differences in lipid reserves among life history morphs of the wing-polymorphic cricket, *Gryllus firmus. Journal of Insect Physiology* 47:1147–1160.

Zera, A. J., J. Potts, and K. Kobus. 1998a. The physiology of life history trade-offs: Experimental analysis of a hormonally-induced life history trade-off in *Gryllus assimilis. American Naturalist* 152:7–23.

Zera, A. J., J. Sall, and K. Grudzinski. 1996a. Flight-muscle polymorphism in the cricket *Gryllus firmus*: Muscle characteristics and their influence on the evolution of flightlessness. *Physiological Zoology* 70:519–529.

Zera, A. J., J. Sall, and R. Schwartz. 1996b. Artificial selection on JHE activity in *Gryllus assimilis*: Nature of activity differences between lines and effect on JH binding and metabolism. *Archives of Insect Biochemistry and Physiology* 32:421–428.

Zera, A. J., T. Sanger, and G. L. Cisper. 1998b. Direct and correlated responses to selection on JHE activity in adult and juvenile *Gryllus assimilis*: Implications for stage-specific evolution of insect endocrine traits. *Heredity* 80:300–309.

Zera, A. J., and S. Tanaka. 1996. The role of juvenile hormone and juvenile hormone esterase in wing morph determination in *Modicogryllus confirmatus. Journal of Insect Physiology* 42:909–915.

Zera, A. J., and K. C. Tiebel. 1989. Differences in juvenile hormone esterase activity between presumptive macropterous and brachypterous *Gryllus rubens*: Implications for the hormonal control of wing polymorphism. *Journal of Insect Physiology* 35:7–17.

Zera, A. J., and C. Zhang. 1995. Direct and correlated responses to selection on hemolymph juvenile hormone esterase activity in *Gryllus assimilis. Genetics* 141:1125–1134.

Zera, A. J., and Z. Zhao. 2003. Life-history evolution and the microevolution of intermediary metabolism: Activities of lipid-metabolizing enzymes in life-history morphs of a wing-dimorphic cricket. *Evolution* 57:568–596.

———. 2006. Intermediary metabolism and life history trade-offs: Differential metabolism of amino acids underlies the dispersal-reproduction trade-off in a wing-polymorphic cricket. *American Naturalist* 167:889–900.

Zera, A. J., Z. Zhao, and K. Kaliseck. 2007b. Hormones in the field: Evolutionary endocrinology of juvenile hormone and ecdysteroids in field populations of the wing-dimorphic cricket *Gryllus firmus. Physiological and Biochemical Zoology* 80:592–606.

Zhao, Z., and A. J. Zera. 2002. Differential lipid biosynthesis underlies a tradeoff between reproduction and flight capability in a wing-polymorphic cricket. *Proceedings of the National Academy of Sciences of the USA* 99:16829–16834.

———. 2004. The hemolymph JH titer exhibits a large-amplitude, morph-dependent, diurnal cycle in the wing-polymorphic cricket, *Gryllus firmus. Journal of Insect Physiology* 50:93–102.

Zijlstra, W., M. Steigenga, P. Koch, B. Zwan, and P. Brakefield. 2004. Butterfly selected lines explore the hormonal basis of interactions between life histories and morphology. *American Naturalist* 163:E76–E87.

Zwann, B., R. Bijlsma, and R. F. Hoekstra. 1995. Direct selection on the life span in *Drosophila melanogaster. Evolution* 49:649–659.

# 11

# BEHAVIOR AND NEUROBIOLOGY

Justin S. Rhodes and Tadeusz J. Kawecki

The tree of life is decorated with an extraordinary diversity of animal behavior (figure 11.1). Such behaviors as foraging, reproducing, moving through the environment, and avoiding predators are all clearly major determinants of survival and reproductive success and hence are thought to be under relatively strong natural and sexual selection. Although some behaviors are culturally transmitted, the vast majority evolve by genetic mechanisms. One of the earliest pieces of direct evidence that behavior can be shaped by evolutionary processes was domestication of wolves into dogs, which is thought to have

FIGURE 11.1

Massive behavioral diversity in feeding and home range size among vertebrate species. Some species stalk, attack, and kill other animals for food (A), whereas others forage entirely on plant material (B). Among the predators, some sit and wait for their prey to come to them (C), whereas others actively pursue their prey (D). Some animals spend their entire life within few square meters of space (E), whereas others roam for miles in the open ocean (F). Presumably these behavioral shifts are mediated by changes in the brain that evolved through structural modifications of the genome.

occurred as far back as fifteen thousand years ago (Savolainen et al. 2002). Since then, a variety of animals have been domesticated, thus providing ample evidence that selective breeding can alter behavior (see also Barnett and Smart 1975; Simões et al. this volume). Domestication also demonstrates that genes can influence behavior and that behavior can evolve rapidly (Garland 2003; Greenspan 2003; Robinson 2004). In recent decades, a number of natural genetic polymorphisms that affect behavior have been identified, and some progress has been made toward understanding how changes in DNA alter gene expression and/or protein structure, nervous system development, and neural physiology to produce differences in behavior (Ross and Keller 1998; Keller and Parker 2002; Greenspan 2004; Fitzpatrick et al. 2005).

Selection experiments and experimental evolution approaches offer powerful tools for elucidating the origin and mechanisms of behavioral diversity. The discipline is useful to establish basic knowledge about nature, but it also has powerful applications for biomedicine. For example, advances in understanding how genes influence behavior could provide insights into the etiology of drug addiction, obesity, or attention-deficit/ hyperactivity disorder (ADHD) (Rhodes et al. 2005).

Here, we review some of the methods in experimental evolution that can be used to study the evolution of behavior (see also Fry this volume; Swallow et al. this volume; Zera and Harshman this volume). We illustrate how these methods can be applied toward understanding the origin and mechanisms of behavioral diversity using examples from our own work and from the literature.

## BEHAVIOR EVOLVES FIRST

A long-standing idea in evolutionary biology is that "behavior evolves first" (Mayr 1958; Blomberg et al. 2003). For example, natural selection will only favor physiological adaptations to a novel host plant species in herbivorous insects after the females have begun utilizing the new host for oviposition. Similarly, before the ancestors of whales evolved fins, they probably evolved a brain that made them want to spend time in the water. Providing support for this idea are fossil whales without fins (Gingerich et al. 2001) and living species that display evidence of a recent behavioral transition but without corresponding morphological adaptations (Fryer and Iles 1972). For example, the speciose family of freshwater tropical fish, cichlids, display an extraordinary diversity in feeding habits with clear evidence for genetic adaptations in dentition (Ruber and Adams 2001). However, a few species show no evidence for specialized dentition. *Cyrtocara moorii*, for instance, has large, irregularly shaped teeth that are inserted on the jaws in an uneven fashion. This cichlid feeds on food particles stirred up by other fish that do not require large dentition. Therefore, it has been proposed that recent ancestors of this fish displayed a different feeding behavior and that at some point in the recent past these fish changed their feeding habits, which relaxed selection on the original form of dentition and allowed the dentition to become irregular. Another cichlid species, *Haplochromis acidens*, displays large teeth that are typically seen in piscivorous fish, but this species feeds nearly entirely on plants. It has been proposed that *H. acidens* recently changed its feeding habits from a carnivore to an herbivore, and that the piscivorous dentition has apparently not greatly impaired the ability of the animals to eat plants and hence has yet remained unchanged (Fryer and Iles 1972).

The implication of "behavior evolves first" is that many adaptations in morphology or whole-animal physiology originate from or are constrained by behavioral shifts (see also the later section on "Testing Adaptive Hypotheses"). Thus, in the whale example, genetic changes in the physiology of the brain were necessary for the appearance of such nonbehavior morphological or physiological traits as fins, blubber, cardiovascular, and breath-holding abilities. In the cichlid example, genetic changes in feeding behavior lead the way for adaptations in dentition.

Hence, behaviors have a unique ecological significance. They often allow the animal to compensate for deficiencies in morphology or physiology (see also Oufiero and Garland 2007; Gibbs and Gefen this volume), facilitating invasion of novel habitats and opening new adaptive zones. Behavior can also be a key aspect of speciation (Fry this volume). The evolution of behavior, especially the higher forms associated with learning and intelligence, also has a unique relevance for understanding human nature because a major difference between human beings and the rest of animals is intelligent behavior.

## THE EVOLUTION OF BEHAVIOR

One goal of studying the evolution of behavior is to understand the underlying changes at different levels of biological organization, such as neural physiology, cell biology, or molecular biology (e.g., molecules and genes). At the level of behavioral phenotypes, one can also explore how a specific behavior changes as a consequence of selection on another behavior (or on organismal traits that have at least some behavioral component; see, e.g., Fry this volume; Gibbs and Gefen this volume; Huey and Rosenzweig this volume; Swallow et al. this volume; Zera and Harshman this volume). For example, selective breeding for high levels of voluntary wheel running in house mice *(Mus domesticus)* resulted in a corresponding decrease in thermoregulatory nest-building behavior (Carter et al. 2000). The reverse was also true: selection for small nests resulted in increased voluntary wheel running (Bult et al. 1993). This consistency implies that nest building and wheel running share a common genetic and/or neural basis (and hence are genetically correlated; e.g., see Rauser et al. this volume; Roff and Fairbairn this volume), even though the specific genes and neural circuits are not yet known.

The prospect of understanding how the brain changes at the level of neural physiology to produce a corresponding shift in behavior is intriguing and has the potential for making important contributions in biomedicine as well as evolutionary biology (discussed later). The nervous system is composed of an extraordinary number of components, any or all of which might play important roles in the evolution of behavior. For example, changes in the density of neurotransmitter receptor proteins, synthesis of neurotransmitters, and/or the structure or quantity of signaling molecules or transcription factors downstream of receptors might underlie evolutionary changes in behavior. Alternatively, changes in the development or connectivity of neurons, neuron numbers, morphology of neurons, electrical properties of neurons, properties of glial cells, or blood vessels in the brain might be subject to change by the evolutionary processes of selection and random genetic drift. We provide some examples of neurophysiological changes associated with genetic adaptation of behavior. In each case, presumably a change in DNA sequence somewhere in the genome underlies the physiological changes that lead to alterations in behavior. However, the genetic architecture of the natural heritable variation underlying evolutionary change in behavior has rarely been traced all the way to the structure of DNA. One rare example where great progress has been made in vertebrates is pair-bonding behavior in vole species. In these animals, investigators have identified a specific pattern of DNA (microsatellite) upstream of a gene that causes differential expression and neuroanatomical distribution of a neuropeptide (arginine vasopressin) receptor protein that appears to have a strong influence over pair-bonding behavior (Young et al. 1999; Hammock and Young 2005) (see later discussion).

The paucity of such examples will likely change soon, owing to recent technological advances for rapid genotyping and profiling of gene expression. The obvious question to ask is, Which genes change to facilitate the evolution of a specific behavior? But that may be somewhat misleading because the changes in DNA that underlie behavioral evolution may not occur directly in sequences of protein-coding genes (exons or introns), but rather in regulatory regions, as in the vole example (Young et al. 1999; Hammock and Young 2005). Such changes may indirectly affect gene expression by altering the folding properties of DNA, patterns of methylation, binding sites for transcription factors, and/or noncoding RNA products that regulate transcription (Lindahl 1981; Castillo-Davis 2005). Thus, rather than say the search is for "the genes," it might be conceptually more appropriate to search for "the structural variation in the genome" that leads to behavioral evolution. Whether those changes happen to occur in protein-coding sequences or not is an empirical question of considerable interest in the study of evolution, and it has enormous biomedical importance. It has been argued that changes in the genome that affect gene expression as opposed to those that affect the structure of proteins are particularly important for the evolution of behavior, though this issue is far from resolved (Whitehead and Crawford 2006).

Other interesting questions to be addressed at the level of genes include testing whether the type of genetic variation that is important for behavioral evolution consists of single nucleotide polymorphisms (SNPs), as compared with microsatellites or tandem repeats upstream of genes (Hammock and Young 2005), and whether behavioral evolution occurs from many small structural changes in DNA that cause many small changes in gene expression or from a few large structural changes or a few large changes in gene expression. Research in these areas has only begun, and the field is wide open for exploration.

## EXPERIMENTAL METHODS

Compared with morphology or life history (see Zera and Harshman this volume), behavior displays some unique features that bring specific challenges (see also Boake 1994). For example, behavior is highly sensitive to small and often uncontrollable environmental influences, as well as the animal's physiological and motivational states. For example, the response of parasitoid wasps to plant volatile chemicals is strongly affected by atmospheric pressure (Steinberg et al. 1992). Furthermore, behavior of most animals may be influenced by learning—that is, by the memory of past experience, which will depend on past environments. But which environments have been encountered in the past may in turn be affected by past behaviors. Thus, behaviors are expected to show complex patterns of genotype-by-environment interactions. The low repeatability of behavior and its dependence on past experience makes reliable measurements of behavior particularly challenging. On the one hand, the low repeatability would require multiple assays on the same individuals; on the other, the experience of being assayed once is likely to affect the animal's behavior in subsequent assays. For example, see Dohm et al. (1996) on differential heritability of sequential measures of sprint-running speed in mice.

To illustrate the complexity of this issue, consider the following example. One of us is currently in the process of establishing pilot data for a large-scale selective breeding experiment for increased physical activity in the home cage of house mice and has encountered an interesting problem. We are using sixteen video cameras mounted to the ceiling that feed into two computers to track the movement of singly-housed animals. Under the cameras, sixty-four mice can be housed simultaneously in modified cages (with food and water delivered from the side, and clear tops). Data are collected continuously in the light and dark (under red light) and are analyzed online with software (TopScan from Clever Sys., Inc., Reston, VA) designed to track the movement of the center of mass of the animals. This gives very precise measurement of total distance traveled at any desired increment, seconds, minutes, days, weeks, and so forth.

We have begun examining mice from the outbred stock Hsd:ICR strain (Harlan-Sprague-Dawley, Indianapolis, IN), the same strain used in another selection experiment for increased levels of voluntary wheel-running behavior (Swallow et al. 1998; Swallow et al. this volume). After measuring hundreds of animals over periods up to four weeks, what we have discovered is that the distance covered by an individual animal in their home cage is extremely repeatable between consecutive days: $R^2$ values are on the order of 0.75 or higher, with lower values occurring occasionally because of outliers. However, the correlation between daily values separated by two weeks is much lower, on the order of 0.25, and after four weeks is near zero! This poses a problem because we were hoping to be able to measure each animal over a period of six days and use the total distance on days 5 and 6 (as in Swallow et al. 1998) to capture a feature of the individual related to their physical activity level. But, now it would seem that the value for an individual would strongly depend on which week that individual happened to be measured.

It does not seem to be related to the age of the animals or their level of habituation to a cage, because some animals at the same age start out inactive the first two weeks and then become active, while others do the reverse. It seems to be related to some spontaneous property within the animal, or perhaps it is influenced by the other animals around them.

The home cages described here were designed so that adjacent animals can interact with each other (lick, smell, or groom each other) through wire mesh separating their cages in one corner. We have also experimented with housing animals without physical contact, using clear plastic inserts instead of the mesh, and have preliminary evidence that repeatability is higher with the plastic inserts. This suggests that the level of physical activity displayed by one animal depends on who its partners are in adjacent cages and that this effect is more potent when animals can interact via smell and touch.

In accord with this discussion, our estimates of narrow-sense heritability from midparent offspring regression of home cage activity (distance traveled on days 5 and 6 in the home cages that allow social contact via the wire grid) was zero, and no response to within-family selection was observed in one line (composed of twelve male-female pairs of mice) after four generations of selection. The take-home message from this example is that repeatability of behavior is difficult to assess and may be lower than predicted because it can change dramatically on different time scales and can be influenced by spontaneous events or events that are not easily measurable.

Another challenge is determining an appropriate assay for quantifying behaviors. Such assays usually aim to reduce the (presumably) high complexity of an animal's behavior under natural conditions to a manageable number of aspects that are measurable under standard conditions. This is illustrated, for example, by the controversy about how to measure host preference in herbivorous insects (e.g., Singer 2000). If the goal is to carry out an evolutionary experiment and breed for a particular behavior, then choosing an appropriate selection criterion is often difficult, especially if the target is a complex, high-level, cognitive behavior, such as locomotor activity, aggression, or learning ability, which can be measured in a variety of ways. One strategy is to screen animals on a variety of tests and then use some form of index selection (e.g., based on the first principal component) (Falconer and Mackay 1996). To our knowledge, this has rarely been done. Index selection may have its own problems if factor loadings change over generations. Moreover, it is not clear if sufficient selection could ever be levied against any single component variable to effect changes in allele frequencies. The problem with choosing one measurement is that the response may not be as generalizable as desired. For example, in the 1940s, in one of the most famous selective breeding experiments in psychology, Robert Tryon and Edward Tolman bred two separate lines of rats based on their performance on a maze-learning task. The selection criterion was total number of errors on a maze. Their goal was to breed "bright" and "dull" rats. However, later it was discovered that the "bright" rats were actually quite limited in their cognitive approach to solving mazes in that they had a strong preference for using spatial strategies (remembering where to move relative to distal landmarks in the room) rather than response strategies (e.g., turn left twice, then right twice, etc.; no need for remembering distal landmarks) (Innis 1992). Similarly, another line of rats selected to learn to avoid auditory and visual stimuli with electric shock turned out to be not better in learning but more fearful (Brush 2003).

Another feature of behavior and of life-history (Zera and Harshman this volume) and whole-organism performance traits (Swallow et al. this volume) is that the underlying genetic architecture is expected to be extremely complex (e.g., Leamy et al. 2008). For example, whereas a difference in coat color may be a (relatively) simple consequence of differences in the activity of an enzyme involved in pigment synthesis (e.g., Hoekstra et al. 2006), for behavior, the pathway from genes to phenotype is likely to be much longer. Changes at a variety of levels, including biochemistry of intracellular signaling and cell-cell communication, development and physiology of nervous and sensory system, endocrine regulation, and so forth, may all underlie an evolutionary change in behavior. Consider, for example, the complexity in how natural polymorphism at a single locus affects foraging behavior in *Drosophila*. It appears to influence behavior by affecting biology at a number of different of levels of organization, including biochemical activity of the enzyme it encodes (Osborne et al. 1997), physiology of synapses at neuromuscular junction (Renger et al. 1999), larval and adult foraging behavior (Pereira and Sokolowski 1993; Osborne et al. 1997), responsiveness to sugar (Scheiner et al. 2004), food intake (Kaun et al. 2007), associative learning (Mery et al. 2007), and success in competition (Fitzpatrick et al. 2007). Similarly, selection for a tonic response to threat (feigning death) in a beetle indicated that this antipredator behavior is negatively genetically correlated with locomotor activity, the correlation apparently being mediated by the levels of dopamine (Miyatake et al. 2008). In another such example, selection for increased voluntary wheel running in mice (Swallow et al. 1998) entailed changes in dopamine signaling (Rhodes et al. 2005), and these changes may in turn account for increased predatory aggression (toward crickets) in the selected lines (Gammie et al. 2003).

Depending on the question of interest, the study organism, the focal behavior, and technical limitations, a researcher intending to do an evolutionary experiment on a behavior faces a choice between three basic approaches (with the distinctions somewhat blurred), which we refer to here as artificial selection, mass selection, and laboratory natural selection (see also Rose and Garland this volume; Futuyma and Bennett this volume). These methods allow the investigator to produce an evolutionary change in a behavior. The result is an animal model that can be used to explore the genetic architecture of the response and the physiological bases for the behavioral shift by use of such additional tools as line-cross analysis, gene mapping (e.g., QTL analysis), and genetic engineering (discussed later).

## ARTIFICIAL SELECTION

In artificial selection, a target behavior is quantified for a number of individuals, and some top or bottom fraction is selected as breeders to produce the next generation (Garland 2003). For example, when selecting for preference for a particular odor, each individual may be repeatedly given a choice between the focal odor and a number of other odors; those choosing most consistently the focal odor would then be selected. Obtaining reliable individual measurements reduces noise, increases repeatability, and thus effectively increases the narrow-sense heritability of the focal behavior (the proportion of phenotypic variance in the trait that is due to additive effects of genes). This is because when noise is reduced, total variance is reduced without changing the genetic contribution, so the proportion that is genetic is larger. This is important because larger values for narrow-sense heritability increase the rate of response to selection. The effectiveness of selection can be further increased by controlling the mating by pairing specific individuals among the selected cohort (Falconer and Mackay 1996). Artificial selection also allows a direct measurement of selection differential or intensity, as well as realized heritability (Falconer and Mackay 1996).

Artificial selection is the approach used in most evolutionary studies of behavior, including virtually all experiments on vertebrates (see also Garland 2003; Eisen 2005; Swallow and Garland 2005; Swallow et al. this volume), where the population size and the number of generations are limited by other considerations (e.g., cost, low reproductive rate, long generation time). Breeding for desired behavioral characteristics has also been practiced for millennia in the process of domestication of animals, long before being applied to scientifically motivated study of the evolution of a behavior. One interesting observation from domestication is that unintentional changes (i.e., correlated responses) in morphological characters are remarkably similar across domestication events and across species of vertebrates (Belyaev 1979). For example, selection for tameness in the Russian fox, *Vulpes vulpes*, found patterned changes in pigment in the skin and fur in the shape of a star on the face (common in dogs, *Canis lupus familiaris*), floppy ears (common to dogs, goats, and sheep), and rolled tails (common in dogs and pigs) (Belyaev 1979; Belyaev et al. 1981; Trut 1999). Many other changes were noted in behavioral and physiological traits, such as the onset of hormonally driven fear and aggression responses during early postnatal development, and changes in serotonin metabolism in the brain (Hare et al. 2005). Although a great deal has been learned from these domestication events, they are not scientific experiments (e.g., variables are not always controlled, lines are not replicated). (For a review of experimental domestication studies in *Drosophila*, see Simoes et al. this volume.) Here, we restrict our attention to artificial selection applied to behaviors in the scientific context. A comprehensive review is beyond the scope of this chapter, but we mention a number of examples to illustrate the variety of experiments that have been performed.

*Dietary and Other Preference Traits*   For example, many different lines of mice and rats have been bred for their preference for ethanol (Mardones and Segovia-Riquelme 1983; Hilakivi et al. 1984; Crabbe et al. 1994; Grahame et al. 1999; Murphy et al. 2002). Honeybees were selected to specialize in foraging for pollen versus nectar, resulting in sixfold difference in the amount of pollen hoarded between colonies of the high and low line after only five generations of selection (Page et al. 1995). Outside of the dietary context, quail were selected for color preference (Kovach et al. 1981).

*Activity Amount and Patterns*   For example, mice were selected for increased levels of voluntary wheel running (Swallow et al. 1998), Japanese quail for dust-bathing activity (Gerken and Petersen 1992), and *Drosophila* for emergence from the pupa at a particular time of the day (Clayton and Paietta 1972). In another avian example, blackcaps showed a strong response in both directions to selection on migratory restlessness, a behavior thought to indicate motivation for seasonal migration (Berthold et al. 1990).

*Courtship and Mating Traits*   For instance, in *Drosophila*, selective breeding was imposed on the interval between courtship song pulses (Ritchie and Kyriacou 1996) and the rate at which males lick female genitalia during courtship (Welbergen and Vandijken 1992). Interestingly, in both cases the response occurred only in the direction of lower performance (i.e., greater interpulse interval and lower licking rate), suggesting that sexual selection keeps these traits at their maxima. Readiness to mate was successfully targeted by artificial selection in *Drosophila* (Manning 1961; Spuhler et al. 1978) and in stalk-eyed flies (Rogers et al. 2005).

*Anxiety and Aggression*   Mice were selectively bred for high and low activity in an open-field arena, which is considered a measure of exploratory behavior, response to novelty, and/or anxiety (DeFries et al. 1978). Mice have also been selected for various types of aggression (e.g., Gammie et al. 2006; references therein). More recently, genetic analysis of *Drosophila* lines bred for high intermale aggression led to the discovery of several genes that affect aggression (Dierick and Greenspan 2006; Edwards et al. 2006).

*Higher Cognitive Abilities, Such as Learning Ability and Memory*   Rats were selected for their learning performance in mazes (Tryon 1940) and for moving to a different compartment in response to an acoustic or visual signal previously associated with electric shock (Brush and Sakellaris 1973). Several species of insects (*Drosophila*, blowflies, honeybees) were selected for performance in various versions of the so-called proboscis extension reflex (i.e., the tendency to extend the mouthparts in response to a stimulus associated with a food reward) (McGuire and Tully 1987; Brandes et al. 1988; Lofdahl et al. 1992).

The main aim of most artificial selection experiments has been to demonstrate heritable variation for a particular behavioral response and to obtain genetically diverged lines. These lines are then used to discover the underlying genetic bases of the response through analysis of line crosses, quantitative trait loci (QTL) mapping, and verification of candidate genes with quantitative complementation tests or genetic engineering (e.g., McGuire and Tully 1987; Chandra et al. 2001; Dierick and Greenspan 2006; Edwards et al. 2006). They can also be used to study the underlying physiological, neural, and molecular mechanisms of the differences in behavior between selected and control lines, or between divergently selected lines.

However, reliable assays of individual behavior are time-consuming and labor-intensive, so artificial selection experiments on behavior face an acute trade-off between the precision of individual measurements and the number of individuals assayed. The latter limits the population sizes. For example, when selectively breeding *Drosophila* for a characteristic of the courtship song (interpulse interval), Ritchie and Kyriacou (1996) were constrained by the workload to a population size of four pairs per selection line, with only one line selected in each direction. Even assuming (rather generously) an effective population size equal to the census size, the selection lines would have lost to drift about one-third of their original heterozygosity within six generations of selection (Hartl and Clark 1997). It is thus not surprising that the response decelerated sharply within just six generations (Ritchie and Kyriacou 1996). With small population sizes, the response to selection will usually underestimate the evolutionary potential of the original base population and may be confounded by drift and inbreeding. This is particularly important for interpreting correlated responses to selection. Rather than being due to pleiotropic effects of alleles favored by selection, they can reflect fortuitous fixation of alleles at loci unrelated to the targeted phenotype. As we discuss later, one way to deal with this problem is to increase the scale of the experiment. Large artificial selection experiments have been performed on behavior, where multiple lines and reasonable population sizes are maintained. This reduces inbreeding depression, and the replicate selected and control lines allow a way to account for drift as an alternative explanation for correlated responses.

To summarize, artificial selection is a powerful tool with which to explore the question of how behavior evolves (i.e., the underlying proximate mechanisms), but it is less informative concerning the adaptive significance (i.e., costs and benefits) of behavior. However, consider the following unusual artificial selection experiment conducted on largemouth bass that has direct adaptive significance because it was conducted in the wild. Between 1977 and 1998 in experimental ponds in central Illinois, bass were caught and released by sport fishers, except the fish were tagged and records were kept about how many times the fish were caught in a season. At the end of a season, the ponds were drained and lines were transplanted into new ponds based on whether they were caught four or more times in a single season or fewer than one time. After, three generations of this type of selection, bass from high- and low-vulnerability lines were stocked into a common-garden pond and permitted to grow for three years. The authors discovered that fish from high-vulnerability lines displayed higher metabolic rates, greater food consumption, and greater investment in parental care as compared with the low lines. The authors concluded that if fish are selectively harvested based on vulnerability, then the remaining fish in the population may be less effective in providing parental care and potentially reduce reproductive output (Cooke et al. 2007). For a recent review of the evolutionary consequences of fishing with respect to salmonids, see Hard et al. (2008).

MASS SELECTION

A second approach, mass selection, relies on an experimental setup that sorts individuals into groups depending on a particular behavior. For example, mass selection on odor preference could be applied by running large numbers of individuals through a Y-maze with the focal odor coming from one arm and another odor from the other arm, and breeding the next generation en masse from those that chose the arm with the focal odor. This approach allows for greater population sizes, thus alleviating the problem of inbreeding. However, such a binary behavioral response will usually have a large random component—even an indifferent individual will have a 50 percent chance of being selected, and so selection imposed this way will be rather weak. Furthermore, the distribution of the underlying preference trait (which could be defined as the probability of choosing the focal odor) remains unknown, so the selection differential cannot be estimated. This problem is less acute if the individuals are sorted into multiple categories that reflect a degree of a particular behavioral tendency. For example, in probably the longest experimental evolution study on behavior in a eukaryote (see also Travisano this volume), *Drosophila* were selected for over five hundred generations for geotaxis, using a simple but ingenious setup that sorted flies according to their geotaxis score on the scale from 1 to 9 (figure 11.2; Ricker and Hirsch 1988). Unfortunately, not all behaviors are amenable to such automatic sorting.

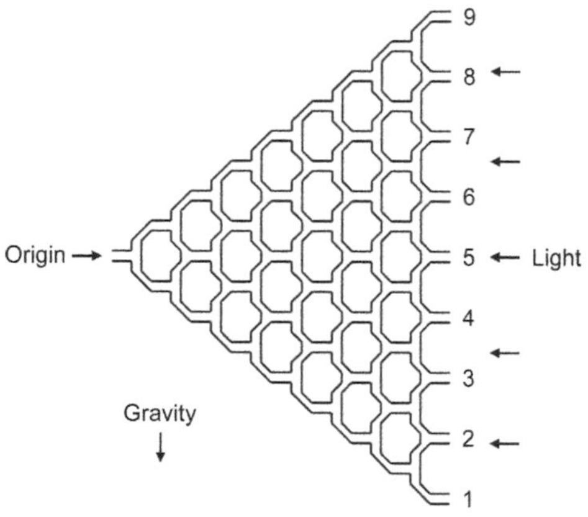

FIGURE 11.2

The apparatus for fractionating flies according to geotaxis used in the five-hundred-generation mass selection experiment of Hirsch and coworkers (Hirsch 1959; Ricker and Hirsch 1988). The flies are released at the origin and move to the right attracted by light, having to choose at each fork whether they move up or down. Traps at the end collect flies according to their geotaxis score, from 1 (strong positive geotaxis) to 9 (strong negative geotaxis). From Toma et al. (2002).

The power of mass selection on behavior is maybe best illustrated by an experiment in which *Drosophila* were selected for increased flying speed in a wind tunnel (Weber 1996). An automatic setup sorted fifteen thousand flies in a single run, according to how far they were able to travel in a wind tunnel with increasing wind speeds from the beginning of the tunnel to the end. The two thousand fastest adults out of about fifty thousand were chosen as breeders each generation. Within one hundred generations, the average speed increased almost 100-fold, from about 0.02 meter/second to over 1.7 meters/second (Weber 1996). In other experiments, *Drosophila* have been successfully selected for the tendency to move along or against a moving striped pattern ("optomotor behavior") (Gotz 1970), or for the height of the pupation site above the medium (Singh and Pandey 1993). However, not all mass selection experiments on behavior were successful. For example, a parasitoid wasp did not respond to selection for a shift in preference of plants used by potential hosts (Rutledge and Wiedenmann 2003). Similarly, in one of our labs, we failed to obtain a response to mass selection on the ability of *Drosophila* to avoid an odor previously associated with mechanical shock, a form of associative learning (Kolss and Kawecki 2008).

To summarize, mass selection provides a way to implement selective breeding with large population sizes. This reduces the problem of inbreeding, but the types of behaviors amenable for this method require creative designs for sorting individuals en masse, such as the geotaxis device or wind tunnel. Mass selection is also less informative than laboratory natural selection or field experiments (Irschick and Reznick this volume) concerning the adaptive significance (i.e., costs and benefits) of behavior.

## LABORATORY NATURAL SELECTION

In both approaches just described, selection is imposed by allowing only a defined subset of individuals to breed. This aspect is absent from a third approach to experimental evolution of behavior, which we refer to as *laboratory natural selection* (see Huey and Rosenzweig this volume; Garland 2003). Here, rather than explicitly excluding some individuals from breeding, experimental populations are maintained under a husbandry regimen in which individuals with particular characteristics are expected to contribute more offspring to the next generations. To the best of our knowledge, behavior has never been the intentional "target" of any field introduction or laboratory natural selection experiment. However, field experiments have been carried out with guppies (see Irschick and Reznick this volume), where guppies are introduced between sections of streams in Trinidad below or above waterfalls with or without predatory fish, and scientists interested in the evolution of behavior have taken advantage of these experiments. For example, Magurran et al. (1992) examined schooling behavior and reaction to the presence of a mock predator in guppies from low- and high-predation sites in the laboratory. Fish were born and reared under the same conditions, and still those from high predation sites displayed stronger schooling behavior and greater avoidance of the mock predator.

Another interesting experiment that has yet to be done would be to introduce fish into replicate streams or aquaria with varying flow speeds to study the evolution of swimming behavior and performance.

To summarize, similar to artificial and mass selection experiments, such natural selection experiments can be used to obtain lines with genetically divergent behavior for genetic analysis or to study constraints on evolution. However, the response is expected to be slower because effective selection on any intended focal trait will be weaker than would be possible under direct artificial selection. Moreover, the results will be less predictable because no particular trait is being selected directly by the investigator. Depending on the goals of the experiment (e.g., if one wishes to discover "multiple solutions"), this may be an advantage or a disadvantage (Garland 2003). The main general advantage of natural selection experiments is that they allow testing adaptive hypotheses about ecological factors thought to favor particular phenotypes (Garland and Carter 1994; Gibbs and Gefen this volume; Huey and Rosenzweig this volume). We return to this application toward the end of the chapter.

## METHODOLOGICAL CONSIDERATIONS IN EVOLUTIONARY EXPERIMENTS

*Choice of Species*   Obviously, the experimental question will, to some extent, dictate the species (e.g., swimming in fish vs. walking in land animals). As a general rule, though, it is desirable to choose a model species with a short generation time so that results can be acquired in a reasonable amount of time. Consider that it might take five or more generations to observe a response, depending on narrow-sense heritability and the intensity of selection. For example, if the goal is to study the behavior of mammals, small-bodied rodents are a good choice because generation time is on the order of three months. In contrast, chimpanzees have a generation time of approximately seventeen years, so it would take more than the lifetime of the researcher to observe a response to selection. Other desirable features are that the animals are easy to breed, produce large litters, are small in body size, and are easy to keep alive (and healthy) in the laboratory. Another consideration if the goal is to carry out an artificial selection experiment in vertebrates where individuals are paired for breeding is that the species must be amenable to investigator-imposed matchmaking. For example, zebra finches do not always pair bond when placed in the same cage, even though they seem to meet all the other criteria nicely.

*Replication*   In a selective breeding experiment, the unit of replication is the line, not the individual (Henderson 1997). This is because even in the absence of selection, lines of finite population size will diverge in gene frequencies, and hence in the mean value for various traits, because of random genetic drift. Thus, without replication (e.g., if only two lines are maintained, one high-selected line and one nonselected line as a control), then it is impossible to know whether the difference in a trait between the two lines is due to selection or drift. Note that the degrees of freedom for the statistical test of the effect of selection on a phenotype are based on the number of replicate lines, not the number of individuals. In a two-line comparison, degrees of freedom are zero. In an eight-line comparison (four high and four control), degrees of freedom are six (Swallow et al. 1998; Garland 2003). This argument has been described in detail elsewhere and in other chapters of this book (e.g., Swallow et al. this volume), so we will not elaborate further here. An empirical example from our own work demonstrating the importance of replication is shown in figure 11.3.

*Line Types* In addition to the selected lines, it is important to maintain unselected lines to serve as controls. This can be done with a random or quasi-random breeding design where pairing of siblings or close relatives is avoided. Sometimes it may be of interest to include lines bred for both high and low values for the trait. However, even when this is done, it is important to maintain control lines. It should not be assumed that the same genes (i.e., different alleles at the same locus) underlie behavioral shifts up versus down because different suites of genes may be involved. For example, high levels of locomotor activity may involve genes that make the organisms perceive the physical activity as pleasurable or rewarding, whereas low levels of activity might involve genes that promote fear or shyness. Figure 11.4 illustrates this point with an empirical example taken from the literature.

*Base Population* For a response to selection to occur, the population must be composed of genetically variable individuals. Depending on the species, this could be accomplished by collecting individuals from a wild population or by use of a deliberately outbred captive stock. If a wild-derived population is to be used as the base for beginning a selection experiment, then one must consider whether it is better to use a population that has experienced only a few generations of potential "domestication" or has had substantial time to adapt to the new (artificial) environment (see Fry this volume; Rauser et al. this volume; Simões et al. this volume). Outbred stocks are relatively large populations that include hundreds or thousands of individuals in each generation. Various outbred strains of rodents are maintained by commercial vendors (e.g., Harlan-Sprague-Dawley), and typically individuals are bred randomly except that siblings are never mated together. The wheel-running experiment of Swallow et al. (1998) began with outbred Hsd:ICR mice. Another option is to start with a population derived from a cross of inbred strains that has subsequently been outbred for two or more generations. This option is commonly used in mouse behavior genetics (Crabbe and Phillips 1993). For example, Lynch (1980, 1994) used HS/Ibg mice, originally created by an eight-way cross of inbred strains.

One advantage of crossing inbred strains is that a population can be produced in which all the alleles at a given locus are represented with approximately equal frequency. This reduces the chance that a given allele in the starting population will be lost due to random genetic drift (Falconer and Mackay 1996). Moreover, distantly related progenitor strains can be crossed, potentially including those derived from separate subspecies,

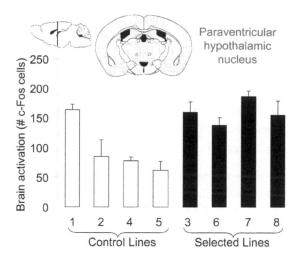

FIGURE 11.3

Statistical evidence for correlated evolution of neurophysiology and behavior requires replication (data from Rhodes et al. 2003). Data from eight separate lines of mice are shown. After initial establishment, four lines were exposed to twenty-nine generations of selective breeding for increased voluntary wheel-running behavior. These are referred to as the "Selected Lines" in the figure and are numbered 3, 6, 7, and 8, based on initial random assignment of the eight lines before selection was applied (Swallow et al. 1998). Four other lines were bred randomly with respect to wheel running for the same number of generations, and they serve as the controls. These are referred to as the "Control Lines" and are numbered 1, 2, 4, and 5. Each bar represents the mean ($\pm$ SEM) value of the trait (described later) measured in three separate animals (i.e., $n = 3$ per bar). Each animal was placed with access to a running wheel for six days. On day 7, access to wheels was blocked by placing a slate between the wheel-access tunnel and the cage. Animals were sacrificed on day 7 at a time when they would normally be running at peak levels. The purpose was to sample the animals at a time when they would be in a psychological state of high motivation to run without actually running. The brains were removed, sectioned, and stained for c-Fos protein, which is a transcription factor that is transiently expressed in the nucleus of brain cells after they are stimulated. The number of nuclei that stained positive for c-Fos were counted in the paraventricular hypothalamic nucleus (PVN), a brain region whose projecting neurons secrete corticotrophin-releasing hormone into the blood and hence initiate a pathway that results in secretion of corticosterone from the adrenal glands. A comparison of the trait values for control lines 1 and 2 shows that sampling error and/or the random evolutionary processes of genetic drift (and, to a lesser extent, founder effects) are capable of producing large differences in this trait that are of a similar magnitude as those produced by selection. Only when all the replicate lines are considered together is it possible to infer, with a reasonable level of statistical significance, that this trait is associated with past selection for high running ($F_{1,6} = 14.7$, $p = 0.009$; see table 2 in Rhodes et al. 2003). The implication with respect to the mechanism of how the behavior evolved is that increased activation in an area of the brain that initiates the stress response appears necessary for the evolution of increased physical activity, but that this activation alone is not sufficient for causing the high wheel running, as suggested by control line 1 (i.e., other mechanisms are also involved).

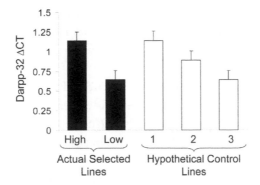

FIGURE 11.4

Control lines are necessary for interpreting results of a selection experiment. Data above the label "Actual Selected Lines" were redrawn from figure 2 in Palmer et al. (2005). Their experiment consisted of two lines of mice. One line was selectively bred for increased locomotor stimulation following administration of methamphetamine (High) and the other was bred for low stimulation in response to methamphetamine (Low) (Palmer et al. 2005). DARPP-32 gene expression was higher in the Low line than in the High line in the nucleus accumbens (a key brain region in the natural reward circuit; see figure 11.5). Note that delta CT stands for change in cycle time and is inversely proportional to transcript abundance. DARPP-32 is an important signaling molecule that responds to dopamine and other neurotransmitters when they bind to their receptors. This result is intriguing because methamphetamine increases dopamine in extracellular spaces and, hence, DARPP-32 seems like a plausible mechanism for the differential response to methamphetamine. However, as no control line was measured, it is impossible to know whether DARPP 32 was associated with low stimulation or high stimulation or both. The hypothetical control lines (1–3) illustrate how the interpretation would change for three possible outcomes: (1) DARPP-32 is associated with low stimulation but not high stimulation; (2) DARPP-32 is associated with both low and high stimulation; (3) DARPP-32 is associated with high but not low stimulation.

such as *Mus spretus, Mus domesticus,* and *Mus musculus,* which increases the number of polymorphic loci. On the other hand, populations composed in this way are derived from strains that were each formerly stripped of their genetic diversity through the process of inbreeding (and some may have been directionally selected for one or more traits prior to the intentional inbreeding). Thus, the number of possible alleles at a given locus is limited to the number of inbred strains in the cross. This may be an important consideration, especially if highly polymorphic regions of the genome, such as microsatellites or tandem repeats, are important for the evolution of behavior, which appears to be the case (see later discussion Hammock and Young 2005). Moreover, the process of removing genetic variation by inbreeding and then piecing it back together by crossing inbred strains introduces an artificiality that may not be desirable if the goal is to develop a model for the evolution of behavior in nature. Although it could be argued that any model derived from laboratory-adapted animals has limited relevance for the natural world, the structure of genetic variation in outbred laboratory mice has been empirically shown to be similar to wild populations of mice (Carter et al. 1999 and references therein). Hence, outbred laboratory stocks may respond to selection and drift in a manner more similar to wild populations than would populations derived from crosses of inbred strains.

Using a cross of inbred strains has some practical advantages, however. One is that progenitor strains can be used that have already been genotyped or entirely sequenced. This makes it easy to find genetic polymorphisms (such as SNPs) that are needed for QTL mapping. Note, however, that this is less of an advantage today than it was a few years ago due to the recent advances in genotyping, which makes it relatively easy to genotype large numbers of animals over a short period of time. In fact, databases are now publicly available that provide information on genetic polymorphisms for a variety of commercially available outbred stocks (see the Mouse SNP detector: http://gscan.well.ox.ac.uk/gs/strains.cgi).

Another practical advantage of using a cross of inbred strains is that once a QTL has been identified, it is possible to develop congenic strains for fine mapping (Silver 2005). Congenic strains are produced by taking two inbred strains that differ for the trait of interest, and backcrossing one on the other many times until a small region of the genome (near the QTL) from one inbred strain is placed on an entire background of the other. This requires genotyping individuals at each generation in the region of the QTL and only backcrossing those that contain the donor DNA in the region. The congenic strains can then be tested for the trait of interest. If they show a trait value similar to the donor strain whose small region was transferred to the background strain, then that constitutes strong evidence that the QTL is located in the small transferred region (Crabbe et al. 1994). However, fine mapping to the level of finding the gene or genes underlying the QTL with congenic strains takes a very long time (ten years or more; see Shirley et al. 2004). Recent advances in molecular genetics offer a new approach that promises to be much faster (see later discussion).

## LINE-CROSS ANALYSIS

A response to selection certifies that the trait targeted by selection has heritable variation. This variation can be quantified in terms of realized heritability or additive genetic variance (Falconer and Mackay 1996). However, additional methods are required to characterize the genetic architecture of the experimental evolutionary change. Some insights into the nature of genetic variation underlying the response to selection (e.g., the prevailing patterns of dominance, overall magnitude of epistatic effects, and the number of loci involved) can be gained from the analysis of crosses of directionally selected lines. For example, line-cross analysis of the response of *Drosophila* to selection on interpulse interval of the courtship song indicated that it was based on many loci spread throughout genome, with no strong pattern of dominance or epistasis (Ritchie and Kyriacou 1996). In turn, an analysis of blowfly lines *(Fromia regina)* subjected to fourteen generations of selective breeding for appetitive learning using the proboscis extension reflex showed a strong contribution of epistasis (McGuire and Tully 1987). It is also useful to examine crosses between parallel (replicate) selection lines within a selection regime. This may help detect confounding effects due to inbreeding as well as additional novel insights. For example, Kawecki and Mery (2006) analyzed crosses between pairs of replicate lines selected for improved olfactory learning ability. While F1 crosses between some pairs of lines performed as well as the parents, F1 crosses between other pairs of lines lost all their response (i.e., they performed as poorly as the unselected controls; the performance of F2 and backcrosses was intermediate). This indicates that the response of replicate lines to selection had different genetic bases, even though the lines were derived from the same base population (Kawecki and Mery 2006).

GENE MAPPING: NEW TECHNOLOGY

Identification of genes responsible for the response to selection can be a starting point for trying to understand how the behavioral change is mediated at the molecular, physiological, and neural level (Wehner et al. 2001). Gene mapping has historically been difficult and time-consuming, with only a few successful attempts at identifying the gene or genes underlying a QTL (Flint and Mott 2001). However, state-of-the-art technology brings new promise. Now it is possible to rapidly genotype and monitor expression of nearly every gene in the genome using microarray technology. If the same animals that are measured for a behavioral trait of interest are also genotyped throughout the genome and assessed for gene expression, then this puts the investigator in an excellent position to find evidence that particular genes are involved in a behavior. This is because the microarray data will likely identify genes whose expression is correlated with the behavior. If these genes happen to be physically located in the QTL interval, then that constitutes strong evidence that those genes probably underlie the QTL (Jansen and Nap 2001; Chesler et al. 2005). Their involvement can be verified with quantitative complementation tests (Mackay and Fry 1996) or with an analysis of mutants or transgenics (e.g., Toma et al. 2002). The complementation test is a tool developed in *Drosophila* genetics where a mutant "null" allele (which does not transcribe the hypothesized candidate gene) is crossed onto a variety of backgrounds with different naturally occurring alleles in the QTL interval. If the heterozygote shows the mutant phenotype, then it is said to have failed the complementation test (because the alternative allele was not able to compensate for the mutation), and it provides evidence that natural variation in the candidate gene underlies the QTL. This strategy has recently been applied for the first time in mammals, to identify genes underlying anxiety in outbred mice (Yalcin et al. 2004).

Another advantage of the new technology is that it is possible to treat expression of each gene on the microarray as a quantitative trait for QTL mapping. A QTL for expression of a gene is referred to as an *eQTL* (with this terminology, *bQTL* is used to specify that the QTL refers to the behavior). Thus, if expression of a gene is correlated with the behavioral trait, and the eQTL for that gene maps to the same location as the bQTL, then that constitutes strong evidence for involvement of the gene in the behavior, even if the physical location of the gene is not in the eQTL interval (i.e., if the gene is *trans* regulated as opposed to *cis* regulated). In the case of *trans* regulation, the structural variation in the genome that underlies evolution of the behavior may be located far from the genes it influences (Doerge 2002; Schliekelman 2008). Whether evolution of behavior tends to involve more *trans*-regulated genes than *cis*-regulated genes is not known (Hoekstra and Coyne 2007). This gene-mapping technology is very new, and to the best of our knowledge, nothing has been published using this method on any vertebrate model of behavioral evolution. However, the tool is currently being used to explore the genetic architecture underlying the evolution of increased voluntary wheel running in house mice (Swallow et al. 1998) and has been used successfully to implicate particular genes in underlying behavioral variation in biomedical animal models of alcoholism, obesity, and hypertension (Dumas et al. 2000; Chesler et al. 2005; Schadt et al. 2005).

Several factors limit the usefulness of expression profiling to study responses to selection. First, the method assumes that changes in gene expression underlie the changes in behavior and ignores the possibility that different alleles might also, or instead, affect the structure of proteins rather than quantity of message RNA. Second, because analysis of microarray data amounts to thousands of statistical comparisons, it faces an unavoidable trade-off between a large number of false positives and false negatives. Third, because of hitchhiking and genetic drift, changes in allele frequencies are likely to occur in selection lines at loci unrelated to the phenotype under selection (although this limitation can be overcome by studying replicate selected and control lines). Nonetheless, microarray expression data are useful to identify candidate loci and metabolic pathways, which can then be verified with other means. For example, Toma et al. (2002) combined microarray technology with quantitative complementation tests to identify a gene involved in the response to selection on geotaxis in *Drosophila*. Hence, empirical data generally support the idea that heritable variation in gene expression is strongly associated with behavioral variation (Chesler et al. 2004; Mery et al. 2007).

We conclude this section by noting an additional tool that has been used widely by behavior geneticists in the biomedical sciences, but not as often by biologists interested in evolutionary mechanisms (but see Hughes and Leips 2006). The approach is to use panels of isogenic strains for gene discovery. Isogenic strains are strains in which same-sex members are genetically identical. These strains are typically produced by repeated inbreeding (breeding of siblings). Two different inbred strains can also be crossed to produce an F1 that is also isogenic but different from an inbred strain in that it is heterozygous (as opposed to homozygous) at all loci where the two strains differ. Examples of useful isogenic strains include sets of standard or recombinant inbred strains of mice (Chesler et al. 2004) or flies (Jordan et al. 2006). Recombinant inbred strains are panels of inbred strains derived from the same outbred population, usually an F2 or higher cross of two or more inbred strains (Silver 2005). The advantage of using isogenic strains is that because same-sex individuals within a strain are all genetically identical, data on each genotype can be collected in different individuals over a period of many years by different investigators, and the data are all cumulative. Hence, once the animals are genotyped at a large number of loci and measured for gene expression in a variety of tissues and time points after different experimental manipulations, the data can be cataloged so that all the investigator needs to do is measure the trait of interest in the panel, and automatically, access is available to a dense genetic map along with gene expression values. Such information is publicly available in the form of the Mouse Phenome Database (http://phenome.jax.org/pub-cgi/phenome/mpdcgi?rtn=docs/home) for approximately one hundred standard inbred strains and WebQTL (www.genenetwork.org/) for approximately eighty-eight recombinant inbred lines derived from a cross of two standard inbred strains, C57BL/6J and DBA/2J. WebQTL also includes tools for gene mapping online. A large effort, entitled the Collaborative Cross, is now underway to produce hundreds of recombinant inbred mouse lines derived from a cross of eight inbred strains that were chosen for biomedical relevance and for genetic diversity (three different subspecies are represented, *Mus musculus*, *Mus domesticus*, and *Mus castaneus*) (Churchill et al. 2004). Note that, at the present time, it is unclear how relevant genes discovered via the Collaborative Cross will prove to be for understanding phenotypic evolution in any one species of *Mus*, whether in experimental settings or in the wild.

Because of inbreeding depression (i.e., low fecundity, low performance), some have questioned the use of inbred strains for studying the evolution of behavior (Fonio et al. 2006). One way to address this issue is to study F1 hybrids of inbred strains, which are isogenic but also heterozygous at all loci that differ between the parental strains. We view the Mouse Phenome Database, WebQTL, and the Collaborative Cross as powerful resources for exploring molecular mechanisms of behavioral evolution. We wish to remind skeptical readers that panels of inbred strains or F1 hybrids are regarded as animal models, and although they may lack certain ecological relevance, they may still be useful to explore features of behavioral evolution—namely, molecular genetic mechanisms.

## GENETIC ENGINEERING

Great advances have been made in recent years in genetic engineering. It is now possible to render a gene nonfunctional (null mutant, or knockout), engineer a gene so that it is over- or underexpressed, and even change the anatomical pattern of expression of the gene in the brain (Wells and Carter 2001). Among the null mutants, it is now possible to control expression of the gene depending on the presence of a chemical (inducible knockout) (Olausson et al. 2006). These tools can be used to directly test whether a gene is necessary for a behavior, but each gene must be evaluated separately. This poses a substantial limitation in the context of studying the evolution of behavior—or, indeed, any complex trait—which presumably involves changes in many genes that interact with each other to coordinate complex changes in physiological pathways (Swallow and Garland 2005). Nonetheless, genetic engineering has its place. For example, later we describe a research program in which transgenic engineering technology was used in an extremely creative way to dissect the genetic mechanism underlying the evolution of pair bonding in voles.

## EXAMPLES OF CORRELATED RESPONSES TO SELECTION ON BEHAVIOR

One of the most interesting aspects of the evolution of voluntary behavior is that presumably it requires changes in the brain related to motivation. This is because voluntary behaviors require motivation (an emotional state of urgency to engage in the behavior, as opposed to doing something else), and motivation is expected to vary among individuals depending on experience, personality, and genetics (Horn et al. 1976). Hence, by studying the evolution of a voluntary behavior, we may gain insight into how the brain processes decisions related to wanting or liking a particular experience. One of the best examples of this phenomenon involves an artificial selection experiment for increased voluntary wheel running in house mice (Rhodes et al. 2005). The experiment is currently in its fifty-first generation. The model consists of four replicate high-runner (HR) lines and four control lines. The base population was derived from an outbred stock (Hsd:ICR strain; Harlan-Sprague-Dawley, Indianapolis, IN. Each generation, mice were given access to running wheels for six days. In the selected lines, mice were chosen as breeders for the next generation based on their total number of revolutions run on days 5 and 6. In the control lines, mice were bred without regard to their wheel running. By generation 15, selected lines ran approximately 13 kilometers/day, whereas control lines ran 5 kilometers/day. The differential (expressed as a ratio) has remained relatively stable since then, although large fluctuations in actual mean distances run by both control and selected lines have occurred from one generation to the next (Garland 2003).

Selection on wheel running has resulted in the correlated evolution of other behaviors. For example, when the animals are placed in an operant situation where a lever press is required to unlock a wheel and allow it to freely rotate for a set duration, the HR mice showed fewer lever presses than control mice when running duration was short, set to ninety seconds, but similar when duration was lengthened to thirty minutes (Belke and Garland 2007). This led Belke and Garland (2007) to hypothesize that there may be an inherent trade-off in the motivational system for activities of short versus long duration. The HR mice are also more active in their cages without wheels (Rhodes et al. 2001; Malisch et al. 2008), build smaller thermoregulatory nests (Carter et al. 2000), and show greater predatory aggression toward crickets (Gammie et al. 2003). Note, however, that HR mice are not more active than control mice when placed in an open-field arena for a three-minute test (Bronikowski et al. 2001). This suggests that locomotor activity in a habituated environment is a different trait (i.e., influenced by different genes and neural pathways) than locomotor activity in a novel environment. Locomotor activity in the open-field test is probably more related to novelty seeking, exploratory behavior or anxiety-related emotions, rather than motivation or physiological capacity for exercise (Kliethermes and Crabbe 2006).

Some of the most striking discoveries in the mice bred for increased wheel running have been changes in the brain that appear to underlie increased motivation to run. For example, dopamine is widely known for its role in motivation, reward, and reinforcement, and the HR lines respond entirely differently than controls to psychoactive drugs that increase dopamine signaling, such as apomorphine, cocaine, methylphenidate, and GBR 12909 (Rhodes et al. 2001; Rhodes and Garland 2003). Moreover, areas of the brain that comprise the natural reward circuit (figure 11.5) show differential levels of activation in HR versus control lines when the animals are prevented from running and presumably are in a state of high motivation or "withdrawal" from running (see figure 11.3) (Rhodes et al. 2003). Presumably, structural changes in the genome underlie these differences in neurophysiology and pharmacology, but the identity and location of those changes in the genome are not known at present. An eQTL project is underway.

The discovery of the involvement of the natural reward circuit in the evolution of running behavior may have implications for how voluntary behaviors evolve, in general. It seems intuitive that motivation should need to evolve to shift levels of a voluntary behavior, whatever the behavior. Given that motivation is influenced by neural activity in the natural reward circuit, it seems likely that this will be a target of change. Consistent with this hypothesis is the evolution of an entirely different behavior, pair bonding in voles, which appears to involve a change in the distribution of arginine vasopressin receptors in the natural reward circuit (discussed later) (Young and Wang 2004). Thus, perhaps increasing motivation for one behavior versus another requires adaptations in specific aspects of the natural reward circuit. So, for wheel running in mice, a signaling molecule downstream of a dopamine receptor may be a target (Rhodes et al. 2005); whereas for pair bonding in voles, distribution of arginine vasopression receptors may be a mechanism (Young and Wang 2004).

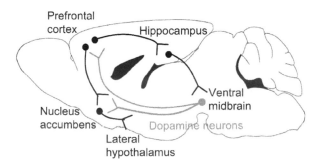

FIGURE 11.5
The natural reward circuit. A drawing of a mouse brain (saggital view) showing connections among a few key brain regions involved in motivation and reinforcement of appetitive behaviors.

FIGURE 11.6

Dramatic evolution of nest size has been accomplished using bidirectional selection for thermoregulatory nest building behavior. A representative mouse from a high line (left) versus a low line (right). The experiment included six lines, two high-selected, two low-selected, and two maintained as non-selected control lines (Lynch 1980, 1994). Photo by Abel Bult-Ito.

A replicated selective breeding experiment for thermoregulatory nest-building behavior has provided another intriguing model with which to explore the evolution of motivation for behavior (Lynch 1980, 1994). This model consists of two high, two low, and two replicate control lines. The selection criterion was the total mass of cotton used to build a nest over four consecutive daily trials. The starting population was derived from a cross of eight standard inbred strains. By generation 15, the high-selected lines were building nests with approximately fifty grams of cotton, whereas the low lines were using only five grams and control lines, fifteen (figure 11.6). Recall from the experiment on wheel running that nest building evolved as a correlated response (Carter et al. 2000). The nest-building experiment provides further support for the hypothesis that these seemingly different behaviors are jointly affected by some of the same genes because the small-nest builders ran more than the controls or high-nest builders (Bult et al. 1993).

Surprisingly little work explored how the brain has changed in the divergent nest-building lines. Two reports found an increase in the number of arginine-vasopressin neurons in the suprachiasmatic hypothalamic nucleus of a low line relative to a high or a control line, but results for the other replicate lines were not reported, and to the best of our knowledge, no one has looked to see whether alterations have taken place in the natural reward circuit (Bult et al. 1992; Bult et al. 1993).

## THE EVOLUTION OF MATING SYSTEMS IN VOLES:
## FROM GENES TO BEHAVIOR

One limitation of experimental evolution is that it is unclear whether behavioral perfor-mance scores targeted by artificial selection will correspond to traits shaped by natural selection. An alternative approach for identifying mechanisms of behavioral evolution is to use experimental methods to explore genetics and physiology of real behavioral shifts that occurred among populations or species in nature. In this section, we describe an in-teresting example of how laboratory experimental tools, such as genetic engineering and pharmacology, were used to discover the evolution of mating systems in voles. It is one of the most complete stories of discovery of a real evolutionary shift all the way from the genes to physiology to behavior in a vertebrate.

Voles of the genus *Microtus* are one of the most speciose genera of mammals (Jaarola et al. 2004). They also display an extraordinary diversity in social behavior. Some species, such as *M. orchrogaster* (prairie vole) and *M. pinetorum* are socially monogamous (i.e., they develop pair bonds for life and display biparental care); whereas others, such as *M. montanus* and *M. pennsylvanicus*, are solitary and nonmonogamous and do not typically display biparental care (figure 11.7). Thus, the *Microtus* genus provides a useful model with which to explore the neurophysiology and molecular-genetic mechanisms underlying the evolution of pair bonding (Young and Wang 2004).

One of the first hypotheses for the mechanism of the evolution of pair bonding was generated from the discovery that central administration of arginine vasopressin increases social behavior in voles and other animals, especially in males (Winslow et al. 1993). Hence, it was hypothesized that the evolution of monogamy would involve alter-ations in expression of the arginine vasopressin receptor in the brain. Consistent with this hypothesis, a comparative study of the four species of voles described here showed strik-ing differences in the pattern of distribution of the arginine vasopressin receptor (V1aR) in the brain (of both sexes), depending on whether the species was monogamous. For ex-ample, male monogamous voles showed more V1aR in the ventral pallidum and less V1aR in the lateral septum than nonmonogamous species (Young et al. 1999). This result has been replicated numerous times (figure 11.7) (for a review, see Young and Wang 2004).

On the surface, this seems like strong evidence that V1aR plays a role in pair boding in voles, but note the severe limitation of this four-species comparison in light of the phylogeny (figure 11.7). Differences in social behavior are completely confounded with genetic relatedness. Hence, this analysis gets dangerously close to a two-species compar-ison with zero degrees of freedom (Garland and Adolph 1994). Even in the absence of a role in pair bonding, V1aR would be expected to show similarities within and differences between groups *orchrogaster-pinetorum* versus *montanus-pennsylvanicus* because of phylo-genetic relatedness.

It is interesting that despite weaknesses in the comparative data, direct evidence sup-ports the hypothesis that V1aR receptor distribution is associated with social behavior in

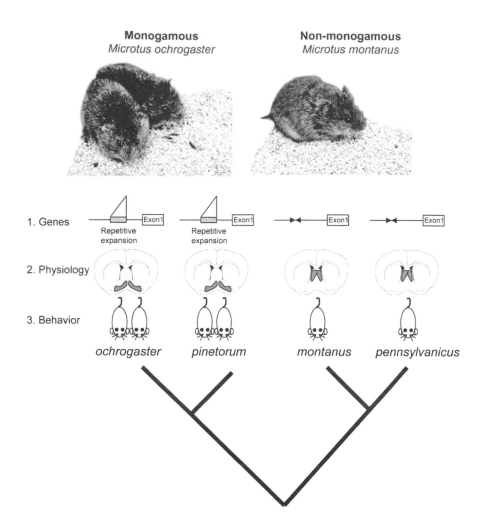

FIGURE 11.7

Strong evidence for the evolution of pair bonding in voles from genes to behavior in spite of
weaknesses in the comparative data. In the speciose genus *Microtus*, some species are monogamous,
whereas others are nonmonogamous. In the four species shown, a repetitive expansion in a
regulatory region upstream of the gene for arginine vasopressin receptor (V1aR) is associated with
a change in distribution of this receptor in the brain that parallels behavior (Young et al. 1999).
Note the weakness in these data in light of the phylogeny. An association between genes, physiology,
and behavior is expected even in the absence of pair bonding because the animals are related to
each other in a hierarchical way. Nonetheless, extensive direct evidence has established a causal
role for the repetitive expansion on V1aR distribution and behavior using a combination of state-of-
the-art transgenic technology, viral gene transfer, and pharmacology (see text). Photos by Larry
Young.

these animals. First, in an impressive feat of genetic engineering, a small segment of DNA containing the V1aR promoter region from the monogamous prairie vole was inserted into the genome of a mouse. Remarkably, the neuroanatomical distribution of V1aR in the genetically engineered mice was similar to that in the prairie vole. The mice were then injected with arginine vasopressin. Male transgenic mice expressing the prairie vole V1aR promoter displayed more affiliative behavior (as measured by olfactory investigation and grooming) than control mice (Young et al. 1999).

The authors later provided additional support for the V1aR hypothesis. They overexpressed V1aR in the ventral pallidum of the polygamous meadow vole, *Microtus pennsylvanicus*, using locally administered viral gene transfer, and discovered that overexpression of this single gene substantially increased partner preference formation. Then they blocked V1aR signaling by injecting V1aR antagonists into the ventral pallidum in male prairie voles, and they discovered that this inhibited partner preference formation. Taken together, these data provide strong support for the hypothesis that V1aR signaling in the ventral pallidum (along with other regions of the natural reward circuit) mediates pair-bond formation in *M. orchrogaster* (prairie vole) (Lim et al. 2004).

A molecular-genetic mechanism underlying the differential pattern of V1aR expression in the brains of monogamous and polygamous voles was hypothesized to involve an expansion and contraction of a microsatellite in the *cis*-regulatory region of the V1aR gene. This microsatellite is greatly expanded in the two monogamous species shown in figure 11.7 as compared with the nonmonogamous species (Young et al. 1999). The mechanism for the expansion was identified as several repeat blocks interspersed with nonrepetitive sequences. This expansion was later determined to have a functional effect on gene expression in cell culture (Hammock and Young 2004).

Despite all the positive evidence, a recent phylogenetic analysis failed to find the predicted association between expansion in this microsatellite region and affiliative behavior in a larger sample of twenty-three species of voles (Fink et al. 2006). All twenty-three species showed the expanded version of the microsatellite except for the two species chosen as representative nonmonogamous voles (figure 11.7). Because many vole species are nonmonogamous, these data demonstrate that expansion of the microsatellite is not sufficient to produce monogamy, but it still might be necessary in the monogamous species shown in figure 11.7.

At least in the prairie vole, the evidence is strong that expansion of the microsatellite region increases affiliative behavior. Within prairie voles, individual alleles of the microsatellite predict both individual differences in receptor distribution patterns and affiliative behavior. Moreover, in three independent cell culture experiments, the longer alleles significantly changed transcriptional activity relative to the shorter alleles (Hammock and Young 2005).

Taken together, these data provide strong evidence that an evolutionary shift in affiliative behavior in voles was caused, in part, by a change in a regulatory region of a gene that resulted in altered distribution and expression of a neurotransmitter receptor in the brain, and that this alteration is important for the behavior. To the best of our knowledge, no one has attempted to test this hypothesis using a selection experiment to increase affiliative behavior in voles (or any other animal), and then see whether the microsatellite locus responds in the same way. This would constitute a very strong alternative test of the primary hypothesis.

## TESTING ADAPTIVE HYPOTHESES

Much research on behavior in recent decades has been done within the framework of an adaptationist research program, looking at behavior as a product of evolution by natural selection and aiming to explain how and when particular forms of behavior contribute to Darwinian fitness. Although adaptive hypotheses about behavior may be easy to formulate, testing them is often difficult (e.g., see Garland and Adolph 1994; Garland and Carter 1994; Garland et al. 2005). Experimental evolution under experimentally imposed natural selection is a direct way to test hypotheses about factors (environmental, social, etc.) thought to favor the evolution of particular behaviors. For example, Joshi and Mueller (1993) showed that high (low) larval density result in natural selection for fast (slow) feeding in *Drosophila* larvae, but does not impose differential selection on pupation height (see also Mueller this volume). In another *Drosophila* study, experimental removal of sexual selection (by enforcing monogamy) led to the evolution of lower courtship rate by males, in addition to having various effects on the physiological aspects of male-female conflict (Holland and Rice 1999).

Experimental natural selection may also help to clarify which traits have the greatest potential to respond to a selection regime that favors several complementary or alternative adaptations. For example, by maintaining bean weevil populations on a mixture of two host seeds (A and B) either alone or together with a competitor specializing on host A, Taper (1990) tested the prediction of ecological character displacement. As predicted, the target species became physiologically better adapted to host B in the presence of specialized competitor but, surprisingly, did not evolve a greater preference for host B. Thus, behavioral traits do not always evolve faster than physiological ones.

Finally, experimental evolution can be used to test hypotheses about the consequences of behavioral evolution for evolutionary processes in general. For example, Mery and Kawecki (2004) used experimental evolution in *Drosophila* to show that, depending on circumstances, an opportunity to learn may either accelerate or slow down behavioral adaptation to a novel environment. This was the first direct experimental test of a century-old idea (known as the Baldwin effect) that learning may accelerate evolution (Baldwin 1896). In a similar vein, Rice and Salt (1987) showed that ecological reproductive isolation can evolve as a by-product of divergent selection on behavioral ("habitat") preferences (a combination of geotaxis, phototaxis, and odor preference). These studies illustrate the potential of experimental evolution for testing hypotheses of evolutionary adaptations.

## RELEVANCE FOR PUBLIC HEALTH

The vast majority of molecular-genetic and neurobiological analyses of behavioral traits has been directed by biomedical interests where the goal is to understand the etiology of mental illness, how to improve the health of the mind, or, more generally, to understand human behavior. Experimental methods in evolution can be useful for studying the etiology of mental illness to the extent that mental disease can be modeled as extreme forms of behavior or neurobiology, symptomatic of the disorder. For example, a great deal of work has explored the neurobiological and genetic bases of motivation for drugs of abuse using animals genetically predisposed to drink intoxicating quantities of ethanol (Mardones and Segovia-Riquelme 1983; Hilakivi et al. 1984; Crabbe et al. 1994; Grahame et al. 1999; Murphy et al. 2002; Kamdar et al. 2007). In principle, the methods described herein hold promise for exploring such mental disorders as addiction, obesity, ADHD (e.g., see Rhodes et al. 2005), mania, learning disorders, stress-related disorders, and depression. It is important to note here that the real challenge is not how to apply the genetic and neurobiological methods, but rather how to represent these illnesses in animal models via behavioral assays that are actually relevant to the human phenomenon (Dole and Gentry 1984; Kamdar et al. 2007).

## CONCLUSIONS AND FUTURE DIRECTIONS

Experimental evolution offers powerful tools to identify the origin and mechanisms of behavioral variation at different levels of biological organization, from structural changes in DNA to expression of genes to physiological pathways to behavior. However, some important practical issues should be considered before embarking on an evolutionary experiment on behavior, such as the difficulty in developing an assay to capture the essence of a complex behavior and the sensitivity to subtle, and often uncontrollable, environmental factors. Behaviors are often influenced by multiple genes with complex gene-by-gene, gene-by-environment, and environment-by-environment interactions. This is one reason, for example, that single-gene mutants are relatively uninformative (see also Rauser et al. this volume), though we described a case in which such mutants were useful for exploring mechanisms underlying the evolution of mating systems in voles.

One feature consistent across many different vertebrate models of behavioral evolution is the involvement of the natural reward circuit in the brain. This circuit has many components that play a role in motivation and reinforcement. Specific alterations in signaling between particular regions could change motivation for one behavior versus another. We discussed an example with V1aR in the ventral pallidum of voles that caused increased affiliative social behavior (Young and Wang 2004), and another example in which dopamine signaling is altered in mice bred for increased wheel running (Rhodes et al. 2005). Exactly which substrates are necessary and how they alter motivation to shift behavior has not been worked out for any vertebrate species.

Technology in genotyping, measuring expression of genes (microarray), gene mapping, genetic engineering, and an understanding of how to combine multiple sources of information (bioinformatics) have greatly advanced recently. These tools offer promise for finding genes or regulatory sites on chromosomes, and for testing mechanisms of behavioral evolution.

When the question changes from trying to understand broad correlations across species to testing individual mechanisms within species or among a handful of species, experimental evolution offers a strategy that nicely complements the comparative approach. The time is right for evolutionary biologists interested in behavior and neurobiology to take full advantage of the new and old technology that the molecular and behavior geneticists have brought to the table.

## SUMMARY

Animal behavior (e.g., foraging, mating, parental care, aggression, territorial defense, moving through the environment, learning, escaping from predators) has obvious evolutionary significance and is amenable for experimental evolution, provided some creative thought is given to surmount certain methodological obstacles. One unique feature about behavior, as compared with morphological or life-history traits, is that behavior is highly sensitive to small and often uncontrollable environmental influences, as well as the animal's motivational states. This sensitivity reduces reliability or repeatability, and it makes measuring or quantifying behavioral traits relatively difficult. Nonetheless, many successful selection experiments have targeted behavior in species ranging from flies to mice, including behaviors ranging from geotaxis to nest building. This chapter reviews some of the methods used and problems encountered. Several conclusions can be drawn. One is that behavior often evolves rapidly and that the changes in DNA that lead to variation in behavior occur in both coding and noncoding, regulatory regions. These changes ultimately lead to developmental changes in the nervous system at all levels of biological organization (molecules, cells, physiology, morphology). In vertebrates, the natural reward circuit in the brain appears to be a major target where alterations take place to shift motivation that underlies voluntary behavior. For example, both the evolution of increased wheel-running behavior in house mice and increased pair-bonding behavior in voles appears to have resulted from specific changes in this circuit. For wheel running, altered dopamine signaling from the ventral midbrain to forebrain areas was implicated in increased motivation for physical activity; whereas for pair bonding, distribution of arginine vasopressin and oxytocin receptors in various regions of the circuit affected reinforcement of social contact. The technology for genotyping, measuring global expression of genes, and mapping location of genes that influence behavioral traits is rapidly advancing, and it is expected that within the next ten to twenty years, the connection between specific polymorphisms in DNA and variation in behavior will be elucidated for many behavioral traits in a variety of organisms. Results of such studies hold great promise for biomedicine because finding genes and pathways that regulate behavior can provide useful targets for therapeutic manipulation of maladaptive behavior and mental illness.

## REFERENCES

Baldwin, J. M. 1896. A new factor in evolution. *American Naturalist* 30:441–451, 536–553.

Barnett, S. A., and J. L. Smart. 1975. The movements of wild and domestic house mice in an artificial environment. *Behavioral Biology* 15:85–93.

Belke, T. W., and T. Garland, Jr. 2007. A brief opportunity to run does not function as a reinforcer for mice selected for high daily wheel-running rates. *Journal of the Experimental Analysis of Behavior* 88:199–213.

Belyaev, D. K. 1979. Destabilizing selection as a factor in domestication. *Journal of Heredity* 70:301–308.

Belyaev, D. K., A. O. Ruvinsky, and L. N. Trut. 1981. Inherited activation-inactivation of the star gene in foxes. *Journal of Heredity* 72:267–274.

Berthold, P., G. Mohr, and U. Querner. 1990. Control and evolutionary potential of obligate partial migration: Results of a 2-way selective breeding experiment with the blackcap (*Sylvia atricapilla*). *Journal für Ornithologie* 131:33–45.

Blomberg, S. P., T. Garland, Jr., and A. R. Ives. 2003. Testing for phylogenetic signal in comparative data: Behavioral traits are more labile. *Evolution* 57:717–745.

Boake, C. R. B., ed. 1994. *Quantitative Genetic Studies of Behavioral Evolution*. Chicago: University of Chicago Press.

Brandes, C., B. Frisch, and R. Menzel. 1988. Time-course of memory formation differs in honey bee lines selected for good and poor learning. *Animal Behaviour* 36:981–985.

Bronikowski, A. M., P. A. Carter, J. G. Swallow, I. A. Girard, J. S. Rhodes, and T. Garland, Jr. 2001. Open-field behavior of house mice selectively bred for high voluntary wheel-running. *Behavior Genetics* 31:309–316.

Brush, F. R. 2003. Selection for differences in avoidance learning: The Syracuse strains differ in anxiety, not learning ability. *Behavior Genetics* 33:677–696.

Brush, F. R., and P. C. Sakellaris. 1973. Bidirectional genetic selection for shuttlebox avoidance learning in rat. *Behavior Genetics* 3:396–397.

Bult, A., L. Hiestand, E. A. Van der Zee, and C. B. Lynch. 1993. Circadian rhythms differ between selected mouse lines: A model to study the role of vasopressin neurons in the suprachiasmatic nuclei. *Brain Research Bulletin* 32:623–627.

Bult, A., E. A. van der Zee, J. C. Compaan, and C. B. Lynch. 1992. Differences in the number of arginine-vasopressin-immunoreactive neurons exist in the suprachiasmatic nuclei of house mice selected for differences in nest-building behavior. *Brain Research* 578:335–338.

Carter, P. A., T. Garland, Jr., M. R. Dohm, and J. P. Hayes. 1999. Genetic variation and correlations between genotype and locomotor physiology in outbred laboratory house mice (*Mus domesticus*). *Comparative Biochemistry and Physiology A Molecular and Integrative Physiology* 123:155–162.

Carter, P. A., J. G. Swallow, S. J. Davis, and T. Garland, Jr. 2000. Nesting behavior of house mice (*Mus domesticus*) selected for increased wheel-running activity. *Behavior Genetics* 30:85–94.

Castillo-Davis, C. I. 2005. The evolution of noncoding DNA: How much junk, how much func? *Trends in Genetics* 21:533–536.

Chandra, S. B. C., G. J. Hunt, S. Cobey, and B. H. Smith. 2001. Quantitative trait loci associated with reversal learning and latent inhibition in honeybees *(Apis mellifera). Behavior Genetics* 31:275–285.

Chesler, E. J., L. Lu, S. Shou, Y. Qu, J. Gu, J. Wang, H. C. Hsu, et al. 2005. Complex trait analysis of gene expression uncovers polygenic and pleiotropic networks that modulate nervous system function. *Nature Genetics* 37:233–242.

Chesler, E. J., L. Lu, J. Wang, R. W. Williams, and K. F. Manly. 2004. WebQTL: Rapid exploratory analysis of gene expression and genetic networks for brain and behavior. *Nature Neuroscience* 7:485–486.

Churchill, G. A., D. C. Airey, H. Allayee, J. M. Angel, A. D. Attie, J. Beatty, W. D. Beavis, et al. 2004. The Collaborative Cross, a community resource for the genetic analysis of complex traits. *Nature Genetics* 36:1133–1137.

Clayton, D. L., and J. V. Paietta. 1972. Selection for circadian eclosion time in *Drosophila melanogaster. Science* 178:994.

Cooke, S. J., C. D. Suski, K. G. Ostrand, D. H. Wahl, and D. P. Philipp. 2007. Physiological and behavioral consequences of long-term artificial selection for vulnerability to recreational angling in a teleost fish. *Physiological and Biochemical Zoology* 80:480–490.

Crabbe, J. C., J. K. Belknap, and K. J. Buck. 1994. Genetic animal models of alcohol and drug abuse. *Science* 264:1715–1723.

Crabbe, J. C., and T. J. Phillips. 1993. Selective breeding for alcohol withdrawal severity. *Behavior Genetics* 23:171–177.

DeFries, J. C., M. C. Gervais, and E. A. Thomas. 1978. Response to 30 generations of selection for open-field activity in laboratory mice. *Behavior Genetics* 8:3–13.

Dierick, H. A., and R. J. Greenspan. 2006. Molecular analysis of flies selected for aggressive behavior. *Nature Genetics* 38:1023–1031.

Doerge, R. W. 2002. Mapping and analysis of quantitative trait loci in experimental populations. *Nature Reviews Genetics* 3:43–52.

Dohm, M. R., J. P. Hayes, and T. Garland, Jr. 1996. Quantitative genetics of sprint running speed and swimming endurance in laboratory house mice *(Mus domesticus). Evolution* 50:1688–1701.

Dole, V. P., and R. T. Gentry. 1984. Toward an analogue of alcoholism in mice: Scale factors in the model. *Proceedings of the National Academy of Sciences of the USA* 81:3543–3546.

Dumas, P., Y. Sun, G. Corbeil, S. Tremblay, Z. Pausova, V. Kren, D. Krenova, et al. 2000. Mapping of quantitative trait loci (QTL) of differential stress gene expression in rat recombinant inbred strains. *Journal of Hypertension* 18:545–551.

Edwards, A. C., S. M. Rollmann, T. J. Morgan, and T. F. C. Mackay. 2006. Quantitative genomics of aggressive behavior in *Drosophila melanogaster. PLoS Genetics* 2:1386–1395.

Eisen, E. J., ed. 2005. *The Mouse in Animal Genetics and Breeding Research.* London: Imperial College Press.

Falconer, D. S., and T. F. C. Mackay. 1996. *Introduction to Quantitative Genetics.* 4th ed. Essex, U.K.: Longman.

Fink, S., L. Excoffier, and G. Heckel. 2006. Mammalian monogamy is not controlled by a single gene. *Proceedings of the National Academy of Sciences of the USA* 103:10956–10960.

Fitzpatrick, M. J., Y. Ben-Shahar, H. M. Smid, L. E. M. Vet, G. E. Robinson, and M. B. Sokolowski. 2005. Candidate genes for behavioral ecology. *Trends in Ecology & Evolution* 20:96–104.

Fitzpatrick, M. J., E. Feder, L. Rowe, and M. B. Sokolowski. 2007. Maintaining a behaviour polymorphism by frequency-dependent selection on a single gene. *Nature* 447:210–212.

Flint, J., and R. Mott. 2001. Finding the molecular basis of quantitative traits: Successes and pitfalls. *Nature Reviews Genetics* 2:437–445.

Fonio, E., Y. Benjamini, A. Sakov, and I. Golani. 2006. Wild mouse open field behavior is embedded within the multidimensional data space spanned by laboratory inbred strains. *Genes, Brain, and Behavior* 5:380–388.

Fryer, G., and T. D. Iles. 1972. *The Chichlid Fishes of the Great Lakes of Africa*. Neptune City, NJ: t.f.h. publications.

Gammie, S. C., T. Garland, Jr., and S. A. Stevenson. 2006. Artificial selection for increased maternal defense behavior in mice. *Behavior Genetics* 36:713–722.

Gammie, S. C., N. S. Hasen, J. S. Rhodes, I. Girard, and T. Garland, Jr. 2003. Predatory aggression, but not maternal or intermale aggression, is associated with high voluntary wheel-running behavior in mice. *Hormones and Behavior* 44:209–221.

Garland, T., Jr. 2003. Selection experiments: An underutilized tool in biomechanics and organismal biology. Pages 23–56 *in* V. Bels, J. Gasc, and A. Casinos, eds. *Biomechanics and Evolution*. Oxford: BIOS Scientific.

Garland, T., Jr., and S. C. Adolph. 1994. Why not to do two-species comparative studies: Limitations on inferring adaptation. *Physiological Zoology* 67:797–828.

Garland, T., Jr., and P. A. Carter. 1994. Evolutionary physiology. *Annual Review of Physiology* 56:579–621.

Garland, T., Jr., A. F. Bennett, and E. L. Rezende. 2005. Phylogenetic approaches in comparative physiology. *Journal of Experimental Biology* 208:3015–3035.

Gerken, M., and J. Petersen. 1992. Direct and correlated responses to bidirectional selection for dustbathing activity in Japanese quail *(Coturnix coturnix japonica)*. *Behavior Genetics* 22:601–612.

Gingerich, P. D., M. Haq, I. S. Zalmout, I. H. Khan, and M. S. Malkani. 2001. Origin of whales from early artiodactyls: Hands and feet of Eocene Protocetidae from Pakistan. *Science* 293:2239–2242.

Gotz, K. G. 1970. Fractionation of *Drosophila* populations according to optomotor traits. *Journal of Experimental Biology* 52:419.

Grahame, N. J., T. K. Li, and L. Lumeng. 1999. Selective breeding for high and low alcohol preference in mice. *Behavior Genetics* 29:47–57.

Greenspan, R. J. 2003. The varieties of selectional experience in behavioral genetics. *Journal of Neurogenetics* 17:241–270.

———. 2004. *E pluribus unum, ex uno plura*: Quantitative and single-gene perspectives on the study of behavior. *Annual Review of Neuroscience* 27:79–105.

Hammock, E. A., and L. J. Young. 2004. Functional microsatellite polymorphism associated with divergent social structure in vole species. *Molecular Biology and Evolution* 21:1057–1063.

———. 2005. Microsatellite instability generates diversity in brain and sociobehavioral traits. *Science* 308:1630–1634.

Hard, J. J., M. R. Gross, M. Heino, R. Hilborn, R. G. Kope, R. Law, and J. D. Reynolds. 2008. Evolutionary consequences of fishing and their implications for salmon. *Evolutionary Applications* 1:388–408.

Hare, B., I. Plyusnina, N. Ignacio, O. Schepina, A. Stepika, R. Wrangham, and L. Trut. 2005. Social cognitive evolution in captive foxes is a correlated by-product of experimental domestication. *Current Biology* 15:226–230.

Hartl, D. L., and A. G. Clark. 1997. *Principles of Population Genetics.* Sunderland, MA: Sinauer.

Henderson, N. 1997. Spurious association in unreplicated selected lines. *Behavior Genetics* 27:145–154.

Hilakivi, L., C. J. Eriksson, M. Sarviharju, and J. D. Sinclair. 1984. Revitalization of the AA and ANA rat lines: Effects on some line characteristics. *Alcohol* 1:71–75.

Hirsch, J. 1959. Studies in experimental behavior genetics. *Journal of Comparative Physiology* 52:304–308.

Hoekstra, H. E., and J. A. Coyne. 2007. The locus of evolution: Evo devo and the genetics of adaptation. *Evolution* 61:995–1016.

Hoekstra, H. E., R. J. Hirschmann, R. A. Bundey, P. A. Insel, and J. P. Crossland. 2006. A single amino acid mutation contributes to adaptive beach mouse color pattern. *Science* 313:101–104.

Holland, B., and W. R. Rice. 1999. Experimental removal of sexual selection reverses inter-sexual antagonistic coevolution and removes a reproductive load. *Proceedings of the National Academy of Sciences of the USA* 96:5083–5088.

Horn, J. M., R. Plomin, and R. Rosenman. 1976. Heritability of personality traits in adult male twins. *Behavior Genetics* 6:17–30.

Hughes, K. A., and J. Leips. 2006. Quantitative trait locus analysis of male mating success and sperm competition in *Drosophila melanogaster*. *Evolution* 60:1427–1434.

Innis, N. K. 1992. Tolman and Tryon: Early research on the inheritance of the ability to learn. *American Psychologist* 47:190–197.

Jaarola, M., N. Martinkova, I. Gunduz, C. Brunhoff, J. Zima, A. Nadachowski, G. Amori, et al. 2004. Molecular phylogeny of the speciose vole genus *Microtus* (Arvicolinae, Rodentia) inferred from mitochondrial DNA sequences. *Molecular Phylogenetics and Evolution* 33:647–663.

Jansen, R. C., and J. P. Nap. 2001. Genetical genomics: The added value from segregation. *Trends in Genetics* 17:388–391.

Jordan, K. W., T. J. Morgan, and T. F. Mackay. 2006. Quantitative trait loci for locomotor behavior in *Drosophila melanogaster*. *Genetics* 174:271–284.

Joshi, A., and L. D. Mueller. 1993. Directional and stabilizing density-dependent natural selection for pupation height in *Drosophila melanogaster*. *Evolution* 47:176–184.

Kamdar, N., S. Miller, Y. Syed, R. Bhayana, T. Gupta, and J. S. Rhodes. 2007. Acute effects of Naltrexone and GBR 12909 on ethanol drinking-in-the-dark in C57BL/6J mice. *Psychopharmacology* 192:207–217.

Kaun, K. R., T. Hendel, B. Gerber, and M. B. Sokolowski. 2007. Natural variation in *Drosophila* larval reward learning and memory due to a cGMP-dependent protein kinase. *Learning & Memory* 14:342–349.

Kawecki, T. J., and F. Mery. 2006. Genetically idiosyncratic responses of *Drosophila melanogaster* populations to selection for improved learning ability. *Journal of Evolutionary Biology* 19:1265–1274.

Keller, L., and J. D. Parker. 2002. Behavioral genetics: A gene for supersociality. *Current Biology* 12:R180–R181.

Kliethermes, C. L., and J. C. Crabbe. 2006. Genetic independence of mouse measures of some aspects of novelty seeking. *Proceedings of the National Academy of Sciences of the USA* 103:5018–5023.

Kolss, M., and T. J. Kawecki. 2008. Reduced learning ability as a consequence of evolutionary adaptation to nutritional stress in *Drosophila melanogaster*. *Ecological Entomology* 33:583–588.

Kovach, J. K., F. R. Yeatman, and G. Wilson. 1981. Perception or preference: Mediation of gene effects in the color choices of naive quail chicks *(C. coturnix japonica)*. *Animal Behaviour* 29:760–770.

Leamy, L. J., D. Pomp, and J. T. Lightfoot. 2008. An epistatic genetic basis for physical activity traits in mice. *Journal of Heredity* 99:639–646.

Lim, M. M., Z. Wang, D. E. Olazabal, X. Ren, E. F. Terwilliger, and L. J. Young. 2004. Enhanced partner preference in a promiscuous species by manipulating the expression of a single gene. *Nature* 429:754–757.

Lindahl, T. 1981. DNA methylation and control of gene expression. *Nature* 290:363–364.

Lofdahl, K. L., M. Holliday, and J. Hirsch. 1992. Selection for conditionability in *Drosophila melanogaster*. *Journal of Comparative Psychology* 106:172–183.

Lynch, C. B. 1980. Response to divergent selection for nesting behavior in *Mus musculus*. *Genetics* 96:757–765.

———. 1994. Evolutionary inferences from genetic analyses of cold adaptation in laboratory and wild populations of the house mouse. Pages 278–301 *in* C. R. B. Boake, ed. *Quantitative Genetic Studies of Behavioral Evolution*. Chicago: University of Chicago Press.

Mackay, T. F. C., and J. D. Fry. 1996. Polygenic mutation in *Drosophila melanogaster*: Genetic interactions between selection lines and candidate quantitative trait loci. *Genetics* 144:671–688.

Magurran, A., B. Seghers, G. Carvalho, and P. Shaw. 1992. Behavioral consequences of an artificial introduction of guppies *(Poecilia reticulata)* in Trinidad: Evidence for the evolution of anti-predator behavior in the wild. *Proceedings Biological Sciences* 248:117–122.

Malisch, J. L., C. W. Breuner, F. R. Gomes, M. A. Chappell, and T. Garland, Jr. 2008. Circadian pattern of total and free corticosterone concentrations, corticosteroid-binding globulin, and physical activity in mice selectively bred for high voluntary wheel-running behavior. *General and Comparative Endocrinology* 156:210–217.

Manning, A. 1961. The effect of artificial selection for mating speed in *Drosophila melanogaster*. *Animal Behaviour* 9:82–92.

Mardones, J., and N. Segovia-Riquelme. 1983. Thirty-two years of selection of rats by ethanol preference: UChA and UChB strains. *Neurobehavioral Toxicology and Teratology* 5:171–178.

Mayr, E. 1958. Behavior and systematics. *In* A. Roe and G. G. Simpson, eds. *Behavior and Evolution*. New Haven, CT: Yale University Press.

McGuire, T. R., and T. Tully. 1987. Characterisation of genes involved with classical conditioning that produces differences between bidirectionally selected strains of the blow fly *Phormia regina*. *Behavior Genetics* 17:97–107.

Mery, F., A. T. Belay, A. K. So, M. B. Sokolowski, and T. J. Kawecki. 2007. Natural polymorphism affecting learning and memory in *Drosophila*. *Proceedings of the National Academy of Sciences of the USA* 104:13051–13055.

Mery, F., and T. J. Kawecki. 2004. The effect of learning on experimental evolution of resource preference in *Drosophila melanogaster*. *Evolution* 58:757–767.

Miyatake, T., K. Tabuchi, S. K., K. Okada, K. Katayama, and S. Moriya. 2008. Pleiotropic antipredator strategies, fleeing and feigning death, correlated with dopamine levels in *Tribolium castaneum*. *Animal Behaviour* 75:113–121.

Murphy, J. M., R. B. Stewart, R. L. Bell, N. E. Badia-Elder, L. G. Carr, W. J. McBride, L. Lumeng, et al. 2002. Phenotypic and genotypic characterization of the Indiana University rat lines selectively bred for high and low alcohol preference. *Behavior Genetics* 32:363–388.

Olausson, P., J. D. Jentsch, N. Tronson, R. L. Neve, E. J. Nestler, and J. R. Taylor. 2006. DeltaFosB in the nucleus accumbens regulates food-reinforced instrumental behavior and motivation. *Journal of Neuroscience* 26:9196–9204.

Osborne, K. A., A. Robichon, E. Burgess, S. Butland, R. A. Shaw, A. Coulthard, H. S. Pereira, et al. 1997. Natural behavior polymorphism due to a cGMP-dependent protein kinase of *Drosophila*. *Science* 277:834–836.

Oufiero, C., and T. Garland, Jr. 2007. Evaluating performance costs of sexually selected traits. *Functional Ecology* 21:676–689.

Page, R. E., K. D. Waddington, G. J. Hunt, and M. K. Fondrk. 1995. Genetic determinants of honey bee foraging behaviour. *Animal Behaviour* 50:1617–1625.

Palmer, A. A., M. Verbitsky, R. Suresh, H. M. Kamens, C. L. Reed, N. Li, S. Burkhart-Kasch, et al. 2005. Gene expression differences in mice divergently selected for methamphetamine sensitivity. *Mammalian Genome* 16:291–305.

Pereira, H. S., and M. B. Sokolowski. 1993. Mutations in the larval foraging gene affect adult locomotory behavior after feeding in *Drosophila melanogaster*. *Proceedings of the National Academy of Sciences of the USA* 90:5044–5046.

Renger, J. J., W. D. Yao, M. B. Sokolowski, and C. F. Wu. 1999. Neuronal polymorphism among natural alleles of a cGMP-dependent kinase gene, foraging, in *Drosophila*. *Journal of Neuroscience* 19:RC28 (21–28).

Rhodes, J. S., S. C. Gammie, and T. Garland, Jr. 2005. Neurobiology of mice selected for high voluntary wheel running activity. *Integrative and Comparative Biology* 45:438–455.

Rhodes, J. S., and T. Garland, Jr. 2003. Differential sensitivity to acute administration of Ritalin, apomorphine, SCH 23390, but not raclopride in mice selectively bred for hyperactive wheel-running behavior. *Psychopharmacology* 167:242–250.

Rhodes, J. S., T. Garland, Jr., and S. C. Gammie. 2003. Patterns of brain activity associated with variation in voluntary wheel-running behavior. *Behavioral Neuroscience* 117:1243–1256.

Rhodes, J. S., G. R. Hosack, I. Girard, A. E. Kelley, G. S. Mitchell, and T. Garland, Jr. 2001. Differential sensitivity to acute administration of cocaine, GBR 12909, and fluoxetine in mice selectively bred for hyperactive wheel-running behavior. *Psychopharmacology* 158:120–131.

Rice, W. R. 1987. Speciation via habitat specialization: the evolution of reproductive isolation as a correlated character. *Evolutionary Ecology* 1:301–314.

Ricker, J. P., and J. Hirsch. 1988. Genetic changes occurring over 500 generations in lines of *Drosophila melanogaster* selected divergently for geotaxis. *Behavior Genetics* 18:13–25.

Ritchie, M. G., and C. P. Kyriacou. 1996. Artificial selection for a courtship signal in *Drosophila melanogaster*. *Animal Behaviour* 52:603–611.

Robinson, G. E. 2004. Genomics. Beyond nature and nurture. *Science* 304:397–399.

Rogers, D. W., R. H. Baker, T. Chapman, M. Denniff, A. Pomiankowski, and K. Fowler. 2005. Direct and correlated responses to artificial selection on male mating frequency in the stalk-eyed fly *Cyrtodiopsis dalmanni*. *Journal of Evolutionary Biology* 18:642–650.

Ross, K. G., and L. Keller. 1998. Genetic control of social organization in an ant. *Proceedings of the National Academy of Sciences of the USA* 95:14232–14237.

Ruber, L., and D. Adams. 2001. Evolutionary convergence of body shape and trophic morphology in cichlids from Lake Tanganyika. *Journal of Evolutionary Biology* 14:325–332.

Rutledge, C. E., and R. N. Wiedenmann. 2003. An attempt to change habitat preference of a parasitoid, *Cotesia sesamiae* (Hymenoptera: Braconidae), through artificial selection. *Journal of Entomological Science* 38:93–103.

Savolainen, P., Y. P. Zhang, J. Luo, J. Lundeberg, and T. Leitner. 2002. Genetic evidence for an East Asian origin of domestic dogs. *Science* 298:1610–1613.

Schadt, E. E., J. Lamb, X. Yang, J. Zhu, S. Edwards, D. Guhathakurta, S. K. Sieberts, et al. 2005. An integrative genomics approach to infer causal associations between gene expression and disease. *Nature Genetics* 37:710–717.

Scheiner, R., M. B. Sokolowski, and J. Erber. 2004. Activity of cGMP-dependent protein kinase (PKG) affects sucrose responsiveness and habituation in *Drosophila melanogaster*. *Learning & Memory* 11:303–311.

Schliekelman, P. D. 2008. Statistical power of expression QTLs for mapping of complex trait loci in natural populations. *Genetics*. Available: www.genetics.org/cgi/rapidpdf/genetics.107.076687v1.

Shirley, R. L., N. A. Walter, M. T. Reilly, C. Fehr, and K. J. Buck. 2004. Mpdz is a quantitative trait gene for drug withdrawal seizures. *Nature Neuroscience* 7:699–700.

Silver, L. M. 1995. *Mouse Genetics: Concepts and Applications*. New York: Oxford University Press. Available: www.informatics.jax.org/silverbook/.

Singer, M. C. 2000. Reducing ambiguity in describing plant-insect interactions: "Preference," acceptability" and "electivity." *Ecology Letters* 3:159–162.

Singh, B. N., and M. B. Pandey. 1993. Selection for high and low pupation height in *Drosophila ananassae*. *Behavior Genetics* 23:239–243.

Spuhler, K. P., D. W. Crumpacker, J. S. Williams, and B. P. Bradley. 1978. Response to selection for mating speed and changes in gene arrangement frequencies in descendants from a single population of *Drosophila pseudoobscura*. *Genetics* 89:729–749.

Steinberg, S., M. Dicke, L. E. M. Vet, and R. Wanningen. 1992. Response of the braconid parasitoid *Cotesia* (=*Apanteles*) *glomerata* to volatile infochemicals: Effects of bioassay set-up, parasitoid age and experience and barometric flux. *Entomologia Experimentalis et Applicata* 63:163–175.

Swallow, J. G., and T. Garland, Jr. 2005. Selection experiments as a tool in evolutionary and comparative physiology: Insights into complex traits: An introduction to the symposium. *Integrative and Comparative Biology* 45:387–390.

Swallow, J. G., P. A. Carter, and T. Garland, Jr. 1998. Artificial selection for increased wheel-running behavior in house mice. *Behavior Genetics* 28:227–237.

Taper, M. L. 1990. Experimental character displacement in the adzuki bean weevil, *Callosobruchus chinensis*. Pages 289–301 *in* K. Fujii, ed. *Bruchids and Legumes: Economics, Ecology and Coevolution*. Dordrecht: Kluwer Academic.

Toma, D. P., K. P. White, J. Hirsch, and R. J. Greenspan. 2002. Identification of genes involved in *Drosophila melanogaster* geotaxis, a complex behavioral trait. *Nature Genetics* 31:349–353.

Trut, L. N. 1999. Early canid domestication: The farm-fox experiment. *American Scientist* 87:160–169.

Tryon, R. C. 1940. Genetic differences in maze learning in rats. *Yearbook of the National Society for the Study of Education* 39:111–119.

Weber, K. E. 1996. Large genetic change at small fitness cost in large populations of *Drosophila melanogaster* selected for wind tunnel flight: Rethinking fitness surfaces. *Genetics* 144:205–213.

Wehner, J. M., R. A. Radcliffe, and B. J. Bowers. 2001. Quantitative genetics and mouse behavior. *Annual Review of Neuroscience* 24:845–867.

Welbergen, P., and F. R. Vandijken. 1992. Asymmetric response to directional selection for licking behavior of *Drosophila melanogaster* males. *Behavior Genetics* 22:113–124.

Wells, T., and D. A. Carter. 2001. Genetic engineering of neural function in transgenic rodents: Towards a comprehensive strategy? *Journal of Neuroscience Methods* 108:111–130.

Whitehead, A., and D. L. Crawford. 2006. Variation within and among species in gene expression: Raw material for evolution. *Molecular Ecology* 15:1197–1211.

Winslow, J. T., N. Hastings, C. S. Carter, C. R. Harbaugh, and T. R. Insel. 1993. A role for central vasopressin in pair bonding in monogamous prairie voles. *Nature* 365:545–548.

Yalcin, B., S. A. Willis-Owen, J. Fullerton, A. Meesaq, R. M. Deacon, J. N. Rawlins, R. R. Copley, A. P. Morris, J. Flint, and R. Mott. 2004. Genetic dissection of a behavioral quantitative trait locus shows that Rgs2 modulates anxiety in mice. *Nature Genetics* 36:1197–1202.

Young, L. J., R. Nilsen, K. G. Waymire, G. R. MacGregor, and T. R. Insel. 1999. Increased affiliative response to vasopressin in mice expressing the V1a receptor from a monogamous vole. *Nature* 400:766–768.

Young, L. J., and Z. Wang. 2004. The neurobiology of pair bonding. *Nature Neuroscience* 7:1048–1054.

# 12

# SELECTION EXPERIMENTS AND EXPERIMENTAL EVOLUTION OF PERFORMANCE AND PHYSIOLOGY

John G. Swallow, Jack P. Hayes, Pawel Koteja, and Theodore Garland, Jr.

Since a seminal paper by Arnold (1983), direct measurement of whole-organism performance has become central to functional evolutionary biology (e.g., Arnold 2003; Ghalambor et al. 2003; Kingsolver and Huey 2003). In this context, "performance" can be most easily defined by example. Assuming that individuals can be fully motivated (e.g., see Swallow et al. 1998a; Harris and Steudel 2002; Losos et al. 2002; Tobalske et al. 2004), it is relatively easy to measure maximal sprint running speed of small mammals and lizards on photocell-timed racetracks or high-speed treadmills (e.g., Calsbeek and Irschick 2007; Chappell et al. 2007). Moving down one level of biological organization to give another example, sprint speed is the product of stride length and stride frequency. In turn, stride length is affected by both limb length and gait, including possible pelvic rotation (e.g., in bipedal lizards). Limb length is the summation of the lengths of individual limb segments and is affected by how fully the limb is extended. Some gaits involve phases in which all limbs leave the ground while the animal continues to move forward, thus further lengthening the stride. Stride frequency results from neural transmission and muscular contractions that cause bones to move as force is transmitted through tendons and ligaments. And, of course, muscular contractions involve numerous structural, ultrastructural, physiological, and biochemical traits.

The example of maximal sprint speed should make it clear that measures of organismal performance constitute "complex traits" (Bennett 1989; Ghalambor et al. 2003; Swallow and Garland 2005; Miles et al. 2007; Oufiero and Garland 2007). As such, their evolution must entail changes in numerous subordinate traits (figure 12.1). If we define *behavior* as anything an animal does (or, in some cases, fails to do) and *performance* as the ability of an animal to do something when maximally motivated (see also Garland 1994), then the evolution of performance ability may have immediate impacts on behavioral repertoires, frequencies, and/or intensities in situations where the animal is maximally motivated. For example, when chased by a predator, an animal cannot choose to sprint faster than its body will allow. Indeed, several workers have emphasized that natural and sexual selection are likely to act on combinations of behavioral and performance traits via "correlational selection" (e.g., Brodie 1992; Garland 1994; Sinervo and Clobert 2003; Ohno and Miyatake 2007). Returning to the example of sprint speed, selection would seem likely to favor individuals that have *both* a behavioral propensity to sprint after prey or away from a predator *and* high physical abilities for sprinting, because such individuals would fare better than those that had either the propensity or the ability, but not both. As discussed in this chapter, depending on the genetic correlation between traits under correlational selection, their joint evolution may be either "constrained" or "facilitated."

Although organismal performance abilities clearly set boundaries within which normal behavior must occur (Bennett 1989; Sears et al. 2006), animals do not necessarily behave in ways that tax their maximal abilities (Hayes 1989a, 1989b; Hayes and O'Connor 1999; Irschick and Garland 2001; Husak 2006). Accordingly, Arnold's (1983) original morphology → performance → fitness paradigm has been extended to include the point that behavior often acts as a "filter" between performance and Darwinian fitness

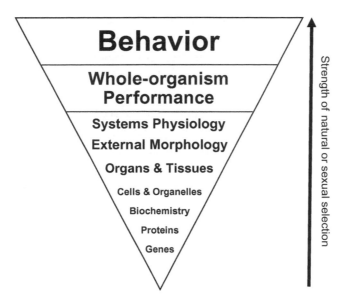

FIGURE **12.1**

Complex traits, such as behavior and performance, comprise hierarchical suites of interacting subordinate traits at lower levels of biological organization. In general, selection is thought to act most strongly on phenotypic variation in traits at higher levels of biological organization, such as behavior (see also Rhodes and Kawecki this volume) and performance, as indicated by the width of the inverted triangle (and the direction of the arrow). In other words, behavior and performance have strong effects on major components of Darwinian fitness, such as survivorship and fecundity. Lower-level traits include a wide range of suborganismal morphological, physiological, and biochemical phenotypes. These lower-level traits, directly and indirectly, influence aspects of whole-organism performance ability that are crucial for survival and reproduction.

(Garland et al. 1990; Garland 1994; Garland and Carter 1994; Garland and Losos 1994; Husak 2006).

For example, when an animal detects a predator, it may or may not "decide" to flee at its maximal sprint speed. If it never runs away at top speed, then its maximal sprinting abilities become moot. Alternatively, animals with low performance abilities—either innately or because of short-term conditions, such as injury, a full stomach, or pregnancy—might modify their behavior in ways that limit their exposure to life-or-death situations where performance is crucial (e.g., Bauwens and Thoen 1981; Clobert et al. 2000). Although many workers have argued that "both natural and sexual selection are expected to operate on performance first, and secondarily on other aspects (e.g., morphology)" (Lailvaux and Irschick 2006, 264), a recent review of the limited available field data does not provide evidence supporting this proposition (Irschick et al. 2008), perhaps because behavior often intervenes between selection and performance and can often compensate for changes in environmental conditions (Huey et al. 2003; Huey and Rosenzweig this volume). Behavior, of course, is highly plastic, but plasticity can be an important component of the response to selection at any level of biological organization (Garland and Kelly 2006). (For an interesting discussion of how performance abilities may constrain the evolution of bird song, see Podos et al. 2004.)

It is easy to conceive—and often to implement (but see Losos et al. 2002)—measures of maximal performance in the context of locomotion; indeed, most studies of performance in the sense used here do involve locomotion of some type. However, many other measures of performance, such as growth rate, thermal tolerance (Futuyma and Bennett this volume; Huey and Rosenzweig this volume), osmotic regulation, desiccation resistance (Gibbs and Gefen this volume), and rate of heat production in response to cold (Hayes and O'Connor 1999; see later sections), are ecologically and evolutionarily important. Nevertheless, to keep this contribution to a manageable size, we focus primarily on locomotor performance. This "dynamical, whole-organism" definition of performance, as recently articulated by Lailvaux and Irschick (2006), is consistent with the majority of performance measures in the literature.

The purpose of this chapter is to review selection experiments that have targeted organismal performance traits, with an emphasis on those involving locomotion or metabolism. Selection experiments are a powerful tool for investigating the linkages between biochemical, morphological, and physiological traits that influence physiological performance, as well as investigating how performance capacities may constrain or facilitate behavioral evolution (Garland and Carter 1994; Garland 1994; Lynch 1994; Rose et al. 1996; Bradley and Zamer 1999; Gibbs 1999; Feder et al. 2000; Bennett 2003; Garland 2003; Bradley and Folk 2004; Swallow and Garland 2005; Fuller et al. 2005; Gibbs and Gefen this volume; Zera and Harshman this volume). Rhodes and Kawecki (this volume) discuss selection experiments that have targeted behavior per se, several of these experiments having led to insights concerning correlated evolution of performance capacities. Both the Rhodes and Kawecki chapter and the present one provide some discussion of a selection experiment that targeted high voluntary wheel running of mice, because increases in this behavior have turned out to involve substantial components of physiology and performance (Swallow et al. 1998a; Garland 2003; Middleton et al. 2008).

Numerous selection experiments have targeted growth rates or body size, as well as such life-history traits as litter size, all of which have major "physiological" components, but these are not reviewed in the present chapter (see references in Falconer 1992; Falconer and Mackay 1996; Bunger et al. 2001; Garland 2003; Renne et al. 2003; Eisen 2005; Bell 2008). Zera and Harshman (this volume) consider the physiological underpinnings of life-history traits in insects from the perspective of selection experiments (see also Zera et al. 2007, which includes vertebrate examples). Other selection experiments have targeted physiological traits below the level of the whole organism, such as circulating corticosterone levels (Roberts et al. 2007), serum cholesterol (Dunnington et al. 1981a, 1981b), sensitivity to acute ethanol administration (Draski et al. 1992), core body temperature (Gordon and Rezvani 2001), and hypothermic responses to specific pharmacological agents (Overstreet 2002). Again, those studies are not reviewed here.

Selection experiments have a long history and come in many types (Falconer and Mackay 1996; Garland 2003; Bell 2008; Futuyma and Bennett this volume; Irschick and Reznick this volume; Rhodes and Kawecki this volume; Rose and Garland this volume). Most of the experiments that we review have used artificial selection rather than laboratory natural selection and have involved rodents. This is for two main reasons. First, measures of whole-organism performance are physically easier with rodents than with insects, let alone microorganisms. Second, laboratory natural selection, in which an entire population must be subjected to altered environmental or husbandry conditions, is impractical or unethical with rodents. Although laboratory mice have been used most commonly, which is not surprising given the long history of genetic and breeding research (Silver 1995; Falconer and Mackay 1996; Eisen 2005), laboratory rats as well as wild rodents have also been employed. So far as we are aware, only three vertebrate studies, two selecting on body size in mice (Bunger et al. 2001; Renne et al. 2003; and Wirth-Dzięciolowska et al. 2000, 2005; see also Rosochacki et al. 2005) and the other selecting for litter size in mice (Holt et al. 2004, 2005), have gone as many as one hundred generations, hence crossing into the category of "long-term" selection experiments (Laurie et al. 2004; Travisano this volume). Surprisingly, with one short-term exception (Shikano et al. 1998), small fishes (e.g., guppies, zebra fish, mosquito fish) have yet to be the subject of a selection experiment designed to alter whole-animal performance (see also Garland 2003).

## THE IMPORTANCE OF REPLICATION

Properly designing and analyzing a selection experiment is nontrivial (e.g., see Robertson 1980; Falconer and Mackay 1996; Rose et al. 1996; Roff 1997; Fuller et al. 2005; Bell 2008; Roff and Fairbairn this volume). In addition to population size, the exact details of how a phenotype will be scored, or how the environment will be altered in an attempt to induce selection, careful attention needs to be paid to the choice of base population and also to the degree of replication (Fry this volume; Huey and Rosenzweig this volume; Rauser et al. this volume; Rhodes and Kawecki this volume; Simões et al. this volume). Unfortunately, a number of interesting experiments that targeted performance or physiological traits are difficult to interpret because of the base population used, lack of replication, and/or lack of a control line (discussed later). With respect to replication, the basic problem is that any finite population will undergo genetic changes due to the stochastic processes of mutation and random genetic drift (founder effects may also be important in selection experiments if a line is begun with very few individuals; e.g., see example calculations in Garland et al., 2002; see also Fry this volume on speciation experiments). Genetic changes caused by random mutation and drift will, in turn, lead to phenotypic changes. These changes may occur for any trait, including one that is directly targeted by artificial selection. Hence, without replication of the selected lines (and maintenance of nonselected control or oppositely selected lines), it is difficult to know if any observed change in the mean phenotype of a single selected line is really because of the selection that was intentionally imposed (for a similar discussion, but of "unreplicated" comparative studies, see Garland and Adolph 1994). The same would be true, but even more so, for any other traits that change across generations in a selected line (i.e., traits other than the one intentionally targeted for selection). So-called correlated responses to selection are notoriously more variable than the direct response to selection (Hill 1978; Robertson 1980; Falconer and Mackay 1996; Roff 1997; Bell 2008).

The accepted way to study both direct and correlated responses to selection is by comparing all of the replicate selected lines with all of the replicate control (and/or oppositely selected) lines by a "mixed-model" nested analysis of variance (ANOVA). The effect of selection is considered a "fixed effect" and is tested relative to the variation among the replicate lines, which is viewed as a "random effect" in statistical terms. Thus, if we had four replicate lines that had been selected for, say, high wheel running (discussed later) and four nonselected lines, then the degrees of freedom for testing the effect of selection would be one and six, regardless of how many individual mice might be measured for a particular trait. Other approaches are possible for analyzing selection experiments that lack replication (Henderson 1989, 1997; Konarzewski et al. 2005), but they are no substitute for proper replication (see also Eisen and Pomp 1990; Rhodes and Kawecki this volume).

Replication also allows tests for "multiple solutions" in response to a particular selective regime (Garland 2003). Selection that acts at a relatively high level of biological organization (figure 12.1)—such as voluntary behavior, whole-organism performance, or major components of life history (e.g., age at first reproduction)—is particularly likely to result in different adaptive responses both at the phenotypic level of subordinate traits and at the genetic level. At the phenotypic level, detailed studies of physiology, morphology, and biochemistry can elucidate whether a higher-level trait has evolved via changes in different subordinate traits. At the genetic level, a first-pass "black box" approach to determine whether different genes underlie the response to selection in replicate lines is to cross those lines and examine the traits of interest in the F1, F2, and/or backcross populations (see also Rhodes and Kawecki this volume). For example, if two replicate lines selected in the same direction for a given trait have different alleles at some of the loci influencing that trait, then a cross should produce an F1 whose mean trait value exceeds that of the two lines (e.g., Eisen and Pomp 1990; Bult and Lynch 1996). Moreover, crosses of replicate selected lines can be useful for renewed selection to break through selection limits (Bult and Lynch 2000).

## EXPERIMENTAL EVOLUTION OF MICE
## IN DIFFERENT THERMAL ENVIRONMENTS

Although temperature is often the environmental factor of choice in laboratory natural selection studies with *Drosophila* and microbes (Gibbs and Gefen this volume; Huey and Rosenzweig this volume), we are aware of only one such study with rodents (see also

Garland 2003). Barnett and Dickson (1984a, 1984b, 1989, and references therein) allowed wild-captured house mice to breed for many generations in either cold (~3°C) or warm (~23°C) environments. They chronicled the changes in size, anatomy, life history, and physiology that resulted, and they performed several common-environment experiments to distinguish genetic changes that resulted from the multigenerational exposure to cold and warm environments from the short-term acclimation effects of temperature. That is, they transplanted mice from cold environments to warm ones and vice versa, so that four groups of mice could be compared. Those groups were (1) evolved at 3°C and tested at 3°C; (2) evolved at 3°C, then acclimated to and tested at 23°C; (3) evolved at 23°C and tested at 23°C; and (4) evolved at 23°C and tested at 3°C. By acclimating mice to temperatures other than the one at which they evolved, short-term acclimation effects of temperatures were controlled, so that genetically determined changes could be demonstrated. Multigenerational cold exposure resulted in increased body mass and fat content of mice. In addition, the population in the cold became more fertile, evolving an earlier age of first reproduction and shorter tails. Mice in the cold evolved faster growth rates—an important measure of physiological performance and one with clear potential links to ecological success. Likewise, to support the faster growth rates of their young, females from populations with multigenerational exposure to the cold environment produced milk with higher fat and protein content than females that were transferred to the cold environment from populations with multigenerational exposure to the warm environment (Barnett and Dickson 1984b).

## WIND TUNNEL FLIGHT IN *DROSOPHILA*

Weber (1996) developed an ingenious apparatus and procedure that allowed him to select for wind tunnel flight in *Drosophila* with very large population sizes; estimated effective sizes in his experiments were $500 \le N_e \le 1,000$. The apparatus was essentially a 1.5-meter-long wind tunnel with forty equal internal compartments pierced by 4-centimeter circular holes, providing a cylindrical potential flight path. Flies were released into the downwind end and would instinctively fly toward a bright light at the upwind end. Flies moving toward the light faced wind speeds that increased in regular increments from zero in the first compartment to a maximum in the last. Walls were coated with a dry, slippery coat of Fluon such that flies could only advance by flying. Trials were ended by $CO_2$ anesthesia that then allowed the phenotypic distribution of the compartment reached to be scored. Up to fifteen thousand flies were fractionated in a single run, and the top few percent were selected as breeders for the next generation. According to Weber (1996, 206), "Various tests and observations show that the trait is actually a composite of phototaxis, activity level, flying speed, and aerial maneuvering skill." In other words, a mixture of behavior and performance was being positively selected.

The mean apparent flight speed in each of two replicate lines increased from about 2 to 170 centimeters/second and did not reach a plateau in one hundred generations.

Competitive fitness tests conducted at generations 50 and 85 indicated little or no loss of fitness in the selected lines as compared with two control lines (see Weber 1996 for details).

Marden et al. (1997) used a computerized system for three-dimensional tracking of large numbers of individual free-flying flies to assess performance at generation 160. The selected lines showed significant increases in mean flight velocity, decreases in angular trajectory, and a significant change in the interaction between velocity and angular trajectory. However, maximal flight velocity differed little between the selected and control lines, although individuals from the selected lines attained maximal performance levels much more frequently. The authors concluded that although the selection regimen had changed the relative effort and/or the frequency of phenotypes capable of attaining a preexisting maximal performance level, it was not able to increase maximal flight performance over entire populations. These results are in sharp contrast to an experiment that selected for high voluntary wheel running in mice (Swallow et al. 1998a; Garland 2003; see also later discussion in this chapter and Rhodes and Kawecki this volume), which found that both mean and maximal voluntary speeds increased significantly (Girard et al. 2001; Koteja and Garland 2001; view movie at www.biology.ucr.edu/faculty/Garland/ Girard01.mov). This difference is important because an increase in the maximal speed or intensity of a behavior is presumably more likely to entail changes in underlying performance capacities.

## ENDURANCE RUNNING AND STRESS-INDUCED ANALGESIA IN MICE

As a performance measure, maximum endurance capacity has received considerable attention. The ability of an organism to sustain physical activity is determined by multiple biochemical and physiological factors, and by such environmental factors as temperature. The evolution of increased endurance capacity has particularly important ramifications for organisms that are wide ranging. For example, selection for higher aerobic capacity may have played an important role in the evolution of endothermy (Hayes and Garland 1995). Endurance capacity may constrain many behaviors, including foraging, mate searching or courtship, territorial defense, and migration. Endurance was important in human evolution (Bramble and Lieberman 2004), and endurance capacity strongly influences the outcome of sporting events, such as marathons. Furthermore, endurance capacity has been linked to disease susceptibility and longevity (Wisløff et al. 2005; Koch and Britton 2007; Koch et al. 2007). Until relatively recently, the heritability of endurance capacity in human populations has been unclear, with estimates of "heritability" (not narrow sense) ranging from essentially zero (Perusse et al. 1987) to over 0.9 (Klissouras 1971). Research with animal models has established that genetic factors explain a significant amount of variation in both exercise capacity in an untrained state (Koch and Britton 2001) and in the physiological responses to training regimens (Troxell et al. 2003).

Bunger et al. (1994) reported the results of sixty generations of selecting laboratory mice for an index combining high body weight and high "stress resistance," where the latter denoted the distance to exhaustion on a treadmill. To our knowledge, this was the first attempt to select on a measure of locomotor performance in a rodent. Unfortunately, this selection experiment was conducted in the former East Germany at a time when publications in English were little supported (L. Bunger, personal communication). Although some interesting results have emerged from this work (e.g., those of Falkenberg et al. 2000 on blood enzymes and substrates), they seem to have been underappreciated in the larger scientific community.

It may seem that estimating an animal's endurance capacity during locomotor activity is a relatively straightforward task, which can be easily achieved by forcing the exercise on a treadmill. However, exposure to threatening, emergency conditions elicits a transient decrease of pain sensitivity—so-called stress-induced analgesia (SIA)—which also affects thermoregulatory mechanisms, locomotor performance, and metabolism (Sadowski and Konarzewski 1999; Lapo et al. 2003a, 2003b, 2003c). The result can be a counterintuitive depression (rather than mobilization) of metabolism in response to acute stressful stimuli, such as those applied during the measurements of performance traits (e.g., shocks from electric grid during forced running or cold exposure during the measurement of thermogenic capacity). Therefore, results obtained from such measurements may to some extent reflect an animal's propensity to develop SIA. Indeed, a comparison of mice from lines selected for high and low propensity to develop SIA (Panocka et al. 1986) showed that the maximum thermoregulatory rate of oxygen consumption (achieved during either swimming in cold water or exposure to $HeO_2$ atmosphere at $-2.5°C$) was lower in the high-SIA than in the low-SIA lines (Sadowski and Konarzewski 1999; Lapo et al. 2003b).

Thus, a response to selection for performance traits measured in conjunction with the application of stressful stimuli could be partly attributable to selection on propensity to develop SIA, rather than mechanisms limiting physiological performance per se. Consequently, selection for performance traits measured with protocols based on forced activity (e.g., endurance running on treadmill) may yield results different from those based on voluntary activity (e.g., voluntary wheel running—see later discussion).

## ENDURANCE RUNNING IN RATS AND VOLUNTARY WHEEL RUNNING IN MICE

Treadmill endurance capacity has also been the subject of a within-family selection experiment that used rats (Koch and Britton 2001). Selection for low and high endurance capacity was based on distance run to exhaustion on an inclined motorized treadmill (with electrical stimulation grid) whose speed was gradually increased. The starting velocity was 10 meters/minute and was increased 1 meter/minute every 2 minutes. The selected lines were not replicated, and no unselected control lines were maintained, which is not an ideal design (see also later discussion of work by Konarzewski and colleagues). The selected lines diverged rapidly, and after six generations the low-capacity runners (LCR) and high-capacity runners (HCR) differed by 2.7-fold in treadmill endurance capacity, with the majority of the change (relative to earlier measurements obtained from the starting population) occurring in the up-selected line (Koch and Britton 2001). By generation 11, the ratio between the average phenotypes of the two selected lines was 4.5 (Wisløff et al. 2005).

In such an experiment, it is possible that lines will diverge in their willingness to run, via changes in brain circuitry related to motivation, reward, or pain sensitivity (Li et al. 2004; Rhodes et al. 2005; Keeney et al. 2008). If so, then a difference in endurance measured during forced exercise might emerge even in the absence of changes in actual physiological capacities for exercise. Thus, it is important that it was found that, by generation 7, maximum oxygen consumption ($VO_2$max)—a key determinant of maximum aerobic speed and hence of endurance capacity per se—differed by 12 percent between the HCR and LCR lines (Henderson et al. 2002).

This rat experiment forms an interesting counterpart to an experiment with mice that used within-family selection to increase voluntary wheel running in four replicate lines while also maintaining four unselected lines as controls (figure 12.2: Swallow et al. 1998a; Garland 2003; Middleton et al. 2008; Rhodes and Kawecki this volume). Mice from the high-runner (HR) lines exhibit an elevated $VO_2$max (Swallow et al. 1998b; Rezende et al. 2006a, 2006b), as well as increased endurance during forced treadmill

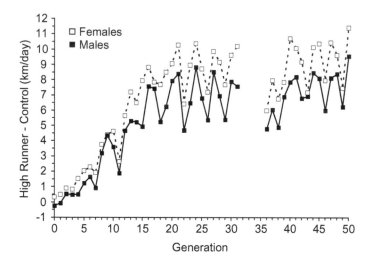

FIGURE 12.2

Average difference in voluntary wheel running on days 5 and 6 of a six-day exposure to wheels (1.12-m circumference) between four High Runner (HR) lines and their four nonselected control lines (Swallow et al. 1998a; Garland 2003; Middleton et al. 2008). After generation 31, mice were moved from the University of Wisconsin–Madison to the University of California–Riverside, and wheel running was not recorded (nor was selection applied) for four generations. For both the HR and C lines, about a third of the total revolutions (and hence distance) per day occur during "coasting" (Koteja et al. 1999; Girard et al. 2001). Although the absolute difference between the HR and control lines is somewhat greater for females, the fold difference is identical for the sexes (e.g., see Swallow et al. 1998a; Garland 2001, 2003; Rezende et al. 2006b). Note the substantial environmental variation across generations.

exercise (Meek et al. 2007). Similarly, the HCR rats also ran longer distances on wheels as compared with LCR rats (Waters et al. 2008). HCR rats ran for more time each day and ran faster on wheels. Striatal dopaminergic responses to the wheel running trial differed between HCR and LCR rats, suggesting a divergence in the central control of locomotor behaviors in these lines (Waters et al. 2008). Alterations in dopamine function have also been strongly implicated in the HR lines of mice (Rhodes et al. 2005; Rhodes and Kawecki this volume). The general consistency of results between these mouse and rat experiments suggests a fairly general relationship between physiological capacity for aerobic performance and the propensity to engage in endurance running, at least in rodents, if not mammals generally. An important area for future research in both experiments will be elucidating sex-specific differences in the lower-level traits that underlie the response to selection (e.g., see Swallow et al. 2005; Keeney et al. 2008).

Another point of consistency between the mouse wheel-running and rat treadmill endurance experiments is that in both cases, the high-selected lines became smaller in body mass. In the mouse experiment, the reduction in body mass may be partly a function of elevated circulating corticosterone levels, both at rest and during wheel running (for details, see Girard and Garland 2002; Malisch et al. 2007, 2008). However, the HCR and LCR rats do not differ in baseline corticosterone levels (Waters et al. 2008). In the mice, line variation in circulating corticosterone levels is inversely related to the ability to clear a parasitic nematode infection (Malisch et al. 2009).

Rats selected for endurance capacity diverged in control of aerobic capacity, including central mechanisms involved in the delivery of $O_2$ to the skeletal muscle as well as peripheral mechanisms involved in the delivery, use, and conversion of $O_2$ within the skeletal muscles. Interestingly, by generation 7 of selection, when maximum oxygen uptake ($VO_2$max) differed by 12 percent between the lines, only divergence in peripheral mechanisms were detected between the HCR and LCR rats. Higher $O_2$ delivery resulted from improved $O_2$ transport from tissue to cells, increased capillary density (Henderson et al. 2002), and increased oxidative enzyme capacity (Howlett et al. 2003)—with no significant changes in pulmonary or cardiovascular function (Henderson et al. 2002). By generation 15 of selection, when $VO_2$max had diverged by 50 percent between the HCR and LCR, the lines no longer differed significantly in $O_2$ extraction. This seeming reversal, however, is not a function of a loss of the peripheral changes (e.g., increased capillary density) observed in generation 7 but, instead, is a function of gains in cardiac output and stroke volumes, with a resultant reduction in blood transit time across the capillary beds in the HCR rats (Gonzales et al. 2006). Taken together, these results suggest that the mechanistically subordinate traits underlying a particular phenotype under selection may evolve at different rates (see also Rezende et al. 2006c; Archer et al. 2007; Guderley et al. 2008).

The rat lines selected for high or low treadmill endurance were developed to test the "metabolic syndrome" hypothesis—that many complex diseases are caused by defects in the metabolic pathways of energy acquisition, deposition, oxidation, and detoxification (Koch and Britton 2005). A comparison at generation 11 showed that the two rat lines had diverged in a number of cardiovascular risk factors. Not only are LCR rats larger, but they have significantly higher total body fat (J. G. Swallow, unpublished data), higher visceral body fat, and higher circulating free fatty acids and triglycerides. (Similarly, mice from the HR lines have reduced body fat as compared with their control lines; see Girard et al. 2007; Vaanholt et al. 2007.) LCR rats had higher abdominal aortic blood pressure measured across multiple time periods. Furthermore, stroke volume and isolated myocyte shortening, standard measures of cardiac function, were impaired in the LCR rats as compared to HCR rats. Taken together, these data constitute evidence for the emergence of "metabolic syndrome" in the LCR line (Koch and Britton 2005; Wisløff et al. 2005). Hence, these lines are becoming an important model for biomedical research.

An unexpected result of the mouse selection for high wheel running was the discovery and increase in frequency (in two of the four HR lines) of a Mendelian recessive allele that, when present in the homozygous condition, causes a 50 percent reduction in hind limb muscle mass (Garland et al. 2002). This reduced muscle mass is related to dramatic differences in muscle fiber type composition (Guderley et al. 2008), and minimuscle individuals demonstrate that the allele has numerous other pleiotropic effects, many of which seem conducive to supporting high levels of endurance exercise (e.g., increases in mass-specific cellular aerobic capacity, heart size, and hindlimb bone lengths; review in Middleton et al. 2008). (Interestingly, however, minimuscle individuals seem to have an elevated cost of transport; Dlugosz et al. 2007.) This gene of major effect has been localized to a 2.6335-megabase interval on MMU11, a region that harbors about one hundred expressed or predicted genes, many of which have known roles in muscle development and/or function (Hartmann et al. 2008). Identification of the genetic variant that underlies minimuscle could elucidate both normal muscle function and the dysregulation of muscle physiology that leads to disease. Beyond this, QTL mapping of the HR lines, which is currently underway, has the potential to uncover "anti–couch potato" genes that positively affect voluntary activity levels, which again could have important biomedical implications.

The experiments that have increased forced treadmill endurance running in rats, while also apparently increasing voluntary wheel running; and those experiments that have increased voluntary wheel running in mice, while also increasing forced treadmill endurance, raise interesting issues concerning phenotypic plasticity (Garland and Kelly 2006). In the mouse experiment, home cage activity is also increased, as compared with control lines, when wheels are not provided (Rhodes et al. 2005; Malisch 2007; Malisch et al. 2008). Whether the same is true for the high-endurance rats is unknown. However, if activity levels are increased, then they may contribute to the physical fitness and performance abilities of the high-selected lines in question. Swallow et al. (2005) coined the term "self-induced adaptive plasticity" to refer to situations in which an organism engages in a behavior (e.g., locomotion) that in turn positively affects its ability to further engage in that behavior, notwithstanding the somewhat paradoxical use of this term to refer to a pattern of plasticity that arises as a result of selection.

A related point that has been emphasized by some exercise physiologists (e.g. Booth et al. 2002a, 2002b) is that human patterns of gene expression may be adapted to function "normally" when we engage in substantially higher levels of daily physical activity, as during the hunter-gatherer stage of our ancestors. In their view, "our current genome is maladapted, resulting in abnormal gene expression, which in turn frequently manifests itself as clinically overt disease" (Booth et al. 2002b, 399). In other words, what might be termed the ancestral physical-activity environment cannot be separated from "normal" physiological function. Or, in their words, "in sedentary cultures, daily physical activity normalizes gene expression towards patterns established to maintain the survival in the Late Palaeolithic era" (Booth et al. 2002b, 399). The rat and mouse selection experiments, or variants on them, may be well suited to address these sorts of hypotheses.

The replicate high-running selected lines of mice show a number of interesting differences that can be viewed as multiple solutions. For example, the components of wheel running (minutes/day and mean speed) differ significantly among the four replicate selected lines (Swallow et al. 1998a; T. Garland, unpublished results). And, as noted, the mini-muscle phenotype occurs in only two of the four selected lines. Pleiotropic effects of the mini-muscle allele in homozygotes (Hannon et al. 2008) also result in line differences for the affected traits. The extent to which various traits differ among the four selected lines after adjusting statistically for the effects of mini-muscle is currently under investigation.

The mouse wheel-running experiment also provides an interesting example of a possible "adaptive" genetic correlation (*sensu* Lynch 1992, 1994) between two behavioral traits. Although they do not differ from control lines in intermale or maternal aggression, mice from the HR lines show elevated predatory aggression when tested with live crickets (Gammie et al. 2003). Even when not fasted, they attack, kill, and eat crickets more rapidly as compared with the nonselected control lines (Gammie et al. 2003; T. Garland, unpublished results). The correlated response to selection by predatory aggression provides evidence that it is positively genetically correlated with locomotor activity. If this correlation is a fairly general feature of mammals, then it has implications for the evolution of an active, predatory lifestyle. Mammals such as wolverines are characterized by high activity levels (e.g., large home ranges and long daily movement distances: Garland 1983; Goszczynski 1986) and, of course, by a tendency to attack and kill suitable prey items when encountered. During the evolution of such a mammal, from an ancestor that was both less active and had lower predatory tendencies, one presumes that correlational selection would favor individuals with both relatively high activity and predatory tendency. If so, then the rate of bivariate evolution would be increased if the two traits were positively genetically correlated: the genetic architecture would "facilitate" what selection was "trying" to do (for a similar argument concerning locomotor speed and endurance in garter snakes, see Garland 1994; for a general argument that is related, see Schluter 1996). Why might locomotor activity and predatory aggression be positively related? Gammie et al. (2003) suggest that both behaviors share similar neuronal substrates, with dopamine being a likely candidate (see also Ohno and Miyatake 2007 and Rhodes and Kawecki this volume).

## EVOLUTION OF THE RATE OF ENERGY METABOLISM IN RODENTS

The performance of vital animal functions, such as food acquisition, locomotion, osmotic and thermal regulation, and reproduction, depends on a complicated network of interacting physiological processes. However, each of those processes involves—directly or indirectly—conversion of energy. Therefore, the rate of energy metabolism can serve as a unifying quantitative measure of organismal functioning (McNab 2002). In most animals, energy is eventually obtained from aerobic oxidation of organic substrates, so the rate of energy metabolism can be measured indirectly as the rate of oxygen consumption. Although describing the complicated network of physiological processes by a single value—the rate of energy metabolism or oxygen consumption—is a gross oversimplification, the question of why animals vary in metabolic rate has been a primary motivation for countless comparative and experimental studies.

Three obvious sources of variation in metabolic rate are body size, body temperature, and physical activity. Larger animals of a given kind must, on average, have higher total metabolic rates than small ones, and the exact way metabolic rate scales in relation to body size (see also Frankino et al. this volume) has important functional implications (e.g., see Garland 1983; Carbone et al. 2005). But even if we compare animals of a particular size under resting conditions and at the same body temperature, their metabolic rates may still differ by more than an order of magnitude (e.g., see Kozłowski and Weiner 1997; Kozłowski et al. 2003; McKechnie et al. 2006; White et al. 2006). Analysis of experiments in which the rate of metabolism is either directly selected or expected to change in response to selection on other traits provides a powerful tool with which to uncover underlying mechanisms (see also Hayes and Garland 1995; Garland and Carter 1994; Garland 2003; Książek et al. 2003, 2004, 2007; Konarzewski et al. 2005; Swallow and Garland 2005; Rezende et al. 2006; Brzęk et al. 2007; Kane et al. 2008; Sadowska et al. 2008).

## SELECTION ON BASAL METABOLIC RATE

*Basal metabolic rate* (BMR) is defined as the minimum rate of metabolism measured in a normothermic, resting, postabsorptive individual under thermally neutral conditions (McNab 2002), and it is widely used as a standard for inter- and intraspecific comparisons in studies on animal energetics (Frappell and Butler 2004, and references cited herewith). We are aware of only one successful experiment in which BMR was the trait directly and purposefully selected (Książek et al. 2003, 2004; Gębczyński 2005; Konarzewski et al. 2005; Brzęk et al. 2007; figure 12.3). Konarzewski and colleagues selected for high and low mass-independent BMR (i.e., residuals from regression of BMR on body mass) in laboratory house mice (Swiss Webster outbred strain). Unfortunately, the experiment had only one line selected for each of the directions and no unselected control line, which undermined its power and forced the authors to use indirect methods to test the statistical

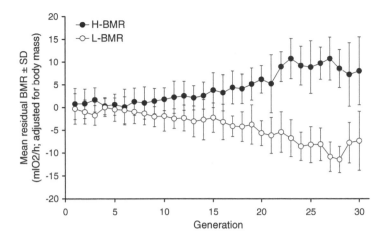

FIGURE 12.3

Changes of basal metabolic rate (BMR) in male laboratory mice selected for high (H-BMR) and low (L-BMR) values of mass-independent BMR (residuals from regression models; Książek et al. 2003, 2004; Gębczyński 2005; Konarzewski et al. 2005; Brzęk et al. 2007). The selection experiment had only one line for each direction, and therefore Henderson's (1989, 1997) approach (see text) has been used to test for the effects of selection (the means and standard deviations were calculated from family means). The data for this figure have been kindly provided by A. Gębczyński and M. Konarzewski.

significance of both the direct and correlated responses to selection (Henderson 1989, 1997; Konarzewski et al. 2005). Nevertheless, the experiment provided important results concerning variation in BMR. First, the selection was effective, and after seven generations, the lines clearly diverged (Książek et al. 2004; figure 12.3). At generation 19, the difference in BMR between the high line (59.0 ml $O_2$/h) and the low line (50.1 ml $O_2$/h) was equivalent to 2.3 phenotypic standard deviations (SD). At generation 22, the difference had increased to 10.8 ml $O_2$/h, or 3.3 phenotypic SD (Brzęk et al. 2007), and it appears that around generation 25 a selection limit was achieved (figure 12.3).

The results of this selection experiment were qualitatively consistent with the narrow-sense heritability ($h^2 = 0.38$, $p = 0.02$) estimated in the base population by means of parent-offspring regression (Konarzewski et al. 2005). This is an important finding, because previous studies in both laboratory mice (Lacy and Lynch 1979; Dohm et al. 2001) and a wild rodent, the leaf-eared mouse *Phyllotis darwini* (Nespolo et al. 2003; Bacigalupe et al. 2004), indicated very low or even negligible heritability of BMR. On the other hand, in a more recent study based on a much larger sample of about one thousand individuals, Sadowska et al. (2005) found a relatively high and significant heritability of mass-adjusted BMR in the bank vole *Myodes* (*Clethrionomys*) *glareolus* ($h^2 = 0.4$). Similarly, Johnson and Speakman (2007) reported high heritability ($h^2 = 0.45$, estimated from mean parent-offspring regression) of resting metabolic rate in short-tailed field voles (*Microtus agrestis*). Thus, it seems that at least in some populations of rodents, both laboratory and wild, there is a substantial additive genetic component to the variation of BMR. Thus, a search for genetic correlations with other traits is not hopeless.

As this selection experiment was performed on mass-independent BMR, mice from the high- and low-BMR lines did not diverge in body mass (Książek et al. 2004; Brzęk et al. 2007). Therefore, these selected lines are particularly suitable material with which to address questions concerning proximate factors underling differences in BMR (Hulbert and Else 2004, 2005). One of the plausible causes of BMR differences is a difference in the size of metabolically active organs, especially brain, heart, liver, kidneys, and the small intestine (Garland and Else 1987; Schmidt-Nielsen 1990; Hulbert and Else 2005). Indeed, the internal organs are larger in mammals than in reptiles of the same size (see Karasov and Diamond 1985; Hulbert and Else 2004, 2005), but several studies of intraspecific correlations between BMR and the internal organ size within birds or mammals have produced ambiguous results (e.g., Daan et al. 1990; Koteja 1996; Speakman and McQueenie 1996; Meerlo et al. 1996; Piersma and Lindström 1997; Chappell et al. 2007).

Results obtained from the BMR selected lines were unequivocal: mice from the high-BMR line had larger hearts, livers, kidneys, and small intestines, yet they were leaner than mice from the low-BMR line (Książek et al. 2004; Brzęk et al. 2007). Standardized differences between the organ sizes in high and low lines were 1.1–1.9 SD in generation 19, which increased to 1.4–2.2 SD in generation 22, significantly larger than a divergence that could appear as a result of random genetic drift alone, as estimated by Henderson's (1997) approach. This corroborated an earlier study, which showed a positive correlation between BMR and the mass of the internal organs across four inbred lines of laboratory mice (Konarzewski and Diamond 1995).

In addition, daily food consumption was also significantly higher in the high-BMR line than in the low-BMR line (Książek et al. 2004). These results, too, are complementary to those of another artificial selection experiment, which showed that laboratory mice from lines selected for high rate of food consumption had both larger sizes of the internal organs and a higher resting metabolic rate, compared with mice from lines selected for low rate of food consumption (Selman et al. 2001a, 2001b).

Thus, studies of these BMR-selected mice provided strong evidence that the level of BMR is associated both with the mass of the central organs involved in whole-animal metabolism and with the total rate of energy turnover in animals. This concurs with the results of studies that have shown an association between BMR and field metabolic rate (FMR)—that is, average daily metabolic rates measured in free-living animals by means of the doubly labeled water technique (Daan et al. 1990; Koteja 1991; Lindström and Kvist 1995; Ricklefs et al. 1996; Speakman et al. 2003).

Surprisingly, however, the mass of interscapular brown adipose tissue (IBAT; see also Lynch 1992, 1994)—the major site of heat production through nonshivering thermogenesis—was lower in the high- than in the low-BMR mice (Książek et al. 2004). Moreover, the maximum rate of oxygen consumption achieved during cold exposure did not differ significantly between the lines, and mice from the high-BMR lines showed slightly deeper hypothermia after cold exposure than those from the low-BMR lines, a difference that was significant when tested with ordinary ANCOVA, but not significant when tested using Henderson's approach. Furthermore, mice from the high-BMR lines did not maintain higher body temperatures and were not able to increase body temperatures faster, after exposure to mildly cold ambient temperatures, compared with mice from the low-BMR lines (Gębczyński 2004). Thus, thermogenic capacity of the mice appeared to be genetically independent from BMR. The same conclusion emerged from a quantitative genetic analysis in the bank vole, which showed no additive genetic correlation between BMR and the maximum cold-induced rate of oxygen consumption, even though both of the traits were heritable in the narrow sense ($h^2 \approx 0.4$; Sadowska et al. 2005). These results are apparently at odds with the common assumption that evolution of high BMR in mammals and birds was driven by selection favoring better thermoregulatory capabilities (Heinrich 1977; Crompton et al. 1978; Pörtner 2004).

The "thermoregulatory" model for the evolution of endothermy had been already challenged long before any information about genetic correlations between BMR and thermogenic capacity was known (Bennett and Ruben 1979). The major concern of its critics was that the supposed benefits from a slightly elevated temperature appear to be vague when compared to immediate and severe energy costs associated with an increase in resting metabolic heat production. This problem was clearly demonstrated in an experiment on varanid lizards, which showed that doubling the rate of energy metabolism resulted in a body temperature elevation of less than 1°C (Bennett et al. 2000). Therefore, over the last three decades, alternative hypotheses have been proposed to explain how an increased level of BMR and endothermy could have evolved as a correlated response to natural selection acting on other traits (see Kemp 2006 for a recent review). Among several proposed hypotheses, the most attention has been given to the aerobic capacity model, according to which an increase BMR and endothermy evolved as a correlated response to selection for high, sustained locomotor activity, supported by aerobic metabolism (Bennett and Ruben 1979).

The crucial assumption of the aerobic capacity model is that BMR and maximal aerobic capacity (i.e., the maximum rate of oxygen consumption achieved during locomotor activity—$VO_2max$) are strongly connected due to shared underlying biochemical processes, physiological functions, or anatomical structures, and hence they cannot evolve independently. The model has been appealing not only because it generally fits the paleontological record (Ruben 1995), but also because the major assumption could be translated into a simple hypothesis of a generally positive genetic correlation between BMR and $VO_2max$ and tested against readily available empirical data. Not surprisingly, several comparative and intraspecific analyses have been performed to test the hypothesis, and they provide mixed support for such a positive correlation at the comparative level (Hayes and Garland 1995; Angilletta and Sears 2003; Gomes et al. 2004; Rezende et al. 2004; Kemp 2006; Wiersma et al. 2007).

However, the first determined attempts to test the hypotheses at the level of genetic variation have been made only recently. Dohm et al. (2001) found limited support for a positive genetic correlation between BMR and $VO_2max$ in outbred laboratory house mice, whereas Sadowska et al. (2005) have shown a significant additive genetic correlation between BMR and $VO_2max$ measured during swimming in the bank vole. The latter study also showed that the factorial aerobic scope (i.e., the ratio of $VO_2max/BMR$) is heritable ($h^2 \approx 0.2$), which indicates that simultaneous selection for an increased $VO_2max$ while maintaining a low BMR should also be effective. Thus, quantitative genetic analyses with rodents to date offer partial support for the key assumption of the aerobic capacity model.

The aerobic capacity model would be ideally tested in a selection experiment in which locomotor activity or $VO_2max$ is the directly selected trait, and BMR is tested for a correlated response. As noted, selection for high voluntary activity levels in house mice (Swallow et al. 1998a; Garland 2003) has indeed resulted in increased maximal oxygen consumption during forced exercise (Swallow et al. 1998b; Rezende et al. 2006a, 2006b), but it has not led to an increase in basal (Kane et al. 2008) or resting (Rezende et al. 2006b) metabolic rate. These results are, thus, also inconsistent with the aerobic capacity model.

However, if the assumption of a tight functional link between BMR and $VO_2max$ holds, the latter should be lower in the low-BMR than in the high-BMR mice from Konarzewski's experiment. To test this possibility, Książek et al. (2004) and Brzęk et al. (2007) have measured the maximum rate of oxygen consumption achieved by the selected mice during swimming in cool water (25°C). Contrary to the expectation, $VO_2max$ during swimming tended to be lower in the high-BMR than in low-BMR line, though the difference was significant when tested with normal ANCOVA, but not when tested with Henderson's approach (Książek et al. 2004; Brzęk et al. 2007). However, the mice became severely hypothermic during the test, more so than after the cold-exposure trials. Again, mice from the high-BMR line showed a significantly deeper hypothermia after the test than those from the low-BMR line. The hypothermia observed at the conclusion of the swimming trials indicated that the animals were under substantial cold stress during the trials, and a considerable part of their oxygen consumption must have been associated with heat generation rather than with locomotor activity. Moreover, individuals with lower thermogenic capability could suffer from a deeper hypothermia, and consequently a compromised capacity of the respiratory system and muscles to support locomotor activity. Thus, the $VO_2max$ measured during the test could actually estimate thermogenic capacity (i.e., maximal oxygen consumption elicited via cold exposure) rather than maximal exercise metabolic rate (i.e., maximal oxygen consumption elicited via exercise). Therefore, it might be more appropriate to read the results as contradicting the "thermoregulatory" rather than the "aerobic capacity" model for the evolution of endothermy.

Perhaps the most intriguing result from Konarzewski's experiment comes from a comparison of cell membrane lipids (Brzęk et al. 2007). Else and Hulbert (1987; also

Hulbert and Else 1989, 1990, 2005; Else et al. 2004) found that permeability of the cell membranes to protons is higher in vertebrate endotherms than in ectotherms. Consequently, endotherms should spend more energy to maintain electric potential across membranes, should have a lower efficiency of ATP synthesis in mitochondria, and consequently should have a higher rate of heat dissipation compared with ectotherms. This difference in "leakiness" could be a cost of being an endotherm, but Else and Hulbert (1987) proposed that the "leaky membranes" could also have adaptive value as a mechanism for heat production per se. The permeability of cell membranes to protons and other ions depends on their composition of fatty acids: the more unsaturated the fatty acid chains, the higher the "leakiness" of cell membranes (e.g., Brookes et al. 1988). Based on their results, Hulbert and Else (1999, 2005) proposed the "membrane pacemaker hypothesis," according to which manipulation of membrane composition, and hence permeability, serves as a major mechanism controlling the overall level of metabolism. Farmer (2000, 2003) incorporated this hypothesis in a "parental care" model of evolution of endothermy, according to which the "leaky membranes" served as a mechanism of thermogenesis associated with incubation of eggs and young.

The lines of mice selected for high and low BMR provided a unique model for testing the assumption of the membrane pacemaker hypothesis. Contrary to the expectation derived from the membrane pacemaker hypothesis, in generation 22 of the experiment, the proportion of unsaturated bonds (unsaturation index) was higher in livers of mice from the low-BMR lines compared with the high-BMR lines, though no difference was observed in kidneys (Brzęk et al. 2007). It is easy to imagine that effective selection for high BMR could be realized through several different mechanisms, so a lack of a difference in composition of cell membranes does not necessarily falsify the "membrane pacemaker" model. However, the finding of a significant difference in the opposite direction should be treated as a serious challenge to the model, one which cannot be ignored in future research.

Finally, Konarzewski and colleagues (Książek et al. 2003, 2007) also investigated how selection for high and low BMR in mice affected their immune competence—which can also be considered an important performance trait. Immune function was measured as a response to injection of sheep red blood cells (SRBC). Mice from the high-BMR line showed a significantly lower response than those from low-BMR lines (Książek et al. 2003). This effect was even more evident when the mice were also exposed to a low ambient temperature—that is, when their energy budgets were simultaneously challenged by thermoregulation and immune response (Książek et al. 2007). As mentioned earlier, among the four control and four high-running lines of mice, variation in circulating corticosterone levels is inversely related to the ability to clear a parasitic nematode infection (Malisch et al. 2009); however, selection lines that differ in BMR have not been studied in this respect.

The scarcity of selection experiments focused on BMR is not surprising, because most of the selection experiments on vertebrates that have involved physiological traits were performed in the context of agricultural animal production. From that perspective, the important questions concern the actual energy cost of animal maintenance rather than a minimum metabolic rate measured under conditions (e.g., fasted) that do not occur under normal housing conditions. In particular, agricultural selection experiments have been designed to ascertain whether it is possible to select for low energetic costs of maintenance, because that would result in a decreased feed intake, and perhaps also would allow conversion of more of the food eaten into useful tissues—both of which would increase economic efficiency of animal production. Although the ultimate objective of animal production science (maximization of financial profit) differs from the objectives of evolutionary biology, several of those experiments have provided results that are informative in an evolutionary context.

Nielsen et al. (1997b) selected laboratory mice for high and low heat loss, measured in adult males during a fifteen-hour assay using a direct calorimetry system. Three replicate lines each were selected for high (MH) and for low (ML) heat loss, and three unselected lines were maintained as controls (MC). To control for the effect of body mass, (heat loss)/(body mass)$^{0.75}$ was used as the selection criterion (in units of kcal $\times$ kg$^{-0.75} \times$ day$^{-1}$). Because the actual slope of the relation between heat loss and body mass of the mice may differ from 0.75, such a correction does not ensure that the trait is mass-independent (Hayes and Shonkwiler 1996). However, the results showed that, in effect, there was no unintentional selection for large or small body mass. The selection on the heat loss character was effective: at generation 15 the trait value was about 33 percent higher in the MH lines and about 20 percent lower in ML lines, compared with the control lines (Nielsen et al. 1997b). Realized heritability, calculated from divergence between MH and ML lines, was about 0.28 $\pm$ 0.01 and tended to be higher when calculated from the divergence between MH and MC lines (0.31 $\pm$ 0.01) than when calculated from divergence between ML and MC lines (0.26 $\pm$ 0.01; Nielsen et al. 1997b).

The authors intended to measure just the costs of maintenance, but, as they admitted, the difference could be related to differences in locomotor activity during the test, which could in turn have been inflated by differences in behavioral response to the calorimetric chamber, isolated from external signals (Nielsen et al. 1997b). However, other measurements showed clearly that the selection affected overall metabolism of the animals under normal housing conditions as well. Mice from the MH lines had a higher rate of food consumption, higher body temperature, and higher locomotor activity, as well as having larger metabolically active organs (liver, heart, spleen) and being leaner than mice from ML lines (mice from control lines were intermediate; Nielsen et al. 1997a; Moody et al. 1997; Mousel et al. 2001; Kgwatalala and Nielsen 2004). Kgwatalala et al. (2004) also found that corticosterone levels were higher in the MH compared to ML lines, which was also consistent with observations that MH mice were more susceptible, and ML less susceptible, to restraint stress. Results for thyroid hormones were surprising: the level of T4 in blood serum tended to be higher in ML than in MH lines, and no clear difference was observed for T3. Thus, contrary to some expectations (Denckla and Marcum 1973; but see Hulbert and Else 2004), the differences in the rate of metabolism between the lines were apparently not mediated by an altered level of thyroid hormones.

Very interestingly, litter size at birth was higher in mice from MH than from ML lines (again, mice from MC lines had intermediate litter sizes; Nielsen et al. 1997a; McDonald and Nielsen 2006). This finding clearly shows the problem of poor exchange of information and ideas between researchers working in distinct areas. Nielsen et al. (1997a, p. 1475) noted that "why selection for higher or lower maintenance energy causes a correlated change in ovulation rate is not clear" and concluded that "the positive correlated responses in number born represent a very undesirable relationship between energy for maintenance and number born." The gloomy comment is understandable in the original context of the research, which was focused on the possibility of lowering costs of maintenance of livestock: the results showed that it indeed had been possible, but at the cost of lowering fecundity, which obviously was not good news. Apparently, the authors were not aware that their results might be a valuable contribution to the debate concerning a hypothetical link between metabolic rate and life-history traits, which has been carried on among comparative, evolutionary, and ecological physiologists.

In a seminal paper, McNab (1980) proposed that individual growth rate, reproductive rate, and the Malthusian population parameter, r, are positively correlated with the basal level of metabolism. These hypotheses have been tested in several comparative and within-species studies. The results of these studies have not provided unequivocal evidence for such associations, but there has been continued interest in this hypothesis nonetheless (e.g., McNab 1980, 2002; Hennemann 1983; Hayssen 1984; Padley 1985; Trevelyan et al. 1990; Harvey et al. 1991; Hayes et al. 1992; Koteja 2000; White and Seymour 2004; Johnston et at. 2007). Results from the selection experiment of Nielsen and colleagues provide important evidence that seems to support the hypothesis of a link between the metabolic rate and reproductive performance. Similarly blind were earlier evolutionary physiologists, who did not refer to these results. For example, Farmer (2000, 2003) and Koteja (2000) proposed that endothermy and high BMR in birds and mammals evolved in connection with evolution of an increased parental effort in reproduction, but they did not refer to results from this selection experiment, even though the results could be read as supporting their hypothesis.

Nielsen and collaborators subsequently put much effort into identification of the genes underlying the selected-line difference in heat loss, using two distinct methods. First, they attempted to identify quantitative trait loci (QTL) associated with the difference in heat loss (Moody et al. 1999), finding several loci influencing heat loss and other related traits such as total body mass, amount of brown adipose tissue, and gonadal fat tissue. Though it was not possible to unambiguously link any of the QTLs to a particular gene, the authors pointed out that some of the identified regions contain genes that are potential candidates for heat loss regulators, such as genes encoding the $\beta$-subunit of the thyroid stimulating hormone, neuropeptide Y receptor Y2, or uncoupling proteins UCP2 and UCP3 (Moody et al. 1997).

The second method used by this group was to investigate differences in gene expression between the lines selected for high and low heat loss. Allan et al. (2000) performed a genome-wide scan of genes differently expressed in hypothalamus and brown adipose tissue (BAT) in mice from inbred lines derived from the MH and ML lines, followed by Northern blot analysis that enabled identification of those genes that showed different expression. These analyses revealed that ribosomal protein L3 (RPL3) mRNA was expressed at a higher level in the ML than MH mice. This protein forms a central channel through which newly synthesized peptides emerge, and, therefore, its function potentially affects all aspects of metabolism. In an extension of the project, Wesolowski et al. (2003) applied microarray analyses followed by real-time PCR and Northern blotting to investigate differences in the expression of several plausible genes in the hypothalamus. They found two genes that had expression significantly different between the inbred mice derived from MH and ML lines: oxytocin and a tissue inhibitor of metalloproteinase 2 (Timp-2), both of which had a higher expression in mice with high heat loss. The role of the latter in regulating heat loss is unclear, but an increased expression of the oxytocin gene is consistent with a higher level of locomotor activity and a higher body temperature of the mice from MH lines (Wesolowski et al. 2003).

In the next study, McDaneld et al. (2002) applied opposite tactics: instead of scanning the entire genome for candidate genes, the authors specifically examined whether the differences between MH and ML lines could be attributed to altered expression or functionality of uncoupling protein UCP1 in BAT (the protein which plays a central role in heat generation in BAT). They compared UCP1 expression in MH and ML lines and also compared heat loss in two new lines, derived from inbred MH and ML lines, but which also had a knockout UCP1 gene. The expression of UCP1 was higher in the mice from ML than MH lines, and heat loss did not differ between mice with functional and knockout UCP1 gene. Thus, the difference in energy expenditure between the lines is not related to UCP1-mediated thermogenesis in BAT (McDaneld et al. 2002). We should also add that the lower expression of UCP-1 in the MH lines is consistent with a lower mass of BAT and a lower thermogenic capacity observed in mice selected for high BMR in Konarzewski's experiment described earlier.

## RATE OF METABOLISM AS A HYPOTHETICAL CORRELATED RESPONSE

In the two selection projects described so far, the basal or average rate of energy metabolism were targets of selection and the selected lines were used in attempts to identify proximate morphophysiological and genetic factors underlying the observed differences.

Although success in identifying the mechanisms involved is, so far, limited, the rapid progress of molecular techniques warrants continued effort. However, evolutionary physiologists would also like to know whether selection for some traits possibly related to Darwinian fitness of free-living animals could trigger evolution of the rate of metabolism. In other words, the question is whether an increased or decreased rate of metabolism is likely to arise as a correlated response to selection for ecologically relevant traits.

Although animals from any selection experiment could reasonably be used to test for a correlated response in the basal or standard metabolic rate, few experiments have been designed specifically for that purpose. Recently, one of us (P.K.) has launched a large-scale selection experiment on a nonlaboratory rodent, the bank vole *Myodes (Clethrionomys) glareolus*. The voles are selected for three characters that are thought to have played an important role in the mammalian adaptive radiation (Eisenberg 1981; Kemp 2007): increased aerobic capacity, the ability to grow on a low-quality herbivorous diet (Karasov and Martinez del Rio 2007), and the intensity of predatory behavior. All three traits are adjusted for body mass during selection (figure 12.4). (For a recent discussion of adaptive radiation in the wild, see Seehausen 2006. For an example of experimental adaptive radiation in a bacterium, see MacLean and Bell 2002.) To produce satisfactory power for statistical tests, four replicated lines are maintained in each of the three selection directions, and four unselected, control lines (Sadowska et al. 2008).

After three generations of within-family selection, the maximum rate of oxygen consumption achieved during voluntary swimming was 15 percent higher in lines selected for high aerobic capacity than in the control lines (252 vs. 219 ml $O_2$/h, $p = 0.0001$). When fed a low-quality diet made of dried grass, voles from lines selected for the ability to cope with an herbivorous diet lost about 0.7 gram less mass than voles from the control lines ($-2.44$ vs. $-3.16$ g/4 days, $p = 0.008$). In lines selected for predatory behavior toward crickets, the proportion of "predatory" individuals was higher than in the selected lines (43.6 percent vs. 31.9 percent, $p = 0.045$), but "time to capture" calculated for the successful trials did not differ between the lines (Sadowska et al. 2008).

In the next generation, basal metabolic rate was significantly higher in the lines selected for high aerobic capacity than in the control lines (unpublished results), as expected from the positive additive genetic correlation between the traits estimated in the base population (Sadowska et al. 2005). This result fits the hypothesis that selection favoring high locomotor activity powered by aerobic metabolism could have been an important factor in the evolution of high BMR in mammals. BMRs of the other selection lines ("herbivorous" and "predatory") did not differ from those of their control lines, but the direct effect of selection was also less profound in these lines than in the high-aerobic lines. The experiment is continuing, and the selected lines of voles will provide a useful case for testing hypotheses concerning the correlated evolution of complex traits that may relate to adaptive radiation of rodents.

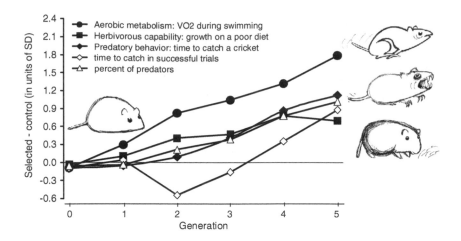

FIGURE 12.4

Multidirectional selection experiment on bank voles *(Myodes glareolus)* intended to provide a laboratory model of adaptive radiation (Sadowska et al. 2008): a "generic," omnivorous rodent species (generation 0), is selected to create highly aerobic athletes (right, top), aggressive predators (middle), and efficient herbivores, capable of growing on low-quality diet (bottom). Four replicate lines are maintained in each of the four line types (three selected plus unselected control). The effects of five generations of selection are shown as differences between mean values of the selected traits, expressed in units of phenotypic standard deviation (filled symbols). For the "predatory" lines, changes of proportion of predatory individuals and their performance in successful trials are also shown (open symbols). The SD is a mean of standard deviations in pooled control and pooled selected lines, and it has been calculated separately for each generation. See text for further explanation. The drawings of voles have been kindly provided by January Weiner.

## SELECTION ON VO₂Max AND THE CORRELATION BETWEEN BMR AND VO₂Max

Another selection experiment (J. P. Hayes, B. Wone, M. K. Labocha, E. R. Donovan, and M. W. Sears) is taking a two-pronged approach to testing the aerobic capacity model. This experiment has used Ibg/HS (i.e., heterogeneous stock) laboratory mice as the base population. Four replicate lines are being selected for each of three treatments. The first treatment is no selection (i.e., control lines). The second treatment is selection to increase mass-independent $VO_2$max elicited during forced treadmill exercise. The third treatment is selection to achieve a negative relationship between mass-independent $VO_2$max and mass-independent BMR. More specifically, the third treatment chooses mice with the most negative cross-product of mass-independent $VO_2$max and mass-independent BMR, with the additional restriction that selected mice have positive $VO_2$max residuals (see also Roff and Fairbairn this volume, and Rhodes and Kawecki this volume, on index selection). Because a correlation is calculated as the covariance of two traits divided by the product of their standard deviations, selecting for the most negative cross-product should also select for negative correlations between mass-independent VO$_2$max and mass-independent BMR.

This selection experiment is currently in its sixth generation, and initial results will soon be forthcoming. If selection for increased mass-independent VO$_2$max is achieved, then a correlated increase in BMR in those lines would support the assumption of the aerobic capacity model. An increase in mass-independent VO$_2$max with no increase in mass-independent BMR would argue against the assumption. However, responses to the third treatment may be the most revealing. If it is possible to select for lines of mice with a negative correlation between BMR and VO$_2$max, then that result would suggest that the physiological machinery of terrestrial vertebrates does not produce an inescapable link between these two metabolic traits. Such a result would argue against the aerobic capacity model. However, the vitality of such laboratory populations would also have to be assessed, because although selection for populations with a negative correlation between BMR and VO$_2$max might be possible in a benign laboratory environment, those populations might suffer fitness losses in comparison with wild populations in natural environments. Ultimately, it will be of considerable interest to compare the results of this experiments with the results of other research groups selecting for voluntary running, endurance capacity, BMR, and VO$_2$max (see earlier discussion).

This selection experiment by Hayes and colleagues attempts to address a specific "macroevolutionary" event, the evolution of mammalian endothermy, using a microevolutionary test. It is certainly possible that this type of microevolutionary study will be conducted on a scale (e.g., too few generations) that is insufficient to test such a macroevolutionary hypothesis. However, some macroevolutionary hypotheses make very specific and testable predictions. In the case of the aerobic capacity model, the specific prediction is that resting metabolism and aerobic capacity are inescapably positively related, at least in the vertebrate lineages that led to birds and mammals (Bennett and Ruben 1989; Hayes and Garland 1995). If this prediction is correct, then it can be argued that resting metabolism and aerobic capacity should be positively related in all extant birds and mammals, except perhaps for species that are extremely specialized. More specifically, these two characters should exhibit a strongly positive additive genetic correlation (for more on the importance of genetic correlations, see Rauser et al. this volume; Roff and Fairbairn this volume). Falsification of this prediction for any species of bird or mammal would undermine the model, although it could always be argued post hoc that the ancestral forms in which endothermy evolved were somehow fundamentally different from modern birds and mammals. Thus, at least in some circumstances, selection experiments may be useful for attempts to falsify a macroevolutionary hypothesis. Indeed, this example suggests that whenever a macroevolutionary hypothesis incorporates a broad prediction about necessitous features of physiological machinery, then selection experiments may be one appropriate way to test this type of hypothesis. Ultimately, understanding the evolution of any "macroevolutionary" feature of physiology or performance will likely benefit from a variety of approaches (Garland and Carter 1994; Feder et al. 2000; Garland 2001; Kemp 2006; Futuyma and Bennett this volume; Kerr this volume).

## SEXUAL SELECTION: EFFECTS OF ORNAMENTS ON PERFORMANCE

Morphology can affect whole-organism performance, and selection on performance can affect morphology. Sexual selection can lead to the very rapid evolution of exaggerated morphological and behavioral characters when individuals of a given sex, usually males, experience a mating advantage. Two processes are thought to drive the elaboration of secondary sexual characters: intrasexual selection (typically male-male competition) and intersexual selection (typically female choice). Thus, secondary sexual characters function as weapons in the context of intrasexual conflict, as attractive advertisements in the context of intersexual choice, or as both if the character serves dual purposes (Andersson 1994). The consequences of sexual selection, and rapid changes in morphology, on whole-organism performance are inadequately addressed in the sexual selection literature.

Extravagant sexually selected structures are notable for their tremendous variation in size and form. Moreover, some structures, such as the long necks of giraffe and of sauropod dinosaurs, traditionally thought to have been elaborated by natural selection, are now thought to have been shaped by sexual selection to a significant extent (Senter 2007, references therein). The assumption that elaborate secondary sexual characteristics are costly to produce and to maintain, and thus will be opposed by natural selection, is central to all models of sexual selection (Kotiaho 2001). Direct and indirect costs as well as trade-offs associated with producing and maintaining elaborate secondary structures are becoming increasingly well documented (Andersson and Simmons 2006).

As a result, the functional or performance consequences of trait elaboration are just beginning to receive wider attention among comparative and evolutionary physiologists (Lailvaux and Irschick 2006; Oufiero and Garland 2007). For example, Emlen (2001) has suggested that developmental trade-offs associated with ornament production may entail performance costs and thus have major, but largely overlooked, implications for evolutionary diversification (see also Frankino et al. this volume). Even in the absence of developmental trade-offs, elaborated sexual ornaments may generate physiological or biomechanical constraints and impair locomotor performance, as has been suggested for stalk-eyed flies (Swallow et al. 2000; Ribak and Swallow 2007).

If sexually selected traits result in evolutionarily significant performance and physiological trade-offs, then patterns of correlated evolution should be reflected in the phylogenetic history during which the ornament evolved and diversified. Such patterns could and should be explored in a comparative context (Garland et al. 2005; Lavin et al. 2008; Frankino et al. this volume). It also follows that if a performance decrement generated by the ornament is selectively important, then compensatory morphological and behavioral changes should arise to mitigate performance trade-offs (Oufiero and Garland 2007; see also Huey and Rosenzweig this volume). To date, however, locomotor trade-offs with secondary sexual characters have yet to be rigorously tested or demonstrated (see also Roff and Fairbairn this volume, on trade-offs between flight capability and reproductive investment in crickets).

One definitive means of determining the mechanistic and adaptive significance of sexually selected traits is to experimentally alter them via artificial selection and then compare the relative performance or fitness attributes of selected and unselected individuals (Garland and Carter 1994; Feder et al. 2000). Relatively few artificial selection studies have targeted sexually selected ornaments per se (e.g., Wilkinson 1993; Houde 1994; Emlen 1996; Frankino et al. this volume). In this section, we discuss two model systems to illustrate how selection on secondary sexual traits overlaps with, and should be informed by, the performance literature and by selection experiments.

## GUPPIES

Guppies have been a widely used model organism for studies of natural and sexual selection both because they are amenable to field observations and because they can be reared with relative ease in the laboratory (Houde 1997). Their short generation time also makes them attractive as a genetic model, permitting analysis of the genetic underpinnings of many of the traits of interest (e.g., Brooks and Endler 2001a, 2001b). In guppies, measures of male ornamentation and behavioral displays that influence attractiveness show substantial additive genetic variation (Brooks and Endler 2001a). Similarly, female mating preference is heritable and is genetically correlated with male color patterns (Houde 1988, 1997; Houde and Endler 1990). Although the pattern of mate preference is more nuanced than we will delineate here, studies of individual and population-level variation indicate that female guppies generally prefer male ornaments with bright orange and black spots and a high degree of contrast (Houde 1997; Brooks and Endler 2001b).

In natural populations where predation pressure is low, males tend to be brightly colored. In populations where predation pressure is high, males are generally dull. Brightly colored males experienced a predation cost, because they are more conspicuous (Houde 1997; see also Irschick and Reznick this volume). Natural selection in high-predation areas had a variety of other effects in addition to changes in coloration, including effects on a range of life-history traits (Reznick et al. 1997) and behavioral antipredator traits (Magurran et al. 1993), including escape performance ability (O'Steen et al. 2002).

In this case, sexual selection clearly has an indirect effect on performance. These indirect effects are manifest in high-predation environments, and they have far-ranging consequences on the phenotype. However, in the absence of predation, there is no inherent biomechanical reason why reduced locomotor performance has to be associated with bright coloration. Unless coloration is condition-dependent and linked with other systems, then selection on coloration alone should not result in any change in swimming performance. However, the latter possibility is not without merit. If the ability to bear and express ornaments are indicators of increased viability, as suggested by "good genes" models of sexual selection (Anderson 1994), then individuals that bear larger or brighter ornaments may actually display greater performance (see Lailvaux and Irschick 2006 for a review). Artificial selection could be used to directly test the connection between ornaments and performance, free of the confounding selection for antipredation abilities. Although Houde (1994) created lines of guppies divergent in coloration to test a variety of hypotheses about mate choice and sensory drive, to our knowledge no one has tested these lines for differences in locomotor performance.

## STALK-EYED FLIES

Stalk-eyed flies are a particularly good model for investigating trade-offs between sexually selected traits and locomotor performance. Stalk-eyed flies (Diopsidae) have eyes placed laterally away from the head on elongated peduncles, and the degree of hypercephaly and sexual dimorphism varies dramatically species (Wilkinson and Dodson 1997; Warren and Smith 2007; see also Frankino et al. this volume). In sexually dimorphic species, exaggerated eye span is driven by intense directional intrasexual and intersexual selection on male eye span. Males with longer eye stalks usually win contests of aggression over resources (Burkhardt et al. 1994), and females prefer to mate with males bearing longer eye stalks (Burkhardt and de la Motte 1988; Wilkinson and Reillo 1994; Wilkinson et al. 1998). Furthermore, both eye stalk length and female preference are genetically heritable and thus capable of responding to selection (Wilkinson 1993; Wilkinson and Reillo 1994).

Because laterally protruding eyes seem like an inefficient arrangement for flight, trade-offs between these secondary sexual traits and aerial performance might be expected. Currently, however, comparative data are insufficient to fully address the relationship between eye span and flight performance. A comparative analysis of the wing beat frequency of thirteen species of stalk-eyed flies in relation to eye span elongation illustrates some of the compensatory mechanisms (see also Oufiero and Garland 2007) that may mitigate the costs of bearing this ornament (Swallow unpublished). As predicted by scaling laws, variation in the wing beat frequency scales negatively with body mass. Large insects oscillate their wings more slowly. Eye stalk elongation, indexed as the ratio of eye span to body length, was correlated with a reduction in wing beat frequency. Furthermore, within the dimorphic species in which males have longer eye span, wing beat frequency was lower in males than females. Mechanistically, these results can largely be explained by correlated compensatory increases in the length and area of the wing, which serve to increase the length of the oscillating arm, and in the thorax of males of the dimorphic species. How these compensatory changes translate into differences in flight behavior and performance is currently under investigation. For example, in comparisons between species and sexes that bear longer eyestalks, there is some evidence for reduced aerial performance and maneuverability (Swallow et al. 2000). However, the reduction in performance is smaller in magnitude than a biomechanical analysis of the consequences of eye span elongation for moment of inertia would imply, providing further evidence of correlated compensatory evolution among other traits (Ribak and Swallow 2007).

One definitive means of determining the mechanistic and adaptive significance of eye stalks is to experimentally alter them via artificial selection and then compare the relative performance or fitness differences between individuals from selected and unselected lines. After more than sixty generations of bidirectional artificial selection on male *Cyrtodiopsis dalmanni*, Wilkinson and colleagues have created replicate lines of stalk-eyed flies that differ dramatically (~2 mm, 20 percent) in relative eye span (Wilkinson 1993; Wilkinson et al. 2005). Male fighting performance and likelihood of winning contests of aggression evolved with eye span (Panhuis and Wilkinson 1999). These selected lines are currently being used to investigate the consequences of eye stalk elaboration outside of the context of mate competition.

## "EXPERIMENTS" WITHOUT PRECISELY DEFINED SELECTION CRITERIA

Besides the formal selection experiments described in this volume, there are many interesting examples of artificial selection and experimental evolution that are less formal but still informative. For example, the process of domestication no doubt was associated with strong selection for a wide range of functional characters (Hemmer 1990; Bradley and Cunningham 1999; Price 2002; see also Simões et al. this volume). Related to the process of domestication is selection on the physiological performance of already-domesticated animals. An excellent example of this is selection on racing performance.

Humans race camels, greyhounds, horses, ostriches, and pigeons. Horse racing has a long history, dating back at least to chariot racing by the Greeks and Romans, and a wide range of racing events for horses are still practiced throughout the world. Many people enjoy betting on racing events, and owners of racing animals are interested in increasing the earnings of their animals. Consequently, informal selection on racing performance is widespread, at least for horses and greyhounds. It is interesting to compare the informal selection on horses and greyhounds, and ask what the prospects are for improving racing performance via further selection.

Horse racing is a diverse enterprise. Races vary by distance, nature of the running surface (dirt or turf), whether there are jumps (steeplechase), and whether the jockey is pulled on a cart (trotters) or rides on the horse's back. Racehorse populations are separated by breed (e.g., standard-bred trotters, thoroughbreds, and Arabians), and they also are segregated into somewhat distinct populations (e.g., British, Brazilian, Japanese, North American, Swedish, Polish, Spanish, Turkish, Tunisian, and other populations). For those of us who study the evolution of performance, selection for racing in these populations constitutes an interesting informal experiment.

Among the most interesting aspects of selection on racing performance are (1) the rate of improvement in racing speeds and performance, (2) the rate of increase in breeding values for those traits, and (3) the amount of genetic variation that remains after many years of selection on performance. An analysis of the winning time for three major thoroughbred races over several decades in Britain suggested that winning time did not appear to be improving (Gaffney and Cunningham 1988). In contrast, winning and average race times got faster in Swedish standard-bred trotters (Árnason 2001). A possible explanation for this discrepancy is that inbreeding and a long history of selection led to the exhaustion of genetic variation for performance in thoroughbreds. Overall, it appears that heritabilities for racing performance are low (usually ~0.2 or lower), but sufficient genetic variation is present for continued genetic progress to be made (Villela et al. 2002; Bokor et al. 2005, 2007; Chico 1994; Ekiz and Kocak 2005; Oki 1995; Sobczynska 2006). The data of horse pedigrees and racing performance have led to investigations of (1) the distribution of allelic effects and scale (additive vs. multiplicative) effects, (2) the possible effects of assortative mating, and (3) the distinctions between speed and other measures of performance. To date, these studies suggest that while fastest race times show little improvement, average racing speeds do appear to be getting faster. Why this is the case deserves further study, and the results could be useful to those interested in the evolution of performance. Another interesting finding is that the informal breeding schemes used by horse breeders result in the selection of sires and dams that do not produce the most rapid genetic progress for performance (Hill 1988; Williamson and Beilharz 1999).

GREYHOUND RACING

In contrast to the long history of horse racing, the racing of dogs (specifically greyhounds) is relatively recent. The racing performance and genetics of greyhounds have not been studied as carefully as for horses, so this description is based on a single recent report on greyhounds in Ireland (Taubert et al. 2007). From 2000 to 2003, average race time over 480 meters has declined by roughly 0.5 second (30.3 to 29.8 seconds). Analysis of the greyhound data shows that heritable variation exists and that selection is increasing performance (i.e., race times are getting shorter, and the dogs are getting faster). As is the case for horses, breeding values (with higher breeding values indicating faster race time) are increasing, but the increase is less than could be achieved by the use of optimal selection criteria.

One of the flaws in the informal breeding design used for Irish greyhounds is that the dogs chosen as sires do not have the highest breeding values (Taubert et al. 2007). Breeding values for race time were predicted for 51,332 dogs. Only 793 of the dogs were used as sires over the time period studied, and 18 of those sires produced 43.7 percent of all offspring. Sires were not optimally chosen from the perspective of improving racing time because the sires that produced the largest number of offspring were not always those with the highest breeding values. For instance, one sire that produced 991 offspring had a slightly negative breeding value. As has been suggested for horses, breeders of racing dogs would likely benefit from working more closely with animal geneticists (Hill 1988; Taubert et al. 2007).

## PHYSIOLOGICAL DIFFERENCES AMONG STRAINS OF MICE AND BREEDS OF DOG

Another rich source of information for students of experimental evolution is the examination of strains or breeds of animals. These populations are fertile ground for study because it is straightforward to identify phenotypic differences that are genetically based (Silver 1995; Crawley et al. 1997; Eisen 2005; Fuller et al. 2005). The genetic differences among strains or breeds of mice, rats, dogs, and other animals may have resulted from deliberate selection for the traits of interest (e.g., blood pH, sensitivity or resistance to drugs, infection, tumor formation, estrus cycle timing, or stress response: Webster 1933a, 1933b; Weir and Clark 1955; Nobunaga 1973; Nagasawa 1976; Liang and Blizard 1978; Caslet et al. 1997; Kemp et al. 2005). Alternatively, physiological differences may be the accidental by-product of selection for other characters, or they may be attributable to founder effects or random genetic drift, with subsequent inbreeding (see also Simões et al. this volume). Thanks to Professor Michael Festing and the Jackson Laboratory, information on the origins of standard mouse strains is easily available online. Phenotypic and genetic differences among mouse strains have been extensively characterized, particularly by biomedical researchers (e.g., Barbato et al. 1998; Biesiadecki et al. 1999; Svenson et al. 2007), but we think that the genetic resources represented by these many strains have been underutilized by evolutionary biologists. The literature on strain difference is vast, and here we report only a few examples to illustrate how the outcomes of the informal or formal selection used to produce laboratory strains may be of interest to comparative, ecological, or evolutionary physiologists.

The ability to cope with hypoxia is physiologically, ecologically, and evolutionarily important (Monge and Leonvelarde 1991; Dudley 1998; Hochachka et al. 1998; Iyer et al. 1998; Hayes and O'Connor 1999; Semenza 1999). Mouse strains vary in their ventilatory response to hypoxia, with some strains coping primarily by altering tidal volume and others having greater changes in breathing frequency (Tankersley et al. 1994). Moderate hypoxia engenders physiological responses to cope with the changed environment, but extreme hypoxia can be immediately life threatening. Inbred C57BL/6J mice survived extreme hypoxia (4.5 percent $O_2$) better than outbred CD-1 mice (Zwemer et al. 2007). In addition, more detailed physiological study suggested that differences in ketone metabolism are one factor that contributes to the differences in hypoxic tolerance. Another factor that may be involved in hypoxic tolerance is angiogenesis. Angiogenesis and the up-regulation of proteins involved in stimulating angiogenesis varied markedly across four strains of mice subjected to severe hypoxia (10 percent $O_2$; Ward et al. 2007). Likewise, the effects of severe hypoxia (10 percent $O_2$) on arterial blood pressure and heart rate depend in large measure on the strain studied (Campen et al. 2004). Lastly, physiological differences in how mice respond to hypoxia might also be affected by their underlying morphology, and strains of mice have genetically based differences in alveolar anatomy (Soutiere et al. 2004). The point here is that the informal or formal selection programs that lead to the production of genetically distinct strains and breeds can be helpful to evolutionary physiologists seeking to understand how integrated physiological phenotypes evolve.

Locomotor performance is one of the themes of this chapter, and several studies have measured mouse running duration during incremental step tests on a treadmill or some other measure of aerobic capacity during exercise. Aerobic capacity is higher in DBA/2 and BALB than in C57BL mice (Lerman et al. 2002), and significant differences were found in aerobic capacity among ten inbred strains that included A, AKR, Balb, C57BL, and DBA/2, among others (Lightfoot et al. 2001). In the latter study, Balb mice had the greatest endurance of any of the strains, and A mice had the lowest endurance. Related to exercise endurance is the ability of an animal to respond to training, because training can substantially increase exercise endurance. Not only does aerobic capacity vary across strains of mice, but strains also differ in the magnitude of their response to exercise training, in how far they run voluntarily, how fast they run voluntarily, and in how much time they run each day (Lightfoot et al. 2004; Massett and Berk 2005). Strains also differ in the critical speed at which lactate production reaches the point that long-term endurance is limited (Billat et al. 2005). As some of the preceding studies demonstrate, differences among strains are fertile ground for probing the morphological, biochemical, genetic, and genomic bases of variation in performance. Such studies have much to offer evolutionary physiologists.

Another underutilized resource for evolutionary physiologists is the genetic variation present among and within breeds of dogs (Parker et al. 2004). To anyone familiar with greyhounds and pit bulls, it is no surprise that these dogs have been selected for very different measures of physiological performance. Pit bulls were bred to fight, and greyhounds were bred to run. The strength of their bones reflects those specializations (Kemp et al. 2005). Likewise, the distribution of muscle and the ability of the musculoskeletal system differ markedly between pit bulls and greyhounds (Pasi and Carrier 2003).

## CONCLUSION

"Survival of the fittest" is the fundamental public metaphor for symbolizing evolution by natural selection. Unfortunately, this public perception conflates physical fitness with Darwinian fitness (Wassersug and Wassersug 1986), which, although complicated, roughly equates with net lifetime reproductive success (McGraw and Caswell 1996; Mueller this volume). Nonetheless, the terminological ambiguity carries more than a grain of truth, because organismal performance traits that involve strength, speed, or stamina may often have strong effects on such major components of Darwinian fitness as survivorship and fecundity. Measures of locomotor performance are now routinely studied in comparative biology, and it has been claimed that "[l]ocomotion, movement through the environment, is the behavior that most dictates the morphology and physiology of animals" (Dickinson et al. 2000, 100).

Overall, the present volume shows that selection experiments of various types are key tools for disentangling the mechanistic basis of evolutionary diversification in all sorts of complex phenotypes (see also Swallow and Garland 2005), and this chapter reveals that performance and physiology are no exception. Selection allows researchers to create populations that explore the limits of physiological performance, while simultaneously facilitating the elucidation of genetic architecture and genes of major effect (e.g., Garland et al. 2002; Hartmann et al. 2008). By studying responses to selection on complex performance traits, biologists can determine both how such traits are integrated phenotypically and the precise mechanisms by which performance evolves (see also Zera and Harshman this volume). These responses also can show whether complex traits (e.g., locomotion, oxygen transport, and delivery: Swallow et al. 1998b, 2005; Rezende et al. 2006c) evolve by correlated progression of all component subordinate traits simultaneously (e.g., stroke volume, heart rate, hematocrit, Bohr shift, mitochondrial density), as might be posited by proponents of symmorphosis (Garland 1998 and references therein), or whether responses reflect mosaic evolution of just one or a few key traits (Garland and Carter 1994; Kemp 2006; Archer et al. 2007). Similarly, correlated responses to selection may reveal both genetic architecture and functional constraints associated with the focal trait being selected. In addition, even when inbreeding effects are minimized, selection that is intense and drastically alters traits potentially may lead to correlated responses that reduce individual Darwinian fitness and hence population viability, at least in some environments. Selected populations that evolve correlated responses rendering them inviable in nature may also reveal ecological constraints. As a hypothetical example, animals selected for high metabolic rate might also evolve as a correlated response immune systems that are too weak to allow survival in nature (see also Malisch et al. 2009).

One additional point to keep in mind regarding the design of selection experiments that target organismal performance traits (including some aspects of behavior [Rhodes and Kawecki this volume] and of life history [Nunney 1996; Zera and Harshman this volume]) is that selection for low values of performance may tend to increase the frequency of any deleterious recessive alleles that are segregating in the population. This may result in "sickly" organisms, and the observed response to selection for low performance may have little to do with the evolutionary potential of the species in nature (Nunney 1996). Indeed, selection for low treadmill endurance in rats has led to animals that are unhealthy, as compared with their high-selected counterparts (Koch and Britton 2005; Wisløff et al. 2005). In addition, selection for low values of performance may not result in a trait that is "the same" but at the opposite end of a continuum as compared with selection for high values. For example, selection for low values of voluntary wheel running (Swallow et al. 1998a) might lead to an increase in fear of entering the wheels, rather than reduced aerobic capacity (see discussion in Garland 2003).

As concern about global warming intensifies, experimental evolution of animals in warm or cold environments could be used to test how populations evolve when faced with climate change. These mesocosm or laboratory studies would not fully reflect ecologically reality, but they would offer a great starting point (see also Huey and Rosenzweig this volume). For example, it would be interesting to know whether mammals, lizards, and birds tend to respond primarily physiologically or behaviorally when housed for multiple generations at different temperatures (see also Gibbs and Gefen this volume; Huey and Rosenzweig this volume). One intriguing possibility for experimental evolution studies would be to consider using different species for replication, an approach that has already been used with *Drosophila* species experimentally evolved with later reproduction (Deckert-Cruz et al. 2004). Similarly, in terms of response to global changes in the ozone layer, one could study the evolutionary responses to increased ultraviolet radiation in multiple species of amphibians.

In conclusion, biologists interested in physiology and performance should consider the unique knowledge that can be gained by selection experiments. In particular, these studies are highly appropriate for understanding (1) the mechanistic bases of adaptation, (2) the extent to which organismal "design" may constrain evolutionary outcomes, and (3) phenotypic integration. We encourage biologists to apply these approaches to a broader range of animal species.

## SUMMARY

In the last twenty-five years, direct measurement of whole-organism performance has become central to those fields of biology that explicitly focus on those physiological, biomechanical, or molecular mechanisms that underlie variation in whole-organism traits. Measures of performance, such as maximal sprint running speed, are presumed to be more direct targets of selection in nature relative to traits below the organismal level. Moreover, voluntary behavior occurs within an envelope of possibilities circumscribed by performance limits; and, conversely, behavioral choices made by animals can shield performance capacities from the direct effects of selection. Both behavioral and performance traits are complex, consisting of numerous subordinate traits and affected by many genes and environmental factors. In spite of the increased attention paid to whole-animal performance, relatively few experiments have intentionally imposed selection at this level, and most that have involve locomotion or metabolism of rodents. Such experiments with rodents have successfully targeted treadmill endurance-running capacities, voluntary locomotion, and metabolic rates. Some of these experiments have found interesting correlated changes in the behavior of the selected animals, as well as identifying mechanistic underpinnings of divergence at the organismal level. Experiments involving sexual selection have altered guppies and stalk-eyed flies in ways that should impinge on performance abilities, but this has yet to be tested. Finally, besides the planned selection experiments emphasized in this volume, there are many interesting examples of animal breeding that are less formal but still informative, including breeds of horse, dog, and mouse. Potentially useful model organisms for selection on target performance have yet to be exploited (e.g., small fishes). Finally, only three vertebrate studies (two selecting for increased body size in mice, the other for increased litter size) have gone as many as one hundred generations, hence crossing into the category of "long-term" selection experiments.

## ACKNOWLEDGMENTS

The work was funded by National Science Foundation grants IOB-0448060 to J.G.S, IOS-0344994 to J.P.H., and IOB-0543429 to T.G. P.K. was funded by Polish MNiSW grant 2752/B/P01/2007/33 and UJ/DS/757. We are grateful to J. Weiner for his drawings and to M. Konarzewski and A. Gębczyński for providing data and helpful comments.

## REFERENCES

Allan, M. F., M. K. Nielsen, and D. Pomp. 2000. Gene expression in hypothalamus and brown adipose tissue of mice divergently selected for heat loss. *Physiological Genomics* 3:149–156.

Andersson, M. 1994. *Sexual Selection*. Princeton, NJ: Princeton University Press.

Andersson, M., and L. M. W. Simmons. 2006. Sexual selection and mate choice. *Trends in Ecology & Evolution* 2:296–302.

Angilletta, M. J., and M. W. Sears. 2003. Is parental care the key to understanding endothermy in birds and mammals? *American Naturalist* 162:821–825.

Archer, M. A., T. J. Bradley, L. D. Mueller, and M. R. Rose. 2007. Using experimental evolution to study the physiological mechanisms of desiccation resistance in *Drosophila melanogaster*. *Physiological and Biochemical Zoology* 80:386–398.

Arnold, S. J. 1983. Morphology, performance and fitness. *American Zoologist* 23:347–361.

———. 2003. Performance surfaces and adaptive landscapes. *Integrative and Comparative Biology* 43:367–375.

Bacigalupe, L. D., R. F. Nespolo, D. M. Bustamante, and F. Bozinovic. 2004. The quantitative genetics of sustained energy budget in a wild mouse. *Evolution* 58:421–429.

Barbato, J. C., L. G. Koch, A. Darvish, G. T. Ciclia, P. J. Metting, and S. L. Britton. 1998. Spectrum of aerobic endurance running performance in eleven inbred strains of rats. *Journal of Applied Physiology* 85:530–536.

Barnett, S. A., and R. G. Dickson. 1984a. Changes among wild house mice *(Mus musculus)* bred for 10 generations in cold environment, and their evolutionary implications. *Journal of Zoology* 203:163–180.

———. 1984b. Milk-production and consumption and growth of young of wild mice after 10 generations in a cold environment. *Journal of Physiology (London)* 346:409–417.

———. 1989. Wild mice in the cold: Some findings on adaptation. *Biological Reviews of the Cambridge Philosophical Society* 64:317–340.

Bauwens, D., and C. Thoen. 1981. Escape tactics and vulnerability to predation associated with reproduction in the lizard *Lacerta vivipara. Journal of Animal Ecology* 50:733–743.

Bell, G. 2008. *Selection: The Mechanism of Evolution.* 2nd ed. Oxford: Oxford University Press.

Bennett, A. F. 1989. Integrated studies of locomotor performance. Pages 191–202 *in* D. B. Wake and G. Roth, eds. *Complex Organismal Functions: Integration and Evolution in Vertebrates.* Chichester, U.K.: Wiley.

Bennett, A. F., J. W. Hicks, and A. J. Cullum. 2000. An experimental test of the thermoregulatory hypothesis for the evolution of endothermy. *Evolution* 54:1768–1773.

Bennett, A. F., and J. A. Ruben. 1979. Endothermy and activity in vertebrates. *Science* 206:649–654.

Biesiadecki, B. J., P. H. Brand, L. G. Koch, P. J. Metting, and S. L. Britton. 1999. Phenotypic variation in sensorimotor performance among eleven inbred rat strains. *American Journal of Physiology* 276 (Regulatory Integrative and Comparative Physiology 45):R1383–R1389.

Billat, V. L., E. Mouisel, N. Roblot, and J. Melki. 2005. Inter- and intrastrain variation in mouse critical running speed. *Journal of Applied Physiology* 98:1258–1263.

Booth, F. W., M. V. Chakravathy, S. E. Gordon, and E. E. Spangenburg. 2002a. Waging war on physical inactivity: Using modern molecular ammunition against an ancient enemy. *Journal of Applied Physiology* 93:3–30.

Booth, F. W., M. V. Chakravathy, and E. E. Spangenburg. 2002b. Exercise and gene expression: Physiological regulation of the human genome through physical activity. *Journal of Physiology* 543:399–411.

Bradley, D. G., and E. P. Cunningham. 1999. Genetic aspects of domestication. Pages 15–31 *in* R. Fries and A. Ruvinsky, eds. *The Genetics of Cattle.* New York: CABI.

Bramble, D. M., and D. E. Lieberman. 2004. Endurance running and the evolution of *Homo. Nature* 432:345–352.

Brodie, E. D., III. 1992. Correlational selection for color pattern and antipredator behavior in the garter snake *Thamnophis ordinoides. Evolution* 46:1284–1298.

Brookes, P. S., J. A. Buckingham, A. M. Tenreiro, A. J. Hulbert, and M. D. Brand. 1998. The proton permeability of the inner membrane of liver mitochondria from ectothermic and endothermic vertebrates and from obese rats: Correlations with standard metabolic rate and phospholipid fatty acid composition. *Comparative Biochemistry and Physiology B* 119:325–334.

Brooks, R., and J. A. Endler. 2001a. Direct and indirect sexual selection and quantitative genetics of male traits in guppies *(Poecilia reticulate). Evolution* 55:1002–1015.

———. 2001b. Female guppies agree to differ: Phenotypic and genetic variation in mate-choice behavior and the consequences for sexual selection. *Evolution* 55:1644–1655.

Brzęk, P., K. Bielawska, A. Książek, and M. Konarzewski. 2007. Anatomic and molecular correlates of divergent selection for basal metabolic rate in laboratory mice. *Physiological and Biochemical Zoology* 80:491–499.

Bult, A., and C. B. Lynch. 1996. Multiple selection responses in house mice bidirectionally selected for thermoregulatory nest-building behavior: Crosses of replicate lines. *Behavior Genetics* 26:439–446.

———. 2000. Breaking through artificial selection limits of an adaptive behavior in mice and the consequences for correlated responses. *Behavior Genetics* 30:193–206.

Bunger, L., A. Laidlaw, G. Bulfield, E. J. Eisen, J. F. Medrano, G. E. Bradford, F. Pirchner, U. Renne, W. Schlote, and W. G. Hill. 2001. Inbred lines of mice derived from long-term growth selected lines: Unique resources for mapping growth genes. *Mammalian Genome* 12:678–686.

Bunger, L., U. Renne, and G. Dietl. 1994. Sixty generations selection for an index combining high body weight and high stress resistance in laboratory mice. *Proceedings of the 5th World Congress on Genetics Applied to Livestock Production, Guelph* 19:16–19.

Burkhardt, D., I. de la Motte, and K. Lunau. 1994. Signalling fitness: Larger males sire more offspring: Studies of the stalk-eyed fly *Cyrtodiopsis whitei* (Diopsidae, Diptera). *Journal of Comparative Physiology A* 174:61–64.

Calsbeek, R., and D. J. Irschick. 2007. The quick and the dead: Locomotor performance and natural selection in island lizards. *Evolution* 61:2493–2503.

Campen, M. J., Y. Tagaito, J. G. Li, A. Balbir, C. G. Tankersley, P. Smith, A. Schwartz, and C. P. O'Donnell. 2004. Phenotypic variation in cardiovascular responses to acute hypoxic and hypercapnic exposure in mice. *Physiological Genomics* 20:15–20.

Carbone, C., G. Cowlishaw, N. J. B. Isaac, and J. M. Rowcliffe. 2005. How far do animals go? Determinants of day range in mammals. *American Naturalist* 165:290–297.

Casley, W. L., J. A. Menzies, N. Mousseau, M. Girard, T. W. Moon, and L. W. Whitehouse. 1997. Increased basal expression of hepatic Cyp1a1 and Cyp1a2 genes in inbred mice selected for susceptibility to acetaminophen-induced hepatotoxicity. *Pharmacogenetics* 7:283–293.

Chappell, M. A., T. Garland, Jr., G. F. Robertson, and W. Saltzman. 2007. Relationships among running performance, aerobic physiology, and organ mass in male Mongolian gerbils. *Journal of Experimental Biology* 210:4179–4197.

Clobert, J., A. Oppliger, G. Sorci, B. Ernande, J. G. Swallow, and T. Garland, Jr. 2000. Trade-offs in phenotypic traits: Endurance at birth, growth, survival, predation, and susceptibility to parasitism in a lizard, *Lacerta vivipara*. *Functional Ecology* 4:675–684.

Connor, J. K. 2003. Artificial selection: A powerful tool for ecologists. *Ecology* 84:1650–1660.

Crompton, A. W., C. R. Taylor, and J. A. Jagger. 1978. Evolution of homeothermy in mammals. *Nature* 272:333–336.

Crawley, J. N., J. K. Belknap, A. Collins, J. C. Crabbe, W. Frankel, N. Henderson, R. J. Hitzemann, S. C. Maxson, L. L. Miner, A. J. Silva, J. M. Wehner, A. WynshawBoris, and R. Paylor. 1997. Behavioral phenotypes of inbred mouse strains: Implications and recommendations for molecular studies. *Psychopharmacology* 132:107–124.

Daan, S., D. Masman, and A. Groenewold. 1990. Avian basal metabolic rates: Their association with body composition and energy expenditure in nature. *American Journal of Physiology* 259:R333–R340.

Deckert-Cruz, D. J., L. M. Matzkin, J. L. Graves, and M. R. Rose. 2004. Electrophoretic analysis of Methuselah flies from multiple species. Pages 237–248 *in* M. R. Rose, H. B. Passananti, and M. Matos, eds. *Methuselah Flies: A Case Study in the Evolution of Aging.* Singapore: World Scientific.

Denckla, W. D., and E. Marcum. 1973. Minimal $O_2$ consumption as an index of thyroid status: Standardization of method. *Endocrinology* 93:61–73.

Dickinson, M. H., C. T. Farley, R. J. Full, M. A. R. Koehl, R. Kram, and S. Lehman. 2000. How animals move: An integrative view. *Science* 288:100–106.

Dlugosz, E. M., M. A. Chappell, and T. Garland, Jr. 2007. Locomotor constraints in mice selected for high voluntary wheel running. *Integrative and Comparative Biology* 47:e29.

Dohm, M. R., J. P. Hayes, and T. Garland, Jr. 2001. The quantitative genetics of maximal and basal rates of oxygen consumption in mice. *Genetics* 159:267–277.

Draski, L. J., K. P. Spuhler, V. G. Erwin, R. C. Baker, and R. A. Deitrich. 1992. Selective breeding of rats differing in sensitivity to the effects of acute ethanol administration. *Alcoholism: Clinical and Experimental Research* 16:48–54.

Dudley, R. 1998. Atmospheric oxygen, giant Paleozoic insects and the evolution of aerial locomotor performance. *Journal of Experimental Biology* 201:1043–1050.

Dunnington, E. A., J. M. White, and W. E. Vinson. 1981a. Selection for serum cholesterol, voluntary physical activity, 56-day weight and feed intake in randombred mice. I. Direct responses. *Canadian Journal of Genetics and Cytology* 23:533–543.

———. 1981b. Selection for serum cholesterol, voluntary physical activity, 56-day weight and feed intake in randombred mice. II. Correlated responses. *Canadian Journal of Genetics and Cytology* 23:545–555.

Eisen, E. J., ed. 2005. *The Mouse in Animal Genetics and Breeding Research.* London: Imperial College Press.

Eisen, E. J., and D. Pomp. 1990. Replicate differences in lines of mice selected for body composition. *Genome* 33:294–301.

Eisenberg, J. F. 1981. *The Mammalian Radiations.* Chicago: University of Chicago Press.

Else, P. L., and A. J. Hulbert. 1987. Evolution of mammalian endothermic metabolism: "Leaky" membranes as a source of heat. *American Journal of Physiology* 253:R1–R7.

Else, P. L., N. Turner, and A. J. Hulbert. 2004. The evolution of endothermy: Role for membranes and molecular activity. *Physiological and Biochemical Zoology* 77:950–958.

Emlen, D. J. 1996. Artificial selection on horn length body size allometry in the horned beetle *Onthophagus acuminatus* (Coleoptera: Scarabaeidae). *Evolution* 50:1219–1230.

———. 2001. Costs and the diversification of exaggerated animal structures. *Science* 291:1534–1536.

Falconer, D. S. 1992. Early selection experiments. *Annual Review of Genetics* 26:1–14.

Falconer, D. S., and T. F. C. Mackay. 1996. *Introduction to Quantitative Genetics.* 4th ed. Essex, U.K.: Longman.

Falkenberg, H., M. Langhammer, and U. Renne. 2000. Comparison of biochemical blood traits After long-term selection on high or low locomotory activity in mice. *Archiv für Tierzucht* 43:513–522.

Farmer, C. G. 2000 Parental care: The key to understanding endothermy and other convergent features in birds and mammals. *American Naturalist* 155:326–334.

———. 2003. Reproduction: The adaptive significance of endothermy. *American Naturalist* 162:826–840.

Feder, M. E., A. F. Bennett, and R. B. Huey. 2000. Evolutionary physiology. *Annual Review of Ecology and Systematics* 31:415–341.

Frappell, P. B., and P. J. Butler. 2004. Minimal metabolic rate: What it is, its usefulness, and its relationship to the evolution of endothermy: A brief synopsis. *Physiological and Biochemical Zoology* 77:865–868.

Fuller, R. C., C. F. Baer, and J. Travis. 2005. How and when selection experiments might actually be useful. *Integrative and Comparative Biology* 45:391–404.

Fuller, T., S. Sarkar, and D. Crews. 2005. The use of norms of reaction to analyze genotypic and environmental influences on behavior in mice and rats. *Neuroscience and Biobehavioral Reviews* 29:445–456.

Gammie, S. C., N. S. Hasen, J. S. Rhodes, I. Girard, and T. Garland, Jr. 2003. Predatory aggression, but not maternal or intermale aggression, is associated with high voluntary wheel-running behavior in mice. *Hormones and Behavior* 44:209–221.

Garland, T., Jr. 1983. Scaling the ecological cost of transport to body mass in terrestrial mammals. *American Naturalist* 121:571–587.

———. 1994. Quantitative genetics of locomotor behavior and physiology in a garter snake. Pages 251–277 *in* C. R. B. Boake, ed. *Quantitative Genetic Studies of Behavioral Evolution.* Chicago: University of Chicago Press.

———. 1998. Testing the predictions of symmorphosis: conceptual and methodological issues. Pages 40–47 *in* E. R. Weibel, L. Bolis, and C. R. Taylor, eds. *Principles of Animal Design: The Optimization and Symmorphosis Debate.* New York: Cambridge University Press.

———. 2001. Phylogenetic comparison and artificial selection: Two approaches in evolutionary physiology. Pages 107–132 *in* R. C. Roach, P. D. Wagner, and P. H. Hackett, eds. *Hypoxia: From Genes to the Bedside. Advances in Experimental Biology and Medicine,* vol. 502. New York: Kluwer Academic/Plenum.

———. 2003. Selection experiments: An under-utilized tool in biomechanics and organismal biology. Pages 23–56 *in* V. L. Bels, J.-P. Gasc, and A. Casinos, eds. *Vertebrate Biomechanics and Evolution.* Oxford: BIOS Scientific.

Garland, T., Jr., and S. C. Adolph. 1994. Why not to do two-species comparative studies: Limitations on inferring adaptation. *Physiological Zoology* 67:797–828.

Garland, T., Jr., A. F. Bennett, and C. B. Daniels. 1990. Heritability of locomotor performance and its correlates in a natural population. *Experientia* 46:530–533.

Garland, T., Jr., A. F. Bennett, and E. L. Rezende. 2005. Phylogenetic approaches in comparative physiology. *Journal of Experimental Biology* 208:3015–3035.

Garland, T., Jr., and P. A. Carter. 1994. Evolutionary physiology. *Annual Review of Physiology* 56:579–621.

Garland, T., Jr., and P. L. Else. 1987. Seasonal, sexual, and individual variation in endurance and activity metabolism in lizards. *American Journal of Physiology* 252 (Regulatory Integrative and Comparative Physiology 21):R439–R449.

Garland, T., Jr., and S. A. Kelly. 2006. Phenotypic plasticity and experimental evolution. *Journal of Experimental Biology* 209:2234–2261.

Garland, T., Jr., and J. B. Losos. 1994. Ecological morphology of locomotor performance in squamate reptiles. Pages 240–302 *in* P. C. Wainwright and S. M. Reilly, eds. *Ecological Morphology: Integrative Organismal Biology*. Chicago: University of Chicago Press.

Garland, T., Jr., M. T. Morgan, J. G. Swallow, J. S. Rhodes, I. Girard, J. G. Belter, and P. A. Carter. 2002. Evolution of a small-muscle polymorphism in lines of house mice selected for high activity levels. *Evolution* 56:1267–1275.

Gębczyński, A. 2005. Daily variation of thermoregulatory costs in laboratory mice selected for high and low basal metabolic rate. *Journal of Thermal Biology* 30:187–193.

Ghalambor, C. K., J. A. Walker, and R. N. Reznick. 2003. Multi-trait selection, adaptation, and constraints on the evolution of burst swimming performance. *Integrative and Comparative Biology* 43:431–438.

Girard, I., and T. Garland, Jr. 2002. Plasma corticosterone response to acute and chronic voluntary exercise in female house mice. *Journal of Applied Physiology* 92:1553–1561.

Girard, I., M. W. McAleer, J. S. Rhodes, and T. Garland, Jr. 2001. Selection for high voluntary wheel running increases intermittency in house mice *(Mus domesticus)*. *Journal of Experimental Biology* 204:4311–4320. Movie: www.biology.ucr.edu/faculty/Garland/Girard01.mov.

Girard, I., E. L. Rezende, and T. Garland, Jr. 2007. Leptin levels and body composition of mice selectively bred for high voluntary activity. *Physiological and Biochemical Zoology* 80:568–579.

Gomes, F. R., J. G. Chauí-Berlinck, J. E. P. W. Bicudo, and C. A. Navas. 2004. Intraspecific relationships between resting and activity metabolism in anuran amphibians: Influence of ecology and behavior. *Physiological and Biochemical Zoology* 77:197–208.

Gonzalez, N. C., S. D Kirkton, R. A. Howlett, S. L. Britton, L. G. Koch, H. E. Wagner, and P. D. Wagner. 2006. Continued divergence in VO$_2$max of rats artificially selected for running endurance is mediated by greater convective blood O$_2$ delivery. *Journal of Applied Physiology* 101:1288–1296.

Gordon, C. J., and A. H. Rezvani. 2001. Genetic selection of rats with high and low body temperatures. *Journal of Thermal Biology* 26:223–229.

Goszczynski, J. 1986. Locomotor activity of terrestrial predators and its consequences. *Acta Theriologica* 31:79–95.

Guderley, H., D. R. Joanisse, S. Mokas, G. M. Bilodeau, and T. Garland, Jr. 2008. Altered fiber types in gastrocnemius muscle of high wheel-running selected mice with mini muscle phenotypes. *Comparative Biochemistry and Physiology B* 149:490–500.

Hannon, R. M., S. A. Kelly, K. M. Middleton, E. M. Kolb, D. Pomp, and T. Garland, Jr. 2008. Phenotypic effects of the "mini-muscle" allele in a large HR × C57BL/6J mouse backcross. *Journal of Heredity* 99:349–354.

Harris, M. A., and K. Steudel. 2002. The relationship between maximum jumping performance and hind limb morphology/physiology in domestic cats *(Felis silvestris catus)*. *Journal of Experimental Biology* 205:3877–3889.

Hartmann, J., T. Garland, Jr., R. M. Hannon, S. A. Kelly, G. Muñoz, and D. Pomp. 2008. Fine mapping of "Mini-Muscle," a recessive mutation causing reduced hindlimb muscle mass in mice. *Journal of Heredity* 99:679–687.

Harvey, P. H., M. D. Pagel, and J. A. Rees. 1991. Mammalian metabolism and life histories. *American Naturalist* 137:556–566.

Hayes, J. P., and T. Garland, Jr. 1995. The evolution of endothermy: Testing the aerobic capacity model. *Evolution* 49:836–847.

Hayes, J. P., T. Garland, Jr., and M. R. Dohm. 1992. Metabolic rates and reproduction of *Mus*: Are energetics and life history linked? *Functional Ecology* 6:5–14.

Hayes, J. P., and C. S. O'Connor. 1999. Natural selection on thermogenic capacity of high-altitude deer mice. *Evolution* 53:1280–1287.

Hayes, J. P., and J. S. Shonkwiler. 1996. Analyzing mass-independent data. *Physiological Zoology* 69:974–980.

Hayssen, V. 1984. Basal metabolic rate and the intrinsic rate of increase: An empirical and theoretical reexamination. *Oecologia* 64:419–421.

Heinrich, B. 1977. Why have some species evolved to regulate a high body temperature? *American Naturalist* 111:623–640.

Hemmer, H. 1990. *Domestication: The Decline of Environmental Adaptation*. Cambridge: Cambridge University Press.

Henderson, K. K., H. Wagner, F. Favret, S. L. Britton, L. G. Koch, P. D. Wagner, and N. C. Gonzalez. 2002. Determinants of maximal $O_2$ uptake in rats selectively bred for endurance running capacity. *Journal of Applied Physiology* 93:1265–1274.

Henderson, N. D. 1989. Interpreting studies that compare high- and low-selected lines on new characters. *Behavior Genetics* 19:473–503.

———. 1997. Spurious associations in unreplicated selected lines. *Behavior Genetics* 27:145–154.

Hennemann, W. W., III. 1983. Relationship among body mass, metabolic rate and the intrinsic rate of natural increase in mammals. *Oecologia* 56:104–108.

Hill, W. G. 1978. Design of selection experiments for comparing alternative testing regimes. *Heredity* 41:371–376.

Hochachka, P. W., H. C. Gunga, and K. Kirsch. 1998. Our ancestral physiological phenotype: An adaptation for hypoxia tolerance and for endurance performance? *Proceedings of the National Academy of Sciences of the USA* 95:1915–1920.

Holt, M., T. Meuwissen, and O. Vangen. 2005. Long-term responses, changes in genetic variances and inbreeding depression from 122 generations of selection on increased litter size in mice. *Journal of Animal Breeding and Genetics* 122:199–209.

Holt, M., O. Vangen, and W. Farstad. 2004. Components of litter size in mice after 110 generations of selection. *Reproduction* 127:587–592.

Houde, A. E. 1988. Genetic differences in mating preferences between guppy populations. *Animal Behaviour* 36:510–516.

———. 1994. Effect of artificial selection on male colour patterns on mating preference of female guppies. *Proceedings of the Royal Society of London B, Biological Sciences* 256:125–130.

————. 1997. *Sexual Selection and Mate Choice in Guppies.* Monographs in Behavioral Ecology. Princeton, NJ: Princeton University Press.

Houde, A. E., and J. A. Endler 1990. Correlated evolution of female mating preference and male color patterns in the guppy *Poecilia reticulata. Science* 248:1405–1408.

Howlett, R. A., N. C. Gonzalez, H. E. Wagner, Z. X. Fu, S. L. Britton, L. G. Koch, and P. D. Wagner. 2003. Genetic models in applied physiology: Selected contribution: Skeletal muscle capillarity and enzyme activity in rats selectively bred for running endurance. *Journal of Applied Physiology* 94:682–1688.

Huey, R. B., P. E. Hertz, and B. Sinervo. 2003. Behavioral drive versus behavioral inertia in evolution: a null model approach. *American Naturalist* 161:357–366.

Hulbert, A. J., and P. L. Else. 1989. The evolution of endothermic metabolism: mitochondrial activity and changes in cellular composition. *American Journal of Physiology* 256: R1200–R1208.

————. 1990. The cellular basis of endothermic metabolism: A role for "leaky" membranes? *News in Physiological Sciences* 5:25–28.

————. 1999. Membranes as possible pacemakers of metabolism. *Journal of Theoretical Biology* 199:257–274.

————. 2004. Basal metabolic rate: History, composition, regulation, and usefulness. *Physiological and Biochemical Zoology* 77:869–876.

————. 2005. Membranes and the setting of energy demand. *Journal of Experimental Biology* 208:1593–1599.

Husak, J. F. 2006. Does survival depend on how fast you *can* run or how fast you *do* run? *Functional Ecology* 20:1080–1086.

Irschick, D. J., and T. Garland, Jr. 2001. Integrating function and ecology in studies of adaptation: Investigations of locomotor capacity as a model system. *Annual Review of Ecology and Systematics* 32:367–396.

Irschick, D. J., J. J. Meyers, J. F. Husak, and J. F. Le Galliard. 2008. How does selection operate on whole-organism functional performance capacities? A review and synthesis. *Evolutionary Ecology Research* 10:1–20.

Iyer, N. V., S. W. Leung, and G. L. Semenza. 1998. The human hypoxia-inducible factor 1 alpha gene: HIF1A structure and evolutionary conservation. *Genomics* 52:159–165.

Johnson, M. S., and J. R. Speakman. 2007. Heritability of resting metabolic rate in short-tailed field voles, *Microtus agrestis. Comparative Biochemistry and Physiology A* 148:S21–S22.

Johnston, S. L., D. M. Souter, S. S. Erwin, B. Tolkamp, J. M. Yearsley, I. Gordon, A. W. Illius, I. Kyriazakis, and J. R. Speakman. 2007. Associations between basal metabolic rate and reproductive performance in C57BL/6J mice. *Journal of Experimental Biology* 210:65–74.

Kane, S. L., T. Garland, Jr., and P. A. Carter. 2008. Basal metabolic rate of aged mice is affected by random genetic drift but not by selective breeding for high early-age locomotor activity or chronic wheel access. *Physiological and Biochemical Zoology* 81:288–300.

Karasov, W. H., and J. Diamond. 1985. Digestive adaptations for fueling the cost of endothermy. *Science* 228:202–204.

Karasov, W. H., and C. Martinez del Rio. 2007. *Physiological Ecology: How Animals Process Energy, Nutrients, and Toxins.* Princeton, NJ: Princeton University Press.

Keeney, B. K., D. A. Raichlen, T. H. Meek, R. S. Wijeratne, K. M. Middleton, G. L. Gerdeman, and T. Garland, Jr. 2008. Differential response to a selective cannabinoid receptor antagonist (SR141716: rimonabant) in female mice from lines selectively bred for high voluntary wheel-running behavior. *Behavioural Pharmacology* 19:812–820.

Kemp, T. J., K. N. Bachus, J. A. Nairn, and D. R. Carrier. 2005. Functional trade-offs in the limb bones of dogs selected for running versus fighting. *Journal of Experimental Biology* 208:3475–3482.

Kemp, T. S. 2006. The origin of mammalian endothermy: A paradigm for the evolution of complex biological structure. *Zoological Journal of the Linnean Society* 147:473–488.

———. 2007. *The Origin and Evolution of Mammals*. Oxford: Oxford University Press.

Kgwatalala, P. M., J. L. DeRoin, and M. K. Nielsen. 2004. Performance of mouse lines divergently selected for heat loss when exposed to different environmental temperatures. I. Reproductive performance, pup survival, and metabolic hormones. *Journal of Animal Science* 82:2876–2883.

Kgwatalala, P. M., and M. K. Nielsen. 2004. Performance of mouse lines divergently selected for heat loss when exposed to different environmental temperatures. II. Feed intake, growth, fatness, and body organs. *Journal of Animal Science* 82:2884–2891.

Kingsolver, J. G., and R. B. Huey. 2003. Introduction: The evolution of morphology, performance, and fitness. *Integrative and Comparative Biology* 43:361–366.

Klissouras, V. 1971. Heritability of adaptive variation. *Journal of Applied Physiology* 31:338–344.

Koch, L. G., and S. L. Britton. 2001. Artificial selection for intrinsic aerobic endurance running capacity in rats. *Physiological Genomics* 5:45–52.

———. 2005. Divergent selection for aerobic capacity in rats as a model for complex disease. *Integrative and Comparative Biology* 45:405–415.

———. 2007. Evolution, atmospheric oxygen, and complex disease. *Physiological Genomics* 30:205–208.

Koch, L. G., U. Wisløff, J. E. Wilkinson, and S. L. Britton. 2007. Rat models of intrinsic aerobic capacity differ in longevity. *FASEB Journal* 21:A1401.

Konarzewski, M., and J. Diamond. 1995. Evolution of basal metabolic rate and organ masses in laboratory mice. *Evolution* 49:1239–2148.

Konarzewski, M., A. Książek, and I. B. Lapo. 2005. Artificial selection on metabolic rates and related traits in rodents. *Integrative and Comparative Biology* 45:416–425.

Koteja, P. 1991. On the relation between basal and field metabolic rates in birds and mammals. *Functional Ecology* 5:56–64.

———. 1996. Limits to the energy budget in a rodent, *Peromyscus maniculatus*: Does gut capacity set the limit? *Physiological Zoology* 69:994–1020.

———. 2000. Energy assimilation, parental care and the evolution of endothermy. *Proceedings of the Royal Society of London B, Biological Sciences* 267:479–484.

Koteja, P., T. Garland, Jr., J. K. Sax, J. G. Swallow, and P. A. Carter. 1999. Behaviour of house mice artificially selected for high levels of voluntary wheel running. *Animal Behaviour* 58:1307–1318.

Kotiaho, J. S. 2001. Costs of sexual traits: A mismatch between theoretical considerations and empirical evidence. *Biological Reviews* 76:365–376.

Kozlowski, J., M. Konarzewski, and A. T. Gawelczyk. 2003. Cell size as a link between noncoding DNA and metabolic rate scaling. *Proceedings of the National Academy of Sciences of the USA* 100:14080–14085.

Kozlowski, J., and J. Weiner. 1997. Interspecific allometries are by-products of body size optimization. *American Naturalist* 149:352–380.

Książek, A., M. Konarzewski, M. Chadzińska, and M. Cichoń. 2003. Costs of immune response in cold-stressed laboratory mice selected for high and low basal metabolism rates. *Proceedings of the Royal Society of London B, Biological Sciences* 270:2025–2031.

———. 2007. Selection for high basal metabolic rate compromises immune response in cold-stressed laboratory mice. *Comparative Biochemistry and Physiology A* 148:S20–S21.

Książek, A., M. Konarzewski, and I. B. Łapo. 2004. Anatomic and energetic correlates of divergent selection for BMR in laboratory mice. *Physiological and Biochemical Zoology* 77:890–899.

Lacy, R. C. and C. B. Lynch. 1979. Quantitative genetic analysis of temperature regulation in *Mus musculus*. I. Partitioning of variance. *Genetics* 91:743–753.

Lailvaux, S., and D. J. Irschick. 2006. A functional perspective on sexual selection: Insights and future prospects. *Animal Behaviour* 72:263–273.

Lapo, I. B., M. Konarzewski, and B. Sadowski. 2003a. Analgesia induced by swim stress: Interaction between analgesic and thermoregulatory mechanisms. *Pflugers Archiv—European Journal of Physiology* 446:463–469.

———. 2003b. Differential metabolic capacity of mice selected for magnitude of swim stress-induced analgesia. *Journal of Applied Physiology* 94:677–684.

———. 2003c. Effect of cold acclimation and repeated swimming on opioid and nonopioid swim stress-induced analgesia in selectively bred mice. *Physiology & Behavior* 78:345–350.

Laurie, C., S. Chasalow, J. Ledeaux, R. Mccarroll, D. Bush, B. Hauge, C. Lai, D. Clark, T. Rocheford, and J. Dudley. 2004. The genetic architecture of response to long-term artificial selection for oil concentration in the maize kernel. *Genetics* 168:2141–2155.

Lavin, S. R., W. H. Karasov, A. R. Ives, K. M. Middleton, and T. Garland, Jr. 2008. Morphometrics of the avian small intestine, compared with non-flying mammals: A phylogenetic perspective. *Physiological and Biochemical Zoology* 81:526–550.

Lerman, I., B. C. Harrison, K. Freeman, T. E. Hewett, D. L. Allen, J. Robbins, and L. A. Leinwand. 2002. Genetic variability in forced and voluntary endurance exercise performance in seven inbred mouse strains. *Journal of Applied Physiology* 92:2245–2255.

Li, G., J. S. Rhodes, I. Girard, S. C. Gammie, and T. Garland, Jr. 2004. Opioid-mediated pain sensitivity in mice bred for high voluntary wheel running. *Physiology & Behavior* 83:515–524.

Liang, B., and D. A. Blizard. 1978. Central and peripheral norepinephrine concentrations in rat strains selectively bred for differences in response to stress: Confirmation and extension. *Pharmacology Biochemistry and Behavior* 8:75–80.

Lightfoot, J. T., M. J. Turner, M. Daves, A. Vordermark, and S. R. Kleeberger. 2004. Genetic influence on daily wheel running activity level. *Physiological Genomics* 19:270–276.

Lightfoot, J. T., M. J. Turner, K. A. Debate, and S. R. Kleeberger. 2001. Interstrain variation in murine aerobic capacity. *Medicine and Science in Sports and Exercise* 33:2053–2057.

Lindström, Å., and A. Kvist. 1995. Maximum energy intake rate is proportional to basal metabolic rate in migrating passerine birds. *Proceedings of the Royal Society of London B, Biological Sciences* 261:337–343.

Losos, J. B., D. A. Creer, and J. A. Schulte II. 2002. Cautionary comments on the measurement of maximum locomotor capabilities. *Journal of Zoology (London)* 258:57–61.

Lynch, C. B. 1992. Clinal variation in cold adaptation in *Mus domesticus*: Verification of predictions from laboratory populations. *American Naturalist* 139:1219–1236.

———. 1994. Evolutionary inferences from genetic analyses of cold adaptation in laboratory and wild populations of the house mouse. Pages 278–301 *in* C. R. B. Boake, ed. *Quantitative Genetic Studies of Behavioral Evolution*. Chicago: University of Chicago Press.

MacLean, R. C., and G. Bell. 2002. Experimental adaptive radiation in Pseudomonas. *American Naturalist* 160:569–581.

Magurran, A. E., B. H. Seghers, G. R. Carvalho, and P. W. Shaw. 1993. Evolution of adaptive variation in antipredator behavior. *Marine Behaviour and Physiology* 23:29–44.

Malisch, J. L. 2007. Micro-evolutionary change in the hypothalamic-pituitary-adrenal axis in house mice selectively bred for high voluntary wheel running. Unpublished PhD diss. University of California, Riverside.

Malisch, J. L., C. W. Breuner, F. R Gomes, M. A. Chappell, and T. Garland, Jr. 2008. Circadian pattern of total and free corticosterone concentrations, corticosteroid-binding globulin, and physical activity in mice selectively bred for high voluntary wheel-running behavior. *General and Comparative Endocrinology* 156:210–217.

Malisch, J. L., S. A. Kelly, A. Bhanvadia, K. M. Blank, R. L. Marsik, E. G. Platzer, and T. Garland, Jr. 2009. Lines of mice with chronically elevated baseline corticosterone are more susceptible to a parasitic nematode infection. *Zoology* 112:316–324.

Malisch, J. L., W. Saltzman, F. R. Gomes, E. L. Rezende, D. R. Jeske, and T. Garland, Jr. 2007. Baseline and stress-induced plasma corticosterone concentrations of mice selectively bred for high voluntary wheel running. *Physiological and Biochemical Zoology* 80:146–156.

Marden, J. H., M. R. Wolf, and K. E. Weber. 1997. Aerial performance of *Drosophila melanogaster* from populations selected for upward flight mobility. *Journal of Experimental Biology* 200:2747–2755.

Massett, M. P., and B. C. Berk. 2005. Strain-dependent differences in responses to exercise training in inbred and hybrid mice. *American Journal of Physiology* 288 (Regulatory Integrative and Comparative Physiology):R1006–R1013.

McDaneld, T. G., M. K. Nielsen, and L. J. Miner. 2002. Uncoupling proteins and energy expenditure in mice divergently selected for heat loss. *Journal of Animal Science* 80:602–608.

McDonald, J. M., and M. K. Nielsen. 2006. Correlated responses in maternal performance following divergent selection for heat loss in mice. *Journal of Animal Science* 84:300–304.

McGraw, J. B., and H. Caswell. 1996. Estimation of individual fitness from life-history data. *American Naturalist* 147:47–64.

McKechnie, A. E., R. P. Freckleton, and W. Jetz. 2006. Phenotypic plasticity in the scaling of avian basal metabolic rate. *Proceedings of the Royal Society of London B, Biological Sciences* 273:931–937.

McNab, B. K. 1980. Food habits, energetics, and the population biology of mammals. *American Naturalist* 116:106–124.

———. 2002. *The Physiological Ecology of Vertebrates*. Ithaca, NY: Cornell University Press.

Meek, T. H., R. M. Hannon, B. Lonquich, R. L. Marsik, R. S. Wijeratne, and T. Garland, Jr. 2007. Endurance capacity of mice selectively bred for high voluntary wheel running. *Integrative and Comparative Biology* 47:e81.

Meerlo, P., L. Bolle, G. H. Visser, D. Masman, and S. Daan. 1997. Basal metabolic rate in relation to body composition and daily energy expenditure in the field vole, *Microtus agrestis*. *Physiological Zoology* 70:362–369.

Middleton, K. M., S. A. Kelly, and T. Garland, Jr. 2008. Selective breeding as a tool to probe skeletal response to high voluntary locomotor activity in mice. *Integrative and Comparative Biology* 48:394–410.

Miles, D. B., R. Calsbeek, and B. Sinervo. 2007. Corticosterone, locomotor performance, and metabolism in side-blotched lizards *(Uta stansburiana)*. *Hormones and Behavior* 51:548–554.

Monge, C., and F. Leonvelarde. 1991. Physiological adaptation to high altitude: Oxygen transport in mammals and birds. *Physiological Reviews* 71:1135–1172.

Moody, D. E., D. Pomp, and M. K. Nielsen. 1997. Variability in metabolic rate, feed intake and fatness among selection and inbred lines of mice. *Genetical Research* 70:225–235.

Moody, D. E., D. Pomp, M. K. Nielsen, and L. D. Van Vleck. 1999. Identification of quantitative trait loci influencing traits related to energy balance in selection and inbred lines of mice. *Genetics* 152:699–711.

Mousel, M. R., W. W. Stroup, and M. K. Nielsen. 2001. Locomotor activity, core body temperature, and circadian rhythms in mice selected for high or low heat loss. *Journal of Animal Science* 79:861–868.

Nagasawa, H., R. Yanai, H. Taniguchi, R. Tokuzen, and W. Nakahara. 1976. Two-way selection of a stock of Swiss albino mice for mammary tumorigenesis: Establishment of 2 new strains (SHN and SLN). *Journal of the National Cancer Institute* 57:425–430.

Nespolo, R. F., L. D. Bacigalupe, and F. Bozinovic. 2003. Heritability of energetics in a wild mammal, the leaf-eared mouse *(Phylottis darwini)*. *Evolution* 57:1679–1688.

Nielsen, M. K., B. A. Freking, L. D. Jones, S. M. Nelson, T. L. Vorderstrasse, and B. A. Hussey. 1997a. Divergent selection for heat loss in mice. 2. Correlated responses in feed intake, body mass, body composition, and number born through fifteen generations. *Journal of Animal Science* 75:1469–1476.

Nielsen, M. K., L. D. Jones, B. A. Freking, and J. A. DeShazer 1997b. Divergent selection for heat loss in mice. 1. Selection applied and direct response through fifteen generations. *Journal of Animal Science* 75:1461–1468.

Nobunaga, T. 1973. Establishment by selective inbreeding of the IVCS strain and related sister strains of the mouse, demonstrating regularly repeated 4-day estrus cycles. *Laboratory Animal Science* 23:803–811.

Nunney, L. 1996. The response to selection for fast larval development in *Drosophila melanogaster* and its effect on adult weight: An example of a fitness trade-off. *Evolution* 50:1193–1204.

Ohno, T., and T. Miyatake. 2007. Drop or fly? Negative genetic correlation between death-feigning intensity and flying ability as alternative anti-predator strategies. Proceedings of the Royal Society B: Biological Sciences 274:555–560.

O'Steen, S., A. J. Cullum, and A. F. Bennett. 2002. Rapid evolution of escape ability in Trinidadian guppies *(Poecilia reticulata)*. *Evolution* 56:776–784.

Oufiero, C. E., and T. Garland, Jr. 2007. Evaluating performance costs of sexually selected traits. *Functional Ecology* 21:676–689.

Overstreet, D. H. 2002. Behavioral characteristics of rat lines selected for differential hypothermic responses to cholinergic or serotonergic agonists. *Behavior Genetics* 32:335–348.

Padley, D. 1985. Do life history parameters of passerines scale to metabolic rate independently of body mass? *Oikos* 45:285–287.

Panhuis, T. M., and G. S. Wilkinson. 1999. Exaggerated male eye span influences contest outcome in stalk-eyed flies. *Behavioral Ecology and Sociobiology* 46:221–227.

Panocka, I., P. Marek, and B. Sadowski 1986. Inheritance of stress-induced analgesia in mice: Selective breeding study. *Brain Research* 397:152–155.

Parker, H. G., L. V. Kim, N. B. Sutter, S. Carlson, T. D. Lorentzen, T. B. Malek, G. S. Johnson, H. B. De France, E. A. Ostrander, and L. Kruglyak. 2004. Genetic structure of the purebred domestic dog. *Science* 304:1160–1164.

Pasi, B. M., and D. R. Carrier. 2003. Functional trade-offs in the limb muscles of dogs selected for running vs. fighting. *Journal of Evolutionary Biology* 16:324–332.

Perusse, L., G. Lortie, C. LeBlanc, A. Tremblay, G. Theriault, and C. Bouchard. Genetic and environmental sources of variation in physical fitness. *Annals of Human Biology* 14:425–434.

Piersma, T., and A. Lindström. 1997. Rapid reversible changes in organ size as a component of adaptive behaviour. *Trends in Ecology & Evolution* 12:134–138.

Pörtner, H. O. 2004. Climate variability and the energetic pathways of evolution: The origin of endothermy in mammals and birds. *Physiological and Biochemical Zoology* 77:959–981.

Price, E. O. 2002. *Animal Domestication and Behavior.* New York: CABI.

Renne, U., M. Langhammer, E. Wytrwat, G. Dietl, and L. Bünger. 2003. Genetic-statistical analysis of growth in selected and unselected mouse lines. *Journal of Experimental Animal Science* 42:218–232.

Rezende, E. L., F. Bozinovic, and T. Garland, Jr. 2004. Climatic adaptation and the evolution of basal and maximum rates of metabolism in rodents. *Evolution* 58:1361–1374.

Rezende, E. L., T. Garland, Jr., M. A. Chappell, J. L. Malisch, and F. R. Gomes. 2006a. Maximum aerobic performance in lines of *Mus* selected for high wheel-running activity: Effects of selection, oxygen availability, and the mini-muscle phenotype. *Journal of Experimental Biology* 209:115–127.

Rezende, E. L., S. A. Kelly, F. R. Gomes, M. A. Chappell, and T. Garland, Jr. 2006b. Effects of size, sex, and voluntary running speeds on costs of locomotion in lines of laboratory mice selectively bred for high wheel-running activity. *Physiological and Biochemical Zoology* 79:83–99.

Rezende, E. L., F. R. Gomes, J. L. Malisch, M. A. Chappell, and T. Garland, Jr. 2006c. Maximal oxygen consumption in relation to subordinate traits in lines of house mice selectively bred for high voluntary wheel running. *Journal of Applied Physiology* 101:477–485.

Reznick, D. N., F. H. Shaw, F. H. Rodd, and R. G. Shaw. 1997. Evaluation of the rate of evolution in natural populations of guppies *(Poecilia reticulata).* *Science* 275:1934–1937.

Rhodes, J. S., S. C. Gammie, and T. Garland, Jr. 2005. Neurobiology of mice selected for high voluntary wheel-running activity. *Integrative and Comparative Biology* 45:438–455.

Ribak, G., and J. G. Swallow. 2007. Free flight maneuvers of stalk-eyed flies: Do eye stalks limit aerial turning behavior? *Journal of Comparative Physiology A* 193:1065–1079.

Ricklefs, R. E., M. Konarzewski, and S. Daan. 1996. The relationship between basal metabolic rate and daily energy expenditure in birds and mammals. *American Naturalist* 147:1047–1071.

Roberts, M. L., K. L. Buchanan, D. Hasselquist, A. T. Bennett, and M. R. Evans. 2007. Physiological, morphological and behavioural effects of selecting zebra finches for divergent levels of corticosterone. *Journal of Experimental Biology* 210:4368–4378.

Robertson, A., ed. 1980. *Selection Experiments in Laboratory and Domestic Animals.* Farnam Royal, Slough, U.K.: Commonwealth Agricultural Bureau.

Roff, D. A. 1997. *Evolutionary Quantitative Genetics.* New York: Chapman and Hall.

Rose, M. R., T. J. Nusbaum, and A. K. Chippindale. 1996. Laboratory evolution: The experimental wonderland and the Cheshire Cat syndrome. Pages 221–241 *in* M. R. Rose and G. V. Lauder, eds. *Adaptation.* San Diego, CA: Academic Press.

Rosochacki, S. J., E. Wirth-Dzięciołowska, M. Zimowska, T. Sakowski, J. Poloszynowicz, E. Juszczuk-Kubiak, and M. Gajewska. 2005. Skeletal muscle and liver protein degradation in mice divergently selected for low and high body weight over 108 generations. *Archives of Animal Breeding* 48:505–517.

Ruben J. A. 1995. The evolution of endothermy in mammals and birds: From physiology to fossils. *Annual Review of Physiology* 57:69–95.

Sadowska, E. T., K. Baliga-Klimczyk, K. M. Chrząścik, and P. Koteja. 2008. Laboratory model of adaptive radiation: A selection experiment in the bank vole. *Physiological and Biochemical Zoology* 81:627–640.

Sadowska, E. T., M. K. Labocha, K. Baliga, A. Stanisz, A. K. Wróblewska, W. Jagusik, and P. Koteja. 2005. Genetic correlations between basal and maximum metabolic rates in a wild rodent: Consequences for evolution of endothermy. *Evolution* 59:672–681.

Sadowski, B., and M. Konarzewski. 1999. Analgesia in selectively bred mice exposed to cold in helium/oxygen atmosphere. *Physiology & Behavior* 66:145–151.

Schluter, D. 1996. Adaptive radiation along genetic lines of least resistance. *Evolution* 50:1766–1774.

Schmidt-Nielsen, K. 1990. *Animal Physiology: Adaptation and Environment.* 4th ed. Cambridge: Cambridge University Press.

Sears, M. W., J. P. Hayes, C. S. O'Connor, K. Geluso, and J. S. Sedinger. 2006. Individual variation in thermogenic capacity affects above-ground activity of high-altitude deer mice. *Functional Ecology* 20:97–104.

Seehausen, O. 2006. African cichlid fish: A model system in adaptive radiation research. *Proceedings of the Royal Society of London B, Biological Sciences* 273:1987–1998.

Selman, C., T. K. Korhonen, L. Bünger, W. G. Hill, and J. R. Speakman. 2001b. Thermoregulatory responses of two mouse *Mus musculus* strains selectively bred for high and low food intake. *Journal of Comparative Physiology B* 171:661–668.

Selman, C., S. Lumsden, L. Bünger, W. G. Hill, and J. R. Speakman. 2001a. Resting metabolic rate and morphology in mice *(Mus musculus)* selected for high and low food intake. *Journal of Experimental Biology* 204:777–784.

Semenza, G. L. 1999. Regulation of mammalian $O_2$ homeostasis by hypoxia-inducible factor 1. *Annual Review of Cell and Developmental Biology* 15:551–578.

Senter, P. 2007. Necks for sex: Sexual selection as an explanation for sauropod dinosaur neck elongation. *Journal of Zoology* 271:45–53.

Shikano, T., E. Arai, and Y. Fujio. 1998. Seawater adaptability, osmoregulatory function, and branchial chloride cells in the strain selected for high salinity tolerance of the guppy *Poecilia reticulata*. *Fisheries Science* 64:240–244.

Silver, L. M. 1995. *Mouse Genetics: Concepts and Applications.* New York: Oxford University Press. Available: www.informatics.jax.org/silverbook/.

Sinervo, B., and J. Clobert. 2003. Morphs, dispersal behavior, genetic similarity, and the evolution of cooperation. *Science* 300:1949–1951.

Soutiere, S. E., C. G. Tankersley, and W. Mitzner. 2004. Differences in alveolar size in inbred mouse strains. *Respiratory Physiology & Neurobiology* 140:283–291.

Speakman, J. R., T. Ergon, R. Cavanagh, K. Reid, D. M. Scantlebury, and X. Lambin. 2003. Resting and daily energy expenditures of free-living field voles are positively correlated but reflect extrinsic rather than intrinsic effects. *Proceedings of the National Academy of Sciences of the USA* 100:14057–14062.

Speakman, J. R., and J. McQueenie. 1996. Limits to sustained metabolic rate: The link between food intake, basal metabolic rate, and morphology in reproducing mice, *Mus musculus*. *Physiological Zoology* 69:746–769.

Svenson, K. L., R. V. Smith, P. A. Magnani, H. R. Suetin, B. Paigen, J. K. Naggert, R. Li, G. A. Churchill, and L. L. Peters. 2007. Multiple trait measurements in 43 inbred mouse strains capture the phenotypic diversity characteristic of human populations. Journal of Applied Physiology 102:2369–2378.

Swallow, J. G., P. A. Carter, and T. Garland, Jr. 1998a. Artificial selection for increased wheel-running behavior in house mice. *Behavior Genetics* 28:227–237.

Swallow, J. G., and T. Garland, Jr. 2005. Selection experiments as a tool in evolutionary and comparative physiology: Insights into complex traits: An introduction to the symposium. *Integrative and Comparative Biology* 45:387–390.

Swallow, J. G., T. Garland, Jr., P. A. Carter, W.-Z. Zhan, and G. C. Sieck. 1998b. Effects of voluntary activity and genetic selection on aerobic capacity in house mice *(Mus domesticus)*. *Journal of Applied Physiology* 84:69–76.

Swallow, J. G., J. S. Rhodes, and T. Garland, Jr. 2005. Phenotypic and evolutionary plasticity of organ masses in response to voluntary exercise in house mice. *Integrative and Comparative Biology* 45:426–437.

Swallow, J. G., G. S. Wilkinson, and J. H. Marden. 2000. Aerial performance of stalk-eyed flies that differ in eye span. *Journal of Comparative Physiology B* 170:481–487.

Tankersley, C. G., R. S. Fitzgerald, and S. R. Kleeberger. 1994. Differential control of ventilation among inbred strains of mice. *American Journal of Physiology* 36 (Regulatory Integrative and Comparative Physiology):R1371–R1377.

Taubert, H., D. Agena, and H. Simianer. 2007. Genetic analysis of racing performance in Irish greyhounds. *Journal of Animal Breeding and Genetics* 124:117–123.

Tobalske, B. W., D. L. Altshuler, and D. R. Powers. 2004. Take-off mechanics in hummingbirds (Trochilidae). *Journal of Experimental Biology* 207:1345–1352.

Trevelyan, R., P. H. Harvey, and M. D. Pagel. 1990. Metabolic rates and life histories in birds. *Functional Ecology* 4:135–141.

Troxell, M. L., S. L. Britton, and L. G. Koch. 2003. Genetic models in applied physiology: Selected contribution: Variability and heritability for the adaptational response to exercise in genetically heterogeneous rats. *Journal of Applied Physiology* 94:1674–1681.

Vaanholt, L. M., P. Meerlo, T. Garland, Jr., G. H. Visser, and G. van Dijk. 2007. Plasma adiponectin is increased in mice selectively bred for high wheel-running activity, but not by wheel running per se. *Hormone and Metabolic Research* 39:377–383.

Ward, N. L., E. Moore, K. Noon, N. Spassil, E. Keenan, T. L. Ivanco, and J. C. LaManna. 2007. Cerebral angiogenic factors, angiogenesis, and physiological response to chronic hypoxia differ among four commonly used mouse strains. *Journal of Applied Physiology* 102:1927–1935.

Warren, I., and H. Smith. 2007. Stalk-eyed flies (Diopsidae): Modeling the evolution and development of an exaggerated sexual trait. *BioEssays* 29:300–307.

Wassersug, J. D., and R. J. Wassersug. 1986. Fitness fallacies. *Natural History* 95:34–37.

Waters, R. P., K. J. Renner, R. B. Pringle, C. H. Summers, S. L. Britton, L. G. Koch, and J. G. Swallow. 2008. Selection for aerobic capacity affects corticosterone, monoamines and wheel-running activity. *Physiology & Behavior* 93:1044–1054.

Weber, K. E. 1996. Large genetic change at small fitness cost in large populations of *D. melanogaster* selected for wind tunnel flight: Rethinking fitness surfaces. *Genetics* 144:205–213.

Webster, L. T. 1933a. Inheritance and acquired factors in resistance to infection. I. Development of resistant and susceptible lines of mice through selective breeding. *Journal of Experimental Medicine* 57:793–817.

———. 1933b. Inheritance and acquired factors in resistance to infection. II. A comparison of mice inherently resistant or susceptible to *Bacillus enteritidis* infection with respect to fertility, weight and susceptibility to various routes and types of infection. *Journal of Experimental Medicine* 57:819–843.

Weir, J. A., and R. D. Clark. 1955. Production of high and low blood-pH lines of mice by selection with inbreeding. *Journal of Heredity* 46:125–132.

Wesolowski, S. R., M. F. Allan, M. K. Nielsen, and D. Pomp. 2003. Evaluation of hypothalamic gene expression in mice divergently selected for heat loss. *Physiological Genomics* 13:129–137.

White, C. R, N. F. Phillips, and R. S. Seymour. 2006. The scaling and temperature dependence of vertebrate metabolism. *Biology Letters* 2:125–127.

White, C. R., and R. S. Seymour. 2004. Does basal metabolic rate contain a useful signal? Mammalian BMR allometry and correlations with a selection of physiological, ecological, and life-history variables. *Physiological and Biochemical Zoology* 77:929–941.

Wiersma, P., A. Munoz-Garcia, A. Walker, and J. B. Williams. 2007. Tropical birds have a slow pace of life. *Proceedings of the National Academy of Sciences of the USA* 104:9340–9345.

Wilkinson, G. S. 1993. Artificial sexual selection alters allometry in the stalk-eyed fly *Cyrtodiopsis dalmanni* (Diptera: Diopsidae). *Genetical Research* 62:213–222.

Wilkinson, G. S., E. G. Amitin, and P. M. Johns. 2005. Sex-linked correlated responses in female reproductive traits to selection on male eye span in stalk-eyed flies. *Integrative and Comparative Biology* 45:500–510.

Wilkinson, G. S., and G. Dodson. 1997. Function and evolution of antlers and eye stalks in flies. Pages 310–328 *in* J. Choe and B. Crespi, eds. *The Evolution of Mating Systems in Insects and Arachnids.* Cambridge: Cambridge University Press.

Wilkinson, G. S., H. Kahler, and R. H. Baker. 1998. Evolution of female mating preferences in stalk-eyed flies. *Behavioral Ecology* 9:525–533.

Wilkinson, G. S., and P. R. Reillo. 1994. Female preference response to artificial selection on an exaggerated male trait in a stalk-eyed fly. *Proceedings of the Royal Society of London B, Biological Sciences* 255:1–6.

Wirth-Dzięciołowska, E., and K. Czumińska. 2000. Longevity and aging of mice from lines divergently selected for body weight for over 90 generations. *Biogerontology* 1:171–178.

Wirth-Dzięciołowska, E., A. Lipska A., and M. Węsierska. 2005. Selection for body weight induces differences in exploratory behavior and learning in mice. *Acta Neurobiologiae Experimentalis* 65:243–253.

Wisloff, U., S. M. Najjar, O. Ellingsen, P. M. Haram, S. Swoap, Q. Al-Share, M. Fernstrom, K. Rezaei, S. J. Lee, L. G. Koch, and S. L. Britton. 2005. Cardiovascular risk factors emerge after artificial selection for low aerobic capacity. *Science* 307:418–420.

Zera, A. J., L. G. Harshman, and T. D. Williams. 2007. Evolutionary endocrinology: The developing synthesis between endocrinology and evolutionary genetics. *Annual Review of Ecology, Evolution, and Systematics* 38:793–817.

Zwemer, C. F., M. Y. Song, K. A. Carello, and L. G. D'Alecy. 2007. Strain differences in response to acute hypoxia: CD-1 versus C57BL/6J mice. *Journal of Applied Physiology* 102:286–293.

# 13

# THROUGH A GLASS, CLEARLY
*Experimental Evolution as a Window on
Adaptive Genome Evolution*

## Frank Rosenzweig and Gavin Sherlock

Genome architecture is defined as the structure, content, and organization of a genome, and it is described in terms of genes, their *cis*-acting regulatory elements, and various noncoding entities that populate intergenic regions. This useful phrase has an unfortunate tendency to promote the view that genomes are static, perhaps because of the intuitive associations conjured up by the term *architecture*. This view has been reinforced by the recent proliferation of whole-genome maps. Such maps, however clever and colorful, are spatially limited to two dimensions and do not represent epigenetic processes that can alter genomes physical structure. (Jaenisch and Bird 2003). Each map represents the genome uniquely characteristic of an individual or clone, or, more specifically, an individual or clone at the time it was sequenced. Such representations capture neither how nor why genomes vary among individuals within populations, much less how genomes change over time within lineages. What is captured, therefore, is not a species' genome architecture, but rather a snapshot of an individual or clone of a particular species at a particular time.

Explaining the dynamic properties of genomes or, more specifically how—and why—genomes behave in certain ways over successive generations is the province of evolutionary biology. Ongoing advances in high-throughput DNA sequencing (e.g., Albert et al. 2005) and methods for whole-genome analyses (Ochmann and Santos 2005) continue to revolutionize how evolutionary biologists can formulate and test hypotheses concerning the forces that shape genome structure. In this enterprise, the comparative and experimental approaches each have their place and offer the prospect of mutual enrichment and validation. For example, detailed understanding of differences in the distribution and abundance of mobile elements in closely related species, or sub-species, can be used to inform the design of experiments to test hypotheses concerning how such patterns arose. Likewise, knowledge of the distribution, abundance, and homology of portable virulence determinants can inform the design of experiments to test mechanisms by which horizontal transfer of such determinants occurs between recipient and probable donor species probable.

In this chapter, we explore recent examples where experimental evolution, empowered by genomic technologies and increasingly informed by systems biology, has fundamentally advanced our understanding of adaptive evolution. We will focus our attention on genomically enabled insights into the following issues: how prevalent adaptive parallelism is; what the relative contributions of structural versus regulatory mutations are to the adaptive process; whether trade-offs inevitably result from pleiotropy; how and why loss of function occurs under strong selection; and how much and how often adaptations arise from large-scale structural changes, such as gene duplications, deletions and translocations. Anticipating further advances in genome technology, especially in whole-genome sequencing and bioinformatics, we offer our perspective on how these technologies have been and can be fruitfully applied in the context of experimental evolution. For example, we are on the verge of precisely understanding the spectra of mutations that we can expect to see generated and propagated under different experimental conditions.

As a recognizable "science," experimental evolution dates back well over a century, but as a planned human activity, experimental evolution brought about the development of agriculture. We contend that the marriage of this ancient technology with the "new genomics" will lead to major advances in medicine, agriculture, and other biology-based industries, as well as a deeper understanding of evolution, the most fundamental biological process of all.

## ADAPTIVE EVOLUTION

Adaptive evolution occurs by natural selection when genetic variants better able to survive and reproduce expand as a fraction of the population, potentially leading to fixation of alleles that confer a fitness advantage. The study of adaptive evolution has been especially fruitful in microbial systems, due to the relative ease with which one can establish and maintain large populations with short generation times, impose specific selective pressures, and preserve a living record of evolutionary change by periodically archiving cells as frozen stocks in glycerol (see figure 13.1). Here, we discuss how recent laboratory evolution experiments, newly enriched with data acquired using genomic technologies, have enlarged our view of adaptive evolution in bacteria and yeast. We also illustrate how these techniques are being extended to the study of real-time evolution in higher eukaryotes, and we conclude by proposing a research agenda for those who wish to harness the power of emerging technologies in genomic science to deepen our understanding of evolutionary processes.

## EXPERIMENTAL EVOLUTIONARY GENOMICS OF BACTERIA

Experimental evolutionary studies using bacteria informed our knowledge of adaptation long before we understood the chemical structure of DNA or the molecular biology of the gene. Foundational work by Luria, Delbruck, and Newcombe establishing the random nature and background frequency of mutations was arguably "experimental evolution" (Luria and Delbruck 1943; Newcombe 1949). Soon after, it was shown that experimental evolution using bacteria provides insight into the mode and tempo of adaptive evolution (Novick and Szilard 1949; Atwood et al. 1951). These pioneering studies also demonstrated the power of using experimental models that could be easily manipulated to test prospectively and retrospectively, and in real time, key predictions of population genetic theory (e.g., clonal replacement as envisioned by Muller 1932). For many years thereafter, experimental studies of bacterial evolution followed parallel paths: one aimed at understanding the relative strength of population genetic forces that guide evolving, and coevolving, populations across adaptive landscapes (e.g., Levin et al. 1977, 1991, 1994), and another focused on identifying specific mechanisms by which novel adaptive

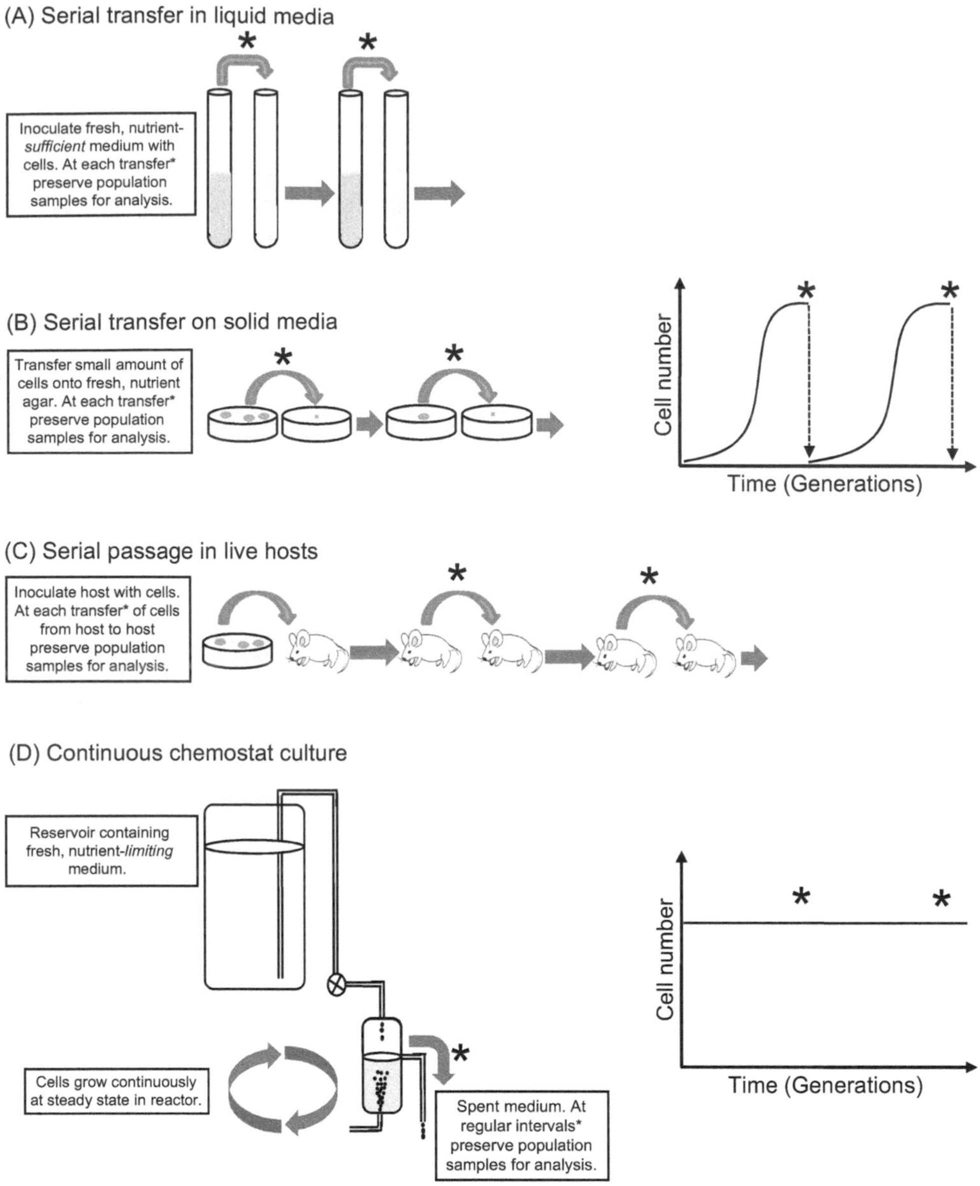

(A) Serial transfer in liquid media

Inoculate fresh, nutrient-*sufficient* medium with cells. At each transfer* preserve population samples for analysis.

(B) Serial transfer on solid media

Transfer small amount of cells onto fresh, nutrient agar. At each transfer* preserve population samples for analysis.

(C) Serial passage in live hosts

Inoculate host with cells. At each transfer* of cells from host to host preserve population samples for analysis.

(D) Continuous chemostat culture

Reservoir containing fresh, nutrient-*limiting* medium.

Cells grow continuously at steady state in reactor.

Spent medium. At regular intervals* preserve population samples for analysis.

FIGURE 13.1

Experimental evolution using microbes can be carried out under very different environmental conditions. Bacterial and eukaryotic microbes can be propagated by serial transfer in liquid (A) or solid media (B), or by serial passage through living hosts (C). Alternatively, microbes can be propagated under continuous resource limitation in chemostats (D). In all these cases, a living record of evolutionary change can be preserved by freezing population samples at time of transfer or at otherwise specified intervals. Populations evolving under regimens A, B, and C experience successive "boom-and-bust" cycles. Following an initial lag phase, population sizes expand rapidly before cells encounter resource limitation. By contrast (and by definition) chemostat populations (D) are propagated at steady state under continuous nutrient limitation. Population size is fixed by the concentration of limiting nutrient, and population growth rate is fixed by the rate at which fresh nutrient is introduced into the reactor.

traits evolve (e.g. Horiuchi et al. 1962; Rigby et al. 1974; Hall and Zuzel 1980). Not surprisingly, population-focused studies commonly underscored the role of stochastic factors in evolution, whereas those focused on specific genes and pathways emphasized deterministic factors.

A hallmark parameter reported in all experimental evolution studies is relative fitness, estimated either by direct pair-wise competition between experimental strains or by measuring strain-specific growth rate and/or survivorship in a "common garden." Whereas population genetic studies typically use relative fitness to plot population trajectories across adaptive landscapes under different modes and intensities of selection, molecular genetic studies tend to use relative fitness as a roadmap to discovering the biochemistry underlying evolved traits. A convergence of these lines of investigation was needed, as understanding how genotypes map onto physiological and fitness phenotypes has long been a "Holy Grail" of evolutionary biology (Koehne et al. 1983). Early examples of this convergence in the bacterial literature include works by Dykhuizen, Dean, and colleagues (Hartl et al. 1985; Dykhuizen et al. 1987), who used the language of metabolic control theory (Kacser and Burns 1973, 1981) to enable experimental evolution to shed light on the selectionist-neutralist controversy (Kimura 1968; Lewontin 1974; see also Dykhuizen and Dean this volume).

Development of high-throughput sequencing and tools for whole-genome analysis have accelerated this convergence of approaches, bringing us closer to the goal of understanding how genetic variation maps onto physiological performance and fitness in bacteria (see figure 13.2). Recent studies integrating experimental bacterial evolution and genomics have provided unprecedented insight into molecular mechanisms that underlie parallelism and convergence (Cooper et al. 2004, 2008; Fong et al. 2005; Woods et al. 2006), the relative contributions of structural and regulatory mutations to the adaptive process (Honish et al. 2004; Pelosi et al. 2006; Herring et al. 2006; Phillipe et al. 2006; Hegreness and Kishnoy 2007; Spencer et al. 2007), how often adaptation occurs by gain (Riehle et al. 2001, 2005) or loss (Maughan et al. 2007) of function, as well as how adaptation may be achieved via major changes in genome structure (Schneider et al. 2000; deVisser et al. 2004; Faure et al. 2004; Zhong et al. 2004; Nilsson et al. 2005). In certain instances (Papin et al. 2003; Fong et al. 2005), global-scale metabolic modeling is providing a framework for interpreting expression data obtained from evolution experiments. And increasingly, experimental evolutionary studies are being informed and sometimes motivated by comparative genomic studies (e.g., Ochman and Santos 2005; Escobar-Paramo et al. 2006; Abby and Daubin 2007).

## HOW WIDESPREAD IS PARALLELISM IN ADAPTATION?

In 1988, Richard Lenski and colleagues embarked on an immensely fruitful, long-term study of bacterial evolution in continuously varying environments (Lenski et al. 1991; Lenski and Travisano 1994; Lenski 2004; see also Travisano this volume). Twelve

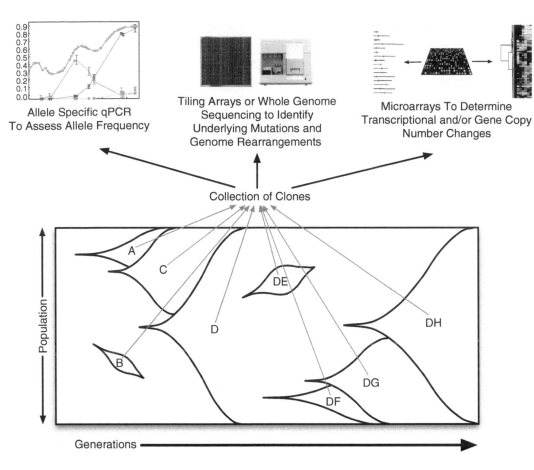

0.9
0.8
0.7
0.6
0.5
0.4
0.3
0.2
0.1
0.0

Allele Specific qPCR
To Assess Allele Frequency

Tiling Arrays or Whole Genome
Sequencing to Identify
Underlying Mutations and
Genome Rearrangements

Microarrays To Determine
Transcriptional and/or Gene Copy
Number Changes

Collection of Clones

A

C

DE

DH

D

B

DG

DF

Population

Generations

FIGURE 13.2

Overview of how molecular and genomic technologies can be used to determine mechanisms underlying adaptation during experimental evolution. Bottom panel: In this case, a large asexually reproducing population is evolving, and various adaptive mutations, A–H, arise. Mutants can easily be recovered, as a living record of evolutionary change is preserved within population samples previously archived as −80°C glycerol stocks (see figure 13.1). Adaptive sweeps occur; the ancestral genotype is replaced by D, which in turn is replaced by DH. Multiple clones may transiently coexist (e.g., D, DE, DF, DG) giving rise to clonal interference, which slows the tempo of evolutionary change. Alternatively, clones may coexist as a stable polymorphism (not shown; but see Rosenzweig et al. 1994; Rozen and Lenski 2002). Upper panels: Through the use of tiling microarrays or whole genome sequencing, the underlying mutations can be identified, and their frequency within the population can then be determined by allele-specific quantitative PCR. Whole-genome sequencing and array CGH can be used to determine gene copy number changes, as well as changes in chromosomal architecture, while expression microarrays can reveal the transcriptional consequences of the adaptive changes that have occurred.

experimental populations were founded by two *Escherichia coli* B strains that differed only in their genotype at the neutral arabinose-resistance marker. Bacteria were propagated by daily serial transfer of cells into fresh glucose-amended (25 $\mu$g/ml) Davis minimal media, so that the evolving populations experienced a regular cycle of nutrient excess and nutrient limitation. Because microbial populations can be cryogenically preserved, and fitness of individual clones assayed by pair-wise competition, these experiments (now at forty thousand generations) have provided unprecedented insight into the tempo and mode of adaptive evolution in asexual organisms, as well as the extent to which this process is convergent and/or subject to historical contingency (e.g., Blount et al. 2008). The advent of techniques for high-throughput sequencing and genomic analysis now make it possible to discover which molecular mechanisms underlie observed differences—and similarities—in fitness and physiology among ancestral and evolved strains.

Cooper et al. (2003) investigated changes in gene expression in these independent lineages, comparing adapted clones from twenty thousand generations to their ancestors. In two of these clones they observed significant and coordinated changes in fifty-nine genes, many of which were in the guanosine tetraphosphate (ppGpp) and cAMP-receptor protein (CRP) regulons (CRP was formerly known as CAP, catabolite activator protein). These observations led them to sequence, in each of their evolved strains, four candidate genes (*spoT* and *relA*; *cyaA* and *crp*) known to influence steady-state abundance of these effectors (see table 13.1 for a description of proteins and protein functions encoded by genes discussed herein). A missense mutation in *spoT* was discovered in one of the evolved clones. When placed in its ancestral background and tested using a laboratory competition experiment, this *spoT* mutation conferred increased fitness and predictably altered expression of several ppGpp-regulated genes. Cooper et al. (2003) further demonstrated that nonidentical, point mutations in *spoT* occurred in eight of twelve experimental populations, illustrating a high degree of genetic parallelism at the locus but not at the nucleotide level. (The generality of this finding has now been extended to *pykF* and *nadR*; Woods et al. 2006.) The observation of similar gene expression patterns in evolved clones that do *not* bear *spoT* mutations points to novel, undiscovered mechanisms for ppGpp regulation. This study illustrates how genomics enables us "to see more clearly through the experimental glass into evolution." Expression profiling reveals candidate loci that can be hypothesized to be under selection; then direct DNA sequencing enables these loci to be screened within and between populations. This approach leads to deeper understanding of the molecular underpinnings of both parallel evolution and ppGpp regulation.

This same general approach has now made it possible to more clearly define the role that *epistasis* plays in adaptive evolution. Cooper et al. (2003) had noted parallel changes in the expression of genes in the CRP regulon, but no change in *crp* coding sequences. Cooper et al. (2008) chose to delete *crp* in the *E. coli* B founder and in clones independently isolated from two lineages at twenty thousand generations, reasoning that the

TABLE 13.1  Glossary of Abbreviation Terms

| Acronym | Meaning of Term | Function | Reference |
|---|---|---|---|
| Genes | | | |
| *spoT* | ppGpp 3'-pyrophosphatase | Dephosphorylates global effector ppGpp to GDP; (p)ppGpp regulates stringent response (see later discussion) | Cooper, Rozen, et al. 2003 |
| *relA* | GDP pyrophosphokinase/GTP pyrophosphokinase | Biosynthesis of (p)ppGpp, regulator of the stringent response. Activated in response to amino acid starvation. | Cooper, Remold, et al. 2003 |
| *cyaA* | Adenylate cyclase | Catalyzes ATP → cAMP, which controls expression of many genes via cAMP receptor protein (CRP) | Cooper et al. 2003 |
| *crp* | CRP-cAMP | When bound to cAMP, binds to specific sites upstream of promoters, activating or repressing transcription | Cooper et al. 2003; Cooper et al. 2008 |
| *pykF* | Pyruvate kinase I | Catalyzes conversion of phosphoenolpyruvate to pyruvate; final step in glycolysis | Woods et al. 2006; Philippe et al. 2007 |
| *nadR* | Transcriptional repressor and kinase/transferase | Multifunctional protein with regulatory and catalytic domains: regulation of NAD biosynthesis, NMN adenylyltransferase, ribosylnicotinamide kinase | Woods et al. 2006 |
| *glpK* | Glycerol kinase | Essential first step in glycerol metabolism | Herring et al. 2006 |
| *rpoBC* | RNA polymerase, $\beta$ and $\beta'$ subunits | DNA-directed RNA polymerase, $\beta$ chain; has "flexible flap" element that contacts $\sigma$ factors | Herring et al. 2006 |
| *rpoS* | Sigma factor, $\sigma^s$ | Starvation-inducible transcription factor; regulates expression of >70 stress resistance genes or operons | Riehle et al. 2001; Maharjan et al. 2006 |

| Gene | Protein | Function | Reference |
|---|---|---|---|
| *rpoD* | Sigma factor, $\sigma^D$ | Regulates transcription of nutrient scavenging genes | Ferenci 2003 |
| *topA* | Topoisomerase I | Global control of DNA supercoiling, which in turn exerts widespread control over gene expression | Mallik et al. 2004 |
| *fis* | DNA-binding protein; histone-like global regulator | Activates ribosomal RNA transcription; prevents DNA replication from OriC | Mallik et al. 2004 |
| *gyrAB* | DNA gyrase subunits A and B | Negatively supercoils closed circular dsDNA in an ATP-dependent manner | Mallik et al. 2004 |
| *matT* | Maltose regulator protein | Transcription factor sensitive to catabolite control; activates several maltose catabolism genes and operons | Pelosi et al. 2006 |
| *rbs* | Ribose ABC transporter | Multimeric proteins that mediate ATP-dependent transport of ribopyranose across plasma membrane | Pelosi et al. 2006 |
| PTS | Phosphotransferase system | Multienzyme phosphoenolpyruvate-dependent system that mediates uptake and phosphorylation of a variety of carbohydrates, including glucose | Rozen et al. 2002; Ostrowski et al. 2005 |
| *lacI* | LacI transcriptional repressor | Regulates several genes involved in lactose catabolism; represses lacZYA operon; binds allolactose autoinducer | Zhong et al. 2004 |
| *lacY* | Lactose MFS transporter | Mediates proton-dependent transport of lactose across cell membrane | Zhong et al. 2004 |
| *galS* | GalSp transcriptional dual regulator | Transcriptional regulator of high-affinity galactose transport; autogenously regulated and CRP-dependent | Zhong et al. 2004 |
| *mglA* | Subunit of galactose ABC transporter | Helps mediate ATP-dependent transport of galactose across cell membrane | Zhong et al. 2004 |
| *pilQ* | Type IV pilus secretin homolog | Secretin homolog required for social motility in *Myxococcus xanthus* | Velicer et al. 2002 |

TABLE 13.1 *(continued)*

| Acronym | Meaning of Term | Function | Reference |
|---|---|---|---|
| *sgaA* | GASP gene created by insertion/inversion event | Activates putative amino acid transport operon *ypeJ-gltJKL-ybeK* by placing *cstA* CRP-binding site upstream | Zinser et al. 2003 |
| *ypeJ-gltJKL-ybeK* | Glutamate ABC transporter (*ybeJ* only) | ATP-dependent transport of glutamate from periplasm to cytosol | Zinser et al. 2003 |
| *cstA* | Oligopeptide permease | Peptide transporter induced by carbon starvation | Zinser et al. 2003 |
| *nlpD* | Putative outer membrane lipoprotein | May be involved in stationary phase survival | Riehle et al. 2001 |
| *pcm* | L-isoaspartate protein carboxyl-methyltransferase Type II | Involved in protein modification and repair | Riehle et al. 2001 |
| *surE* | Broad specificity 5'(3')-nucleotidase and polyphosphatase | Functions in water homeostasis; catalyzses hydrolysis of nucleosides and ribonucleosides | Riehle et al. 2001 |
| *iap* | Alkaline phophatase isozyme conversion protein | Involved in protein modification, including proteolysis | Riehle et al. 2001 |
| *prpC* | Subunit of methylcitrate synthase | Hydrolyses oxaloacetate with propionyl-CoA to form methylcitrate | Zhong et al. 2004 |
| *yedX* | Conserved protein, family of transthyretin-related proteins | Transport function inferred by homology | Zong et al. 2004 |
| *yedY* | Periplasmic molybdoprotein | Periplasmic sulfoxide reductase | Zong et al. 2004 |

| | | | |
|---|---|---|---|
| *proBA* | Subunits of the γ-glutamyl kinase-GP-reductase multienzyme complex | Catalyses ATP-dependent conversions of L-glutamate required for proline biosynthesis | Faure et al. 2004 |
| *crl* | Crlp protein | Helps transcription factor $\sigma^s$ outcompete $\sigma^{70}$ at RNApcl | Faure et al. 2004; Typas et al. 2007 |
| *dinB* | DNApolymerase IV | Nonproofreading DNA polymerase; does not contribute to chromosomal mutation rate under normal conditions; may enhance cell survival during nucleotide starvation | Faure et al. 2004 |
| *acs* | Acetyl CoA synthetase | Low-affinity acetate scavenging enzyme; converts coenzyme A, ATP, and acetate to acetyl-CoA | Rosenzweig et al. 1994; Treves et al. 1998 |
| *hxt6/hxt7* | High-affinity glucose transporters | High-affinity glucose transporters of major facilitator superfamily; hxt6 and hxt7 are tandemly arrayed and encode nearly identical proteins | Brown et al. 1998 |
| *cit1* | Mitochondrial citrate synthase | Catalyzes condensation of acetyl coA and oxaloacetate to form citrate; rate-limiting enzyme of the TCA cycle; nuclear-encoded mitochondrial protein | Dunham et al. 2002 |
| *gal80* | Transcriptional factor | Repression of GAL genes required for galactose utilization in the absence of galactose | Segre et al. 2006 |
| Effector molecules | | | |
| (p)ppGpp | guanosine-5'-(tri)diphosphate-3'-diphosphate | Major effectors of stringent response to amino acid depletion in *E. coli*; repress genes required for ribosome biosynthesis; activate scores of genes needed to cope with stress and starvation | Cashel et al. 1996 |
| CRP | cAMP receptor protein (formerly catabolite activator protein [CAP]) | Activated by cAMP in response to glucose starvation and other stresses; CRP-dependent glucose activation and repression regulates expression of >70genes in *E. coli* | Martinez-Antonio Collado-Vides 2003; Gosset et al. 2004 |

TABLE 13.1 (*continued*)

| Acronym | Meaning of Term | Function | Reference |
|---|---|---|---|
| Techniques | | | |
| a-CGH | Array-based comparative genomic hybridization | DNA microarray-based technique for detecting loss and/or gain of gene copy number by comparing the ratio of signal intensities from differently labeled reference and test samples of total genomic DNA (see http://www.chem.agilent.com/scripts/LiteraturePDF.asp?iWHID=44844) | Dunham et al. 2002 |
| CGS | Comparative genome sequencing | CGS is a two-phase process. First, regions of genomic difference are identified by comparative hybridization of test DNA versus reference DNA on a whole-genome tiling array. Then, identified regions of genomic difference are sequenced to produce a set of fully characterized SNPs (see http://www.nimblegen.com/products/cgr/). | |
| SAM | Significance analysis of microarrays | Permutation-based method that relies on variance information present in measurements obtained from all probes on a microarray. Available as an Excel add-in for identifying genes with significantly differing expression levels between sets of samples; a priori hypothesis is that some genes will have significantly different mean expression levels between different sets of samples (see http://lane.stanford.edu/howto/index.html?id=_1906). | Tusher et al. 2001 |

relative impact of these lesions on the CRP regulatory network would uncover modifiers that had undergone evolutionary change. As expected, both maximum specific growth rate and global expression patterns were profoundly altered in all three deletants. However, these effects were far more pronounced in the ancestor. A Venn diagram of observed changes in gene expression enabled them to distinguish the ancestral regulon (consisting of 171 differentially regulated genes), from a smaller core regulon (consisting of 25 genes altered in expression among *all* three strains), and a much larger meta-regulon (consisting of 1,089 genes altered in *any* of the three strains). These analyses also revealed a group of 117 genes whose change in expression was unique to the evolved clones. Consistent with their previous report of parallelism, 115 of these changes had similar directionality, all being up-regulated. The authors then used a systems biology approach to draw inferences concerning *crp*-dependent changes, placing them in the context of the characterized *E. coli* transcriptional network, a curated database of interactions comprising 1,217 genes, including 135 genes that mediate 2,333 transcriptional interactions. They discovered that 20 of 135 regulons changed independently in *both* of the two evolved strains chosen for detailed analysis, and that of these 20, 14 changed in common and in the same direction, 12 becoming *less* sensitive to *crp* deletion! By combining experimental evolution with genomics and systems biology, Cooper et al. (2008) reveal how genes are recruited to and uncoupled from regulatory networks during the course of adaptive evolution: parallel changes occur not just at the level of individual genes but also epistatically at the level of individual regulons, at least in an experimental system where recombination was forestalled.

Theoretically, at least, the combinatorics of metabolic pathway interactions implies that multiple gene expression states can give rise to a common phenotype (Mahadevan and Schilling 2003). Fong et al. (2005) provide empirical evidence to support this prediction. Seven replicate populations founded by wild-type *E. coli* K12 were experimentally evolved by serial dilution in either glycerol- or lactate-amended minimal media, for six hundred and one thousand generations, respectively (fourteen total populations). In each case, the selection regime resulted in more than a twofold increase in growth rate, in association with dramatic increases in substrate uptake and oxygen consumption. Phenotypic convergence was manifest in low variance among replicates for each of these physiological parameters, among replicates within selection regimes. Nevertheless, replicate populations varied widely in their patterns of gene expression as well as in the adaptive trajectories each took in converging on these phenotypes, as inferred from physiological performance on alternative carbon sources that were not present in the culture environments employed during selection. Another conspicuous feature of evolution on both lactate and glycerol was the observation of early (day 1) widespread changes in gene expression, relative to the profile of glucose-grown K12, followed by what appear to be compensatory changes that reduce the number of significant gene expression differences observed at later time points, relative to glucose-grown cells as well as cells from day 1 grown on the evolutionary substrate. Replicate endpoint clones nevertheless exhibit different patterns of gene expression: excluding compensatory changes, the average number of evolutionarily significant changes was estimated to be 1,109 and 203 for glycerol and lactate evolution experiments, respectively. Of these, few were shared among replicates: seventy genes showed changes in common among six of seven glycerol selection replicates compared to only two in six of seven lactate selection replicates!

Herring et al. (2006) used a novel microarray-based technique for whole-genome resequencing to identify mutations in Fong et al.'s glycerol-selected populations (for a review, see Hegreness and Kishony 2007). Comparative Genome Sequencing (CGS) (Albert et al. 2005) of terminal isolates from five of the seven populations revealed candidate SNPs and probe-level differences. Thirteen of these were confirmed by PCR amplification and Sanger sequencing to be *de novo* mutations that arose during the course of about six hundred generations of experimental evolution. The selective value of these mutations was demonstrated by introducing them into the wild-type ancestor, singly and in tandem. Because no one evolved strain contained more than three of these mutations, it was possible to evaluate systematically the additive effect of these alleles in a common background. Remarkably, evolved strains' growth rates were reproduced in four of five cases of deliberate allele substitution, indicating not only successful mutation recovery but that each recovered allele was adaptive and that adaptive evolution on glycerol occurs in the absence of epistasis. (The one exception contained a 1.3-kb rearrangement, which the authors did not attempt to reconstruct.) Two of five glycerol-evolved clones contained mutations in RNA polymerase $\beta'$ (*rpoC*), and all contained mutations in glycerol kinase (*glpK*), the first step in glycerol metabolism. Although every mutation was unique, partially purified mutant GlpK proteins all showed dramatic increases (51–133 percent) in their $V_{max}$ for glycerol phosphorlyation, changes that likely account for their strong selective value in populations evolving on this nonfermentable carbon source.

Using MALDI-TOF mass spectrometry-based estimates of changes in allele frequencies, Herring et al. (2006) also assayed the evolutionary dynamics of all thirteen mutations in the five populations Fong et al. evolved on glycerol. As might be expected, mutations in *glpK* and *rpoBC* that confer large increases in fitness arose earliest and went quickly to fixation, whereas mutations that provide more incremental increases arose later and took longer to sweep through experimental populations. Interestingly, sequencing of *glpK* and *rpoC* in multiple clones isolated from each experiment on day 15 revealed several alleles that failed to go to fixation, including those where a nucleotide substitution affected the same amino acid in *glpK*. Remarkably, at approximately 180 generations in one 600-generation experiment, Herring et al. (2006) discovered an *rpoC* allele that eventually "won" paired with a *glpK* mutation that did not. Clearly, there are alternative scripts to the "evolutionary play," and it seems likely that epistatic interactions influence which is eventually chosen for production. Another important conclusion we may draw from this study is that there are two general mechanisms by which cells growing under resource limitation obtain large increases in fitness: mutation at so-called structural genes like *glpK* that initiate catabolism of the limiting resource, and mutation at regulatory genes like *rpoB* and *ropC* that promote widespread adaptive change in gene expression. These strategies are not mutually exclusive: epistasis may determine which specific pairs of alleles are favored, but successive structural and regulatory mutations can become fixed within the same lineage. Analogizing to the metazoa, these kinds of structural changes make better teeth, beaks, and claws for resource acquisition, while regulatory changes lead to organ-level improvements in resource processing!

## STRUCTURAL VERSUS REGULATORY MUTATION: WHICH CONTRIBUTES MORE TO ADAPTATION?

The tempo of evolutionary change in bacterial populations can be inferred from changes in the frequency of marker allele(s) determined to be selectively neutral in the experimental environment (Novick and Szilard 1950; Atwood et al. 1951). Using resistance to bacteriophage T5 as such a marker, Adams and Helling (1983) estimate that new adaptive clones arise in glucose-limited *E. coli* populations on the order of once every hundred generations. Adaptive mutations do not necessarily accumulate within one lineage; indeed, they may be segregated within multiple coevolving lineages (Helling et al. 1987; Rosenzweig et al. 1994). Proteomic analyses of one such polymorphic population reveal widespread differences in patterns of gene expression among adaptive strains, relative to one another and to their common ancestor (Kurlandzka et al. 1991). Of approximately 1,200 gene products resolved and identified by two-dimensional PAGE, about 20 percent vary significantly among ancestral and evolved clones isolated after approximately eight hundred generations. The most parsimonious explanation for these observations is that the few (less than ten) adaptive mutations fixed in this population occurred at regulatory rather than at structural loci. In this regard, investigations of chemostat evolution under glucose limitation by Ferenci and colleagues (Notley-McRobb et al. 2002, 2003; Maharjan et al. 2006) demonstrate strong adaptive value associated with mutations in *rpoS*, a global regulator of about 10 percent of the *E. coli* genome (see the later section "Does Loss of Function Result from Selection or Drift?"). An alternative explanation is that structural mutations either cause or relieve metabolic bottlenecks, and that the resulting changes in flux are fed back to the genome via changes in the pool size of effector molecules that can signal changes in the rate and/or specificity of gene expression. Examples of such global effector molecules include pyruvate and glyoxylate (Lorca et al. 2007), glutamine (Magasanik 2000), and the small nucleotide (p)ppGpp (guanosine-5'-diphosphate-3'-diphosphate and guanosine-5'-triphosphate-3'-diphosphate) (Cashel et al. 1996) (see table 13.1).

The Lenski experimental populations have now evolved for thirty years, more than an order of magnitude longer than the longest chemostat studies. Indeed, the Lenski experiments now span in human generations the length of time between the emergence of *Homo erectus* and the present! However, a consideration of the effects of drift and natural selection suggests that no more than one hundred mutations have become fixed in any one adaptive lineage (Lenski 2004; Philippe et al. 2007). Increasing evidence from this long-term study suggests that global regulatory mutations may bring about the greatest overall gains in fitness. Such mutations could be expected to fall into two general classes of genes: those that exert direct control over gene expression (e.g., transcription factors and topoisomerases), and those that exert indirect control by altering pool size(s) of global effector molecules. We have already encountered examples of the latter; genome-enabled analyses indicate that adaptive evolution in the ongoing Lenski experiments is mediated in part by genes that regulate CRP and ppGpp levels (Cooper et al. 2003, 2008; Woods et al. 2006). Regulatory mutations that alter DNA supercoiling also appear to play a key role, presumably because changes in this parameter facilitate transitions between reproduction and starvation (Hatfield and Benham 2002), transitions that are daily forced on populations in serial-dilution batch culture. Ten of the twelve populations of the Lenski populations show increased supercoiling (Crozat et al. 2005), and in each case that they examined, this trait was associated with mutations at either or both *topA* and *fis*. These loci encode, respectively, topoisomerase I and histone-like global regulators, the latter of which can couple transcription to nutrient status either via interaction with *topA* gene product or by repression of *gyrAB* (Mallik et al. 2004).

Structural mutations are also observed in these populations—specifically, deletions in *malT* activator and *rbs* (Pelosi et al. 2006). And fitness benefits may accrue to lineages bearing such deletions because their cells conserve energy and material resources that might otherwise be unproductively allocated to ribose and maltose catabolism. These observations have given rise to a view of evolution in experimental bacterial populations proceeding via changes at "hubs and spokes" (i.e., mutations at global and local regulators). The latter may consist of changes in structural genes whose effects are limited to specific core metabolic processes, such as *pykF* (pyruvate kinase) (Woods et al. 2006; Philippe et al. 2007).

By contrast, when Honisch et al. (2004) and Herring et al. (2006) comprehensively screened replicate *E. coli* populations evolved on glycerol, they observed *glpK* mutants in every lineage. In each case, these structural mutants demonstrated increased $V_{max}$ for glycerol kinase, and markedly elevated growth rates on the evolutionary substrate. This observation is reminiscent of observations in the older literature that structural mutations quickly arise in bacterial populations selected for growth on novel, even xenobiotic, substrates. These mutations may alter either or both of (1) transport of the novel substrate and (2) specificity and activity of gene products that introduce that substrate into central metabolism (e.g., Mortlock 1984 and references therein). Indeed, Herring et al. (2006) partly attribute their findings to the fact that, for all practical purposes, glycerol is an unfamiliar substrate for *E. coli* K12, which has been serially passaged for decades on rich media containing fermentable carbon.

Clearly, while it is tempting to claim that regulatory mutations have greater evolutionary potential, even in prokaryotes, evidence to date from experimental evolutionary genomic studies is equivocal. Depending on the selection pressure and the position of genes in metabolic networks, mutations at so-called structural loci can confer strong adaptive advantages. These findings are consistent with recent analyses indicating that adaptation and speciation are likely driven by a combination of structural and *cis*-regulatory mutations (Hoekstra and Coyne 2007), a finding that is consistent with observation that both types of mutations can go to fixation in the same lineage (Herring et al. 2006).

## DOES ADAPTIVE EVOLUTION REQUIRE TRADE-OFFS?

By necessity, microbial evolution experiments are performed by continuous culture of organisms either in defined solid (Nilsson et al. 2005) or liquid media, or, alternatively, by serial passage within a host organism (Ebert 1998; Nilsson et al. 2004) (see figure 13.1). When experiments are conducted in defined media, selection typically acts via nutrient limitation, imposed either *continuously* (Ferenci 2008) or *discontinuously* (Lenski 2004), in the form of one or, more rarely, two growth-supporting substrates (e.g., Zhong et al. 2004; Spencer et al. 2007). When nutrient limitation is intermittent, its severity (Zinser et al. 2003) or periodicity (de Visser et al. 2004) can also be varied. Additional selective pressures can be superimposed on these experimental designs, by culturing populations at nonoptimal temperatures (Bennett and Lenski 1993; Riehle et al. 2005), or by using conditions where bacteria may or may not be required to retain sporulation competence (Maughan et al. 2007) and social motility (Velicer et al. 2002, 2006). Heritable phenotypic changes that result from evolution in response to such manipulations have enlarged our understanding of the trade-offs and constraints that guide the adaptive process. Genomic analyses now promise to elucidate their mechanistic bases.

When we select on a population for many generations to grow more rapidly or more efficiently on a single resource, we expect to favor lineages that are less able to utilize alternative substrates. In other words, we tend to view specialist and generalist lifestyles as being mutually incompatible. This incompatibility can arise by two very different genetic mechanisms. Functionality of catabolic pathways not under selection may degenerate over time as they gradually accumulate neutral changes *(mutation accumulation)*. Alternatively, key elements of those same pathways may be co-opted to serve in pathways that are under selection *(antagonistic pleiotropy)*. For the latter mechanism to come into play, those elements must lose some degree of their original functionality under positive selection. Cooper and Lenski (2000) tested these two possibilities by assaying the growth of clones from twelve independent evolutionary lineages on sixty-four substrates. Each of these lineages had been subjected to discontinuous glucose limitation for twenty thousand generations. If specialization resulted from the accumulation of neutral alleles in pathways not under selection, then one would expect populations to lose function on different nonevolutionary substrates at different rates. Cooper and Lenski (2000), however, found that total catabolic function on alternative substrates decayed rapidly and in parallel, supporting the view that antagonistic pleiotropy attends adaptation under these experimental conditions.

Although expression profiling points to the importance of pathways encompassed within the ppGpp and CRP regulons (Cooper et al. 2003, 2008), the exact genetic basis for these observations remains obscure. In fact, a candidate gene approach has revealed widespread *agonistic* pleiotropy! Evolved clones that were isolated after fewer than four hundred generations, and that contained no more than one spontaneous beneficial mutation, were individually competed against their common ancestor on five novel carbon sources. In most instances, mutants showed *increased* fitness, as well as substantial heterogeneity in fitness effects on PTS versus non-PTS growth substrates (Rozen et al. 2002; Ostrowski et al. 2005; Ostrowski et al. 2008). (The PTS, or phosphotransferase system in *E. coli*, is a multienzyme complex that mediates the transport and initial phosphorylation of certain sugars, such as glucose; see table 13.1.) It remains to be seen whether these seemingly contradictory results can be explained by differences in the number and types of novel carbon sources assayed, early- versus late-arising mutations, or the imposition of conditions needed to follow clonal interference in the later experiments.

Using an explicitly genomic approach, Zhong et al. (2004) investigated constraints on the evolution of specialists and generalists under continuous carbon limitation. Chemostat cultures were founded by one of two strains differing only in *lac* alleles that confer higher fitness on either lactulose or methyl-galactoside. Populations were cultured for 190 to 598 generations in media containing either lactulose, methyl-galactoside, or a 72:28 mixture of both sugars. Array-based comparative genomic hybridization (a-CGH) revealed that adaptation to one or the other carbon source occurred with a high degree of specificity and repeatability. Evolution on lactulose commonly resulted in *IS-2* mediated gene duplications of variable length around *lac* that increased *lacY* copy number by at least threefold. By contrast, evolution on methyl-galactoside commonly resulted in *IS*-element insertions into *galS* and *mglA*, mutations expected either to promote constitutive expression of the methyl-galactoside-specific transporter or to abolish its function altogether, forcing the sugar through an alternative uptake pathway. Remarkably, from among all populations evolved on mixed substrates Zhong et al. (2004) recovered only one mutant having higher fitness on *both* sugars. In most of their experiments, they isolated pairs of lactulose and methyl-galactoside specialists, each of which was genotypically similar to clones evolved on one or the other pure substrate. Thus, while there is clearly an evolutionary pathway by which generalists can evolve, generalists are nevertheless rare. To explain these findings, Zhong et al. invoke antagonistic pleiotropy, perhaps arising from metabolic problems associated with simultaneous overexpression of *mgl* and *lac*.

## DOES LOSS OF FUNCTION RESULT FROM SELECTION OR DRIFT?

Adaptive evolution may result in loss of function, either because certain traits are selectively disfavored in the experimental environment, or because selection-independent genetic drift causes mutations to accumulate in neutral or nearly neutral genes. Recent theoretical work suggests that antagonistic pleiotropy is more prevalent in large microbial populations with wild-type mutation rates, while mutational degradation is preponderant in populations with small effective population sizes and/or mutation rates (Masel et al. 2007). These predictions are supported by experimental evolution of wild-type *E. coli* under different thermal regimes (Cooper et al. 2001), as well as under conditions where the decay of unused catabolic functions is also monitored (Cooper and Lenski 2000). The number of genes underlying a trait of interest may also determine whether selection or drift leads to loss of function, as highly polygenic traits can be expected to provide a larger mutational target.

A characteristic feature of the life history of gram-positive bacteria is a developmental program, induced by deteriorating environmental conditions, that leads a cell to produce a metabolically quiescent, stress-resistant spore. Experimental evolution of *Bacillus subtilis* in rich medium repeatedly (but not invariably) results in the evolution of cells that cannot sporulate (Maughan et al. 2006). While 5 percent of the *B. subtilis* genome (at least 210 genes) is required to execute this program (Piggot and Losick 2002), transcriptional profiling indicates that more than 20 percent of these genes are also highly expressed under noninducing conditions (Maughan et al. 2007). This phenomenon seemingly raises the possibility that selection will favor gene loss because of antagonistic pleiotropy. Moreover, previous work indicated that there may be trade-offs between maximizing growth rate and/or flux of growth-supporting substrate and retaining sporulation ability (Maughan et al. 2006; Fischer and Sauer 2005). Using a statistical genetic approach, Maughan et al. (2007) conclude that mutational degradation best explains loss of sporulation ability in four of five populations experimental evolved for six thousand generations in sporulation-repressing medium. Even though effective population size ($N_e$) is large, (~$1 \times 10^7$ cells), the highly polygenic nature of this trait offers a large mutational target. Weak mutators can arise in these populations; these are novel variants that have background mutation rates greater than $10^{-8}$ bp per generation, owing to defects in DNA metabolism. Under their influence, Spo+ to Spo− mutation rate increases to about 0.003 (Maughan and Masel 2007). Loss of sporulation ability in one experimental *Bacillus* population *is* therefore attributable to selection, though the exact basis for antagonistic pleiotropy has yet to be established.

Other evidence indicates that experimental evolution of bacteria under benign conditions favors loss of activities required for resistance to stress. In *E. coli*, loss-of-function mutations in *rpoS* rapidly proliferate when cells are evolved under glucose limitation, even though *rpoS⁻* mutants are strongly disadvantaged during prolonged starvation (Ferenci 2008). In this case, the basis for antagonistic pleiotropy apparently relates to alleviating competition between the RpoS sigma factor ($\sigma^s$) required for transcription of stress resistance genes, and the RpoD sigma factor ($\sigma^D$) used to transcribe genes associated with scavenging nutrients to support vegetative growth (Ferenci 2003).

Social bacteria in the genus *Myxococcus* have a complex life history in which one stage is reminiscent of Metazoa and Metaphyta in terms of cellular differentiation and division of labor. In a deteriorating environment, motile *Myxococcus* cells aggregate to form spore-bearing fruiting bodies whose color and morphology are highly species-specific. Velicer et al. (1998) report that when wild-type *Myxococcus xanthus* is experimentally evolved for more than one thousand generations in nutrient-rich media, selection favors mutants that increase growth rate, but ignores or disfavors those that mediate social development leading to stress-resistant myxospores. Adaptive clones repeatedly evolve that have higher growth rates but wholly or partly lack the social and developmental pathways required to produce fruiting bodies. This result suggests that during evolution in a nutrient-rich environment, selection is relaxed on genes that encode for social motility and sporulation, making possible the accumulation of neutral mutations that lead to loss of these functions. Remarkably, certain adaptive clones can "cheat" when placed in mixed starvation cultures with socially competent cells (Velicer et al. 2000); a subset of these regain social proficiency following only six rounds of starvation and growth (Fiegna and Velicer 2003).

Using whole-genome sequencing Velicer et al. (2006) uncovered all mutations in clones isolated from a two-stage evolution experiment: wild-type to "cheater," and "cheater" to "phoenix" (a socially proficient adaptive mutant). Conspicuously absent was evidence for large-scale changes in genome architecture such as transpositions, duplications, and multiple-base deletions. Instead, the first stage one thousand generations of growth in rich media) was marked by fourteen single-base mutations, the second stage (sixty generations alternating starvation and growth) by one. Eleven mutations occurred in protein-coding regions, only one of which was synonymous, suggesting that most substitutions are nonneutral. Nine mutations occurred at annotated loci, notably *pilQ*, which encodes a PilQ secretin homologue that is required for social motility, and whose activity places a decrement on fitness in a resource-nonlimiting environment (Velicer et al. 2002).

Interestingly, the single compensatory mutation that distinguishes "cheater" from "phoenix" occurs in a noncoding region 128 bases upstream of a GNAT family acetyltransferase. The role this gene plays in restoring cooperative behavior is unclear. Global expression profiling indicates that its expression as well as that of many other genes in "phoenix" significantly differs from its progenitor, "cheater," and their common wild-type ancestor (Fiegna et al. 2006). These results not only illustrate how genomic technologies can help us discover the "genes that matter," but also how fine-scale alteration in genome structure can result in widespread changes in gene expression and cell phenotype.

## HOW DO ADAPTATIONS ARISE FROM LARGE-SCALE CHANGES IN GENOME STRUCTURE?

Single base substitutions can exert enormous effects on an organism's fitness; they can even turn a "cheater" into a "phoenix"! However, adaptation in experimentally evolving bacteria can also be mediated by larger-scale changes such as deletions, duplications, and rearrangements driven by mobile elements, especially insertion sequences. Rapid and continuous evolution occurs in prolonged starvation cultures of both bacteria (Finkel and Kolter 1999) and yeast (Coyle and Kroll 2008). In starving *E. coli*, adaptation has been attributed to so-called GASP genes (growth advantage in stationary phase) that when mutated confer the ability to invade wild-type populations in stationary phase. Perhaps not surprisingly, one GASP mutant is an allele of *rpoS* (Zinser and Kolter 2000). However, another GASP mutant, *sgaA*, arises via a two-step genomic rearrangement mediated by the mobile insertion element, *IS5* (see following paragraph). This insertion/inversion event activates the *ybeJ-gltJKL-ybeK* operon by placing upstream a CRP-binding element expropriated from *cstA* encoding oligopeptide permease (Zinser et al. 2003). In this remarkable example, functional trade-offs occur instantaneously: enhanced capacity to scavenge individual amino acids is purchased at the expense of the ability to take up small polypeptides.

Insertion sequence (*IS*) elements are short (700–2,500bp) DNA sequences that usually contain two genes that are flanked upstream and downstream by a pair of inverted repeat (IR) sequences. In an active *IS* element, the two genes encode for transposase activity and regulation of that activity. *IS* elements are a conspicuous feature of most bacterial genomes, and more than five hundred different types of *IS* sequences have been described to date (Schneider and Lenski 2004). Due to their flanking IR sequences, *IS* elements are highly recombinogenic and account for a large proportion of all spontaneous mutations, especially large-scale chromosome rearrangements such as inversions, duplications, and deletions (Blot 1994; Hall 1999). Movement of an *IS* element into a gene's coding region effectively silences that gene. But because *IS* elements contain outwardly directed promoters, insertion of an *IS* element into an upstream regulatory region may also result in gene activation.

It has long been clear that *IS* elements are highly successful genetic parasites. Over the past decade, a plethora of experimental studies has made it equally clear that *IS* elements actively—and sometimes beneficially—reorganize genome architecture under selection. Indeed, when experimental evolutionists seek to uncover the molecular genetic basis for adaptive change, they frequently encounter the ghost of *IS* past. (Of course, they are unlikely to recover deleterious changes caused by these elements.) *IS* activity appears to underlie adaptation to oxygen stress in *Lactococcus lactis* (de Visser et al. 2004), to growth at nonoptimal temperatures in *E. coli* (Riehle et al. 2001), to discontinuous (Schneider et al. 2000) and continuous carbon limitation (Treves et al. 1998; Zhong et al. 2004), as well as to prolonged starvation (Naas et al. 1994). *IS* elements also underlie adaptive remodeling of the *Salmonella typhimurium* genome during serial passage on solid media and through mice (Nilsson et al. 2004, 2005). Mira et al. (2006) hypothesize that the recent expansions of *IS* elements found in human-associated pathogens, but absent from their non-human-associated relatives, arise from selective pressures caused by the post-Neolithic increase in human population size. The comparative genomic evidence is striking. Analysis of paralogous genes across 255 fully sequenced prokaryotic genomes reveals that 69 out of 89 bacteria in which more than 75 percent of paralogues correspond to *IS* elements are microbes uniquely associated with humans.

Because the sequence is known for many *IS* elements, it is relatively straightforward to produce by Southern analysis "*IS*-element-specific fingerprints" of strains isolated during an evolution experiment. Band migration and/or the appearance of new bands reveal *IS* activity and enable direct sequencing of *IS*-associated mutations. Such tried-and-true approaches have been greatly augmented by the advent of DNA microarray-based comparative genomic hybridization (a-CGH) (Dunham et al. 2002; Zhong et al. 2004; Pinkel and Albertson 2005). a-CGH now makes it possible to systematically screen for the deletion or amplification of every gene for which there is a probe on the array, and it can be used in conjunction with expression profiling to gain insight into the possible phenotypic consequences of such changes.

These complementary genome-wide methods were first used together in experimental evolution to illuminate how genome architecture changes as *E. coli* adapt to nonoptimal temperature (Riehle et al. 2001). Beginning with a common ancestor, twelve lines were propagated for two thousand generations by daily serial passage in defined glucose minimal media, six at 41.5°C and six at 37°C. Five duplication and deletion events were detected in three of six high-temperature lines, four of which were consistent with accretion and deletions previously inferred by Pulsed Field Gel Electrophoresis (PFGE) of genomic DNA digested with either *Not*I or *Bln*I (Bergthorsson and Ochman 1999). a-CGH localized three duplications at 2.85 Mb, providing evidence for replicability of specific genome rearrangements. No such changes were observed in lines evolving at 37°C. Candidate genes duplicated during thermal adaptation include *rpoS*, *nlpD*, *pcm*, and *surE*, each of which shows higher levels of expression in duplication-containing lineages than in non-duplication-containing lineages and the common ancestor. Single-gene mutants of these genes perform poorly upon entry into stationary phase and, in certain instances, upon exposure to high temperature. The genome rearrangements described by Riehle et al. (2001) do *not* appear to arise via *IS* elements, even though *IS*186 and *IS*150 are near 2.85 Mb. Direct sequencing of the break points leads these authors to propose an alternative mechanism involving two homologous exchange events within a region of 29-bp repeats downstream of the *iap* locus. Their findings underscore the need to follow up a-CGH results with sequencing before making claims as to specific mechanisms underlying genome rearrangement.

Zhong et al. (2004) used a-CGH to illuminate the genetics underlying specialization of *E. coli* as it evolves in nutrient-limited chemostats. Examining forty-two evolvants from twenty-three populations, they detected eight unique duplications, ranging in size from eighteen to seventy-four genes. Each duplication resulted in amplification of thirteen genes between *prpC* and *lacI*. Zong et al. also detected sixteen unique deletions, ranging in size 18.3 to 45.0 kb, all ending between *yedX* and *yedY*. Direct sequencing of regions flanking large-scale changes revealed *IS* elements in all but one instance; vectorette PCR (vPCR) (Riley et al. 1990) uncovered an additional sixteen insertions. The widespread involvement of *IS* elements reported here contrasts with adaptation to elevated temperatures. This observation may relate to the relative sizes of targets for selection under different experimental regimens: Zhong et al. evolved cells on limiting concentrations of lactulose and/or methyl-galactoside, growth substrates whose transport and catabolism require a very specific set of genes located near *IS* elements, while Riehle et al. challenged cells to adapt to high temperature, the evolutionary response to which is likely to involve many more genes that may or may not be located near *IS* elements.

In addition, a-CGH has been used to investigate evolution of bacterial genome architecture in long-term stab cultures, small tubes containing sterile nutrient agar into which a microbial sample has been introduced using a sterile toothpick or flamed loop. Because cells in stab cultures are assumed to be nondividing or dividing at a negligible rate, they are sometimes used to archive strains. In *Salmonella enterica*, five of thirteen strains tested showed large-scale genome rearrangements relative to the Typhimurium LT2 reference genome; these include three small (one- or two-gene) deletions and two large amplifications, one involving fourteen genes, another a 180-kb fragment between two rRNA clusters (Porwollik et al. 2004). These changes were not attributed to any specific mechanism. Faure et al. (2004) followed up on previous observations of *IS* element activity in resting *E. coli* K12 (Naas et al. 1995), examining clones isolated from stab cultures maintained for thirty years and five years. a-CGH revealed 19.1-kb (Type I) and 23.4 (Type II) deletions in the genome of the five-year-old stab culture, relative to its ancestor. The Type I deletion included *proBA* and *crl*, resulting in proline auxotrophy and loss of Crlp, a regulatory protein that supports $\sigma^S$ (RpoS) in its competition with $\sigma^{70}$ for RNA polymerase during the transition to stationary phase (Typas et al. 2007). They also determined that the proline auxotroph phenotype was selectively favored in twelve independent stabs maintained for 2, 42, and 183 days. Southern analyses indicated that variants of Type I deletions were found in all thirty-eight mutants examined in these evolutionary series. *IS5* consistently flanked the right boundary; the left boundary varied over a 5-kb range, always including *crl* but never extending beyond the *dinB* locus which encodes DNA polymerase IV. Once again, *IS* elements appear to play a key role in genome rearrangement. The authors speculate that variation in deletion size is driven by positive selection both for eliminating the RpoS regulator CrlP, and retaining the key stress response protein, DNA Pol IV. This fascinating example should serve as a cautionary tale for anyone who thinks of stationary phase as evolutionarily uninteresting and/or who expects their stab cultures to remain isogenic!

Finally, our labs are using a-CGH and global gene expression profiling to gain insight into mechanisms underlying the emergence and persistence of stable polymorphisms in *E. coli* evolved under glucose limitation. Specifically, we are investigating changes in genome structure and steady-state mRNA abundance among clones initially described by Helling et al. (1987) and further reported on by Kurlandzka et al. (1991) and Rosenzweig et al. (1994). As predicted from specific activity data, and consistent with the report of Treves et al. (1998), microarray analyses indicate that *acs*, the gene encoding acetyl coA synthestase, is highly expressed in clone CV101, enabling it to exploit acetate excreted by the dominant clone, CV103 (M. Kinnersley, W. Holben and F. Rosenzweig, unpublished manuscript). Remarkably, 94 percent of the transcriptome is unchanged in adapted clones relative to their ancestor. Among those transcripts that do differ significantly, almost twice as many genes are down-regulated than up-regulated. A majority of these are CRP regulated, consistent with the findings of Cooper et al. (2003) described earlier. Equally remarkable, and in marked contrast with several studies noted herein, members of the stable polymorphism are not strongly differentiated with respect to large-scale genome rearrangements. The sole exception is a 29-kb deletion in the dominant clone involving twenty-eight genes (M. Kinnersley, W. Holben, and F. Rosenzweig, unpublished manuscript). As about half of these genes are involved in anaerobic respiration, we contend that they are likely dispensable under aerobic glucose limitation. Of course, the CV103 deletion would be strongly disadvantageous under anoxic conditions.

*Summary*    Bacteria are admirably well suited to experimental evolutionary studies. Much is already known about gene function in their relative small genomes. They are easily propagated and cryopreserved, and we can control their growth rate, population size, and the design of their "ecological theater." Newly empowered by genomic technologies, experimental bacterial evolution is providing unprecedented insight into fundamental issues such as bacterial genome plasticity, functional trade-offs, pleiotropy, epistasis, parallelism, and convergence. It is essential to gain clarity on these issues in bacteria, not just because they constitute the majority of species on Earth, but also because bacteria ultimately control many disease processes and all biogeochemical cycles. However, because bacteria are haploid organisms that lack meiosis and syngamy, there are key questions in evolutionary biology for which they are poor experimental models. Fortunately, we have in the simple Eukaryote, *Saccharomyces cerevisiae*, a model organism that is equally well suited to experimental laboratory evolution and whose genetics and physiology are remarkably similar to our own.

## EXPERIMENTAL EVOLUTIONARY GENOMICS OF YEAST

In a seminal paper, Paquin and Adams (1983) observed fluctuations in neutral marker frequency in adaptively evolving populations of yeast cells, grown in continuous culture under glucose limitation. These fluctuations were taken as evidence for the periodic emergence and fixation of adaptive clones. However, the underlying adaptive mutations were not elucidated, largely because neither the technology nor the genome sequence existed then, both of which are required to identify such mutations. Furthermore, while terminal adaptive clones from these evolutions were characterized from a physiological perspective (Adams et al. 1985), the molecular consequences of the adaptive mutations also remained uncharacterized. A decade and a half later, our first glimpse at the underlying causes of adaptation in these evolved clones was revealed, when Rosenzweig and coworkers (Brown et al. 1998) hypothesized that amplification of hexose transporter genes was a likely candidate for adaptive mutation under glucose-limited conditions. Their experiments determined that there had been recurrent tandem duplication of a chimera involving the upstream regulatory region of the high-affinity glucose transporter *HXT7* and the coding region of its homologue, *HXT6*; these events brought about four copies of the high-affinity transporter, leading to an increase in its expression and an overall enhanced capacity to scavenge the limiting evolutionary substrate, glucose. Not only did this experiment reveal the nature of an adaptive mutation itself, but it also suggested that the underlying mutational event was unequal crossing-over, facilitated by the presence of adjacent paralogous hexose transporter genes. Indeed, the presence of these adjacent hexose transporter genes suggests that this mechanism has already played a role in the evolution of the *S. cerevisiae* genome, and that what Brown et al. discovered was a continuation of a process that had previously occurred.

Three years before Brown et al.'s discovery, the first report of microarrays appeared (Schena et al. 1995), followed two years later by the use of whole yeast genome arrays (DeRisi et al. 1997), that were used to monitor gene expression and also the absence or presence (and, by extension, copy number) of genes within the genome (Lashkari et al. 1997). The advent of microarrays allowed a look in unprecedented detail at the genome and transcriptome of evolved clones. Ferea et al. (1999) profiled the transcriptome of one of the adaptive clones from Paquin and Adams' seminal paper, and they also profiled two independently derived clones, also adaptively evolved under glucose limitation. Remarkably, the independently evolved clones all showed similar changes in gene expression, each showing a reduction in transcript abundance, relative to the parental strain, of genes involved in fermentative metabolism, and a corresponding increase in genes required for respiration and glucose transport. While the underlying changes and the trajectory taken across the evolutionary fitness landscape remain unknown, it is unlikely that each of the clones experienced the same changes in the same order, yet each ended up, at least from a phenotypic perspective (both molecular and gross), in a similar place. That relatively few changes (adaptive mutations arise every fifty to one hundred generations in yeast evolving under these conditions) can result in such large and directed changes to the transcriptome, allowing the balance between fermentation and respiration to shift, is a remarkable result. It is unlikely that the transcriptional network has been entirely reprogrammed with so few changes—elucidation of the underlying mutations will shed further light on how such networks are able to evolve.

Dunham and colleagues (Dunham et al. 2002) went on to characterize genomic changes in eight evolved clones—six from the Paquin and Adams (1983) study and two from the Ferea et al. (1999) study, using a combination of whole genome array CGH and single chromosome array CGH. These studies confirmed the previously identified hexose transporter amplification seen previously in one of the clones, but they also found amplification of the entire chromosome arm on which those transporters reside in two other clones. These data suggest that the adaptive advantage enjoyed by this amplification is considerable, and that this amplification can arise frequently and by more than one mechanism. More remarkable is the fact that three of the clones each shared a breakpoint on chromosome 14. The location of this breakpoint was shown to be just upstream of *CIT1*, which encodes the mitochondrial citrate synthase; *CIT1* is one of the key regulated points in the TCA cycle (Stryer 1995). Given that Ferea et al. had profiled one of these strains with this chromosome 14 breakpoint, Dunham et al. speculated that *CIT1* might be brought into a new promoter context and that its changed regulation might result in changes in the expression of other genes involved in the TCA cycle, to facilitate this observed shift in the balance between fermentation and respiration. In considering all observed breakpoints, Dunham et al. noted that they typically occurred at either transposon sequences or tRNA sequences, which likely form good substrates for rearrangements due to the fact that identical, or almost identical, copies exist in the genome. Dunham et al. (2002) speculate that the arrangement of these sequences may be such that they are in positions that might confer adaptive evolvability, the ability of a population to generate and use genetic variation to respond to natural selection (Colegrave and Collins 2008).

## IDENTIFYING ADAPTIVE MUTATIONS IN YEAST

Initial genomics studies that shed light on adaptive evolution in yeast, like those described here, either focused on the phenotype (in the form of changes in gene expression) or on obvious and large changes to the genome, such as amplifications and/or chromosomal rearrangements. What they were not able to do was observe adaptive events occurring at the single nucleotide level, either in adaptive clones from the end point of an evolutionary series or from those interim adaptive clones along the way, which may or may not have become extinct before the end of the experiment. However, recent advances in microarray technology have enabled millions of probes to be placed on a single microarray; this in turn has allowed interrogation of the genome at the nucleotide resolution, whereby a single nucleotide changes can affect the hybridization of a clone's genomic DNA to several probes that interrogate that base. Two studies (Gresham et al. 2006; Segre et al. 2006) have used yeast tiling microarrays to identify single nucleotide changes that occurred during adaptive evolution (see figure 13.2). Analysis of known SNPs between two different yeast strains suggests that more than 85 percent of true positives can be predicted in a background of a few false positives. Because all predicted mutations (of which there are a handful per clone) can be subsequently confirmed with traditional sequencing, the end result is no false positives, with the majority of true positive changes identified (Gresham et al. 2006). Gresham et al. evolved yeast populations under sulfur limitation in chemostat cultures, and they were able to identify nucleotide changes that were either unique to or shared between independently evolved clones. Segre et al. used this approach to map and identify mutations that had arisen independently in strains that had adapted to a fluctuating glucose-galactose environment. In each of four independently derived strains, one or more missense mutations were observed in the *GAL80* gene, which encodes the repressor of the galactose utilization pathway.

## OPENING WINDOWS ON GENOME EVOLUTION
## IN HIGHER EUKARYOTES

In this review, we have focused our discussion on how genomic tools such as DNA microarrays and ultrahigh-throughput sequencing are revolutionizing experimental evolutionary studies that use microbes. We would be remiss if we failed to point out that truly "whole-genome-scale" experiments were carried out prior to the advent of such tools. We refer to seminal work by Wichman, Bull, and colleagues (1999, 2005, and references therein), who have conducted an elegant series of evolution experiments growing bacteriophage and *E. coli* in two-phase chemostats that simulate infinitely expanding populations. Their ability to rapidly acquire complete viral genome sequence data provided their system with unusual power, enabling them first to describe the molecular basis for parallel evolution, and now to begin unraveling the genetics that underlies a perpetual arms race carried out between a bacteriophage and its host (Wichman et al. 2005).

We would also be remiss if we failed to point out that advances in genome technologies are opening new windows onto real-time evolution experiments using higher Eukaryotes, notably *Drosophila* and *Mus*. We consider here three exemplary studies to illustrate the power and problems associated with these new tools. Mackay et al. (2005) performed gene expression analysis on flies selected for either fast or slow mating speed, and they were able to show heritable changes in gene expression across 21 percent of more than 3,700 probe elements on their microarrays. Expression levels were markedly differentiated with respect to genes in development and behavior, most notably olfaction. As noted in the previous section, the number of elements that can be placed on a microarray is increasing rapidly. Sorenson et al. (2007) used an Affymetrix array containing 13,966 probe sets representing about 13,000 unique genes to investigate heritable patterns of gene expression in *Drosophila* evolved under multiple selective regimens. These included starvation and desiccation, different thermal tolerances, and longevity. Remarkably, initial analyses using the method of Significant Analysis of Microarrays (SAM) (Tusher et al. 2001) showed very few significant changes. Accepting a false discovery rate of 20 percent, however, Sorenson et al. were able to detect 262 genes that differed in expression among treatments. Although genes clustered by treatment, considerable overlap among treatments was nevertheless observed, suggesting a connection among starvation, dessication, and longevity phenotypes previously noted by Hoffman and Harshman 1999 and others.

Expression profiling has also been carried out on mice selected in the laboratory for increased voluntary wheel running (Bronikowski et al. 2004). Gene expression profiles were obtained on hippocampus tissue, as that brain region had previously been shown to undergo marked physiological changes in response to wheel running. Using an

Affymetrix chip that contained 12,422 probe sets representing about 12,000 known or putative genes, and accepting a false discovery rate of 10 percent, Bronikowski et al. identified 53 genes whose expression pattern differed between the selected and control lines. These genes were distributed across multiple functions; the majority (thirty) were involved in transcription and translation, while others fell into diverse categories, including neuronal signaling and inflammatory response.

These studies are but a small sampling of studies in which DNA microarrays are now being used to study adaptive evolution of "higher" species under laboratory selection. These projects illustrate the usefulness of such tools in narrowing the search for genes and pathways that underlie observed phenotypic changes. They also illustrate the challenges of finding the genes that matter in organisms whose genomes are many-fold larger than those of microbes, whose metabolic and regulatory networks are far more intricate, and whose organization has reached the tissue and organ levels. Nevertheless, we hopefully anticipate that the rapid pace of innovation in genome technology will help overcome these difficulties and enable detailed insight into the mechanisms that underlie adaptive evolution in plants and animals.

## FUTURE PROSPECTS

While we have barely scratched the surface, technologies now exist that allow us to completely characterize at a molecular level the changes that occur during adaptive evolution. Using high-resolution microarrays, we can determine the single nucleotide changes that occur as cells undergo adaptive evolution; using array CGH, we can identify gross chromosomal changes (except for inversions); and using microarrays to look at gene expression, we can determine the transcriptional phenotype that results from those changes and begin to infer why such changes may provide an adaptive advantage (figure 13.2). However, it is clear that genomic technologies are further improving, in both the precision and specificity that they can provide, as well as the throughput that is possible. With new sequencing technologies (e.g., Shendure et al. 2005), it is now possible to sequence a yeast genome of 12 Mb at six times the coverage for about the same price as a tiling microarray. In the near future, it is likely that this price will drop dramatically, while the coverage and read lengths will increase, such that in a few years, a fully assembled genome can be derived for a few tens of dollars in a day or so. At this point, it will be straightforward and practical to sequence adaptive clones, both from the end of long-term experiments, as well as those that arose during the course of evolution, providing us with a much more complete picture of the underlying molecular changes that have occurred. It will also enable us to ask new questions: How do the initial events in an adaptively evolving population influence later events? What are the conditions, both environmental and genetic, that favor stable polymorphism over clonal interference? How does the strength of the selective pressure, or the type of selective pressure, determine the spectrum of mutations that are adaptive? Now more than ever, we have the tools available, and the prospect of them becoming readily affordable, so that we can construct deeply detailed pictures of the molecular events underlying adaptive evolution, and the resulting consequences of them.

## SUMMARY

Genome architecture is defined in terms of the gene content and organization uniquely characteristic of a given individual. Populations and species can also be said to have genome architectures that are characteristic in their dynamic features, as well as in their probable manifestation over successive generations. Laboratory experimental evolution provides a powerful tool to test hypotheses concerning factors that shape genome structure at the population level, especially if microbes are used that can be cryogenically preserved then revived for later study. Extrinsic factors acting at the population level can be easily manipulated, including selection, population size, standing genetic variation, and level of recombination permitted among variants. Intrinsic factors such as mutation rate and both the number and distribution of recombinogenic sequences can also be manipulated using standard techniques of molecular genetics. The response of taxon-specific architectures to controlled experimental variation in these factors reveals constraints governing genome evolution not clearly seen using comparative approaches. Such prospective experimental studies complement comparative studies, whose inferences are generated retrospectively. This chapter discusses classic and recent studies that illustrate the unusual power of new genomic technologies coupled with experimental evolution to reveal otherwise hidden features of genome evolution, with a special emphasis on bacteria and yeast.

## ACKNOWLEDGMENTS

We thank T. Garland, Jr. and M. R. Rose for the invitation to participate in this project, and for their encouragement and constructive criticism during the review process. The writing was facilitated by grants from the National Institutes of Health (R01-HG003328-01 and R15 GM79762-01) and NASA (NNX07AJ28G) to G.S. and F.R. We thank E. Kroll and J. Piotrowski for helpful comments.

## REFERENCES

Abby, S., and V. Daubin. 2007. Comparative genomics and the evolution of prokaryotes. *Trends in Microbiology* 15:135–141.

Adams, J. 2004. Microbial evolution in laboratory environments. *Research in Microbiology* 155:311–318.

Adams, J., and R. Helling 1983. Adaptive changes in bacterial populations. Abstract 437. *In Proceedings XV of the International Congress of Genetics*. New Delhi: International Congress of Genetics.

Adams, J., C. Paquin, P. W. Oeller, and L. W. Lee. 1985. Physiological characterization of adaptive clones in evolving populations of the yeast, *Saccharomyces cerevisiae. Genetics* 110:173–185.

Albert, T. J., D. Dailidiene, G. Dailide, G., J. E. Norton, A. Kalia, T. A. Richmond, M. Molla, J. Singh, R. D. Green, and D. E. Berg. 2005. Mutation discovery in bacterial genomes: Metronidazole resistance in *Helicobacter pylori. Nature Methods* 2:951–953.

Atwood, K. C., L. K. Schneider, and F. J. Ryan. 1951. Periodic selection in *Escherichia coli. Proceedings of the National Academy of Sciences of the USA* 37:146–155.

Bennett, A. F., and R. E. Lenski. 1993. Evolutionary adaptation to temperature. II. Thermal niches of experimental lines of *Escherichia coli. Evolution* 47:1–12.

Bergthorsson, U., and H. Ochman. 1999. Chromosomal changes during experimental evolution in laboratory populations of *Escherichia coli. Journal of Bacteriology* 181:1360–1363.

Binnewies, T. T., Y. Motro, P. F. Hallin, O. Lund, D. Dunn, T. La, D. J. Hampson, M. Bellgard, T. M. Wassenaar, and D. W. Ussery. 2006. Ten years of bacterial genome sequencing: Comparative-genomics-based discoveries. *Functional Integrative Genomics* 6:165–185.

Blot, M. 1994. Transposable elements and adaptation of host bacteria. *Genetica* 93:5–12.

Blount, Z. D., C. Z. Borland, and R. E. Lensksi. 2008. Historical contingency and the evolution of a key innovation in an experimental population of *Escherichia coli. Proceedings of the National Academy of Sciences of the USA* 105:7899–7905.

Bronikowski, A. M., J. S. Rhodes, T. Garland, T. A. Prolla, T. A. Awad, and S. C. Gammie. 2004. The evolution of gene expression in the hippocampus in response to selective breeding for increased locomotor activity. *Evolution* 58:2079–2086.

Brown, C. J., K. M. Todd, and R. F. Rosenzweig. 1998. Multiple duplication of yeast hexose transport genes in response to selection in a glucose-limited environment. *Molecular Biology and Evolution* 15:931–942.

Cashel M., D. R. Gentry, V. J. Hernandez, and D. Vinella. 1996. The stringent response. Pages 1458–1496 *in* F. C. Neidhardt et al., eds. Escherichia coli *and* Salmonella: *Cellular and Molecular Biology.* Washington, DC: American Society for Microbiology.

Colegrave, N., and S. Collins. 2008. Experimental evolution and evolvability. *Heredity* 100: 464–470.

Cooper, V. S., and R. E. Lenski. 2000. The population genetics of ecological specialization in evolving *Escherichia coli* populations. *Nature* 407:736–739.

Cooper, V. S., A. F. Bennett, and R. E. Lenski. 2001. Evolution of thermal dependence of growth rate of *Escherichia coli* populations during 20,000 generations in a constant environment. *Evolution* 55:889–896.

Cooper, T. F., D. E. Rozen, and R. E. Lenski. 2003. Parallel changes in gene expression after 20,000 generations of evolution in *Escherichia coli. Proceedings of the National Academy of Sciences of the USA* 100:1072–1077.

———. 2004. Parallel changes in gene expression after 20,000 generations of evolution in *Escherichia coli. Proceedings of the National Academy of Sciences of the USA* 100:1072–1077.

Cooper, T. F., and S. K. Remold, R. E. Lenski, and D. Schneider. 2008. Expression profiles reveal parallel evolution of epistatic interactions involving the CRP regulon in *Escherichia coli. PLoS Genetics* 4:e35 (1–10).

Coyle, S., and E. Kroll. 2008. Starvation induces genomic rearrangements and starvation-resilient phenotypes in yeast. *Molecular Biology and Evolution* 25:301–309.

Crombach, A., and P. Hogeweg. 2007. Chromosome rearrangements and the evolution of genome structuring and adaptability. *Molecular Biology and Evolution* 24:1130–1139.

Crozat, E., N. Philippe, R. E. Lenski, J. Geiselmann, and D. Schneider. 2005. Long-term experimental evolution in *Escherichia coli*. XII. DNA topology as a key target for selection. *Genetics* 169:523–532.

de Visser, J. A. G., A. D. L. Akkermans, R. F. Hoekstra, and W. M. de Vos. 2004. Insertion-sequence-mediated mutations isolated during adaptations to growth and starvation in *Lactococcus lactis*. *Genetics* 168:1145–1157.

DeRisi, J. L., V. R. Iyer, and P. O. Brown. 1997. Exploring the metabolic and genetic control of gene expression on a genomic scale. *Science* 278:680–686.

Dettman, J. R., C. Sirjusingh, L. M. Kohn, and J. B. Anderson. 2007. Incipient speciation by divergent adaptation and antagonistic epistatsis in yeast. *Nature* 447:585–589.

Dobrindt, U., A. Franziska, K. Michalis, A. Janka, C. Buchrieser, M. Samuelson, C. Svanborg, G. Gottschalk, H. Karch, and J. Hacker. 2003. Analysis of genome plasticity in pathogenic and commensal *Escherichia coli* isolates by use of DNA arrays. *Journal of Bacteriology* 185:1831–1840.

Dunham, M. J., H. Badrane, T. Ferea, J. Adams, P. O. Brown, F. Rosenzweig, and D. Botstein. 2002. Characteristic genome rearrangements in experimental evolution of *Saccharomyces cerevisiae*. *Proceedings of the National Academy of Sciences of the USA* 99:16144–16149.

Dykhuizen, D. E., A. M. Dean, and D. L. Hartl. 1987. Metabolic flux and fitness. *Genetics* 115:25–31.

Ebert, D. 1998. Experimental evolution of parasites. *Science* 282:1432–1435.

Escobar-Paramo, P., A. L. Menac'h, T. Le Gall, C. Amorin, S. Gouriou, B. Picard, D. Skurnik, and E. Denamur. 2006. Identification of forces shaping the commensal *Escherichia coli* genetic structure by comparing animal and human isolates. *Environmental Microbiology* 8:1975–1984.

Faure, D., R. Frederick, D. Wloch, P. Portier, M. Blot, and J. Adams. 2004. Genomic changes arising in long-term stab cultures of *Escherichia coli*. *Journal of Bacteriology* 186:6437–6442.

Ferea, T. L., D. Botstein, P. O. Brown, and R. F. Rosenzweig. 1999. Systematic changes in gene expression patterns following adaptive evolution in yeast. *Proceedings of the National Academy of Sciences of the USA* 96:9721–9726.

Ferenci, T. 2003. What is driving the acquisition of mutS and rpoS polymorphisms in *Escherichia coli*? *Trends in Microbiology* 11:457–461.

———. 2008. The spread of a beneficial mutation in experimental bacterial populations: The influence of the environment and genotype on the fixation of rpoS mutations. *Heredity* 100:446–452.

Fiegna, F., Y. T. N. Yu, S. V. Kadam, and G. J. Velicer. 2006. Evolution of an obligate social cheater to a superior cooperator. *Nature* 441:310–314.

Finkel, S. E., and R. Kolter. 1999. Evolution of microbial diversity during prolonged starvation. *Proceedings of the National Academy of Sciences of the USA* 96:4023–4027.

Fisher, E., and U. Sauer. 2005. Large-scale in vivo flux analysis shows rigidity and suboptimal performance of *Bacillus subtilis* metabolism. *Nature Genetics* 37:636–640.

Fong, S. S., A. R. Joyce, and B. O. Palsson. 2005. Parallel adaptive evolution cultures of *Escherichia coli* lead to convergent growth phenotypes with different gene expression states. *Genome Research* 15:1365–1372.

Gresham, D., D. M. Ruderfer, S. C. Pragtt, J. Schacherer, M. J. Dunham, D. Botstein, and L. Kruglyak. 2006. Genome-wide detection of polymorphisms at nucleotide resolution with a single DNA microarray. *Science* 311:1932–1935.

Hall, B. G. 1999. Spectra of spontaneous growth-dependent and adaptive mutations at *ebgR*. *Journal of Bacteriology* 181:1149–1155.

Hall, B. G., and T. Zuzel. 1980. Evolution of a new enzymatic activity by recombination within a gene. *Proceedings of the National Academy of Sciences of the USA* 77:3529–3533.

Hartl, D. L., D. E. Dykhuizen, and A. M. Dean. 1985. Limits of adaptation: The evolution of selective neutrality. *Genetics* 111:655–674.

Hatfield, G. W., and C. J. Benham. 2002. DNA topology-mediated control of gene expression in *Escherichia coli*. *Annual Review Genetics* 36:175–203.

Hegreness, M., and R. Kishnoy. 2007. Analysis of genetic systems using experimental evolution and whole-genome sequencing. *Genome Biology* 8:201.

Helling, R. B., C. Vargas, and J. Adams. 1987. Evolution of *Escherichia coli* during growth in a constant environment. *Genetics* 116:349–358.

Herring, C. D., A. Raghunathan, C. Honisch, T. Patel, M. K. Applebee, A. R. Joyce, et al. 2006. Comparative genome sequencing of *Escherichia coli* allows observation of bacterial evolution of a laboratory timescale. *Nature Genetics* 38:1406–1412.

Hoekstra, H. E., and F. A. Coyne. 2007. The locus of evolution: Evo-devo and the genetics of adaptation. *Evolution* 61:995–1016.

Hoffmann, A. A., and L. G. Harshman. 1999. Desiccation and starvation resistance in *Drosophila*: Patterns of variation at the species, population and intrapopulation levels. *Heredity* 83:637–643.

Holloway, A. K., T. P. Palzkill, and J. J. Bull. 2007. Experimental evolution of gene duplicates in a bacterial plasmid model. *Journal of Molecular Evolution* 64:215–222.

Honisch, C., Raghunathan, C. R. Cantor, B. O. Palsson, and D. van den Bloom. 2004. High throughput mutation detection underlying adaptive evolution of *Escherichia coli*—K12. *Genome Research* 14:2495–2502.

Horiuchi, T, J. I. Tomizawa, and A. Novick. 1962. Isolation and properties of bacteria capable of high rates of β-galactosidease synthesis. *Biochimica et Biophysica Acta* 55:152–163.

Jaenisch, R., and A. Bird. 2003. Epigenetic regulation of gene expression: How the genome integrates intrinsic and environmental signals. *Nature Genetics* 33:245–245.

Kacser, H., and J. A. Burns. 1973. The control of flux. *Symposia of the Society of Experimental Biology* 27:65–104.

———. 1981. The molecular basis of dominance. *Genetics* 97:1149–1160.

Kassen, R., and T. Bataillon. 2006. Distribution of fitness effects among beneficial mutations before selection in experimental populations of bacteria. *Nature Genetics* 38:484–488.

Kimura, M. 1968. Evolutionary rate at the molecular level. *Nature* 217:624–626.

Knight, C. G., N. Zitzmann, S. Prabhakar, R. Antrobuis, R. Dwek, H. Hebestreit, and P. B. Rainey. 2006. Unraveling adaptive evolution: How a single point mutation affects the protein co-regulation network. *Nature Genetics* 38:1015–1022.

Koehn, R. K., A. J. Zera, and J. G. Hall. 1983. Enzyme polymorphism and natural selection. Pages 115–136 *in* M. Nei and R. K. Koehn, eds. *Evolution of Genes and Proteins*. Sunderland, MA: Sinauer.

Kurlandzka, A., R. F. Rosenzweig, and J. Adams. 1991. Identification of adaptive changes in an evolving population of *Escherichia coli*: The role of changes with regulatory and highly pleiotropic effects. *Molecular Biology Evolution* 8:261–281.

Lashkari, D. A., J. L. DeRisi, J. H. McCusker, A. F. Namath, C. Gentile, S. Y. Hwang, P. O. Brown, and R. W. Davis. 1997. Yeast microarrays for genome wide parallel genetic and gene expression analysis. *Proceedings of the National Academy of Sciences of the USA* 94:13057–13062.

Lenski, R. E. 2004. Phenotypic and genomic evolution during a 20,000-generation experiment with the bacterium *Escherichia coli*. *Plant Breeding Reviews* 24:225–265.

Lenski, R. E., M. R. Rose, S. C. Simpson, and S. C. Tadler. 1991. Long-term experimental evolution in *Escherichia coli*. I. Adaptation and divergence during 2,000 generations. *American Naturalist* 138:1315–1341.

Lenski, R. E., and M. Travisano. 1994. Dynamics of adaptation and diversification: A 10,000-generation experiment with bacterial populations. *Proceedings of the National Academy of Sciences of the USA* 91:6808–6814.

Lewontin, R. 1974. *The Genetical Basis of Evolutionary Change*. New York: Columbia University Press.

Lorca, G. L., A. Ezersky, V. V. Lunin, J. R. Walker, S. Altamentova, E. Evdokimova, M. Vedadi, A. Bochkarev, and A. Savchenko. 2007. Glyoxylate and pyruvate are antagonistic effectors of the *Escherichia coli* IclR transcriptional regulator. *Journal of Biological Chemistry* 282:16476–16491.

Luria, S., and M. Delbruck. 1943. Mutations of bacteria from virus sensitivity to virus resistance. *Genetics* 28:491–511.

Mackay, T. F. C., S. L. Heinsohn, R. F. Lyman, A. J. Mochring, T. J. Morgan, and S. M. Rollman. 2005. Genetics and genomics of *Drosophila* mating behavior. *Proceedings of the National Academy of Sciences of the USA* 102:6622–6629.

Magasanik, B. 2000. Global regulation of gene expression. *Proceedings of the National Academy of Sciences of the USA* 97:14044–14045.

Mahadevan, R., and C. H. Schilling. 2003. The effects of alternate optimal solutions in constraint-based genome-scale-metabolic models. *Metabolic Engineering* 5:264–276.

Maharjan, R., S. Seeto, L. Notley-McRobb, and T. Ferenci. 2006. Clonal adaptive radiation in a constant environment. *Science* 313:514–517.

Mallik, P. T., T. S. Pratt, M. B. Beach, M. D. Bradley, J. Undamatla, et al. 2004 Growth phase-dependent regulation and stringent control of *fis* are conserved processes in enteric bacteria and involve a single promoter (*fisP*) in *Escherichia coli*. *Journal of Bacteriology* 186:122–135.

Martinez-Antonio, A., and J. Collado-Vides. 2003 Identifying global regulators in transcriptional regulatory networks in bacteria. *Current Opinion in Microbiology* 6:482–489.

Masel, J., and H. Maughan. 2007. Mutations leading to the loss of sporulation ability in *Bacillus subtilis* are sufficiently frequent to favor genetic canalization. *Genetics* 175:453–457.

Masel, J., O. D. King, and H. Maughan. 2007. The loss of adaptive plasticity during long periods of environmental stasis. *American Naturalist* 169:38–46.

Maughan, H., V. Callicotte, A. Hancock, C. W. Birky Jr., W. L. Nicholson, and J. Masel. 2006. The population genetics of phenotypic deterioration in experimental populations of *Bacillus subtilis*. *Evolution* 60:686–695.

Maughan, H., J. Masel, C. W. Birky, and E. L. Nicholson. 2007. The roles of mutation accumulation and selection in loss of sporulation in experimental populations of *Bacillus subtilis. Genetics* 177:937–948.

Mira, A., R. Pushker, and F. Rodriguez-Valera. 2006. The Neolithic revolution of bacterial genomes. *Trends in Microbiology* 14:200–206.

Mortlock, R. P. 1984. *Microorganisms as Model Systems for Studying Evolution.* New York: Plenum.

Muller, H. J. 1932. Some genetic aspects of sex. *American Naturalist* 66:118–138.

Naas, T., M. Blot, W. M. Fitch, and W. Arber. 1994. Insertion sequence related genetic variation in resting *Escherichia coli* K-12. *Genetics* 136:721–730.

Naas, T., M. Blot, W. M. Fitch, and W. Arber. 1995. Dynamics of *IS*-related genetic rearrangements in resting *Escherichia coli* K-12. *Molecular Biology Evolution* 12:198–207.

Newcombe, H. B. 1949. Origin of bacterial variants. *Nature* 164:150–151.

Nilsson, A. I., E. Kugelberg, O. G. Berg, and D. I. Andersson. 2004. Experimental adaptation of *Salmonella typhimurium* to mice. *Genetics* 168:1119–1130.

Nilsson, A. I., S. Koskiniemi, S. Eriksson, E. Kugelberg, J. C. D. Hinton, and D. I. Andersson. 2005. Bacterial genome size reduction by experimental evolution. *Proceedings of the National Academy of Sciences of the USA* 102:12112–12116.

Notley-McRobb, L. T. King, and T. Ferenci. 2002. *rpoS* mutations and loss of general stress resistance in *Escherichia coli* populations as a consequence of conflict between competing stress responses. *Journal of Bacteriology* 187:806–811.

———. 2003. The influence of cellular physiology on the initiation of mutational pathways in *Escherichia coli* populations. *Proceedings of the Royal Society of London B, Biological Sciences* 270:843–848.

Novick, A., and L. Szilard. 1950. Experiments with the chemostat on spontaneous mutations of bacteria. *Proceedings of the National Academy of Sciences of the USA* 36:708-719.

Ochman, H., and S. R. Santos. 2005. Exploring microbial microevolution with microarrays. Infection. *Genetics and Evolution* 5:103–108.

Ostrowski, E. A., D. E. Rozen, and R. E. Lenski. 2005. Pleiotropic effects of beneficial mutations in *Escherichia coli. Evolution* 59:2343–2352.

Ostrowski, E. A., R. J. Woods, and R. E. Lenski. 2008. The genetic basis of parallel and divergent phenotypic responses in evolving populations of *Escherichia coli. Proceedings of the Royal Society of London B, Biological Sciences* 275:277–284.

Papin, J. A., N. D. Price, S. J. Wilback, D. A. Fell, and B. O. Palsson. 2003. Metabolic pathways in the post-genome era. *Trends in Biochemical Sciences* 28:250–258.

Paquin C., and J. Adams. 1983. Frequency of fixation of adaptive mutations is higher in evolving diploid than haploid yeast populations. *Nature* 30:495–500.

Pelosi, L., L. Kuhn, D. Guetta, J. Garin, J. Geiselmann, R. E. Lenski, and D. Schneider. 2006. Parallel changes in global protein profiles during long-term experimental evolution in *Escherichia coli. Genetics* 173:1851–1869.

Philippe, N., E. Crozat, R. E. Lenski, and D. Schneider. 2007. Evolution of global regulatory networks during a long-term experiment with *Escherichia coli. BioEssays* 29:846–860.

Piggot, P. J., and R. Losick. 2002. Sporulation genes and intercompartmental regulation. Pages 483–517 *in* A. L. Sonenshien, J. A. Hoch, and R. Losick, eds. Bacillus subtilis *and Its Closest Relatives: From Genes to Cells.* Washington, DC: American Society for Microbiology.

Pinkel, D., and D. G. Albertson. 2005. Array comparative genomic hybridization and its application to cancer. *Nature Genetics* 37:S11–S17.

Porwollik, S., R. M.-Y. Wong, R. A. Helm, K. K. Edwards, M. Calcutt, A. Eisenstark, and M. McClelland. 2004. DNA amplification and rearrangements in archival *Salmonella enterica* serovar Typhimurium LT2 cultures. *Journal of Bacteriology* 186:1678–1682.

Proschel, M., Z. Zhang, and J. Parsch. 2006. Widespread adaptive evolution of *Drosophila* genes with sex-biased expression. *Genetics* 174:893–900.

Riehle, M. M., A. F. Bennett, and A. D. Long. 2001. Genetic architecture of thermal adaptation in *Escherichia coli*. *Proceedings of the National Academy of Sciences of the USA* 98:525–530.

———. 2005. Differential patterns of gene expression and gene complement in laboratory-evolved lines of *E. coli*. *Integrative Comparative Biology* 45:532–538.

Rigby, P. W. J., B. D. Burleigh, and B. S. Hartley. 1974. Gene duplication in experimental enzyme evolution. *Nature* 251:200–204.

Riley, J., R. Butler, D. Ogilvie, R. Finniear, D. Jenner, S. Powell, R. Anand, J. C. Smith, and A.F. Markham. 1990. A novel, rapid method for the isolation of terminal sequences from yeast artificial chromosome (YAC) clones. *Nucleic Acids Research* 18:2887–2890.

Rosenzweig, R. F., R. R. Sharp, D. Treves, and J. Adams. 1994. Microbial evolution in a simple unstructured environment: Genetic differentiation in *Escherichia coli*. *Genetics* 137:903–17.

Rozen, D. E., L. deVisser, and P. J. Garrish. 2002. Fitness effects of fixed beneficial mutations in microbial populations. *Current Biology* 12:1040–1045.

Rozen, D. E., and R. E. Lenski. 2000. Long-term experimental evolution in *Escherichia coli*. VIII. Dynamics of a balanced polymorphism. *American Naturalist* 155:24–35.

Schena, M., D. Shalon, R. W. Davis, and P. O. Brown. 1995. Quantitative monitoring of gene expression patterns with complementary DNA microarray. *Science* 270:467–470.

Schneider, D., E. Duperchy, E. Courange, R.E. Lenski, and M. Blot. 2000. Long-term experimental evolution in *Escherichia coli* IX. Characterization of insertion sequence-mediated mutations and rearrangements. *Genetics* 156:477–488.

Schneider, D., and R. E. Lenski. 2004. Dynamics of insertion sequence elements during experimental evolution of bacteria. *Research in Microbiology* 155:211–215.

Segre, A. V., A. W. Murray, and J.-Y. Leu. 2006. High-resolution mutation mapping reveals parallel experimental evolution in yeast. *PLoS Biology* 4:1372–1385.

Shendure, J., G. J. Porreca, N. B. Reppas, X. Lin, J. P. McCutcheon, A. M. Rosenbaum, M. D. Wang, K. Zhang, R. D. Mitra, and G. M. Church. 2005. Accurate multiplex polony sequencing of an evolved bacterial genome. *Science* 309:1728–1732.

Sliwa, P., and R. Korona. 2005. Loss of dispensable genes is not adaptive in yeast. *Proceedings of the National Academy of Sciences of the USA* 102:17670–17674.

Sorensen, J. G., M. M. Nielsen, and V. Loeschke. 2007. Gene expression profile analysis of *Drosophila melanogaster* selected for resistance to environmental stressors. *Journal of Evolutionary Biology* 20:1624–1636.

Spencer, C. C., M. Bertrand, M. Travisano, and M. Doebeli. 2007. Adaptive diversification in genes that regulate resource use in *Escherichia coli*. *PLoS Genetics* 3:83–88.

Springer, B., P. Sander, L. Sedlacek, W.-D. Hardt, V. Mizrahi, P. Schar, and E. C. Bottger. 2004. Lack of mismatch correction facilitates genome evolution in mycobacteria. *Molecular Microbiology* 53:1601–1609.

Stryer, L. 1995. *Biochemistry*. New York: Freeman.

Treves, D. S., S. Manning, and J. Adams. 1998. Repeated evolution of an acetate-crossfeeding polymorphism in long-term populations of *Escherichia coli*. *Molecular Biology Evolution* 15:789–797.

Tusher, V. G., R. Tibshirani, and G. Chu. 2001. Significance analysis of microarrays applied to the ionizing radiation response. *Proceedings of the National Academy of Sciences of the USA* 198:5116–5121.

Typas, A., C. Barembruch, A. Prossling, and R. Hengge. 2007. Stationary phase reorganization of the *Escherichia coli* transcription machinery by Crl protein, a fine tuner of sigmas activity and levels. *Embo Journal* 26:1569–1578.

Ussery, D. W. 2006. Leaner and meaner genomes in *Escherichia coli*. *Genome Biology* 7:237.

Velicer, G. J., L. Kroos, and R. E. Lenski. 1998. Loss of social behaviors by *Myxococcus xanthus* during evolution in an unstructured habitat. *Proceedings of the National Academy of Sciences of the USA* 95:12376–12380.

———. 2000. Developmental cheating in the social bacterium *Myxococcus xanthus*. *Nature* 404:598–601.

Velicer, G. J., R. E. Lenski, and L. Kroos. 2002. Rescue of social motility lost during evolution of *Myxococcus xanthus* in an asocial regime. *Journal of Bacteriology* 184:2719–2727.

Velicer, G. J., G. Raddatz, H. Keller, S. Deiss, C. Lanz, I. Dinkelacker, and S. C. Schuster. 2006. Comprehensive mutation identification in an evolved bacterial co-operator and its cheating ancestor. *Proceedings of the National Academy of Sciences of the USA* 103:8107–8112.

Wichman, H. A., M. R. Badgett, L. A. Scott, C. M. Boulianne, and J. J. Bull. 1999. Different trajectories of parallel evolution during viral adaptation. *Science* 258:422–424.

Wichman, H. A., J. Millstein, and J. J. Bull. 2005. Adaptive molecular evolution for 13,000 phage generations: A possible arms race. *Genetics* 170:19–31.

Woods, R. D. Schneider, C. L. Winkworth, M. A. Riley, and R. E. Lenski. 2006. Tests of parallel molecular evolution in a long-term experiment with *Escherichia coli*. *Proceedings of the National Academy of Sciences of the USA* 103:9107–91127.

Zhong, S., A. Khordursky, D. Dykhuizen, and A. M. Dean. 2004. Evolutionary genomics and ecological specialization. *Proceedings of the National Academy of Sciences of the USA* 101:11719–11724.

Zinser, E. R., D. Schneider, M. Blot, and R. Kolter. 2003. Bacterial evolution through the selective loss of beneficial genes: Trade-offs in expression involving two loci. *Genetics* 164:1271–1277.

# PART IV

# APPLICATIONS OF EXPERIMENTAL EVOLUTION

# 14

# UNDERSTANDING EVOLUTION THROUGH THE PHAGES

Samantha E. Forde and Christine M. Jessup

The primary goal of experimental evolution is to directly test theories of evolution using controlled experiments. Experiments in the laboratory offer particular control over both environmental conditions and the starting populations for evolutionary experimentation. Experiments in the laboratory also allow for a high degree of replication, and thus an increased probability of capturing the distribution of potential mutations, which is particularly important when testing evolutionary theory. Developments in molecular biology now make it possible to pinpoint the mutations responsible for adaptation and to determine how these mutations function. A comprehensive picture of adaptation, one that integrates genetic information with the fitness consequences of mutations, is now emerging.

Microbes have become increasingly popular experimental tools for testing evolutionary theory due to their ease of manipulation, large population sizes and short generation times (see Elena and Lenski 2003 as well as other chapters in this book). The short generation times of such systems facilitate the characterization of many generations in relatively short time periods, enabling the researcher to observe the mutational and fitness changes underlying adaptation. Strains can be archived in a state of suspended animation, allowing for a *posteriori* evaluation of evolutionary patterns. There is also a wealth of genetic and physiological information available for many microbes, enabling a mechanistic understanding of evolutionary dynamics.

In this chapter, we focus on empirical tests of evolutionary theory using viruses, specifically bacteriophage (or "phage" for short). The French-Canadian microbiologist Felix d'Herelle coined the term *phage* from the Greek term *phagein* meaning "to eat," which loosely describes what phage do to their bacterial hosts. Studies of experimental evolution with phage date back to before the 1950s and several fundamental evolutionary processes have been studied with phage. For example, Luria and Delbruck (1943) measured resistance in *Escherichia coli* to the phage T1 to demonstrate that mutations occur randomly, rather than as adaptive responses within individuals. Research with the phage lambda provided some of the first information about the mechanism of recombination (Meselson and Weigle 1961).

Although phage have long been a workhorse of microbiology and have been utilized in some key evolutionary studies, historically there has been a division between evolutionary biologists and phage biologists (Duffy and Turner 2008). Phage biologists were often physically isolated from evolutionary biologists due to the common split between molecular and organismal research in university biology departments (Rouch 1997; Duffy and Turner 2008). Recent discovery of the high abundance of phage in many environments (Breitbart and Rohwer 2005) has renewed interest in using phage to test ecological and evolutionary theory, thus motivating researchers to investigate the ecology and evolution of phage *in situ*. Evolutionary biologists are recognizing the advantages to employing relatively simple and rapidly replicating organisms in their research, while microbiologists are recognizing the importance of ecological and evolutionary theory in their research. Studies of phage evolution are flourishing, and the results of these studies illustrate the ongoing and far-reaching contributions of bacteriophage research to molecular biology, ecology, and evolutionary biology (Davis 2003).

There are numerous benefits to using phage experimental systems for testing evolutionary theory. Phage have short generation times, small genomes, and high mutation rates (particularly single-stranded RNA phages), and they are easily propagated in the laboratory. Because of the small size of phage genomes, it is now feasible to sequence the entire genome of both ancestral and evolved phage strains. The past twenty years have seen advances in molecular biology that facilitate characterizing the nucleotide changes underlying adaptations. This is particularly true for phages (and viruses in general), owing to their small genomes. In fact, 1.8 percent of the phage genome has been observed to change in single evolutionary experiments (Wichman et al. 2005). The small genome size also facilitates site-directed mutagenesis, so that genotypes with known differences in amino acids can be constructed and the evolutionary implications of these genetic differences can be investigated (Pepin et al. 2006). Experimental evolution with phage can be exploited to reveal key relationships between the fundamentals of molecular evolution and models of evolutionary processes (Bull et al. 1993).

Recently, phage have been used in evolution experiments to study questions concerning the evolution of sex (Frankino et al. this volume), as well as levels of selection and altruism (Rauser et al. this volume). Phage are also key components in experimental studies of community interactions and adaptive radiations with microbes (Kerr this volume). In this chapter, we focus on recent research on both microevolutionary processes—specifically, the genetic basis of adaptation—and macroevolutionary patterns, such as tests of phylogenetics, in an effort to illustrate how both evolutionary processes and patterns can be evaluated using experiments with microbes. We focus on studies with phage, but also discuss a few nonbacteriophage examples, to illustrate the use of viruses, in general, for studies of experimental evolution.

We begin by reviewing phage biology. We discuss the phage life cycle and describe laboratory procedures for cultivating phage and assaying fitness. We also provide a brief review of the theoretical models that have been used to describe phage ecology and evolution, in order to emphasize the advantages of using phage experimental systems to test theoretical predictions. Our discussion then turns to recent phage experiments that have investigated the genetic basis and molecular mechanisms underlying adaptation, including such key evolutionary concepts as pleiotropy, epistasis, compensation, evolvability, robustness, and parallel versus divergent evolution. We then discuss research that examines the fitness consequences of adaptation to the abiotic environment (e.g., temperature) and adaptation to the biotic environment (i.e., the host cell), addressing trade-off theory, host shifts, and coevolutionary dynamics. We illustrate the utility of phage experiments for studying the mechanisms that underlie divergence from a common ancestor and for testing the accuracy of phylogenetic methods. Finally, we emphasize that the insights gained from experimental evolution studies with phage have important implications not only for phage ecology and evolution, but also for predator-prey and host-parasite interactions in general.

## THE BIOLOGY OF PHAGES

Bacteriophages (or phage) are viruses that infect bacteria. Thus, the bacterial host is a fundamental component of the phage's environment. Phage particles contain genomic material surrounded by a protein coat. Phage are generally divided into types based on their morphology and genomes. Phage genomes can be ssDNA, dsDNA, ssRNA, or dsRNA, with the nature of the genome having important implications for both the infection process and evolution. For example, RNA viruses exhibit elevated mutation rates compared to DNA viruses, as they lack the proofreading capacity offered by DNA polymerase.

The general life cycle of a phage is illustrated in figure 14.1. Phage first bind to receptors on the surface of a susceptible host cell (adsorb), and then the genetic material enters the cytoplasm. The adsorption step is arguably the most critical, as the outcome of this encounter determines whether or not the infection will proceed. Once the phage genome is inside the host cell, marking the beginning of the latent period, the machinery of the host cell is redirected to synthesize phage nucleic acids and proteins. New phages are assembled and packaged, and then the cell lyses at the end of the latent period, releasing the new phage particles into the surrounding environment. Temperate phage, in contrast to lytic phage, first enter into a state called *lysogeny*, where most phage genes are not expressed. The phage genome (called a *prophage*) is replicated in synchrony with the bacterial genomes. Temperate phage can be passed on to subsequent host cells during cell division, and, under certain conditions, lysogenic cells will spontaneously produce and release new phage by reentering the lytic cycle.

In the laboratory, phages are propagated on permissive hosts. Just as bacteria are serially propagated for evolutionary experiments (see Rosenzweig and Sherlock this volume), phage, too, can be propagated on hosts in liquid culture using serial transfer, and the phage progeny harvested (e.g., via selective filtration or treatment with chloroform) to initiate the next passage. Another method for phage propagation is continuous culture in which phage and bacteria are co-cultured in chemostat vessels where resources are supplied and waste products removed at a constant rate. A variation of such chemostat

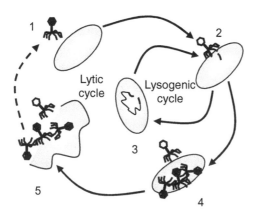

FIGURE **14.1**

The general life cycle of a lytic bateriophage.
1: An infective phage and bacterial host. 2: The phage binds to receptors on the host cell (adsorption) and injects its genome into the cell. 3: Lysogenic cycle—the phage genome integrates into the bacterial chromosome. 4: Lytic cycle—the phage genome is copied, phage proteins are synthesized, and the genome is packaged, all using the host cell's machinery. 5: The host cell lyses and releases new phage, which can then infect another host.

culture which is sometimes used to propagate phage is a two-stage chemostat in which host cells are grown in one chemostat and supplied to a second chemostat containing phage (e.g., Bull et al. 2006). Such culturing enables phage to be presented with a continuous supply of sensitive hosts as the source population of bacteria is näive to the selective pressures of phage. Finally, phage can be serially propagated on a spatially structured solid environment such as an agar plate. Upon incubation, an individual phage particle will form a plaque (or a clearing) in a confluent lawn of bacterial host on an agar plate. With each of these culture methods, the experimenter is able to control various aspects of the abiotic and biotic environment (e.g., dispersal of host and parasite, host genotype, resource level, and/or temperature).

The ability to quantify changes in fitness is central to experimental evolution and is also particularly tractable with bacteriophage. Just as the relative fitness of bacterial strains can be determined through pairwise resource competition (see Rosenzweig and Sherlock this volume and references therein), phage fitness can also be determined through paired growth experiments (e.g., Chao 1990; see also Turner and Chao 1998; Burch and Chao 1999). In such experiments, the growth rate of a test phage is determined relative to an ancestral phage that has a differentiating phenotypic marker, such as a particular plaque morphology. The two phages are combined and allowed to grow on the host of interest. After incubation (typically twenty-four hours), the ratio of test phage in the starting mixture is compared to that in the harvested lysate at the end of the assay in order to compute fitness.

## QUANTITATIVE MODELS OF PHAGE-HOST SYSTEMS

One of the benefits of experimental evolution with phage is that relatively simple but biologically relevant models can be developed and compared to empirical results. Some of the earliest quantitative models of phage-host interactions were developed by Campbell (1961), who presented phage-host systems as ideal for studying predator-prey equilibriums. He also, however, noted that one of the challenges to this application is the rapid evolution of phage-resistant mutants. Levin et al. (1977) and Lenski and Levin (1985) built on this theoretical foundation by further developing the models, supplying them with empirically determined parameter values and, most significantly, testing their predictions in the laboratory. Phage-host interactions in continuous culture have been extensively modeled using variations on this approach, in order to investigate the effects of resource concentration and phage resistance through both theoretical analysis and experimentation (e.g., Bohannan and Lenski 1997; Bohannan and Lenski 1999; Bohannan and Lenski 2000; Bull et al. 2006).

The models just described are fundamentally ecological models, and components of these models have direct analogies to the predator-prey models found in general ecology textbooks. Such ecological models incorporate evolution through the introduction of a new population with a different phenotype, although Bull et al. (2006) address the interaction between ecology and evolution via population density (discussed later). More recently, Weitz and colleagues (2005) developed coupled ecological and evolutionary models for trait adaptation in coevolutionary settings, and in their experimental models, they directly observed diversification of phage and bacteria. In other theoretical work, You and Yin (2002) investigated the interaction between mutational severity and resource environment on epistasis. The challenges of characterizing mutations and their effects on fitness in the field led the authors to develop a model that describes the major molecular process in the phage life cycle. They then studied how simulated mutations interact to affect fitness.

Together, these studies demonstrate the relative ease with which appropriate predator-prey and host-parasite mathematical models can be developed for the case of phage, enabling researchers to develop theoretical predictions, experimentally test these predictions in the lab, and then refine the models. The ability to address all stages of this iterative process over relatively short time scales using well-characterized biological systems is unique to microbial experimental systems in general and studies of bacteriophages in particular. This scientific setting might thus be described as having a "virtuous cycle" linking theoretical and experimental work.

## THE GENETIC BASIS OF EVOLUTIONARY ADAPTATION

Phage experiments provide a powerful system with which to address the genetic basis of adaptation because the entire genome of evolved and ancestral lines can be sequenced. A number of general have emerged from evolutionary experiments with phage, patterns which we outline here. For example, early substitutions in new environments often confer the largest increases in fitness, and then fitness eventually plateaus (Holder and Bull 2001; Silander et al. 2007). Fitness changes over time can be directly related to genetic diversity in these systems: genetic diversity is often purged during a fitness sweep, and then diversity increases over the period of time when the population reaches a fitness plateau (Wahl and Krakauer 2000). Mechanisms underlying the recovery of fitness due to deleterious mutations, such as compensatory mutations, have also been studied in a number of phage systems. Finally, there is often parallel evolution both in genotypes and phenotypes across replicate populations (Bull et al. 1997; Cuevas et al. 2002; Cunningham et al. 1997; Wichman et al. 1999, 2000). The studies outlined here also illustrate that even in simple, well-characterized systems like phage, genetic architecture can have complex effects on phenotypic expression (Pepin et al. 2006).

### THE MOLECULAR SIGNATURE OF ADAPTIVE EVOLUTION OVER TIME

Studies with phage provide a unique opportunity to document the evolutionary dynamics of novel mutations and the fitness effects of these mutations over time. Phage isolates can be sampled over multiple generations from evolving populations, and mutational changes in these isolates can then be documented and linked to measures of fitness. The small size of phage genomes enables the identification of mutations underlying adaptive changes. Documenting the temporal dynamics of mutations and their fitness effects are key to understanding evolution, but this is logistically difficult for most organisms.

Experiments with phage have revealed that the greatest increase in fitness often occurs early in adaptive evolution experiments, and genetic diversity is often low during this period (Wahl and Krakauer 2000). This was demonstrated with the pair of phages, ΦX174 and G4, for example (Holder and Bull 2001). The overall greatest fitness effects per substitution occurred during the early stages of adaptation when the phage were grown at an elevated temperature on *E. coli*. In addition, polymorphisms were seen only briefly in the early part of the experiment but were more persistent in the later period of the experiment for both phage. Related work by Wichman and colleagues on ΦX174 characterized the adaptation of the phage to a continuous supply of sensitive hosts over 13,000 generations (180 days) (Wichman et al. 2005). The authors found that substitutions accumulated at a low rate for the first 20 days of the experiment, and then in a clocklike fashion for the remainder of the study. The rates of substitution of silent and missense mutations varied over time and among genes, but these changes were probably related to adaptation.

Several studies of experimental evolution have demonstrated that during the process of adaptation in a constant environment, fitness does not increase indefinitely. Rather, fitness plateaus are observed in adapting populations (e.g., de Visser and Lenski 2002; Elena and Lenski 2003; Lenski and Travisano 1994; Travisano et al. 1995; Moore et al. 2000). These plateaus may occur because new beneficial mutations become rare relative to deleterious mutations and/or because new beneficial mutations become smaller in effect. Although increases in fitness during experimental evolution are well studied, the dynamics of fitness declines have received less attention. Through experimental evolution with bacteriophage, Silander et al. (2007) demonstrated that fitness also *declines* to a plateau when drift overwhelms selection (figure 14.2). The authors further determined that this was due to increases in the rate of beneficial mutations as fitness decreases (the size of the mutational effect changed little), indicative of compensatory epistasis. Compensatory epistasis occurs when the mean effect of each mutational class stays constant, but the fraction of beneficial mutations increases as fitness decreases (Silander et al. 2007).

## THE LINK BETWEEN GENOTYPE AND PHENOTYPE IS RARELY SIMPLE: PLEIOTROPY AND EPISTASIS

The relationship between genotype and fitness components can be complicated by genetic effects, such as pleiotropy and epistasis. Pleiotropy arises when a single gene affects multiple phenotypic traits. Interactions between the effects of allelic differences arising from distinct genetic loci are commonly referred to as epistatic interactions. Thus, both pleiotropic and epistatic effects can influence phenotypic expression. Furthermore,

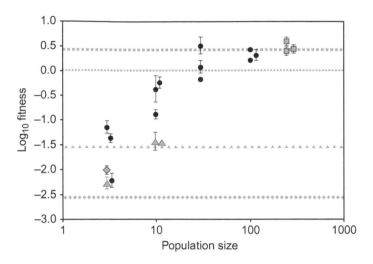

FIGURE 14.2

Changes in mean population fitness. Dotted lines indicate the fitness of ancestral clones: the high-fitness clones (black circles), lower-fitness clones (triangles and diamonds), and the evolved highest-fitness ancestor (squares); this ancestor was derived from the population having the highest fitness after ninety transfers. Each point indicates the average fitness of an evolved population, with error bars indicating one standard error. Some populations are slightly displaced on the *x*-axis for clarity. The population bottleneck sizes were 3, 10, 30, 100, and 250. Five lineages were propagated at bottleneck sizes 3 and 10, and three lineages were propagated for all other bottleneck sizes. The shape of each point indicates the ancestral clone from which that population was derived. The fitness of the high-fitness ancestor was set equal to zero on a $\log_{10}$ scale, and all other fitness values are relative. From Silander et al. (2007).

theoretical work has demonstrated that epistatic effects can have a significant impact on a number of evolutionary processes, such as the evolution of recombination, mutation load, and speciation (Barton and Charlesworth 1998; Charlesworth et al. 1993; Muller 1964; Orr 1995). Experiments with phage have proven to be particularly powerful for understanding the evolutionary implications of both pleiotropy and epistasis.

Two recent studies with phage highlight the subtle but important effects of pleiotropic interactions. First, Duffy et al. (2006) demonstrated that niche expansion, or the ability to attack novel hosts, can confer a cost due to antagonistic pleiotropy. The researchers isolated ten spontaneous host-range mutants of the phage Φ6 on three novel hosts and found that each mutant had one of nine nonsynonymous mutations in a gene important for host attachment. They then went on to demonstrate that seven of these mutations were costly in the original host, confirming the existence of antagonistic pleiotropy (figure 14.3). Second, Pepin et al. (2006) constructed ΦX174 mutants that differed by only one or two mutations in the genome and exhibited variation in host attachment ability. The authors then examined whether the genotype and the environment (different hosts) affected the predictability of host fitness. Three of the mutants exhibited rapid

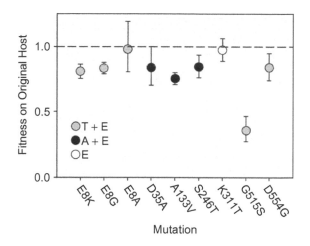

FIGURE 14.3

Frequent antagonistic pleiotropy of expanded host range mutation. The relative fitness of thirty host range mutants of Φ6 on the original host, grouped by their mutations. The fitness of the common ancestor ATCC Φ6 is shown by the dashed line. Each point represents the grand mean fitness (± 95 percent confidence interval) of the collected mutants bearing the indicated mutation. The killing spectrum of each mutation across the novel hosts is indicated as follows: solid circles, *P. syringae* pv. *atrofaciens* and *P. pseudoalcaligenes* ERA; shaded circles, *P. syringae* pv. *tomato* and *P. pseudoalcaligenes* ERA; open circle, *P. pseudoalcaligenes* ERA. From Duffy et al. (2006).

attachment but low fitness on most of the hosts. Thus, some mutations had pleiotropic effects on a fitness component other than attachment rate. The authors also found that the pleiotropic effects depended on the type of host.

Epistasis has also received much attention in phage experimental systems. Epistasis can be detected by measuring the interaction between specific mutations in determining phenotypes. Recent work by Pepin and Wichman (2007) examined variation in the epistatic effects of mutations on the host recognition sites of ΦX174. In this work, the authors constructed two single mutants and their corresponding double mutants and measured fitness on six different hosts. They showed that epistatic effects differed in degree, sign, and variability across hosts. In experiments with the RNA bacteriophage Φ6, Burch and Chao (2004) demonstrated positive epistasis, where the fitness of organisms bearing multiple mutations is higher than expected based on each mutation alone. The authors estimated the effects of deleterious mutations by performing mutation-accumulation experiments. Mutation-accumulation experiments involve subjecting genetic lineages to repeated population bottlenecks of a single individual. Previous work in this system demonstrated that sequential bottleneck passages of Φ6 lead to the accumulation of deleterious mutations (Burch and Chao 1999; Chao 1990). Burch and Chao (2004) found that mutational effects were smaller in low-fitness phage genotypes than in high-fitness types, and they concluded that the effect of deleterious mutations decreases with increasing mutation number.

When organisms suffer reductions in fitness due to adverse mutations, further mutations can occur that recover lost fitness. Phage have been used in several experiments that document the processes and mechanisms underlying fitness recovery. Compensatory mutations, which restore the fitness of evolved types to similar levels to that of the wild-type strain, are commonly found. Alternatively, organisms can recover lost fitness by reversions to the original mutations.

Bull et al. (2003) carried out a two-phase experiment to determine whether a heavily mutated phage could recover lost fitness, and to identify the types of mutations involved in this fitness recovery. Two lines of T7 were handled so as to fix mutations randomly using a combination of population bottlenecks and mutagenesis. In the first phase, consisting of growth of the phage under mutagenic conditions with frequent bottlenecks, the goal was to fix mutations at a high rate in an effort to construct an experimental phylogeny (see discussion under "Phylogenetics"; Hillis et al. 1992). In the second phase of the experiment, heavily mutated phage were maintained at large population sizes to evolve higher fitness. The authors found that substitution levels during the second phase of the experiment were less than 6 percent of those during the first phase, and only two of the changes during the recovery phase were reversions to the original mutations.

Rokyta and colleagues (2002) used site-directed mutagenesis to introduce deleterious mutations into the T7 genome and then investigate subsequent compensation. One of the first-round mutations resulted in a deletion of the viral ligase gene, which drastically reduced fitness. However, during subsequent population growth, compensatory changes occurred in genes involved in DNA metabolism resulting in the recovery of much of the lost fitness. Poon and Chao (2005) used site-directed mutagenesis in a study with $\Phi$X174 to generate twenty-one missense mutations with deleterious effects. The deleterious mutations that were most severe were more likely to be compensated for, while the overall frequency of compensatory mutations was approximately 70 percent.

In another study, lysozyme, which helps a phage lyse the host cell, was artificially deleted to investigate compensatory mutations restoring lysis in T7 (Heineman et al. 2005). Initially, phage suffered greatly reduced fitness and delayed lysis. However, after many generations of subsequent evolution, the phage approached wild-type fitness values. Mutations involved in fitness recovery were determined to be compensatory based on sequences of T7 genome (Heineman et al. 2005).

Recent work by Bull and colleagues (2007) utilized an engineered phage to understand the nature of compensatory changes in T7. The authors replaced the T7 RNA polymerase (RNAP) gene, a central component of the phage developmental pathway, with that of the T3 phage RNAP gene. This engineered phage was then selected for rapid growth, and fitness evolved from five doublings per hour to thirty-three doublings per hour, close to wild-type fitness measures. More than thirty compensatory mutations were found in the evolved genome.

Host shifts can also confer fitness costs, and the organisms must recover lost fitness if they are to persist on novel hosts. Crill et al. (2000) examined viral adaptation to novel hosts in the phage ΦX174 to understand the genetic mechanisms underlying host shifts. Adaptation of the phage to a *Salmonella* host was associated with reduced ability to then grow on the traditional *E. coli* host, whereas adaptation to *E. coli* did not affect subsequent growth on *Salmonella*. Fitness recovery on *E. coli* occurred predominantly by reversions at two to three sites on the major capsid gene, rather than by compensatory mutations elsewhere in the genome.

In addition to understanding what kind of mutations underlie fitness recovery, the time it takes an organism to recover lost fitness is also of interest. To directly examine the time course of fitness recovery in the phage Qβ, Schuppli et al. (1997) used a phage that was adapted to a host strain with an inactivated host factor gene that is required for genome replication. The authors found that, on this deficient host, initially the phage produced approximately 10,000-fold lower titer than did phage on the wild type host. However, after approximately twelve generations, the fitness of the phage returned to levels near those that it had on the wild-type host.

The fitness effects of compensatory mutations can be small, or they can be large enough to restore fitness to that of the ancestral genotype (Burch and Chao 1999). Fisher's model of adaptive evolution argues that evolution should proceed by the substitution of many mutations of small effect. As natural selection moves a population toward an adaptive peak, Fisher supposed, the process should be by small steps because advantageous mutations of small effect should be more common than advantageous mutations of large effect. Mutations of large effect are more likely to be deleterious. Furthermore, in large populations, one expects evolution to proceed by large steps. This is because although the population can harbor advantageous mutations of large and small effect, large effect mutations with beneficial effects have a higher probability of fixation. By contrast, in small populations, only common mutations appear. Thus, Fisher's model also makes the prediction that the step size should be smaller in smaller populations under selection, but empirical tests of this hypothesis have been lacking. Burch and Chao (1999) used the RNA virus Φ6 and its bacterial host *Pseudomonas* to test this idea. RNA viruses, such as Φ6, exhibit short generation times and particularly high genomic mutation rates. The authors subjected the phage to intensified genetic drift in small populations and caused fitness to decline through the accumulation of deleterious mutations. Phage fitness was then allowed to recover evolutionarily in larger population sizes. Although fitness declined in one large step, it was usually recovered in smaller steps. Furthermore, the sizes of the recovery steps were smaller in smaller recovery populations than in larger recovery populations. Thus, Fisher's main predictions were supported: advantageous mutations of small effect were more common than advantageous mutations with large effect.

One of the most important features of biology is the ability of organisms to persist in the face of change (Lenski et al. 2006). Thus, organisms must balance evolvability and robustness; in other words, they must balance resisting and allowing evolutionary change. The evolvability of an organism can be defined as the ability to generate adaptive variation. Phage are capable of rapid adaptation and thus display significant evolvability (Duffy and Turner 2009). Burch and Chao (2000) found that two Φ6 populations that were derived from the same ancestor repeatedly evolved at different rates and toward different fitness maxima. Fitness measurements showed that the fitness distributions of mutants also differed between the two populations. Thus, the evolvability of Φ6 was determined by the mutational neighborhood of each strain (Duffy and Turner 2008).

Hayashi and colleagues (2003) also investigated evolvability in phage using artificial selection. The authors asked whether an arbitrary sequence could evolutionarily acquire function in the fd-tet phage. The evolvability of the mutant was investigated by means of iterative mutation selection, where the clone with the highest infectivity was selected to be the parent of the next generation. At the end of the experiment, the clone that was produced by this evolutionary process showed a 240-fold increase in infectivity, indicating that an arbitrary sequence can indeed evolve toward acquiring function, in this case infectivity.

Mutational or genetic robustness can be defined as phenotypic constancy in the face of mutational changes in the genome (Montville et al. 2005). Robustness should be favored when mutation rates are high, which is a common feature of some RNA viruses, including the phage Φ6. However, selection for robustness may be relaxed under phage co-infection, because many phage overcome their own mutational deficiencies by co-opting proteins that are produced by more fit phage co-infecting the same host. Montville et al. (2005) hypothesized that selection for genetic robustness in Φ6 should decrease with increasing frequency of co-infection. The authors subjected sixty lineages of Φ6 to a mutation accumulation experiment. In such an experimental design, phage populations are passaged through bottlenecks—in this case, single individuals from isolated plaques were used to start the next round of population growth. Intense drift allows the fixation of nonlethal mutations, the majority of which are presumed to be deleterious. Montville and colleagues found greater variance in fitness change for the high co-infection lineages compared to the low co-infection lineages, supporting the hypothesis that selection for mutational robustness is stronger in the absence of co-infection.

More recently, Sanjuan et al. (2007) used the vesicular stomatitis RNA virus (not a bacteriophage) to test the hypothesis that at high mutation rates, slow-replicating genotypes can potentially outcompete faster counterparts if they benefit from a higher robustness. The authors analyzed the fitness distributions and genetic variability of two populations of the virus during a mutation accumulation experiment. At artificially enhanced mutation rates, the more robust viral population outcompeted the other population, despite having a lower replication rate, providing indirect evidence of natural selection for mutational robustness.

Finally, research by McBride and colleagues (2009) directly tested the relationship between evolvability and robustness with Φ6. The authors exposed robust and brittle (nonrobust) strains of the phage to high temperature and demonstrated that lineages founded by robust strains evolved greater resistance to heat shock than those founded by brittle strains. This work provides some of the first experimental evidence for a positive relationship between evolvability and robustness.

## PARALLEL EVOLUTION

Populations of closely related organisms adapting to the same environmental challenge can evolve via either parallel or divergent evolution. Experiments with phage have often exhibited parallel evolution, such as the appearance of identical evolutionary changes among lineages evolving on the same host (Wahl and Krakauer 2000). Cunningham and colleagues were some of the first researchers to document parallel evolution at the DNA sequence level (Cunningham et al. 1997). The authors serially propagated six lineages of T7 in the presence of a mutagen, and found parallel evolution of both deletions and nonsense mutations (figure 14.4).

In a more recent study, Bull et al. (2006) documented the cumulative nucleotide changes during chemostat evolution of ΦX174 on two different hosts, *S. enterica and E. coli*, in order to understand the molecular basis of the feedback between adaptation and population size. The authors found that rates of parallel evolution were extraordinarily high in the study: fifty-one of the changes that they found were identical in at least one other replicate line.

Wichman et al. (1999) adapted replicate lineages of ΦX174 to both high temperature and a novel host and found evidence of parallel evolution. Fifty percent of the changes that occurred in one lineage also occurred in the other, and all but one of the twenty-five amino acid substitutions appeared to be adaptive. However, the order of these changes differed between lineages, and parallel substitutions did not begin with alleles having the largest beneficial effects. This suggests that stochastic factors, such as the nature of the first mutation to sweep through a population or the order in which mutations arise, likely influence patterns of adaptation, even in systems where parallel evolution appears to be common (Wichman et al. 1999).

## ADAPTATION TO ABIOTIC CONDITIONS: GENOTYPE-BY-ENVIRONMENT INTERACTIONS

How organisms adapt to changes in their abiotic environment is a central problem in evolutionary biology, and one that is particularly relevant as organisms face anthropogenically induced environmental change. Several studies have addressed phage

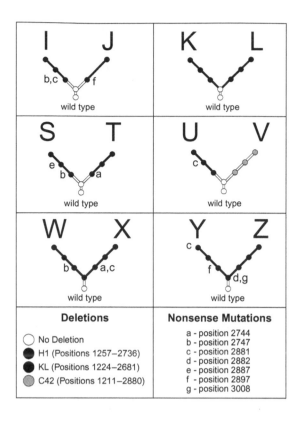

FIGURE 14.4

Parallel evolution of deletions and nonsense mutations in six bifurcating phylogenies bacteriophage T7 propagated in the presence of the mutagen nitrosoguanidine. The number of phage passages from lysate to lysate was in excess of one hundred thousand, except at each of the nodes (designated by a circle), where lineages were bottlenecked to a single plaque. DNA sequences were obtained at every node (designated by circles) except for one node in the J lineage, which was lost. In every lineage, a deletion removed between 3.6 and 4.2 percent of the genome. Each deletion created a fusion protein joining the 0.3 and 0.7 genes. Two of the deletions (KL and C42) are out of frame, and the frameshifts introduce stop codons soon after the break point (three and ten codons after, respectively). The most common deletion (H1) joins the 0.3 and 0.7 genes in frame, forming a fully translated fusion protein. Although nonsense mutations were never observed in the 0.3 portion, the 0.7 portion of the H1 fusion protein was always rapidly truncated by nonsense mutations (a–g). Only nonsense mutations appearing in the H1 lineage are shown. All position numbers are given as in the published sequence of bacteriophage T7 (Dunne and Studier 1983, cited in Cunningham et al. 1997). The number of lytic cycles between each node is fifty, except for the branch between the wild-type ancestor and the bifurcation (ten lytic cycles for U and KL, twenty for ST and UV, thirty for WX and YZ.). From Cunningham et al. (1997).

adaptation to abiotic conditions. While most studies have focused on adaptation to thermal changes (and consequently such studies comprise the bulk of our discussion), some have focused on other aspects of the environment. For example, over the course of twenty generations, Gupta et al. (1995) adapted phage T7 to tolerate inhibitory concentrations urea, a protein-denaturing agent.

Experimental adaptation to thermal environments has long served as a model for adaptation to novel environments. Temperature is relatively easy to control and measure in the laboratory and often has transparent effects on organismal growth or performance, through its effects on biochemical interactions (Knies et al. 2006). As already discussed, Holder and Bull (2001) demonstrated that phage populations evolved higher growth rates as they adapted to a high and inhibitory temperature. They also showed that fitness gains per substitution were highest in the early stages of adaptation. Furthermore, the authors observed that the fitness gains observed at high temperature were not accompanied by fitness losses at the original temperature.

The traits exhibited by organisms are the product of complex interactions between genotype and environment. One approach to characterize such interactions is the norm of reaction. A reaction norm describes the phenotypic expression of a particular genotype across a range of environments. Analyzing reaction norms in natural populations has been challenging because it requires measuring the phenotype of a particular genotype in a number of different environments. During the process of adaptation, reaction norms may shift, but *how* they shift has not been well characterized for any biological system. Taking advantage of the archivable nature of microbial experimental systems, Knies et al. (2006) revisited studies of Holder and Bull (2001) in which the phage G4 adapted to high temperatures. Several shifts in reaction norms could have led to the observed adaptation: (1) fitness across temperatures may increase, resulting in a vertical shift of the reaction norm; (2) the optimum temperature may increase, reflected as a horizontal shift; or (3) broadening of niche width (evolving from a specialist to a generalist). The authors sought to determine whether adaptation of G4 to inhibitory temperatures was due to changes in average performance, optimal temperature, or niche width. To address this question, they determined correlated responses in growth rates across a range of temperatures for genotypes isolated from several time points in the Holder and Bull (2001) evolving population. The authors observed that a horizontal shift in optimal temperature explained the majority (47.38 percent) of shape variation in the data (figure 14.5). Niche width and vertical shifts contributed only 12.65 percent and 11.75 percent, respectively. The authors were also able to associate reaction norm shifts with particular nucleotide changes.

## ADAPTATION TO BIOTIC CONDITIONS

Phage depend on their hosts for growth, reproduction, and survival. Therefore, adaptation to hosts is critical to fitness. Several areas of research are relevant to this phenomenon, such as the evolution of niche breadth, evolution of host range mutants, and diversification. Understanding the conditions that favor host range extensions in viruses also has important implications for emerging viral diseases, many of which arise as host range variants (Nichol et al. 2000).

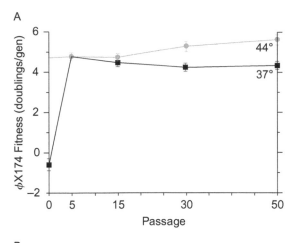

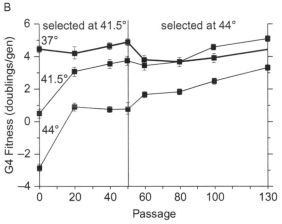

FIGURE 14.5
Fitness at multiple temperatures. a, ΦX174 at 37° and 44°; b, G4 at 37°, 41.5°, and 44°. From Holder and Bull (2001).

## TRADE-OFFS AND THE COSTS OF ADAPTATION

Most theories that describe adaptation and diversification, or the evolution of specialists and generalists, rely on the assumption that adaptation to a new environment confers an associated fitness cost. Several experiments with phage have sought to identify and measure such costs of adaptation, investigate specialist-generalist trade-offs, and identify the mechanisms underlying trade-offs.

Chao et al. (1977) demonstrated that coevolution between phage T7 and *E. coli* B yielded phage-resistant strains of bacteria and subsequent host range variants of T7 that could infect these strains. The first-order extended host range phage, $T7_1$, was capable of infecting both the ancestral *E. coli* B and the T7-resistant strain. Competition assays between $T7_1$ and the ancestral T7 on the ancestral *E. coli* B host showed that the evolved phage, $T7_1$, was less fit on the ancestral host than the narrow host range T7 (Chao et al. 1977).

Studies with the bacteriophage Ox2 have revealed the molecular mechanisms underlying phage-host receptor recognition and also provided evidence for a cost of generalism. Derivatives of Ox2 with extended host ranges were capable of infecting strains deficient in OmpA, the wild-type receptor. These host range mutants evolved to use OmpC on their new hosts and exhibit reduced affinity for OmpA on the wild-type host (Henning 1984; Morona et al. 1985; Drexler et al. 1989, 1991), suggesting a trade-off between host range and infectivity for phage Ox2. Drexler et al. (1989) sequenced the tail fiber gene (gene 38) for each sequentially isolated host range variants; for each shift, they observed primarily single point mutations yielding amino acid substitutions within the hypervariable region. The researchers found that a duplication event yielded an additional glycine residue, in the glycine-rich conserved region of the tail fiber protein, which resulted in improved adsorption to the novel host in conjunction with reduced adsorption to the original host. Based on structural modeling of the observed protein sequences, the authors suggest that the glycine-rich conserved regions offer flexibility to the protein in the form of a "wobble" (Riede et al. 1987; Drexler et al. 1989).

A number of recent studies have explicitly investigated the issue of specialist-generalist trade-offs. For example, Turner and Elena (2000) evolved replicate populations of an RNA virus on novel hosts using a single novel host or alternating novel hosts. The authors observed improvements in fitness on the novel hosts. In most cases, improvement on the selected host was associated with reduced fitness on the ancestral host. Furthermore, adaptation to one novel host did not correlate with improved performance on an unselected novel host. Those lineages that were evolved on alternating novel hosts showed improvements in fitness on both novel hosts as well as a fitness cost on the ancestral host. Interestingly, the fitness of phage lineages evolved in fluctuating novel environments was comparable to those evolved on single hosts: the generalists that evolved in the fluctuating environments did not suffer a fitness cost compared to evolved specialist phage, suggesting that simultaneous adaptation to multiple novel hosts does not constrain adaptation (Turner and Elena 2000). In different study, Crill et al. (2000) experimentally evolved phage ΦX174 to alternately infect *Salmonella* and *E. coli* by switching hosts every eleven days. When lines evolved to grow on *E. coli*, after having evolved on *Salmonella*, they maintained the ability to infect *Salmonella* with no reduction in fitness.

If "costs of adaptation" to hosts are observed, they could be the result of different evolutionary processes—specifically, antagonistic pleiotropy or mutation accumulation. As discussed previously, antagonistic pleiotropy occurs when a particular allele that is beneficial in one environment is detrimental in another environment. Alternatively, as organisms adapt to novel environments, they may accumulate mutations that are detrimental in another environment which they are not exposed to. To distinguish between these processes, Duffy et al. (2006) isolated spontaneous host range mutants on novel hosts and found that each mutant had a single nonsynonymous mutation in the gene important for host attachment. Several of these mutations were costly on the original host, consistent with antagonistic pleiotropy arising immediately, rather than mutation accumulation.

## ADAPTATION TO A CHANGING BIOTIC ENVIRONMENT: COEVOLUTION

Just as phage may gain mutations that confer the ability to infect new hosts, bacteria are also evolving and may acquire mutations that confer phage resistance. Such reciprocal evolutionary change underlies the process of coevolution. However, direct evidence of coevolution is often challenging to document for most organisms, because it requires demonstrating reciprocal evolutionary change in both species involved in the interaction; species which often have prohibitively long generation times for experimental assays. The interaction between phages and their bacterial hosts offers experimentally convenient systems for understanding the dynamics of coevolutionary interactions. In fact, one of the earliest experimental evolution studies involving bacteriophage documented coevolutionary change in T7 and *E. coli*, as described previously (Chao et al. 1977). More recently, there have been a number of studies that use phage-host systems to further understand the dynamics of coevolution (Buckling and Rainey 2002; Forde et al. 2004, 2007; Buckling et al. 2006).

Upon observing long-term coexistence of both phage and bacteria in laboratory cultures, Buckling and Rainey (2002) suspected that these experimental systems exhibited coevolutionary dynamics. They demonstrated long-term antagonistic coevolution between the *Pseudomonas fluorescence* bacteria and its naturally associated DNA phage, SBW25. The authors observed continual cycles of defense and counterdefense that were consistent with directional selection. The bacteria that had evolved in these cultures were resistant to co-occurring phage populations as well as ancestral phage, while phage samples isolated from all time points of the evolutionary process were capable of infecting ancestral bacteria. Thus, they observed local adaptation of bacteria: bacteria were resistant to contemporary phage (determined by streaking bacterial isolates against phage isolates from past, contemporary, and subsequent cultures). Although not the first to demonstrate phage-host coevolution (see Chao et al. 1977; Levin et al. 1977), this was the first study to demonstrate the persistence of coevolutionary change over three hundred bacterial generations.

In order to test the hypothesis that the presence of parasites may result in more rapid purging of deleterious mutations from host populations, Buckling et al. (2006) explored the effects of coevolution in the presence of different levels of deleterious mutations. The authors coevolved a phage parasite with bacteria harboring high mutational loads and with bacteria harboring low mutational loads. Through the course of coevolution with phage, bacterial population fitness decreased more rapidly in bacteria with higher mutation loads, an example of synergistic epistasis. Resistance was more costly in the presence of deleterious mutations. Furthermore, directional coevolution proceeded more slowly in the presence of deleterious host mutations, and, on average, bacteria with high mutational loads exhibited lower levels of resistance to sympatric phage populations, suggesting an advantage to the phage population in this coevolutionary process.

Over long time scales, coevolutionary interactions can form the basis for diversification, as was demonstrated by Buckling and Rainey (2002). In the absence of phage, the authors had previously observed rapid diversification of bacteria into a variety of niche morphotypes, resulting in high levels of sympatric diversity. In their 2002 study, the authors propagated cultures of *Pseudomonas fluorescens* with and without phage and observed that sympatric diversity was significantly lower in the presence of phage. However, allopatric diversity (between replicate flasks) was significantly higher in the presence of phage, with different morphotypes dominating different populations. The authors argued that the presence of phage decreased sympatric diversity by reducing population density and reducing competition for resources. Furthermore, they argued that increases in allopatric diversity were observed because of chance—different resistance-conferring mutations in different populations initiated divergent evolutionary trajectories.

Many have suggested that the relative number of generations that a host or parasite undergoes determines whether the host or parasite is locally adapted or maladapted (Gandon et al. 2002; Hoeksema and Forde 2008). To test this hypothesis, Morgan and Buckling (2006) indirectly manipulated the relative number of generations by separately determining the dispersal rates of host and parasites. Surprisingly, they observed that the number of generations had little effect on local adaptation over time. They concluded that, over time, resistance and infectivity evolution may proceed more slowly due to fitness costs associated with both bacterial resistance and phage infectivity. Furthermore, selection may act on other traits affecting bacterial and phage fitness in addition to the assayed traits of resistance and infectivity later on in the course of coevolution. Thus, the number of generations may be less important than factors such as migration (discussed later), which also acts to increase genetic variation. These experimental results are consistent with a recent meta-analysis of coevolutionary host-parasite interactions, which showed that generation times did not predict variation in local adaptation; instead, rate of gene flow was a significant predictor of local adaptation (Hoeksema and Forde 2008).

A number of recent studies with phage and bacteria have explicitly tested the role of migration, or gene flow, in local adaptation. Forde et al. (2004) demonstrated that gene flow across a heterogeneous landscape affects coevolution. The authors manipulated resource input into chemostats containing communities of T7 and *E. coli*. Phage and bacteria were dispersed together from high to intermediate to low resource communities in a stepping-stone fashion. Another set of chemostats in which bacteria and phage were not dispersed, but which differed in resource input, served as controls. The authors found that differences in resource input resulted in differences in levels of local adaptation of the phage to the bacteria. On average, adaptation was greatest in the high-resource communities. The authors also demonstrated that local adaptation of the phage to the bacteria was higher in communities that were open to dispersal, suggesting that gene flow offered beneficial mutations. In contrast, several of the closed communities exhibited evidence of maladaptation. In another study, Morgan et al. (2005) separately manipulated the dispersal rates of coevolving hosts and parasites. The authors found that phage migration had a large effect on phage local adaptation, but bacterial migration had little effect. Overall, this body of work provides further evidence that, at relatively low rates, migration provides gene flow and evolutionary potential.

In contrast to the beneficial effects of migration at relatively low rates, at high rates of migration, local adaptation should be reduced, or organisms may even be maladapted to their environment. This can occur because locally beneficial genotypes may be swamped by genotypes that are not locally adapted. Brockhurst et al. (2007) tested this hypothesis by measuring the rate of coevolution under different migration regimes. They found that local adaptation of bacteria to the phage was highest at intermediate rates of migration. High immigration rates favored the phage, because they increased the relative number of susceptible hosts. This resulted in maximal sympatric resistance at intermediate rates of migration (figure 14.6). Furthermore, although coevolution was accelerated at low and high migration rates, at intermediate migration rates bacteria were locally adapted, slowing the rate of evolution in the phage, which in turn affected the rate of evolution in the bacteria.

The complex effects of migration on subpopulations within a metapopulation were studied by Kerr et al. (2006). Metapopulation theory has shown that extinction-prone subpopulations can be recolonized by migration, enabling global persistence. However, too little migration prevents recolonization, and too much migration synchronizes population dynamics and increases the risk of global extinction. Both the magnitude of migration and the pattern of migration can influence evolutionary dynamics. Kerr et al. (2006) demonstrated that different patterns of migration (spatially restricted vs.

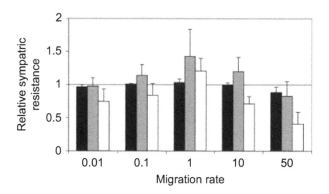

FIGURE 14.6

The effect of migration rate on evolution of bacterial resistance ranges through time. Bars show mean (± 1 SEM) proportion of bacteria resistant to phages from all migration treatments from the same time point. Black bars are transfer (time point) 4; light gray bars, transfer 8; dark gray bars, transfer 12. The system is directional, with bacteria becoming more resistant to a wider range of genotypes, so a higher proportion of resistant bacteria indicates that populations have evolved more rapidly. From Brockhurst et al. (2007).

unrestricted) select for different pathogen strategies. Highly connected dispersal favors the evolution of virulent, more rapacious phage. In contrast, more restricted migration favors the more prudent use of a shared resource, averting a tragedy of the commons. This research is discussed further in Rauser et al. (this volume), from the standpoint of levels of selection.

## PHYLOGENETICS

Understanding patterns of diversity is central to both ecology and evolution. These patterns are the product of adaptation and evolution over long time periods, and characterizing the resulting macroevolutionary patterns hinges on phylogenetic analysis. Phylogenetics is the study of the evolutionary relatedness of organisms. The ancestral states in most phylogenies are rarely known with certainty, and the processes underlying patterns of relatedness of macro-organisms can often only be inferred indirectly. However, experimentally generated phylogenies provide the rare situation in which the actual ancestors are known, and such phylogenetic experiments offer the opportunity to understand the evolutionary processes that underlie patterns of relatedness (Hillis et al. 1992; Oakley and Cunningham 2000; Oakley this volume).

Evolution experiments with phage have provided powerful tests of the accuracy of molecular phylogenies because the phylogenetic history is known in these systems. Hillis et al. (1992) exposed T7 to a mutagen to create the first completely known phylogeny based on restriction site maps. The authors used five different methods to reconstruct the branching pattern and they found that all methods predicted the correct topology but varied in their predictions of branch lengths. Based on the most parsimonious phylogeny, Bull and colleagues (1993) went on to produce a model that described the rates at which restriction sites were gained and lost. This model was used to determine the rates of molecular divergence and convergence in the experimental phylogeny. Hillis and colleagues (1994) then extended tests of phylogenetic accuracy in experimental lineages of T7 by comparing the observed amount of changes in nucleotide sequences to the number inferred from parsimony methods. As in the previous study (Hillis et al. 1992), all the methods tested were successful at recovering the known phylogeny but the analyses differed in their ability to estimate correctly the branch lengths of the phylogeny (figure 14.7).

Further research using phage T7 evaluated phylogenetic reconstruction methods based on continuous phenotypic characters in the phage (Oakley and Cunningham 2000). The authors quantified different measures of growth rate of ancestral and terminal types from the known T7 phylogeny, and then used the character values of the terminal types to reconstruct the ancestral character values. In this case, the authors found that, aside from one slowly evolving character, the estimated ancestral states were highly inaccurate, even when including a known ancestor in the root of the tree. However, Oakley and Cunningham (2000) demonstrated that although explicit ancestor reconstruction

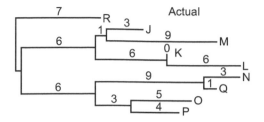

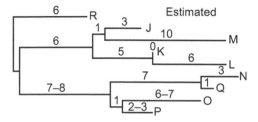

FIGURE 14.7
Comparison of an observed phylogeny of viruses derived from bacteriophage T7 with an estimated phylogeny from the parsimony method, on the basis of analysis of the terminal sequences (J–R). The numbers above the branches indicate the actual or estimated number of substitutions that occurred along the respective lineages. The actual number of substitutions were determined by sequencing the ancestral viruses. Ranges of values on the estimated tree indicate that multiple, equally parsimonious reconstructions of character states are possible. From Hillis et al. (1994).

failed, independent contrasts accurately estimated the correlation among characters. Thus, these experimentally generated phage phylogenies enabled the first direct tests of the methods used to reconstruct character evolution, and they showed that independent contrasts should be used when possible.

## IMPLICATIONS OF PHAGE EVOLUTION EXPERIMENTS
### MEDICAL APPLICATIONS

Experimental evolution studies with phage offer powerful systems for investigating the process of adaptation, particularly because they readily extend from identifiable molecular changes to their fitness effects. Such experiments with phage also offer important insights for treating emerging infectious diseases, phage therapy, and vaccine development. Understanding the conditions that favor hosts and parasites in the coevolutionary process is central to understanding disease evolution (Woolhouse et al. 2002). In the host range study by Duffy and colleagues (2006) described earlier, the authors found some evidence for antagonistic pleiotropy, but they also observed some point mutations responsible for host range extension that exhibited little or no cost. Antagonistic pleiotropy should limit the conditions favoring host range extension, so the observation that some mutations do not exhibit antagonistic pleiotropy would actually broaden the conditions under which expanded host range might evolve. Furthermore, the authors characterized some host range mutations that also conferred the ability to infect unselected novel hosts, which is particularly alarming in light of the potential for zoonotic RNA viruses to rapidly emerge.

Experiments with phage may also help guide the development of attenuated viral vaccines. Typically attenuated viruses are developed through in vitro passage experiments in order to reduce virulence. However, a better understanding of fitness changes during adaptation would aid in determining the duration of propagation required for attenuation, and an understanding of the genetics of adaptation would provide more reasonable expectations for the number of substitutions required for reversion to virulence (Holder and Bull 2001).

Not only are phage-bacterial systems important models of host-parasite interactions and disease evolution, but phage can also serve as useful clinical tools. The use of phage to treat bacterial infections is not new (Merril 1996; Ho 2001), however the increased incidence of bacterial infections that defy antibiotic treatment has prompted renewed interest in using phages as therapeutic tools. The extent to which phage therapy offers a means to eliminate or control the target pathogen in a reliable manner depends on how these organisms coevolve. Recently, researchers investigated coevolution between the pathogenic *E. coli* O157:H7 and bacteriophage PP01 (Mizoguchi et al. 2003). The addition of phage PP01 to *E. coli* O157:H7 in chemostat culture initially resulted in cell lysis. Soon, however, phage-resistant isolates were observed as well as mutant phage capable of infecting these new mutant cells. With an improved understanding of the mechanisms underlying coevolution, phage therapy may offer a complement or alternative to traditional antibiotic therapy.

## PHAGES IN NATURE

It is estimated that there $10^{31}$ viruses on earth, and the majority of these are phages that infect bacteria (Breitbart and Rohwer 2005). As obligate parasites of bacteria, phages impact the biosphere by affecting cycling of nutrients in the marine environment (Breitbart and Rohwer 2005). The study of viruses and bacteriophage in nature has typically fallen within the discipline of environmental microbiology; consequently, it has been fairly removed from the laboratories of experimental evolutionists. However, this, too, has begun to change. Several recent laboratory experimental studies have directly addressed the evolution of phage in nature. For example, in studying the genes responsible for thermal adaptation discussed above, Knies et al. (2006) identified mutations associated with shifts in reaction norms for their laboratory strains. They then went on to show that one particular mutation which was favored in the course of laboratory evolution was also found in natural G4-like phage populations that had recently been isolated. They found that this mutation was associated with increased growth rates at high temperatures, but not at low temperatures, consistent with laboratory evolution results.

Collaborations between researchers characterizing ecological patterns in microbial diversity and function in the field and those conducting phage evolution experiments in the laboratory is a key direction for future research. Such an interdisciplinary approach to testing evolutionary theory will greatly advance the understanding of evolution across all scales of biology-from molecular genetics to fitness to ecosystem effects.

## CONCLUSION

The relative ease of evolution experiments with microbes, as well as the associated genetic and physiological information available for many microbes, provides a powerful platform for testing evolutionary theories. However, performing useful experiments with phage requires researchers to understand the field of evolutionary biology in conjunction with the physiology and genetics of the phage and the host bacterium (Duffy and Turner 2008). We caution that although microbial systems are touted for their simplicity, there are considerable intricacies that cannot be ignored and that can often lead to the same complex and often confusing results found in experiments with macroorganisms (which is discussed in Irschick and Reznick this volume).

We have discussed several studies that illustrate the utility of phage experimental studies in understanding relationships between the fundamentals of molecular machinery and models of evolutionary processes (Bull et al. 1993). Experiments with phage can be used to directly link genotypes to measures of fitness, they can be used to understand the processes that lead to the branching patterns of molecular evolution, and they can advance our understanding of the generation and maintenance of biological diversity.

## SUMMARY

The primary goal of experimental evolution is to directly test theories of evolution in a controlled setting. Microbes have become increasingly popular tools to test evolutionary theory due to their ease of manipulation, large population sizes, and short generation times. Although bacteriophage have long been a workhorse of microbiology, historically there has been a division between evolutionary biologists and phage biologists. Evolutionary biologists are now recognizing the advantages of employing relatively simple and rapidly replicating organisms in their research, while microbiologists are appreciating the importance of evolutionary theory in their work. This chapter highlights recent research with phage to illustrate how evolutionary processes and patterns can be evaluated using experiments with microbes. Studies with phage provide a rare opportunity to directly link the genotype to the phenotype, and to ultimately understand the processes that lead to patterns of divergence from a common ancestor.

## REFERENCES

Barton, N. H., and B. Charlesworth. 1998. Why sex and recombination? *Science* 281:1986–1990.

Bohannan, B. J. M., and R. E. Lenski. 1997. Effect of resource enrichment on a chemostat community of bacteria and bacteriophage. *Ecology* 78:2303–2315.

———. 1999. Effect of prey heterogeneity on the response of a model food chain to resource enrichment. *American Naturalist* 153:73–82.

———. 2000. The relative importance of competition and predation varies with productivity in a model community. *American Naturalist* 156:329–340.

Breitbart, M., and F. Rohwer. 2005. Here a virus, there a virus, everywhere the same virus? *Trends in Microbiology* 13:278–284.

Brockhurst, M. A., A. Buckling, V. Poullain, and M. E. Hochberg. 2007. The impact of migration from parasite-free patches on antagonistic host-parasite coevolution. *Evolution: International Journal of Organic Evolution* 61:1238–1243.

Buckling, A., and P. B. Rainey. 2002. Antagonistic coevolution between a bacterium and a bacteriophage. *Proceedings of the Royal Society of London B, Biological Sciences* 269:931–936.

Buckling, A., Y. Wei, R. C. Massey, M. A. Brockhurst, and M. E. Hochberg. 2006. Antagonistic coevolution with parasites increases the cost of host deleterious mutations. *Proceedings of the Royal Society of London B, Biological Sciences* 273:45–49.

Bull, J. J., M. R. Badgett, D. Rokyta, and I. J. Molineux. 2003. Experimental evolution yields hundreds of mutations in a functional viral genome. *Journal of Molecular Evolution* 57:241–248.

Bull, J. J., M. R. Badgett, H. A. Wichman, J. P. Huelsenbeck, D. M. Hillis, A. Gulati, C. Ho, et al. 1997. Exceptional convergent evolution in a virus. *Genetics* 147:1497–1507.

Bull, J. J., C. W. Cunningham, I. J. Molineux, M. R. Badgett, and D. M. Hillis. 1993. Experimental molecular evolution of bacteriophage T7. *Evolution* 47:993–1007.

Bull, J. J., J. Millstein, J. Orcutt, and H. A. Wichman. 2006. Evolutionary feedback mediated through population density, illustrated with viruses in chemostats. *American Naturalist* 167:E39–E51.

Bull, J. J., R. Springman, and I. J. Molineux. 2007. Compensatory evolution in response to a novel RNA polymerase: Orthologous replacement of a central network gene. *Molecular Biology and Evolution* 24:900–908.

Burch, C. L., and L. Chao. 1999. Evolution by small steps and rugged landscapes in the RNA virus phi6. *Genetics* 151:921–927.

———. 2000. Evolvability of an RNA virus is determined by its mutational neighbourhood. *Nature* 406:625–628.

———. 2004. Epistasis and its relationship to canalization in the RNA virus phi 6. *Genetics* 167:559–567.

Campbell, A. 1961. Sensitive mutants of bacteriophage lambda. *Virology* 14:22.

Chao, L. 1990. Fitness of RNA virus decreased by Muller's ratchet. *Nature* 348:454–455.

Chao, L., B. R. Levin, and F. M. Stewart. 1977. A complex community in a simple habitat: An experimental study with bacteria and phage. *Ecology* 58:369–378.

Charlesworth, D., M. T. Morgan, and B. Charlesworth. 1993. Mutation accumulation in finite outbreeding and inbreeding populations. *Genetical Research* 61:39–56.

Crill, W. D., H. A. Wichman, and J. J. Bull. 2000. Evolutionary reversals during viral adaptation to alternating hosts. *Genetics* 154:27–37.

Cuevas, J. M., S. F. Elena, and A. Moya. 2002. Molecular basis of adaptive convergence in experimental populations of RNA viruses. *Genetics* 162:533–542.

Cunningham, C. W., K. Jeng, J. Husti, M. Badgett, I. J. Molineux, D. M. Hillis, and J. J. Bull. 1997. Parallel molecular evolution of deletions and nonsense mutations in bacteriophage T7. *Molecular Biology and Evolution* 14:113–116.

Davis, R. H. 2003. *The Microbial Models of Molecular Biology: From Genes to Genomes.* New York: Oxford University Press.

de Visser, J. A., and R. E. Lenski. 2002. Long-term experimental evolution in *Escherichia coli*. XI. Rejection of non-transitive interactions as cause of declining rate of adaptation. *BMC Evolutionary Biology* 2:19.

Drexler, K., J. Dannull, I. Hindennach, B. Mutschler, and U. Henning. 1991. Single mutations in a gene for a tail fiber component of an *Escherichia coli* phage can cause an extension from a protein to a carbohydrate as a receptor. *Journal of Molecular Biology* 219:655–663.

Drexler, K., I. Riede, D. Montag, M. L. Eschbach, and U. Henning. 1989. Receptor Specificity of the *Escherichia-coli* T-even type phage-Ox2: Mutational alterations in host range mutants. *Journal of Molecular Biology* 207:797–803.

Duffy, S., and P. E. Turner. 2008. Phage evolutionary biology. *In* S. T. Abedon, ed. *Bacteriophage Ecology: Population Growth, Evolution, and Impact of Bacterial Viruses*. Cambridge, Cambridge University Press.

Duffy, S., P. E. Turner, and C. L. Burch. 2006. Pleiotropic costs of niche expansion in the RNA bacteriophage phi 6. *Genetics* 172:751–757.

Elena, S. F., and R. E. Lenski. 2003. Evolution experiments with microorganisms: The dynamics and genetic bases of adaptation. *Nature Reviews Genetics* 4:457–469.

Forde, S. E., J. N. Thompson, and B. J. Bohannan. 2007. Gene flow reverses an adaptive cline in a coevolving host-parasitoid interaction. *American Naturalist* 169:794–801.

Forde, S. E., J. N. Thompson, and B. J. M. Bohannan. 2004. Adaptation varies through space and time in a coevolving host-parasitoid interaction. *Nature* 431:841–844.

Gandon, S., P. Agnew, and Y. Michalakis. 2002. Coevolution between parasite virulence and host life-history traits. *American Naturalist* 160:374–388.

Gupta, K., Y. Lee, and J. Yin. 1995. Extremo-phage: In vitro selection of tolerance to a hostile environment. *Journal of Molecular Evolution* 41:113–114.

Hayashi, Y., H. Sakata, Y. Makino, I. Urabe, and T. Yomo. 2003. Can an arbitrary sequence evolve towards acquiring a biological function? *Journal of Molecular Evolution* 56:162–168.

Heineman, R. H., I. J. Molineux, and J. J. Bull. 2005. Evolutionary robustness of an optimal phenotype: re-evolution of lysis in a bacteriophage deleted for its lysin gene. *Journal of Molecular Evolution* 61:181–191.

Hillis, D. M., J. J. Bull, M. E. White, M. R. Badgett, and I. J. Molineux. 1992. Experimental phylogenetics: generation of a known phylogeny. *Science* 255:589–592.

Hillis, D. M., J. P. Huelsenbeck, and C. W. Cunningham. 1994. Application and accuracy of molecular phylogenies. *Science* 264:671–677.

Ho, K. 2001. Bacteriophage therapy for bacterial infections: Rekindling a memory from the pre-antibiotics era. 44:1–16.

Hoeksema, J. D., and S. E. Forde. 2008. A meta-analysis of factors affecting local adaptation between interacting species. *American Naturalist* 171:275–290.

Holder, K. K., and J. J. Bull. 2001. Profiles of adaptation in two similar viruses. *Genetics* 159:1393–1404.

Kerr, B., C. Neuhauser, B. J. Bohannan, and A. M. Dean. 2006. Local migration promotes competitive restraint in a host-pathogen "tragedy of the commons." *Nature* 442:75–78.

Knies, J. L., R. Izem, K. L. Supler, J. G. Kingsolver, and C. L. Burch. 2006. The genetic basis of thermal reaction norm evolution in lab and natural phage populations. *PLoS Biology* 4:e201.

Lenski, R. E., J. E. Barrick, and C. Ofria. 2006. Balancing robustness and evolvability. *PLoS Biology* 4:e428.

Lenski, R. E., and B. R. Levin. 1985. Constraints on the coevolution of bacteria and virulent phage: A model, some experiments, and predictions for natural communities. *American Naturalist* 125:585–602.

Lenski, R. E., and M. Travisano. 1994. Dynamics of adaptation and diversification: A 10 ;000-generation experiment with bacterial-populations. *Proceedings of the National Academy of Sciences of the USA* 91:6808–6814.

Levin, B. R., F. M. Stewart, and L. Chao. 1977. Resource-limited growth, competition, and predation: A model and experimental studies with bacteria and bacteriophage. *American Naturalist* 111:3–24.

Luria, S. E., and M. Delbruck. 1943. Mutations of bacteria from virus sensitivity to virus resistance. *Genetics* 28:491–511.

McBride R. C. and P. E. Turner. 2009. Genetic robustness and evolvability. *Microbe* 9:409–415.

Merril, C. R. 1996. Long-circulating bacteriophage as antibacterial agents. *Proceedings of the National Academy of Sciences of the USA* 93:3188–3192.

Meselson, M., and J. J. Weigle. 1961. Chromosome breakage accompanying genetic recombination in bacteriophage. *Proceedings of the Royal Society of London B, Biological Sciences* 47:857–868.

Mizoguchi, K., M. Morita, C. R. Fischer, M. Yoichi, Y. Tanji, and H. Unno. 2003. Coevolution of bacteriophage PP01 and Escherichia coli O157:H7 in continuous culture. *Applied and Environmental Microbiology* 69:170–176.

Montville, R., R. Froissart, S. K. Remold, O. Tenaillon, and P. E. Turner. 2005. Evolution of mutational robustness in an RNA virus. *PLoS Biology* 3:e381.

Moore, F. B., D. E. Rozen, and R. E. Lenski. 2000. Pervasive compensatory adaptation in Escherichia coli. *Proceedings of the Royal Society of London B, Biological Sciences* 267:515–522.

Morgan, A. D., and A. Buckling. 2006. Relative number of generations of hosts and parasites does not influence parasite local adaptation in coevolving populations of bacteria and phages. *Journal of Evolutionary Biology* 19:1956–1963.

Morgan, A. D., S. Gandon, and A. Buckling. 2005. The effect of migration on local adaptation in a coevolving host-parasite system. *Nature* 437:253–256.

Morona, R., and U. Henning. 1984. Host range mutants of bacteriophage Ox2 can use two different outer-membrane proteins of *Escherichia coli* K12 as receptors. *Journal of Bacteriology* 159:579–582.

Morona, R., C. Kramer, and U. Henning. 1985. Bacteriophage receptor area of outer-membrane protein ompA of *Escherichia coli* K12. *Journal of Bacteriology* 164:539–543.

Muller, H. J. 1964. The relation of recombination to mutational advance. *Mutational Research* 1:2–9.

Nichol, S. T., J. Arikawa, and Y. Kawaoka. 2000. Emerging viral diseases. *Proceedings of the National Academy of Sciences of the USA* 97:12411–12412.

Oakley, T. H., and C. W. Cunningham. 2000. Independent contrasts succeed where ancestor reconstruction fails in a known bacteriophage phylogeny. *Evolution: International Journal of Organic Evolution* 54:397–405.

Orr, H. A. 1995. The population genetics of speciation: the evolution of hybrid incompatibilities. *Genetics* 139:1805–1813.

Pepin, K. M., M. A. Samuel, and H. A. Wichman. 2006. Variable pleiotropic effects from mutations at the same locus hamper prediction of fitness from a fitness component. *Genetics* 172:2047–2056.

Pepin, K. M., and H. A. Wichman. 2007. Variable epistatic effects between mutations at host recognition sites in phiX174 bacteriophage. *Evolution: International Journal of Organic Evolution* 61:1710–1724.

Poon, A., and L. Chao. 2005. The rate of compensatory mutation in the DNA bacteriophage phiX174. *Genetics* 170:989–999.

Riede, I., K. Drexler, M. L. Eschbach, and U. Henning. 1987. DNA sequence of genes 38 encoding a receptor-recognizing protein of bacteriophage T2, bacteriophage K3 and of bacteriophage K3 host range mutants. *Journal of Molecular Biology* 194:31–39.

Rokyta, D., M. R. Badgett, I. J. Molineux, and J. J. Bull. 2002. Experimental genomic evolution: extensive compensation for loss of DNA ligase activity in a virus. *Molecular Biology and Evolution* 19:230–238.

Rouch, W. 1997. Biology departments restructure. *Science* 275:1556–1558.

Sanjuan, R., J. M. Cuevas, V. Furio, E. C. Holmes, and A. Moya. 2007. Selection for robustness in mutagenized RNA viruses. *PLoS Genetics* 3:e93.

Schuppli, D., G. Miranda, H. C. Tsui, M. E. Winkler, J. M. Sogo, and H. Weber. 1997. Altered 3'-terminal RNA structure in phage Qb adapted to host factor-less *Escherichia coli*. *Proceedings of the National Academy of Sciences of the USA* 94:10239–10242.

Silander, O. K., O. Tenaillon, and L. Chao. 2007. Understanding the evolutionary fate of finite populations: the dynamics of mutational effects. *PLoS Biology* 5:e94.

Travisano, M., J. A. Mongold, A. F. Bennett, and R. E. Lenski. 1995. Experimental tests of the roles of adaptation, chance, and history in evolution. *Science* 267:87–90.

Turner, P. E., and L. Chao. 1998. Sex and the evolution of intrahost competition in RNA virus phi6. *Genetics* 150:523–532.

Turner, P. E., and S. F. Elena. 2000. Cost of host radiation in an RNA virus. *Genetics* 156:1465–1470.

Wahl, L. M., and D. C. Krakauer. 2000. Models of experimental evolution: the role of genetic chance and selective necessity. *Genetics* 156:1437–1448.

Weitz, J. S., H. Hartman, and S. A. Levin. 2005. Coevolutionary arms races between bacteria and bacteriophage. *Proceedings of the National Academy of Sciences of the USA* 102:9535–9540.

Wichman, H. A., M. R. Badgett, L. A. Scott, C. M. Boulianne, and J. J. Bull. 1999. Different trajectories of parallel evolution during viral adaptation. *Science* 285:422–424.

Wichman, H. A., J. Millstein, and J. J. Bull. 2005. Adaptive molecular evolution for 13,000 phage generations: a possible arms race. *Genetics* 170:19–31.

Wichman, H. A., L. A. Scott, C. D. Yarber, and J. J. Bull. 2000. Experimental evolution recapitulates natural evolution. *Philosophical Transactions of the Royal Society of London B, Biological Sciences* 355:1677–1684.

Woolhouse, M. E., J. P. Webster, E. Domingo, B. Charlesworth, and B. R. Levin. 2002. Biological and biomedical implications of the co-evolution of pathogens and their hosts. *Nature Genetics* 32:569–577.

You, L., and J. Yin. 2002. Dependence of epistasis on environment and mutation severity as revealed by in silico mutagenesis of phage T7. *Genetics* 160:1273–1281.

# 15

# EXPERIMENTAL APPROACHES TO STUDYING THE EVOLUTION OF ANIMAL FORM
## The Shape of Things to Come

### W. Anthony Frankino, Douglas J. Emlen, and Alexander W. Shingleton

A systematic study of the effects of selecting for different kinds of proportional change in body parts would provide valuable evidence on the relative constancy or liability of the growth patterns which determine morphology. Indeed certain questions can be answered only this way. Selection is a useful tool which remains largely unexploited by students of physiology and development. Planned changes in growth and form can provide a wide range of differences for comparison and analysis.

F. W. ROBERTSON (1962)

Morphology most often evolves not through the appearance of new or "novel" traits, but through changes in the shape of existing structures (figure 15.1). Thompson presented shape variation as deformations in the dimensions and relative size of body parts (Thompson 1917); his approach captures differences in organismal shape across groups by

FIGURE 15.1

Morphological diversity as generated through changes in shape and relative size of existing structures. Morphological variation among primates, shown in the first two rows, results from changes in the orbital fossa, teeth, maxilla and mandibles. Diversity in these same structures is even more extreme among other mammals, as shown in the bottom two rows. From left top to right, primate skulls are from an owl monkey, human, orangutan, gorilla, gelada baboon, and mandrill. Other mammal skulls belong to a walrus, babirusa, wombat, rough-toothed dolphin, beaver, and wolverine. Photography by David Littschwager with the California Academy of Sciences; all images used with permission.

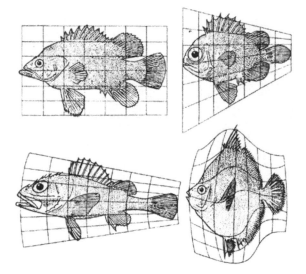

FIGURE 15.2
Transformation of fish morphology represented as deformation
of the Cartesian coordinates of a
generic fish shape (upper left) to
produce the changes in the relative size of eyes, fins, mouths,
and body dimensions of other
species (after Thompson 1917).

compressing, stretching, or bending a reference image, such as a common ancestral form or the mean shape calculated from lineages exhibiting shape variation (figure 15.2). This approach illustrates intuitively how disproportionate changes in dimensions across a generic, reference shape can produce both subtle variations on morphological themes and dramatically different, specialized morphologies. Such changes in animal shape are responsible for much of the gross-level morphological diversity in multicellular life, contributing to variation at all taxonomic levels: among orders, families, genera, species, and populations and even between sexes or among alternative morphotypes within a sex.

Variation in shape is most obvious when morphological proportions reach extremes. Species that attain extraordinary, disproportionate trait sizes are often the focus of great attention by hobbyists, agriculturalists, and scientists. In an effort to produce attractive or more profitable forms, hobbyists and agriculturalists have used artificial selection to generate varieties with extreme—even grotesque—morphologies, such as many domestic breeds of pigeon or dog (Darwin 1859; figure 15.3A–B), and livestock with larger profitable parts (e.g., disproportionately large breast meat weights in domestic chickens; see Le Bihan-Duval et al. 1999). Not all extreme morphologies are artificially produced, however; animals of many species naturally exhibit extreme trait proportions. Fiddler crabs, swordtail fish, spoonbills (figure 15.3C), stalk-eyed flies (see also Swallow et al. this volume), and horned beetles are just a few examples of organisms that derive their common names from exaggerations in the relative size of particular body parts. Such extreme morphologies often are associated with the evolution of highly derived behaviors or ecological specialization, and sexual selection in particular has yielded a tremendous diversity of extreme animal shapes (Darwin 1859; Andersson 1982).

Even when it is not extreme, shape variation is important. Proper scaling of morphological traits is so ubiquitous that it typically goes unnoticed. We expect that, within

FIGURE 15.3

Evolved diversity in animal shape as produced through changes in the relative trait size. Such changes can result from artificial or natural selection, as shown by the sequences of (A) pigeons and (B) dogs. In each sequence, the upper left panel shows the ancestral species of the derived, domesticated varieties that follow in subsequent panels. Domestic pigeons are selected winners from the 2007 National Pigeon Association Grand National Competition. The domestic dogs have been identified by the American Kennel Club as exemplar specimens of various breeds. The bottom row (C) shows a spoonbill, fiddler crab, and swordtail fish, all named for their naturally occurring extreme morphologies. Pigeon images kindly provided by the National Pigeon Association; dog images provided by the American Kennel Club, copyright M. Bloom; spoonbill provided by G. Coe, Wild Images Florida; crab by H. Jones, and swordtail by I. Langworth. All images used with permission.

species, individuals with larger bodies will tend to have larger limbs, larger teeth, and larger internal organs, whereas smaller individuals will tend to have parts of reduced size. Such scaling of body parts makes intuitive sense; ecological performance requires that functionally related suites of morphological traits be scaled properly to each other and to the size of the body. For example, efficient flight requires that wings be the correct size relative to the body, cursorial locomotion similarly requires that limbs be scaled appropriately to one another and to the body, and feeding efficiency benefits from jaws that scale properly to one another and to the size of the head.

In sum, the scaling of body parts may be the quintessential feature of animal morphology (Calder 1984; Schmidt-Nielsen 1984). Individual deviations in the relative sizes of body parts comprise one of the most pervasive and biologically relevant sources of phenotypic variation in natural (and artificial) populations, and changes in the underlying developmental mechanisms responsible for this variation likely have contributed to the majority of historical transformations in animal form. Although shape variation is somewhat easy to quantify (e.g., see Bookstein 1997; Dryden and Mardia 2002; Zelditch et al. 2004), the developmental variation that gives rise to it is not. Consequently, we know a great deal more about the phylogenetic patterns in morphological diversity than the developmental changes that have produced these patterns.

Here, we briefly review traditional and newer approaches to the empirical study of biological form, highlighting exciting ways that experimental evolution studies are contributing to an improved understanding of why and how particular transformations in animal shape occur. We first review different mathematical methods for quantifying and comparing organismal shape. We focus on one methodology in particular: the use of the scaling relationship between two traits. This function describes shape as the size relationship between any two structures within any biological group. For example, scaling relationships can describe how the size of an appendage, such as a wing, is related to body size among species. They can also describe how the relative size of two appendages, such as a forewing and a hind wing, change within or among populations of a single species. Scaling relationships can also be used to describe how the changes in the dimensions of a single structure, like the width and span of a wing, are related among individuals of a single population. After discussing the utility, limitations, and complications of using scaling relationships to describe shape variation, we examine the biology of shape expression and evolution. We then review case studies where experimental evolution has been used to explore various aspects of scaling relationship evolution. As we review these research programs, our goal is to identify what phenotype-based (i.e., "top-down") and developmentally based (i.e., "bottom-up"; see also Dykhuizen and Dean this volume) approaches have revealed about the expression and diversification of morphology. We close the chapter with some generalities regarding the power experimental evolution offers for the future of studying the proximate and ultimate factors shaping the evolution of form.

## MATHEMATICAL DESCRIPTIONS OF SHAPE VARIATION

Since Thompson's deformations, a variety of mathematical techniques have been developed to describe shape variation within biological groups. Perhaps the most widely used method is to calculate the scaling relationship between two morphological traits. Developed by Huxley, this approach models organismal shape as a power equation with the form $y = bx^\alpha$, where $x$ is the size (length or mass) of the body and $y$ is the size of some trait (Huxley 1924; Huxley and Teissier 1936a, 1936b; see also Lapicque 1907; Teissier 1926). Log-transforming this equation yields $\log(y) = \alpha\log(x) + \log(b)$, and log-log plots of the size of different traits among individuals of the same species often reveal linear scaling relationships with an intercept of $b$ and a slope of $\alpha$ (figure 15.4A).

Frequently, the allometric coefficient, $\alpha$, is close to 1, such that the relative size of the two traits is constant. This reflects a condition where all individuals have the same shape, regardless of size; such geometric scaling is called *isometry*. However, the size of $y$ may change disproportionally with $x$, producing a condition called *allometry* where shape changes across body sizes. More specifically, hyperallometries occur when $\alpha > 1$, whereas hypoallometries occur when $\alpha$ is $<1$. The allometric equation, therefore, describes the scaling between the size of morphological structures and the body as a single complex trait that reflects the covariation between size and shape (figure 15.4B).

Hyperallometric traits are often greatly exaggerated in size and represent some of the best-studied cases of morphological scaling relationships (e.g., Huxley 1932; Eberhard

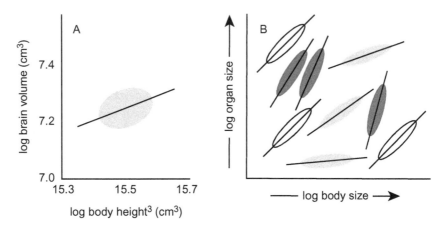

FIGURE 15.4

Scaling relationships as descriptors of shape variation. A, Shape variation within a group of organisms is shown by a line fit to a data cloud representing the size of two traits for a group of organisms. Shape variation within a group of organisms is shown by a line (dark line) fit to a data cloud (gray elipse) representing the size of two traits for a group of organisms, in this case the brain-body size relationship in humans (data from Koh 2005). B, Scaling relationships are divided into three classes based on the pattern of variation they describe. Here, hypothetical populations are represented by ellipses and each is fit with a scaling relationship. Isometric scaling ($\alpha = 1$) is shown by the white ellipses, hyperallometry ($\alpha > 1$) by the dark populations, and hypoallometry ($\alpha < 1$) by the gray populations.

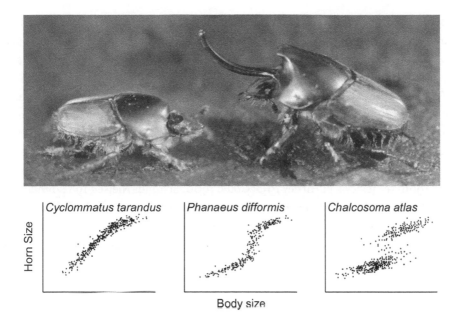

FIGURE 15.5

Nonlinear scaling relationships. Like many horned beetles, *Onthophagus nigriventris* exists as two alternative, size-dependent morphs. "Majors" (right beetle) possess large bodies and develop exaggerated horns on their bodies or heads, whereas "minors" (left beetle) are smaller and have small or no horns. The scaling relationships between horn and body size is often complex in such systems, as illustrated in the bottom panels. Photo: D. Emlen; data are from Emlen and Nijhout 2000.

and Gutierrez 1991; Emlen and Nijhout 2000; figure 15.5). Traits exhibiting hyperallo-metric relationships can rapidly convey honest information about individual size or quality (Nur and Hasson 1984; Zeh and Zeh 1988; Wilkinson and Dodson 1997a; Bonduriansky and Day 2003; Kodric-Brown et al. 2006; Bonduriansky 2007b) and are therefore often the product of sexual selection (see also Swallow et al. this volume), where the degree of exaggeration exhibited the sexually selected trait is a function of individual condition (Rowe and Houle 1996). The potential benefits of exaggerated sexual traits are thought to be greater for larger individuals (Petrie 1988; Petrie 1992; Price 1993; Schluter and Price 1993) because hyperallometries are more sensitive to body size or condition than traits that are isometric or hypoallometric to the body (e.g., male fiddler crab claws increase disproportionally with body size (Huxley 1932); the antlers of male Irish elk have an allometric coefficient of 2.6! (Gould 1974, 1977)). Social status signaling may select for trait exaggeration similarly (Kraaijeveld et al. 2007). Hypoallometries are less well studied but are of interest because they indicate instances where selection presumably favors a fixed or intermediate trait size, regardless of overall body size (e.g., Palestrini et al. 2000; Eberhard et al. 1998; Hosken and Stockley 2003; Kawano 2004; Shingleton et al. 2007). It is important to note, however, that recent evidence suggests difficulties when inferring the pattern of selection from the pattern of trait scaling (Bertin and Fairbairn 2007; Bonduriansky and Day 2003; Bonduriansky 2007b), and so the presumptive selective forces hypothesized to act on individual scaling relationships need to be tested directly.

Perhaps because of the relative simplicity of calculation and standard set of parameters it generates, historically, the allometric equation has been the tool of choice for quantifying organismal shape. Biologists have used it to describe the scaling relationships for numerous morphological traits among countless groups of animals in an effort to identify patterns of morphological variation within and among biological groups (Cock 1966; see reviews in Huxley 1932; Gould 1966; Schmidt-Nielsen 1984; Emlen and Nijhout 2000). The totality of this work has revealed that changes in the parameter values of the allometric equation have contributed to an astonishing breadth of morphological diversity. Closely related populations or species often differ dramatically in the intercepts or slopes of their morphological scaling relationships, revealing profound shifts in the relative proportions of the involved body parts and changes in how these parts scale with body size. Given this, it is not surprising that the allometric equation has become a central, essential instrument in the study of animal form.

Part of the attractiveness of Huxley's allometric equation lies in its simplicity—it distills shape variation down to just two parameters that can be used to compare shape among biological groups. But this simplicity can be deceiving, and care must be taken when estimating and interpreting allometries. In addition to the sensitivity of scaling relationships to metrics and scale (described later), the intercept and slope of the relationship will be affected by the methodology used to fit the scaling relationship itself. For example, for all natural linear allometries, there is error in both the $x$ and the $y$ trait, and so major axis regression, reduced major axis regression, or a more sophisticated measurement error model should be used when fitting a linear scaling relationship to empirical data. The issue becomes even more complex when fitting nonlinear scaling relationships (e.g., figure 15.5). Conceptual and mathematical development of line-fitting approaches remains an active area of research (e.g., Smith 1980; Harvey 1982; Rayner 1985; Long et al. 2006; Warton et al. 2006; Ives et al. 2007).

Despite the utility of scaling relationships, other methodologies exist for describing shape mathematically. These are particularly useful when more than two traits are required to describe shape or when shape analysis is not concerned with relative size. In particular, advances in geometric morphometrics have generated something of a cottage industry in the study of complex morphologies (e.g., Bookstein 1985; Dryden and Mardia 2002; Zelditch et al. 2004; Claude 2009). Here, the shapes of complex structures are estimated through the use of multiple landmarks (e.g., Gabriel and Sokal 1969; Fiorello and German 1997; Langlade et al. 2005) or continuous mathematical functions (e.g., Liu et al. 1996; Houle et al. 2003). Principal component analysis or other statistical approaches (e.g., thin plate splines) are then often used to construct composite shape variables for further analysis. Although such multivariate approaches offer a powerful means to quantify and compare shape variation (e.g., Fink and Zelditch 1995; Arnold and Phillips 1999; Klingenberg and Leamy 2001), they are sensitive to the number of landmarks or parameters used. Moreover, the composite variables that summarize shape variation are potentially difficult to interpret biologically and not necessarily comparable across studies or biological groups. Hence, development of such techniques to handle these morphometric data is also an active area of research (e.g., Adams and Rosenberg 1998; Rohlf 1998; Zelditch et al. 1998; Houle et al. 2002; Mezey and Houle 2003). Central aims of all these new methods are the development of techniques for the inclusion of different sources of variation (e.g., various components of measurement error, phylogenetic nonindependence among compared groups [see also Blomberg et al. 2003], etc.) and the development of approaches for transforming data to increase biological signal in the estimated components of scaling relationships. The ultimate goal of these efforts, of course, is to increase our ability to accurately describe shape and to enhance our ability to compare morphologies across biological groups.

## A COMMENT ON USE OF THE TERM *ALLOMETRY*

In the strictest sense, *allometry* refers only to scaling relationships where the allometric coefficient differs from one (i.e., hypo- and hyperallometric relationships; figure 15.4). Under such conditions, changes in shape accompany change in size, which is implicit in the etymology of the term. However, over time and through use, the originally narrow definition of allometry has been modified and expanded. No longer applied solely to identify nonisometric scaling between a morphological trait and body size on a log-log scale, allometry is now commonly used to refer to any biological scaling relationship, on any scale. The inadvertent and planned revisions of the meaning of allometry have many causes, some of which can be gleaned from historical reviews on this topic (e.g., Gould 1966; Smith 1980; Blackstone 1987; Gayton 2000). Here, we highlight the three most relevant to this chapter.

First, there are issues relating to linearity. Many scaling relationships are linear without transformation; consequently, log-transformation is not always desirable (Smith 1980; Cheverud 1982b). Furthermore, not all scaling relationships are linear on any scale (Long et al. 2006). They can be sigmoidal or discontinuous, depending on the trait, the species, and the unit of measurement (e.g., Eberhard 1982; Garland and Else 1987; Eberhard and Gutierrez 1991; Kawano 1995; Emlen and Nijhout 2000). Nevertheless, these scaling relationships are often referred to as allometries or "nonlinear allometries" (e.g., Fairbairn and Preziosi 1994; Emlen 1996; Shingleton et al. 2007), which is, strictly speaking, an oxymoron.

Second, parameter values of the allometric equation are extremely sensitive to the unit and dimension of measurement. Consequently, both the metric and quantification procedure used can affect the intercepts and slopes estimated for the scaling relationships between traits (Moczek and Cruickshank 2006). Such methodological artifacts can result in an estimated scaling relationship that is formally an allometry *(sensu stricto)* in one study but is not an allometry in another. Broadening the definition of allometry avoids this confusion.

Third, interest in scaling relationships is not limited to those between morphological traits and the body. For example, large literatures are devoted to describing the scaling between morphological structures or how body size scales with ecological performance or physiological traits such metabolic rate or digestive efficiency (e.g., Calder 1984; Schmidt-Nielsen 1984; Garland and Else 1987; Garland and Carter 1994; Long et al. 2006). As the study of scaling relationships has expanded, the term allometry has been used to describe the scaling between nearly any two biological traits, be they the dimensions of the same structure, the size of two structures, or the scaling between performance and body size.

It worth noting that, in an effort to refine the term, allometry has been appended to describe different kinds of biological variation (Cock 1966; Cheverud 1982b; Klingenberg and Zimmermann 1992; Schlichting and Pigliucci 1998; Shingleton et al. 2007). When $x$ and $y$ are measured in a single individual through developmental time, the relationship is called an *ontogenetic allometry*; this describes the relative growth of the two structures over time. When they are measured in different conspecific individuals at the same developmental stage, the relationship is referred to as a *static allometry*. Finally, when traits are measured in differentiated populations or species, the relationship is an *evolutionary allometry*.

For better or worse, examples can be drawn from across the literature where allometry is used to refer to nearly any kind of biological scaling relationship. This means that care must be taken when interpreting studies of allometries as the term may be used to refer to different things across studies. Also, it suggests caution when applying the allometric equation to one's own data on scaling; strict adherence to the allometric approach (e.g., mandatory log-transformation), for example, may not always be desirable. Ultimately, such thoughtful interpretation and application of the term will maintain a useful distinction between the methodology used to describe scaling or shape and the biology underlying the scaling relationships themselves (Smith 1980; Blackstone 1987; Shingleton et al. 2007).

## DEVELOPMENT, SELECTION, AND THE EVOLUTION OF SCALING RELATIONSHIPS

Regardless of the traits considered or the scale on which they are plotted, placing the static allometries for different biological groups on the same figure yields a visual representation of the distribution of phenotypes in morphological space (figures 15.6, 15.7A). Three trends become apparent from such figures. First, there is considerable variation among groups in the intercept, slope, and shape of their scaling relationships. Second, despite this variation, much of the available morphological space is unoccupied. Groups

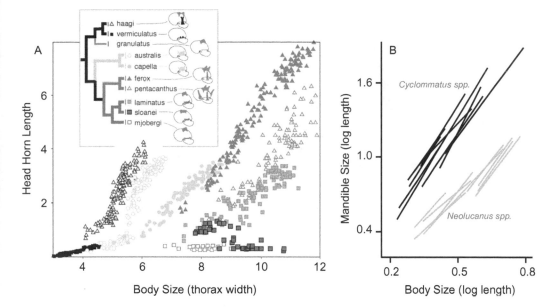

FIGURE 15.6

Patterns in scaling relationship distribution. Within groups, individuals cling tightly to specific scaling relationships, whereas diversity in the slopes and intercepts among groups has phylogenetic pattern. Closely related groups cluster, leaving much of the morphological space unoccupied. A, Horn length–body size scaling relationships shown for a monophyletic clade of nine Australian species of *Onthophagus* beetles. From Emlen et al. (2005). B, Scaling relationships describing the mandible size–body size scaling for several species of two genera of stag beetles, shown in black and gray, respectively (modified from Kawano 2000). See also figure 15.7A.

tend to cluster within regions of the overall morphospace, and these clusters typically represent higher taxonomic groups. For example, all the species within a genus may show little variation in the intercepts or slopes of their respective static allometries, but collectively these scaling relationships may be quite different from those of species within sibling genera. In other words, there is pattern in the distribution of evolutionary allometries that reflects ancestry among biological groups. Third, within a group, individuals tend to exhibit low variation about the static allometry. In sum, these trends reflect a pattern of low intra- and high intergroup variation in shape. These patterns are common for many complex biological traits, and different general hypotheses have been proposed to explain them (see Blomberg and Garland 2002; Blomberg et al. 2003). They may reflect constraints resulting from the developmental mechanisms that regulate and integrate the growth of traits. Alternatively, the pattern may result from drift or from external natural or sexual selection. Before examining in detail these hypotheses addressing how drift or selection may generate such patterns and discussing the roles experimental evolution can play in testing these hypotheses, we briefly introduce theoretical morphology, a computational technique that can be used to in experimental evolutionary research programs focused on the evolution of shape.

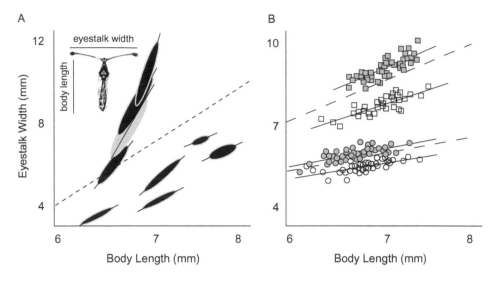

FIGURE 15.7

Natural and experimental evolution of scaling relationships in stalk-eyed flies. A, Distribution of
eyestalk–body size in morphological space for males from several species of stalk-eyed flies (photo inset).
Distribution of each species is shown by an ellipse and fit with a line illustrating the species-specific scaling relationship. The filled gray ellipse is *Teleopsis dalmanni*, the species subject to artificial selection on relative eyestalk size shown in the next panel. The dashed line illustrates isometry for the traits on this scale.
B, Results after ten generations of artificial selection to increase (filled symbols) or decrease (open symbols) relative eyestalk size in male (square) *C. dalmanni*. Circles show the correlated response of females.
Dashed lines represent the scaling relationships of control lineages. Figures modified from (A) Burkhardt
and de la Motte (1985) and (B) Wilkinson 1993. Photo inset in (A) from Baker and Wilkinson (2001).

## THEORETICAL MORPHOLOGY: PUTTING SHAPE IN CONTEXT

Theoretical morphology is a mathematical approach that seeks to determine what morphologies are geometrically possible using a (typically small) set of set of shape parameters (Raup and Michelson 1965; Raup 1966). Here, a model describing the shape of
some structure is constructed, and the values of the shape parameters are varied to
explore how combinations of values produce divergent or convergent morphologies.
Extreme parameter values will generate extreme phenotypes, which define the borders
of the morphospace containing all geometrically possible morphologies under the current
model. This total morphospace need not be continuous, solid or contiguous; some combinations of parameter values generate morphologies that are geometrically impossible
under the current model. In essence, theoretical morphology reveals the complex, total
morphospace of geometrically permissible shapes. Entering parameter values for observed
biological groups (extant or extinct) in this total morphospace yields the distribution of
real shapes within the total theoretical morphospace. Although this approach has been
used extensively by paleontologist, it has been perhaps underexploited as a tool for
studying shape in other fields (McGee 1999).

This is unfortunate, as theoretical morphological approaches have utility beyond simply describing the range of possible phenotypes under a given shape model. The pattern of occupation of actual forms within the total morphospace can be used to develop hypotheses regarding how particular phenotype distributions came to be—hypotheses that can be tested with experimental evolution using both empirical and computational approaches. In particular, the theoretical morphology methodologies can be applied to simulated populations to evaluate the roles of drift, selection, or developmental constraints in producing shape distributions within and among simulated biological groups. These resultant phenotype distributions may serve as null models for the interpretation of real distributions from empirical studies (e.g., see, for early examples, Raup et al. 1973; Raup and Gould 1974; see also Niklas 1999; Rasskin-Gutman and Izpisua-Belmonte 2004), including those generated by experimental evolutionary approaches. Exactly how such patterns of morphospace occupation can be compared statistically between simulated and empirically derived distributions, however, is a topic in need of further development (Hutchinson 1999).

As described earlier, only a few parameters are needed to describe biological shape as a scaling relationship between traits; hence, the scaling relationship is a shape descriptor well suited to the approaches of theoretical morphology. In the following sections, we explore how experimental evolution, paired with theoretical morphology, can be used to explore the potential roles of developmental constraints, drift, and natural selection in shape evolution.

## DEVELOPMENTAL CONSTRAINTS AND SCALING RELATIONSHIP EVOLUTION

Developmental constraints (*sensu* Maynard Smith et al. 1985) may limit the evolution of scaling relationships in several ways. Low genetic or phenotypic variation for a given aspect of a scaling relationship, be it intercept, slope, or shape, will restrict the response to selection on that component. Pleiotropy in the underlying regulators of these components may also limit the evolution of a scaling relationship through counter selection on other aspects of the regulator's functions. Finally, developmental constraints can result from physical processes during ontogeny. In fully metamorphic insects, for example, the growth of most adult structures occurs during the larval and pupal stages. This potentially constrains the time and physical space in which a given morphological structure can grow, thereby limiting the size of some traits or restricting particular combinations of trait sizes (see also Zera and Harshman this volume). In sum, such developmental constraints may limit the evolution of animal shape, even when such shapes are favored by selection.

Whereas empirical demonstration of developmental constraint can be challenging, demonstrating the absence of constraint is somewhat easier. The work of agriculturalists and hobbyists shows that selective breeding can dramatically alter shape through changes in the relative size of morphological traits (figure 15.3). Indeed, it is tempting to interpret the production of such extreme morphologies as indicative of an absence of absolute developmental constraint limiting the evolution of animal form.

Constraints need not be absolute, however. Developmental processes might constrain the evolution of animal shape by introducing biases in the expression of phenotypes or in the immediate direction and speed of the response to selection. Because development structures the relationship between genotype and phenotype, development can bias shape evolution in a few different ways. Development determines phenotype variability, or the tendency to produce phenotypic variation in the face of new mutations (Wagner and Altenberg 1996). Hence, a robust regulatory structure will dampen or channel the phenotypic effects of mutation, making some trait values or combinations of trait values difficult to achieve even as mutations accumulate. Mutational effects on phenotype expression can be determined by conducting mutation induction and transgenic experiments (e.g., Camara et al. 2000; Monteiro et al. 2003; Ramos and Monteiro 2007) or by measuring the effects of mutations, insertions, or deletions on patterns of trait covariation (e.g., Trotta et al. 2005; Breuker et al. 2006; Dworkin and Gibson 2006; Hallgrimsson et al. 2006; Willmore et al. 2006; Hallgrimsson et al. 2007). In the context of studying shape, such experiments could establish how development may bias the effects of mutations on patterns of trait covariation that will affect the intercept or slope of morphological scaling relationships (see Cock 1966 for an early example).

Development can also bias the response to selection on the standing variation in a population, affecting the likelihood of which alternative evolutionary trajectory a population will follow as it evolves. Experimental evolution offers a very powerful approach to test this hypothesis, especially when combined with a clear definition of the putative constraint in question (Beldade and Brakefield 2003). Here, effective utilization of experimental evolution entails the application of artificial selection in contrasting directions that are predicted to differ a priori in the extent to which they are constrained by development. Artificial selection can be focused on morphological shape itself (i.e., on components of the scaling relationship) or applied to specific developmental mechanisms believed to generate a putative constraint. Similar rates of response in the presumptively constrained evolutionary trajectory relative to alternative evolutionary trajectories provide compelling evidence for an absence of developmental constraints. Conversely, the magnitude of a developmental constraint on shape evolution is quantified by the degree to which lineages differ in their rates of response to artificial selection or in the degree to which they fail to follow their predicted evolutionary trajectories. Care must be taken, however, in the design and interpretation of such experiments as the effects of such bias may be subtle over the short term, and even mild drift for a brief period can affect the pattern of covariation among traits and thereby alter the response to selection (Phillips et al. 2001; Whitlock et al. 2002; see later discussion). Roff and Fairbairn (this volume) discuss modeling experimental evolution in a way that facilitates the construction of experimental designs with adequate statistical power.

Although theoretical morphology was originally developed with only minimal consideration given to how the shapes were generated developmentally (Raup and Michelson 1965; Raup 1966), some recent theoretical morphology models explicitly consider shape as a product of the growth trajectories of the parts of the structure (e.g., Lovtrup and Lovtrup 1988; Rasskin-Gutman and Izpisua-Belmonte 2004; in particular see Rice 1998a). Using both traditional (i.e., nondevelopmental) and newer, developmentally derived shape models, theoretical morphology could be used to determine how development might define or restrict the accessible regions of morphospace; absolute developmental constraints would be indicated in areas of morphospace rendered inaccessible under the developmentally derived models but accessible under traditional, development-independent shape models. (Conversely, expanded morphospace under the development-based model means that the development-independent model may not describe shape adequately; Rice 1998a). Moreover, simulations might be performed to predict evolutionary trajectories and final distributions in the morphospace produced under nonconstrained shape models (examples of this approach using development-independent models are in Raup et al. 1973; Raup and Gould 1974). Differences in the results from models using developmentally derived morphologies and morphologies derived independently of development would be indicative of the role of developmental constraint or bias in the evolution of form. The results of these simulation tests could also serve as a baseline for comparison for empirical experimental evolution experiments designed to look for developmental constraints or bias as described earlier.

## DRIFT, NATURAL SELECTION, AND SCALING RELATIONSHIP EVOLUTION

The observed within- and among-group patterns in scaling relationships (figures 15.6, 15.7A) could result from random genetic drift or could be the product of selection favoring specific animal shapes. Drift is predicted to produce proportional changes in the pattern of covariance among traits, whereas selection is predicted to alter the covariance structure (Lande 1980; Lofsvold 1988; Roff 2000). Comparing covariance structures has, therefore, been proposed as a methodology for distinguishing between drift- versus selection-based divergence (e.g., Weaver et al. 2007). In terms of scaling relationships, this means that drift will increase size variation proportionally for each trait, but it will not alter the covariation between the traits. Hence, the range of phenotypic variation is predicted to increase under drift, but the intercepts and slopes of the scaling relationships between traits will not change. Contrary to this prediction, however, drift can affect trait covariances, particularly in small or inbred populations (Phillips et al. 2001). Such changes can occur rapidly and can persist long after a population recovers from a bottleneck, potentially affecting the response to selection (Whitlock et al. 2002). Consequently, drift could alter the components of scaling relationships among traits and thereby affect the evolutionary trajectories of populations, producing or at least contributing to the observed among-group patterns in scaling relationships.

Experimental evolution can be used to assess the role of drift in shape evolution in at least two ways. First, drift experiments (e.g., bottleneck, mutation accumulation, or inbreeding experiments) can be used to see if, at what rate, and through what changes (i.e., shifts in intercepts or slopes of scaling functions, increases in variance) populations will move within the available morphospace. In addition to testing for the effects of drift, such experiments can provide data on divergence rates and repeatabilities that can be used as a null model for comparison to artificially selected lineages as described below. Second, computer simulations can be used in a theoretical morphology framework to explore the role of drift in producing particular scaling relationships within and among groups. Such experiments will also generate a null model for comparison to the observed empirical within- and among-group distributions in total morphological space (e.g., Raup et al. 1973; Raup and Gould 1974). We return to this idea again at the close of the chapter.

Instead of moving randomly into different regions via drift, however, biological groups may come to occupy specific areas of morphospace under the effects of selection. Under this scenario, groups occupy regions of morphospace conveying high fitness, whereas empty areas correspond to morphologies with lower fitness. Many patterns of selection could produce the observed pattern of low intra- and high intergroup variation in scaling relationships. Ecological performance requires that functionally related traits be properly scaled and shaped to perform a given task. Traits that are mismatched in size or misshapen will reduce ecological performance, which in turn reduces reproductive success. Because selection on the relative size or shape of traits can be strong, even over the short term (e.g., Boag and Grant 1981; Kingsolver 1999), phenotypes that deviate from the favored scaling relationship will be removed by selection.

Although this line of argument seems intuitive, it begs the question as to why some traits exhibit extreme exaggeration of shape that would appear to inhibit some ecological functions. As already discussed, the expression of extreme morphologies is favored by sexual selection in some cases (e.g., earwig forceps (Simmons and Tomkins 1996), stag beetle mandibles (Kawano 2006), and manikin or widowbird bird tails (Cuervo and Moller 2001; Arevalo and Heeb 2005)). In other cases, the extreme morphology may actually enhance ecological function, by allowing feeding on an otherwise unexploitable resource (e.g., the elongated digit of aye-ayes (Jouffroy et al. 1991; Lemelin and Jungers 2007)).

Tests of these selective hypotheses for particular animal shapes requires measuring the relative ecological performance or fitness of individuals with divergent trait proportions. Such tests can be difficult to perform because within-group patterns of trait scaling are typically extremely tight. Consequently, shape variation is usually low about the group mean, hindering informative measures of natural or sexual selection; selection cannot be measured on trait values that do not exist.

One solution to this problem is provided by experimental evolution, which can be used to generate novel phenotypes, dramatically expanding the range of phenotypic variation available for estimating the pattern and strength of selection. In this way, hypotheses regarding how selection acts to determine specific values for allometric components can be tested. Experimental evolution can be a particularly powerful tool to generate phenotypes because it can alter the natural pattern of covariation among traits, changing the components (e.g., slope or intercept of the scaling relationship) describing shape. This allows isolation of the contributions of different components of shape to ecological performance or fitness. In sum, experimental evolution can be used to test rigorously for the roles of both developmental constraints and natural selection in the evolution of organismal shape.

## INCORPORATING DEVELOPMENT INTO STUDIES OF SHAPE EVOLUTION

The examples of naturally occurring and artificially produced extreme scaling relationships cited here demonstrate that highly derived morphologies can be produced when favored by selection. But how do evolving developmental processes generate variation in animal shape? And how do these developmental processes interact with different patterns of selection to affect the evolution of form? These questions can be addressed empirically using experimental evolution. However, a generalizable theoretical framework is needed to guide empirical research and to give it context. Some of the most exciting recent advances in this regard incorporate aspects of development into models of the evolution of complex traits such as scaling relationships.

Theoretical approaches to the evolution of complex traits can be divided into two general categories. The first category is comprised of a "top-down" approach, using statistical estimates of trait integration derived from breeding designs or artificial selection experiments to infer the degree to which the components of a scaling relationship can respond to selection. This approach includes quantitative genetic models of multivariate trait evolution (e.g., Cowley and Atchley 1990; Atchley et al. 1992; Roff and Fairbairn this volume) and newer "phenotype landscape" models (e.g., Rice 1998b).

In the second category of theoretical approaches, information regarding the proximate mechanisms regulating and integrating the growth of traits is used to explicitly model shape evolution from the "bottom up" (e.g., Nijhout and Wheeler 1996; Stern and Emlen 1999; Emlen and Nijhout 2000; Emlen and Allen 2004; Shingleton et al. 2007; Shingleton et al. 2008; see also Dykhuizen and Dean this volume). These models attempt to predict the most likely developmental changes producing diversification of complex morphologies based on the specific dynamics of the developmental mechanisms regulating and integrating trait growth. Both classes of models provide a rich framework for experimental studies of shape evolution, and in the following sections, we briefly review these in the context of studying and predicting the evolution of scaling relationships. After reviewing these two classes of model, we turn to examining case studies that have used experimental evolution to examine the evolution of animal form using these approaches.

Quantitative genetic models can be used to predict how complex traits—such as relative trait sizes or scaling relationships—will evolve under genetic drift or different patterns of selection. These models predict how the slope, intercept, or shape of the scaling relationship can evolve in different selective environments based on estimates of the additive genetic variation underlying the traits comprising a scaling relationship, the patterns of genetic covariation among traits or the strength of the genetic correlations among traits. A strong genetic correlation is typically interpreted as being reflective of trait pairs where growth is tightly coupled, biasing the scaling relationship to evolve along the primary axis of covariation (Schluter 1996), thereby inhibiting evolution of the scaling relationship intercept or slope. These models usually make necessary simplifying assumptions that genetic correlations underlying phenotypic covariance are constant over evolutionary time and result from gene pleiotropy or (less likely) linkage (e.g., Lande 1979; Lande and Arnold 1983; Cheverud et al. 1983, 1997; Cheverud 1984, 1988, 1996; Cheverud and Routman 1995; Klingenberg et al. 2001b). These models also generally assume that most evolutionary change occurs primarily through modification of many genes with small, additive effects (Falconer and Mackay 1996; reviewed in Via et al. 1995; Wolf et al. 2001, 2004). This means that the evolution of scaling relationship intercepts or slopes will generally be restricted to small, gradual steps.

Typically, these approaches entail estimating the heritabilities for individual focal traits and genetic (co)variance among them. The model is often tested by applying artificial selection to alter specific components of the scaling relationship in question. As we review later in this chapter, several studies have used experimental evolutionary approaches to explore the evolution of morphological scaling relationships within this framework. These studies have revealed the surprising ease with which the components of scaling relationships can evolve in the short term.

Nevertheless, the findings of these studies should be interpreted with caution. Perhaps in part because of the biology of the model systems used and the simplifying assumptions required for particular studies, top-down research programs can run the risk of overlooking or confounding different sources of developmental variation contributing to the growth of traits and, ultimately, to trait size (see discussions in Zelditch et al. 2004, 2006, and Mitteroecker and Bookstein 2007). This is particularly true in cases where the size or shape of a single morphological trait has multiple parts, each of which may have their own developmental trajectories. This can decrease the accuracy of predictions regarding how a complex trait, such as morphological shape, will respond to selection. Moreover, the general view that strong genetic correlations reflect absolute constraints on the independent evolution of traits (and therefore of their scaling relationships) can produce predictions that underestimate the evolutionary lability of morphological shape. Genetic correlations may influence the ability of traits to evolve independently, but they do not reflect the parameter values of the allometric equation. Correlations standardize variances so that all traits have a mean of zero and a standard deviation of one, so they do not capture information about the intercept and slope of a scaling relationship. Rather, genetic correlations give information about the strength of the relationship between traits. This is in contrast to genetic covariances, which describe the extent to which genetic variation in one trait is accompanied by genetic variation in another. Hence, genetic covariances are essentially a description of linear scaling relationships—in fact, the first principle component of a covariance matrix for multiple traits describes their multivariate allometric relationship (Cheverud 1982a). Consequently, changes in the slope of a scaling relationship will necessarily alter the covariance of two traits, but it need not influence their correlation.

The development and testing of multivariate approaches that address these potential shortcomings is an area of active research. One method entails using quantitative genetic tools to model variation exhibited in a complex morphological structure based on the patterns of (co)variation present among the components of the structure though ontogeny. The patterns of covariation among these components are used to explain and predict how the complex morphological structure will evolve (e.g., Atchley and Rutledge 1980; Atchley et al. 1981, 1994; Atchley and Hall 1991; Cowley and Atchley 1992; see also Zelditch et al 2004, 2006). Although these models have provided important insights into the evolution of complex traits, the components of the complex traits they measure can always be divided into ever-smaller constituent developmental parts, each with their own patterns of covariation. Hence, these models still risk confounding or missing sources of evolutionarily important developmental variation. Moreover, at some level these models are system-specific, as they are derived from empirical data on the growth and development of components of the specific morphological traits in particular model systems.

Another generalized approach models the evolution of complex traits as functions of the developmental processes that underlie their expression. Because these approaches model general classes of developmental interactions (e.g., additive or nonadditive interactions among developmental modules) and are not derived from empirical studies regarding how a particular biological system works, they represent a generalizable top-down approach. Here, complex, multivariate traits such as animal shape are expressed as a function of interactions among any number of underlying developmental variables that contribute to trait variation (Rice 1998b, 2000, 2002, 2004a). These "phenotypic landscape" models hold great promise, because instead of avoiding the complexity of development through necessary simplifying assumptions, they explicitly include the developmental bases of trait expression and evolution (Wolf 2002). The phenotype landscape approach has been extended to connect with existing quantitative genetic treatments of multivariate evolution, yielding an emergent theory exploring how developmental integration, or "entanglement," among traits affects the symmetry and rates of trait evolution; the evolution of heritabilities; the impact of genetic correlations on evolutionary trajectories across different time scales; the evolutionary relationships among trait means, variances, and covariances; and the distribution of traits in phenotypic space (Wolf et al. 2001, 2004; Rice 2004b, 2008). Particularly relevant for the evolution of scaling relationships, these models have revealed that the developmental basis of genetic correlations (e.g., the degree to which a given genetic correlation results from additive or nonadditive epistatic interactions among traits) can profoundly affect the evolutionary malleability of the correlation, trait covariation, and the evolutionary trajectory of the complex phenotype (Wolf et al. 2001, 2004; Rice 2002, 2004a, 2008). This means that it is the developmental basis of trait integration, not simply the strength of the genetic correlations and observable patterns of covariation among traits, that will affect how components of a scaling relationship can evolve.

Although these powerful phenotype landscape models have generated important insights into the evolution of complex traits such as scaling relationships, they are difficult to test empirically (see Rice 2008). This is primarily because the number of developmental parameters that might contribute to variation in most complex traits is large, the degree of precision with which they can be measured can be relatively low, and knowledge regarding the adaptive landscape is limited (Wolf et al. 2004). Despite these difficulties, the models indicate that the details of how morphological traits are developmentally integrated is critically important for predicting how complex phenotypes evolve. This finding has reinforced the importance of understanding the proximate basis of trait covariation. The second approach to modeling shape evolution shares this perspective, building from the developmental mechanisms that control and integrate the growth of traits. Consequently, these bottom-up models are a promising complement to the top-down models.

## "BOTTOM-UP" MODELS FOR THE EVOLUTION OF ANIMAL SHAPE

Bottom-up models take a direct developmentally and physiologically based framework to elucidate how the regulation and integration of trait growth affect the expression and evolution of scaling relationships. Some of the earliest developmental models of scaling relationship expression focused on holometabolous insects, and the insect-based models are currently the most refined in this category.

Fully metamorphic (holometabolous) insects are appealing biological models for the study of shape evolution because the physiology and genetics of trait and body growth are tractable in these systems. In holometabolous insects, all body growth occurs during the larval stages, and most adult tissues develop and grow within the larvae and pupae as discrete structures, the imaginal discs. Within each disc, patterns and rates of cell proliferation, the axes of cell divisions, and patterns of programmed cell death determine the final size and shape of adult traits. All discs are likely exposed to the same levels of circulating nutrition, growth factors, and endocrine cues, coordinating growth and development among discs. This developmental framework provides a useful starting point for exploring the proximate basis of scaling relationship expression and thus for considering bottom-up models of shape evolution. Cowley and Atchley (1990) used proximity of grow-ing traits within the body to make explicit a priori predictions regarding how they ought to be developmentally integrated. These predictions proved surprisingly effective at ex-plaining the phenotypic and genetic correlation structure among traits in adult *Drosophila* (Cowley and Atchley 1990). Similarly, Nijhout and Wheeler (1996) used these same mechanistic principles to model disc-disc interactions and showed that competition between discs for limiting "growth factors" could, in theory, influence disc growth and alter the final shapes of the adult insect. Their hypothesis was supported by subsequent manipulative developmental experiments although the presumptive factors or resources for which the growing traits compete has not yet been identified (e.g., Klingenberg and Nijhout 1998; Nijhout and Emlen 1998; Moczek and Nijhout 2004).

Recent advances in our understanding of the mechanisms that regulate and integrate trait growth have permitted the development of a detailed set of bottom-up models of scaling relationship expression in holometabolous insects. In such insects, larvae grow until they achieve a species-characteristic "critical size" or "critical weight." Attainment of critical size begins a hormonal cascade that initiates the termination of disc and body growth and culminates in metamorphosis. The period between achievement of the crit-ical size and termination of disc and body growth is called the *terminal growth period* (TGP). Although the termination of growth is ultimately a response to the hormonal cas-cade initiated at critical size, discs have different sensitivities to the hormones, primarily ecdysone and juvenile hormone, that constitute this cascade. Consequently, the body and individual discs vary in when they stop growing and thus, individual discs and the body have their own TGPs. At the same time, all growing tissues are exposed to circulat-ing insulin and other growth factors that regulate and coordinate disc and body growth rates during the TGP (Shingleton et al. 2007). Under this model, the final size of an adult trait is determined by the size of a structure at the critical size, the length of the structure's TGP, and the growth rate of the structure during its TGP. Imaginal discs that have relatively short TGPs or have relatively weak responses to insulin (i.e., low growth rates) compared to the body will produce structures that are hypoallometric to body size, whereas discs with longer TGPs or high insulin activity signaling will result in traits that are hyperallometric to body size. Although these models (Shingleton et al. 2007; Shingleton et al. 2008) were conceived to explain the scaling exhibited by a single genotype in different nutritional environments, it seems likely that genetic variation underlying these physiological components could respond to selection on the intercept or slope of the scaling relationship. Hence, these models identify plausible combinations of candidate mechanisms that may underlie the evolution of divergent, or convergent, morphological scaling relationships.

The explanatory and predictive power of these models will likely be enhanced as more major developmental parameters are added to them. For example, programmed cell death can substantially reshape and reduce the ultimate size of adult structures (e.g., Dohrmann and Nijhout 1988; Sameshima et al. 2004; Moczek 2006b). Furthermore, genes responsible for patterning developing discs also influence disc size and therefore may be important in regulating scaling relationships (Emlen and Allen 2004). Finally, direct interactions between growing traits are likely to affect the size and shape of the final structures (e.g., Simpson et al. 1980; Klingenburg and Nijhout 1998; Nijhout and Emlen 1998; Moczek and Nijhout 2004; Zelditch et al. 2004, 2006; Willimore et al. 2006). Although they are incomplete, the current generation of models provide a useful starting point for bottom-up study of scaling relationship expression and evolution.

## EXPERIMENTAL EVOLUTION OF SCALING RELATIONSHIPS

The previous sections detail some of the mathematical, computational, and experimental approaches used to study shape evolution. In the next sections, we review a few case studies where investigators have taken an experimental evolutionary approach to the topic, and we discuss how they fit with the top-down and bottom-up approaches just described. Our goals are to identify what these research programs have revealed about the evolution of the components of scaling relationships and to highlight some of the questions these studies raise or leave unanswered.

Many of the studies we review differ in their application of the term *allometry*. Consequently, we necessarily employ an inclusive definition of allometry, using it as a blanket term describing the scaling relationship between any two morphological traits within a defined biological group, regardless of the metric used to quantify the traits or the scale on which the scaling relationship has been determined. However, we restrict application of the term to instances where we discuss the components of the scaling relationship—namely, the slope or intercept. This broad definition has utility and biological meaning because it is consistent with the vocabulary used in the original publications on which we draw, and because it enables us to meaningfully compare, with caution, the intercepts or slopes of scaling relationships among studies that use similar methodologies.

### THE EVOLUTION OF EYE STALK SPAN–BODY SIZE SCALING RELATIONSHIPS IN DIOPSID FLIES

One of the more extraordinary examples of scaling relationship evolution is seen in the stalk-eyed flies, Diopsidae (see also Swallow et al. this volume). All the members of this monophyletic family exhibit lateral extensions of the head so that the eye and antennae are located at the end of long stalks (e.g., inset, figure 15.7A). In basal diopsids, both males and females posses these stalks, and eye-stalk length is monomorphic (Baker and Wilkinson 2001). This suggests that holding the eyes out from the head initially evolved as a response to ecological factors, possibly to increase field of view (Chapman, 2003). However, among more derived groups within the Diopsidae, eye stalks have become involved in both inter- and intraspecific signaling (Burkhardt and de la Motte 1987; Wilkinson et al. 1998; Baker and Wilkinson 2001; Hingle et al. 2001). This is marked by the evolution of sexual dimorphism for eye span, with exaggerated hypercephaly in males, which has evolved at least four times in the group (Baker and Wilkinson 2001).

One hypothesis explaining exaggerated male traits is that the trait is an indicator of male condition (Schluter and Price 1993; Rowe and Houle 1996). If the size of the trait indicates male quality, hyperallometry of that trait is thought to ensure that it is a more sensitive advertisement than traits that are isometric or hypoallometric to body size (Wilkinson and Dodson 1997b). Hypoallometry is also predicted to evolve if the relative advantage of an exaggerated trait is greater for larger individuals (Bonduriansky and Day 2003). Phylogenetic evidence indicates that among the diopsids, the evolution of sexual dimorphism is associated with hyperallometry of male eye span. The greater the sexual dimorphism in a species, the greater the slope of the eye span to body size allometry in males (Burkhardt and de la Motte 1987; Baker and Wilkinson 2001); and females in these dimorphic species (but not in the monomorphic species) use eye span as a basis for mate choice (e.g., Burkhardt and de la Motte 1983; Cotton et al. 2006).

These data suggest that the eyestalk span–body size scaling should be evolutionary labile within the diopsids, and this appears to be the case (see figure 15.7). Among the stalk-eyed flies, species show considerable difference in their intraspecific phenotypic covariance matrices (Baker & Wilkinson, 2003), a mark of interspecific variation in scaling relationships (Cheverud, 1982a). Perhaps surprisingly, the same is not true for the phenotypic *correlation* matrices, which do not differ significantly across diopsid species. The evolutionary stability of the phenotypic correlation and instability of phenotypic covariance in the diopsids suggests that while scaling relationships are evolutionarily labile, the underlying developmental and functional relationships between traits remains stable across species.

The evolutionary labiality of eyestalk span–body size scaling in the Diopsidae is echoed by evidence for standing genetic variation in the slope of the scaling relationship within a species. In *Teleopsis dalmunni* (formerly known as *Cyrtodiopsis dalmanni*), there is considerable genetic variation in how male eyestalk span–body size ratio responds to changes in nutrition (David et al. 2000). For most genotypes, as food quality falls and body size decreases, relative eye span decreases disproportionately, reflecting the hyperallometric relationship between eye span and body length (David et al. 1998). However, some genotypes are able to maintain a constant relative eye span size across a range of nutritional conditions, suggesting that eye span is isometric to body size in these genotypes. The existence of standing genetic variation for scaling relationship slope within a population suggest that the eye span allometry in *C. dalmanni* could respond to artificial selection.

Wilkinson (1993) tested these predictions by performing experimental evolution with *C. dalmanni*. He artificially selected on the intercept, but not the slope, of the eye span to body length scaling relationship by selecting males with a high or a low eyestalk span–body size ratio over ten generations. Despite a strong genetic correlation between body size and eye span in male stalk-eyed flies (Baker and Wilkinson 2003), the allometry intercept evolved easily; relative eye span deviated ~6% in each selected direction from unselected control lineages (figure 17.7B), indicating that genes affecting relative eye span are as likely to increase as decrease expression of the trait. There was a tight correlation between the relative eye span of fathers and sons, suggesting that the genes controlling eye span allometry are likely to be numerous and of small effect. Interestingly, body lengths were the same across lineages at the end of the experiment, with the exception that body size was significantly larger in one of the lineages selected for increased relative eyespan. This means that, among lineages, relative eye span evolved primarily through changes in eye span alone.

Selection on relative eye span in males produced a correlated, although less extreme, response in female morphology (figure 15.7B), altering the degree of sexual dimorphism in a manner consistent with the phylogenetic pattern exhibited between male hypercephaly and sexual dimorphism across species.

Although Wilkinson (1993) only selected on the intercept of the eyestalk span–body size allometry, the selection regime also changed the slope of the relationship (figure 15.7B); lineages selected for greater relative eye span evolved a more hypoallometric relationship between eye span and body size (figure 15.7B). Thus, the phenotypic covariation between traits was also altered easily through selection, consistent with its phylogenetic instability (lability). Furthermore, the correlated change of the allometry slope to selection on its intercept suggests that the developmental mechanisms that signal male quality to potential mates may be the same, or genetically or functionally linked, to those ensuring that the signal is a sensitive advertisement of quality.

Using individuals from the lineages artificially selected for extreme relative eye spans revealed that wild type females prefer to associate with males possessing large relative eye spans—including those even more exaggerated than the eye spans possessed by wild type males. Females from the artificially selected lineages, however, preferred males with the derived phenotypes from their respective selected directions. Thus, using individuals from the lineages selected for extreme eye span-body size scaling relationships in preference trials elucidated the pattern and strength of intersexual selection acting on this allometry in nature and revealed a genetic correlation between derived male phenotype and female preference (Wilkinson and Riello 1994).

The maintenance of additive genetic variation for male relative eye span and sexual dimorphism in the presence of strong sexual selection suggests that there are selective factors opposing female preference for this exaggerated male trait. Flies from lineages selected for increased relative eye span exhibited increased developmental time, possibly because larger eyes take longer to grow. Competition among larvae for food or an increased likelihood of ingesting parasites while feeding could mean that slower developing larvae feed for longer periods of time and thereby suffer fitness decrements (Wilkinson 1993). Hence, this suggests the testable hypothesis that sexual selection for increased relative eye span is countered by fitness costs incurred through increased development time.

Collectively, the data from these selection experiments support the conclusions of the top-down phylogenetic analyses of eye span allometry evolution—namely, that (1) hypercephally in male diopsids is evolutionarily labile; (2) male hypercephally is associated with hyperallometry of eye span; and (3) there is an association between the hyperallometry of male eye span and sexual dimorphism in relative eye span. However, the experimental evolution data provide two additional insights into the developmental mechanisms that may regulate this scaling relationship in stalk-eyed flies, and thus may be informative for bottom-up models of the scaling relationship evolution.

First, despite the tight genetic correlation between eye span and body length within populations of stalk-eyed flies, it is relatively easy to change the size of these traits independently. Thus, the developmental mechanisms regulating eye span are at least partially independent of those regulating body size. Second, although relative eye span evolved through different combinations of changes in eye span and body size, most change occurred through alteration of eye span. This is perhaps not surprising. Wilkinson measured body size as distance from the face of a fly to the tip of its wings. Hence, his estimate of body size is the sum of multiple parts (head, thorax, and wing), which are derived from multiple imaginal discs. Changes in body length may therefore necessarily involve developmental mechanisms that affect many of the imaginal discs throughout the body. In contrast, the eyes and eyestalks are derived from a single pair of imaginal discs. Therefore, changes in eye span are likely to involve developmental mechanisms that influence the eye-antennal discs alone. Consequently, because developmental regulators of body size are likely to have more pleiotropic effects than the developmental regulators of eye span, it is perhaps not surprising that selection on relative eye span affects the latter more than the former.

To test hypotheses addressing the proximate basis of the response to artificial selection on the relative size of these traits, it is necessary to understand how eye span and body length are developmentally regulated in stalk-eyed flies. The model of the regulation body and organ size presented earlier provides a focus for future research. For example, the evolved increased relative eye span could result from a larger eye-antennal disc at the critical size, an extended disc TGP, or a faster disc growth rate during the TGP. Although an area of active research, unfortunately little is known of the mechanisms that regulate the development of eyestalks in the diopsids. Several studies have looked at the eye-antennal imaginal discs to elucidate the developmental origins of different parts of the adult head, in particular the eyestalks, with mixed results (Hurley et al. 2001, 2002). Surprisingly, these studies indicate that structures separated by the eyestalks are derived from adjacent parts of the imaginal discs; conversely, widely separate parts of the eye-antennal discs give rise to adjacent structures in the adult. More importantly, the parts of the disc that turn into the eyestalk are unknown.

In the case of the diopsids, experimental evolution has been productive in confirming the findings of top-down models of scaling relationship evolution. However, while these studies reveal important trends in the evolution of eyestalk span–body size scaling—for example, the evolutionary lability of the scaling relationship slope but stability of trait pleiotropy—they do not address the functional developmental mechanisms that underlie these trends. To begin to elucidate these mechanisms, a bottom-up approach may prove fruitful. Information regarding the imaginal structures that give rise to the eyestalks, the extent to which these structures' growth rates are influenced by factors that also influence body size, the duration of their TGPs, and other issues, is essential. Once these basic developmental parameters are established, artificial selection studies can explore which, if any, of these parameters change in response to selection on aspects of trait scaling. Thus, in this case experimental evolution has the exciting potential to link top-down and bottom-up models of scaling relationship evolution. For the diopsids, our limited understanding of the developmental mechanisms that regulate variation in eye span prohibits such analysis at this time. In contrast, considerably more is known of the developmental mechanisms that regulate trait size in *Drosophila*, setting the stage for a more direct means of linking observed phenotypic change through experimental evolution to the developmental mechanisms and genetic loci of this change.

## THE EVOLUTION OF WING SCALING RELATIONSHIPS
## IN *DROSOPHILA MELANOGASTER*

One of the most intensively studied morphological traits of any metazoan is the wing of *Drosophila*. Wings are well suited to the study of morphological scaling relationships for several reasons. First, at a gross level, wings are two-dimensional structures. Their relative flatness and transparency, combined with an abundance of landmarks, facilitates precise quantification of wing shape. Ease of measurement has led to a substantial literature on many aspects of wing shape, including the discovery of interesting phylogenetic and geographic patterns of variation in *Drosophila* wings. Wing shape, absolute wing sizes, and relative wing sizes all differ among geographic populations of the same species, and several convergent and divergent clines in these variables have been described (e.g., Starmer and Wolf 1989; Barker and Krebs 1995; Azevedo et al. 1998; Huey et al. 2000; Hoffmann and Shirriffs 2002).

Despite this variation, wing shape (as estimated by landmarks internal to the wing plus landmarks along the wing margin) is generally conserved across *Drosophila* populations and species relative to other Dipterans (Houle et al. 2003). Presumably, these patterns result primarily from strong natural selection on wing shape. Wing shape and relative wing size likely affect flight performance; in fact, wing aspect ratio (the span of a wing squared divided by wing area) and wing loading (body mass divided by the wing area) are central components of flight models (Dickinson et al. 1999; Dudley 2000). However, male *Drosophila* also use their wings during courtship song production; consequently, female choice (Ritchie et al. 1998) may act on wing shape or relative wing size if these characters affect aspects of this male mating behavior (Ewing 1964).

*Drosophila* wings have also served as models for studying appendage development and for elucidating the effects of the developmental environment on trait size and shape (e.g., Debat et al. 2003; Breuker et al. 2006). There is a considerable literature regarding the developmental genetics of growth in fly wings (e.g., Bier 2000; de Celis 2003; de Celis and Diaz-Benjumea 2003; Hafen and Stocker 2003). This includes mechanisms affecting wing shape or size, such as those controlling cell growth and proliferation (e.g., Partridge et al. 1994; Zwaan et al. 2000) and wing vein placement (e.g., Stark et al. 1999; Bier 2000; Birdsall et al. 2000; de Celis 2003; de Celis and Diaz-Benjumea 2003). And of course, *Drosophila* became a model system because it is amenable to genetic study (e.g., Morgan 1911) and because of the many aspects of its biology that make it well suited for population-level studies in the laboratory (e.g., Dobzhansky and Pavlovsky 1957; Weber and Diggins 1990). It is not surprising, then, that the fly wing has been the subject of extensive experimental evolutionary studies of morphological scaling relationships. Although differences in methodology and data presentation among studies preclude a detailed synthesis of these results, some interesting generalities seem clear. We briefly describe these here, focusing on experimental evolution studies of two research programs, the first involving the wing size–body size allometry and the second the scaling of the dimensions, or shape, of the wing itself.

## THE EVOLUTION OF THE WING SIZE–BODY SIZE ALLOMETRY IN *DROSOPHILA*: BODY SHAPE AND WING LOADING

The wing–body size scaling relationships among *Drosophila* species is typically isometric, although some species exhibit hyperallometry (Starmer and Wolf 1989). Nevertheless, while the slope of the wing size–body size scaling relationship is fairly constant, the intercept shows considerable genetic variation among geographic populations along latitudinal and altitudinal clines (Starmer and Wolf 1989; Azevedo et al. 1998; Gilchrist et al. 2000). The cooler the average temperature of the source population, the larger the wing relative to the body (i.e., the lower the wing loading). This change in relative wing size is produced primarily through changes in cell number in the wings (Zwaan et al. 2000). The wing size–body size scaling relationship also exhibits phenotypic plasticity in response to temperature; flies reared at lower temperatures tend to have relatively larger wings than flies from the same population reared at warmer temperatures (Barker and Krebs 1995; Gilchrist and Huey 2004). In contrast to the genetically based variation in wing loading among different fly populations, however, the plastic response of wing size to temperature is produced primarily through and changes in cell size. Nevertheless, both the genetic and plastic responses of wing loading to temperature are thought to be an adaptive response to increase flight performance at low temperatures (e.g., Stalker 1980; Frazier et al. 2008). The genetic correlation between wing size and thorax size (a proxy for body size) has a broad range, being estimated at 0.4–0.8 (Robertson and Reeve 1952; Petavy et al. 2004), and there is abundant genetic variation underlying both traits (e.g., Houle et al. 2003; Petavy et al. 2004). Consequently, it appears that the evolution of wing loading in flies is not restricted by developmental constraints, at least in the short term. Furthermore, these data suggest that wing loading may potentially be altered through a number of different developmental mechanisms.

In an early artificial selection study of wing loading in flies, Robertson (1962) selected on the ratio of wing size to body size (thorax size in this experiment) in each direction (increased or decreased wing loading). Although his methodology imposed very strong selection to small experimental populations, and many results are presented pooling males and females (which differ significantly in wing loading), the study produced several interesting and important results. First, wing loading diverged fairly steadily and symmetrically, producing mean wing size–body size ratios that differed from the wild-type controls by about 15 percent in each direction. He notes that the change in the mean value of this ratio is associated with, but distinct from, changes in the slope of the scaling between wing size and body size. Second, evolution of the ratio occurred through changes in wing size only; thorax size remained the same as wing loading evolved. These first two results indicate that the shift in the mean wing size–body size ratio of the lineages can be cautiously interpreted as a response of both the scaling relationship intercept (in this context, the mean ratio) and the slope of the wing size–body size allometry. Third, these changes in relative wing size occurred through alteration of wing cell number, mimicking the basis of genetic differentiation in wing loading among natural populations. Fourth, segments of the leg, which grow from imaginal discs adjacent to the wing discs, exhibited a positive correlated response to selection on relative wing size. This is somewhat surprising based on the hypothesis discussed earlier that competition among growing discs for some circulating factor will produce a trade-off in relative trait size. Differences in the timing of wing and leg disc growth could reduce or ameliorate competition between these structures (e.g., Moczek and Nijhout 2004). This, coupled with genetic changes in the shared mechanisms regulating wing and leg (and perhaps other, unmeasured) imaginal disc growth, could explain the correlated response of the legs to selection on relative wing size. Fifth, crosses among the derived experimental populations suggested substantial nonadditivity in the genetic basis of the response to artificial selection. Finally, when larvae from these lineages were subjected to variation in temperature or crowding, the average morphological response differed among lineages, presumably yielding lineage-by-environment interactions for wing loading. However, these environmental effects were achieved through alteration of wing cell size, mimicking the mechanism of wing-loading plasticity in nature.

The primary insight offered by this early work is that although wing loading evolved easily in these experimental populations, it involved changes in wing size only; body size did not change. This suggests the presence of developmental bias in the evolution of wing loading. It is unlikely that such a bias results from differences between wings and bodies in their genetic variation for size, as both exhibit similar levels of genetic variation. Such bias could be caused by genetic correlations among unmeasured traits. For example, fecundity selection on body size may act strongly against selection for relatively small bodies when both traits are free to vary. This is perhaps particularly true in lab populations such as these where such fecundity selection could be strong and where selection imposed on relative wing size through its effect on locomotor performance is presumably reduced. Alternatively, as was the case with the stalk-eyed flies discussed earlier, evolution of a scaling relationship that involves changes in the size of an appendage but not the body may result from the ease of changing the growth of a single trait (the wing or eyestalk) relative to the multitude of changes that are involved when body size evolves. The genetic or developmental basis of the response to selection and the degree to which wing shape changes underlie evolution of the wing–body size ratio remain open and interesting questions. Addressing these would allow us to build bottom-up models for the evolution of the wing size–body size scaling relationship and perhaps deepen our understanding of the biogeographic patterns in wing loading exhibited by *Drosophila*.

Understanding and predicting the course of wing size–body size scaling relationship evolution must, of course, take into account the different selective factors acting on wing loading. This is a complicated undertaking, as the fitness contributions of the absolute and relative size of the wings and body must be determined. Again, experimental evolution can be used to address these topics. Males selected for large absolute body size spend less time vibrating their wings during courtship and enjoy increased mating success relative to males from lineages selected for small body size (Ewing 1961). Hence, large absolute wing size and body size convey a fitness advantage, but the mechanism by which this occurs is unclear. Selection for changes in wing loading produced a correlated response in both male courtship song performance and mating success (Ewing 1964); flies from lineages selected to have lower wing loading (i.e., large wings relative to the body) spent less time vibrating their wings and had greater mating success as compared to males from lineages selected to have higher wing loading (i.e., selected to have small relative wings). Moreover, this pattern held in wild-type males that had their wing loading altered experimentally through manipulation of rearing temperature or by physically clipping the wings.

Interestingly however, direct artificial selection on the vibratory components of male song in *Drosophila* produced the opposite relationship between morphology, performance, and fitness (McDonald and Crossley 1982). Flies from lineages selected for increased rates of wing vibration evolved disproportionally larger wings and lower wing loadings, whereas flies from lineages selected for decreased rates of wing vibration evolved in the opposite morphological direction. It is noteworthy that, as in the previous examples, these correlated evolutionary responses of wing loading occurred through changes in wing size only; thorax size did not evolve. In competitive mating trials, females again preferred males with lower rates of wing vibration, although this time it was the males with the proportionally smaller wings that had the preferred courtship phenotype. These results point to a complex, albeit interesting, pattern of sexual selection acting on male wing loading. Combined with the effects of wing loading on flight performance (e.g., Frazier et al. 2008) and dispersal (e.g., Hoffmann et al. 2007), this research suggests that selection on wing loading is very complex. More generally, these studies indicate that an understanding of scaling relationship evolution requires not only consideration of the relative size of the traits in question or the developmental and physiological mechanisms that regulate size, but also the relationship between relative trait size and ecological performance. In this case, the pattern of selection acting on these complex wing traits is likely to be complicated further by considerations of wing shape.

## THE EVOLUTION OF WING DIMENSION ALLOMETRY
## IN *DROSOPHILA*: WING SHAPE

Perhaps in part because body size in flies is difficult to measure, and in part because of the relative ease with which fly wing shape can be quantified, many experimental evolutionary studies on fly wing scaling relationships have not focused on wing loading but rather have been concerned with the scaling between pairs of wing dimensions (e.g., Weber 1990, 1992). Despite the relative constancy of wing shape across populations and species, quantitative genetic studies reveal high levels of genetic variation in wing shape and low genetic covariation among landmarks (e.g., Mezey and Houle 2005); hence, wing shape is predicted to be highly malleable evolutionarily.

Using an elegant mathematical approach, Weber (1990) selected individuals based on the deviations of their wing dimensions from the wing dimension allometry for the population (figure 15.8). By choosing the individuals possessing the most extreme deviations in the relative size of the wing dimensions, his approach sought to alter wing shape by selecting to change the intercept of the allometry between various pairs of wing dimensions. By design, the dimensions of each trait were free to vary as the allometry intercept evolved. In addition to allowing flexibility in the morphological basis of the response to selection, this approach has the advantage of selecting individuals equally from along the entire range of wing dimension sizes (i.e., the full range of wing size), which is correlated with body size. Hence, selection on wing shape is independent of wing size and body size under this experimental design.

Four of the five artificially selected dimension pairs exhibited remarkably strong, smooth, and usually symmetrical responses to selection; after just fifteen generations of artificial selection, wing shapes had diverged on average about 14.5 SD from the wild-type control populations to produce totally distinct populations inhabiting previously unoccupied regions of morphospace (figure 15.8). In all but one case, shifts in the allometry intercepts involved changes in both dimensions, although often to different degrees. Using a method that estimated shape as a set of continuous functions and removed the effects of wing size differently than Weber's (1990) approach, Houle found a similar rates and patterns of response to artificial selection on wing shape (Houle et al. 2003).

These results demonstrate that wing shape, as described by the scaling relationships between wing dimensions, can respond rapidly to selection against their primary axis of

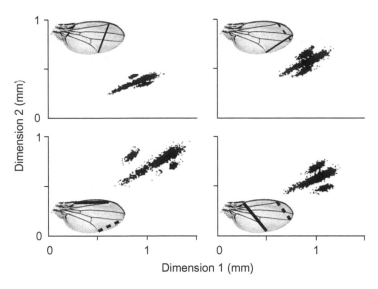

FIGURE 15.8

Natural and artificially selected variation in *Drosophila* wing shape. Each panel shows individuals selected for fifteen generations to produce increases or decreases in Wing Dimension 1 (solid line) relative to Wing Dimension 2 (dashed line), shown on inset images. Control distribution was taken from wild-type flies reared at several temperatures. After Weber (1990).

covariation. In Weber's (1990) experiments, however, some dimensions evolved more easily than others. Many of the wing dimensions span regions of the wing that fall into distinct developmental, morphogenic fields defined by morphogenic gradients (e.g., *dpp*; Lecuit and Le Goff 2007). The degree to which members of these pairs of dimensions can respond independently to selection may be affected by whether the wing dimensions share a single field, are in different fields, or extend across several fields.

The rapidity and degree of independence in the response to selection on minute aspects of wing shape support the hypothesis of myriad genes of small, additive effect contributing to wing shape variation in *Drosophila*. This is indeed the case; genetic analyses of Weber's experimentally derived lineages indicated that wing shape variation is determined by tens of segregating loci, each affecting very small regions of the wing. These data have subsequently been confirmed by additional experimental evolution studies where he selected to alter the relative size of very small (0.5–1.0 mm) wing dimensions (Weber 1992). Four studies have used a genome-wide approach to infer the relationship between genetic variation and interpopulation wing shape variation in *Drosophila* (Weber et al. 1999, 2001; Zimmerman et al. 2000; Mezey et al. 2005). These studies describe a highly complex genetic architecture for *D. melanogaster* wing shape and suggest that interpopulation variation in wing shape is a result of diverse genetic changes. Thus, variation in wing shape, both within and between fruit fly populations, appears to be controlled by a large number of genes, each with small and diverse effects. This raises the question of what the actual genetics of wing shape determination are and whether these same genes are involved in the evolutionary response to selection on wing loading.

One solution to describing this complex genetic architecture more completely may be to break wing shape down into its constituent developmental parts. This is ostensibly done in studies that estimate wing shape using internal wing landmarks, typically the location of vein intersections. Although an interesting morphometric trait, the locations of wing vein intersections do not contribute to wing shape per se, which is defined perhaps more intuitively by the shape of the wing margin. Hence, such studies may confound the genetics of wing shape (i.e., the wing silhouette) with the genetics of cross vein location, potentially influencing the results of association studies. Some studies analyze landmark configurations from two major developmental wing regions, the anterior and posterior compartments, separately (e.g., Pezzoli et al. 1997). We suggest that further incorporation what is known about wing shape determination into such analyses—that is, the bottom-up approach—may be fruitful.

A great deal is known about how pattern is laid down in developing *Drosophila* wings, including which genes are involved and how these genes interact (e.g., see Lecuit and Le Goff 2007). These studies provide the raw material for a developmental model of wing growth that would provide a means for generating a priori predictions regarding the covariation structure among landmarks that define developmentally meaningful units. Specifically, by using information about wing development, it should be possible to assign landmarks to well-known subregions within the developing wing imaginal disc (e.g., anterior/posterior compartments). Consequently, landmarks could be placed in such a way to track movement within and among, say, regions of the wing defined by different morphogenic fields. This general approach has been used successfully to study the covariances between wing vein landmarks in bumblebees (Klingenberg et al. 2001a) and modularity in adult (e.g., Mezey et al. 2000; Klingenburg et al. 2004) and developing mammal jaws (e.g., Cheverud et al. 1997; Badyaev and Foresman 2000) and skulls (e.g., Zelditch et al. 2004, 2006). The one study that has applied this methodology to wing shape variation in *Drosophila* found no evidence reflecting developmental integration of shape across the anterior/posterior patterning boundary of the wing (Klingenberg and Zaklan 2000). Although surprising, this result is consistent with the findings that many genes of small effect can act on a very local scale to affect wing shape. Nevertheless, Klingenberg and Zaklan (2000) used offspring from crosses between genetically differentiated *D. melanogaster* populations; study of lineages with more extreme wing shape phenotypes may prove more fruitful.

Despite numerous studies investigating the genetic basis for wing shape variation in *Drosophila*, there are few data on the role natural selection plays in wing shape evolution. One area that has been well studied is the occurrence of convergent altitudinal and latitudinal clines in wing shape and wing loading (e.g., Huey et al. 2000; Gilchrist et al. 2001). Clearly, there is substantial genetic variation underling many aspects of wing shape, and wing shape phenotypes can change radically due to drift (Phillips et al. 2001;

Whitlock et al. 2002), natural selection, and, of course, artificial selection—yet *Drosophila* wing shape remains relatively conserved (Weber 1990; Houle et al. 2003; Mezey and Houle 2005). It is likely that variation in the shape of the wing margin and placement of the internal wing veins affects performance with respect to courtship and locomotion, both of which are probably broad targets of selection. It seems that the genetics, development, and evolutionary ecology of wing shape variation in *Drosophila* is a rich topic well suited to further exploration using an experimental evolution approach.

## THE EVOLUTION OF WING SCALING RELATIONSHIPS IN THE LITTLE BROWN BUTTERFLY, *BICYCLUS ANYNANA*

*Bicyclus anynana* is a small African butterfly that has been used in a variety of experimental evolutionary studies. Generally, the goals of these studies have been to examine experimentally the role of various kinds of developmental constraints in the evolution of complex suites of correlated morphological, physiological, and life-historical traits, and then to dissect the proximate basis of integration or divergence of these complex phenotypes (e.g., Brakefield et al. 1996; Koch et al. 1996; Zijlstra et al. 2004; Pijpe et al. 2008). In particular, artificial selection experiments have been used to explore evolution of the color composition, shape, and absolute and relative size of eyespots—conspicuous color patterns on the wing (e.g., Holloway and Brakefield 1995; Brakefield et al. 1996; Monterio et al. 1997; Beldade et al. 2002; Allen et al. 2008; see reviews in Brakefield 2003; Brakefield and Frankino 2009). The eyespots are interesting ecologically because they exhibit two different alternative seasonal morphs—a conspicuous wet season form and a cryptic dry season morph. Eyespots function in mate choice, signaling, and predator evasion (Brakefield and Larsen 1984; Robertson and Monteiro 2005; Monteiro and Costanzo 2007; Brakefield and Frankino 2009; Stevens et al. 2008). Although not typically described in these terms, many of the studies that use artificial selection to alter eyespot characters can be viewed as focusing on the scaling relationship between eyespot size and wing size (e.g., figure 15.9). Because this body of work has been reviewed recently and extensively (e.g., McMillan et al. 2002; Brakefield 2003; Frankino and Raff 2004; Beldade et al. 2005; Brakefield and Frankino 2009; Zera and Harshman this volume), and because physiology of wing color pattern determination involves many developmental pathways distinct from those regulating trait growth in our model, here we do not dwell on these wing pattern studies. Nevertheless, this work has revealed a remarkable lack of constraint in the evolution of relative eyespot size (but see Allen et al. 2008 for an important exception) and has indicated relationships between eyespot characters and the endocrine signals that determine the molting cycle and the seasonal morph.

Experimental evolution and mesocosm studies have been used to determine the relative roles of developmental constraints and natural selection in producing the forewing size–hind wing size and the forewing size–body size scaling relationships in *B. anynana*

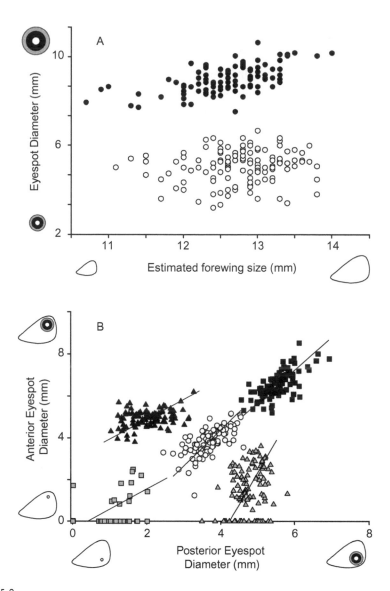

FIGURE 15.9

Artificial selection changes the scaling of butterfly eyespot characters. A, The ratio of eyespot diameter to wing size was selected in alternate directions, producing distinct populations with novel eyespot size–wing size scaling relationship intercepts. Data are shown for females from lineages subjected to selection in each direction for five generations (data are from Monteiro et al. 1994). B, Individuals from a control, stock population (open circles) were selected based on the relative size of their anterior and posterior forewing eyespots. Populations were established to shift the size of the eyespots in the same direction increasing (black squares) or decreasing (gray squares) the size of both eyespots. Selection was also performed antagonistically, increasing (black triangles) or decreasing (gray triangles) the size of the anterior eyespot relative to the posterior eyespot. Points are shown for females from selected lineages after twenty generations of selection, and each is fit with its own scaling relationship using reduced major axis regression. Note that the anterior eyespot responded so strongly in some lineages that it became essentially nonexistent in most individuals (data are from Beldade et al. 2002). Artificial selection methods are given in Monteiro et al. (1994) and Beldade et al. (2002).

(Frankino et al. 2005, 2007). The scaling relationships between the size of the body and the wings in Lepidoptera are good models to study evolution of morphological scaling for both developmental and ecological reasons. First, on the developmental front, fore- and hind wing imaginal discs are developmental homologues (Carroll et al. 2001) that exhibit congruent patterns of developmental gene expression (Carroll et al. 1994; Keys 1999). This suggests that intrinsic changes in wing disc growth should be shared between fore- and hind wings, and that all wings should respond similarly to changes in circulating factors that regulate growth. Second, final wing size is not only determined by wing imaginal disc growth but also by substantial programmed cell death (Dohrmann and Nijhout 1988; Nijhout 1991; Nitsu 2001; Viloria et al. 2003). This means that wing discs often grow to a size much larger than the final adult wing, perhaps exacerbating competitive interactions among adjacently growing discs (e.g., Nijhout and Emlen 1998). Their shared growth regulation and developmental proximity suggest a priori that developmental constraint will resist or bias evolution of the forewing–hind wing scaling relationship. Finally, many Lepidoptera larvae are large, allowing detailed exami- nation and manipulation of the physiology of growth in a manner less challenging than in smaller insects.

In addition to these developmental reasons, the ecology of wing scaling in Lepi- doptera make these interesting models for study. Natural selection on locomotor perfor- mance (Dudley 2000; see also Swallow et al. this volume) and mate acquisition (Wickman 1992) may favor particular within-group scaling relationships, which may account for the near-isometric relationship exhibited for wing loading and between fore- and hind wing size across butterfly species, and the high among-lineage diversity in relative wing size (Strauss 1990, 1992). In sum, the patterns of forewing–hind wing and wing size–body size scaling of Lepidoptera are ecologically relevant, developmentally generalizable, and likely to be subject to both substantial developmental constraint and strong natural selection.

The size of the forewing scales with the size of the hind wing and with body mass in near perfect isometry in *B. anynana* (Frankino et al. 2005, 2007). Using an approach similar to Weber's (1990) described earlier, Frankino and colleagues used artificial selec- tion to determine if the moderate to strong genetic correlations (0.7–1.0) underlying the forewing size–hind wing size or the forewing size–body size scaling relationships would constrain the evolution of these allometry intercepts (Frankino et al. 2005, 2007). After thirteen generations, the allometry intercepts of the artificially selected populations had evolved approximately two standard deviations from the intercept of the wild-type scaling relationship, with roughly one half of the experimental populations evolving to occupy novel morphospace (figure 15.10). Furthermore, there was considerable variation in how individual traits contributed to evolution of the scaling relationship intercept. In some lin- eages, the intercept evolved through changes in wing size alone, while in others there was change in both wing size and in body size. (For further discussion of "multiple solutions" in response to selection, see Swallow et al. this volume.) Moreover, changes in wing

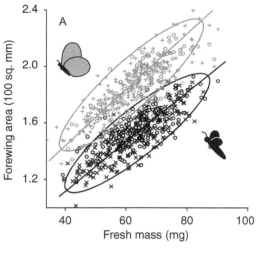

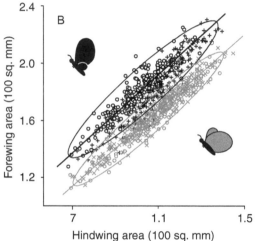

FIGURE 15.10

Phenotype distributions after twelve generations of selection for changes in wing
scaling relationships intercepts in *Bicyclus
anynana*. Each selected population is
shown as a different symbol, replicates of
a selection direction share colors. The scaling relationships of each selected direction
(replicates combined) are shown and individuals from each direction are enclosed
by 95 percent confidence ellipses.
Cartoons show the selected phenotypes in
the appropriate area of morphospace.
A, Populations selected for shifts in wing
loading. B, Populations selected for
changes in the forewing–hind wing scaling. After Frankino et al. (2005, 2007).

shape, not just simple geometric magnification or reduction of wing size, contributed to
the evolution of the intercepts of these scaling relationships. These results, in combination with the similar rates of evolutionary response to exhibited by absolute wing or body
size relative to the rate of evolutionary response to selection on the scaling relationship
intercepts, indicate that the genetic correlations between trait pairs did not impose short-
term constraints on the evolution of their allometry intercept. However, patterns in the
indirect response of the individual traits to direct selection on their scaling relationship
intercept suggest the possibility of developmental bias. Consistent with the response to
selection on the eyestalk span–body size allometry and wing loading in *Drosophila*, nearly
all of the response to selection on the intercept of the forewing size–body size scaling relationship in *B. anynana* involved changes in appendage size. Interestingly, the degree to

which the forewing was involved in the response to selection on the forewing size–hind wing size allometry was dependent on the direction of selection.

Changes in several developmental mechanisms could underlie evolution of the scaling relationship intercepts in these artificially selected *B. anynana* lineages. Modification in the size of the fore- or hind wing imaginal discs at the critical size, change in the duration of each wing's TGP, alteration in the growth rates during the TGP, or manipulation of the patterns of programmed cell death could all underlie instances of intercept evolution that occurred solely through changes in wing size. In the single case where both wing and body size changed as the scaling relationships evolved, alteration in the growth of both traits are required. Changes in the critical size, coupled with modifications that affect wing size alone, could produce evolutionary shifts in scaling relationship intercept involving both traits. A wide variety of approaches can be used to determine the physiological and genetic basis of scaling relationship intercept evolution in this system.

The relatively large size of the body and imaginal discs of Lepidoptera facilitates investigation of the physiological basis of scaling relationship evolution. For example, wing disc size can be easily assessed (Miner et al. 2000) to estimate the initial size of a disc and the pattern of disc growth (i.e., the ontogenetic allometry for the disc). Discs can also be grown in novel signaling environments to determine which, if any, changes are responsible for alterations in disc growth rate; reciprocal wing disc transplants among lineages can be used to determine if derived disc growth patterns result from changes extrinsic or intrinsic to the discs (Kamimura et al. 1996; Hojyo and Fujiwara 1997), and wing disc culture in defined media can be used to test specific hypotheses regarding how changes in sensitivity to a given signaling factor may underlie a derived disc growth pattern (e.g., Hojyo and Fujiwara 1997). The changes in wing shape that contributed to altered wing size discussed earlier may result from programmed cell death toward the end of wing ontogeny. The timing, degree, and pattern of programmed cell death in the wing can be studied using simple staining procedures (e.g., Dohrmann and Nijhout 1988) to test this hypothesis. Finally, starvation experiments can be used to explore the factors underlying critical size or body size evolution (e.g., D'Amico et al. 2001; Davidowitz et al. 2004; Davidowitz and Nijhout 2004) in cases where changes in the body size contribute to evolution of the intercept for the wing size–body size allometry. Similar to how the relatively large body size of Lepidoptera facilitates execution of these experiments, large size also enables relatively easy study of the evolutionary ecology of morphological scaling relationships in a natural setting.

Using the new range of phenotypic variation produced through artificial selection, Frankino and colleagues (2005, 2007) examined the pattern of selection acting on these scaling relationship intercepts in a large, realistically planted tropical greenhouse. Using males with novel, extreme relative trait sizes as well as males with wild-type phenotypes, they documented strong, consistent, stabilizing natural selection favoring the wild-type scaling relationship intercept; control lineage, wild-type females mated preferentially

FIGURE 15.11

Selection on relative forewing size in male *Bicyclus anynana* in a natural environment. A, Large, naturally planted greenhouse, where (B) males and females interacted as in the wild for two days. C and D, Mating success of competing wild-type and extreme-phenotype males from the (C) wing loading selected lineages or (D) forewing–hind wing size selected males shown in figure 15.9. Columns indicate percentage of recaptured females that mated with males in each male phenotype class and are shown with 95 percent confidence intervals based on a bimodal distribution. Similar shading indicates data from replicate trials. Cartoons represents male phenotype in each class. Although these results show strong stabilizing selection on scaling relationship intercepts, the agents of selection remain unknown. Data are from Frankino et al. (2005, 2007). Photos: (A) W. A. Frankino and (B) M. Joron.

with wild-type males three times more than they did with novel-phenotype males (figure 15.11). Although an interesting and important result, unfortunately, the agents of selection were not identified in these experiments. For example, intrasexual competition among males or sexual selection by females may play a significant role in determining male mating success. Rival *Bicyclus* males interact intensively, engaging in aerial chases in an attempt to monopolize access to, or perhaps to display for, females. Wild-type males may have higher fitness because of greater locomotor performance relative to males with derived relative wing sizes. Alternatively, the higher fitness of wild-type males be an design artifact; only wild type, control lineage females were used in this study, and these strongly preferred to mate with males possessing a phenotype similar to their own (i.e., wild type). Females from the artificially selected lineages may also prefer wild-type males whereas females from the artificially selected lineages may have evolved a preference for males with (novel) phenotypes similar to their own (e.g., see Wilkinson and Riellio 1994). Experiments that pit different combinations of male phenotypes against one another in the presence of females derived from the various selected lineages could be used to distinguish

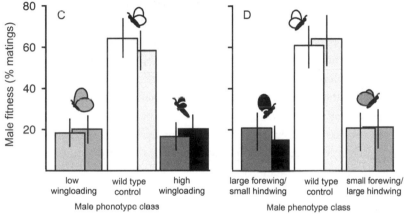

FIGURE 15.11 (*continued*)

among these alternatives. Moreover, the variation among lineages in the morphological basis of scaling relationship intercept evolution—where lineages evolved through various combinations of changes in trait sizes—offers a rich resource for teasing out the independent effects of absolute and relative trait size to male fitness.

In *B. anynana*, the experimental evolution approach indicated a lack of developmental constraint in intercept evolution for wing allometries in the short term, but it also revealed a possible bias in the morphological basis of scaling relationship evolution. Moreover, it allowed estimation of the strong stabilizing natural selection acting on these scaling relationship intercepts—measurements that would not have been possible if only naturally occurring phenotypic variation had been used. In sum, these results suggest that natural selection probably maintains the wild-type scaling relationship intercept in *Bicyclus*, and that ample variation underlying the scaling relationship intercepts allows a rapid evolutionary response to changes in the patterns of selection. It will be interesting to see what future work reveals regarding the proximate basis of scaling relationship evolution in this system.

## SPECIAL OPPORTUNITIES PROVIDED BY POLYPHENIC SPECIES WITH NONLINEAR SCALING RELATIONSHIPS

Species that exhibit alternative, environmentally induced phenotypes (i.e., polyphenisms) hold great promise for the study of shape evolution. In such species, environmental cues such as crowding, day length, temperature, predator presence, or dietary factors determine the phenotype expressed by an individual. Individuals that experience and respond to the inducing cues exhibit different patterns of growth in some traits and therefore produce discrete shapes that differ from those expressed by individuals that do not experience or respond to the cue. Among insects, striking examples include caste differences in head, mandible, and sometimes leg morphology of social insects; winged versus wingless forms in crickets (see also Roff and Fairbairn this volume; Zera and Harshman this volume), plant hoppers, and other insects; and male-dimorphic weapons such as horns or forceps in beetles and earwigs (reviewed in Roff 1996). In polyphenic systems, scaling relationships that describe the within-population variation in shape for these polyphenic traits can be curvilinear or even discontinuous in form (for examples, see figures 15.5 and 15.6, and Emlen and Nijhout 2000).

The maintenance and evolution of polyphenisms has been well studied both theoretically (e.g., Lively 1986; Hazel et al. 1990; Moran 1992; Gross 1996) and empirically (e.g., see reviews in Frankino and Raff 2004; Brakefield and Frankino 2009). Although these studies have revealed much about the evolutionary ecology of polyphenisms, less is known about the role of development in their expression and evolution (Nijhout 1999, 2003; Hartfelder and Emlen 2005). Because they offer the opportunity to induce alternative patterns of trait growth through the manipulation of environmental cues, lineages exhibiting polyphenisms can be used to explore the evolution and development of morphology without the confounding differences present among genetically differentiated populations, species, or other taxa. Thus, polyphenisms offer a uniquely powerful model to dissect the proximate developmental mechanisms underlying trait growth, the expression of scaling relationships, and the roles of development in shape evolution (Gilbert 2001; Emlen and Allen 2004; Frankino and Raff 2004; Hartfelder and Emlen 2005; Brakefield and Frankino 2009). Experimental evolution has been performed on components of polyphenic scaling relationships in at least three different insect systems; the eyespot-wing diphenism in *B. anynana* discussed earlier, a horn–body size diphenism in beetles *(Onthophagus taurus)*, and a wing–body size diphenism in crickets *(Gryllus firmus)*. Because the top-down, bottom-up, and experimental evolution studies of these systems have been reviewed recently elsewhere (e.g., Fairbairn and Roff 1999; Zera 1999; McMillan et al. 2002; Brakefield 2003; Frankino and Raff 2004; Beldade et al. 2005; Emlen et al. 2005; Moczek et al. 2007; Brakefield and Frankino 2009; Zera and Harshman this volume), we discuss only two of them here briefly in order to draw some more general conclusions.

Like many other species in the genus, male *Onthophagus acuminatus* and *O. taursus* beetles exist as two diet-induced morphs (Emlen and Nijhout 2000; Emlen 2001). Small males are hornless and called "minors," whereas larger males possess horns and are referred to as "majors" (figure 15.5; Eberhard 1982). Populations differ in the threshold body size at which individuals will express horns (Moczek et al. 2002). Although neither body size nor horn size had detectable heritabilities in either species (Emlen 1994; Moczek and Emlen 1999), the scaling relationship between these traits can evolve rapidly. The body size threshold for horn induction (i.e., the inflection point of the scaling relationship) in *O. acuminatus* responded to artificial selection in just a few generations (Emlen 1996), and in *O. taurus*, population differences in this threshold persisted in a common garden experiment (Moczek et al. 2002). The physiological mechanisms underlying morph expression are understood somewhat (see reviews in Emlen and Allen 2004; Frankino and Raff 2004; Moczek et al. 2007), but the genetic or developmental mechanisms that produce variation in the threshold, in either wild or artificially selected populations, are not yet known. It is noteworthy, however, that the absence of horns in such beetles does not necessarily result from a lack of horn growth during the immature stages. Rather, in some cases horns grow and are then resorbed before eclosion of the adult (Moczek 2006b), similar to the dying-back of butterfly wing margins mentioned earlier. Hence, there are many developmental possibilities that underlie the expression and diversification of these nonlinear scaling relationships, some of which involve negative growth of the horns themselves. This is important, because a strict top-down approach of estimating the evolvability of the horn–body size allometry from breeding experiments (Emlen 1996) apparently missed this evolutionarily important source of variation. Current studies of the ontogeny of these beetles are ongoing (e.g., Emlen et al. 2006; Moczek 2006a; Moczek and Cruickshank 2006), which will enable the development of a richer, more comprehensive bottom-up model for horn expression and evolution.

Wing polyphenisms are common in crickets. Depending on the environments they experience as juveniles, adults express either small rudimentary wings or fully developed wings and flight musculature (Zera and Denno 1997; Zera and Harshman this volume). A large body of work supports the hypothesis that wing morph is determined largely by juvenile hormone (JH) levels during critical periods of the immature stages (reviewed in Zera and Denno 1997; Frankino and Raff 2004); high JH levels during these periods produce short wings, whereas low JH levels during these times permit the growth of larger wings and flight muscles appropriate for dispersal. The propensity to express a given wing morph in the cricket *Gryllus firmus* responds to artificial selection, and it produces a correlated evolutionary response in the level of juvenile hormone esterase activity (JHE), an enzyme that degrades JH. Lineages selected for increased frequency of long-winged morphs showed an increase in JHE activity, whereas lineages selected for decreased frequency of long-winged morphs showed a correlated decrease in JHE activity (Fairbairn and Yadlowski 1997; Roff et al. 1997; Zera and Huang 1999). Interestingly, selection on JHE activity did not affect wing length expression in the monomorphic species *G. assimilis*, but it did affect wing muscle development and duration of the final immature stage (Zera 2006). These results indicate that wing length determination in monomorphic (ancestral) and polyphenic (derived) species may be determined differently, or perhaps that other endocrine factors, such as ecdysone, play a larger role in wing morph determination than previously thought. In any case, the size of the major elements of the flight apparatus—the wings and flight muscles—are apparently not physiologically integrated tightly as would be predicted based on their ecological function. It would be very interesting to see if selection on JHE activity in a wing dimorphic cricket such as *G. firmus* would alter the frequency of wing morph expression.

These research programs illustrate how a bottom-up approach, informed by a knowledge of the developmental basis of trait integration, may reveal much about the evolution of animal form. They also demonstrate how experimental evolution can inform the building of bottom-up models. Clearly, polyphenic systems offer rich opportunities for the study of form; hopefully, they will be incorporated into experimental evolutionary studies of animal shape with increasing frequency.

## CONCLUSIONS AND FUTURE DIRECTIONS

Both Huxley's equation and top-down quantitative-genetic models incorporate important assumptions of growth and development that may strongly—and perhaps incorrectly—influence our understanding of scaling relationship evolution. For example, Huxley's model assumes that the allometric coefficient, $\alpha$, is determined by the differential exponential growth rate of two traits growing simultaneously. Although this has not been tested explicitly, it is certainly not true for the myriad examples where growth of different structures is not synchronous. Moreover, the final size of morphological traits is not solely determined by the pattern of cell proliferation as predicted by Huxley's model; programmed cell death can dramatically affect the final size and shape of a structure (e.g., Dohrmann and Nijhout 1988; Stern and Emlen 1999; Moczek and Nijhout 2004; see also Lecuit and Le Goff 2007). From the very beginning, therefore, our understanding of allometry evolution has been limited largely to descriptions of diversity in the intercepts and slopes of scaling relationships and to predictions of their evolution using necessary—but overly simplistic and largely untested—assumptions of top-down models.

We have shown how experimental evolution has been used to test some of the general predictions of top-down models. Given the diversity of the trait pairs considered in the empirical examples described in this chapter—from the relative dimensions within the same trait to appendage–body size and appendage–appendage size scaling relationships—experimental evolution studies have demonstrated that the components of animal shape typically can evolve rapidly. This is apparently true even when top-down approaches predict that components of the scaling relationship should not respond readily to selection, either because of estimated low genetic variation for a particular trait or because of tight genetic covariation among traits. In some cases, the morphological basis of the response to selection on aspects of shape can be surprisingly variable, even among replicate selection lineages. These studies reveal that animal shape can evolve easily, and multiple evolutionary and developmental trajectories are often taken to produce similar forms. In other cases, there is pattern, or bias, in how different traits evolve in response to artificial selection on overall shape. Together, these case studies offer a window into the importance of the developmental regulation and integration of trait growth in determining the expression and diversification of morphological variation.

The bottom-up research program we have outlined is one attempt to study shape variation in the context of the proximate mechanisms that produce it. Recent advances in our understanding of the mechanisms regulating and integrating trait growth have ushered in a new generation of developmental models to explain and predict the expression and evolution of shape. These models attempt to reveal the biological mechanisms that regulate the slope and intercept of linear scaling relationships (e.g., Shingleton et al. 2008). This effort is in its infancy and currently, as was the case for top-down models, suffers from simplification. Nevertheless, the components of the models are real biological parameters (Long et al. 2006; Shingleton et al. 2007), and variation in them can be determined empirically. Consequently, they have utility for experimental evolutionary studies of shape evolution.

We have shown how experimental evolution can be used to test for developmental constraints in the evolution of shape. The approach also offers a powerful tool for testing the predictions of the existing bottom-up models. More specifically, experiments can be designed to estimate the extent of variation in the developmental parameters believed to contribute to the regulation of the intercept or slope of a scaling relationship, allowing prediction of which parameter changes are likely to respond under different selective scenarios. Artificial selection can then be used to test these predictions to see if the direct and indirect responses to selection follow their predicted courses. Selection can be focused on the components of the scaling relationships or on the developmental mechanisms hypothesized to contribute to shape variation.

None of the case studies we reviewed used the approaches of theoretical morphology to explore or define the totality of geometrically possible shapes or the roles of developmental constraints, drift, or natural selection in producing the observed distribution in morphospace of naturally occurring or experimentally produced phenotypes. As described here, this is unfortunate as the theoretical morphology approach can suggest hypotheses regarding how biological groups came to occupy different regions of morphological space, and when coupled with experimental evolutionary studies—both empirical and computational—it offers a powerful tool for the study of animal shape. Because scaling relationships have few parameters and the biology underlying their expression in insect morphology is reasonably well understood, we feel allometries offer an excellent system to combine the approaches of theoretical morphology and experimental evolution to elucidate the evolution of form in both top-down and bottom-up research programs.

The lineages created through artificial selection and experimental evolution also offer a rich resource for studying how natural or sexual selection acts on organismal shape. Such experimentally derived lineages possess novel patterns of covariation among shape components or express convergent changes in complex shapes through different morphological means, allowing identification of how each shape component contributes to fitness. In some ways, this is comparable to performing physiological or developmental manipulations (i.e., phenotypic engineering (Sinervo and Huey 1990; Ketterson et al. 1992; Sinervo and Basolo 1996) to create novel phenotypes to achieve the same goal. Even if selection on such phenotypes cannot be measured directly, the relationship between trait variation and ecological performance (e.g., locomotor performance, prey handling time) can be measured as a proxy (Arnold 1983).

It is an exciting time to study the evolution of animal shape. Lineages created through selection experiments can be used to describe more completely the performance and fitness landscapes in which complex shapes evolve, addressing long-standing questions about the general shape and coarseness of the adaptive landscape. Moreover, we are on the cusp of developing realistic, generalizable, multivariate models of trait growth and integration. Experimental evolution offers the primary method for testing both top-down inductive models and bottom-up deductive models of shape evolution, and consequently, it has the potential to unify the findings of both approaches. Although this seems a daunting task, it is already underway. Using advances in morphometrics, developmental genetics, and genomics, researchers are exploring the evolution of shape using extreme phenotypes in a diversity of domesticated taxa. In addition to the studies mentioned here, for example, such studies have revealed interesting patterns of shape change associated with size in rabbits (Fiorello and German 1997), the genetic basis of extreme reduction in body size and shape in dogs (figure 15.3; Long et al. 2006; Sutter et al. 2007), and in the mechanistic basis of dramatic wing size reduction in silk moths (Fujiwara and Hojyo 1997; Hojyo and Fujiwara 1997). This kind of research is certain to accelerate as developmental genetic and genomic resources become increasingly available for such nontraditional model systems. Clearly, experimental evolution—directed at morphological phenotypes or the developmental processes that produce them—is central to an interdisciplinary research program seeking to discover generalities in the evolution of animal form.

## SUMMARY

The study of organismal shape has a long and distinguished history in biology; hence, trends in shape evolution are well described. A primary conclusion of interspecific, comparative morphology is that much morphological diversity has been generated through changes in the relative sizes of traits. Beyond such descriptive and comparative work, identifying the selective factors that shape morphological evolution and elucidating the genetic or physiological basis of morphological variation have been central goals in the study of biological diversity. Evolutionary study of biological shape presents at least three challenges. First, shape is an inherently complex, multivariate trait, and so shape quantification can be difficult. Second, the components of shape tend to covary strongly within biological groups (e.g., populations, species). The resultant low phenotypic variation within groups inhibits measurement of natural selection on shape as well as identification of the fitness contributions of individual shape components. Third, until recently, uncovering generalities that underlie the regulation and integration of trait growth had proven elusive, precluding the development of a general theory of shape expression and evolution.

This chapter examines how experimental evolution can be used to meet these challenges. We focus primarily on one widely used method for describing shape, the scaling relationship between morphometric traits. We connect it to new treatments that model shape expression and evolution as functions of developmental mechanisms that affect the components of morphological scaling relationships. We review case studies that have used experimental evolution to alter the components of morphological scaling relationships in a variety of insect systems, discussing their results in two contexts. We use them first to investigate how development influences the response to different patterns of selection and to evaluate predictions generated by the emergent models of scaling relationship expression and evolution. We then discuss how the novel phenotypes produced via experimental evolution have been used to explore natural selection on the parameters of shape. In sum, these studies reveal that components of shape can evolve rapidly and that natural selection on shape components can be very strong. Our review supports the notion that discovering generalities in the evolution of form may require an explicitly developmental approach to study how growth is regulated and integrated across morphological traits. Experimental evolution offers a powerful methodology for enriching our understanding of the expression and diversification of morphology.

## ACKNOWLEDGMENTS

We thank the editors for the invitation to contribute to this volume, Daphne Fairbairn and Jason Wolf for helpful comments on earlier drafts of this chapter, and Antonia Monteiro and Patricia Beldade for providing the data for figure 15.9. figure 15.1 was inspired by a talk presented by Cerisse Allen at the Fifth International Conference on the Biology of Butterflies. The National Science Foundation supported the writing of this chapter through grants to D.J.E. (IOS-0642409) and W.A.F. (DEB-0805818).

## REFERENCES

Adams, D. C., and M. S. Rosenberg. 1998. Partial warps, phylogeny, and ontogeny: A comment on Fink and Zelditch. *Systematic Biology* 47:168–173.

Allen, C. E., P. Beldade, B. J. Zwaan, and P. M. Brakefield. 2008. Differences in the selection response of serially repeated color pattern characters: standing variation, development, and evolution. *BMC Evolutionary Biology* 8:94.

Andersson, M. 1982. Female choice selects for extreme tail length in a widowbird. *Nature* 299:818–820.

Arevalo, J. E., and P. Heeb. 2005. Ontogeny of sexual dimorphism in the long-tailed manakin *Chiroxiphia linearis*: Long maturation of display trait morphology. *Ibis* 147:697–705.

Arnold, S. J. 1983. Morphology, performance and fitness. *American Zoologist* 23:347–361.

Arnold, S. J., and P. C. Phillips. 1999. Hierarchical comparison of genetic variance-covariance matrices II: Costal-inland divergence in the garter snake, *Thamnophis elegans*. *Evolution* 53:1516–1527.

Atchley, W. R., D. E. Cowley, C. Vogl, and T. McLellan. 1992. Evolutionary divergence, shape change, and genetic correlation structure in the rodent mandible. *Systematic Biology* 41:196–221.

Atchley, W. R., and B. K. Hall. 1991. A model for development and evolution of complex morphological structures. *Biological Reviews* 66:101–157.

Atchley, W. R., and J. J. Rutledge. 1980. Genetic components of size and shape. I. Dynamics of components of phenotypic variability and covariability during ontogeny in the laboratory rat. *Evolution* 34:1161–1173.

Atchley, W. R., J. J. Rutledge, and D. E. Crowley. 1981. Genetic components of size and shape. II. Multivariate covariance patterns in the rat and mouse skull. *Evolution* 35:1037–1055.

Atchley, W. R., S. Xu, and C. Vogl. 1994. Developmental quantitative genetic models of evolutionary change. *Developmental Genetics* 15:92–103.

Azevedo, R. B. R., A. C. James, J. McCabe, and L. Partridge. 1998. Latitudinal variation of wing:thorax size ratio and wing-aspect ratio in *Drosophila melanogaster*. *Evolution* 52: 1353–1362.

Badyaev, A. V., and K. R. Foresman. 2000. Extreme environmental change and evolution: Stress-induced morphological variation is strongly concordant with patterns of evolutionary divergence in shrew mandibles. *Proceedings of the Royal Society of London* 267:371–377.

Baker, R. H., and G. S. Wilkinson. 2001. Phylogenetic analysis of sexual dimorphism and eyespan allometry in stalk-eyed flies (Diopsidae). *Evolution* 55:1373–1385.

———. 2003. Phylogenetic analysis of correlation structure in stalk-eyed flies (*Diasemopsis*, Diopsidae). *Evolution* 57:87–103.

Barker, J. S. F., and R. A. Krebs. 1995. Genetic variation and plasticity of thorax length and wing length in *Drosophila aldrichi* and *D. buzzatii*. *Journal of Evolutionary Biology* 8:689–709.

Beldade, P., and P. M. Brakefield. 2003. The difficulty of agreeing about constraints. *Evolution and Development* 5:119–120.

Beldade, P., P. M. Brakefield, and A. D. Long. 2002. Contribution of *Distal-less* to quantitative variation in butterfly eyespots. *Nature* 415:315–318.

Beldade, P., K. Koops, and P. M. Brakefield. 2002. Developmental constraints versus flexibility in morphological evolution. *Nature* 416:844–847.

———. 2005. Generating phenotypic variation: Prospects from "evo-devo" research on *Bicyclus anynana* wing patterns. *Evolution and Development* 7:101–107.

Bertin, A., and D. J. Fairbairn. 2007. The form of sexual selection on male genitalia cannot be inferred from within-population variance and allometry: A case study in *Aquarius remigis*. *Evolution: International Journal of Organic Evolution* 61:825–837.

Bier, E. 2000. Drawing lines in the *Drosophila* wing: Initiation of wing vein development. *Current Opinion in Genetics & Development* 10:393–398.

Birdsall, K., E. Zimmerman, K. Teeter, and G. Gibson. 2000. Genetic variation for the positioning of wing veins in *Drosophilia melanogaster*. *Evolution and Development* 2:16–24.

Blackstone, N. W. 1987. Allometry and relative growth: Pattern and process in evolutionary studies. *Systematic Zoology* 36:76–78.

Blomberg, S. P., and T. Garland, Jr. 2002. Tempo and mode in evolution: Phylogenetic inertia, adaptation, and comparative methods. *Journal of Evolutionary Biology* 15:899–910.

Blomberg, S. P., T. Garland, Jr., and A. R. Ives. 2003. Testing for phylogenetic signal in comparative data: Behavioral traits are more labile. *Evolution* 57:717–745.

Boag, P. T., and P. R. Grant. 1981. Intense natural selection in a population of Darwin's finches (Geospizinae) in the Galapagos. *Science* 214:82–85.

Bookstein, F. L. 1997 *Morphometric Tools for Landmark Data: Geometry and Biology*. Cambridge: Cambridge University Press.

Bonduriansky, R. 2007a. The evolution of condition-dependent sexual dimorphism. *American Naturalist* 169:9–19.

———. 2007b. Sexual selection and allometry: A critical reappraisal of the evidence and ideas. *Evolution* 61:838–849.

Bonduriansky, R., and T. Day. 2003. The evolution of static allometry in sexually selected traits. *Evolution: International Journal of Organic Evolution* 57:2450–2458.

Brakefield, P. M. 2003. The power of evo-devo to explore evolutionary constraints: Experiments with butterfly eyespots. *Zoology* 106:283–290.

Brakefield, P. M., and W. A. Frankino. 2009. Polyphenisms in Lepidoptera: Multidisciplinary approaches to studies of evolution and development. *In* T. N. Ananthakrishnan and D. W. Whitman, eds. *Phenotypic Plasticity in Insects: Mechanisms and Consequences*. Oxford: Oxford University Press.

Brakefield, P. M., J. Gates, D. Keys, F. Kesbeke, P. J. Wijngaarden, A. Monteiro, V. French, et al. 1996. Development, plasticity and evolution of butterfly eyespot patterns. *Nature* 384:236–242.

Brakefield, P. M., and T. B. Larsen. 1984. The evolutionary significance of dry and wet season forms in some tropical butterflies. *Biological Journal of the Linnean Society* 22:1–12.

Breuker, C. J., J. S. Patterson, and C. P. Klingenberg. 2006. A single basis for developing buffering of *Drosophila* wing shape. *PLoS One* 1:1–7.

Burkhardt, D., and I. de la Motte. 1983. How stalk-eyed flies eye stalk-eyed flies: Observations and measurements of the eyes of *Cyrtodiopsis whitei* (Diopsidae, Diptera). *Journal of Comparative Physiology* 151:407–421.

———. 1985. Selective pressures, variability, and sexual dimorphism in stalk-eyed flies (Diopsidae). *Naturwissenschaften* 72:204–206.

———. 1987. Physiological, behavioral, and morphometric data elucidate the evolutive significance of stalked eyes in Diopsidae (Diptera). *Entomologia Generalis* 12:221–233.

Calder, W. A. 1984. *Size, Function and Life History.* Cambridge, MA: Harvard University Press.

Camara, M. D., C. A. Ancell, and M. Pigliucci. 2000. Induced mutations: A novel tool to study phenotypic integration and evolutionary constraints in *Arabidopsis thaliana. Evolutionary Ecology Research* 2:1009–1029.

Carroll, S. B., J. Gates, D. N. Keys, S. W. Paddock, G. E. F. Panganiban, J. E. Selegue, and J. A. Williams. 1994. Pattern formation and eyespot determination in butterfly wings. *Science* 265:109–114.

Carroll, S. B., J. K. Grenier, and S. D. Weatherbee. 2001. *From DNA to Diversity: Molecular Genetics and the Evolution of Animal Design.* Malden, MA: Blackwell Science.

Cheverud, J. M. 1982a. Phenotypic, genetic, and environmental morphological integration in the cranium. *Evolution* 36:499–516.

———. 1982b. Relationships among ontogenetic, static, and evolutionary allometry. *American Journal of Physical Anthropology* 59:139–149.

———. 1984. Quantitative genetics and developmental constraints on evolution by selection. *Journal of Theoretical Biology* 110:155–171.

———. 1988. The evolution of genetic correlation and developmental constraints. Pages 94–101 *in* G. de Jong, ed. *Population Genetics and Evolution.* Berlin: Springer.

———. 1996. Developmental integration and the evolution of pleiotropy. *American Zoologist* 36:44–50.

Cheverud, J. M., and E. J. Routman. 1995. Epistasis and its contribution to genetic variance components. *Genetics* 139.

Cheverud, J. M., E. J. Routman, and E. J. Irschick. 1997. Pleiotropic effects of individual gene loci on mandibular morphology. *Evolution* 51:2006–2016.

Cheverud, J. M., J. J. Rutledge, and W. R. Atchley. 1983. Quantitative genetics of development: Genetic correlations among age-specific trait values and the evolution of ontogeny. *Evolution* 37:895–905.

Claude, J. 2008. Morphometrics with R. Springer, New York.

Cock, A. G. 1966. Genetical aspects of metrical growth and form in animals. *Quarterly Review of Biology* 41:131–190.

Cotton, S., D. W. Rogers, J. Small, A. Pomiankowski, and K. Fowler. 2006. Variation in preference for a male ornament is positively associated with female eyespan in the stalk-eyed fly *Diasemopsis meigenii. Proceedings of the Royal Society of London B, Biological Sciences* 273:1287–1292.

Cowley, D. E., and W. R. Atchley. 1990. Development and quantitative genetics of correlation structure among body parts of *Drosophilia melanogaster. American Naturalist* 135:242–268.

———. 1992. Quantitative genetic models for development, epigenetic selection, and phenotypic evolution. *Evolution* 46:495–518.

Cuervo, J. J., and A. P. Moller. 2001. Components of phenotypic variation in avian ornamental and non-ornamental feathers. *Evolutionary Ecology* 15:53–72.

D'Amico, L. J., G. Davidowitz, and H. F. Nijhout. 2001. The developmental and physiological basis of body size evolution in an insect. *Proceedings of the Royal Society of London* 268:1589–1593.

Darwin, C. 1859. *The Origin of Species by Means of Natural Selection.* London: Murray.

David, P., T. Bjorksten, K. Fowler, and A. Pomiankowski. 2000. Condition-dependent signalling of genetic variation in stalk-eyed flies. *Nature* 406:186–188.

David, P., A. Hingle, D. Greig, A. Rutherford, A. Pomiankowski, and K. Fowler. 1998. Male sexual ornament size but not asymmetry reflects condition in stalk-eyed flies. *Proceedings of the Royal Society of London B, Biological Sciences* 265:2211–2216.

Davidowitz, G., L. J. D'Amico, and H. F. Nijhout. 2004. The effects of environmental variation on a mechanism that controls insect body size. *Evolutionary Ecology Research* 6:49–62.

Davidowitz, G., and H. F. Nijhout. 2004. The physiological basis of reaction norms: The interaction among growth rate, the duration of growth, and body size. *Integrative and Comparative Biology* 44:443–339.

Debat, V., M. Begin, H. Legout, and J. R. David. 2003. Allometric and nonallometric components of *Drosophila* wing shape respond different to developmental temperature. *Evolution* 57:2773–2784.

de Celis, J. F. 2003. Pattern formation in the *Drosophila* wing: The development of the veins. *BioEssays* 25:443–451.

de Celis, J. F., and J. Diaz-Benjumea. 2003. Developmental basis for vein pattern variations in insect wings. *Internal Journal of Developmental Biology* 47:653–663.

Dickinson, M. H., F. Lehmann, and S. P. Sane. 1999. Wing rotation and the aerodynamic basis of insect flight. *Science* 284:1954–1960.

Dobzhansky, T., and O. Pavlovsky. 1957. An experimental study of interaction between genetic drift and natural selection. *Evolution* 11:311–319.

Dohrmann, C. E., and H. F. Nijhout. 1988. Development of the wing margin in *Precis coenia* (Lepidoptera: Nymphalidae). *Journal for Research on the Lepidoptera* 27:151–159.

Dryden, I. L., and K. V. Mardia. 2002. *Statistical Shape Analysis.* West Sussex, U.K.: Wiley.

Dudley, R. 2000. *The Biomechanics of Insect Flight: Form, Function, Evolution.* Princeton, NJ: Princeton University Press.

Dworkin, I., and G. Gibson. 2006. EGF-R and TGF-ß signalling contributes to variation for wing shape in *Drosophila melanogaster. Genetics* 173.

Eberhard, W. G. 1982. Beetle horn dimorphism: Making the best of a bad lot. *American Naturalist* 119:420–426.

Eberhard, W. G., and E. E. Gutierrez. 1991. Male dimorphisms in beetles and earwigs and the question of developmental constraints. *Evolution* 45:18–28.

Eberhard, W. G., B. A. Huber, R. L. Rodriguez S, R. D. Briceno, I. Salas, and V. Rodriguez. 1998. One size fits all? Relationships between the size and degree of variation in genitalia and other body parts in twenty species of insects and spiders. *Evolution* 52:415–431.

Emlen, D. J. 1994. Environmental control of horn length dimorphism in the beetle *Onthophagus acuminatus* (Coleoptera: Scarabaeidae). *Proceedings of the Royal Society of London* 256:131–136.

———. 1996. Artificial selection on horn length–body size allometry in the horned beetle *Onthophagus acuminatus* (Coleoptera: Scarabaeidae). *Evolution* 50:1219–1230.

———. 2001. Costs and the diversification of exaggerated animal structures. *Science* 291:1534–1536.

Emlen, D. J., and C. E. Allen. 2004. Genotype to phenotype: Physiological control of trait size and scaling in insects. *Integrative and Comparative Biology* 43:617–634.

Emlen, D. J., J. Hunt, and L. W. Simmons. 2005. Evolution of sexual dimorphism and male dimorphism in the expression of beetle horns: Phylogenetic evidence for modularity, evolutionary lability, and constraint. *American Naturalist* 166:S42–S68.

Emlen, D. J., L. C. Lavine, and B. Ewen-Campen. 2007. On the origin and evolutionary diversification of beetle horns. *Proceedings of the National Academy of Sciences of the USA* 104:8661–8668.

Emlen, D. J., and H. F. Nijhout. 2000. The development and evolution of exaggerated morphologies in insects. *Annual Review of Entomology* 45:661–708.

Emlen, D. J., and T. K. Philips. 2006. Phylogenetic evidence for an association between tunneling behavior and the evolution of horns in dung beetles (Coleoptera: Scarabaeidae: Scarabaeinae). *Coleopterists Society Monographs* 5:47–56.

Emlen, D. J., S. Szafran, L. Corley, and I. Dworkin. 2006. Candidate genes for the development and evolution of beetle horns. *Heredity* 97:179–191.

Ewing, A. W. 1964. The influence of wing area on the courtship behaviour of *Drosophila melanogaster*. *Animal Behaviour* 12:316–320.

Fairbairn, D. J., and R. F. Preziosi. 1994. Selection and the evolution of allometry for sexual size dimorphism in the water strider, *Aquarius remigis*. *American Naturalist* 144:101–118.

Fairbairn, D. J., and D. A. Roff. 1999. The endocrine genetics of wing polymorphism in *Gryllus*. A response to Zera. *Evolution* 53:977–979.

Fairbairn, D. J., and D. E. Yadlowski. 1997. Coevolution of traits determining migratory tendency: Correlated response of a critical enzyme, juvenile hormone esterase, to selection on wing morphology. *Journal of Evolutionary Biology* 10:495–513.

Falconer, D. S., and T. F. C. Mackay. 1996. *Introduction to Quantitative Genetics*. 4th ed. Essex, U.K.: Longman.

Fink, W. L., and M. L. Zelditch. 1995. Phylogenetic analysis of ontogenetic shape transformations: A reassessment of the piranha genus *Pygocentrus* (Teleostei). *Systematic Biology* 44:343–360.

Fiorello, C. V., and R. Z. German. 1997. Heterochrony within species: Craniofacial growth in giant, standard, and dwarf rabbits. *Evolution* 51:250–261.

Frankino, W. A., and R. A. Raff. 2004. Evolutionary importance and pattern of phenotypic plasticity: Insights gained from development. Pages 64–81 *in* T. J. DeWitt and S. M. Scheiner, eds. *Phenotypic Plasticity, Functional and Conceptual Approaches*. New York: Oxford University Press.

Frankino, W. A., B. J. Zwaan, D. L. Stern, and P. M. Brakefield. 2005. Natural selection and developmental constraints in the evolution of allometries. *Science* 307:718–720.

———. 2007. Developmental constraints and natural selection in the evolution of a morphological allometry. *Evolution* 61:2958–2970.

Frazier, M. R., J. F. Harrison, S. D. Kirkton, and S. P. Roberts. 2008. Cold rearing improves cold-flight performance in *Drosophila* via changes in wing morphology. *Journal of Experimental Biology* 211:2116–2122.

Fujiwara, H., and T. Hojyo. 1997. Developmental profiles of wing imaginal discs of *flugellos* (*fl*), a wingless mutant of the silkworm, *Bombyx mori*. *Developmental Genes and Evolution* 207:12–18.

Gabriel, K. R., and R. R. Sokal. 1969. A new statistical approach to geographic variation analysis. *Systematic Zoology* 18:259–270.

Garland, T., Jr., and P. A. Carter. 1994. Evolutionary physiology. *Annual Review of Physiology* 56:579–621.

Garland, T., Jr., and P. L. Else. 1987. Seasonal, sexual, and individual variation in endurance and activity metabolism in lizards. *American Journal of Physiology* 252 (Regulatory Integrative and Comparative Physiology 21):R439–R449.

Gayton, J. 2000. History of the concept of allometry. *American Zoologist* 40:748–758.

Gilbert, S. F. 2001. Ecological developmental biology: Developmental biology meets the real world. *Developmental Biology* 233:1–12.

Gilchrist, A. S., R. B. R. Azevedo, L. Partridge, and P. O. O'Higgins. 2000. Adaptation and constraint in the evolution of *Drosophila melanogaster* wing shape. *Evolution and Development* 2:114–124.

Gilchrist, G. W., and R. B. Huey. 2004. Plastic and genetic variation in wing loading as a function of temperature within and among parallel clines in *Drosophila subobscura*. *Integrative and Comparative Biology* 44:461–470.

Gilchrist, G. W., R. B. Huey, and L. Serra. 2001. Rapid evolution of wing size clines in *Drosophila subobscura*. *Genetica* 112–113:273–286.

Gould, S. J. 1966. Allometry and size in ontogeny and phylogeny. *Biological Reviews* 41:587–640.

———. 1973. Positive allometry of antlers in the "Irish Elk" *Megaloceros giganteus*. *Nature* 244:375–376.

———. 1974. The origin and function of "bizarre" structures: Antler size and skull size in the "Irish Elk," *Megaloceros giganteus*. *Evolution* 28:191–220.

Gross, M. R. 1996. Alternative reproductive strategies and tactics: Diversity within sexes. *Trends in Ecology & Evolution* 11:92–98.

Hafen, E., and H. Stocker. 2003. How are the sizes of cells, organs, and bodies controlled? *Public Library of Science* 1:319–323.

Hallgrimsson, B., J. J. Y. Brown, A. F. Ford-Hutchinson, H. D. Sheets, M. L. Zelditch, and F. R. Jirik. 2006. The brachymorph mouse and the developmental-genetic basis for canalization and morphological integration. *Evolution and Development* 8:61–73.

Hallgrimsson, B., D. E. Lieberman, W. Liu, A. F. Ford-Hutchinson, and F. R. Jirik. 2007. Epigenetic interactions and the structure of phenotypic variation in the cranium. *Evolution and Development* 9:76–91.

Hartfelder, K., and D. J. Emlen. 2005. Endocrine control of insect polyphenism. *In* L. I. Gilbert and K. Iatrou, eds. *Comprehensive Insect Molecular Science*, vol. 3. Oxford: Elsevier Science.

Harvey, P. H. 1982. On rethinking allometry. *Journal of Theoretical Biology* 95:37–41.

Hazel, W. N., R. Smock, and M. D. Johnson. 1990. A polygenic model for the evolution and maintenance of conditional strategies. *Proceedings of the Royal Society of London B, Biological Sciences* 242:181–187.

Hingle, A., K. Fowler, and A. Pomiankowski. 2001. Size-dependent mate preference in the stalk-eyed fly *Cyrtodiopsis dalmanni*. *Animal Behaviour* 61:589–595.

Hoffmann, A. A., E. Ratna, C. M. Sgro, M. Barton, M. Blacket, R. Hallas, S. De Garis, et al. 2007. Antagonistic selection between adult thorax and wing size in field released *Drosophila melanogaster* independent of thermal conditions. *Journal of Evolutionary Biology* 20:2219–2227.

Hoffmann, A. A., and J. Shirriffs. 2002. Geographic variation for wing shape in *Drosophila serrata*. *Evolution* 56:1068–1073.

Hojyo, T., and H. Fujiwara. 1997. Reciprocal transplantation of wing discs between a wing deficient mutant (*fl*) and wild type of the silkworm, *Bombyx mori*. *Development, Growth, & Differentiation* 39:599–606.

Holloway, G. J., and P. M. Brakefield. 1995. Artificial selection of reaction norms of wing pattern elements in *Bicyclus anynana*. *Heredity* 74:91–99.

Hosken, D. J., and P. Stockley. 2003. Sexual selection and genital evolution. *Trends in Ecology & Evolution* 19:87–92.

Houle, D., J. Mezey, and P. Galpern. 2002. Interpretation of the results of common principal components analyses. *Evolution* 56:433–440.

Houle, D., J. G. Mezey, P. Galpern, and A. Carter. 2003. Automated measurement of *Drosophila* wings. *BMC Evolutionary Biology* 3:25.

Huey, R. B., G. W. Gilchrist, M. L. Carlson, D. Berrigan, and L. Serra. 2000. Rapid evolution of a geographic cline in size in an introduced fly. *Science* 287:308–309.

Hurley, I., K. Fowler, A. Pomiankowski, and H. Smith. 2001. Conservation of the expression of DII, en, and wg in the eye-antennal imaginal disc of stalk-eyed flies. *Evolution and Development* 3:408–414.

Hurley, I., A. Pomiankowski, K. Fowler, and H. Smith. 2002. Fate map of the eye-antennal imaginal disc in the stalk-eyed fly *Cyrtodiopsis dalmanni*. *Development Genes and Evolution* 212:38–42.

Hutchinson, J. M. C. 1999. But which morphospace to choose? *Trends in Ecology & Evolution* 14:414.

Huxley, J. S. 1924. Constant differential growth-ratios and their significance. *Nature* 114: 895–896.

———. 1932. *Problems of Relative Growth*. London: Methuen.

Huxley, J. S., and G. Teissier. 1936a. Terminologie et notation dans las description de la croissance relative. *Competes Rendus des Seances de la Societe de Biologie* 121:934–937.

———. 1936b. Terminology of relative growth. *Nature* 137:780–781.

Ives, A. R., P. E. Midford, and T. Garland, Jr. 2007. Within-species variation and measurement error in phylogenetic comparative methods. *Systematic Biology* 56:252–270.

Jouffroy, F. K., M. Godinot, and Y. Nakano. 1991. Biometrical characteristics of primate hands. *Human Evolution* 6:269–306.

Kamimura, M., S. Tomita, and H. Fujiwara. 1996. Molecular cloning of an ecdysone receptor (B1 isoform) homologue from the silkworm, *Bombyx mori*, and its mRNA expression during wing disc development. *Comparative Biochemistry and Physiology* 113B:341–347.

Kawano, K. 1995. Horn and wing allometry and male dimorphism in giant rhinoceros beetles (Coleoptera: Scarabacidae) of tropical Asia and America. *Annals of the Entomological Society of America* 88:92–99.

———. 2004. Developmental stability and adaptive variability of male genitalia in sexually dimorphic beetles. *American Naturalist* 163:1–15.

———. 2006. Sexual dimorphism and the making of oversized male characters in beetles (Coleoptera). *Annals of the Entomological Society of America* 99:327–341.

Ketterson, E. D., V. Nolan, M. J. Cawthorn, P. Parker, and C. Ziegenfus. 1992. Phenotypic engineering: Using hormones to explore the mechanistic and functional bases of phenotypic variation in nature. *Ibis* 138:70–86.

Keys, D. N., et al. 1999. Recruitment of a *hedgehog* regulatory circuit in butterfly eyespot evolution. *Science* 283:532–534.

Kingsolver, J. G. 1999. Experimental analyses of wing size, flight, and survival in the western white butterfly. *Evolution* 53:1479–1490.

Klingenberg, C. P., A. V. Badyaev, S. M. Sowry, and N. J. Beckwith. 2001a. Inferring developmental modularity from morphological integration: Analysis of individual variation and asymmetry in bumblebee wings. *American Naturalist* 157:11–23.

Klingenberg, C. P., and L. J. Leamy. 2001. Quantitative genetics of geometric shape in the mouse mandible. *Evolution* 55:2342–2352.

Klingenberg, C. P., L. J. Leamy, and J. M. Cheverud. 2004. Integration and modularity of quantitative trait locus effects on geometric shape in the mouse mandible. *Genetics* 66:1909–1921.

Klingenberg, C. P., L. J. Leamy, E. J. Routman, and J. M. Cheverud. 2001b. Genetic architecture of mandible shape in mice: Effects of quantitative trait loci analyzed by geometric morphometrics. *Genetics* 157:785–802.

Klingenberg, C. P., and H. F. Nijhout. 1998. Competition among growing organs and developmental control of morphological asymmetry. *Proceedings of the Royal Society of London B, Biological Sciences* 265:1135–1139.

Klingenberg, C. P., and S. D. Zaklan. 2000. Morphological integration between developmental compartments in the *Drosophila* wing. *Evolution* 54:1273–1285.

Klingenberg, C. P., and M. Zimmermann. 1992. Static, ontogenetic, and evolutionary allometry: A multivariate comparison in nine species of water striders. *American Naturalist* 140:601–620.

Koch, P. B., P. M. Brakefield, and R. Kesbeke. 1996. Ecdysteroids control eyespot size and wing color patter in the polyphenic butterfly *Bicyclus anynana*. *Journal of Insect Physiology* 42:223–230.

Kodric-Brown, A., R. M. Sibly, and J. H. Brown. 2006. The allometry of ornaments and weapons. *Proceedings of the National Academy of Sciences of the USA* 23:8733–8738.

Koh, I., M. Lee, N. J. Lee, K. W. Park, K. H. Kim, H. Kim, I. J. Rhyu. Body size effect on brain volume in Korean youth. *Neuroreport* 16, 2029–32 (2005).

Kraaijeveld, K., F. J. L. Kraaijeveld-Smit, and J. Komdeur. 2007. The evolution of mutual ornamentation. *Animal Behaviour* 74:657–677.

Lande, R. 1979. Quantitative genetic analysis of multivariate evolution, applied to brain:body size allometry. *Evolution* 33:402–416.

———. 1980. Genetic variation and phenotypic evolution during allopatric speciation. *American Naturalist* 116:463–479.

Lande, R., and S. J. Arnold. 1983. The measurement of selection on correlated characters. *Evolution* 37:1210–1226.

Langlade, N. B., X. Feng, T. Dransfield, L. Copsey, A. I. Hanna, C. Thebaud, A. Bangham, et al. 2005. Evolution through genetically controlled allometry space. *Proceedings of the National Academy of Sciences of the USA* 102:10221–10226.

Lapicque, L. 1907. Tableau général des poids somatiques et encéphaliques dans les espèces animales. *Bulletins Société Anthropologie* 9:248–269.

Le Bihan-Duval, E., S. Mignon-Grasteau, N. Millet, and C. Beaumont. 1999. Genetic analysis of a selection experiment on increased body weight and breast muscle as well as on limited abdominal fat weight. *British Poultry Science* 39:346–353.

Lecuit, T., and L. Le Goff. 2007. Orchestrating size and shape during morphogenesis. *Nature* 450:189–192.

Lemelin, P., and W. L. Jungers. 2007. Body size and scaling of the hands and feet of prosimian primates. *American Journal of Physical Anthropology* 133:828–840.

Liu, J. L., J. M. Mercer, L. F. Stam, G. C. Gibson, Z. Zeng, and C. C. Laurie. 1996. Genetic analysis of a morphological shape difference in the male genitalia of *Drosophila simulans* and *D. mauritiana*. *Genetics* 142:1129–1145.

Lively, C. M. 1986. Canalization versus developmental conversion in a spatially variable environment. *American Naturalist* 128:561–572.

Lofsvold, D. 1988. Quantitative genetics of morphological differentiation in *Peromyscus*. II. Analysis of selection and drift. *Evolution* 42:54–67.

Long, F., Y. Q. Chen, J. M. Cheverud, and R. Wu. 2006. Genetic mapping of allometric scaling laws. *Genetical Research* 87:207–216.

Lovtrup, S., and M. Lovtrup. 1988. The morphogenesis of molluscan shells: A mathematical account using biological parameters. *Journal of Morphology* 197:53–62.

Maynard Smith, J., R. Burian, S. Kauffman, P. Alberch, J. Campbell, B. Goodwin, R. Lande, et al. 1985. Developmental constraints and evolution. *Quarterly Review of Biology* 60: 265–287.

McDonald, J., and S. Crossley. 1982. Behavioral analysis of lines selected for wing vibration in *Drosophila melanogaster*. *Animal Behaviour* 30:802–810.

McGee, G. R. 1999. *Theoretical Morphology: The Concept and Its Applications*. New York: Columbia University Press.

McMillan, W. O., A. Monteiro, and D. D. Kapan. 2002. Development and evolution on the wing. *Trends in Ecology & Evolution* 17:125–133.

Mezey, J. G., J. M. Cheverud, and G. P. Wagner. 2000. Is the genotype-phenotype map modular? A statistical approach using mouse quantitative trait loci data. *Genetics* 156:305–311.

Mezey, J. G., and D. Houle. 2003. Comparing G matrices: Are common principal components informative? *Evolution* 165:411–425.

———. 2005. The dimensionality of genetic variation for wing shape in *Drosophila melanogaster Evolution* 59:1027–1038.

Mezey, J. G., D. Houle, and S. V. Nuzhdin. 2005. Naturally segregating quantitative trait loci affecting wing shape of *Drosophila melanogaster*. *Genetics* 169:2101–2113.

Miner, A. L., A. J. Rosenberg, and H. F. Nijhout. 2000. Control of growth and differentiation of the wing imaginal disk of *Precis coenia* (Lepidoptera: Nymphalidae). *Journal of Insect Physiology* 46:251–258.

Mitteroecker, P., and F. Bookstein. 2007. The conceptual and statistical relationship between modularity and morphological integration. *Systematic Biology* 56:818–836.

Moczek, A. P. 2006a. A matter of measurements: Challenges and approaches in the comparative analysis of static allometries. *American Naturalist* 167:606–611.

———. 2006b. Pupal remodeling and the development evolution of sexual dimorphism in horned beetles. *American Naturalist* 168:711–729.

Moczek, A. P., J. Andrews, T. Kijimoto, Y. Yerushalmi, and D. Rose. 2007. Emerging model systems in evo-devo: Horned beetles and the origins of diversity. *Evolution and Development* 9:323–328.

Moczek, A. P., and T. Cruickshank. 2006. When ontogeny reveals what phylogeny hides: Gain and loss of horns during development and evolution of horned beetles. *Evolution* 60:2329–2341.

Moczek, A. P., and D. J. Emlen. 1999. Proximate determination of male horn dimorphism in the beetle *Onthophagus taurus* (Coleoptera: Scarabaeidae). *Journal of Evolutionary Biology* 12:27–37.

Moczek, A. P., J. Hunt, D. J. Emlen, and L. W. Simmons. 2002. Threshold evolution in exotic population of a polyphenic beetle. *Evolutionary Ecology Research* 4:587–601.

Moczek, A. P., and H. F. Nijhout. 2004. Trade-offs during the development of primary and secondary sexual traits in a horned beetle. *American Naturalist* 163:184–191.

Monteiro, A., and K. Costanzo. 2007. The use of chemical and visual cues in female choice in the butterfly *Bicyclus anynana*. *Proceedings of the Royal Society of London* 274:845–851.

Monteiro, A., J. Prijs, M. Bax, T. Hakkaart, and P. M. Brakefield. 2003. Mutants highlight the modular control of butterfly eyespot patterns. *Evolution and Development* 5:180–187.

Monteiro, A. F., P. M. Brakefield, and V. French. 1994. The evolutionary genetics and developmental basis of wing pattern variation in the butterfly *Bicyclus anynana*. *Evolution* 48:1147–1157.

———. 1997. Butterfly eyespots: The genetics and development of the color rings. *Evolution* 51:1207–1216.

Moran, N. A. 1992. The evolutionary maintenance of alternative phenotypes. *American Naturalist* 139:971–989.

Morgan, T. H. 1911. The origin of five mutations in eye color in *Drosophila* and their modes of inheritance. *Science* 7:537–538.

Nijhout, H. F. 1991. *The Development and Evolution of Butterfly Wing Patterns*. Washington, DC: Smithsonian Institution Press.

———. 1999. Hormonal control in larval development and evolution: Insects. Pages 217–254 *in* B. K. Hall and M. H. Wake, eds. *The Origin and Evolution of Larval Forms*. San Diego, CA: Academic Press.

———. 2003. Development and evolution of adaptive polyphenisms. *Evolution and Development* 5:9–18.

Nijhout, H. F., and D. J. Emlen. 1998. Competition among body parts in the development and evolution of insect morphology. *Proceedings of the National Academy of Sciences of the USA* 95:3685–3689.

Nijhout, H. F., and D. E. Wheeler. 1996. Growth models of complex allometries in holometabolous insects. *American Naturalist* 148:40–56.

Niklas, K. J. 1999. Evolutionary walks through a land plant morphospace. *Journal of Experimental Botany* 50:39–52.

Nitsu, S. 2001. Wing degeneration due to apoptosis in the female winter moth *Nyssiodes lefuarius* (Lepidoptera, Geometridae). *Entomological Science* 4:1–7.

Nur, N., and O. Hasson. 1984. Phenotypic plasticity and the handicap principle. *Journal of Theoretical Biology* 110:275–298.

Palestrini, C., A. Rolando, and P. Laiolo. 2000. Allometric relationships and character evolution in *Onthophagus taurus* (Coleoptera: Scarabaeidae). *Canadian Journal of Zoology* 78:1199–1206.

Partridge, L., R. Barrie, K. Fowler, and V. French. 1994. Evolution and development of body size and cell size in *Drosophila melanogaster* in response to temperature. *Evolution* 48:1269–1276.

Petavy, G., J. R. David, V. Debat, P. Gibert, and B. Moreteau. 2004. Specific effects-of cycling stressful temperatures upon phenotypic and genetic variability of size traits in *Drosophila melanogaster*. *Evolutionary Ecology Research* 6:873–890.

Petrie, M. 1988. Intraspecific variation in structures that display competitive ability: Large animals invest relatively more. *Animal Behaviour* 36:1174–1179.

———. 1992. Are all sexual display structures positively allometry and if so, why? *Animal Behaviour* 43:173–175.

Pezzoli, M. C., D. Guerra, G. Giorgi, F. Garoia, and S. Cavicchi. 1997. Developmental constraints and wing shape variation in natural populations of *Drosophila melanogaster*. *Heredity* 79:572–577.

Phillips, P. C., M. C. Whitlock, and K. Fowler. 2001. Inbreeding changes the shape of the genetic covariance matrix in *Drosophila melanogaster*. *Genetics* 158:1137–1145.

Pijpe, J., P. M. Brakefield, and B. J. Zwaan. 2008. Increased life span in a polyphenic butterfly artificially selected for starvation resistance. *American Naturalist* 171:81–90.

Price, T., Schluter, D., Heckman, N. E. 1993. Sexual selection when the female directly benefits. *Biological Journal of the Linnean Society* 48:187–211.

Ramos, D. M., and A. Monteiro. 2007. Transgenic approaches to study wing color pattern development in Lepidoptera. *Molecular Biosystems* 3:530–535.

Rasskin-Gutman, D., and J. C. Izpisua-Belmonte. 2004. Theoretical morphology of developmental asymmetries. *BioEssays* 26:405–412.

Raup, D. M. 1966. Geometric analysis of shell coiling: General problems. *Journal of Paleontology* 40:1178–1190.

Raup, D. M., and S. J. Gould. 1974. Stochastic simulation and evolution of morphology: Towards a nomothetic paleontology. *Systematic Zoology* 23:305–322.

Raup, D. M., S. J. Gould, T. J. M. Schopf, and D. S. Simberloff. 1973. Stochastic models of phylogeny and the evolution of diversity. *Journal of Geology*:525–542.

Raup, D. M., and A. Michelson. 1965. Theoretical morphology of the coiled shell. *Science* 147:1294–1295.

Rayner, J. M. V. 1985. Linear relations in biomechanics: The statistics of scaling functions. *Journal of Zoology* 206:415–439.

Rice, S. H. 1998a. The bio-geometry of mollusc shells. *Paleobiology* 24:133–149.

———. 1998b. The evolution of canalization and the breaking of von Baer's laws: Modeling the evolution of development with epistasis. *Evolution* 52:647–656.

———. 2000. The evolution of developmental interactions: epistasis, canalization, and integration. *In* J. B. Wolf, E. D. Brodie, and M. J. Wade, eds. *Epistasis and the Evolutionary Process*. Oxford: Oxford University Press.

———. 2002. A general population genetic theory for the evolution of developmental interactions. *Proceedings of the National Academy of Sciences of the USA* 99:15518–15523.

————. 2004a. Developmental associations between traits: Covariance and beyond. *Genetics* 166:513–526.

————. 2004b. *Evolutionary Theory: Mathematical and Conceptual Foundations.* Sunderland, MA: Sinauer.

————. 2008. Theoretical approaches to the evolution of development and genetic architecture. Pages 1167–1186 *in* C. D. Schlichting, and T. A. Mousseau, eds. *The Year in Evolutionary Biology 2008: Annals of the New York Academy of Sciences.* New York: Wiley-Blackwell.

Ritchie, M. G., R. M. Townhill, and A. Hoikkala. 1998. Female preference for fly song: Playback experiments confirm the targets of sexual selection. *Animal Behaviour* 56:713–717.

Robertson, F. W. 1962. Changing the relative size of the body parts of *Drosophila* by selection. *Genetical Research* 3:169–180.

Robertson, F. W., and E. C. R. Reeve. 1952. Studies in quantitative inheritance. I. The effects of selection of wing and thorax length in *Drosophila melanogaster. Journal of Genetics* 50:414–448.

Robertson, K. A., and A. Monteiro. 2005. Female *Bicyclus anynana* butterflies choose males on the basis of their dorsal UV-reflective eyespot pupils. *Proceedings of the Royal Society of London* 272:1541–1546.

Roff, D. A. 1996. The evolution of threshold traits in animals. *Quarterly Review of Biology* 71:3–35.

————. 2000. The evolution of the G matrix: Selection or drift? *Heredity* 84:135–142.

Roff, D. A., G. Stirling, and D. J. Fairbairn. 1997. The evolution of threshold traits: A quantitative genetic analysis of the physiological and life-historical correlates of wing dimorphism in the sand cricket. *Evolution* 51:1910–1919.

Rohlf, F. J. 1998. On applications of geometric morphometrics to studies of ontogeny and phylogeny. *Systematic Biology* 47:147–158.

Rowe, L., and D. Houle. 1996. The lek paradox and the capture of genetic variance by condition dependent traits. *Proceedings of the Royal Society of London B, Biological Sciences* 263: 1415–1421.

Sameshima, S. Y., T. Miura, and T. Matsumoto. 2004. Wing disc development during caste differentiation in the ant *Pheidole megacephala* (Hymenoptera: Formicidae). *Evolution and Development* 6:336–341.

Schlichting, C. D., and M. Pigliucci. 1998. *Phenotypic Evolution: A Reaction Norm Perspective.* Sunderland, MA: Sinauer.

Schluter, D. 1996. Adaptive radiation along genetic lines of least resistance. *Evolution* 50:1766–1774.

Schluter, D., and T. Price. 1993. Honesty, perception and population divergence in sexually selected traits. *Proceedings of the Royal Society of London B, Biological Sciences* 253:117–122.

Schmidt-Nielsen, K. 1984. *Scaling: Why Is Animal Size So Important?* Cambridge: Cambridge University Press.

Shingleton, A. W., Mirth, C. K. & Bates, P. W. Developmental model of static allometry in holometabolous insects. *Proceedings of the Royal Society of London.* 275, 1875–85 (2008).

Shingleton, A. W., W. A. Frankino, T. Flatt, H. F. Nijhout, and D. J. Emlen. 2007. Size and Shape: The regulation of static allometry in insects. *BioEssays* 29:536–548.

Simpson, P., P. Berreur, and J. Berreur-Bonnenfant. 1980. The initiation of pupariation in *Drosophila*: Dependence on growth of the imaginal discs. *Journal of Embryology and Experimental Morphology* 57:155–165.

Simmons, L. W., and J. L. Tomkins. 1996. Sexual selection and the allometry of earwig forceps. *Evolutionary Ecology* 10:97–104.

Sinervo, B., and A. L. Basolo. 1996. Testing adaptation using phenotypic manipulations. Pages 149–185 *in* M. J. Rose and G. V. Lauder, eds. *Adaptation*. San Diego, CA: Academic Press.

Sinervo, B., and R. B. Huey. 1990. Allometric engineering: An experimental test of the causes of interpopulational differences in locomotor performance. *Science* 248:1106–1109.

Smith, R. J. 1980. Rethinking allometry. *Journal of Theoretical Biology* 87:97–111.

Stalker, H. D. 1980. Chromosome studies in wild populations of *Drosophilia melanogaster*. II. Relationships of inversion frequencies to latitude, season, wing-loading and flight activity. *Genetics* 95:211.

Stark, J., J. Bonacum, J. Remsen, and R. DeSalle. 1999. The evolution and development of dipteran wing veins: A systematic approach. *Annual Review of Entomology* 44:97–129.

Starmer, W. T., and L. L. Wolf. 1989. Causes of variation in wing loading among *Drosophila* species. *Biological Journal of the Linnean Society* 37:247–261.

Stern, D. L., and D. J. Emlen. 1999. The developmental basis of allometry in insects. *Development* 126:1091–1101.

Stevens, M., C. L. Stuggins, and C. J. Hardman. 2008. The anti-predator function of "eyespots" on camouflaged and conspicuous prey. *Behavioral Ecology and Sociobiology* 62:1787–1793.

Strauss, R. E. 1990. Patterns of quantitative variation in lepidopteran wing morphology: The convergent groups heliconiinae and ithomiinae (Papilionoidea: Nymphalidae). *Evolution* 44:86–103.

———. 1992. Lepidopteran wing morphology: The multivariate analysis of size, shape, and allometric scaling, Pages 157–179 *in* J. T. Sorensen and R. Foottit, eds. *Ordination in the Study of Morphology, Evolution, and Systematics of Insects: Applications and Quantitative Genetic Rationales*. Amsterdam: Elsevier Science.

Sutter, N. B., C. D. Bustamante, K. Chase, M. M. Gray, K. Zhao, L. Zhu, B. Padhukasahasram, et al. 2007. A single IGF1 allele is a major determinant of small size in dogs. *Science* 316:112–115.

Teissier, G. 1926. Sur la biométrie de l'œil composé des insectes. *Bulletin de la Société Zoologique de France* 51:510–501.

Thompson, D. W. 1917. *On Growth and Form*. Cambridge: Cambridge University Press.

Trotta, V., F. Garoia, D. Guerra, M. C. Pezzoli, D. Grifoni, and S. Cavicchi. 2005. Developmental instability of the *Drosophila* wing as an index of genomic perturbation and altered cell proliferation. *Evolution and Development* 7:234–243.

Via, S., R. Gomulkiewicz, G. DeJong, S. M. Scheiner, C. D. Schlichting, and P. H. Van Tienderen. 1995. Adaptive phenotypic plasticity: consensus and controversy. *Trends in Ecology & Evolution* 10:212–217.

Viloria, A. L., T. W. Pyrcz, J. Wojtusiak, J. R. Ferrer-Paris, G. W. Beccaloni, K. Sattler, and D. C. Lees. 2003. A brachypterous butterfly? *Proceedings of the Royal Society of London (Suppl.)* 270:S21–S24.

Wagner, G. P., and L. Altenberg. 1996. Perspective: Complex adaptations and the evolution of evolvability. *Evolution* 50:967–976.

Warton, D. I., I. J. Wright, D. S. Falster, and M. Westoby. 2006. Bivariate line-fitting methods for allometry. *Biological Review* 81:259–291.

Weaver, T. D., C. C. Roseman, and C. B. Stringer. 2007. Were neandertal and modern human cranial differences produced by natural selection or genetic drift? *Journal of Human Evolution* 53:135–145.

Weber, K. E. 1990. Selection on wing allometry in *Drosophila melanogaster*. *Genetics* 126:975–989.

———. 1992. How small are the smallest selectable domains of form? *Genetics* 130:345–353.

Weber, K. E., and L. T. Diggins. 1990. Increased selection response in larger populations II. Selection for ethanol vapor resistance in *Drosophila melanogaster* at two population sizes. *Genetics* 125:585–597.

Weber, K. E., R. Eisman, S. Higgins, L. Morey, A. Patty, M. Tausek, and Z. Zeng. 2001. An analysis of polygenes affecting wing shape on chromosome 2 in *Drosophila melanogaster*. *Genetics* 159:1045–1057.

Weber, K. E., R. Eisman, L. Morey, A. Patty, J. Sparks, M. Tausek, and Z. Zeng. 1999. An analysis of polygenes affecting wing shape on chromosome 3 in *Drosophila melanogaster*. *Genetics* 153:773–786.

Whitlock, M. C., P. C. Phillips, and K. Fowler. 2002. Persistence of changes in the genetic covariance matrix after a bottleneck. *Evolution* 56:1968–1975.

Wickman, P. 1992. Sexual selection and butterfly design: A comparative study. *Evolution* 46:1525–1536.

Wilkinson, G. S. 1993. Artificial selection alters allometry in the stalk-eyed fly *Cyrtodiopsis dalmanni* (Diptera: Diopsidae). *Genetical Research* 62:213–222.

Wilkinson, G. S. & Reillo, P. R. Female Choice Response to Artificial Selection on an Exaggerated Male Trait in a Stalk-Eyed Fly. Proceedings of the *Royal Society of London Series B-Biological Sciences* 255, 1–6 (1994).

Wilkinson, G. S., and G. N. Dodson. 1997a. Function and evolution of antlers and eye stalks in flies. Pages 310–328 in J. C. Choe and B. J. Crespi, eds. *Mating Systems in Insects and Arachnids*. Cambridge: Cambridge University Press.

———. 1997b. *Function and Evolution of Antlers and Eye Stalks in Flies*. Cambridge: Cambridge University Press.

Wilkinson, G. S., D. C. Presgraves, and L. Crymes. 1998. Male eye span in stalk-eyed flies indicates genetic quality by meiotic drive suppression. *Nature* 391:276–279.

Willmore, K. E., L. Leamy, and B. Hallgrimsson. 2006. Effects of developmental and functional interaction on mouse cranial variability through late ontogeny. *Evolution and Development* 8:550–567.

Willmore, K. E., M. L. Zelditch, N. Young, A. Ah-Seng, S. Lozanoff, and B. Hallgrimsson. 2006. Canalization and developmental stability in the Brachyrrhine mouse. *Journal of Anatomy* 208:361–372.

Wolf, J. B. 2002. The geometry of phenotypic evolution in developmental hyperspace. *Proceedings of the National Academy of Sciences of the USA* 99:15849–15851.

Wolf, J. B., C. E. Allen, and W. A. Frankino. 2004. Multivariate phenotypic evolution in developmental hyperspace. Pages 366–389 *in* M. Pigliucci and K. Preston, eds. *Phenotypic Integration: Studying the Ecology and Evolution of Complex Phenotypes.* Oxford: Oxford University Press.

Wolf, J. B., W. A. Frankino, A. F. Agrawal, E. D. Brodie III, and A. J. Moore. 2001. Developmental interactions and the constituents of quantitative variation. *Evolution* 55:232–245.

Zeh, D. W., and J. A. Zeh. 1988. Condition-dependent sex ornaments and field tests of sexual-selection theory. *American Naturalist* 132:454–459.

Zelditch, M. L., W. L. Fink, D. L. Swiderski, and B. L. Lundrigan. 1998. On application of geometric morphometrics to studies of ontogeny and phylogeny: A reply to Rohlf. *Systematic Biology* 47:159–167.

Zelditch, M. L., B. L. Lundrigan, and T. Garland, Jr. 2004. Developmental regulation of skull morphology. I. Ontogenetic dynamics of variance. *Evolution and Development* 6:194–206.

Zelditch, M. L., J. Mezey, H. D. Sheets, B. L. Lundrigan, and T. Garland, Jr. 2006. Developmental regulation of skull morphology II. Ontogenetic dynamics of covariance. *Evolution and Development* 8:46–60.

Zelditch, M. L., D. L. Swiderski, H. D. Sheets, and W. L. Fink. 2004. *Geometric Morphometrics for Biologists: A Primer.* San Diego, CA: Elsevier Academic Press.

Zera, A. J. 1999. The endocrine genetics of wind polymorphism in *Gryllus*: Critique of recent studies and state of the art. *Evolution* 53:973–977.

———. 2006. Evolutionary genetics of juvenile hormone and ecdysteroid regulation in *Gryllus*: A case study in the microevolution of endocrine regulation. *Comparative Biochemistry and Physiology Part A* 144:365–379.

Zera, A. J., and R. F. Denno. 1997. Physiology and ecology of dispersal polymorphism in insects. *Annual Review of Entomology* 42:207–230.

Zera, A. J., and Y. Huang. 1999. Evolutionary endocrinology of juvenile hormone esterase: Functional relationship with wing polymorphism in the cricket, *Gryllus firmus*. *Evolution* 53:837–847.

Zijlstra, W., M. Steigenga, P. Koch, B. Zwaan, and P. M. Brakefield. 2004. Butterfly selected lines explore the hormonal basis of interactions between life histories and morphology. *American Naturalist* 163:E76–E87.

Zimmerman, E., A. Palsson, and G. Gibson. 2000. Quantitative trait loci affecting components of wing shape in *Drosophila melanogaster*. *Genetics* 155:671–683.

Zwaan, B., R. B. R. Azevedo, A. C. James, J. van't Land, and L. Partridge. 2000. Cellular basis of wing size variation in *Drosophilia melanogaster*: A comparison of latitudinal clines on two continents. *Heredity* 84:338–347.

# 16

# SEXUAL EXPLOITS IN EXPERIMENTAL EVOLUTION

Paul E. Turner, Robert C. McBride, and Clifford W. Zeyl

An intellectual is a person who has discovered something more interesting than sex.

ALDOUS HUXLEY

The evolution of sex is the hardest problem in evolutionary biology.

JOHN MAYNARD SMITH

*Experimental Evolution: Concepts, Methods, and Applications of Selection Experiments*, edited by Theodore Garland, Jr., and Michael R. Rose. Copyright © by the Regents of the University of California. All rights of reproduction in any form reserved.

Humans tend to be fascinated with sex. This is not surprising, as we are members of an obligate sexual species, composed of males and females that must experience sexual reproduction to make offspring. Thus, many of us are sometimes understandably fixated with actions relating to finding a mate, keeping a mate, and mating itself. Without such interests, we would not persist as a species, and there would be no one around to purchase this book.

Evolutionary biologists have also had a historical fascination with sex. This interest began when macroorganism researchers realized that sexual reproduction is a trait and that not all species of animals, plants, and other eukaryotes feature the trait (Steenstrup 1845). Rather, some eukaryotes reproduce asexually; for example, roughly seventy of the over forty-two thousand named vertebrate species undergo exclusively asexual reproduction by making offspring that have direct copies of the single parental genome (Vrijenhoek 1989; Wilson 1992). Still, it is remarkable that the overwhelming majority of eukaryote species are at least sometimes sexual (Bell 1982; Otto and Lenormand 2002). Why should this be so? For over a century, evolution researchers have expended tremendous energy on theory and experiments addressing this question (de Visser and Elena 2007). Despite this effort, an answer continues to be elusive, causing studies on the evolution of sex to remain a fertile (pun intended) area for the use of experimental evolution. Perhaps the mysteries surrounding the evolution and maintenance of sex persist because its effect is straightforward—increased genetic diversity—but the reasons *why* sex is maintained may vary dramatically and according to the particulars of the sexual mechanism in a biological species or group.

In this chapter, we review many of the varied attempts to study the evolution of sex. We note that our purpose is not to exhaustively review all the ideas and studies put forth for the evolution and maintenance of sex. There are entire books devoted to the subject (Williams 1975; Maynard Smith 1978; Bell 1982; Michod and Levin 1988), as well as several relatively recent reviews (Otto and Lenormand 2002; Xu 2004; de Visser and Elena 2007). Rather, our goal is to convince the reader that a wide range of biological systems have been usefully harnessed to explore questions pertaining to the evolution and maintenance of sex by reviewing a select collection of experimental evolution studies involving viruses, unicellular organisms, and multicellular organisms.

## COSTS AND BENEFITS OF SEX

The near ubiquity of sex in eukaryote species is inherently intriguing because sexuality seems to incur quite obvious costs. From a genetic standpoint, sexual reproduction is less efficient than asexual (clonal) reproduction. A female that reproduces sexually only contributes half of her genetic information to her offspring on average, whereas she would contribute the entirety if she reproduced asexually (figure 16.1). Roughly half of the members of a sexual population are males that are themselves incapable of producing young, suggesting that an asexual population has an intrinsic capacity to grow faster

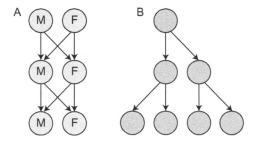

FIGURE 16.1

The twofold cost of sex. If each individual were to contribute to the same number of offspring (two), (a) the sexual population remains the same size each generation, where the (b) asexual population doubles in size each generation. Diagram by Michael Reeve, 2004.

than a typical sexual population with two sexes. All else being equal, this "twofold cost of sex" (Maynard Smith 1978) would favor an asexual population because it could as much as double in size each generation, relative to the size of an otherwise identical sexual population (figure 16.1).

In addition, several other identifiable costs are associated with the specifics of particular sexual reproduction systems. For example, the requirement for separate sexes to produce offspring also necessitates that individuals must find a suitable mate. This need may prove formidable, especially if the ecological habitat causes populations to subsist at very low densities or to thrive under seemingly challenging conditions, such as the total darkness of the deep ocean. For these reasons, some sexual species have evolved elaborate and expensive traits and structures that aid in mate recognition and attraction (Kirkpatrick and Ryan 1991). The act of mating can engender certain costs as well, because copulating individuals may be less able to focus on predator avoidance and copulation places individuals at a greater risk of acquiring sexually transmitted diseases. Last, if the care of young falls more heavily on one member of the sexes, this creates a relative energetic cost that is associated with gender.

Historically, the vast majority of effort exerted to explain the evolution and maintenance of sex has centered on sexually reproducing plants and animals. In such organisms there are distinct genders that have contrasting roles. Females usually contribute almost all the resources with which the progeny begin life, while males add little more than alleles. This disparity in investment is the basis for the twofold cost: asexual females hypothetically could transmit twice as many alleles at the same cost.

In most plants and animals, mates tend to be unrelated, leading to outcrossing. But sex usually also involves the basic process of physical recombination: the breakage and reunion of two different DNA or RNA molecules. Of these two processes, recombination is clearly the more widespread feature of sexual reproduction. A variety of reproductive systems, such as selfing and automixis, involve recombination but not outcrossing. In contrast, relatively few reproductive systems have outcrossing without recombination.

The widespread occurrence of recombination has prompted a broader definition of sex as any natural process that combines genetic material from more than one source into a single individual (Margulis 1986). Thus, although reproduction in viruses and bacteria is strictly asexual, these microbes feature many "parasexual" mechanisms, such as genetic exchange among viral particles, as well as conjugation, transduction, and transformation in prokaryotes. In addition, most microbial eukaryotes reproduce asexually but are occasionally sexual when alternating between haploid and diploid states.

But this broader definition of sex also implies a very general cost. Sex introduces the risk of creating a recombinational load if particular combinations of alleles at different loci are better adapted than alternative combinations (Crow and Kimura 1965). Long ago, Wright (1931) noted that sex may destroy adaptation because "a successful combination of characteristics is attained in individuals only to be broken up in the next generation by the mechanisms of meiosis itself." Similarly, if alleles at different loci were jointly responsible for the production of phenotypes, sex has the potential to break apart coadapted gene complexes, as it moves alleles away from genetic backgrounds where beneficial epistatic interactions have evolved through natural selection.

Why should sex therefore be so common, given the obvious costs? One explanation is that sex is merely the by-product of selection to evolve and maintain physical recombination. This "repair hypothesis" asserts that sexual mechanisms arose due to the selective advantage resulting from recombinational repair of genetic damage (Bernstein et al. 1985). That is, physical recombination can foster survival in the event of DNA damage, and this benefit has nothing to do with allelic recombination.

In contrast, most other hypotheses concerning the evolution of sex invoke the advantage of allelic recombination. This chapter focuses mainly on these ideas, because they have been more often addressed using experimental evolution. The general notion is that sexuality is favored over asexuality because it promotes linkage equilibrium, which is presumed to be useful for evolution by natural selection. Genetic variability is the raw material for evolution by natural selection, and one way to obtain this variation is through mutations. But most evolutionary biologists would agree that random mutations tend to be deleterious. In contrast, increased linkage equilibrium also leads to new genetic combinations, but here the alleles may have already run the gamut of selection. For this reason, sex has the ability to combine useful alleles into the same genetic background more quickly than their sequential fixation in an asexual lineage. It is therefore possible for sex to be advantaged over asexuality because it may speed the rate of adaptation through natural selection (figure 16.2), as independently proposed by Fisher (1930) and Muller (1932). Similarly, the increased linkage equilibrium associated with sex can reduce the frequency of suboptimal allele combinations (Crow and Kimura 1965; Felsenstein 1974). Thus, the overall effect of sex is that it promotes genetic mixis. In many cases, the important underlying assumption in arguments for the evolutionary advantage of sex is that this increased mixis tends to be generally beneficial in evolving populations.

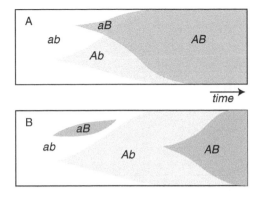

FIGURE 16.2

Adaptive evolution in sexual versus asexual populations of finite size. The *y*-axis represents proportion of the total population carrying a given allele. The *x*-axis represents generation time. Panel (a) shows that sex allows novel mutations arising in independent lineages to be efficiently incorporated into the same genetic background. By contrast, panel (b) shows that this ideal combination of mutations occurs more slowly because the mutations must sequentially fix in the same lineage

In population-genetic terms, sex can affect the distribution of genotype frequencies in a population if linkage disequilibrium exists between alleles at two or more loci; that is, if some combinations, such as *AB* and *ab*, are present in excess and others, such as *Ab* and *aB*, are deficient. For example, in large finite populations experiencing linkage disequilibrium, sexuality can be favored over asexuality because sex brings together the allelic combinations most important for fitness, relative to the time required for them to independently fix in the same genetic background—the so-called Fisher-Muller hypothesis (Crow and Kimura 1965; figure 16.2). Similarly, sex can be selectively favored because it is more efficient at relieving the mutational load brought on by deleterious alleles. Sex brings harmful alleles together into the same genetic background, allowing selection to more efficiently purge them from the population and potentially producing some offspring that are fitter than either parent.

However, the benefit of recombining deleterious mutations may depend on the nature of the epistatic interactions between them. The mutational deterministic hypothesis (Kondrashov 1988) depends partly on this epistasis. If there are synergistic interactions between deleterious mutations in a population, the proportion of individuals that fail to reproduce as a result of selection against mutations, the genetic load (given by $L = 1 - e^{-u}$), may be small in a sexual population, even with a large deleterious mutation rate per genome per generation *(u)* totaled over all loci. The benefit of sex due to a reduced genetic load relative to the genetic load in the absence of sex could be greater than twofold if $u > 1$ (Kondrashov 1988), and thus with these two criteria being fulfilled, it is conceivable that sex could overcome the suggested twofold cost and that sexuality could invade a previously asexual population.

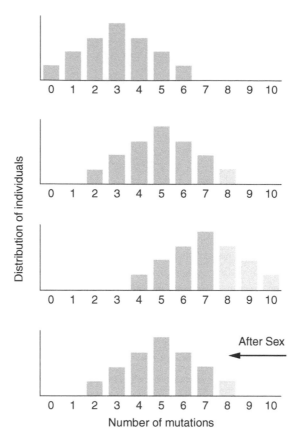

FIGURE 16.3

Over time, the mutational load tends to increase in an asexual population of small size due to the strong effects of genetic drift, and this can create a large fraction of individuals of very low fitness (light shaded bars). *Muller's ratchet* is the term for this process that leads to loss of the least mutated classes of individuals. Sex may be selectively advantaged over asexuality because it can reverse the process, thereby combating the mutational load.

Still other important hypotheses have been proposed to explain why sex might have evolved to combat mutational load when epistasis does not play a prominent role. Muller (1964) envisioned the buildup of mutations as an inevitable process in a small asexual population—a buildup likened to a ratchet tool that clicks forward but cannot move in reverse (figure 16.3). But he also believed sex might have evolved because it is capable of reversing "Muller's ratchet" by allowing the creation of hybrid offspring that harbor fewer mutations than the parent genomes (figure 16.3).

Many other theories have been put forth to explain sex, and while reviewing them here is beyond the scope and space of this chapter, they have been reviewed in some detail by others (Charlesworth 1989; Kondrashov 1993).

Because theories on the evolution of sex were originally formulated around biological systems that feature sexual reproduction, less effort has been made to identify and discuss the costs and benefits of parasexual mechanisms in asexually reproducing microbes. However, it is apparent that microbes should also suffer some of the same identifiable costs of sex as sexually reproducing species. In particular, the possibility of separating coadapted combinations of genes is an expected cost of sex, regardless of reproduction mode. Thus, the expectation is that nearly any model system—microbe or macroorganism—might be used to address many of the hypotheses put forth for the evolution of sex.

However, the ideal approach for testing hypotheses concerning the evolutionary advantages of sex is to use an experimental system where otherwise identical sexual and asexual populations can be created. The problem is that there are very few systems in which this comparison can be simply made. Most often experimenters have needed to resort to artful and ingenious techniques to enable such comparisons. Here we review some of the attempts to empirically test various hypotheses for the evolution of sex, emphasizing the innovative methods and the wide variety of biological systems employed. Our description is not exhaustive, and we refer the interested reader to a more complete list of experimental evolution studies published on the topic, provided in the tables accompanying each of the following sections.

## VIRAL SEX

Viruses are different from other biological entities because they are not composed of cells. But if sex is defined broadly as any exchange of genetic material between individuals, viruses can be considered just as "sexy" as many cellular organisms. Viruses have genomes that are composed of either DNA or RNA. In DNA viruses, sex is promoted usually by recombination (generation of a new nucleotide strand from two or more parental strands; figure 16.4), the same mechanism underlying sexual processes in cellular DNA organisms. In some RNA viruses, sex also involves physical joining of strands of nucleic acids, but in others it does not. Rather, genetic exchange is achieved by splitting the viral genome into several smaller RNA molecules (often termed *segments* or *chromosomes*), and hybrid progeny are produced as random reassortments of segments descending from the co-infecting parent viruses (Chao et al. 1992). Reassortment in this case is analogous to the shuffling of two (or more) hands of playing cards to create a series of novel hands (figure 16.4).

Viruses are made up of largely inert molecules, both protein and nucleic acid, unless they encounter a cell suitable for infection. Thus, it is important to note that viral sex cannot occur unless two or more sexually compatible virus particles happen to co-infect the same host cell. There is clear evidence that viruses can undergo sex in laboratory experiments, so long as researchers create conditions where virus particles are abundant enough to co-infect individual cells (Delbruck and Bailey 1946; Turner and Chao 2003). In addition, data show that sex occurs often enough in natural populations of viruses that hybrid genotypes can sometimes be easily found in the wild (Silander et al. 2005),

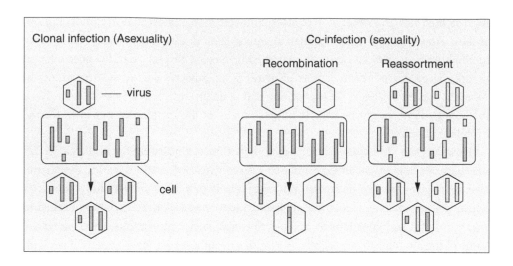

FIGURE 16.4

Clonal infection by a single virus is an asexual process because no genetic mixis is involved. During co-infection, the sexual processes of recombination or reassortment can lead to formation of hybrid genotypes in the viral progeny. Virus particles: hexagons; host cells: rectangles; genetic material (DNA or RNA): light and dark bars.

and recombinant viruses can be isolated from hosts infected by a diverse population of such viral pathogens as influenza virus or HIV (Heeney et al. 2006a, 2006b; Steinhauer and Skehel 2002; Varma and Malathi 2003). For these reasons, it is quite relevant to use viruses as model systems for examining the evolutionary importance of recombination and reassortment, especially the ability of these processes to generate new combinations of alleles among multiple loci on which natural selection can act. Table 16.1 provides an overview of experiments that have been conducted using viruses and bacteria to address similar questions.

## SEX PROMOTES DIRECTIONAL SELECTION IN VIRUS T4

It is noteworthy that viruses were used in the first experimental evolution study to test a hypothesis for the evolution of sex. To do so, Malmberg (1977) employed *Escherichia coli* and one of its viruses, the bacteriophage T4 (a *bacteriophage*, or *phage*, is a virus that specifically attacks bacterial hosts). His experiments addressed two hypotheses for the maintenance of sex. The first prediction derived from the well-known Fisher-Muller hypothesis: Malmberg predicted that beneficial mutations should be incorporated more rapidly in virus T4 lineages undergoing recombination, relative to their nonrecombining counterparts. A second prediction tested was that an absence of recombination would allow for greater epistatic interactions to evolve among genes, owing to the increased likelihood of coadapted gene interactions under linkage disequilibrium.

TABLE 16.1    Summary of Experiments Relevant to the Evolution of Sex Using
Viruses and Bacteria as Model Systems

| Topic/Hypothesis | Conclusion/Summary | Reference |
|---|---|---|
| Fisher-Muller theory | Populations (of T7) evolved at high MOI (with sex) adapt more quickly to a particular environment (proflavine) than populations evolved at a low MOI (without sex). | Malmberg 1977 |
| Mutational deterministic hypothesis | Deleterious mutations do not on average interact synergistically in *E. coli;* rather, they tend to on average act multiplicatively. | Elena and Lenski 1997 |
| Fisher-Muller theory | Sex decreased clonal interference and enabled beneficial mutations to fix more readily, speeding adaptation in rec + (sexual) versus rec−(asexual) lines of *E. coli.* | Cooper 2007 |
| Fisher-Muller theory | Sex did not increase the rate of adaptation of the virus phi 6 to the bacterial host *Pseudomonas phaseolicola.* | Turner and Chao 1998 |
|  | Disadvantages of complementation outweigh the benefits of sex; deleterious load purged faster in the absence of co-infection (sex) in the bacteriophage phi 6. | Froissart et al. 2004 |
| Sex and effective population size | Advantages of sex decreased as effective population size increased in bacteriophage phi 6. | Poon and Chao 2004 |
| Sex, variation, and adaptation | While sex increased variability, it did not increase adaptive evolution in *E. coli.* | Souza, Turner, and Lenski 1997 |
| Muller's ratchet | Fitness of RNA virus decreased by Muller's ratchet | Chao 1990 |
| Muller's ratchet | Provides indirect support for Muller' ratchet as well as showing how nonrecombining populations can recover from fitness loss resulting from the accumulation of deleterious mutations through the acquisition of compensatory mutations | Chao, Tran, and Tran 1997 |
| Sex and clonal interference | Rate of adaptation levels off as population size increased, indicating clonal interference. | Miralles et al. 1999 |
| Sex and clonal interference | Rate of adaptation levels off as population size increased, indicating clonal interference. | Miralles, Moya, and Elena 2000 |
| Muller's ratchet | Advantage to sex was significant for a small number of hybrids, but no advantage of sex averaged over all hybrids. | Chao, Tran, and Mathews 1992 |

Malmberg manipulated the presence or absence of viral sex in a test population, by controlling the degree to which virus T4 particles experienced co-infection. Multiple infection allows for different virus genotypes to enter the same bacterial cell and for the occurrence of recombination (figure 16.4). Thus, if large numbers of viruses are mixed with a relatively small number of bacterial cells, co-infection—and, hence, recombination—should be common. In contrast, as the ratio of viruses to cells becomes less skewed, viruses should more often singly infect cells, and recombination should be less prevalent.

To test the Fisher-Muller hypothesis, Malmberg looked at the effect of recombination level when virus lineages were evolutionarily challenged with environments containing proflavine (3,6-diaminoacridine), a chemical that causes frameshift mutations in viruses such as T4 (Ferguson and Denny 1991). Recombination level was controlled indirectly by manipulating multiplicity of infection (MOI), the ratio of viruses to cells. The high co-infection treatment imposed MOI = 0.2, whereas the low co-infection treatment featured MOI = 0.013. Assuming Poisson sampling (Sokal and Rohlf 1995), the proportion of cells infected with 0, 1, and $\geq 2$ viruses is, respectively, $P(0) = e^{-MOI}$, $P(1) = (e^{-MOI} \times MOI)/1$, and $P(\geq 2) = 1 - P(0) - P(1)$. Thus, only $P(\geq 2)/(1 - P(0))$, or 0.65 percent, of all infected cells contain two or more viruses at an MOI of 0.013, and recombination is very rare. By the same logic, at an MOI = 0.2 recombination is more common because about 10 percent of cells should experience multiple infections. Therefore, in both treatments, the majority of infected cells should contain a single virus, and the expected differences in recombination level offered a somewhat conservative test of Malmberg's prediction.

Malmberg tracked adaptation in his experimental lineages by repeatedly assaying the mean number of offspring produced per parental virus, a measure of fitness, over evolutionary time. This fitness component is similar to the classic microbiology measurement of burst size (virus particles produced per infected cell), but it incorporates other "life-history" stages of the virus such as adsorption rate (viral attachment rate to cells). The results showed that in all cases the lineages undergoing the higher level of recombination experienced a selective advantage in adapting to proflavine, strongly suggesting that greater sexuality in virus populations allowed more rapid fitness improvement (figure 16.5). In turn, these data were consistent with the general prediction that sexuality is favored over asexuality because it increases the rate of adaptation.

To examine the hypothesis that recombination affects evolution because of epistasis, Malmberg isolated multiple viruses that had evolved very strong resistance to high levels of proflavine. He then partitioned the virus T4 genome into eight discrete regions and determined the degree of proflavine resistance associated with each region for every one of the collected viruses. This was achieved by performing genetic crosses between the evolved genotypes and a marked version of the wild-type virus T4. These recombinants were then scrutinized to determine the resistance associated with an individual region of the evolved genome. The pattern of epistasis was inferred by

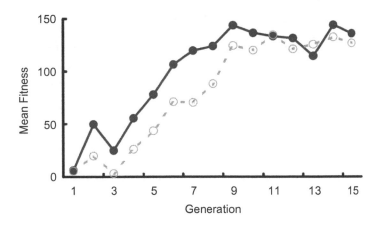

FIGURE 16.5

Sex (recombination) accelerates adaptive change in phage T4. The solid line represents fitness improvement in virus lineages experiencing a relatively higher multiplicity of infection that allowed for sex among co-infecting viruses, compared with less sexual populations (dashed line). The probabilities that the data are best fit by a single curve from a nonlinear regression are 0.025.

measuring whether the resistance of viruses bearing multiple genome regions was greater than that expected based on the resistance of single regions, thereby testing whether there was nonadditive resistance due to positive epistasis among regions. The results showed that greater epistasis existed for genetic regions stemming from viruses isolated from the low-recombination lineages, relative to those drawn from the high-recombination populations. These data were consistent with the idea that the greater linkage equilibrium arising from more frequent recombination hinders the evolution of epistatic interactions.

Although Malmberg's study was truly groundbreaking, it had some shortcomings. One problem is the manner in which Malmberg measured the dynamics of fitness in his test lineages. The results suggesting faster adaptation of sexual lineages relative to asexual populations depended on estimated deviations in fitness of these test lineages from control populations evolving in parallel. Thus, in some cases, the benefit accredited to sex might be better attributed to a benefit associated with changes occurring in the controls, instead of the treatment populations themselves (Rice 2002). Moreover, Malmberg tracked population fitness and not the fate of individual mutations, leaving the specific genetic outcome of the experimental evolution unknown (Rice and Chippindale 2001). Finally, it has been argued that strong laboratory selection imposed on systems of small genome size (in the case of virus T4, ~170 kb) can lead to small standing variation and consequently low background selection, making the effect of recombination on selection response unclear (Rice and Chippindale 2001). Nevertheless, this paper was an excellent first illustration of how experimental evolution could be effectively used to test alternative hypotheses for the evolution of sexual recombination.

As we have already mentioned, one classic theory for the evolution of sex is that genetic mixis evolved to combat Muller's ratchet, the buildup of deleterious mutations in small populations (Muller 1964). In asexual populations of small size there is a tendency for slightly deleterious mutations to accumulate; mutation-free individuals become rare and can be easily lost due to drift (figure 16.3). Asexual populations can only reconstitute mutation-free individuals through reversions or compensatory mutations, which are less likely the smaller the population.

Muller's ratchet was first demonstrated empirically with the RNA phage $\phi6^2$ (Chao 1990). When these phages are grown at low densities on agar containing a superabundant amount (a "lawn") of *Pseudomonas phaseolicola* bacteria, an individual phage infects a cell, and the exiting progeny infect neighboring cells in the lawn. Over many hours, this continued process yields a visible hole in the lawn termed a *plaque*, which can contain millions of phage progeny. But because each plaque is initiated by a single virus particle, an experimenter who propagates a phage population using plaque-to-plaque serial transfers repeatedly forces the virus population through bottlenecks consisting of a single individual ($N = 1$). This propagation scheme should lead to intensified genetic drift in the small virus populations; because the effects of drift are stronger than those of natural selection, the lineages should fix mutations virtually at random. Thus, it is expected that repeated bottleneck passaging of viruses should cause rapid mutation accumulation that leads to a fitness decline, owing to a large proportion of random mutations being deleterious.

Chao (1990) passaged twenty replicate lineages of $\phi6$ through forty such bottlenecks, and measured the fitness (using growth rate as the phenotypic assay) of an evolved virus lineage relative to its unevolved ancestor; fitness above or below 1.0 indicated that the virus was, respectively, more or less fit than its ancestor. Although the genomic mutation rate in virus $\phi6$ was unknown when Chao conducted his experiments, this rate is now gauged to be 0.067 deleterious mutations per generation (Burch and Chao 2004). Therefore, we can now estimate that nearly three mutations on average were fixed in each of his virus lineages (i.e., 0.067 × 40 bottleneck events ≈ 2.7 mutations), assuming that the majority of spontaneous mutations are deleterious. All twenty of the bottlenecked virus lineages featured fitness values much less than 1.0 (showing drops in fitness of 6 to 71 percent), indicating that deleterious mutations had appeared spontaneously, fixed, and consequently severely limited ability of the viruses to grow on the host bacterium. Clearly, drift rather than natural selection drove the fitness changes. In addition, these results showed that deleterious mutations were clearly more common than both reverse and compensatory mutations.

Chao et al. (1997) then examined whether sex could reverse the negative fitness consequences associated with Muller's ratchet. This was possible because phage $\phi6$ is a segmented RNA virus, whose genome is split up into three RNA segments per virus particle. When multiple $\phi6$ genotypes co-infect the same bacterial cell, the segments from the co-infecting parent viruses can be shuffled to create different combinations that are inherited by the virus progeny. This form of sex is called *segment reassortment* (figure 16.4), and it is noteworthy that physical recombination within chromosome segments does not occur at measurable rates in $\phi 6$, allowing each multigene segment to be envisioned as a single locus. As with Malmberg's (1977) experiments with virus T4, it is possible to manipulate the MOI of a virus $\phi 6$ population and to thus control the likelihood it will undergo sex via segment reassortment.

To determine whether segment reassortment combats Muller's ratchet, Chao et al. (1997) used the viruses from Chao's mutation accumulation experiments and conducted genetic crosses of these viruses with one another. Sex is advantageous if crosses tend to yield hybrids (or "reassortants") whose fitness is greater than that of either of the two mated parents. One difficulty of this experimental design is that locating such a hybrid in a genetically crossed population is difficult; the reassortants that were produced in these experiments are only expected to be at a low frequency in the population, perhaps roughly 5 to 20 percent. To overcome this problem, the researchers enriched for such genotypes by subjecting the hybrid population to mass selection. This was achieved by passaging the population through several growth cycles featuring a relatively large bottleneck size of about $10^3$ phages. With this larger bottleneck, selection operates with minimal drift, and hybrid populations containing higher-fitness reassortants should evolve a higher fitness, yielding a measure of the improvement due to sex ($I_S$). To control for the possibility that a hybrid population improves because of back and/or compensatory mutations, the parent strains were "self-crossed" and subjected to mass selection, yielding a measure of the improvement due to such back or compensatory mutation ($I_M$).

Figure 16.6 shows a typical result from these experiments: sex led to the formation of hybrid genotypes featuring a greater fitness than that of either parent. Improvement

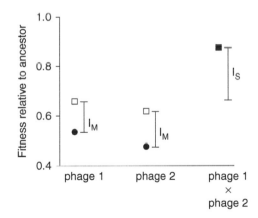

FIGURE 16.6

A hybrid cross between two phages of low fitness demonstrates that sex can combat the mutational load; $I_M$ is the improvement due to mutation in a self-cross; $I_S$ is the improvement due to sex in a hybrid-cross. After Chao et al. (1997).

$(I_S > 0)$ could only have occurred if deleterious mutations had appeared on separate segments; that is, if harmful mutations were confined to a single segment (a hot spot), then segment reassortment would have provided no advantage. Because these data confirm that reassortment can slow or reverse the debilitating fitness effects of Muller's ratchet, they suggest that reassortment might have originally evolved as a mechanism to combat the buildup of harmful mutations. However, Muller's ratchet is only expected to operate in small populations, motivating later experiments where researchers examined whether reassortment was also beneficial in phage $\phi6$ populations of large size (Turner and Chao 1998; Froissart et al. 2004).

## COMPLEMENTATION OVERWHELMS THE ADVANTAGE OF SEX FOR PURGING VIRUS MUTATIONS

Malmberg (1977) examined the advantage of recombination for producing genetic variation that enhanced the response to directional selection in virus T4. Purifying selection is the other general process sometimes proposed to select for sexual lifestyles over asexual ones (e.g., Felsenstein 1974). Here sex is expected to provide a benefit in promoting selection's efficiency at removing deleterious mutations from populations of large size. In such populations, selection will cause harmful mutations to decrease in frequency over time because the genotypes harboring them are of low fitness relative to others in the population. Sex has the ability to combine such deleterious alleles into the same genetic background, allowing selection to be more efficient.

When the harmful effects of mutations are enhanced through negative epistasis, sex should provide an even greater advantage over asexuality. Negative epistasis between deleterious alleles makes combinations of these alleles in the same genetic background more harmful, relative to the sum of their fitness effects when present alone (figure 16.7). Theoretical analysis shows that, if negative epistasis exists between deleterious alleles, then selection should be especially efficient at purging deleterious alleles because genomes bearing an increasing number of mutations are of very low fitness (Eshel and Feldman 1970; Kondrashov 1993). Sex has the ability to readily bring these mutations together in the same genetic background, and thus sex coupled with negative epistasis is expected to provide a substantial advantage for evolving populations.

Froissart et al. (2004) examined this proposed advantage of sex using experimental evolution of virus $\phi6$. They began by identifying three mutations, each of which was unique to one of the three RNA segments of the virus. Using genetic crosses, they determined that the mutations interacted through negative epistasis; that is, double and triple mutants generated through reassortment were of lower than expected fitness (figure 16.7). Consider the fate of these three mutations if they were present in a virus population containing the fitter wild type. If the virus population were asexual, selection would cause the mutations to be purged through time such that the population approaches the known fitness optimum achievable by fixation of the wild type. In

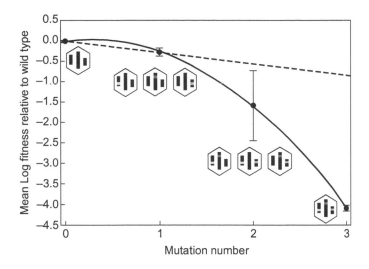

FIGURE 16.7

Negative synergistic epistasis is associated with the decrease in log fitness as three mutations are added to independent RNA segments of the phage $\phi6$ genome. Each point (± standard error) is the mean of replicate measures versus the wild-type virus, and a curvilinear regression (solid line) provides the best fit to the data. Dashed line is the expected decrease in fitness if the mutations affected fitness independently. Virus particles: hexagons; RNA segments: black bars; shaded squares: deleterious mutations.

contrast, co-infecting populations might allow the optimum to be approached more rapidly, because co-infection allows for sex (specifically, reassortment) that creates double and triple mutants of extremely low fitness. This benefit of sex should be especially substantial due to the known negative epistasis. Thus, one might expect that sexual (i.e., co-infecting) populations should purge deleterious mutations faster than their asexual counterparts.

However, Froissart et al. (2004) recognized that genetic complementation could also operate in populations of phage $\phi6$, and that this phenomenon might lead to an opposite result. During co-infection, virus genotypes of low fitness can have their deleterious alleles masked *in trans* by other genotypes bearing superior alleles. In the parlance of Mendelian genetics, single virus infections are strictly haploid, whereas co-infected cells are subject to transient virus polyploidy. Such polyploidy allows genetic dominance that masks one allele's effects by another's. With phage, this dominance is observed when deleterious mutations are complemented *in trans* by unmutated proteins of a co-infecting virus. Complementation occurs because during infection phage gene products, such as replication enzymes and structural proteins, diffuse within the cell to create a resource pool that is commonly available to all co-infecting genotypes (Nee 2000; Nee and Maynard Smith 1990; Turner and Chao 1999, 2003). As a consequence, inferior and genetically recessive phage genotypes can potentially benefit from genetically dominant superior products circulating in the pool. This buffering potential of complementation can thus weaken the effects of selection, causing the virus population to retain harmful alleles for longer than expected.

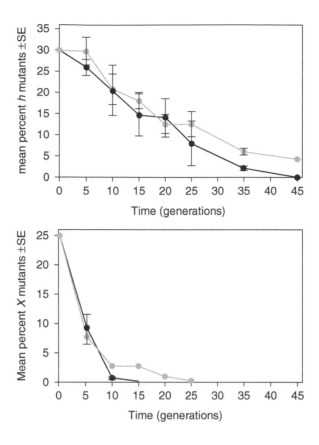

FIGURE 16.8

Loss through time of virus mutants in populations propagated in the presence and absence of co-infection. Each point is the back-transformed least-square-regressed mean frequency (± 95 percent confidence interval) of marked viruses in three replicate populations evolved at MOI = 2 (grey circles) or MOI = 0.0001 (black circles). Populations started with wt $\phi$6 and three marked viruses at an initial ratio of ~1:1:1:1. A, Trajectory for mutants bearing the $h$ (host-range) mutation on the medium segment; B, trajectory for mutants bearing the $X$ (beta-gal gene insertion) on the large segment. In all cases, mutations disappear faster in the absence of co-infection.

The experiment by Froissart et al. (2004) consequently determined which of two intracellular forces, negative epistasis versus complementation, was more important in the evolution of phages experiencing multiple infections. If the effects of genetic mixis and negative epistasis were more important than complementation, then the deleterious mutations would disappear faster from the co-infecting (sexual) lineages. However, if complementation was the more potent force, this masking effect would weaken selection against the harmful alleles during co-infection, causing them to disappear faster in the non-co-infecting (asexual) lineages.

To test these two alternatives, Froissart et al. (2004) created large populations of phage $\phi$6 containing equal mixtures of four virus genotypes: the wild-type plus three mutants that differed by a single unique deleterious allele on one of each of the three $\phi6$ segments. These lineages were then serially transferred in the presence and absence of co-infection (MOI of 2 and 0.0001, respectively). Because the study was short-duration (fewer than ten days, fifty generations of virus evolution), the importance of new spontaneous mutations in the populations was minimized. Thus, regardless of reproductive mode, the expectation was that the deleterious mutations would be purged from each test population, leading to fixation of the wild-type variant (the known equilibrium) that represents the genotype of highest fitness. Each day, samples from the test populations were scrutinized to estimate the frequency of marked genotypes that remained. The results showed that the deleterious alleles disappeared faster or slower from all test populations, depending on whether they strongly or weakly affected virus fitness, respectively. But regardless of the magnitude of their harmful effect, the mutations were retained for longer periods of time on average in the populations undergoing relatively more frequent genetic mixis (figure 16.8). Thus, the data indicated that the effects of complementation overwhelmed the expected advantage of sex arising from its enhancement of purifying selection.

## BACTERIAL SEX

There are evolutionary theories for sex that cannot be tested using simple virus systems but that can be tested in bacteria rather than eukaryotes. The tractability and short generation times of bacteria foster their usefulness as model systems with which to address many evolutionary questions. Bacterial sex occurs through at least three mechanisms: conjugation, transformation, and transduction. All of these mechanisms are asymmetrical; that is, the transfer of genetic material occurs unidirectionally. Conjugation allows DNA to be passed from one cell to another via a pore that forms during a cell-cell junction (Thomas and Nielsen 2005), and the conjugative process is typically mediated by an F plasmid or other similar, accessory, genetic element of bacteria. Transformation occurs when extracellular DNA is brought into the cell and then integrated into the bacterial chromosome (Thomas and Nielsen 2005). Transduction is mediated by a phage that does not complete its infection cycle within the cell but instead causes foreign DNA to recombine with the bacterial cell's chromosome. Here we describe two studies that harness bacteria to test the mutational deterministic hypothesis and the Fisher-Muller hypothesis.

### NEGATIVELY EPISTATIC MUTATIONS ARE NOT COMMON IN *E. COLI* BACTERIA

The mutational deterministic hypothesis of Kondrashov (1988) describes two necessary conditions for sexuality to be favored over asexuality. First, the rate at which deleterious mutations occur should be at least one per genome per generation. Second, deleterious mutations should tend to have negative synergistic effects on fitness (Kondrashov 1982).

An experiment testing this second condition was conducted using *E. coli* populations by Elena and Lenski (1997).

Elena and Lenski used *E. coli* as their model system to examine epistatic interactions, because they had ready access to twelve bacterial strains that had been selected for over ten thousand generations in the laboratory (Lenski and Travisano 1994) and were thus presumably at or nearing some sort of selective optimum in the particular environment of minimal glucose medium. They began by obtaining three different mini-transposons, each bearing a different antibiotic marker. These genetic elements were inserted randomly into the *E. coli* genome, allowing the construction of 225 mutated genotypes of *E. coli* derived from the long-adapted bacterial populations. The engineered genotypes carried one, two, or three mutations per bacterial genome. Populations founded from these genotypes then underwent selection in a nutritionally rich growth medium (agar supplemented with the relevant antibiotic) in order to ensure the survival of unbiased samples of mutants. Elena and Lenski then measured the fitness (using relative growth rate as their assay phenotype) of these mutants using competition assays against an unmutated common competitor in minimal glucose medium, the original selective environment. If mutations interact synergistically with negative epistasis, the expectation is that the relationship between log fitness and increasing mutation number should be curvilinear and concave downward. Since this relationship was shown to be statistically significant when fitted using a simple linear regression model, the researchers concluded that the interactions between deleterious mutations were on average strictly additive, as indeed visual inspection of figure 16.9 clearly suggests. The results from this experiment did not rule out the possibility that deleterious

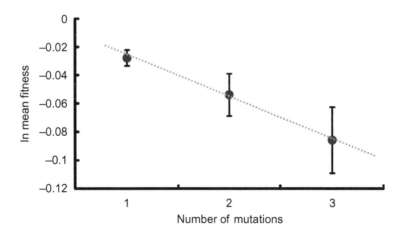

FIGURE 16.9

The observed effect of increasing deleterious mutation number on fitness of *E. coli*. The points represent the average fitness of seventy-five genotypes carrying one, two, or three insertion mutations. The error bars are standard errors generated using the jackknife method. The solid line represents the best fit of a log-linear (multiplicative) model to the data.

mutations might act epistatically, in some way or other, in particular cases; rather, it provided support for the assertion that, on average, the interactions between deleterious mutations were predominantly neither negative nor positive in synergism.

This first experiment by Elena and Lenski added successive mutations to the bacterial genome at random. While this is convenient, greater power is gained in comparing the fitnesses of genomes with single mutations versus the fitnesses of genomes with two mutations. To investigate this phenomenon, Elena and Lenski constructed twenty-seven recombinant *E. coli* genotypes bearing pairs of mutations. These mutations were taken from the collection of individual mutations generated in the first experiment and that spanned a large range of fitness values. They combined these mutations into pairs whose separate and combined effects on fitness were determined. They then compared the actual fitness of the combined mutants to the predicted fitness generated by multiplying the fitness of the individual mutants. They found that in seven cases, the actual fitness of the combined mutants was significantly lower than their predicted combined fitness, demonstrating that these mutations exhibited negative epistasis. However, there were also pairs of mutations that exhibited positive epistasis, and on average, the interaction between the pairs was additive.

These experiments did not find support in *E. coli* for one of the key assumptions of the mutational deterministic hypothesis, preponderant negatively synergistic epistasis. But as these bacteria do not undergo sexual reproduction, this result may not seem that damning a critique of the mutational deterministic hypothesis. However, together with the rather prescriptive requirement that the deleterious mutation rate be greater than 1.0, and data from other studies showing no negative epistasis in organisms that have eukaryotic sexual reproduction (Peters and Keightley 2000), the results that we now have suggest that the mutational deterministic hypothesis is not a compelling general explanation for the maintenance of sex in biological populations.

## RECOMBINATION REDUCES COMPETITION AMONG BENEFICIAL MUTATIONS IN *E. COLI*

Using the same model system as Elena and Lenski, Cooper (2007) tested the Fisher-Muller model in a novel way that circumvented many of the problems associated with prior tests of the hypothesis. Cooper sought direct evidence for the Fisher-Muller hypothesis by examining whether recombination enhanced the rate of fitness improvement in a population where the supply of beneficial mutations (and hence the degree of competition among beneficial alleles undergoing substitution—or "clonal interference") varied. In addition, he examined how clonal interference in the absence of recombination affected the spread of a beneficial mutation.

Problems with previous attempts to test the Fisher-Muller hypothesis can be attributed to the difficulty of testing the underlying assumptions of the model, and those of incorrectly testing the formal hypothesis. One key requirement for a good experimental test of this hypothesis is the need to establish actual competition among beneficial mutations, such that their individual fixation in an asexual population is appreciably slowed. While many studies have shown this to be true (de Visser et al. 1999; Imhof and Schlotterer 2001; Rozen et al. 2002), they have not eliminated confounding factors between treatments such as differences between treatments with respect to the strength of selection for the beneficial mutations arising from differences in population size. Many studies of the Fisher-Muller hypothesis have shown that recombination is beneficial, but these experiments were potentially confounded by treatment regimes that involved differences in environmental or life-history parameters.

Cooper attempted to overcome these problems by building on a previous experiment by de Visser et al. (1999) that identified conditions in which *E. coli* strains containing mutated DNA repair systems experience elevated mutation rates. Using one of these strains, mutS, Cooper determined that he could alter the supply of mutations experienced by a lineage by roughly thirty-fold, relative to the isogenic nonmutated strain. Thus, he could control the amount of competition occurring among mutations as they attempted to fix in the population. Cooper standardized all aspects of the evolutionary process except the mutation rate, which was either high or low. Presence and absence of recombination were controlled by making strains recombination proficient through the introduction of the rec+ and rec– F plasmids into the high- and low-mutation strains. He evolved eight populations under each of these four treatments (rec+, high mutation; rec+, low mutation; rec–, high mutation; rec–, low mutation) for one thousand bacterial generations in minimal glucose medium, and then he estimated the overall adaptation of each lineage by monitoring how fitness changed over time.

In agreement with the Fisher-Muller hypothesis, Cooper found that the lines undergoing recombination experienced greater increases in fitness than those not undergoing mixis. The mechanism driving this result was unclear, however. The results are consistent with a mechanism of decreased interference with substitution among beneficial mutations in the recombining lineages. But the data could not be used to rule out an alternative benefit to recombination: increased linkage equilibrium that tended to separate beneficial mutations from linked deleterious alleles. To explore this second possibility, Cooper compared the dynamics of a focal beneficial mutation in both the rec+ and rec– lines. He found that in the absence of recombination, the focal beneficial mutation took longer to fix (figure 16.10). Additionally, he measured the fitness of this beneficial mutation at different points in its spread through both the rec+ and rec– populations. In the rec– lines, the fitness of the beneficial mutation decreased relative to common competitors as evolutionary time progressed. This suggests that over the course of the experiment, other beneficial mutants arose, which decreased the *relative* fitness of genotypes bearing the focal beneficial mutation, providing support for the idea that competition among beneficial mutations slows their spread in a nonrecombining population. By contrast, in the rec+ lines, there was no change in allelic fitness relative to common competitors over the duration of the selective sweep, suggesting that sex enabled the more rapid spread of the beneficial mutant by reducing its competition with other beneficial mutants.

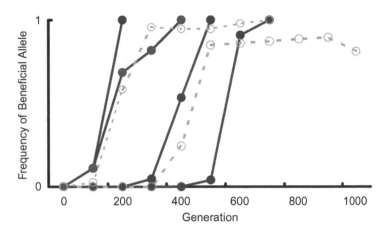

FIGURE 16.10

The effect of recombination in *E. coli* on fixation dynamics of a beneficial allele (spoTEv). The solid lines are rec+ populations, whereas the dashed lines are rec–. Data show that the beneficial allele spreads significantly faster in the rec+ lines than in the rec– lines, suggesting a benefit to sex.

## EUKARYOTIC SEX

Although genetic exchange and recombination are widespread and important processes in viral and bacterial evolution, the cycle of mating and meiosis in eukaryotes has been the context for most evolutionary theory to date and a good deal of the experimentation. It is to anisogamous eukaryotes, those with contrasting male and female functions in sexual reproduction, that the twofold cost of sex applies. In most plants and animals, sex is a necessary component of reproduction, and the question for evolutionary biologists is why reproductive mechanisms have evolved that way. In one of the experiments described next, evolutionary geneticists have nevertheless devised a way to compare evolution with and without recombination in the obligately sexual fruit fly. But most evolution-of-sex experiments with eukaryotes have featured facultatively sexual microbes, in particular the unicellular fungus *Saccharomyces cerevisiae* (henceforth "yeast," although the closely related *S. paradoxus* is emerging as a valuable subject that, unlike *S. cerevisiae*, is thought to be ecologically untainted by domestication) and the unicellular alga *Chlamydomonas reinhardtii*. In addition to offering many of the same advantages as bacteria and viruses—easy maintenance, immense population sizes, short generation times, cryopreservation, sequenced and readily manipulated genomes—these model systems have sexual cycles that are readily manipulated, so that otherwise identical experimental populations can be propagated with or without sex.

Compared to plants and especially animals, many microbial eukaryotes pose different challenges for evolutionary biologists to explain the evolution and maintenance of sex. The canonical twofold cost of sex does not apply to yeast or *C. reinhardtii* because rather than functionally contrasting genders and gametes with their disparity of contributions to sexual progeny, these microbes have mating types. Apart from functions directly involved in sex, these mating types are physiologically and ecologically very similar.

But a different challenge arises from the fact that sex in these species is optional. Rather than being essential for reproduction, sex in yeast and *C. reinhardtii* could be considered a digression from reproduction, which occurs by mitosis and budding or cytokinesis. The ability to propagate these populations indefinitely without sex is a major advantage in experiments on the evolution of sex, because it facilitates straightforward comparisons between otherwise identical sexual and asexual populations.

But more important, the dispensable role of sex in the population growth of these species suggests that sex must provide some selective advantage other than reproduction. A complex trait remains functional only if there is continual selection against the random mutations that otherwise accumulate and degrade it. This "use it or lose it" principle has been observed in the erosion of mating and meiotic competence in lab populations after hundreds of mitotic generations in the fungus *Cryptococcus neoformans* by Xu (2002), in yeast by Zeyl and colleagues (2005)—although Hill and Otto (2007) observed no such trend—and in *C. reinhardtii* by DaSilva and Bell (1992). The fact that newly isolated yeast and *Chlamydomonas* strains are very proficient in mating and meiosis is evidence that sex is maintained in natural habitats by selection.

In both yeast and *C. reinhardtii*, sex is connected with an immediate physiological advantage: the sexual cycle includes the formation of a dormant stage that is highly resistant to desiccation and other stresses. But from an evolutionary perspective, this is not a satisfying explanation for sex, as the ability to form resistant spores without mating and meiosis could plausibly evolve with the substitution of just a few regulatory mutations.

Breaking down linkage disequilibrium by recombination is adaptive only when there is negative linkage disequilibrium for fitness—that is, when alleles of high fitness are frequently linked to low-fitness alleles. Recent theory on the evolution of sex has focused on two potential causes of negative linkage disequilibrium: (1) interaction between finite population size (chance) and selection and (2) epistasis. The potential role of epistasis alone increasingly seems limited, both by theoretical restrictions on the range of epistatic effects that can provide an advantage for recombination and by growing evidence that eukaryote genomes do not consistently generate the necessary weakly negative epistatic interactions (Gandon and Otto 2007; Kouyos et al. 2006; Otto and Feldman 1997). We describe three experiments that tested for hypothesized acceleration of adaptation in sexual populations that were placed under strong directional selection, by novel environments in the first two and by artificial selection in the third. First, a yeast experiment tested for a general advantage of sex in populations introduced into a harsh environment that demanded rapid adaptation. The second experiment used populations of *C. reinhardtii* to test for the advantage inferred by Cooper (2007) for *E. coli*: the recombination of adaptive mutations that otherwise compete. The third experiment looked for an advantage of recombination resulting from the separation of deleterious mutations from an eye color allele artificially selected in *Drosophila melanogaster* populations.

Before we summarize the microbial eukaryote experiments, a brief explanation of the yeast and algal sexual cycles is in order. Both *S. cerevisiae* and *C. reinhardtii* are isogamous, meaning that instead of large female gametes and much smaller male gametes, there are mating types that produce gametes with equal roles in the sexual cycle (figures 16.11, 16.12). This eliminates the twofold cost of sex that complicates anisogamous systems such as the familiar male-female dichotomy of most plants and animals. In isogamous mating systems, where mating types contribute equally to zygotes, the likely disadvantages of sex should be less drastic. One cost, however, is particular to facultatively sexual microbes: mating and meiosis take much longer than asexual reproduction by cell division or budding. In both yeast and *C. reinhardtii*, the sexual cycle is driven by

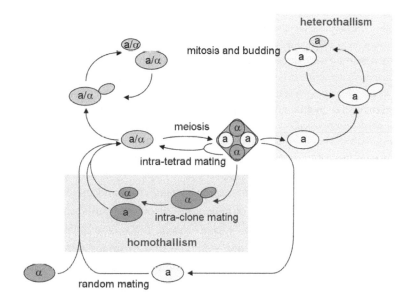

FIGURE 16.11

The life cycle of *Saccharomyces cerevisiae*. Reproduction is by mitosis and budding, which continue indefinitely as long as adequate nutrients permit. Mating occurs spontaneously when MATa and MATα cells encounter each other. Diploids (heterozygous at the MAT locus) reproduce asexually by mitosis and budding, until induced to sporulate and undergo meiosis by starvation for nitrogen and restriction to carbon sources that must be respired. Sporulation produces four haploid ascospores, two of each mating type, which have thick walls and are resistant to stresses including desiccation and enzymatic treatments that destroy the walls of vegetative cells. Upon the return of environmental conditions that permit growth and budding, spores germinate. Mating often occurs almost immediately because of the proximity of germinated spores of complementary mating types, either from the same ascus or from different asci. Most wild isolates are *homothallic*, switching mating types after budding off daughter cells. Such strains are usually highly inbred, because of matings between mother and daughter cells. Lab strains are often converted to heterothallism (see text) so that haploid strains can be maintained asexually.

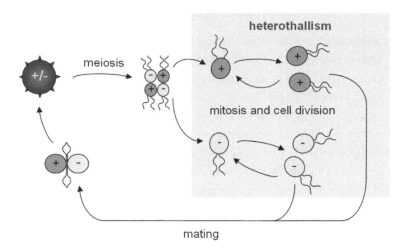

FIGURE 16.12

The life cycle of *Chlamydomonas reinhardtii*. Vegetative cells are normally haploid and reproduce asexually by mitosis. Natural isolates are heterothallic, so populations established with a single mating type remain asexual. Starvation for nitrogen induces gametogenesis and, if both mating types are present, mating, which normally occurs in liquid cultures. Cultures containing both mating types remain asexual as long as adequate nitrogen is supplied, but one mating type may be lost due to selective sweeps or, in long-term cultures, by drift. After mating, thick-walled zygotes are formed during a maturation process that in lab culture usually occurs in the dark. After four to five days, zygotes can be returned to vegetative growth and reproduction by restoring light and adequate nitrogen. Meiosis occurs spontaneously, and haploid spores emerge to resume growth and division.

nitrogen starvation, which induces mating in *C. reinhardtii* and meiosis in yeast. This response may reflect an advantage of genetically diversifying offspring in a harsh or deteriorating environment, or this may have evolved as a way to obtain the benefits of sex at a time when asexual reproduction would be slow or impossible anyway due to a shortage of nutrients. As with other sexual systems, a possible cost of sex is the breakup of adaptive combinations of alleles. But the goal of evolution experiments using *C. reinhardtii* and yeast is not to explain the specific sexuality of these organisms; rather, the aim is to apply these species as models for testing general hypotheses concerning the evolution of sex in eukaryotes.

For species in which the haploid products of meiosis are capable of mitotic (asexual) reproduction, there is a broad distinction between homothallic systems, in which mitotic lineages produce both mating types, and heterothallic systems, in which a clone produces only the mating type of its founder (figure 16.11). The difference has implications for both the genetic effects of sex and for the design of experiments on the evolution of sex. Homothallic sexual cycles have the potential for high levels of inbreeding, by permitting mating within a haploid clone between cells that differ only in whatever mutations they have accumulated since the meiotic division by their common ancestor. In the most extreme case, a daughter cell may mate with its mother cell. In a heterothallic population, the closest inbreeding that can occur is between descendants of the same meiotic division. Homothallism tends to be a nuisance in the lab, making it impossible to maintain haploid asexual populations. Fortunately, at least in yeast, it is typically straightforward to switch the homothallic system usually found in nature to heterothallism, by knocking out a single gene. The potential for vegetative heterothallic populations to remain haploid for many generations is another potentially significant difference from homothallism, as great differences between asexual haploid and diploid populations in rates of adaptation have been observed (Zeyl et al. 2003).

## FASTER ADAPTATION TO A HARSH ENVIRONMENT BY SEXUAL YEAST POPULATIONS

The tractability of laboratory research and the availability of molecular tools in *S. cerevisiae* reflect its development as a leading model organism in molecular and cellular biology; these same advantages can also be exploited in experimental evolution, especially when sex and recombination are of interest (table 16.2). Yeast offer the typical microbial features, such as the potential for thousands of mitotic generations per year and cryopreservation of ancestors and samples of evolving populations (Zeyl 2000), as well as standard tools and methods for precise genetic manipulation. Genetic crossover occurs at a high rate per kilobyte (roughly ten times that of humans (Mortimer et al. 1991)), which makes the deletion and insertion of genes and markers very efficient. There is also the increasing allure of a parallel natural model in the closely related *S. paradoxus*, which is easier to find in natural habitats and probably has a population structure that hasn't been directly altered by human domestication (Johnson et al. 2004).

Most wild yeast strains are homothallic, but this inconvenience is easily remedied. Knocking out homothallism relies on the HO gene, which encodes an endonuclease that makes a double-stranded cut in the MAT locus on chromosome III, where the active mating type gene resides. At each end of that chromosome is a repressed copy of a MAT sequence, MATa and MATα, which are termed idiomorphs rather than alleles because they are not homologues (Butler et al. 2004). Double-stranded cuts in the active copy of the MAT locus are repaired using one of the inactive telomeric copies, chosen at random. This switches the mating type of a parent cell after budding off a daughter cell by mitosis. A homothallic strain can be converted to heterothallism simply by replacing the HO gene with one of three widely used antibiotic markers, which fortuitously has no detectable effect on mitotic fitness (Goldstein and McCusker 1999), thereby preventing switching of mating type.

Rigorous comparisons of adaptation by sexual and asexual yeast populations are complicated by the fact that nitrogen starvation and forced respiration are required to induce sex. A resulting flaw of early yeast experiments (e.g., Zeyl and Bell 1997) is that sexual and asexual populations experience different environments. This problem is now routinely solved by deleting regulatory genes that are required to convert the

TABLE 16.2 Summary of Experiments Relevant to the Evolution of Sex Using Budding Yeast,
*Saccharomyces cerevisiae*

| Topic/Hypothesis | Conclusion/Summary | Reference |
|---|---|---|
| Fitness effects of recombination | One sexual cycle increased competitive fitness of strains with abundant heterozygosity over isogenic diploids that were asexual due to homozygosity at theMAT locus; no such advantage was observed in homozygotes. The effect of recombination was confounded with overdominance at MAT (competitive advantage in the absence of meiosis). | Birdsell and Wills 1996 |
| Fisher-Muller theory | Sexual populations were fitter than asexuals after ~600 generations on familiar glucose carbon source but not after ~400 generations of adaptation to novel galactose. Inferred advantage was clearance of deleterious mutations. | Zeyl and Bell 1997 |
| Fisher-Muller theory | One round of meiosis initially reduced mean fitness of both homozygous and heterozygous populations in a thermally stressful (37°) environment. During 500 generations of ensuing selection, recombined heterozygotes but not homozygotes were competitively superior. Results are consistent with recombination reducing mean fitness and increasing its variance. | Greig, Borts, and Louis 1998 |
| Sexual transmission of parasitic DNA | 2 Mu-M plasmid reduces fitness by ~1% and invades only outcrossing populations. | Futcher, Reid, and Hickey 1988 |
| Sexual transmission of parasitic DNA | Retrotransposon Ty3 invaded sexual populations more frequently than asexuals. | Zeyl, Bell, and Green 1996 |
| Deterministic mutation hypothesis | Mean epistasis among 620 pairs of deletion mutations is weakly positive. | Jasnos and Korona 2007 |
| Epistasis between deleterious mutations | Among mutations induced by the mutagen EMS and by deletion of open reading frames, the direction of epistasis varies, with no tendency to be negative (synergistic). | Wloch et al. 2001; Jasnos and Korona 2007 |
| Evolution of mate choice | Strong mating preferences evolved within 36 sexual cycles of strong selection for assortative mating. | Leu and Murray 2006 |
| Dispersal and outcrossing by spores | When ingested and excreted by *Drosophila*, vegetative cells were killed, ascus walls were broken down, and random mating occurred. | Reuter, Bell, and Greig 2007 |
| Sex and natural variation | Sex increases efficacy of natural selection in a new harsh environment. | Goddard, Godfray, and Burt 2005 |
| Outcrossing and the spread of a selfish gene | Outcrossing speeds the spread of a selfish gene relative to inbreeding | Goddard, Greig, and Burt 2001 |
| Sex and adaptation to changing environments | Sex advantageous in one of two environments examined | Grimberg and Zeyl 2005 |

signal that initiates meiosis upon exposure to a stressful environment. One option is to delete *IME1*, the master regulator of meiosis. Strains lacking *IME1* (*ime1Δ*) simply wait out nitrogen starvation, while wild types undergo sporulation and meiosis. The deletion has no detectable effect on mitotic fitness (Greig et al. 1998; M. Ostasiewski, unpublished data), suggesting that the cost of not producing four meiotic offspring is balanced by not losing any time to mating afterward. A second option is to delete both *SPO11* and *SPO13*. *SPO11* encodes the endonuclease that makes the double-stranded breaks to begin crossing over in meiosis. Without it, recombination does not occur, and chromosome segregation is erratic. *SPO13* is required for the first meiotic division. When both are deleted, the result is a viable cell that sporulates as usual, but with neither a reductional meiotic division nor recombination (Klapholz et al. 1985; Steele et al. 1991).

Mutation rates in wild-type yeast are low, with estimates of $U$ ranging from $9 \times 10^{-5}$ to $1.1 \times 10^{-3}$ (Wloch et al. 2001; Zeyl et al. 2001), but they can be increased by about two orders of magnitude by deleting the MSH2 gene, which encodes a protein involved in DNA mismatch repair. The added mutations are predominantly single-nucleotide substitutions and deletions (Marsischky et al. 1996). The resulting mutation rate is still well below the threshold of 1.0 required by the deterministic mutational hypothesis to explain anisogamous sex, but it can cause major fitness losses in small populations (Wloch et al. 2001; Zeyl and DeVisser 2001; Joseph and Hall 2004).

The idea that sex accelerates adaptation by increasing genetic variation is about as old as the field of evolutionary genetics itself (Weismann et al. 1904). But for decades, one of the most striking gaps in evolutionary biology was the absence of experimental tests of this idea, despite its immediate intuitive appeal. A recent comparison of adaptation by sexual and asexual yeast populations (Goddard et al. 2005) is one of the clearest and most direct demonstrations to date of Weismann's principle in action.

The principle, later developed by Fisher and Muller and introduced near the beginning of this chapter, is that recombination reduces negative linkage disequilibrium at loci that are under selection. Negative linkage equilibrium is a tendency for the fittest alleles to be linked to alleles of lower fitness at other loci. An adaptive allele can be dragged to extinction by an otherwise inferior genome, or its increase in frequency in response to selection can be slowed by competition with genomes that carry fitter alleles at other loci.

In the Fisher-Muller scenario, negative linkage disequilibrium results from directional selection in a new environment: new adaptive mutations arise in different lineages, so they compete, each burdened by linkage to inferior alleles at the other polymorphic loci. This competition is another form of clonal interference. A straightforward way to test the Fisher-Muller hypothesis is to manipulate the level of clonal interference and compare rates of adaptation by sexual and asexual populations. If the advantage of recombination is that it relieves clonal interference, then increasing clonal interference should magnify the advantage of recombination. Microbial populations are ideally suited to such experiments because they offer several independent ways of increasing clonal interference, in each case by increasing the abundance of spontaneous adaptive mutations. A novel and challenging environment puts multiple loci under selection for new mutations, while increasing population size or mutation rate increases the input of adaptive mutations at each locus.

Goddard and colleagues tested the prediction of faster adaptation by sexual than asexual populations in a harsh environment that imposed strong directional selection. The same ancestral populations were also placed in a benign environment to which they were already well adapted, anticipating that, with little potential for further adaptation, adaptive mutations would be too few and far between to interfere with one others' selection.

The experiment was run in chemostats: culture vessels where microbes undergo continuous growth. The benign environment was a standard minimal defined medium with the glucose concentration reduced to 0.08 percent, making it the limiting nutrient, at least for the ancestral populations. For the harsh environment, the chemostat temperature was turned up from the standard 30°C to 37°C and 0.2 M NaCl was added. Eight replicate sexual populations and eight asexual populations were propagated in each environment. The only difference between asexual and sexual populations was the deletion of *SPO11* and *SPO13* in the asexuals, causing them to produce two clones of the diploid parent cell during sporulation, instead of four recombined haploids. Because the asexuals still sporulated as efficiently as the sexuals, the environmental cues used to stimulate meiosis in the sexuals could be applied to asexual populations too.

All populations were propagated through a cycle that alternated asexual reproduction and selection with sporulation. A major technical challenge in this experiment was enforcing random mating. After meiosis, yeast spores are enclosed in an ascus, so that when nutrients return and they germinate, the closest compatible mating types are likely to come from the same ascus (this is generally thought to lead to high levels of inbreeding). To randomize encounters between potential mates, the ascus wall can be digested with the enzyme zymolyase and the spores dissociated by shaking with tiny glass beads or by sonication. (This has the added utility of destroying the cells that failed to sporulate, enabling the experimenter to enforce completion of the sexual cycle.) For each experimental cycle, Goddard et al. had to replace the benign or harsh selective media in the chemostats with sporulation medium, stop inflow and outflow for a week while sporulation occurred, remove each population from its chemostat to digest the ascus walls, and finally resume selection back in the chemostats. The experiment ran for about three hundred mitotic generations, with ten sporulation cycles in the harsh environment and five in the benign environment, because faster growth in the latter added up to more mitotic generations per cycle.

In the hot salty environment, the sexuals enjoyed a clear evolutionary advantage. From the fourth sporulation (one hundred generations) on, mean fitness was about 10 percent higher in sexual populations, corresponding to a growth rate about 14 percent higher. By contrast, there was no detectable effect of sex in the benign environment—where there was also no detectable fitness change and therefore no detectable adaptive mutations to recombine. Recombination clearly accelerated adaptation, although the mechanism—the cause of negative linkage disequilibrium—could not be specified. The next experiment that we discuss was designed to focus on a mechanism for this advantage.

## SEX RELIEVES CLONAL INTERFERENCE IN *CHLAMYDOMONAS*

*Chlamydomonas*, a genus of unicellular Chlorophyte algae, possesses many of the experimentally convenient traits of yeast, in a radically different ecology. For simplicity and brevity we focus on *Chlamydomonas reinhardtii*, although research that is highly relevant to the evolution of sex has also been carried out with *C. moewusii* (see table 16.3). *C. reinhardtii* has been isolated from moist soil, particularly where enriched and disturbed by agriculture (Sack et al. 1994). It uniquely offers the experimenter a choice between autotrophic and heterotrophic populations. Given sufficient light it photosynthesizes, swimming toward the light source in liquid cultures using flagellae. *C. reinhardtii* can also use carbon sources supplied in the growth medium, and darkness has been used as a novel environment that imposes both strong directional selection for more efficient heterotrophy (Colegrave et al. 2002; Kaltz and Bell 2002) and purifying selection against mutations that are effectively neutral when carbon can be obtained by photosynthesis (Lee et al. 2007).

As with yeast, the sexual cycle of *C. reinhardtii* is driven by nitrogen availability, but it is gametogenesis and mating (rather than meiosis) that are induced by starvation for nitrogen (figure 16.12). A population containing both mating types (*mt+* and *mt–*) can therefore be kept asexual, simply by supplying adequate nitrogen. Mating is usually restricted to liquid cultures. Zygotes are transferred to agar plates and stored in the dark for about five days, where they mature into a thick-walled, dormant stage that is resistant to desiccation and freezing—and, fortuitously, chloroform fumes. Chloroform kills vegetative cells, so it can be used to select for zygotes produced by syngamy. When nitrogen and light are restored, the haploid spores are released, so unlike yeast, *C. reinhardtii* does not offer the option of vegetative diploid populations. There is evidence of trade-offs between vegetative and sexual fitness (DaSilva and Bell 1992) as noted earlier, and some fascinating responses to sexual selection on *C. reinhardtii* have been observed (Bell 2005).

The results of the yeast experiment by Goddard et al. (2005) unambiguously showed faster adaptation by sexual populations than by asexuals. The observed benefit of recombination indicates that selection imposed by a harsh new environment must have generated negative linkage disequilibrium that impeded adaptation by asexuals. But that experiment did not distinguish which of two possible mechanisms was responsible for linkage disequilibrium: negative epistasis or clonal interference. If epistatic interactions between adaptive mutations reduce their fitness benefits when they are combined, then genotypes with multiple adaptive mutations will be underrepresented.

TABLE 16.3   Summary of Experiments Relevant to the Evolution of Sex Using the Unicellular Alga
*Chlamydomonas reinhardtii*

| Topic/Hypothesis | Conclusion/Summary | Reference |
|---|---|---|
| Sexual selection | Evolution of a homothallic mating system from the ancestral heterothallic system, and of spontaneous gametogenesis before nitrogen depletion | Bell 2005 |
| Genotype-by- environment interaction for fitness | Complex pattern of strong genotype-by- environment interaction, consistent with an advantage of genetically diverse offspring due to environmental heterogeneity (the "Tangled Bank") | Bell 1991 |
| Deterministic mutation hypothesis | Increase in variance for fitness in one of three crosses, but no effect of sex on mean fitness<br>Results suggest no epistasis among deleterious mutations (and thus a random distribution of mutations among clonal lines before sex), very low mutation rate, or low statistical power. | DaSilva, Bell, and Ve 1996 |
| Deterministic mutation hypothesis | Sex reduced mean log fitness in crosses between *C. moewusii* strains that had been mutagenized with UV or allowed to accumulate random mutations, but not in control crosses.<br>Interpreted as evidence of synergistic epistasis | de Visser, Hoekstra, and van den Ende 1997 |
| Epistasis and recombination | Neither growth rates nor competitive fitness differed between asexuals isolated and assayed after 1,000 generations and sexuals after 113 sexual cycles | DaSilva, Bell, and Ve 1996 |
| Fisher-Muller hypothesis | One sexual cycle increased fitness variance and reduced mean; during subsequent selection in darkness on novel carbon sources, sexuals were temporarily fitter but after 50 generations, their advantage had disappeared. | Colegrave, Kaltz, and Bell 2002 |
| Fisher-Muller hypothesis | Sex initially increased fitness variation and reduced mean, but subsequent adaptation to growth in the dark on carbon sources supplied in the medium was greater in sexual than asexual populations, particularly on multiple carbon sources.<br>Interpreted as support for Fisher-Muller, with temporary reduction of mean fitness after sex possibly caused by synergistic epistasis between adaptive mutations | Kaltz and Bell 2002 |
| Mutation and recombination | Various stresses, including starvation such as that used to induce gametogenesis, heritably reduced mean fitness and increased variance.<br>Interpreted as an increase in mutation rates due to stress | Goho and Bell 2000 |

Unlike clonal interference, epistasis is a deterministic mechanism, predicted to be unaffected by population size. Colegrave (2002) compared the effects of sex on fitness in *C. reinhardtii* populations that varied in size by more than three orders of magnitude. A similar advantage of sex at all population sizes would imply negative epistasis. By contrast, because the abundance of adaptive mutations would increase with population size, an advantage of sex due to clonal interference would be greater in larger populations.

Standard lab techniques allow the experimenter to vary the population size of tractable microbes over several orders of magnitude, simply by manipulating culture volumes or in serial transfer experiments by varying the numbers of cells sampled at each transfer. For serially transferred evolving populations of microbes, the effective population size ($N_e$) with respect to the selection of adaptive mutations is a simple function of the number of cells transferred and the number of generations between transfers. But manipulating either of these features of an experimental design runs the risk of confounding effects of population size with unintended differences in selection pressures that could result from different culture volumes or cell densities. Colegrave (2002) varied the size of the bottleneck at each transfer, to arrive at $N_e$ values of $1.5 \times 10^4$, $8.5 \times 10^5$, and $5.5 \times 10^6$. The varying bottleneck sizes also resulted in a threefold range in the numbers of generations between transfers. To minimize the risk that selective pressures might differ at different densities, he chose transfer regimes that kept evolving populations from reaching the carrying capacity of their ten-milliliter cultures, reasoning that if the populations spent the selection phase of the experiment in exponential growth, exponential growth rate would be the direct target of selection. The growth rates of evolved populations at the two extremes of densities that were used during the experiment were highly correlated, suggesting that density effects were not important factors in adaptation.

The algal model system for this experiment required a design and analysis that were less straightforward than that of Goddard et al. (2005). Vegetative cultures of *C. reinhardtii* remain haploid until gametogenesis is induced by nitrogen starvation, and one of the mating types in an initially mixed population may be extinct after many asexual generations. Colegrave avoided this complication by applying selection to four asexual *mt*₁ populations (each founded by a random isolate from a variable population) at each $N_e$ for 150 generations, and then comparing the effects of recombination on their response to a further 50 generations of selection. Prior to recombination, varying $N_e$ had the expected result: for all four ancestors, larger populations showed larger fitness gains, but by less and less as $N_e$ increased, just the kind of diminishing-returns effect that clonal interference would produce. Each population was then crossed with the same *mt*– partner, and seven more cycles of mating and meiosis were performed to allow the recombination of adaptive mutations that had been segregating in a population after 150 generations of adaptation. The recombined populations, and control asexual populations that were derived from the same evolved populations but never mated, were returned to the same selective regime as before, except that the same bottleneck of $10^4$ was applied to all populations to prevent any confounding interactions of recombination with subsequent $N_e$.

Sex led to fitness increases that clearly scaled with population size: the fitness advantage of recombined populations ranged from a barely significant about 2 percent for the lowest $N_e$ to approximately 20 percent for the largest. Because the number of adaptive mutations segregating after 150 generations would, on average, be greater in larger populations, it is in the largest populations that clonal interference should be most intense and therefore that recombination should provide the greatest benefit. The strong effect of $N_e$ argues against epistasis among adaptive mutations as a deterministic cause of the linkage disequilibrium that accumulated over the first 150 generations of Colegrave's experiment.

The competition between adaptive mutations that is implied by this result is just one way in which random processes interacting with selection can generate negative linkage disequilibrium. By placing initially homogeneous populations of large size in novel environments that imposed strong directional selection, Goddard et al. and Colegrave maximized the role of competition between new mutations. But it seems unlikely that abiotic environments keep most natural populations in this situation most of the time, so clonal interference probably does not fully account for the ubiquity of sex. In the longer term, another form of negative linkage disequilibrium is likely to arise: adaptive mutations trapped in otherwise low-fitness genomes. The initial association between a new mutation, whether adaptive or deleterious, and the fitness of the genome in which it occurs is random, but the associations that persist the longest and therefore contribute the most to linkage disequilibrium are genomes in which adaptive and maladaptive alleles are combined. Adaptive mutations that occur in fitter-than-average genomes are fixed relatively quickly, and deleterious ones in genomes that already had low fitness accelerate their extinction. The next experiment that we discuss was designed to place $D. melanogaster$ populations in just such a predicament.

## SEPARATING THE RED FROM THE BAD IN DROSOPHILA MELANOGASTER

Although the fruit fly has been a mainstay of genetic research for roughly a century, its obligate sexuality might seem to limit its utility for experiments on the evolution of sex. But a Dipteran idiosyncrasy provides a starting point: there is virtually no crossing over in males. In addition, it is possible to block recombination from specific and extensive regions of the genome using chromosome rearrangements. In heterozygotes for large inversions, recombination within the inverted region leads to further chromosome rearrangements that are lethal. In heterozygotes for reciprocal translocations (nonhomologous chromosomes that have exchanged large fragments), it is independent assortment that is prevented: gametes that receive just one of the translocated chromosomes are not viable because they lack part of one chromosome and have a duplicate copy of another (figure 16.13). The translocations used by geneticists also carry phenotypic markers so that progeny that inherit both normal chromosomes can be distinguished from those that receive the translocated pair. Both crossing over and independent segregation can therefore be prevented in males.

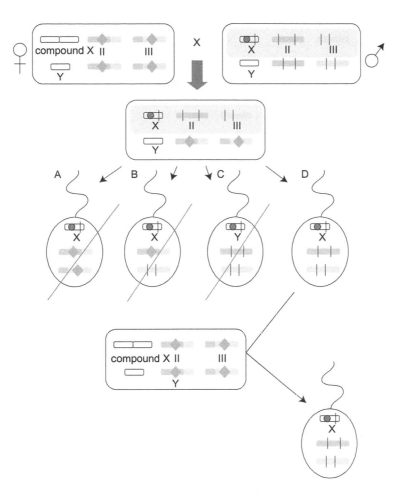

FIGURE 16.13

The genetic strategy for propagating paternally transmitted haplotypes without recombination in *Drosophila melanogaster* (Rice and Chippindale 2001). A male haplotype carrying the artificially selected $w^+$ allele (shaded circle) on a randomly sampled X chromosome is cloned from a variable base population by crossing a normal red-eyed male with a female that is homozygous for a translocation of chromosomes II and III, and carries a compound X chromosome as well as a Y chromosome. The compound X provides the 2:2 dosage ratio of X chromosomes to autosomes that makes the flies female despite the Y chromosome, which does not affect sex determination. Vertical lines on chromosomes represent deleterious mutations segregating in the base population. Diamonds on autosomes represent phenotypic markers. Sons from this cross are backcrossed to the same female genotype. The only sperm that contribute to the next generation of the experimental population are those that carry the artificially selected X chromosome bearing $w^+$ and the two autosomes making up the rest of the clonally transmitted haplotype. A, Sperm carrying the translocated autosomes are selected against using phenotypic markers on the translocations. B, Sperm carrying one translocated autosome and a normal autosome are inviable due to a large duplication and a large deletion. C, Sperm carrying a Y chromosome produce inviable zygotes with two Y chromosomes and no X, or compound X/Y zygotes which are female; only sons from this cross are candidates to reproduce. In each subsequent generation, this same haplotype is transmitted paternally through crosses with the same stock of "clone-generating" females. Paternal transmission prevents recombination by crossing over, which does not occur in male *Drosophila*. Crossing with the clone-generating female stock prevents recombination by segregation because inheritance of one translocated autosome but not the other is lethal.

This still leaves the problem of female meiosis, which necessarily involves crossing over. But that, too, can be avoided, using further cytogenetic ingenuity and the fact that the mechanism of sex determination in *Drosophila* is entirely different from that in mammals. Instead of a dominant gene for maleness on the Y chromosome, it is the ratio of X chromosomes to autosomes that determines gender. The 2:2 ratio of XX females and the 1:2 ratio in XY males produce different ratios of regulatory proteins encoded by X-linked and autosomal genes. Those regulatory genes in turn cause transcripts of the regulatory *Sex-lethal (Sxl)* gene to be spliced differently in males and females, which begins the process of sexual differentiation. A fly with two X chromosomes can therefore carry a Y and still be a fertile female, leading to a paradoxical sex chromosome system in which males inherit X chromosomes from their fathers (figure 16.13).

Rice and Chippindale (2001) used a combination of these genetic techniques to test the hypothesis that recombination would permit a faster and more consistent response to artificial selection, by freeing the favored allele from the low-fitness genetic backgrounds in which chance would often introduce it to the population. To be selectively favored in an asexual population, an allele must raise the fitness of its genome to exceed the fittest genotype in the population. Background selection in a genetically variable population can thus eliminate many alleles that would be favored if they were not linked to deleterious alleles. Mutations whose fitness effects are too small to overcome drift will not be helped by recombination; mutations with large fitness advantages relative to the standing variation can make any genotype the fittest in the population and do not need any help. Rice and Chippindale tracked the fate of an allele artificially given an intermediate fitness advantage in recombining and clonal *Drosophila* haplotypes.

In the yeast and algal experiments discussed earlier, the founding populations were clones of a single genotype, and adaptation occurred by the selection of new mutations that were never identified. Rice and Chippindale used a very different approach, seeding genetically diverse populations with twenty copies of the allele that would play the role of an adaptive mutation: the familiar $w^+$ allele encoding wild-type red eyes by which Thomas Hunt Morgan discovered sex linkage about a century earlier (Morgan 1910). This allele was given a 10 percent selective advantage, calculated to fall within the range of fitness effects that would make an adaptive mutation vulnerable to background selection given the known genetic variation in the starting populations.

In the nonrecombining treatment, the $w^+$ allele was trapped in a single genetic background consisting of the X chromosome and chromosomes II and III, which constitute over 99 percent of the genome (the minute fourth chromosome was disregarded). To prevent recombination, this haplotype was transmitted strictly through meiosis in males, using the cytogenetic trickery already described (see figure 16.13). The experiment was replicated seventeen times, in each case $w^+$ introduced in a unique haplotype sampled randomly from the variable base population. Its competitors were eighty $w^-$ alleles on diverse haplotypes. By starting the experiment with the favored allele already well established, Rice and Chippindale avoided the quick stochastic extinction that befalls most new mutations, since this would fail to test the hypothesis of interest and would have required a prohibitively high number of trials.

In the recombination treatment, the genetic backgrounds of selected $w^+$ alleles were varied in two ways: first, in each of the seventeen replicates, the twenty copies of $w^+$ were introduced on a variety of X chromosomes representing the diversity of the base population; and second, in each generation the selected males were crossed with females who had the altered sex chromosome system causing paternal transmission of X chromosomes, but normal autosomes sampled from the diverse base population (figure 16.14). After the initial twenty varying X chromosomes carrying the favored $w^+$ allele were chosen, they were paternally transmitted and so had no further opportunity to recombine;

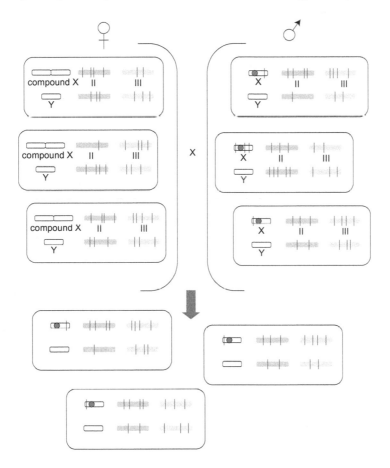

FIGURE 16.14

Artificial selection for $w^+$ in recombining haplotypes. Each recombining population is established with $w^+$ on a random sample of twenty X chromosomes from the base population, graphically represented here as three male genotypes with differing deleterious mutations (vertical lines). During the experiment, males are crossed with females whose compound X/Y sex chromosomes cause paternal transmission of X chromosomes, as in the previous figure, but whose normal autosomes allow segregation between autosomes of paternal origin and the diverse autosomes in the base population.

but over the course of the experiment, the rest of the paternally transmitted haplotypes were selected from the base population, potentially freeing the $w^+$ alleles from autosomal deleterious mutations in the starting haplotypes.

Selection for $w^+$ was applied by brute force: in each generation, one hundred males were picked to reproduce, preserving the ratio of red-eyed to white-eyed males in the whole population, and then the number of red-eyed males was augmented by 10 percent. Background selection occurred spontaneously, as the net effect of standing genetic variation for viability and fecundity, although quite significantly mating was allowed only in the dark so that the blindness of white-eyed males was not a handicap.

For six generations the mean frequency of $w^+$, contrived to be an adaptive mutation, increased steadily in both recombining and non-recombining populations. The trajectories varied widely, reflecting variation among replicate populations in the X chromosomes on which $w^+$ originated. But over the following five generations, the mean $w^+$ frequency in nonrecombining populations leveled off at about 0.3, while its rate of increase in recombining populations accelerated slightly, such that the frequency of the $w^+$ allele reached about 0.4. The final increase in mean $w^+$ frequency in the recombining populations was double that in the absence of recombination and was also more consistent; the favored allele went extinct in two nonrecombining populations and was perilously close to disappearing from several others when the experiment concluded. The prediction that recombination would not affect the fate of a mutation whose fitness effect is large relative to existing fitness variation was not tested, but in a population as variable as the base population for this experiment (the standard deviation of fitness was estimated as 0.11), such a mutation would need to have a tremendous fitness advantage, greater than 50 percent, to overcome background selection.

This fly experiment thus illustrates an effect of background selection similar to the clonal interference found in the *C. reinhardtii* experiment: without recombination, a smaller subset of adaptive mutations can be fixed. An important difference is that clonal interference, and the Fisher-Muller hypothesis in general, applies only to populations under strong directional selection with a combination of size and mutation rate that generates an excess of adaptive mutations. It seems unlikely that this is the usual state of natural populations. Background selection occurs in the more general context of sporadic adaptive mutations in variable populations, requiring neither the perpetually changing Fisher-Muller environment nor the permanently fixed environment of Muller's ratchet or the deterministic mutation hypothesis.

## WHAT HAVE WE LEARNED FROM SEXUAL EXPLORATIONS USING EXPERIMENTAL EVOLUTION?

Most of the model organisms harnessed to conduct experimental evolution also feature sexual systems that are readily manipulated. Given that the evolution and maintenance of sex remain major puzzles for evolutionary biologists, it is therefore unsurprising that these model systems have figured prominently in evolution experiments. What have we learned from these efforts thus far?

Beyond a doubt, experimental evolution has established that allelic recombination has the potential to accelerate adaptation. Faster adaptation in recombining populations relative to their clonal counterparts has been observed in all of the prominent microbial models: viruses, *E. coli*, yeast, and *Chlamydomonas*. These observations indicate that, at least in the laboratory, strong directional selection generates negative linkage disequilibrium. Theoretical work has emphasized two possible causes: (1) negative epistasis among adaptive mutations and (2) clonal interference (the Fisher-Muller hypothesis). The yeast experiment by Goddard et al. (2005) does not discriminate between these two causes, but the *Chlamydomonas* experiment by Colegrave et al. (2002) clearly indicates that sex relieved clonal interference that was generating negative linkage disequilibrium. Similarly, Cooper (2007) showed that *E. coli* recombination speeds adaptation by reducing clonal interference.

The experiments summarized in tables 16.2 and 16.3 are more equivocal. Recombination is expected to increase variance in fitness regardless, and this is consistently observed in yeast and *Chlamydomonas* experiments. Negative epistasis is predicted also to lower the mean fitness of recombinants, which occurred in some experiments but not others.

The microbial evolution experiments that we have reviewed here began with populations that had little or no standing genetic variation. However, this condition is probably not representative of most natural populations. By contrast, the experimental fly populations of Rice and Chippindale (2001) began with substantial fitness variation. Instead of competition between multiple new mutations, the impediment to their adaptation was that most "adaptive mutations" were trapped in mediocre genomes. But the principle is similar: adaptation is facilitated by allowing beneficial mutations to be detached from most of the genome in which they originate.

## SUMMARY

Since the time of Darwin, evolutionary biologists have been preoccupied with elucidating why sex (genetic exchange) evolved and is maintained in biological populations. Various hypotheses have been formulated to explain the maintenance of sex, but the majority of these ideas hinge on the potential for sex to promote linkage equilibrium and on the presumed advantage of such mixis for enhancing the response to either directional or purifying selection. One useful approach to this question is to employ laboratory populations of micro- or macroorganisms, manipulating them to create sexual and asexual lineages that are experimentally evolved to test whether sex is advantageous. This chapter showcases how myriad biological systems have been successfully put to this task, systems ranging from viruses and bacteria, to unicellular and multicellular eukaryotes. We defer from exhaustively reviewing all such efforts, but rather focus on detailed descriptions of a few key experiments that span a wide variety of study systems and theoretical predictions. In so doing, we highlight the advantages, disadvantages, and empirical obstacles these diverse systems present in testing theories for the evolution and maintenance of sex, as well as the general strengths and shortcomings of the experimental evolution approach for this line of inquiry.

## REFERENCES

Bell, G. 1982. *The Masterpiece of Nature: The Evolution and Genetics of Sexuality.* Berkeley: University of California Press.

———. 1991. The ecology and genetics of fitness in *Chlamydomonas*. 3. Genotype-by-environment interaction within strains. *Evolution* 45:668–679.

———. 2005. Experimental sexual selection in *Chlamydomonas*. *Journal of Evolutionary Biology* 18:722–734.

Bernstein, H., H. C. Byerly, F. A. Hopf, and R. E. Michod. 1985. The evolutionary role of recombinational repair and sex. *International Review of Cytology: A Survey of Cell Biology* 96:1–28.

Birdsell, J., and C. Wills. 1996. Significant competitive advantage conferred by meiosis and syngamy in the yeast *Saccharomyces cerevisiae*. *Proceedings of the National Academy of Sciences of the USA* 93:908–912.

Burch, C. L., and L. Chao. 2004. Epistasis and its relationship to canalization in the RNA virus phi 6. *Genetics* 167:559–567.

Butler, G., C. Kenny, A. Fagan, C. Kurischko, C. Gaillardin, and K. H. Wolfe. 2004. Evolution of the MAT locus and its Ho endonuclease in yeast species. *Proceedings of the National Academy of Sciences of the USA* 101:1632–1637.

Chao, L. 1990. Fitness of RNA virus decreased by Muller ratchet. *Nature* 348:454–455.

Chao, L., T. R. Tran, and C. Matthews. 1992. Muller ratchet and the advantage of sex in the RNA virus-phi-6. *Evolution* 46:289–299.

Chao, L., T. T. Tran, and T. T. Tran. 1997. The advantage of sex in the RNA virus phi 6. *Genetics* 147:953–959.

Charlesworth, B. 1989. The evolution of sex and recombination. *Trends in Ecology & Evolution* 4:264–267.

Charlesworth, B., M. T. Morgan, and D. Charlesworth. 1993. The effect of deleterious mutations on neutral molecular variation. *Genetics* 134:1289–1303.

Colegrave, N., O. Kaltz, and G. Bell. 2002. The ecology and genetics of fitness in *Chlamydomonas*. VIII. The dynamics of adaptation to novel environments after a single episode of sex. *Evolution: International Journal of Organic Evolution* 56:14–21.

Cooper, T. F. 2007. Recombination speeds adaptation by reducing competition between beneficial mutations in populations of *Escherichia coli*. *PLoS Biology* 5:1899–1905.

Crow, J. F., and M. Kimura. 1965. Evolution in sexual and asexual populations *Drosophila*. *American Naturalist* 99:439–450.

Cutter, A. D. 2005. Mutation and the experimental evolution of outcrossing in *Caenorhabditis elegans*. *Journal of Evolutionary Biology* 18:27–34.

DaSilva, J., and G. Bell. 1992. The ecology and genetics of fitness in *Chlamydomonas*. 6. Antagonism between natural-selection and sexual selection. *Proceedings of the Royal Society of London B, Biological Sciences* 249:227–233.

DaSilva, J., G. Bell, and Ve. 1996. The ecology and genetics of fitness in *Chlamydomonas*.7. The effect of sex on the variance in fitness and mean fitness. *Evolution* 50:1705–1713.

Davies, E. K., A. D. Peters, and P. D. Keightley. 1999. High frequency of cryptic deleterious mutations in *Caenorhabditis elegans*. *Science* 285:1748–1751.

Decaestecker, E., S. Gaba, J. A. M. Raeymaekers, R. Stoks, L. Van Kerckhoven, D. Ebert, and L. De Meester. 2007. Host-parasite "Red Queen" dynamics archived in pond sediment. *Nature* 450:870–U16.

Delbruck, M., and W. T. Bailey. 1946. Induced mutations in bacterial viruses. *Cold Spring Harbor Symposia on Quantitative Biology* 11:33–37.

de Visser, J., and S. F. Elena. 2007. The evolution of sex: Empirical insights into the roles of epistasis and drift. *Nature Reviews Genetics* 8:139–149.

de Visser, J. A., R. F. Hoekstra, and H. van den Ende. 1997. An experimental test for synergistic epistasis and its application in *Chlamydomonas*. *Genetics* 145:815–819.

de Visser, J., C. W. Zeyl, P. J. Gerrish, J. L. Blanchard, and R. E. Lenski. 1999. Diminishing returns from mutation supply rate in asexual populations. *Science* 283:404–406.

Drake, J. W. 1999. The distribution of rates of spontaneous mutation over viruses, prokaryotes, and eukaryotes. *Annals of the New York Academy of Sciences* 870:100–107.

Elena, S. F., and R. E. Lenski. 1997. Test of synergistic interactions among deleterious mutations in bacteria. *Nature* 390:395–398.

Eshel, I., and M. W. Feldman. 1970. On the evolutionary effect of recombination. *Theoretical Population Biology* 1:88–100.

Felsenstein, J. 1974. Evolutionary advantage of recombination. *Genetics* 78:737–756.

Ferguson, L. R., and W. A. Denny. 1991. The genetic toxicology of acridines. *Mutation Research* 258:123–160.

Fisher, R. A. 1930. *The Genetical Theory of Natural Selection*. Oxford: Clarendon.

Froissart, R., C. O. Wilke, R. Montville, S. K. Remold, L. Chao, and P. E. Turner. 2004. Co-infection weakens selection against epistatic mutations in RNA viruses. *Genetics* 168:919.

Futcher, B., E. Reid, and D. A. Hickey. 1988. Maintenance of the 2-Mu-M circle plasmid of *Saccharomyces cerevisiae* by sexual transmission: An example of a selfish DNA. *Genetics* 118:411–415.

Gandon, S., and S. P. Otto. 2007. The evolution of sex and recombination in response to abiotic or coevolutionary fluctuations in epistasis. *Genetics* 175:1835–1853.

Goddard, M. R., H. C. Godfray, and A. Burt. 2005. Sex increases the efficacy of natural selection in experimental yeast populations. *Nature* 434:636–640.

Goddard, M. R., D. Greig, and A. Burt. 2001. Outcrossed sex allows a selfish gene to invade yeast populations. *Proceedings of the Royal Society of London B, Biological Sciences* 268:25372542.

Goho, S., and G. Bell. 2000. The ecology and genetics of fitness in *Chlamydomonas*. IX. The rate of accumulation of variation of fitness under selection. *Evolution* 54:416–424.

Goldstein, A. L., and J. H. McCusker. 1999. Three new dominant drug resistance cassettes for gene disruption in *Saccharomyces cerevisiae*. *Yeast* 15:1541–1553.

Greig, D., R. H. Borts, and E. J. Louis. 1998. The effect of sex on adaptation to high temperature in heterozygous and homozygous yeast. *Proceedings of the Royal Society of London B, Biological Sciences* 265:1017–1023.

Grimberg, B., and C. Zeyl. 2005. The effects of sex and mutation rate on adaptation in test tubes and to mouse hosts by *Saccharomyces cerevisiae. Evolution* 59:431–438.

Haag-Liautard, C., M. Dorris, X. Maside, S. Macaskill, D. L. Halligan, B. Charlesworth, and P. D. Keightley. 2007. Direct estimation of per nucleotide and genomic deleterious mutation rates in *Drosophila. Nature* 445:82–85.

Hamilton, W. D., R. Axelrod, and R. Tanese. 1990. Sexual reproduction as an adaptation to resist parasites (a review). *Proceedings of the National Academy of Sciences of the USA* 87:3566–3573.

Heeney, J. L., A. G. Dalgleish, and R. A. Weiss. 2006a. Origins of HIV and the evolution of resistance to AIDS. *Science* 313:462–466.

Heeney, J. L., E. Rutjens, E. J. Verschoor, H. Niphuis, P. ten Haaft, S. Rouse, H. McClure, et al. 2006b. Transmission of simian immunodeficiency virus SIVcpz and the evolution of infection in the presence and absence of concurrent human immunodeficiency virus type 1 infection in chimpanzees. *Journal of Virology* 80:7208–7218.

Hill, J. A., and S. P. Otto. 2007. The role of pleiotropy in the maintenance of sex in yeast. *Genetics* 175:1419–1427.

Imhof, M., and C. Schlotterer. 2001. Fitness effects of advantageous mutations in evolving *Escherichia coli* populations. *Proceedings of the National Academy of Sciences of the USA* 98:1113–1117.

Jasnos, L., and R. Korona. 2007. Epistatic buffering of fitness loss in yeast double deletion strains. *Nature Genetics* 39:550–554.

Johnson, L. J., V. Koufopanou, M. R. Goddard, R. Hetherington, S. M. Schafer, and A. Burt. 2004. Population genetics of the wild yeast *Saccharomyces paradoxus. Genetics* 166:43–52.

Joseph, S. B., and D. W. Hall. 2004. Spontaneous mutations in diploid *Saccharomyces cerevisiae*: More beneficial than expected. *Genetics* 168:1817–1825.

Kaltz, O., and G. Bell. 2002. The ecology and genetics of fitness in *Chlamydomonas*. XII. Repeated sexual episodes increase rates of adaptation to novel environments. *Evolution: International Journal of Organic Evolution* 56:1743–1753.

Keightley, P. D., and S. P. Otto. 2006. Interference among deleterious mutations favours sex and recombination in finite populations. *Nature* 443:89–92.

Kirkpatrick, M., and M. J. Ryan. 1991. The evolution of mating preferences and the paradox of the lek. *Nature* 350:33–38.

Klapholz, S., C. S. Waddell, and R. E. Esposito. 1985. The role of the spo11 gene in meiotic recombination in yeast. *Genetics* 110:187–216.

Kondrashov, A. S. 1982. Selection against harmful mutations in large sexual and asexual populations. *Genetical Research* 40:325–332.

———. 1988. Deleterious mutations and the evolution of sexual reproduction. *Nature* 336:435–440.

———. 1993. Classification of hypotheses on the advantage of Amphimixis. *Journal of Heredity* 84:372–387.

Kouyos, R. D., S. P. Otto, and S. Bonhoeffer. 2006. Effect of varying epistasis on the evolution of recombination. *Genetics* 173:589–597.

Kouyos, R. D., O. K. Silander, and S. Bonhoeffer. 2007. Epistasis between deleterious mutations and the evolution of recombination. *Trends in Ecology & Evolution* 22:308–315.

Lee, J. H., S. Waffenschmidt, L. Small, and U. Goodenough. 2007. Between-species analysis of short-repeat modules in cell wall and sex-related hydroxyproline-rich glycoproteins of *Chlamydomonas* (1). *Plant Physiology* 144:1813–1826.

Lenski, R. E., and M. Travisano. 1994. Dynamics of adaptation and diversification: A 10,000-generation experiment with bacterial-populations. *Proceedings of the National Academy of Sciences of the USA* 91:6808–6814.

Leu, J. Y., and A. W. Murray. 2006. Experimental evolution of mating discrimination in budding yeast. *Current Biology* 16:280–286.

Malmberg, R. L. 1977. Evolution of epistasis and advantage of recombination in populations of bacteriophage-T4. *Genetics* 86:607–621.

Margulis, L., and D. Sagan. 1986. *Origins of Sex: Three Billion Years of Genetic Recombination.* New Haven, CT: Yale University Press.

Marsischky, G. T., N. Filosi, M. F. Kane, and R. Kolodner. 1996. Redundancy of *Saccharomyces cerevisiae* MSH3 and MSH6 in MSH2-dependent mismatch repair. *Genes & Development* 10:407–420.

Maynard Smith, J. 1978. *The Evolution of Sex.* Cambridge: Cambridge University Press.

Michod, R. E., and B. R. Levin. 1988. *The Evolution of Sex: An Examination of Current Ideas.* Sunderland, MA: Sinauer.

Miralles, R., P. J. Gerrish, A. Moya, and S. F. Elena. 1999. Clonal interference and the evolution of RNA viruses. *Science* 285:1745–1747.

Miralles, R., A. Moya, and S. F. Elena. 2000. Diminishing returns of population size in the rate of RNA virus adaptation. *Journal of Virology* 74:3566–3571.

Morgan, T. H. 1910. Sex limited inheritance in *Drosophila. Science* 32:120–122.

Mortimer, R. K., D. Schild, C. R. Contopoulou, and J. A. Kans. 1991. Genetic and physical maps of Saccharomyces cerevisiae. *Methods in Enzymology* 194:827–863.

Muller, H. J. 1932. Some genetic aspects of sex. *American Naturalist* 66:118–138.

———. 1964. The relation of recombination to mutational advance. *Mutation Research* 1:2–9.

Nee, S. 2000. Mutualism, parasitism and competition in the evolution of coviruses. *Philosophical Transactions of the Royal Society of London B, Biological Sciences* 355:1607–1613.

Nee, S., and J. Maynard Smith. 1990. The evolutionary biology of molecular parasites. *Parasitology* 100:S5–S18.

Otto, S. P., and N. H. Barton. 2001. Selection for recombination in small populations. *Evolution* 55:1921–1931.

Otto, S. P., and M. W. Feldman. 1997. Deleterious mutations, variable epistatic interactions, and the evolution of recombination. *Theoretical Population Biology* 51:134–147.

Otto, S. P., and T. Lenormand. 2002. Resolving the paradox of sex and recombination. *Nature Reviews Genetics* 3:252–261.

Palsson, S. 2002. Selection on a modifier of recombination rate due to linked deleterious mutations. *Journal of Heredity* 93:22–26.

Peters, A. D., and P. D. Keightley. 2000. A test for epistasis among induced mutations in *Caenorhabditis elegans*. *Genetics* 156:1635–1647.

Poon, A., and L. Chao. 2004. Drift increases the advantage of sex in RNA bacteriophage phi 6. *Genetics* 166:19–24.

Reuter, M., G. Bell, and D. Greig. 2007. Increased outbreeding in yeast in response to dispersal by an insect vector. *Current Biology* 17:R81–R83.

Rice, W. R. 2002. Experimental tests of the adaptive significance of sexual recombination. *Nature Reviews Genetics* 3:241–251.

Rice, W. R., and A. K. Chippindale. 2001. Sexual recombination and the power of natural selection. *Science* 294:555–559.

Rozen, D. E., J. A. G. M. de Visser, and P. J. Gerrish. 2002. Fitness effects of fixed beneficial mutations in microbial populations. *Current Biology* 12:1040–1045.

Sack, L., C. Zeyl, G. Bell, T. Sharbel, X. Reboud, T. Bernhardt, and H. Koelewyn. 1994. Isolation of 4 new strains of *Chlamydomonas reinhardtii* (Chlorophyta) from soil samples. *Journal of Phycology* 30:770–773.

Silander, O. K., D. M. Weinreich, K. M. Wright, K. J. O'Keefe, C. U. Rang, P. E. Turner, and L. Chao. 2005. Widespread genetic exchange among terrestrial bacteriophages. *Proceedings of the National Academy of Sciences of the USA* 102:19009–19014.

Sokal, R. R., and F. J. Rohlf. 1995. *Biometry: The Principles and Practice of Statistics in Biological Research.* New York: Freeman.

Souza, V., P. E. Turner, and R. E. Lenski. 1997. Long-term experimental evolution in *Escherichia coli.* 5. Effects of recombination with immigrant genotypes on the rate of bacterial evolution. *Journal of Evolutionary Biology* 10:743–769.

Steele, D. F., M. E. Morris, and S. Jinksrobertson. 1991. Allelic and ectopic interactions in recombination-defective yeast strains. *Genetics* 127:53–60.

Steenstrup, J. J. 1845. *On the Alternation of Generations or the Propagation and Development of Animals through Alternate Generations.* Trans. George Busk. Ray Society London.

Steinhauer, D. A., and J. J. Skehel. 2002. Genetics of influenza viruses. *Annual Review of Genetics* 36:305–332.

Szafraniec, K., D. M. Wloch, P. Sliwa, R. H. Borts, and R. Korona. 2003. Small fitness effects and weak genetic interactions between deleterious mutations in heterozygous loci of the yeast *Saccharomyces cerevisiae. Genetical Research* 82:19–31.

Thomas, C. M., and K. M. Nielsen. 2005. Mechanisms of, and barriers to, horizontal gene transfer between bacteria. *Nature Reviews Microbiology* 3:711–721.

Turner, P. E., and L. Chao. 1998. Sex and the evolution of intrahost competition in RNA virus phi 6. *Genetics* 150:523–532.

———. 1999. Prisoner's dilemma in an RNA virus. *Nature* 398:441–443.

———. 2003. Escape from Prisoner's Dilemma in RNA phage phi 6. *American Naturalist* 161:497–505.

Varma, A., and V. G. Malathi. 2003. Emerging geminivirus problems: A serious threat to crop production. *Annals of Applied Biology* 142:145–164.

Vrijenhoek, R. C., R. M. Dawley, C. J. Cole, and J. P. Bogart. 1989. A list of known unisexual vertebrates. Pages 19–23 in R. Dawley and J. Bogart, eds. *Evolution and Ecology of Unisexual Vertebrates.*

Weismann, A., J. A. Thomson, and M. R. Thomson. 1904. *The Evolution Theory*. London: Arnold.

Williams, G. C. 1975. *Sex and Evolution*. Monographs in Population Biology, 8. Princeton, NJ: Princeton University Press.

Wilson, E. O. 1992. *The Diversity of Life: Questions of Science*. Cambridge, MA: Belknap Press of Harvard University Press.

Wloch, D. M., K. Szafraniec, R. H. Borts, and R. Korona. 2001. Direct estimate of the mutation rate and the distribution of fitness effects in the yeast *Saccharomyces cerevisiae*. *Genetics* 159:441–452.

Wright, S. J. 1931. p. 145 In Evolution in Mendelian Populations. *Genetics* 16:97–159.

Xu, J. 2002. Estimating the spontaneous mutation rate of loss of sex in the human pathogenic fungus *Cryptococcus neoformans*. *Genetics* 162:1157–1167.

Zeyl, C. 2000. Budding yeast as a model organism for population genetics. *Yeast* 16:773–784.

Zeyl, C., and G. Bell. 1997. The advantage of sex in evolving yeast populations. *Nature* 388:465–468.

Zeyl, C., G. Bell, and D. M. Green. 1996. Sex and the spread of retrotransposon Ty3 in experimental populations of *Saccharomyces cerevisiae*. *Genetics* 143:1567–1577.

Zeyl, C., C. Curtin, K. Karnap, and E. Beauchamp. 2005. Antagonism between sexual and natural selection in experimental populations of *Saccharomyces cerevisiae*. *Evolution* 59:2109–2115.

Zeyl, C., and J. A. de Visser. 2001. Estimates of the rate and distribution of fitness effects of spontaneous mutation in *Saccharomyces cerevisiae*. *Genetics* 157:53–61.

Zeyl, C., M. Mizesko, and J. A. de Visser. 2001. Mutational meltdown in laboratory yeast populations. *Evolution: International Journal of Organic Evolution* 55:909–917.

Zeyl, C., T. Vanderford, and M. Carter. 2003. An evolutionary advantage of haploidy in large yeast populations. *Science* 299:555–558.

# 17

# PHYSIOLOGICAL ADAPTATION IN LABORATORY ENVIRONMENTS

## Allen G. Gibbs and Eran Gefen

## ENVIRONMENTAL STRESS IN THE LABORATORY

Almost any study in experimental evolution requires an altered environment in which selection is expected to occur. Sometimes the environmental variable can be biological (e.g., predators, potential mates). Often, however, it is the abiotic environment that is changed. In nature, thermodynamic variables such as temperature, pressure, and chemical activity (i.e., the concentration of salts, hydrogen ions, etc.) differ across habitats. Life itself requires input of raw materials from the environment (nutrients, water, ions, etc.) that can then be used to drive physiological processes and make more organisms.

We consider here two categories of environmental variables that have been used as selective agents in laboratory natural selection experiments. Temperature is the most important and common *physical* variable affecting the distribution and abundance of organisms in nature, as a 10°C increase in temperature causes most biochemical reactions to increase in rate two- to threefold. Typical physiological temperatures span 0°–40°C, although more extreme limits are well known (e.g., overwintering plants and insects, hot springs bacteria). Thus, selection experiments using temperature may be highly relevant to the real world. For aquatic organisms, the osmotic strength of the surrounding medium is an important environmental variable. Hypotonic or hypertonic surroundings can cause water to leak in or out, respectively, and a few selection studies have considered the effects of medium concentration.

The second category is *resource* variables, chemicals organisms need for survival and reproduction, energy sources and water being the most fundamental. Heterotrophic organisms also need chemicals such as essential amino acids, polyunsaturated fatty acids in the case of animals, oxygen in organisms using aerobic respiration, and so forth. One reason to distinguish resource from physical variables is the time course of imposed stress. The physical environment can be altered very rapidly—for example by an abrupt temperature shift or transfer into a new medium—whereas resource stress generally requires longer time scales. It takes time for an organism to consume resources, so phenotypic responses that reduce consumption rates may be part of the evolutionary response (Garland and Kelly 2006). In addition, acquisition of resources before selection is imposed (i.e., behavioral or life-history differences contributing to resource storage) further increases the time frame for evolutionary responses.

## TEMPERATURE STRESS IN *DROSOPHILA*

Temperature has frequently been used as an agent of selection in laboratory experiments (see also Estes and Teotonio this volume; Huey and Rosenzweig this volume; Swallow et al. this volume). Although surprisingly little is known about the natural thermal environment of *Drosophila*, numerous studies have showed that flies respond to temperature selection (reviewed by Hoffmann et al. 2003). The experimental approaches used to study the evolution of thermal resistance have varied widely (Hoffmann et al. 2003, table 3 therein), including acute selection for performance at extreme temperatures and laboratory natural selection of populations maintained continuously at moderately high or low temperatures. Despite the different methodologies, selection experiments highlight the ample genetic variation for thermal resistance.

Direct lab selection for increased heat survival can be difficult, as surviving flies are often sterile. Siblings of the stressed flies can be used to rear the next generation, but because siblings are only 50 percent identical genetically, this procedure provides a slower selective response. Alternative methods have therefore been developed. For example, selection for high "knockdown temperature" ($T_{KD}$), at which insects lose their ability to cling to vertical surfaces, results in a rapid increase in heat tolerance in *D. melanogaster* (Huey et al. 1991). In a modification of the method, knockdown times for *D. melanogaster* at 39°C increased fourfold after eighteen generations of selection (McColl et al. 1996). Similarly, flight ability following exposure to high temperature stress also responds rapidly to selection (Krebs and Thompson 2006). Perhaps the most interesting finding from thermal knockdown experiments is that $T_{KD}$ is bimodally distributed in natural populations of *D. melanogaster* (Gilchrist and Huey 1999; Folk et al. 2006), probably due to polymorphism in a gene with major effect. Selection for high or low $T_{KD}$ removes the low or high mode, respectively, from the population.

Measures of thermal tolerance usually exhibit consistent patterns across different assays (Hoffmann et al. 2003). For example, in natural populations of *D. buzzatii*, thirteen of nineteen assayed traits relevant to thermal adaptation were found to be significantly correlated with climatic variables (Sarup et al. 2006). However, selection for increased knockdown resistance does not always result in correlated increases in other thermoresistance traits, such as survival, recovery time, or critical thermal maximum, suggesting different genetic bases for these traits (Hoffmann et al. 1997; Bubliy et al. 1998; Folk et al. 2007). Microarray data on stress-selected populations support this conclusion. Many genes exhibit higher or lower expression in populations selected in different ways for heat tolerance (heat shock, $T_{KD}$, or chronic high temperatures; Sorensen et al. 2007). However, there is very little overlap between these sets of genes, suggesting there is fundamental variation in how organisms respond to different types of thermal stress.

It is not clear which of the different traits often assayed for comparing selected or natural populations is the most relevant for Darwinian fitness, and therefore the evolution of thermal resistance in nature. While $T_{KD}$ may be an ecologically relevant indicator of adaptation to high temperature (an animal that cannot move cannot forage or evade predators), heat shock survival may not be, as extreme high temperatures are unlikely to be reached without prior gradual warming that induces the heat shock response (Sorensen et al. 2001, 2005). Indeed, a potentially important problem in selection experiments is phenotypic plasticity (Garland and Kelly 2006). Even a brief experimental treatment will cause changes in neuronal, hormonal, and intracellular signaling, gene expression, membrane properties, and so forth. Organisms respond rapidly to changes in their environment, and these responses can themselves be targets of selection.

Many thermal selection experiments using *Drosophila* have subjected populations to constant high or low temperatures (Kilias and Alahiotis 1985; Huey et al. 1991; Loeschke and Krebs 1996). Phenotypic responses to temperature change are then of less concern, but the simplicity of these experiments reduces their relevance to the real world. In nature, flies may experience the mean temperature only twice a day, as temperatures increase in the morning and decline in the evening (Feder 1997). Variable thermal regimes are clearly more ecologically relevant. Designing experiments with ecologically relevant thermal conditions is complicated by the scarcity of information regarding the temperatures *Drosophila* experience in the field. Adults can potentially avoid high temperature be seeking cooler microclimates, although these may not necessarily be available (Gibbs et al. 2003b). Immobile life stages (eggs, pupae, and to some extent larvae) are more likely to be targets for thermal adaptation (Sarup et al. 2006), but these stages have not been the subject of laboratory evolution experiments.

Survival is not the only fitness trait affected during adaptation to changing thermal environments. Life-history traits (see also Zera and Harshman this volume) also respond to temperature selection. Comparison of *D. melanogaster* populations adapted in the lab to 16.5° and 25°C showed that females had higher fertility and fecundity at their selection temperature in comparison with females adapted to the other (Partridge et al. 1995). Likewise, a comparison of *D. melanogaster* adapted to 18° and 25°C for ten years in lab culture found a higher mating success with control females for males at their selection temperature compared with males selected at a different temperature (Dolgin et al. 2006).

## TEMPERATURE SELECTION IN MICROBES

Studies of temperature selection in bacteria and other microorganisms have primarily been motivated by interest in testing theoretical predictions regarding trade-offs in adaptation (see also Futuyma and Bennett this volume). Does evolutionary adaptation at high temperatures, for example, result in reduced fitness at low temperatures? Does evolution at a single temperature result in reduced fitness at other temperatures (i.e., a narrower thermal niche) or reduced fitness in variable environments? Does adaptation to one type of stress improve bacterial strains' ability to survive other stresses, suggesting that there are general mechanisms of stress resistance? A fundamental advantage of *Escherichia coli* and most other microbes is the ability to store stocks, including the ancestral strain, in a frozen condition, then revive them. Thus, fitness changes can be directly assessed by competition experiments using ancestral and evolved strains. (More details on how these experiments are performed can be found in Estes and Teotonio this volume; Futuyma and Bennett this volume; Travisano this volume.)

Several early studies by Bennett, Lenski, and colleagues suggested that predictions of trade-offs were true. Replicated strains of *E. coli* that had evolved at 42°C for two thousand generations had higher fitness than their ancestor at this temperature, but they had reduced relative fitness at the ancestral temperature of 37°C (Lenski and Bennett 1993). Surprisingly, the upper thermal limit for growth did not increase very much. Mutants that did achieve higher thermal tolerance lost fitness at lower temperatures (Mongold et al. 1999), also consistent with the existence of trade-offs. After twenty thousand generations at 37°C, fitness declined at extreme temperatures (Cooper et al. 2001), indicating narrowing of the thermal niche. Subsequent work reveals that trade-offs in performance at different temperatures, although common, are not universal (Bennett and Lenski 2007).

Of course, natural thermal habitats are not static. Microbes have also been selected for performance in variable habitats, either predictable (Leroi et al. 1994) or randomly changing (Ketola et al. 2004). In the former study, lines that were alternated between 32° and 42°C showed improved fitness in the variable environment, as well as at each of these temperatures when held under constant conditions. Surprisingly, they actually showed decreased fitness, relative to the ancestor, during the transition between temperatures (Leroi et al. 1994). Thus, just when phenotypic responses to changing conditions would seem to be most critical, lines selected in variable thermal regimes did not improve.

## LABORATORY EVOLUTION IN AQUATIC ENVIRONMENTS: OSMOTIC STRESS

Temperature is not the only physical variable. Besides hydrostatic pressure, for which we are unaware of any experimental evolution studies, the osmotic strength of the medium is an important factor for aquatic organisms (Evans 1993). Only a handful of studies have examined laboratory selection for osmotic resistance in animals. For example, one of us (A.G.G.) once received a set of salt-selected *Drosophila* populations from J. S. F. Barker. One line developed and lived its adult life on media containing 9.5 percent NaCl (nearly three times the strength of seawater). We eventually disposed of these unreplicated lines (because they grew so vigorously that they threatened to invade other lab stocks), and Barker apparently never published anything on them, so we know nothing about their mechanisms of adaptation. Clearly, however, salt selection works in *Drosophila*.

More relevant to the real world are selection experiments using aquatic organisms. Marine teleost fishes are hypo-osmotic relative to their surroundings and therefore must ingest seawater to balance the osmotic loss of water. As the teleost fish kidney cannot produce hyperosmotic urine, mitochondria-rich cells (MRCs) in the gills, often referred to as "chloride cells," are a major site for osmoregulation through excretion of $Na^+$ and $Cl^-$. Euryhaline teleosts transferred from freshwater to seawater exhibit correlated changes in both number and structure of MRCs (Evans et al. 2005). In addition to the plasticity in this trait, levels of proliferation and hypertrophy of MRCs (and the ensuing osmoregulatory capacity) also appear to have a genetic component. Three generations of selection for high salinity tolerance in the guppy *Poecilia reticulata* resulted in an increase in both size and number of chloride cells (Shikano et al. 1998). This was correlated with increased seawater tolerance and enhanced osmoregulatory function, in comparison with controls from the same original fish stock.

The Baltic copepod, *Eurytemora affinis*, has invaded numerous aquatic ecosystems in the United States over the past century (Lee 1999). This species is normally intolerant of fresh water, but invasive populations have colonized several river drainages. Freshwater and saline populations have diverged in their osmoregulatory capabilities and survival at low salinities. When reared at an intermediate salinity for six generations, freshwater populations survive better at low salinity, but their survival is the same at high salinity (Lee et al. 2007). Recent work suggests that the physiological differences that evolved in the laboratory mimic those observed in natural populations (C. E. Lee, personal communication).

## RESOURCE LIMITATION IN INSECTS: SURVIVAL OF THE FATTEST?

Because of the Second Law of Thermodynamics (entropy always increases in an isolated system), organisms need an input of energy simply to maintain themselves. Starvation stress presents an interesting range of potential survival strategies. Organisms can acquire energy and store it in anticipation of food limitation, as do many animals in nature as winter approaches, or decrease energy consumption (i.e., metabolic rate) when food is unavailable. The first option is well documented in *Drosophila*. Starvation-selected populations accumulate large quantities of lipids, the most energy-dense storage form (Chippindale et al. 1998). Differences in energy content can explain almost all of the variation in starvation resistance in *Drosophila* populations selected for various physiological and life-history traits (Chippindale et al. 1996; Bradley and Folk 2004; Zera and Harshman this volume).

Starvation selection experiments serve as a good example of the possible effect of different experimental protocols on the results and their interpretation. Experiments in *Drosophila* usually involve exposure of selected populations to stressful conditions as young adults, typically about fourteen days after egg collection, corresponding to an adult age of about four days posteclosion (Chippindale et al. 1996, 1998; Djawdan et al. 1997). This selection procedure resulted in a twofold increase in resistance after twenty generations (Rose et al. 1992). A different experimental approach, selecting on newly eclosed flies, resulted in a less profound increase in resistance following twenty generations of selection, possibly by omitting adult feeding from the overall response to selection (Baldal et al. 2006). Both larval and adult derived energy are important sources of adult nutrition, even in unselected lines. Wild-type flies actually become less starvation resistant in the first few days of adult life, even as they are feeding (Aguila et al. 2007). In contrast, starvation-selected populations show increased starvation resistance in early adulthood (Chippindale et al. 1996).

The effects of starvation selection on metabolic rate are not clear. Hoffmann and Parsons (1989) reported that mass-specific metabolic rates were lower in stress-selected populations. A potential analysis problem arises in animals that store very large quantities of energy as lipids or glycogen. This storage may come at a low energy cost, so expressing metabolic rates per unit mass may bias estimates downward. Should glycogen and lipid sitting in a cell be counted as part of the metabolizing animal? Djawdan et al. (1997) argued that it should not and that a better measure of mass is the lipid- and carbohydrate-free mass. When they subtracted energy stores from the total mass, starvation-selected flies did not have lower mass-specific metabolic rates. Harshman and Schmid (1998) used a third approach, the per-animal metabolic rate, and also found no reduction in metabolism. In summary, it appears that lower metabolic rates can evolve, but the primary selection response is increased energy storage before starvation is imposed.

Nutritional selection need not require complete removal of food. Low food quality has been used in selection experiments in a few cases. Harshman et al. (1999) reared replicate populations on lemons (a poor larval substrate) or standard *Drosophila* medium. Many of the same physiological differences appeared as observed in the studies described here. Lemon-selected flies had greater energy stores and lower metabolic rates (quantified on a mass-specific basis). (See also Garland and Kelly 2006 for a discussion of evolved differences in plasticity in these lines.) Lemon-reared larvae developed slowly, and their slower development remained when reared on bananas. In a non-*Drosophila* example, Warbrick-Smith et al. (2006) varied the dietary protein/carbohydrate ratio of *Plutella xylostella* caterpillars and demonstrated changes in lipid storage after six generations of selection. Another example of nutritional selection concerns *Drosophila* reared for ninety generations in the presence of ethanol (Fry 2001). Ethanol is a potential energy source for all life stages in nature but potentially toxic at high levels. In fact, adult *D. mojavensis* can actually gain dry mass with ethanol vapor as its sole food source (Etges 1989). In the presence of ethanol, selected lines developed faster and survived better to adulthood than controls, without any disadvantage (i.e., trade-off) in the absence of ethanol (Fry 2001).

The rapid response to starvation selection in laboratory experiments illustrates the substantial genetic variability in natural populations, which should result in differences between natural populations exposed to different environmental conditions. Several studies report negative correlations between latitude and starvation resistance in *Drosophila* species (Karan and Parkash 1998; Karan et al. 1998; Parkash and Munjal 2000; Hoffmann et al. 2001), but these correlations are inconsistent (Harshman and Hoffmann 2000; Griffiths et al. 2005). Furthermore, Hoffmann et al. (2001) found that the latitudinal cline effect on starvation resistance in Australian populations of *D. melanogaster* was weaker than the variation within populations. Indeed, conflicting results led Karan et al. (1998) to argue that, because flies always survive starvation longer than desiccation, flies will die of starvation in nature only in humid areas.

Variation in starvation resistance to among natural populations could also result from a correlated response to selection for resistance other stresses. For example, mechanisms associated with resistance to starvation and cold stresses are antagonistic (Hoffmann et al. 2005). Latitude is generally negatively correlated with winter temperature; thus, decreased starvation resistance with latitude may be associated with increased cold resistance. Cold tolerance is also affected by altitudinal differences (Collinge et al. 2006), and therefore clines for starvation resistance could be affected by altitudinal differences along the latitudinal gradient. Furthermore, high- and low-altitude sites are often separated by short distances thus facilitating gene flow (Blanckenhorn 1997), which may result in deviations from clines. Opposite clines and deviations from clines highlight the need to further explore the causes for evolution of stress resistance in natural populations. Furthermore, they emphasize the extent to which "laboratory natural selection experiments" oversimplify selection in nature (see also Huey and Rosenzweig this volume).

## NUTRITIONAL SELECTION IN MICROBIAL POPULATIONS

In bacteria, selection using alternative energy sources has been performed on numerous occasions, using progenitor strains that have been grown on a defined diet for two thousand generations or more. When exposed to novel energy sources, replicate lines diverge markedly in their fitness (Travisano et al. 1995; Travisano and Lenski 1996). This is especially true for carbon sources that do not share the same uptake mechanism as glucose, the ancestral carbon source.

Experiments using serial dilution expose bacteria to fluctuating nutritional conditions: scramble competition for resources when cultures are diluted into fresh medium starvation, followed by resource limitation when food runs out and bacteria enter the stationary phase. The growth phase is especially important for selection. The ability to enter the exponential phase rapidly and grow quickly is critical to relative fitness (Vasi et al. 1994), but indirect evidence suggests that the ability to tolerate nutritional stress is also important. For example, thirty-six lines adapted to serial dilution versus chemostat conditions (chronic poor nutrition) exhibited little sign of trade-offs, as indicated by reduced fitness in the alternative regime (Velicer and Lenski 1999). These results would not be anticipated if rapid growth were the only target of selection in serial dilution experiments. In a different experiment, multiple lines selected for high temperature fitness had similar gene duplication events in a region containing several genes known to affect stress and starvation resistance (Riehle et al. 2001).

In most serial dilution experiments, nutritional stress is relatively brief, less than twenty-four hours. An interesting example of an alternative starvation selection regime comes from *E. coli* subjected to long-term starvation (Vasi and Lenski 1999). Replicated lines were maintained in stationary phase cultures for forty-nine days, long after energy from the media had been consumed, and the density of viable cells had decreased by 99.99 percent. In an example of artificial selection, colonies were grown from the survivors, and only those visibly different from progenitor-strain colonies were selected for experiments. Out of five mutants, three died less rapidly when grown in pure culture, exactly what one might expect. Three, but not the same three, died less rapidly in competition with the progenitor strain, in a frequency-specific manner (figure 17.1). This suggests that

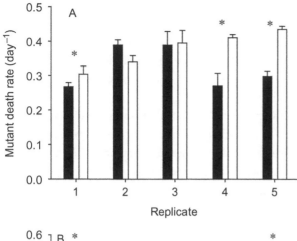

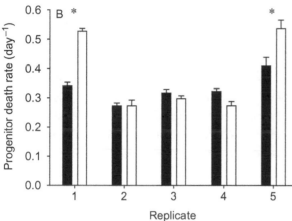

FIGURE 17.1
Death rates of stationary-phase
selected and progenitor strains of
*E. coli* in competition experiments.
A, Death rates of selected lines at
low (10 percent) and high
(90 percent) relative density.
Lower death rates at low density
suggest that these strains are acquiring resources from the progenitor, possibly by cannibalism. B,
Death rates of progenitor strains at
low (10 percent) and high (90 percent) relative density. Higher death
rates at low density may indicate
allelopathy (bacteriocide) by the
selected strain. Data redrawn from
Vasi and Lenski (1999).

they were deriving nutrition from live or dead progenitor cells. In two cases, the progenitor died more rapidly when the mutant was more abundant, suggesting that the evolution
of allelopathy by the mutant.

The study by Vasi and Lenski (1999) shows that bacteria can evolve to kill and eat
the competition. What about in *Drosophila*? Huey et al. (2004) tested and rejected the
hypothesis that adult *Drosophila* consume their conspecifics when food is scarce. Larvae
are not so fastidious. We have seen active larvae with red guts, apparently from consum-
ing the eye pigments of dead adults in their vials. Typical starvation selection experiments
involve population cages containing large numbers of fly corpses, which theoretically
could serve as a food source. The potential for evolution of cannibalism in starvation-
selected adult flies has not been investigated. Indirect cannibalism, by feeding on
microbes colonizing dead conspecifics, is another possibility.

Many bacteria can survive starvation stress by entering a dormant spore state.
Hypometabolic states (hibernation, diapause, anhydrobiosis) in animals are well documented, but they have not been observed in laboratory selection studies. This may reflect

the design of the experiments. In most cases, selection continues until a certain proportion of the population appears dead. Food is restored, and those animals that can feed immediately have an advantage. In the case of bacteria, hypometabolic mutants will be flushed out of a chemostat. In experiments involving serial dilution, competition for resources will select for genotypes that can acquire resources quickly, and emergence from dormancy takes time (Vasi et al. 1994).

One selection experiment was designed with hypometabolism in mind. Maughn and Nicholson (2004) selected replicated strains of *Bacillus subtilis* for the ability to form heat-resistant spores under nutritional stress for five thousand generations. The progenitor strain sporulated 70 percent of the time under these conditions. Despite the fact that other strains exhibit varying degrees of sporulation efficiency and increasing mutation rates as selection continued, sporulation efficiency did not respond to selection. The proposed explanation for this surprising result concerns stochasticity in the decision to sporulate. *Bacillus subtilis* can take alternative developmental pathways under stress, including becoming competent for DNA uptake, growth form shifts, adaptive mutagenesis (Robleto et al. 2007), or spore formation (Maughan and Nicholson 2004). If each cell has an independent probability of taking one of these pathways, this bet-hedging strategy can allow survival of the genotype under a variety of stressful conditions. In this example, different strains may differ in which route occurs most often, but the "decision" at the individual level is stochastic. It remains to be seen whether this is a common phenomenon or whether some species or strains can evolve increased sporulation efficiency.

### *DROSOPHILA* IN LABORATORY DESERTS

Water is perhaps the most important resource determining where organisms actually live in terrestrial ecosystems. Two-thirds of the mass of a typical animal is water, and deviations impose severe physiological and cellular stress. Several labs have subjected *Drosophila* to selection for the ability to resist desiccation stress. Desiccation selection generally results in a rapid and significant increase in desiccation resistance of adult flies (reviewed by Hoffmann and Harshman 1999; also see Bubliy and Loeschcke 2005; Gefen et al. 2006). Figure 17.2 shows the response of nine populations of *D. melanogaster* (three source populations) to desiccation selection, as indicated by the time necessary to reach 80–85 percent mortality each generation. Despite similar selection responses in different experiments, response patterns vary because of differences in genetic variation for desiccation resistance (Hoffmann and Parsons 1993; Hoffmann et al 2003). For example, three lines derived from one source population in figure 17.2 responded more rapidly to selection and evolved significantly greater desiccation resistance than the other six populations.

Because desiccation selection has been performed several times, in several labs, it provides an opportunity to examine the repeatability of evolution in different genetic backgrounds. This must be done cautiously, however, as the details of selection, measurements,

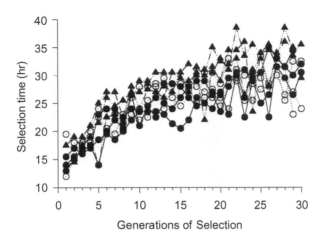

FIGURE 17.2

Desiccation resistance responds rapidly to selection. Data represent the amount of time required for about 85 percent of each cage's population to die in dry air. Three starting populations represented by different symbols were used, with three replicates of each. Note that replicate populations indicated by the triangles consistently outperformed the other two sets of replicates. Data are from A. G. Gibbs and C. H. Vanier (unpublished).

and analytical procedures can differ greatly. A few consistent patterns are evident. Desiccation resistance can evolve through one or more of the following mechanisms: an increase in body water content, reduced rate of water loss to the surrounding environment, and the ability to tolerate reduced body water content as a result of dehydration (dehydration tolerance). Desiccation-selected *D. melanogaster* generally have increased body mass, carbohydrate levels and water contents. In short-term selection experiments, the percentage of body mass composed of water does not change (Hoffmann and Harshman 1999; Folk et al. 2001; Gefen et al. 2006), whereas long-term selection (more than one hundred generations) can result in a differential increase in water content (Gibbs et al. 1997; Folk et al. 2001).

Greater water storage may be related to the high-carbohydrate contents in desiccation-selected flies (Hoffmann and Parsons 1993; Gibbs et al. 1997; Chippindale et al. 1998; Gefen et al. 2006). Glycogen binds three to five times its own mass in water (Schmidt-Nielsen 1997), and therefore accumulation of glycogen could increase body water storage capabilities. Gibbs et al. (1997) estimated that 47 percent of the additional water in selected flies could be bound to intracellular glycogen stores. This water cannot be released unless glycogen is metabolized. Djawdan et al (1997) reported that *D. melanogaster* shift to carbohydrate catabolism under desiccation stress, thereby making free water available to replace that which has been lost. Similarly, carbohydrate catabolism during desiccation stress was significantly higher than during starvation in four out of five *Drosophila* species (Marron et al. 2003). The second location for water storage is the hemolymph. Folk et al. (2001) estimated that selected flies had a

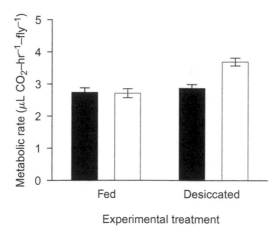

FIGURE 17.3

Metabolic rates of desiccation-selected and control flies. Selected flies (filled bars) do not exhibit an increase in metabolic rate when exposed to dry air, whereas control flies (open bars) do. Carbon dioxide production was measured using flow-through respirometry, in the presence or absence of a food and water source. Data are from A. G. Gibbs and C. H. Vanier (unpublished).

hemolymph volume four times greater than controls, equivalent to 68 percent of their extra water content.

Water loss rates are significantly lower in desiccation-selected flies than in controls (Hoffmann and Parsons 1993; Gibbs et al. 1997; Williams et al. 1997). This may be the result of reduced respiratory water loss, as selected lines are less active (Williams et al. 2004) and exert better spiracular control than control populations (Williams et al. 1997, 1998). This may, in turn, be caused by a reduced behavioral response to desiccation stress (Williams et al. 2004). For convenience, metabolic rates are typically measured in dry air; when flies have access to food and water, the differences in metabolic rate can disappear (figure 17.3). In this example, reduced plasticity in activity and metabolic rate are part of the overall selection response.

Behavioral responses to desiccation are very apparent in natural populations of *Drosophila* and differ according to habitat. Desiccation-sensitive species become very active shortly after exposure to dry air, whereas desert-dwelling species remain inactive for fifteen hours or more (Gibbs et al. 2003a). We note that experimenters have generally measured water loss and metabolic rates early in desiccation stress, when behavioral differences between species or selected populations and their controls are most evident. Thus, some of the reported differences may be artifacts of when measurements were made. Both desert-adapted and desiccation-selected flies may exhibit the same behavioral and physiological responses as mesic or control populations, but these may be delayed until the flies actually "feel" stressed.

Despite consistent changes in water loss rates, carbohydrate storage, and behavior, different studies can yield what appear to be conflicting results. Both age and sex of experimental flies are potential sources of discrepancy. Desiccation resistance of *Drosophila* species decreases with age after the first week of adult life (Lamb 1984; Nghiem et al. 2000; Gibbs and Markow 2001). In contrast, Chippindale et al. (1998) reported that resistance of desiccation-selected males decreased during the first four days posteclosion, but that of females remained unchanged. Female flies generally tend

TABLE 17.1 Water Contents of Control and Desiccation-Selected *Drosophila*

| Line | Initial Water Content (mg) | % Water Lost | Net Water Lost (mg) | Net Water Remaining at Death (mg) |
|---|---|---|---|---|
| C1 | 1.087 | 47.36 | 0.515 | 0.572 |
| C2 | 1.058 | 50.43 | 0.534 | 0.524 |
| D1 | 1.290 | 60.39 | 0.779 | 0.511 |
| D2 | 1.178 | 58.22 | 0.698 | 0.480 |

NOTE: Flies were weighed before the experiment and immediately after dying from desiccation stress. Masses are milligrams per individual. Data calculated from Telonis-Scott et al. 2006.

to be more desiccation resistant than males, but Gefen et al. (2006) reported only minor differences between newly eclosed desiccation-selected males and females. Large sex-related differences in older flies would then reflect different patterns of water accumulation in adult males and females.

Apparently conflicting results between lab selection experiments may also stem from different terminology. In contrast to Gibbs et al. (1997), Hoffmann and Parsons (1993) reported that D flies had increased dehydration tolerance compared to their controls. However, this discrepancy may result from different definitions of dehydration tolerance. Hoffmann and Parsons (1993) calculated dehydration tolerance as the total water loss prior to death. Other authors have used similar definitions (Telonis-Scott et al. 2006; Archer et al. 2007), whereas Gibbs et al. (1997) described dehydration tolerance as the body water content at time of death. Table 17.1 uses data from Telonis-Scott et al. (2006) to illustrate how these different definitions can result in different interpretations. Desiccation-selected flies (D1, D2) have higher body water contents than controls (C1, C2) and therefore have more water available to lose. Selected flies lose a greater fraction of their initial water content before they die, but the actual amount remaining in their bodies when death occurs does not differ (table 17.1; Gibbs et al. 1997; Gefen et al. 2006), so it appears these seemingly conflicting reports may actually be in agreement. The higher initial bulk water stores of desiccation-selected flies allow higher total water loss prior to death, which may occur at a similar body water content regardless of selection treatment.

The major routes for water loss from *Drosophila* are cuticular transpiration and respiratory water loss (Gibbs et al. 2003a). Cuticular waxes (the main barrier to water loss) do not differ between selected and control populations (Gibbs et al. 1997), suggesting that desiccation selection had not resulted in a less permeable cuticle. Another possibility that has not been tested in experimental evolution studies is an increase in melanization. In interspecific comparisons, population comparisons, and studies of body color mutants, darker adult *Drosophila* lose water less rapidly than lighter ones and are more resistant to desiccation (Rajpurohit et al. 2008).

If desiccation selection does not affect cuticular permeability, by default respiratory water loss must be reduced. This hypothesis is very difficult to test in such small insects, but reduced metabolic rates in these lines are consistent with a reduced need to open the spiracles to allow gas exchange (Hoffmann and Harshman 1999; Gibbs 2002; but see Williams and Bradley 1998). Comparative studies of water balance yield similar conclusions; desert species have lower metabolic rates for their size than mesic ones (Gibbs et al. 2003a).

Desiccation selection experiments are generally designed to press the physiological limits of organisms, but that does not preclude behavioral responses. Desiccation-selected *Drosophila* are less active than control populations when water is available and do not become more active when desiccated (Hoffmann and Parsons 1993; Williams et al. 2004). These differences are supported by measurements of metabolic rate in the presence and absence of water (figure 17.3). Control populations have higher metabolic rates when desiccated, in contrast to selected populations. In this case, selected lines have lost the plastic response of increasing metabolic rate, at least during the early stages of desiccation. We have also recently noticed an interesting phenomenon in desiccation-selected flies in our lab. Flies are reared in thirty-five-milliliter vials during preadult

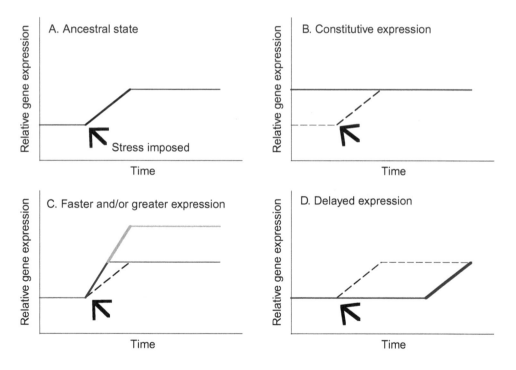

FIGURE 17.4

Potential effects of selection on gene expression. A, In the ancestor, expression of a given gene increases when stress is imposed (arrow). B, Stress selection results in high constitutive expression of the gene. In this and following panels, the dashed indicates the ancestral condition; the solid line, the evolved pattern. C, Selection results in faster induction and/or higher expression under stress (see also Garland and Kelly 2006). D, Induction is delayed, relative to the ancestor, when stress is imposed. This pattern may occur in resource selection, in which increased storage of resources delays stress responses.

stages and the first few days of adult life, then are transferred into population cages (Gefen et al. 2006). Because not all populations can be transferred simultaneously, flies that are transferred first are provided with media until all populations have been dumped. Desiccation-selected flies congregate and feed on the medium, while control populations remain dispersed around their cages. It appears that we have selected for flies that respond to the physical trauma of being dumped into a population cage, by using the last-minute opportunity to acquire additional water and nutrition that they will need to survive the selection bout. Alternatively, they may simply spend more time feeding, both before and after transfer in cages.

Because some *Drosophila* species inhabit desert environments, desiccation selection provides an opportunity to compare evolutionary responses in the lab to those in nature (Gibbs 2002). Desert flies are subjected to highly variable temperatures as well as low humidity (Gibbs et al. 2003b), so comparisons need to be made with caution. Some of the same differences appear in lab-selected and desert *Drosophila*, particularly reduced water loss rates. Desert flies are not more tolerant of dehydration (after correction for phylogeny; Gibbs and Matzkin 2001), nor do desert-adapted flies store more water and carbohydrate than mesic ones. One possible reason for the lack of water and carbohydrate storage in natural populations is that food and water sources are the same in nature. Carrying extra mass in the form of water or energy might compromise flight and the ability to avoid predators. Preliminary work from our lab suggests that, despite their larger mass, desiccation-selected flies do not have larger wings, which results in greater wing loading and potentially diminished flight performance. This is of little consequence in a predator-free, confined ecosystem like a population cage, but potentially important in nature.

## WHAT'S THE GENE?

An ultimate goal of evolutionary studies is to understand the nuts and bolts of the evolutionary process. What are the "genes that matter," in nature or in the laboratory? Experimental evolutionists have taken two general approaches to this issue (see also Dykhuizen and Dean this volume; Rosenzweig and Sherlock this volume). The first is to investigate candidate genes that have been independently identified by researchers in other disciplines. A prime example is heat shock genes. HSP70 is a heat-inducible chaperone that helps to minimize cellular damage by binding and enabling refolding of heat-damaged proteins (Feder and Hoffman 1999). These genes are expressed in response to numerous stresses, and most of them are expressed constitutively at a basal level (Hoffmann et al. 2003).

In nature, HSP expression and the threshold for expression often correlate with the levels of stress to which species are exposed naturally and their stress resistance (Feder and Hofmann 1999). In the laboratory, long-term maintenance of *Drosophila* at constant temperatures results in lower basal HSP70 expression (Krebs and Loeschcke 1994;

Cavicchi et al. 1995; Bettencourt et al. 1999), perhaps because it has deleterious effects below the threshold temperature, while HSP expression is higher in lines selected in variable thermal environments (Sorenson 1999). A similar pattern is seen in the protozoan, *Tetrahymena thermophila*, with higher induction of *Hsp90* in lines selected in the most variable conditions (Ketola et al. 2004). Changes in basal HSP level can evolve independently of an increase in HSP induction (Bettencourt et al. 1999).

Naturally occurring protein polymorphisms have been associated with performance differences in stress resistance or clinal variation in natural populations (Dahlhoff and Rank 2000, 2007). These provide candidate genes to examine in laboratory-selected populations. One can also use candidate physiological processes to identify candidate genes. For example, oxidative damage induced by free radicals has been implicated in aging-related loss of function, responses to hyperoxia, reoxygenation damage following hypoxia, salinity stress, desiccation stress, and so forth. It stands to reason that selection to survive such stresses might involve increased expression of proteins that reduce oxidative damage—for example, superoxide dismutase. Even if these efforts are successful (e.g., Deckert-Cruz et al. 1997), however, observations of allele frequency shifts in the lab need to be followed up with more experiments, such as measuring survival of different genotypes or manipulating gene expression in model organisms (Sun et al. 2002).

Despite occasional successes, the use of information from other fields has not been very successful in finding genes that actually are under selection in laboratory experiments. In experimental evolution, many genes will respond to selection, most with relatively minor effect. Even obvious candidates are components of pathways; existing genetic variation in ancestral populations may favor selection on other pathway members. One exception may be $T_{KD}$ in *Drosophila*, where the bimodal distribution of knockdown temperatures is consistent with polymorphism for a gene of major effect existing in natural populations (Gilchrist and Huey 1999; Folk et al. 2006).

Genomic methods provide a second avenue to identify genes under selection (see also Rosenzweig and Sherlock this volume). Whole-genome microarrays allow rapid screening of differences in gene expression, as well as identification of sequence variation between selected and control lines. The first such study used yeast (Ferea et al. 1998), but bacteria provide the most detailed information to date. Microarray experiments have revealed parallel changes in gene expression in replicate lines (Riehle et al. 2001; Cooper et al. 2001, 2003; Crozat et al. 2005). Genes that show up in microarray analyses include heat shock genes, PIMT (a protein repair gene), topoisomerase, and spoT, which is involved in metabolism of the important signaling molecule, ppGpp. Gene duplication and loss also affect thermal tolerance (Riehle et al. 2003). It is encouraging to see that molecular evolution is repeatable to some extent, although the specific mutations that occur usually differ among replicates (Woods et al. 2006). One could just as easily imagine that, for example, different subunits of multiunit proteins, or different enzymes in metabolic pathways, would mutate and have similar effects on fitness. Protein expression experiments also indicate parallel evolution (Pelosi et al. 2006).

As of this writing, one *Drosophila* microarray study using laboratory-selected populations has been published (Sorensen et al. 2007). The study includes lines selected for resistance to high and low temperatures, desiccation and starvation stress, and aging. The source populations used to found the selection lines are problematic, including an eclectic mix of outbred populations, isofemale lines, and even populations that had previously been selected for stress resistance. Clustering analysis revealed that selection replicates had similar expression patterns, again indicating the repeatability of molecular evolution. There was relatively high overlap in the functional categories of genes responding to selection (e.g., metabolic enzymes), but relatively little overlap in specific genes. Thus, the situation in *Drosophila* appears more complicated than in bacteria, with selection apparently acting on the same stress response pathways, but not necessarily the same genes. Of course, the differences between microbial and insect responses could also reflect differences in population size (thousands vs. millions) and number of generations under selection (twenty vs. twenty thousand).

The study by Sorensen et al. (2007) used flies that had been selected for stress resistance but had not been directly exposed to stress. It is likely that selection will affect induction of gene expression as well (i.e., expression plasticity). For example, genes involved in thermal responses may be turned on more quickly or to a higher level in heat-selected lines. Alternatively, if populations have been subjected to resource selection, they may have slower responses to stress. Desiccation-selected *Drosophila* may not mount a response until they have been desiccated for several hours, when controls may already be dead. It is interesting, then, to compare the results of Sorensen et al. (2007; selected lines not exposed to stress) with those in which flies have been directly exposed to stress. Harbison et al. (2005) exposed flies to severe starvation stress (estimated to cause 50 percent mortality) and quantified gene expression across the genome. Over 3,400 genes were differentially expressed during starvation, with nearly half having higher expression. In contrast, Sorensen et al. (2007) found that 230 genes were downregulated in starvation-selected lines, and none were up-regulated. Only one gene was common to these data sets. Thus, the selection response and the phenotypic response to starvation stress are dramatically different.

The lack of concordance between these studies extends to the functional categories of genes. Starvation selection affects energy acquisition and storage before starvation actually occurs, and it is not surprising that a few carbohydrate metabolism genes are downregulated in selected lines. However, genes involved in transcriptional regulation are represented much more heavily (more than seventy-five genes out of approximately two hundred annotated genes; Sorensen et al. 2007). In contrast, most of the genes that are differentially regulated during starvation stress are involved in biosynthesis and protein metabolism (Harbison et al. 2005). It will be very interesting to see if selection affects the regulation of these same genes, and certainly more genomic comparisons of selection responses and plastic responses to stress are needed. Figure 17.4 illustrates the types of changes in gene expression one might expect. Genes that are normally turned on (or off) by stress might evolve differences in constitutive expression, or the speed, degree or timing of the stress response might change (see also Garland and Kelly 2006). To understand which changes are most common, what is needed is a time series of gene expression measurements, in replicate selected and control populations. This would be a large undertaking, but certainly within the scope of current technology and bioinformatics.

Identification of candidate genes is only the first step to demonstrating that these genes actually have an effect on fitness in laboratory populations. A few such experiments have been performed. Sun et al. (2002) overexpressed superoxide dismutase in *D. melanogaster* to confirm its hypothesized role in aging. In *E. coli*, Crozat et al. (2005) transferred candidate genes into the ancestral genetic background and demonstrated that these could account for one-third of the gain in fitness. These studies are clearly just the start of efforts to investigate the function of target genes identified by experimental evolution studies.

## UNINTENDED SELECTION IN LABORATORY ENVIRONMENTS

In nature, environmental variables are often highly correlated. For example, high-pressure deep-sea habitats are generally cold, hydrothermal vents being an extremely rare (but extremely interesting) exception. The solubility of oxygen in water is negatively related to temperature; thus, even oxygen-saturated aquatic environments can have less available oxygen than colder, subsaturated regions. In terrestrial environments, the saturating vapor pressure of water increases dramatically with temperature, so that a parcel of air containing the same absolute quantity of water vapor will have a lower relative humidity as it warms. The combination of high temperatures and low humidity in deserts, therefore, provides a double stress. A potential interaction between cold and desiccation stress also exists. Long-tem exposure to subfreezing temperatures can lead to organismal and cellular dehydration (Ring and Danks 1994). Thus, selection for resistance to either cold or desiccation stress might cause a correlated response of resistance to the other.

Physiological ecologists have been well aware of and particularly interested in interactions between environmental variables for decades, but studies in experimental evolution have generally not considered these. We discuss here one hypothetical example. Consider a laboratory natural selection experiment using *Drosophila* populations exposed to high temperatures throughout the life cycle. The larval stages are semiaquatic, usually feeding in the upper one to two centimeters of the medium. At high temperatures, the solubility of oxygen in the media declines. Although larvae are generally considered to respire through the spiracles, they spend much of their time foraging below the surface. Their cuticle is very permeable to water, and oxygen fluxes across the surface are certainly possible.

Oxygen availability affects larval growth rate, developmental time, survival, and adult body size (Loudon 1989; Greenberg and Ar 1996; Frazier et al. 2001; Harrison et al. 2006). In a selection experiment, Henry and Harrison (2004) reared populations of *D. melanogaster* in atmospheric oxygen levels ranging from 10 to 40 percent. After six generations, larvae from hypoxic populations had wider tracheae than larvae from normoxic and hyperoxic populations, when all were reared in the same conditions (Henry and Harrison 2004). Wider tracheae allow more rapid diffusion of oxygen to the tissues. Interestingly, neither hypoxia nor hyperoxia significantly affected the body mass of third-instar larvae. Instead, adult mass was correlated with rearing oxygen levels, suggesting an important effect of oxygen during pupal development, perhaps related to the reorganization of the tracheal system during metamorphosis.

*Drosophila* larvae respond to hypoxia by moving to the upper regions of the medium or even out of the medium entirely. This can affect growth rates and body size in two ways. First, larvae that exit the medium cannot feed, so their development will be delayed. Eventually, they must return to the medium, but surface crowding may occur. Food quality under these conditions will be compromised by the accumulation of ammonia (Borash et al. 1998) and other waste products, also contributing to slow development and smaller adult body size. Thus, a seemingly simple temperature selection experiment might inadvertently reduce larval oxygen availability and thereby indirectly select for resistance to hypoxia, larval starvation, and resistance to toxins.

Interactions between the environmental variables of temperature and food quality were examined in one experiment using *Drosophila* (Bochdanovits and de Jong 2003). Larvae were reared at high and low temperatures and on high- and poor-quality food, with adults being maintained at an intermediate temperature. Larvae from the cold selection lines produced consistently larger adults, especially when reared at high temperature. They also had greater survivorship, though development times did not differ. Larvae selected on poor-quality food were not smaller, even though they had lower feeding rates. Significant interactions between selection temperature and food quality appeared for all traits assayed. These complicated results suggest that low temperatures select for efficient nutrient processing. As the authors note, the findings highlight the potential importance of food quality in nature for this widespread species.

Unintentional selection has been well documented in the case of life-history evolution (Chippindale 2006). It may also be common in environmental selection experiments (see also Garland 2003). The example we have discussed is purely hypothetical (so far), but perhaps one example is already known from desiccation selection. In these experiments, selection is imposed by removing flies' access to food, which is their source of water. Thus, the control treatment is usually to have water accessible but not food—in other words, starvation is imposed (Gibbs et al. 1997). Control lines, therefore, undergo starvation stress, and they respond by storing more lipid than selected populations (Chippindale et al. 1996; Djawdan et al. 1997). This makes sense, as starved flies normally metabolize lipids (Marron et al. 2003). Carbohydrate accumulation in desiccation-selected flies (Djawdan et al. 1997) also makes sense, as it is metabolized when flies are desiccated (Marron et al. 2003). Thus, the control treatment undergoes starvation selection that is relatively mild, but enough to generate a significant physiological response. Is it then a proper control, or would fed flies (Telonis-Scott et al. 2006) be better? One way to address this issue would be to have two control treatments, both fed and starved.

In summary, even seemingly simple selection regimes may contain hidden complexity that can result in unintended selection. Laboratory experimenters need to consider the entire ecology of their model environments, including potential interactions between environmental variables, and even control environments can be problematic. This does not mean selection experiments are not useful, only that they are not as easy to interpret as one would hope.

## CONCLUSION

Selection experiments have demonstrated that responses to physical and resource variables are different. Physical variables generally induce direct responses to a change in environmental conditions. This is evident in the improved fitness of bacterial lines selected under variable temperature regimes. Selection in constant physical conditions tends to foster specialization (niche narrowing). In contrast, in insects, selection for survival of resource limitation results in prestress resource accumulation, as well as conservation of existing resources. Thus, preparation for stress is an important component of resource adaptation. Resource selection experiments using microbes have not examined resource acquisition, but other mechanisms have been shown to evolve, such as shorter lag phase and faster growth rates when food is available. In extreme circumstances, allelopathy and cannibalism can evolve.

To what extent does evolution in "simple" laboratory environments mimic nature? The jury is still out (see also Huey and Rosenzweig this volume). Similar physiological differences are often found in stress-selected laboratory populations, natural populations along environmental clines, and species living in different habitats. However, natural environments are far more variable in time and space, making it difficult to identify which environmental variables are most important in shaping populations. That is the great advantage of laboratory selection, but lab environments can have their own complexity. This is well known in the context of life-history evolution (Chippindale 2006; Zera and Harshman this volume), but environmental physiologists also need to keep this in mind.

Finally, high-throughput molecular techniques are beginning to change our understanding of the details of evolution (see also Rosenzweig and Sherlock this volume). These studies are certain to change our understanding of basic biology. Candidate genes identified in genomic analyses of stress-selected populations provide testable hypotheses for their function in stress resistance and under non-stressful conditions. Thus, selection experiments have the potential to inform mechanistic biology (and for the reverse perspective, see Dykhuizen and Dean this volume).

## SUMMARY

Environments change on time scales ranging from seconds and minutes to millions of years. Natural environments are complex, sometimes making it difficult to identify general principles of adaptation. In principle, laboratory models simplify the environment to the point where environmental variables can be manipulated independently, thereby providing a complementary approach to interspecific comparative analyses. The types of environmental variables can be broadly separated into physical ones, such as temperature, and resources, such as food and water. One potential advantage of laboratory selection experiments using model organisms is that genomic approaches can provide unparalleled insight into the mechanisms of adaptation. These experiments are in their infancy, but they will eventually allow deeper understanding of both evolutionary and mechanistic biology. Laboratory environments can also contain unintended complexity. This has been recognized in the context of life-history evolution as, for example, resources necessary to survive stress may be acquired well before they are actually needed. Interactions among environmental factors have not generally been considered in experimental evolution, despite their recognition by ecological physiologists outside the laboratory. The relevance of laboratory studies to nature may therefore be problematic, and indeed the complex interactions found in nature may also arise in laboratory settings.

## ACKNOWLEDGMENTS

We thank Drs. Garland and Rose for inviting us to write this chapter, two reviewers for their excellent suggestions to improve an early draft, and the National Science Foundation for financial support of our research.

## REFERENCES

Aguila, J. R., J. Suszko, A. G. Gibbs, and D. K. Hoshizaki. 2007. The role of larval fat cells in adult *Drosophila melanogaster*. *Journal of Experimental Biology* 210:956–963.

Archer, M. A., T. J. Bradley, L. D. Mueller, and M. R. Rose. 2007. Using experimental evolution to study the physiological mechanisms of desiccation resistance in *Drosophila melanogaster*. *Physiological and Biochemical Zoology* 80:386–398.

Baldal, E. A., P. M. Brakefield, and B. J. Zwaan. 2006. Multitrait evolution in lines of *Drosophila melanogaster* selected for increased starvation resistance: The role of metabolic rate and implications for the evolution of longevity. *Evolution* 60:1435–1444.

Bennett, A. F., and R. E. Lenski. 2007. An experimental test of evolutionary trade-offs during temperature adaptation. *Proceedings of the National Academy of Sciences of the USA* 104:8649–8654.

Bettencourt, B. R., M. E. Feder, and S. Cavicchi. 1999. Experimental evolution of Hsp70 expression and thermotolerance in *Drosophila melanogaster*. *Evolution* 53: 484–492.

Blanckenhorn, W. U. 1997. Altitudinal life history variation in the dung flies *Scathophaga stercoraria* and *Sepsis cynipsea*. *Oecologia* 109:342–352.

Bochdanovits, Z., and G. de Jong. 2003. Experimental evolution in *Drosophila melanogaster*. Interaction of temperature and food quality selection regimes. *Evolution* 57:1829–1836.

Borash, D. J., A. G. Gibbs, A. Joshi, and L. D. Mueller 1998. A genetic polymorphism maintained by natural selection in a changing environment. *American Naturalist* 151:148–156.

Bradley, T. J., and D. G. Folk. 2004. Analyses of physiological evolutionary response. *Physiological and Biochemical Zoology* 77:1–9.

Bubliy, O. A., A. G. Imasheva, and V. Loeschcke. 1998. Selection for knockdown resistance to heat in *Drosophila melanogaster* at high and low larval densities. *Evolution* 52:619–625.

Bubliy, O. A., and V. Loeschcke. 2005. Correlated responses to selection for stress resistance and longevity in a laboratory population of *Drosophila melanogaster*. *Journal of Evolutionary Biology* 18:789–803.

Cavicchi, S., D. Guerra, V. Latorre, and R. B. Huey. 1995. Chromosomal analysis of heat shock tolerance in *Drosophila melanogaster* evolving at different temperatures in the laboratory. *Evolution* 49:676–684.

Chippindale, A. K. 2006. Experimental evolution. *In* C. Fox and J. Wolf, eds. *Evolutionary Genetics* New York: Oxford University Press.

Chippindale, A. K., T. J. F. Chu, and M. R. Rose. 1996. Complex trade-offs and the evolution of starvation resistance in *Drosophila melanogaster*. *Evolution* 50:753–766.

Chippindale, A. K., A. G. Gibbs, M. Sheik, K. Yee, M. Djawdan, T. J. Bradley, and M. R. Rose. 1998. Resource acquisition and the evolution of stress resistance in *Drosophila melanogaster*. *Evolution* 52:1342–1352.

Collinge, J. E., A. A. Hoffmann, and S. W. McKechnie. 2006. Altitudinal patterns for latitudinally varying traits and polymorphic markers in *Drosophila melanogaster* from eastern Australia. *Journal of Evolutionary Biology* 19:473–482.

Cooper, V. S., A. F. Bennett, and R. E. Lenski. 2001. Evolution of thermal dependence of growth rate of *Escherichia coli* during 20,000 generations in a constant environment. *Evolution* 55:889–896.

Cooper, T. F., D. E. Rozen and R. E. Lenski. 2003. Parallel changes in gene expression after 20,000 generations of evolution in *Escherichia coli*. *Proceedings of the National Academy of Sciences of the USA* 100:1072–1077.

Crozat, E., N. Phillippe, R. E. Lenski, J. Geiselmann, and D. Scheider. 2005. Long-term experimental evolution in *Escherichia coli*. XII. DNA topology as a key target of selection. *Genetics* 189:523–532.

Dahlhoff, E. P., and N. E. Rank. 2000. Functional and physiological consequences of genetic variation at phosphoglucose isomerase: Heat shock protein expression is related to enzyme genotype in a montane beetle. *Proceedings of the National Academy of Sciences of the USA* 97:10056–10061.

———. 2007. The role of stress proteins in responses of a montane willow leaf beetle to environmental temperature variation. *Journal of Biosciences* 32:477–488.

Deckert-Cruz, D. J., R. H. Tyler, J. E. Landmesser, and M. R. Rose. 1997. Allozymic differentiation in response to laboratory demographic selection of *Drosophila melanogaster*. *Evolution* 51:865–872.

Djawdan, M., M. R. Rose, and T. J. Bradley. 1997. Does selection for stress resistance lower metabolic rate? *Ecology* 78:828–837.

Dolgin, E. S., M. C. Whitlock, and A. F. Agrawal. 2006. Male *Drosophila melanogaster* have higher mating success when adapted to their thermal environment. *Journal of Evolutionary Biology* 19:1894–1900.

Etges, W. J. 1989. Influences of atmospheric ethanol on adult *Drosophila mojavensis*: Altered metabolic rates and increases in fitness among populations. *Physiological Zoology* 62:170–193.

Evans, D. E. 1993. Osmotic and ionic regulation. Pages 315–341 *in* D. E. Evans, ed. *The Physiology of Fishes*. Boca Raton, FL: CRC Press.

Evans, D. H., P. M. Piermarini, and K. P. Choe. 2005. The multifunctional fish gill: Dominant site of gas exchange, osmoregulation, acid-base regulation, and excretion of nitrogenous waste. *Physiological Reviews* 85:97–177.

Feder, M. E. 1997. Necrotic fruit: A novel model system for thermal ecologists. *Journal of Thermal Biology* 22:1–9.

Feder, M. E., and G. E. Hofmann. 1999. Heat-shock proteins, molecular chaperones, and the stress response: Evolutionary and ecological physiology. *Annual Reviews of Physiology* 61:243–282.

Ferea, T. L., D. Botstein, P. O. Brown, and R. F. Rosenzweig. 1999. Systematic changes in gene expression patterns following adaptive evolution in yeast. *Proceedings of the National Academy of Sciences of the USA* 96:9721–9726.

Folk, D. G., and T. J. Bradley. 2004. Evolved patterns and rates of water loss and ion regulation in laboratory-selected populations of *Drosophila melanogaster*. *Journal of Experimental Biology* 206:2779–2786.

Folk, D. G., C. Han, and T. J. Bradley. 2001. Water acquisition and partitioning in *Drosophila melanogaster*: Effects of selection for desiccation-resistance. *Journal of Experimental Biology* 204:3323–3331.

Folk, D. G., L. A. Hoekstra, and G. W. Gilchrist. 2007. Critical thermal maxima in knockdown-selected *Drosophila*: Are thermal endpoints correlated? *Journal of Experimental Biology* 210:2649–2656.

Folk, D. G., P. Zwollo, D. M. Rand, and G. W. Gilchrist. 2006. Selection on knockdown performance in *Drosophila melanogaster* impacts thermotolerance and heat-shock response differently in females and males. *Journal of Experimental Biology* 209:3964–3973.

Frazier, M. R., H. A. Woods, and J. F. Harrison. 2001. Interactive effects of rearing temperature and oxygen on the development of *Drosophila melanogaster*. *Physiological and Biochemical Zoology* 74:641–650.

Fry, J. D. 2001. Direct and correlated responses to selection for larval ethanol tolerance in *Drosophila melanogaster*. *Journal of Evolutionary Biology* 14:296–309.

Garland, T., Jr. 2003. Selection experiments: An under-utilized tool in biomechanics and organismal biology. Pages 23–56 *in* V. L. Bels, J.-P. Gasc, and A. Casinos, eds. *Vertebrate Biomechanics and Evolution*. Oxford: BIOS Scientific.

Garland, T., Jr., and S. A. Kelly. 2006. Phenotypic plasticity and experimental evolution. *Journal of Experimental Biology* 209:2344–2361.

Gefen, E., A. J. Marlon, and A. G. Gibbs. 2006. Selection for desiccation resistance in adult *Drosophila melanogaster* affects larval development and metabolite accumulation. *Journal of Experimental Biology* 209:3293–3300.

Gibbs, A. G. 2002. Water balance in desert *Drosophila*: lessons from non-charismatic micro-fauna. *Comparative Biochemistry and Physiology A* 133:781–789.

Gibbs, A. G., A. K. Chippindale, and M. R. Rose. 1997. Physiological mechanisms of evolved desiccation resistance in *Drosophila melanogaster*. *Journal of Experimental Biology* 200:1821–1832.

Gibbs, A. G., F. Fukuzato, and L. M. Matzkin. 2003a. Evolution of water conservation mechanisms in *Drosophila*. *Journal of Experimental Biology* 206:1183–1192.

Gibbs, A. G., and T. A. Markow. 2001. Effects of age on water balance in *Drosophila* species. *Physiological and Biochemical Zoology* 74:520–530.

Gibbs, A. G., and L. M. Matzkin. 2001. Evolution of water balance in the genus *Drosophila*. *Journal of Experimental Biology* 204:2331–2338.

Gibbs, A. G., M. C. Perkins, and T. A. Markow. 2003b. No place to hide: Microclimates of Sonoran desert *Drosophila*. *Journal of Thermal Biology* 28:353–362.

Gilchrist, G. W., and R. B. Huey. 1999. The direct response of *Drosophila melanogaster* to selection on knockdown temperature. *Heredity* 83:15–29.

Greenberg, S., and A. Ar. 1996. Effects of chronic hypoxia, normoxia and hyperoxia on larval development in the beetle *Tenebrio molitor*. *Journal of Insect Physiology* 42:991–996.

Griffiths, J. A., M. Schiffer, and A. A. Hoffmann. 2005. Clinal variation and laboratory adaptation in the rainforest species *Drosophila birchii* for stress resistance, wing size, wing shape and development time. *Journal of Evolutionary Biology* 18:213–222.

Harbison, S. T., S. Chang, K. P Kamdar, and T. F. C. Mackay. 2005. Quantitative genomics of starvation stress resistance in *Drosophila*. *Genome Biology* 6:R36.

Harrison, J., M. R., Frazier, J. R., Henry, A. Kaiser, C. J. Klok, and B. Rascon. 2006. Responses of terrestrial insects to hypoxia or hyperoxia. *Respiration Physiology and Neurobiology* 154:4–17.

Harshman, L. G. and A. A. Hoffmann. 2000. Laboratory selection experiments using *Drosophila*: What do they really tell us? *Trends in Ecology & Evolution* 15:32–36.

Harshman, L. G., A. A. Hoffmann, and A. G. Clark. 1999. Selection for starvation resistance: Physiological correlates, enzyme activities and multiple stress responses. *Journal of Evolutionary Biology* 12:370–379.

Harshman, L. G., and J. L. Schmid. 1998. Evolution of starvation resistance in *Drosophila melanogaster*: Aspects of metabolism and counter-impact selection. *Evolution* 52:1679–1685.

Henry, J. R., and J. F. Harrison. 2004. Plastic and evolved responses of larval tracheae and mass to varying atmospheric oxygen content in *Drosophila melanogaster*. *Journal of Experimental Biology* 207:3559–3567.

Hoffmann, A. A., and L. G. Harshman. 1999. Desiccation and starvation resistance in *Drosophila*: Patterns of variation at the species, population and intrapopulation levels. *Heredity* 83:637–643.

Hoffmann, A. A., H. Dagher, M. Hercus, and D. Berrigan. 1997. Comparing different measures of heat resistance in selected lines of *Drosophila melanogaster*. *Journal of Insect Physiology* 43:393–405.

Hoffmann, A. A., R. Hallas, A. R. Anderson, and M. Telonis-Scott. 2005. Evidence for a robust sex-specific trade-off between cold resistance and starvation resistance in *Drosophila melanogaster*. *Journal of Evolutionary Biology* 18:804–810.

Hoffmann, A. A., R. Hallas, C. Sinclair, and P. Mitrovski. 2001. Levels of variation in stress resistance in *Drosophila* among strains, local populations, and geographic regions: Patterns for desiccation, starvation, cold resistance, and associated traits. *Evolution* 55:1621–1630.

Hoffmann, A. A., and P. A. Parsons. 1989. An integrated approach to environmental stress tolerance and life history variation: Desiccation tolerance in *Drosophila*. *Biological Journal of the Linnean Society* 37:117–136.

———. 1993. Direct and correlated responses to selection for desiccation resistance: A comparison of *Drosophila melanogaster* and *D. simulans*. *Journal of Evolutionary Biology* 6:643–657.

Hoffmann, A. A., J. G. Sorensen, and V. Loeschcke. 2003. Adaptation of *Drosophila* to temperature extremes: Bringing together quantitative and molecular approaches. *Journal of Thermal Biology* 28:175–216.

Huey, R. B., L. Partridge, and K. Fowler. 1991. Thermal sensitivity of *Drosophila melanogaster* responds rapidly to laboratory natural election. *Evolution* 45:751–756.

Huey, R. B., J. Suess, H. Hamilton, and G. W. Gilchrist. 2004. Starvation resistance in *Drosophila melanogaster*: Testing for a possible "cannibalism" bias. *Functional Ecology* 18:952–954.

Karan, D., N. Dahiya, A. K. Munjal, P. Gibert, B. Moreteau, R. Parkash, and J. R. David. 1998. Desiccation and starvation tolerance of adult *Drosophila*: Opposite latitudinal clines in natural populations of three different species. *Evolution* 52:825–831.

Karan, D., and R. Parkash. 1998. Desiccation tolerance and starvation resistance exhibit opposite latitudinal clines in Indian geographical populations of *Drosophila kikkawai*. *Ecological Entomology* 23:391–396.

Ketola, T., J. Laakso, V. Kaitala, and S. Airaksinen. 2004. Evolution of Hsp90 expression in *Tetrahymena thermophila* (Protozoa, Ciliata) populations exposed to thermally variable environments. *Evolution* 58:741–748.

Kilias, G., and S. N. Alahiotis. 1985. Indirect thermal selection in *Drosophila melanogaster* and adaptive consequences. *Theoretical and Applied Genetics* 69:645–650.

Krebs, R. A., and K. A. Thompson. 2006. Direct and correlated effects of selection on flight after exposure to thermal stress in *Drosophila melanogaster*. *Genetica* 128:217 225.

Krebs, R. A., and V. Loeschcke. 1994. Costs and benefits of activation of the heat-shock response in *Drosophila melanogaster*. *Functional Ecology* 8:730–737.

Lamb, M. J. 1984. Age-related changes in the rate of water loss and survival time in dry air of active *Drosophila melanogaster*. *Journal of Insect Physiology* 30:967–973.

Lee, C. E. 1999. Rapid and repeated invasions of fresh water by the copepod *Eurytemora affinis*. *Evolution* 53:1423–1434.

Lee, C. E., J. L. Remfert, and Y. M. Chang. 2007. Response to selection and evolvability of invasive populations. *Genetica* 129:179–192.

Lenski, R. E., and A. F. Bennett. 1993. Evolutionary response of *Escherichia coli* to thermal-stress. *American Naturalist* 142:S47–S64.

Leroi, A. M., R. E. Lenski, and A. F. Bennett. 1994. Evolutionary adaptation to temperature. III. Adaptation of *Escherichia coli* to a temporally varying environment. *Evolution* 48:1222–1229.

Loeschcke, V., and R. A. Krebs. 1996. Selection for heat-shock resistance in larval and in adult *Drosophila buzzatii*: Comparing direct and indirect responses. *Evolution* 50:2354–2359.

Loudon, C. 1989. Tracheal hypertrophy in mealworms: Design and plasticity in oxygen supply systems. *Journal of Experimental Biology* 147:217–235.

Marron, M. T., T. A., Markow, K. J. Kain, and A. G. Gibbs. 2003. Effects of starvation and desiccation on energy metabolism in desert and mesic *Drosophila*. *Journal of Insect Physiology* 49:261–270.

Maughan, H., and W. L. Nicholson. 2004. Stochastic processes influence stationary-phase decisions in *Bacillus subtilis*. *Journal of Bacteriology* 186:2212–2214.

McColl, G., A. A. Hoffmann, and S. W. McKechnie. 1996. Response of two heat shock genes to selection for knockdown heat resistance in *Drosophila melanogaster*. *Genetics* 143: 1615–1627.

Mongold, J. A., A. F. Bennett, and R. E. Lenski. 1999. Evolutionary adaptation to temperature. VII. Extension of the upper thermal limit of *Escherichia coli*. *Evolution* 53:386–394.

Nghiem, D., A. G. Gibbs, M. R. Rose, and T. J. Bradley. 2000. Postponed aging and desiccation resistance in *Drosophila melanogaster*. *Experimental Gerontology* 35:957–969.

Parkash, R., and A. K. Munjal. 2000. Evidence of independent climatic selection for desiccation and starvation tolerance in Indian tropical populations of *Drosophila melanogaster*. *Evolutionary Ecology Research* 2:685–699.

Partridge, L., B. Barrie, N. H. Barton, K. Fowler, and V. French. 1995. Rapid laboratory evolution of adult life history traits in *Drosophila melanogaster* in response to temperature. *Evolution* 49:538–544.

Pelosi, L., L. Kuhn, D. Guetta, J. Garin, J. Geiselmann, R. E. Lenski, and D. Schneider. 2006. Parallel changes in global protein profiles during long-term experimental evolution in *Escherichia coli*. *Genetics* 173:1851–1869.

Rajpurohit, S., R. Parkash, and S. Ramniwas. 2008. Body melanization and its adaptive role in thermoregulation and tolerance against desiccating conditions in drosophilids. *Entomological Research* 38:49–60.

Riehle, M. M., A. F. Bennett, R. E. Lenski, and A. D. Long. 2003. Evolutionary changes in heat-inducible gene expression in lines of *Escherichia coli* adapted to high temperature. *Physiological Genomics* 14:47–58.

Riehle, M. M., A. F. Bennett, and A. D. Long. 2001. Genetic architecture of thermal adaptation in *Escherichia coli*. *Proceedings of the National Academy of Sciences of the USA* 98:525–530.

Ring, R. A., and H. V. Danks. 1994. Desiccation and cryoprotection: Overlapping adaptations. *Cryo-Letters* 15:181–190.

Robleto, E. A., C., Ross, R. Yasbin, and M. Pedraza-Reyes. 2007. Stationary phase mutagenesis in *Bacillus subtilis*: A paradigm to study genetic diversity programs in cells under stress. *Critical Reviews in Biochemistry and Molecular Biology* 42:327–339.

Rose, M. R., L. N. Vu, S. U. Park, and J. L. Graves. 1992. Selection on stress resistance increases longevity in *Drosophila melanogaster*. *Experimental Gerontology* 27:241–250.

Sarup, P., J. G. Sorensen, K. Dimitrov, J. S. F. Barker, and V. Loeschcke. 2006. Climatic adaptation of *Drosophila buzzatii* populations in southeast Australia. *Heredity* 96:479–486.

Schmidt-Nielsen, K. 1997. *Animal Physiology: Adaptation and Environment.* 5th ed. Cambridge: Cambridge University Press.

Shikano, T., E. Arai, and Y. Fujio. 1998. Seawater adaptability, osmoregulatory function, and branchial chloride cells in the strain selected for high salinity tolerance of the guppy *Poecilia reticulata*. *Fisheries Science* 64:240–244.

Sorensen, J. G., J. Dahlgaard, and V. Loeschcke. 2001. Genetic variation in thermal tolerance among natural populations of *Drosophila buzzatii*: Down regulation of Hsp70 expression and variation in heat stress resistance traits. *Functional Ecology* 15:289–296.

Sorensen, J. G., P. Michalat, J. Justesen, and V. Loeschke. 1999. Expression of the heat-shock protein HSP70 in *Drosophila buzzatii* lines selected for thermal resistance. *Hereditas.* 131:155–164.

Sorensen, J. G., M. M. Nielsen, and V. Loeschke. 2007. Gene expression profile analysis of *Drosophila melanogaster* selected for resistance to environmental stressors. *Journal of Evolutionary Biology* 20:1824–1838.

Sorensen, J. G., F. M. Norry, A. C. Scannapieco, and V. Loeschcke. 2005. Altitudinal variation for stress resistance traits and thermal adaptation in adult *Drosophila buzzatii* from the New World. *Journal of Evolutionary Biology* 18:829–837.

Sun, J., D. Folk, T.J. Bradley, and J. Tower. 2002. Induced overexpression of mitochondrial Mn-superoxide dismutase extends the life span of adult *Drosophila melanogaster*. *Genetics* 161:661–672.

Telonis-Scott, M., K. M. Guthridge, and A. A. Hoffmann. 2006. A new set of laboratory-selected *Drosophila melanogaster* lines for the analysis of desiccation resistance: Response to selection, physiology and correlated responses. *Journal of Experimental Biology* 209:1837–1847.

Travisano, M., and R. E. Lenski. 1996. Long-term experimental evolution in *Escherichia coli*. IV. Targets of selection and the specificity of adaptation. *Genetics* 143:13–26.

Travisano, M., F. Vasi, and R. E. Lenski. 1995. Long-term experimental evolution in *Escherichia coli*. III. Variation among replicate populations in correlated responses to novel environments. *Evolution* 49:189–200.

Vasi, F. K., and R. E. Lenski. 1999. Ecological strategies and fitness tradeoffs in *Escherichia coli* mutants adapted to prolonged starvation. *Journal of Genetics* 78:43–49.

Vasi, F., M. Travisano, and R. E. Lenski. 1994. Long-term experimental evolution in *Escherichia coli*. II. Changes in life-history traits during adaptation to a seasonal environment. *American Naturalist* 144:432–456.

Velicer, G. V., and R. E. Lenski. 1999. Evolutionary trade-offs under conditions of resource abundance and scarcity: Experiments with bacteria. *Ecology* 80:1168–1179.

Warbrick-Smith, S. T. Behmer, K. P. Lee, D. Raubenheimer, and S. J. Simpson. 2006. Evolving resistance to obesity in an insect. *Proceedings of the National Academy of Sciences of the USA* 103:14045–14049.

Williams, A. E., and T. J. Bradley. 1998. The effect of respiratory pattern on water loss in desiccation-resistant *Drosophila melanogaster*. *Journal of Experimental Biology* 201:2953–2959.

Williams, A. E., M. R. Rose, and T. J. Bradley. 1997. $CO_2$ release patterns in *Drosophila melanogaster*: The effect of selection for desiccation resistance. *Journal of Experimental Biology* 200:615–624

———. 1998. Using laboratory selection for desiccation resistance to examine the relationship between respiratory pattern and water loss in insects. *Journal of Experimental Biology* 201:2945–2952.

―――. 2004. The respiratory pattern in *Drosophila melanogaster* selected for desiccation resistance is not associated with the observed evolution of decreased locomotory activity. *Physiological and Biochemical Zoology* 77:10–17.

Woods, R. D. Schneider, C. L. Winkworth, M. A. Riley, and R. E. Lenski. 2006. Tests of parallel molecular evolution in a long-term experiment with *Escherichia coli. Proceedings of the National Academy of Sciences of the USA* 103:9107–9112.

# 18

# EVOLUTION OF AGING AND LATE LIFE

Casandra L. Rauser, Laurence D. Mueller,
Michael Travisano, and Michael R. Rose

Aging, like all biological characters, evolves. However, unlike many other biological characters, the evolution of aging is not primarily shaped by powerful natural selection balanced against such mitigating factors as clonal interference, linked deleterious alleles, directional mutation, and the like. Instead, aging is evolutionarily distinctive because it arises from the progressive fall in Hamilton's forces of natural selection acting on the survival and reproduction of the somata of ovigerous species (Hamilton 1966; Rose et al. 2007). Therefore, the evolutionary mechanisms underlying aging are different from those underlying most other phenotypes. That is, aging does not have a function or purpose, nor is it adaptive.

Thus, aging must be studied from a different theoretical and experimental perspective than other biological characteristics. From an evolutionary point of view, aging can be defined as a sustained age-specific decline of fitness-related characteristics, such as survival probability and fertility, that is not due to external environmental factors like disease, predation, or climate. Indeed, experimental evolutionary approaches to the problem of aging have mainly focused on manipulating and measuring age-specific survival probability and age-specific fecundity.

Experimental evolution has played an important role in aging research over the last few decades. However, the success of the experimental evolution approach in aging research is in large part due to the availability of a well-developed body of evolutionary theory that concerns aging (see also Mueller this volume). This strong theoretical foundation has provided experimental evolutionary biologists with relatively straightforward predictions to test. Nonetheless, the evolutionary biology of aging would not have progressed as it has over the last few decades without the relatively clear results obtained from some remarkably repetitious, if not dreary, experimental work.

In this chapter, we discuss the evolutionary theories of aging and the experimental studies that have tested and, in a few cases, challenged these theories. We also discuss the postaging phase of life, referred to here as "late life," and the experimental and theoretical work surrounding this phenomenon. The discovery, explanation, and manipulation of late life have served to put the evolution of aging into a strikingly different context. However, we will also discuss a variety of dissenting views, ranging from biomedically motivated research with large-effect mutants to anti-Hamiltonian research by demographers. As we will explain in detail, we regard much of this work as positively misleading with respect to how and why aging evolves.

## EVOLUTIONARY THEORIES OF AGING

Although the evolution of aging was discussed in the nineteenth century by Alfred Russel Wallace and August Weismann, it was not until the advent of theoretical population genetics in the twentieth century that the evolutionary analysis of aging was made relatively coherent (Rose 1991). In early discussions of aging, a few evolutionary biologists erroneously proposed that aging evolved to eliminate older individuals to make way for the young (e.g., Weismann 1891). This is an idea that is still repeated to this day, at least by some authors without training in evolutionary theory. The key mistake common to all these proposals is that they presume that, without aging, there would be an abundance of frail older organisms in natural populations. However, for the vast majority of species in the wild, this is not likely to be the case because disease, starvation, predation, accident, and bad weather normally remove most individuals from wild populations before significant aging has occurred (Medawar 1952), the most important exceptions being species that have benefited from human interventions, including humans themselves, our pets, zoo animals, laboratory model organisms, and species kept in ecological preserves from which their normal predators and other hazards have been wholly or largely removed.

By the middle of the twentieth century, R. A. Fisher (1930), J. B. S. Haldane (1941), P. B. Medawar (1946, 1952), and G. C. Williams (1957) had developed a different kind of evolutionary theory of aging. Their proposals were based on the idea that natural selection should operate with less effectiveness on later fitness components because these characters would be expressed less often, per lifetime, than early fitness components, simply because the probability of survival from birth to any age falls with age. Haldane (1941) first intuited the fall in the force of natural selection acting on mortality when he proposed that the human genetic syndrome of Huntington's disease was common relative to many other genetic diseases, despite its lethality and allelic dominance, because it was not expressed until middle age, after most of its carriers would have reproduced. Verbal hypotheses like these set the stage for the development of the formal evolutionary theory of aging, which was to follow in the last third of the twentieth century.

## HAMILTON'S FORCES OF NATURAL SELECTION

The basic elements of the mathematical population genetics of aging were developed primarily by Hamilton (1966) and Charlesworth (e.g., 1980, 1994). Hamilton (1966) derived the result that the force of natural selection acting on mortality is given by $s(x)/T$, where $x$ is chronological age and $T$ is a measure of generation length. The function $s$ at age $x$ is given by

$$s(x) = \sum_{y=x+1} e^{-ry} l(y)m(y),$$

where $r$ is the Malthusian parameter, or the growth rate of the population, associated with the specified $l(y)$ survivorship and $m(y)$ fecundity functions. The $s(x)$ function represents the relative fitness impact of an individual's future reproduction, after age $x$. Note that, before the first age of reproduction, $s$ is always equal to 1 once reproduction has ended, $s$ is equal to 0 forever after and during the reproductive period, $s(x)$ progressively falls (figure 18.1). This result shows that the *force of natural selection acting on survival falls with adult age*, at least when the assumptions of this analysis are met.

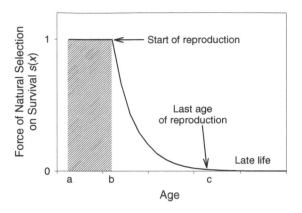

FIGURE 18.1

The age-specific force of natural selection acting on survival, as a percentage of its full force, from Hamilton's (1966) equations. Natural selection is strong at earlier ages (a) until the start of reproduction (b), after which it starts to decline until the last age of reproduction (c) in a population's evolutionary history. Note that the force of natural selection becomes weaker with age until it reaches, and remains at, zero after the last age of reproduction (c). The onset of late life also occurs after this point (c), when the force of natural selection is negligible.

Recently, the generality of Hamilton's results have been challenged (Baudisch 2005, 2008). The function $s(x)$ was derived by Hamilton by implicitly differentiating the Lotka equation

$$\sum_{y=0}^{\infty} e^{-ry} l(y) m(y) = 1$$

and finding the first partial differential

$$\frac{dr}{d \ln p_a},$$

where $p_a$ is the survival from age $a$ to age $a + 1$. A new and different approach to the problem of aging was suggested by Baudisch (2008:22), who proposed that "equally reasonable, alternative forms would have been $dr/dp_a$, $dr/dq_a$, $dr/d\ln q_a$ or $dr/d\ln \bar{u}_a$," where $q_a = 1\text{-}p_a$, and $\ln \bar{u}_a = -\ln p_a$. If by "reasonable" Baudisch means "easy to compute," then her suggestion might be acceptable. However, there is more known about the role of $s(x)$ in the evolution of age-structured populations today than there was when Hamilton first wrote his paper. Explicit population genetic models for survival show that the fate of alleles at a single locus are dependent on the genotypic equivalent of $s(x)$ (Charlesworth 1980:207–208). We do not have equivalent results for the other proposed measures of Baudisch, and therefore, her proposed fitness measures do not have equal standing with Hamilton's original measure. Put another way, in terms of explicit population genetics, they have no well-founded basis, unlike Hamilton's measures, which repeatedly crop up as terms in the equations of theoretical population genetics (e.g., Rose 1985).

A similar analysis for effects on age-specific fecundity gives a comparable "scaling equation," as follows:

$$s'(x) = e^{-rx}l(x).$$

All variables in this equation have the same definitions as those in the equation for the force of natural selection acting on survival. In this case, if population growth is not negative, then the force of natural selection once again declines with age. But if population growth is strongly negative, then the force of natural selection acting on age-specific fecundity may increase with age for a time. This arises when a population is declining rapidly, so that two offspring produced when the population size is six hundred are "worth" less to natural selection than two offspring produced some time later when the population size is sixty. In the latter instance, the offspring are a larger proportion of the population. However, these cases will normally be rare, since populations that are declining rapidly will often become extinct, rendering them unobservable. The typical pattern will remain one in which the force of natural selection acting on fecundity will decline with age and, most important, will converge on zero after the last age of survival in the population's evolutionary history.

The application of Hamilton's (1966) forces of natural selection to the evolution of aging has some tricky features when population genetics are introduced theoretically (for examples, see Charlesworth and Williamson 1975; Charlesworth 1980; Rose 1985). But the effect of the pronounced decline in these functions is so marked that Hamilton's (1966) original verbal interpretation of these functions remains salient to this day (Rose et al. 2007), with one striking exception that we will discuss in detail later.

From these theoretical results, evolutionary biologists have been able to explain the evolution of declining age-specific fitness components among adults, or aging, in very general terms. In particular, the evolutionary theory of aging does not require or predict the action of any particular mechanism of physiological deterioration. Instead, aging is expected to be a pervasive failure of adaptation across most, if not all, of the physiological mechanisms that sustain survival and reproduction among young individuals. For this reason, evolutionary biologists have generally been skeptical of proposals that attribute "the cause of aging" to any one physiological mechanism or gene for aging or programmed death. Although common genetic pathways might be identified that contribute to aging among a variety of organisms (cf. Guarante and Kenyon 2000), a possibility that the Hamiltonian theory of aging does not preclude, the cause of aging in any one species is expected to involve a diversity of physiological mechanisms and genes, with additional diversity of mechanisms among different species (Rose 1991). Leaving aside specific physiological mechanisms, the forces of natural selection acting on mortality and fecundity are quantitatively similar in their effects at later ages, and they thus will shape both age-specific mortality rates and fecundity within populations in a comparable manner. Specifically, age-specific mortality rates should tend to increase with organismal age, while fecundity should decrease with age, whenever Hamilton's forces of natural selection decline.

The Hamiltonian theory of aging predicts that all species that exhibit a well-defined separation of germ line from soma will age, while all species with strictly symmetrical fissile reproduction will be free of aging (Rose 1991; Rose et al. 2007). A significant complication, and a source of some misunderstanding, is that many species exhibit neither a well-defined soma nor fissile reproduction. Still other species combine fissile with sexual reproduction.

Aging is expected to be universal in very large phylogenetic groups that meet the requirements of Hamiltonian theory. The most common cases are to be found in the ovigerous metazoa that lack the sort of vegetative reproduction that is found among coelenterates and many plants. In obligately ovigerous animals, an egg develops into an immature animal and ultimately into a reproductively mature animal. Both insects and vertebrates exhibit this type of life cycle.

Insects and vertebrates are two of the most intensively studied groups of species. Actuarial and physiological aging are hard to detect in most wild populations of these organisms. For example, several fish species appear to attain great age in the wild without signs of physiological aging, with prominent examples being sturgeon and rockfish. However, when cohorts of fish are maintained in the laboratory free of contagious disease until all die, they show demographic aging (see Comfort 1979). For mammals, the evidence for the ubiquity of aging is best, with respect to both the numbers of species studied and the quality of their care. In general, in insects and vertebrates, there are no well-attested laboratory refutations of the Hamiltonian prediction that all these species must age (Comfort 1979; Rose 1991). This is a notable comparative finding, particularly in view of the prediction of some non-Hamiltonian demographers (e.g., Baudisch 2008) that nonaging species can evolve under such conditions. However, as noted previously, these theoretical predictions of nonaging come from models that use indicators of evolution that are different than those that naturally follow from population genetic models.

At the other end of the spectrum with respect to the expectations of the Hamiltonian theory of aging are the symmetrically fissile species. It is important to understand that the concept of fissile here does not mean merely asexual reproduction or budding. Obviously asymmetrical fission in organisms like budding yeast, the asexual protozoan *Tokophyra*, and the bacterium *Caulobacter crescentus* are all known to show observable demographic aging in laboratory cohorts (Rose 1991; Ackermann et al. 2003). What has been surprising recently is that less obviously asymmetrical fissile species, such as *Schizosaccharomyces pombe* and even the humble *Escherichia coli*, also, at least sometimes, show both asymmetrical fission and demographically measurable aging (Barker and Walmsley 1999; Stewart et al. 2005).

The crux of Hamiltonian expectations for the evolution of aging is whether or not asexual reproduction involves a kind of "adult" producing "juvenile" offspring, in which the juvenile is produced relatively intact, while the adult accumulates damage (cf. Stewart et al. 2005; Ackermann et al. 2007). When this occurs in a way that can give rise to differential genetic effects on such "adults" and "juveniles," then aging is expected to evolve in Hamiltonian theory. The cases where aging cannot, according to Hamiltonian theory, evolve are those with strict symmetry between the products of fission. In these cases, aging would extinguish all the descendant lineages, wiping out any such lineage.

## SPECIFIC POPULATION GENETIC HYPOTHESES FOR AGING

Subordinate to basic Hamiltonian theory for aging are alternative population genetic hypotheses for the evolution of aging. These hypotheses are not incompatible with one another; they could be simultaneously valid. At present, there are two population genetic mechanisms of primary interest: antagonistic pleiotropy and mutation accumulation.

### ANTAGONISTIC PLEIOTROPY

Antagonistic pleiotropy arises when alleles that have beneficial effects on one set of fitness components also have deleterious effects on other fitness components, a longstanding concept in evolutionary theory (Rose 1982). Both Medawar (1952) and Williams (1957) verbally argued for the importance of this population genetic mechanism in the evolution of aging during the 1950s. The underlying concept is one of trade-offs, such that alleles with early beneficial effects in some way produce bad side effects later in life. Genes having such actions remain in populations because the forces of natural selection are typically high at earlier ages when these genes have beneficial effects and low at later ages when these same genes have detrimental effects, and thus are not selected against.

Charlesworth (1980) and Rose (1985) analyzed the action of antagonistic pleiotropy mathematically using explicit population genetics and showed that the declining force of natural selection would lead to a tendency for selection to fix alleles that have early beneficial effects but later deleterious effects. It is a noteworthy result of these analyses that Hamilton's forces of natural selection reappear as basic terms in the equations determining the outcome of natural selection when the Malthusian parameter is fitness, contrary to some of the speculations of Baudisch (2005, 2008). In addition, antagonistic pleiotropy may lead to the maintenance of genetic variability for aging and related characters, which is of great experimental significance. It should be noted that antagonistic pleiotropy does not always maintain genetic variability within populations, as it is possible to have genes with antagonistic pleiotropic effects that are fixed in the population but that still contribute to aging.

Other theories similar to the antagonistic pleiotropy theory have independently been proposed to explain the evolution of aging. One such theory is the "disposable soma" theory, which more specifically assumes trade-offs between somatic maintenance and reproduction (Kirkwood 1977; Kirkwood and Holliday 1979). According to this theory, investment in reproduction or somatic maintenance at earlier ages should result in a decrease in an individual's ability to maintain somatic tissue at later ages. This theory is more specific in its form than the antagonistic pleiotropy theory, which allows for gene actions that involve trade-offs between different ages with respect to fecundity alone, among other possibilities (Kirkwood and Rose 1991).

## MUTATION ACCUMULATION

The other cogent population genetic mechanism for the evolution of aging is mutation accumulation. Mutation accumulation arises when the force of natural selection has declined to a point where it has little impact on recurrent deleterious mutations with effects confined to late-life. Medawar (1946, 1952) was the main advocate of the importance of this mechanism in the evolution of aging. Charlesworth (e.g., 1980, 1994, 2001) analyzed the population genetics of mutation accumulation mathematically, showing that the frequency of deleterious mutations can rise with adult age because of the declining force of natural selection at late ages, although Baudisch (2005, 2008) has strongly argued against the importance of mutation accumulation in the evolution of aging. That is, alleles that differ only with respect to detrimental effects expressed only at later ages will tend to remain and accumulate within a population over time because the force of natural selection is too weak to eliminate them. This population genetic mechanism also can maintain genetic variability for aging, like antagonistic pleiotropy.

## LARGE-EFFECT MUTANTS AND THE GENETICS OF AGING

One approach that has become increasingly common in the characterization of the genetics of aging is to isolate aging mutants, usually from mutagenesis experiments, and then to determine the mechanistic basis for the unusual life span in the mutants. This approach has led to the discovery of genes that can enhance (e.g., Maynard Smith 1958; Lin et al. 1988; reviewed in Guarente and Kenyon 2000, Kim 2007) or reduce life span (e.g., Pearl and Parker 1922). Most of the large-effect mutants affecting aging decrease longevity or fecundity, but it has been possible to use fairly ingenious protocols to find mutants that increase longevity, particularly in *Drosophila* (e.g., Lin et al. 1988), nematodes (e.g., Lin et al. 2001), and yeast (e.g., Kennedy and Guarente 1996). The obvious value of this approach has been the direct confirmation of the genetic control of aging.

Unfortunately, it only provides identification of genes that have demonstrable effect as major mutants. As such, it is a relatively indirect approach that requires tremendous effort to determine the evolutionary biology involved (Toivonen et al. 2007; Giannakou et al. 2008). Of greater concern, however, is the difficulty of reproducing the beneficial effects on life span of some of these mutations (e.g., Khazaeli et al. 2005). Among other problems, genotype-by-environment interactions arise for such mutants, such that they survive well under the conditions of specific, sometimes highly artificial, laboratory screening protocols, but not otherwise (Van Voorhies et al. 2006). Within the context of experimental evolutionary research, particularly in its applications to the problem of aging, inbreeding and genotype-by-environment interactions have long been of concern, both in general (e.g., Rose 1991) and in specific hypothesis-testing experiments (e.g., Service and Rose 1985). Studying the genetics of aging using inbred stocks that have not been given sufficient opportunity to adapt to laboratory conditions (cf. Simões et al. 2007, 2008) is expected to generate a variety of artifacts. This makes the failures of replication that have arisen with mutant stocks (e.g., Khazaeli et al. 2005) unsurprising.

This criticism does not necessarily imply that there is never any value in such genetic studies. For one thing, there is little doubt that the long tradition of large-effect "longevity" mutants shows that such mutants predictably, if not universally, entail fitness costs that will be either obvious (e.g., Maynard Smith 1958) or somewhat recondite (Van Voorhies et al. 2006). In particular, it is at least plausible that many of the "longevity mutants" supply interesting worked-out examples of antagonistic pleiotropy, often with much physiological detail.

Unfortunately, it is very difficult to proceed from the pleiotropic effects of alleles that have probably never risen to high frequency in natural populations to the evolution of aging in general. For example, it would be an inappropriate conclusion from such work, though one that is common enough (e.g., Baudisch 2008), to conclude that such genetic research constitutes support for the importance of antagonistic pleiotropy, relative to mutation accumulation, in the evolution of aging. Yet this is not to deny that the loci that are revealed by such mutants may indeed have other, moderately hypomorphic alleles that do indeed play a role in the evolution of aging. They might. But, as we will now show, there are much better methods for studying the evolution of aging than research using "longevity mutants."

## THE ROLE OF EXPERIMENTAL EVOLUTION IN TESTING HAMILTONIAN THEORIES OF AGING
### LABORATORY EVOLUTION OF AGING

Unlike many theories in biology, the Hamiltonian analysis of aging provides fairly obvious strategies for experimentally testing its validity. Two straightforward predictions can be extracted from theoretical work on the evolution of aging: (1) natural selection should accelerate aging in populations with relatively earlier ages of reproduction, and (2) natural selection should slow aging in populations with relatively later ages of reproduction. One of the most elegant experimental approaches in gerontology is the manipulation of the force of natural selection to shape the evolution of aging patterns. This experimental strategy was first proposed by Edney and Gill (1968). The design of these experiments depends on the manipulation of the age of reproduction.

For example, normal laboratory fruit fly culture involves reproduction at fourteen days of age, when cultured at 25°C, with just a few hours for egg laying allowed. This focuses the force of natural selection on that, relatively early, age. Genetic effects expressed at later ages, much after fourteen days, are subject to negligible natural selection. This type of experimental regime allows the evolution of accelerated aging. Sokal (1970) in *Tribolium castaneum*, Mueller (1987) using *Drosophila melanogaster*, and Passananti et al. (2004) also using *D. melanogaster* all found more rapid aging after selection for early fertility.

An alternative experimental regime is to keep adult flies alive for some time before they are allowed to contribute offspring to the next generation. This is done by discarding any eggs that they lay until they have reached the age allowed for reproduction, which can be as late as ten weeks from emergence of the larva. Note that this procedure does not require that the fruit flies be kept virgin; mating is allowed, just not successful reproduction. This regime is expected to lead to the evolution of relatively later aging. Wattiaux (1968) and Rose and Charlesworth (1980, 1981b) found evidence of enhanced later-age fertility and longevity, with depressed early fertility, when looking at *Drosophila* populations selected for later ages of reproduction without replication.

Properly replicated experiments using this second experimental approach were not performed until the 1980s, particularly by Rose (1984) and Luckinbill et al. (1984). Rose (1984) analyzed longevity and fecundity differences between three populations selected for earlier reproduction and three populations selected for increasingly later reproduction. These early and late reproducing populations were derived from the same outbred laboratory population of *D. melanogaster*, but they had been separated and selected for their relative ages of reproduction for more than fifteen generations at the time of their first assays. Significant differences were observed in mean longevity between the early and late reproducers, with the late reproducers having an increased mean longevity (figure 18.2). Earlier fecundity was also found to be higher in the earlier-reproducing populations at earlier ages compared to the later-reproducing populations, which provides support for the predictions made by the antagonistic pleiotropy theory of aging. Luckinbill et al. (1984) found essentially the same results, further demonstrating that selection on age of reproduction can alter longevity in ways consistent with the antagonistic pleiotropy mechanism for the evolution of aging.

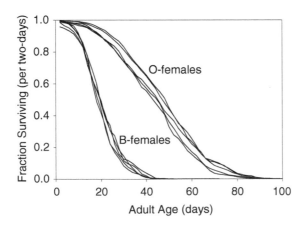

FIGURE 18.2
Fraction of females surviving during adulthood in the B and O populations of Rose (1984). Data are taken from Rose et al. (2002), but newly plotted.

Experiments using one or the other of these basic designs are now routine, often using fruit fly species of the genus *Drosophila*, but sometimes other species are used (see, e.g., Nagai et al. 1995; Reed and Bryant 2000).

One concern that has been raised about this experimental strategy involves the nature of the starting populations (e.g., Linnen et al. 2001). In this respect, the field of experimental evolution as a whole is fortunate that the Matos laboratory (see Simões et al. this volume) has devoted some years to sorting out the pattern and reproducibility of initial laboratory adaptation. Without rehearsing the findings of Matos and colleagues at length here, it is clear that *Drosophila* samples taken from nature, if they are even moderately outbred during laboratory culture, then rapidly adapt to laboratory conditions, with significant changes in life-history characters, particularly early fecundity. Thus, in the case of the much-studied populations of Rose (1984; Rose et al. 2004), it is relevant that these populations were maintained for more than one hundred generations in the laboratory prior to the start of experimental evolution work in which the age of first reproduction was progressively delayed. The findings of Simões et al. (this volume; see also Simões et al. 2007, 2008) show that these populations had probably largely adapted to laboratory conditions by that time. This in turn raises the question as to whether or not populations like this are "representative" of populations in nature. The answer is obvious. They *are not* representative of the populations from which they were derived. It is pellucidly clear that aging and other life-history characters rapidly evolve in laboratory populations of reasonable size, even without deliberate selection, as decades of work in the Rose (Rose et al. 2004) and Matos (Simões et al. this volume) laboratories have illustrated. Instead, such laboratory populations are like particles accelerated to great speeds in the experiments of high-energy physics: particular contrived cases in which evolutionary mechanisms that might act in nature are deliberately exaggerated to extreme levels. This is a general principle that affects all of experimental evolution (see also Futuyma and Bennett this volume; Huey and Rosenzweig this volume); there is nothing special about experimental evolution research on aging in this regard.

The results of changing the ages at which selection is strong have been striking for aging research with *Drosophila*. Mean and maximum life span dramatically increase when the force of natural selection is experimentally strengthened at later ages, as the evolutionary theory predicts. In addition, female fecundity, male virility, flight endurance, locomotion, stress resistance, among a variety of functional characters, are also enhanced (Rose et al. 2004). These results show that aging is not a by-product, accidental or otherwise, of an unmodifiable biochemical process. Rather, aging is an easily modified product of evolution. Rose and colleagues have further gone on to spend some years working out the physiological particulars that underlie this evolutionary transformation in the case of *Drosophila*, much of this work being compiled in Rose et al. (2004), but such research is not the particular concern of this article. (This type of research is considered in more detail in Harshman and Zera this volume.) However, it might at least be argued that such evolutionary physiology research is a preferable alternative to the longevity mutant research, given all the problems for evolutionary interpretation that attend the latter (Van Voorhies et al. 2006).

EXPERIMENTAL POPULATION GENETICS OF AGING

Given the preeminence of the strength of natural selection in determining patterns of aging, the next major experimental question is, What are the population genetic mechanisms that underlie the evolution of aging? As described earlier, there are two main hypotheses: antagonistic pleiotropy and mutation accumulation. The experimental evidence that has been brought to bear on these hypotheses has been of two main kinds: (1) correlation between relatives and (2) indirect responses to selection. More recently, researchers have also investigated the effects of changing extrinsic mortality rates both in the lab and in nature (e.g., Stearns et al. 2000). In addition, there is the issue of the number of genetic loci involved in the evolution of aging. Several different methods have been used to address this last question.

*CORRELATION BETWEEN RELATIVES*

One of the classic techniques in quantitative genetics is the study of the correlations between relatives for characters that are, in part, inherited. Correlated inheritance of two characters indicates the pattern of pleiotropy, with positive correlations indicating that alleles affect two characters in the same direction, and negative correlations indicating that alleles affect two characters in opposite directions, on average. Linkage disequilibrium can, in principle, also generate genetic correlations, but the likelihood that it will systematically generate strong genetic correlations of one sign or another, when many segregating loci affect a quantitative character, is generally discounted. The antagonistic pleiotropy mechanism for the evolution of aging requires that some early and late characters exhibit negative genetic correlations with respect to each other in populations with abundant segregating genetic variation. This pattern has been found in a few cases, notably between early fecundity and longevity in fruit flies (Rose and Charlesworth 1981a). However, a positive correlation has been found in many more (reviewed in Rose et al. 2005, 2007). Artifacts may be responsible for some of these results, particularly inbreeding and novel environment effects, both of which are expected on theoretical grounds to bias genetic correlations toward positive values (Rose 1991) and which have been experimentally demonstrated (e.g., Service and Rose 1985), as already discussed.

The mutation accumulation hypothesis for the evolution of senescence requires that genetic correlations between early and late characters be approximately zero. The problem with this hypothesis is that these genetic correlations tend to be strongly positive among newly occurring mutations, and sometimes negative in populations with established genetic variation. Mutation accumulation also requires that heritable genetic variation increase with age, but this has not been shown to occur in some experiments

(Rose and Charlesworth 1980; Promislow et al. 1996; Shaw et al. 1999), although other experiments have found this effect (Hughes and Charlesworth 1994; Shaw et al. 1999). Shaw et al. (1999) went to considerable trouble to unravel the many difficulties with experiments of this kind, in particular and in general. One basic conclusion that has been offered is that evidence relating age-specific genetic variance to the evolution of aging is arduous to collect and ambiguous in net import (Rose et al. 2007). In particular, since both antagonistic pleiotropy and mutation accumulation can generate changes in age-specific genetic variance, and their joint action would allow almost any pattern of age dependence in the quantitative genetics of age-specific mortality and fecundity, it is doubtful that experiments of this kind are worthwhile from the standpoint of general scientific inference.

The antagonistic pleiotropy and mutation accumulation hypotheses have both received weak and inconsistent experimental support with regard to genetic correlations (Rose et al. 2005, 2007). An important factor to note, however, is that both these population genetic mechanisms may act simultaneously and, in so doing, cancel out the predicted effects of *both* mechanisms on correlations between relatives when they are acting jointly. This is like the difficulties facing experiments that study age dependence of genetic variances. Despite the years of work, some of it our own, that have gone into measuring these quantitative-genetic parameters, it is doubtful that their estimation is the most efficient way to address the population genetics that underlie the evolution of aging.

## INDIRECT RESPONSES TO SELECTION

To some extent, selection experiments are better able to detect the simultaneous action of antagonistic pleiotropy and mutation accumulation. When, for example, postponed aging has evolved as a result of later reproduction in fruit flies or as a result of direct selection on longevity (e.g., Zwaan et al. 1995), it is often inferred that antagonistic pleiotropy causes a reduction in early fecundity (Rose and Charlesworth 1980, 1981b; Rose 1984; Clare and Luckinbill 1985; Luckinbill et al. 1987; Partridge et al. 1999). Sometimes this reduction in early fecundity is coupled with enhanced later fecundity, sometimes with a decrease in subsequent mortality, and sometimes with both enhanced later fecundity and decreased mortality. With antagonistic pleiotropy, it is expected that some early, functional characters, if not fecundity, will be depressed by the laboratory evolution of postponed aging (see Partridge and Fowler 1992). The observation of this pattern in a number of instances indicates that antagonistic pleiotropy is, in some cases, an important genetic mechanism in the evolution of aging.

The effects of mutation accumulation can also be observed in laboratory evolution experiments. For example, when later reproductive opportunities are denied, it is expected that after many generations later fecundity should be reduced, due to the accumulation of alleles with late-acting deleterious effects, while hybrids should show recovery of such reduced fecundity. This result was obtained by Mueller (1987) in fruit flies denied any opportunity for later reproduction for more than one hundred generations. Similarly, Borash et al. (2007) found evidence of greater mutation accumulation at later adult ages in crosses of long-established experimental evolution lines of *Drosophila* tested for male virility. Mutation accumulation evidently can act over hundreds of generations to undermine functional characters, when the force of selection is reduced, making it an important mechanism for the evolution of aging, particularly in smaller populations.

Although it is ultimately difficult to determine the relative contribution that mutation accumulation makes to overall aging compared to antagonistic pleiotropy, we do not agree with Baudisch's (2008) strong dismissal of mutation accumulation. For instance, Baudisch (2008) relies heavily on the results of Pletcher et al. (1998) in formulating her arguments against the importance of mutation accumulation. This study, like that of Yampolsky et al. (2001), relies on lines of *Drosophila* that were allowed to accumulate mutations. The experiments are then designed to investigate the properties of these new mutants. An experimental difficulty with these types of experiments as tests of the population genetic mechanisms underlying aging are the number of new mutants one can reasonable expect to accumulate. In the Yampolsky et al (2001) study, there were originally two hundred parents; and in each generation, two hundred new parents were chosen—thus four hundred chromosomes chosen each generation. They used techniques to make $N_e$ as large as possible and to avoid inbreeding effects. Now if we assume a mutation rate of $10^{-6}$ at loci that affect age-specific survival at later ages, and perhaps one thousand such loci, then over thirty generations the total number of newly arising mutants would be $(400) \times (30) \times (10^{-6}) \times (1,000) = 12$. (Note that these hypothetical assumptions are biased so as to favor the use of this experimental design to test for mutation accumulation.) At best, these new mutants are neutral—given the culture techniques—and therefore, if they arise as a single copy, there is a 25 percent chance of the newly arisen mutant being lost in the next generation. So at best we might be expecting around three new mutants in the course of this entire experiment. For these reasons, we believe it will be difficult to use these types of studies as a means of gaining detailed understanding of the role of mutation accumulation in the evolution of aging.

If mutation accumulation is contributing to aging in existing populations of *Drosophila*, then there ought to be many such deleterious alleles at selection-mutation equilibrium in existing populations. Consequently, experiments designed to uncover these existing deleterious alleles provide direct evidence for the importance of mutation accumulation to aging. Two studies not cited by Baudisch have done exactly this (Mueller 1987; Borash et al. 2007), as already mentioned. These studies show that, over hundreds of generations, existing deleterious late-acting alleles can rise to high frequency once they are made effectively neutral. In addition, other evidence supporting the contribution of mutation accumulation to *Drosophila* aging exist that are also not cited by Baudisch (e.g., Kosuda 1985; Hughes et al. 2002). In conclusion, we feel the existing evidence does in fact support the conclusion that mutation accumulation contributes to aging in *Drosophila* and probably other organisms.

Inherent within the evolutionary theories of aging is the idea that an increase in extrinsic mortality rates could, over many generations, lead to increased intrinsic mortality rates, decreased life span, increased development time, and increased earlier reproduction in a population, depending on patterns of pleiotropy and the presence of sufficient genetic variation. This is because with higher extrinsic mortality rates, the force of natural selection will decline more rapidly. Researchers have investigated the evolutionary effects of changing extrinsic mortality rates both in the lab and in nature. Stearns et al. (2000) created two types of *Drosophila* lines in the lab by applying either high or low extrinsic mortality on the populations, using artificially imposed mortality schedules. Over time, they measured various life-history characters and found that lines that had been subjected to higher extrinsic mortality rates evolved as expected.

Reznick and colleagues have similarly manipulated extrinsic mortality rates in populations of guppies in nature by exposing these populations to different predation levels (Reznick et al. 1990, 1997). They have found that development time and early fecundity patterns have increased with increased extrinsic mortality. However, these predicted changes in early-age life-history characters are not coupled with the expected earlier onset of aging in either mortality or fecundity in the same populations (Reznick et al. 2004, 2006). Ackermann et al. (2007) found a similarly paradoxical result for the evolution of aging in asymmetrically fissile bacteria. The latter case is particularly instructive, in that Ackermann et al. were able to resolve some of the evolutionary mechanisms involved in their overall failure to get the result expected from a simple application of Hamiltonian principles, particularly the complexity of the two-morph life cycle of the bacterium that they studied. In addition, Ackermann et al. did observe one case in which Hamiltonian expectations were met, the invasion of a mutant that increased the rate of aging in one of their bacterial lines that had been reproduced predominantly with younger bacterial "mothers." Such examples illustrate the importance of knowing enough about the biology of the experimental evolution system that you are using, in that a complex set of selection mechanisms may be at work, some of which may artifactually undermine Hamiltonian force of natural selection expectations.

These examples serve to illustrate the general point that the more complex designs of experiments that manipulate the level of imposed mortality rates, unlike the simpler procedure of altering the first age of reproduction in a laboratory population, may in turn make these experiments systematically more difficult to interpret. Futuyma and Bennett (this volume) also discuss the merits of simple experimental manipulations.

## THE NUMBER OF GENES AFFECTING AGING

Early evolutionary discussions of aging, such as those by Williams (1957) and Maynard Smith (1966), characteristically concluded that a large number of loci are likely to affect aging. This gave rise to some pessimism among evolutionary biologists concerning the feasibility of postponing aging. However, the success of laboratory evolution experiments in producing organisms with genetically postponed aging forced a reexamination of earlier assumptions concerning the number of loci affecting aging. Various experimental techniques have been used to answer this question, including segregation analysis (Hutchinson and Rose 1990), 2D protein electrophoresis (Fleming et al. 1993), and gene expression microarrays (Pletcher et al. 2002). Given the ambiguities and limitations of large-effect mutant studies of aging, discussed earlier, those publications do not provide very useful evidence with respect to the question of the number of loci that affect aging. At present, the best answer to the question of the number of genes controlling aging is many (Rose and Long 2002), in keeping with the original expectations of evolutionary biologists.

However, studies of the genetics of the experimental evolution of aging are now amenable to the application of genomic methods. Among the possibilities that are likely to transform research in this area are global comparisons of gene expression microarrays, whole-genome SNP association studies, and even large-scale genomic resequencing between populations that have evolved different rates of aging. With results like these in hand, we should be in a much better position to clarify the extent, and possibly the nature, of the genes involved in the evolution of aging and related characters. What will be of particular interest as this research unfolds is the relationships between the loci revealed by such whole-genome methods and the physiological mechanisms that underlie the evolution of life-history (see Harhsman and Zera this volume). Some hints of such relationships are provided by the tentative findings from protein electrophoretic studies of populations in which physiological mechanisms have been partially uncovered. For example, phosphoglucomutase is an enzyme that plays a key role in energetic catabolism; the Rose populations that live longer have both altered energetic metabolism and corresponding changes in the frequencies of alleles at this locus (Deckeret-Cruz et al. 1997; Rose et al. 2004). However, such research was carried out using very primitive technology compared to what is now available. Research over the next decade or so should reveal many points of interest concerning the physiological genetics that underlie the evolution of aging across entire genomes.

## EVOLUTIONARY BIOLOGY OF LATE LIFE
### THE DISCOVERY OF LATE LIFE

Aging has most often been numerically characterized in terms of the following equation for age-specific mortality rates, based on an idea originally proposed by Benjamin Gompertz early in the nineteenth century:

$$\mu(x) = Ae^{\alpha x}, \tag{1}$$

where $x$ is age, $u(x)$ is the age-specific mortality rate, $A$ an age-independent parameter that gives the baseline mortality rate, and $\alpha$ an age-dependent parameter, or the rate of aging. Although this equation works quite well when fitting mortality data obtained from small cohorts, it has long been known to fail when it came to late-age mortality data from very large human cohorts (e.g., Greenwood and Irwin 1939; Comfort 1964; Gavrilov and Gavrilova 1991), which show a slowing or stabilization of age-specific mortality rates at very late ages. However, this quantitative aging pattern in humans was generally ignored until it was also observed in the 1990s in large cohorts from two dipteran species (Carey et al. 1992; Curtsinger et al. 1992; reviewed in Charlesworth and Partridge 1997). In these studies, late ages were characterized by an apparent cessation of age-related deterioration in age-specific survival probabilities (see figure 18.3). That is, mortality rates increased exponentially during midlife in these populations, as expected in aging organisms, but stopped increasing rapidly and crudely "plateaued" at later ages.

At first, the cessation of aging at late ages seemed paradoxical to most biologists who studied aging, because the widely presumed progressive accumulation of molecular and cellular damage with age was expected to increase until mortality rates reached 100 percent, especially given the Gompertz equation. However, since the definitive discovery of this postaging period of life, or "late life," data from a variety of labs has suggested that late life occurs generally, though not universally, among aging organisms (Fukui et al. 1993; Tatar et al. 1993; Brooks et al. 1994; Kannisto et al. 1994; Charlesworth and Partridge 1997; Vaupel et al. 1998; Carey 2003). Without question, mortality patterns in very old organisms from sufficiently large cohorts do not always follow the Gompertz

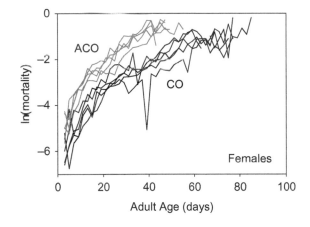

FIGURE 18.3

Two-day mortality rates in females for five replicated populations selected for early reproduction (ACO, gray lines) and five replicated populations selected for later reproduction (CO, black lines). Age-specific mortality rates increase with age until later ages when the rate of increase slows and mortality rates "plateau." Note that the slowing in mortality rate increase occurs at later ages in populations selected for later reproduction (CO) compared to populations selected for early reproduction (ACO). This finding supports the predictions of the evolutionary theory of aging and late life based on the force of natural selection. Data from Rose et al. (2002).

pattern. Late-life mortality rates vary widely among species, but a reliable attribute of late life is the switch from accelerating mortality to a relatively stable age-specific mortality rate, on average.

Mortality levels during the late-life period vary widely. The late-life age-specific mortality rates of some animals, humans being one example, are sometimes very high, relative to the baseline mortality rate, $A$. In other species, such as the medfly, the late-life mortality rate is not as high relative to $A$ (Carey et al. 1992). In such species, it is conceivable that many adult organisms will achieve late life in laboratory cohorts and other protected settings. In any case, regardless of its observability in small cohorts, the distinctive feature of late life is the transition from rapidly accelerating mortality rates to a rough "plateau" of mortality. This is one of the most important findings of aging-related research since 1990, as it indicates that there can be an end to aging among some sufficiently old organisms.

Late life is characterized not only by a deceleration in age-specific mortality rates but, more recently, also by a cessation in the age-specific decline in reproduction (Rauser et al. 2003, 2006b). Together these unexpected observations of mortality and fecundity patterns strongly suggest that late life is as distinct a phase of life history as either development or aging. From an evolutionary standpoint, the onset of late life can be defined as the age at which age-specific fitness components stop deteriorating, on average. Since the first laboratory observations of late life, significant experimental and theoretical attention has been focused on explaining this phenomenon.

## EXPLAINING LATE LIFE WITH HAMILTONIAN THEORY

### PLATEAUS IN THE FORCES OF NATURAL SELECTION

Recall that the Hamiltonian theory for the evolution of aging is based on the a priori analysis of Hamilton (1966), as later refined by Charlesworth (e.g., 1980, 1994). A key feature of this theory is that the force of natural selection acting on survival falls to, and remains at, zero once reproduction has ended (see figure 18.1). In other words, the force of natural selection "plateaus" at zero at very late ages. This suggests that this same theory can explain late-life plateaus in mortality rates, providing only that the decline in age-specific fitness components observed during aging is due to the parallel decline in Hamilton's forces of natural selection, with the eventual plateau in these forces generating the fitness component plateaus of late life. The key to this is that $s$ and $s'$ must both equal zero for all ages after reproduction and survival ceased in the evolutionary history of a population. These plateaus in the forces of natural selection imply that natural selection did not discriminate among genetic effects that act at ages so late that they have no impact on fitness during the evolutionary history of a population.

This intuition was confirmed in explicit numerical simulations of population genetic scenarios for the evolution of age-specific mortality rates, scenarios that included antagonistic pleiotropy, mutation accumulation, or combinations of the two mechanisms

(Mueller and Rose 1996). Charlesworth (2001) supplied analytical solutions of this kind for special cases restricted to mutation accumulation. In both of these studies, it turned out that aging was generated by falling forces of natural selection, with the eventual plateau in these forces leading to a marked deceleration in the deterioration of aging. There have been several critiques of these evolutionary theories of late-life mortality plateaus (Charlesworth and Partridge 1997; Pletcher and Curtsinger 1998; Wachter 1999), and additional theoretical research on these models would no doubt improve them as purely theoretical constructs. Of greater interest here, however, is the success of these theories in tests using experimental evolution.

As with the force of natural selection acting on survival probabilities, Rauser et al. (2003, 2006b) noted that the force of natural selection acting on fecundity, as described earlier, asymptotically declines to zero at late ages. This plateau in the force of natural selection acting on fecundity was also shown to translate into late-life plateaus in fecundity by Rauser et al. (2006b) using computer simulations. The age at which $s'$ declines to zero is dependent on the last age of survival in the environment in which evolution has occurred, rather than the last age of reproduction as with survival. Therefore, age-specific fecundity should stop declining and roughly "plateau" at late ages like the plateau in age-specific mortality rates, because the force of natural selection acting on fecundity is so low at late ages that it cannot distinguish fitness differences in fecundity.

Although it is possible that these late-life plateaus will be at the zero-survival or zero-fecundity levels in some species (cf. Pletcher and Curtsinger 1998), when there are enough alleles that have sufficiently age-independent beneficial effects, it is possible to have positive-valued average survival and average fecundity values (see Charlesworth 2001; Reynolds et al. 2007). Any beneficial effect that is not overly age-dependent will continue to benefit individuals who remain alive after the force of natural selection has converged on zero. And after that age, the intensity of natural selection will no longer depend on chronological age. If there are any age-independent genetic benefits, they will be favored by natural selection acting at early ages, with a pleiotropic echo benefiting later ages.

The expansion of the evolutionary theory of aging to include late life has been significant for the understanding of both aging and late life. Although a more formal mathematical derivation of the late-life theory is necessary, this theory yields testable predictions regarding age-specific fitness components of late life. Now evolutionary biologists must not only be concerned with why aging happens but why, when, and how aging ceases at late ages.

## EXPLAINING LATE LIFE WITH NONEVOLUTIONARY THEORIES

Hamiltonian theory is not the only idea that has been proposed to explain the leveling of mortality rates at late ages. Another set of theories that have attempted to do so are demographic in nature and are based on the hypothesis of lifelong differences in individual robustness within a population. We will call these theories collectively the "lifelong heterogeneity" theories of late life. The common assumption among these theories is that there are significant differences in robustness between individuals that make up a population and that these differences are not age-specific, but are maintained for the entire duration of an individual's life. That is, an individual that is born more robust remains robust until the day it dies, and conversely for those individuals that are less robust. Lifelong heterogeneity can produce rough late-life mortality rate plateaus because the less robust individuals will die off at earlier ages, leaving only the more robust individuals at late ages to define the population's mortality rate pattern.

Vaupel et al. (1979) developed the first lifelong heterogeneity theory that could conceivably explain late-life mortality patterns. Their theoretical analysis assumed that a population was made up of subgroups that were each characterized by unique Gompertz functions. Thus, one subgroup could have a low baseline mortality rate, $A$ in equation (1), compared to other subgroups, but still have the same rate of aging. This would reduce this subgroup's age-specific mortality rate throughout life. Another formulation of this idea allows for variation among subgroups for the aging parameter, $\alpha$, of equation (1) (Pletcher and Curtsinger 2000).

There are no lifelong heterogeneity theories that have been proposed to explain the fecundity plateaus that have been observed at late ages, and that are readily explicable using Hamiltonian theory. However, post hoc explanations could be concocted. For late-life fecundity plateaus, it is conceivable that there are lifelong differences in individual female fecundity, with some females laying a lot of eggs per day and some females laying only a few eggs per day. For example, high-rate egg layers might lay a lot of eggs at early ages resulting in their early death, leaving only the low-rate egg layers at later ages contributing to the population's late-life fecundity plateau. Note that this explanation assumes that there is a trade-off between fecundity and mortality. That is, high-rate egg layers also have high mortality, and vice versa for low-rate egg layers, a natural "reproductive effort" trade-off concept. A number of variations on this same idea are imaginable. However, it should be noted that, to the extent to which such variant life histories are affected by segregating alleles, Hamiltonian theory suggests that there should be selection against slower egg laying in most natural populations, all other things being equal.

## EXPERIMENTALLY TESTING HAMILTONIAN THEORIES OF LATE LIFE

### BASIC PREDICTIONS OF THE HAMILTONIAN THEORY OF LATE LIFE

Because the Hamiltonian late-life theory is an evolutionary theory, late life should evolve in ways predictable from the forces of natural selection, given enough genetic variation and an ample number of generations. Rose et al. (2002) used computer simulations of life-history evolution in which late-life mortality plateaus evolved in response to changes in the timing of reproduction. Recall that the force of natural selection acting on mortality converges on zero sometime after the last age of reproduction. Therefore, Rose et al. (2002) predicted that late-life mortality plateaus should evolve accordingly in the experimental evolution of life history. Their simulation results explicitly demonstrated that the start of mortality rate plateaus should be positively correlated with the last age of reproduction, when Hamilton's forces of natural selection control the evolution of age-specific life-history characters. Similarly, Rauser et al. (2006a) found from numerical simulations that the last age of survival in the evolutionary history of a population should determine the start of late-life plateaus in fecundity. The evolutionary theory of late life based on the force of natural selection thus predicts that late life should evolve according to the timing of the convergence of the forces of natural selection on zero.

Experimental evolution has been used to test both these predictions. Both mortality rate and fecundity patterns in late life have been measured using multiple populations of *D. melanogaster* that have been maintained for many generations with different ages at which the forces of natural selection decline to zero. Thus far, all such experimental studies have strongly corroborated the predictions inferred from evolutionary theory. These studies will now be discussed in some detail.

### MORTALITY

Rose et al. (2002) first tested the Hamiltonian analysis of late life using different types of replicated populations of *D. melanogaster* long selected for different ages of reproduction. They experimentally demonstrated that the start of late-life mortality plateaus evolves according to the last age of reproduction as predicted by the Hamiltonian theory for late life (figure 18.3). This study involved three different independent experimental tests, using a total of twenty-five evolutionary distinct cohorts of fruit flies. Two of these tests compared replicated populations long selected for different ages of reproduction. These tests featured either a fifty-five-day or a twenty-day contrast in last ages of reproduction. Age-specific mortality rates were measured for all populations, and the ages at which mortality rates plateaued were determined and compared between the different groups of populations. They found that late life started later in populations with *later* last ages of reproduction compared to populations that had *earlier* last ages of reproduction.

### FECUNDITY

The Hamiltonian predictions for the evolution of late-life fecundity were experimentally tested by Rauser et al. (2006b)—specifically, whether late-life fecundity evolves as predicted using a comparison analogous to the comparison tests used by Rose et al. (2002) for mortality, except that Rauser et al. measured mid- to late-life fecundity from large cohorts derived from ten populations. Like the study of Rose et al. (2002), Rauser et al. (2006b) found that the start of late-life fecundity plateaus was earlier in the populations that had an earlier last age of survival in their evolutionary history, compared to the populations allowed to survive for longer periods, as predicted by Hamiltonian theory (figure 18.4).

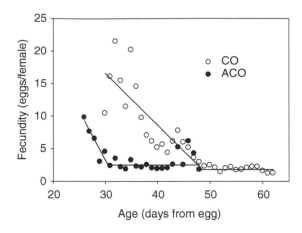

FIGURE 18.4

Mean age-specific fecundity for a later-reproducing CO population (open circles) and an early-reproducing ACO population (solid circles) for mid and late-life ages. This figure represents one of five pairwise comparisons between the CO and ACO populations from Rauser et al. (2006b) and demonstrates, after statistical analyses, that fecundity declines during midlife and stops declining and "plateaus" at late ages. Note that the slowing in the decline in fecundity occurs at later ages in the population selected for later reproduction (CO) compared to the population selected for early reproduction (ACO). This finding was consistent among all five pairwise comparisons and supports the predictions of the evolutionary theory of aging and late life based on the force of natural selection.

## POPULATION GENETIC MECHANISMS OF LATE-LIFE EVOLUTION

Both theoretical and experimental analyses have revealed that late-life patterns of survival and reproduction are greatly affected by plateaus in Hamilton's forces of natural selection. The next question is whether the population genetic mechanisms that underlie the evolution of aging also underlie the evolution of late life.

With antagonistic pleiotropy, genes having a positive effect on early-age fitness components also have a detrimental effect at later ages. Therefore, with antagonistic pleiotropy involving late life, a shift to earlier *last* ages of reproduction and survival should result in a rapid shift in the age at which fitness components stabilize at late ages. Rose et al. (2002) and Rauser et al. (2006b) found just such rapid shifts in the onset of both late-life mortality and fecundity, respectively, in *Drosophila* populations. Rose et al. (2002) used an experimental design that started with replicated populations of fruit flies long selected for late reproduction. They then changed the last age of reproduction to a much earlier age, fifty-six days earlier, for only twenty-four generations, and then they compared the age of mortality plateau onset of the newly derived early reproduced populations to the original late-reproduced populations. The onset of late-life mortality plateaus from populations long having late last ages of reproduction responded quickly to selection for an earlier cessation of reproduction, implicating antagonistic pleiotropy as a genetic mechanism shaping late-life mortality patterns (Rose et al. 2002), since the response was too rapid to be explained by mutation accumulation, given the limits on the quantitative effects of mutation accumulation over a short number of generations, as discussed earlier.

Rauser et al. (2006b) used the same experimental design to test whether antagonistic pleiotropy also works to shape late-life fecundity patterns, except that they started with replicated populations that last reproduced in midlife. Thus, the difference in age of reproduction between the original and newly derived populations with earlier termination of reproduction was only eighteen days. After making mid- and late-life fecundity comparisons and statistically determining the age of onset of late-life plateaus in fecundity, Rauser et al. (2006b) found that late-life fecundity also responds quickly to selection for earlier reproduction, with an earlier last age of survival at which reproduction can occur. Thus, antagonistic pleiotropy is also implicated as a genetic mechanism shaping late-life fecundity patterns.

Rose et al. (2002) tested whether mutation accumulation was contributing to late-life mortality patterns by using five genetically independent populations, long selected for the same age of reproduction, and performing all possible crosses between these populations to create new "hybrid" populations. The age of onset of late-life mortality plateaus was measured in all such hybrid populations and compared to the original populations from which they were derived. However, Rose et al. (2002) did not observe differences between crossed and uncrossed populations in late-life survival probabilities, failing to support the hypothesis that mutation accumulation acts in shaping late-life mortality, unlike the results obtained by Mueller (1987) or Borash et al. (2007) for reproductive characters during aging. Although this experiment from Rose et al. (2002) did not support mutation accumulation as a genetic mechanism in the evolution of late life, it did not refute it. While hybrid superiority supports the involvement of mutation accumulation, failure to detect it does not necessarily show that it isn't involved in the evolution of a life-history character.

Reynolds et al. (2007) used chromosomes extracted from isofemale lines of *D. melanogaster* to test for age specificity in allelic effects on age specificity. Their goal was to determine whether or not such allelic effects were too age-specific for Charlesworth's (2001) explanation for the evolution of late-life plateaus with nonzero values of life-history characters. They found enough width of age specificity to support Charlesworth's hypothesis, in their opinion. However, their experimental analysis suffers from uncertainty about the degree to which chromosomes extracted from isofemale lines are adequately representative of the effects of the genetic variation that predominates among outbred populations, given the artifacts discussed at some length earlier, both inbreeding effects and genotype-by-environment interaction. Nonetheless, their results certainly constitute an example in which a conscientious attempt to critically test Hamiltonian theory for late life failed to refute that theory.

## EXPERIMENTALLY TESTING NONEVOLUTIONARY THEORIES OF LATE LIFE

Although the lifelong heterogeneity theories of late life are heavily emphasized by some scientists (e.g., Vaupel et al. 1979), these theories have not withstood experimental testing. This is unlike the experimental work that has been done to test the alternative Hamiltonian theories of late life.

### MORTALITY

*Inbreeding*  Genetic variation could drive the heterogeneity models. But extensive experimental work has shown that, after removing genetic variation by extensive inbreeding, well-defined plateaus continue to be observed (Curtsinger et al. 1992; Fukui et al. 1993, 1996). In the absence of genetic variation, all variation must be environmental in origin. Detailed studies with *Drosophila* have shown that environmental changes that affect longevity, like dietary restriction, do so by changing the age-independent parameter of the Gompertz equation not the age-dependent parameter (Nusbaum et al. 1995). However, empirical estimates of the levels of variation in the age-independent parameter that would be required to produce plateaus suggest these requirements are far greater than observed levels of variation (Mueller et al. 2003). Consequently, standard lifelong heterogeneity models fail to explain the absence of an effect of inbreeding on the occurrence of late-life mortality plateaus.

*Experimental Manipulation of Environmental Variation*  The experiments establishing the existence of late-life mortality rate plateaus were all carried out under laboratory conditions where environmental variation is purposely minimized, although it is technically impossible to eliminate all environmental variation. Khazaeli et al. (1998) reduced the environmental variation experienced by the preadult, developmental stages of *Drosophila* by collecting subsamples of eggs and pupae that had more similar environmental histories. The mortality of adults from these subsamples was compared to adults that experienced the full range of environments. If the Gompertz demographic parameters are affected by these environmental effects then plateaus should be less prominent or nonexistent in the subsampled populations. There was no difference found in the timing of mortality deceleration, as a result of this heterogeneity-reducing procedure, suggesting that these preadult environments contribute little to the creation of the lifelong heterogeneity in demographic parameters required by anti-Hamiltonian demographers.

*Age-Specific Variance*  Service (2000, 2004) showed that the natural log of age-specific mortality rates should show a unimodal distribution if there is sufficiently large variation in $A$ and $\alpha$ among genetically different populations. We have examined this variance across highly differentiated populations. Despite the fact that these populations had been isolated and had undergone independent evolution for one hundred to five hundred generations, the pattern predicted by Service was not seen (Mueller et al. 2003). Such negative outcomes from experimental tests of lifelong heterogeneity theory do not preclude the possibility that purposeful methods of creating genetic differentiation between populations, like selection or inbreeding, might result in the patterns adduced by Service as a result of lifelong heterogeneity.

*Extreme Age Distribution*   Several studies have examined models that assume there is population variation for $\alpha$ in the Gompertz equation (Pletcher and Curtsinger 2000; Service 2000; Mueller et al. 2003). When Service varied $\alpha$ in order to produce late-life mortality rate plateaus, he generated simulated populations with average longevities of 50 days, which is reasonable for *Drosophila*, but simulated maximum life spans of 365 days, which are unknown for this species. Mueller et al. (2003) followed up on this result by attempting to make parameter estimates of a model with variation in $\alpha$ that was based on observed mortality data from many populations of *Drosophila*. These maximum-likelihood estimates also lead to predictions of more very old individuals than were observed in these populations, if one starts from the assumption that lifelong heterogeneity produces late-life mortality rate plateaus.

The conclusions of Mueller et al. (2003) have been criticized by de Grey (2003). A response to those criticisms have also been published (Mueller and Rose 2004). Without rehashing the details of that response, there are two specific problems with de Grey's (2003) analysis.

The first is that de Grey expresses concerns about the consequences of using the maximum-likelihood techniques and then goes on to develop his own method for estimating model parameters. The scientific onus is on de Grey to show that his technique provides superior estimators of model parameters relative to the maximum-likelihood estimators. This must be done in terms of specific properties of estimators, such as their bias and variance (e.g., see Mueller et al. 1995).

Second, the heterogeneity versions of the Gompertz equation that are at issue are continuous time models with instantaneous mortality functions that we will label $f(t)$. Therefore, in an experiment where observations are made at discrete times, the observed mortality, $\Delta m$, between times $t_1$ and $t_2$ should be equal to $1 - (p(t_2)/p(t_1))$, where $p(t)$ is

$$\exp\left\{ -\int_0^t f(x)dx \right\}.$$

de Grey's parameter estimation scheme equates $\Delta m$ to $f(t)$. Parameters estimated in this fashion have no logical connection to the original model.

*Robust Flies*   Theories that explain late-life mortality deceleration on the basis of a conjectured abundance of lifelong differences in robustness must imply differences in the properties of late life when selection radically improves robustness at early ages. In particular, populations that are much more robust should proceed to late-life mortality stabilization at later ages. Using populations of *Drosophila* successfully selected for much greater starvation resistance and their controls, Drapeau et al. (2000) found no such late-life differences between more and less robust populations. But a reanalysis of these data led Steinsaltz (2005) to different conclusions. However, in a major methodological departure from normal practice in these experiments, Steinsaltz chose to remove from his analysis the mortality data observed in early life. It is hardly surprising that under these conditions the results might differ between the studies of Drapeau et al. (2000) and their reanalysis by Steinsaltz (2005). The process of removing data is always fraught with danger since it is by and large a subjective procedure often guided by a priori expectations that are in fact part of the hypotheses being tested.

*Density Treatments*  Carey has argued (Carey et al. 1995; Carey 2003) that if mortality is increased by increasing the population density, then the age at which a mortality plateau occurs should decline. This follows reasonably enough because, at high density, the less robust groups are being eliminated faster, and thus the age at which only the most robust groups are left, and the age at which pronounced late-life deceleration in mortality occurs, should come sooner. However, in experiments with Mediterranean fruit flies, changing adult density had no detectable effect on the age at which mortality rates leveled off (Carey et al. 1995; Carey 2003). Carey (2003) concludes that "leveling off of mortality is not an artifact of changes in cohort composition."

*Natural Selection and Heterogeneity*  As evolutionary theory predicts late-life plateaus in underlying propensities to die or reproduce, proponents of heterogeneity theories also need to supply a model as to why this type of evolution will not happen. Admittedly, creationists have such a model, but there are some important details left out of creationist theory, such as how creation is supposed to work. A lot of explanatory work is required to justify why evolution does not shape characters like mortality and fecundity, which are integral to the population genetic theory of selection and adaptation.

If environmental variation affects age-specific survival probabilities or age-specific fertility, then it affects fitness. There is significant mathematical theory that shows that under these conditions, natural selection will favor the evolution of biological mechanisms that reduce the environmental variation in components of fitness (Gillespie 1973). Such evolution would again narrow the conditions under which the lifelong heterogeneity theory of late-life plateaus works.

## FECUNDITY

Although the lifelong heterogeneity theories have not been extended to include fecundity, as previously mentioned, there are several post hoc explanations based on the lifelong heterogeneity hypothesis that can be contrived so as to explain the observed late-life fecundity plateau data (Rauser et al. 2003; Rauser et al. 2006b). The scientifically inspiring thing about robustness in fecundity is that it can be measured over the lifetime of an individual, unlike the case of the hypothetical robustness underlying mortality in life-long heterogeneity theories. Thus, the observation of late-life fecundity plateaus provides a platform for more direct tests of the lifelong heterogeneity theories for late life. One way of testing for a connection between individual "robustness" and fecundity is to experimentally measure daily fecundity throughout adult life for all individuals in a *Drosophila* population, and compare the fecundity of those individuals that live to lay eggs in late life with those that do not.

Rauser et al. (2005) undertook the task of measuring the daily fecundity and time of death in three large cohorts of *D. melanogaster* in order to experimentally test lifelong heterogeneity theories of late life. With either of the lifelong heterogeneity theories described earlier for fecundity, early-life fecundity of an individual should indicate the fecundity of that individual in late life: the high-rate egg layers at early ages should either die before late life or be the individuals contributing to late-life fecundity patterns, depending on the lifelong heterogeneity theory you are adhering to. Rauser et al. (2005) could easily test both versions of the lifelong heterogeneity theory for fecundity: the results showed that knowledge of female fecundity in early life does not predict survival to late life. This was yet another refutation of the lifelong heterogeneity theory of late life.

### ARE WE NEEDLESSLY CRUEL?

Were it not for the vast amounts of data required to test lifelong heterogeneity theories for late life properly, it would be an amusing recreation to refute them over and over again. Furthermore, as evolutionary biologists have a perfectly usable, and already corroborated, Hamiltonian alternative, it seems unreasonable for us to continue pummeling this theory. However, we have late-life physiological data forthcoming that will do just that, though we will spare our present readers such gratuitous intellectual bloodshed. For now.

### THE IMPACT OF EXPERIMENTAL EVOLUTION ON AGING RESEARCH

Experimental evolution has had a major impact on aging research over the last few decades. It has corroborated much of the evolutionary theory derived by Hamilton (1966) and Charlesworth (e.g., 1980, 1994) and has plausibly explained why aging occurs. Experimental evolution has also revealed the need for more theoretical work, especially in the newly discovered area of late life.

Although late life was first established as a scientific phenomenon by biologists who were interested in demographic theories appropriate to the "oldest old" (see Vaupel et al. 1998), evidence is accumulating that the phenomenon is not a mere "sampling effect" arising within cohorts. Rather, we contend that late life is an evolutionarily distinct phase of life history, evolving according to strictures very different from those that mold both early life and aging. Late life arises after the strength of natural selection acting on mortality and fecundity have approached zero. From work on both mortality and fecundity in late life, there is at least some pleiotropic connection with early life. When the last age of reproduction is abruptly changed, the timings of the mortality rate and fecundity plateaus shift evolutionarily (Rose et al. 2002; Rauser et al. 2006b). Mortality rate plateaus change by almost a day for each generation of laboratory evolution. This is a remarkable speed of evolution that is only explicable in terms of late-life pleiotropic effects of genetic change arising from strong selection on alleles that have effects early in life.

We have only begun to characterize the evolution of late life. But it is already quite clear that it is as different from aging as aging is from development. These phases of life are connected by pleiotropic gene action, but each phase evolves according to very different rules. Given that late life apparently evolves at least as rapidly as aging itself does, it is apparent that this is a fertile area for the use of experimental evolution, just as aging is.

## SUMMARY

Evolutionary biologists have supplied a formal theory for the evolution of aging, a theory that depends critically on Hamilton's forces of natural selection. This "Hamiltonian" theory has been tested repeatedly using the techniques of both genetics and experimental evolution, particularly using the genus *Drosophila*. Of the two approaches, experimental evolution has given much clearer, readily reproducible, and generally supportive results. However, Hamiltonian evolutionary theory seemed to come under challenge with the discovery of a rough plateauing in age-specific mortality rates during the later ages of very large cohorts of insects. This period during which aging ostensibly slows, or even stops, has been called "late life," particularly in the evolutionary literature. It turned out that, rather than being undermined by the existence of late life, Hamiltonian theory naturally implied the possibility of a late-life deceleration of aging. Furthermore, experimental evolution has corroborated this extension of the original theory, in particular by showing that changing the ages at which Hamilton's forces of natural selection plateau leads to corresponding changes in the ages at which late life starts, as predicted by evolutionary theory. Thus, the case of aging and late life is one of the better illustrations of the value of experimental evolution as a tool for testing and refining evolutionary theories.

## ACKNOWLEDGMENTS

We wish to apologize to any readers who have reached this point in our monstrously long chapter. It was no intention of ours to write a chapter of such length, and indeed our original submission was nothing so elephantine. We would like to give credit where it is due and acknowledge that it was only the diligence of our corresponding editor, T. Garland, that forced us to expatiate so tediously. We are also grateful for the comments and suggestions from one of our referees. Parts of the experimental work discussed in this chapter were supported by a Sigma Xi grant to C.L.R. and a National Science Foundation Doctoral Dissertation Improvement Grant to M.R.R. and C.L.R. C.L.R. was supported by Graduate Assistance in Areas of National Need (GAANN) and American Association of University Women (AAUW) fellowships during the tenure of this work. C.L.R. and M.R.R. would also like to thank the numerous undergraduate students who helped in the collection of much of the mortality and fecundity data from the University of California–Irvine discussed here.

## REFERENCES

Abrams, P. A., and D. Ludwig. 1995. Optimality theory, Gompertz' Law, and the disposable-soma theory of senescence. *Evolution* 49:1055–1066.

Ackermann, M., A. Schauerte, S. Stearns, and U. Jenal. 2007. Experimental evolution of aging in a bacterium. *BMC Evolutionary Biology* 7:126.

Ackermann, M., S. Stearns, and U. Jenal. 2003. Senescence in a bacterium with asymmetric division. *Science* 300:1920.

Barker, M. G., and R. M. Walmsley. 1999. Replicative ageing in the fission yeast *Schizosaccharomyces pombe*. *Yeast* 15:1511–1518.

Baudisch, A. 2005. Hamilton's indicators of the force of natural selection. *Proceedings of the National Academy of Sciences of the USA* 102:8263–8268.

———. 2008. *Inevitable Aging? Contributions to Evolutionary-Demographic Theory*. Berlin: Springer.

Borash, D. J., M. R. Rose, and L. D. Mueller. 2007. Mutation accumulation affects male virility in *Drosophila* selected for later reproduction. *Physiological and Biochemical Zoology* 80:461–72.

Brooks, A., G. J. Lithgow, and T. E. Johnson. 1994. Mortality-rates in a genetically heterogeneous population of *Caenorhabditis elegans*. *Science* 263:668–671.

Carey, J. R. 2003. *Longevity: The Biology and Demography of Life Span*. Princeton, NJ: Princeton University Press.

Carey, J. R., P. Liedo, D. Orozco, and J. W. Vaupel. 1992. Slowing of mortality rates at older ages in large medfly cohorts. *Science* 258:457–461.

Carey, J. R., P. Liedo, and J. W. Vaupel. 1995. Mortality dynamics of density in the Mediterranean fruit fly. *Experimental Gerontology* 30:6–5–29.

Charlesworth, B. 1980. *Evolution in Age-Structured Populations*. Cambridge: Cambridge University Press.

———. 1994. *Evolution in Age-Structured Populations*. 2nd ed. Cambridge: Cambridge University Press.

———. 2001. Patterns of age-specific means and genetic variances of morality rates predicted by the mutation-accumulation theory of ageing. *Journal of Theoretical Biology* 210:47–65.

Charlesworth, B., and L. Partridge. 1997. Ageing: Leveling of the Grim Reaper. *Current Biology* 7:R440–R442.

Charlesworth, B., and J. A. Williamson. 1975. The probability of survival of a mutant gene in an age-structured population and implications for the evolution of life-histories. *Genetical Research* 26: 1–10.

Clare, M. J., and L. S. Luckinbill. 1985. The effect of gene-environment interaction on the expression of longevity. *Heredity* 55:19–29.

Comfort, A. 1964. *Ageing: The Biology of Senescence*. London: Routledge and Kegan Paul.

———. 1979. *The Biology of Senescence*. 3rd ed. Edinburgh: Churchill Livingstone.

Curtsinger, J. W., H. H. Fukui, D. R. Townsend, and J. W. Vaupel. 1992. Demography of genotypes: failure of the limited life span paradigm in *Drosophila melanogaster*. *Science* 258:461–463.

Deckert-Cruz, D. J., R. H. Tyler, J. E. Landmesser, and M.R. Rose. 1997. Allozymic differentiation in response to laboratory demographic selection of *Drosophila melanogaster*. *Evolution* 51: 865–872.

de Grey, A. D. N. J. 2003. Letter to the editor. *Experimental Gerontology* 38:921–923.

Drapeau, M.D., E. K. Gass, M. D. Simison, L. D. Mueller, and M. R. Rose. 2000. Testing the heterogeneity theory of late-life mortality plateaus by using cohorts of *Drosophila melanogaster*. *Experimental Gerontology* 35:71–84.

Edney, E. B., and R. W. Gill. 1968. Evolution of senescence and specific longevity. *Nature* 220:281–282.

Fisher, R. A. 1930. *The Genetical Theory of Natural Selection*. Oxford: Clarendon.

Fleming, J. E., G. S. Spicer, R. C. Garrison, and M. R. Rose. 1993. Two dimensional protein electrophoretic analysis of postponed aging in *Drosophila*. *Genetica* 91:183–198.

Fukui, H. H., L. Ackart, and J. W. Curtsinger. 1996. Deceleration of age-specific mortality rates in chromosomal homozygotes and heterozygotes of *Drosophila melanogaster*. *Experimental Gerontology* 31:517–531.

Fukui, H. H., L. Xiu, and J. W. Curtsinger. 1993. Slowing of age-specific mortality rates in *Drosophila melanogaster*. *Experimental Gerontology* 28:585–599.

Gavrilov, L. A., and N. S. Gavrilova. 1991. *The Biology of Lifespan: A Quantitative Approach*. New York: Harwood Academic.

Giannakou, M., M. Goss, and L. Partridge. 2008. Role of dFOXO in lifespan extension by dietary restriction in *Drosophila melanogaster*: Not required, but its activity modulates the response. *Aging Cell* 7:187–198.

Gillespie, J. H. 1973. Natural selection with varying selection coefficients: A haploid model. *Genetical Research* 21:115–120.

Greenwood, M., and J. O. Irwin. 1939. Biostatistics of senility. *Human Biology* 11:1–23.

Guarente, L., and C. Kenyon. 2000. Genetic pathways that regulate aging in model organisms. *Nature* 408:255–262.

Haldane, J. B. S. 1941. *New Paths in Genetics*. London: Allen and Unwin.

Hamilton, W. D. 1966. The moulding of senescence by natural selection. *Journal of Theoretical Biology* 12:12–45.

Hughes, K. A., and B. Charlesworth. 1994. A genetic analysis of senescence in *Drosophila*. *Nature* 367:64–66.

Hutchinson, E. W., and M. R. Rose. 1990. Quantitative genetic analysis of *Drosophila* stocks with postponed aging. Pages 66–87 *in* D. E. Harrison, ed. *Genetic Effects on Aging II*. Caldwell, NJ: Telford.

Kannisto, V., J. Lauristen, and J. W. Vaupel. 1994. Reduction in mortality at advanced ages: Several decades of evidence from 27 countries. *Population Development Review* 20:793–810.

Kennedy, B. K., and L. Guarente. 1996. Genetic analysis of aging in *Saccharomyces cerevisiae*. *Trends in Genetics* 12:355–359.

Khazaeli, A. A., S. D. Pletcher, and J. W. Curtsinger. 1998. The fractionation experiment: Reducing heterogeneity to investigate age-specific mortality in *Drosophila*. *Mechanics of Ageing and Development* 16:301–317.

Khazaeli, A. A., W. Van Voorhies, and J. W. Curtsinger. 2005. The relationship between life span and adult body size is highly strain-specific in *Drosophila melanogaster*. *Experimental Gerontology* 40:377–85.

Kim, S. K. 2007. Common aging pathways in worms, flies, mice and humans. *Journal of Experimental Biology* 210:1607–1612.

Kirkwood, T. B. L. 1977. Evolution of aging. *Nature* 270:301–304.

Kirkwood, T. B. L., and R. Holliday. 1979. The evolution of ageing and longevity. *Proceedings of the Royal Society of London B, Biological Sciences* 205:531–546.

Kirkwood, T. B. L., and M. R. Rose. 1991. Evolution of senescence: Late survival sacrificed for reproduction. *Philosophical Transactions of the Royal Society of London B, Biological Sciences* 332:15–24.

Kosuda, K. 1985. The aging effect on male mating activity in *Drosophila melanogaster*. *Behavioral Genetics* 15:297–303.

Kowald, A., and T. B. L. Kirkwood. 1993. Explaining fruit fly longevity. *Science* 260:1664–1665.

Lin, K., H. Hsin, N. Libina, and C. Kenyon. 2001. Regulation of the *Caenorhabditis elegans* longevity protein DAF-16 by insulin/IGF-1 and germline signaling. *Nature Genetics* 28:139–145.

Lin, Y. J., L. Seroude, and S. Benzer. 1988. Extended life-span and stress resistance in the *Drosophila* mutant Methusclah. *Science* 282:943–946.

Linnen, C., M. Tatar, and D. Promislow. 2001. Cultural artifacts: A comparison of senescence in natural, laboratory-adapted and artificially selected lines of *Drosophila melanogaster*. *Evolutionary Ecology Research* 3:877–888.

Luckinbill, L. S., R. Arking, M. J. Clare, W. C. Cirocco, and S. A. Buck. 1984. Selection of delayed senescence in *Drosophila melanogaster*. *Evolution* 38:996–1003.

Luckinbill, L. S., M. J. Clare, W. L. Krell, W. C. Cirocco, and P. A. Richards. 1987. Estimating the number of genetic elements that defer senescence in *Drosophila*. *Evolutionary Ecology* 1:37–46.

Maynard Smith, J. 1958. Prolongation of the life of *Drosophila subobscura* by a brief exposure of adults to a high temperature. *Nature* 181:496–97.

———. 1966. Theories of ageing. Pages 1–35 *in* P. L. Krohn, ed. *Topics in the Biology of Aging*. New York: Wiley Interscience.

Medawar, P. B. 1946. Old age and natural death. *Modern Quarterly* 1:30–56.

———. 1952. *An Unsolved Problem of Biology*. London: Lewis.

Mueller, L. D. 1987. Evolution of accelerated senescence in laboratory populations of *Drosophila*. *Proceedings of the National Academy of Sciences of the USA* 84:1974–1977.

Mueller, L. D., M. D. Drapeau, C. S. Adams, C. W. Hammerle, K. M. Doyal, A. J. Jazayeri, T. Ly, S. A. Beguwala, A. R. Mamidi, and M. R. Rose. 2003. Statistical tests of demographic heterogeneity theories. *Experimental Gerontology* 38:373–386.

Mueller, L. D., T. J. Nusbaum and M. R. Rose. 1995. The Gompertz equation as a predictive tool in demography. *Experimental Gerontology* 30:553–569.

Mueller, L. D., C. L. Rauser, and M. R. Rose. 2007. An evolutionary heterogeneity model of late-life fecundity. *Biogerontology* 8:147–161.

Mueller, L. D., and M. R. Rose. 1996. Evolutionary theory predicts late-life mortality plateaus. *Proceedings of the National Academy of Sciences of the USA* 93:15249–15253.

Nagai, J., C. Y. Lin, and M. P. Sabour. 1995. Lines of mice selected for reproductive longevity. *Growth, Development, and Aging* 59:79–91.

Nagylaki, T. 1992. *Introduction to Theoretical Population Genetics.* Berlin: Springer.

Partridge, L., and K. Fowler. 1992. Direct and correlated responses to selection on age at reproduction in *Drosophila melanogaster. Evolution* 46:76–91.

Partridge, L., N. Prowse, and P. Pignatelli. 1999. Another set of responses and correlated responses to selection on age at reproduction in *Drosophila melanogaster. Proceedings of the Royal Society of London B, Biological Sciences* 266:255–261.

Passananti, H. B., D. J. Deckert-Cruz, A. K. Chippindale, B. H. Le, and M. R. Rose. 2004. Reverse evolution of aging in *Drosophila melanogaster.* Pages 296–322 *in* M. R. Rose, H. B. Passananti, and M. Matos, eds. *Methuselah Flies: A Case Study in the Evolution of Aging.* Singapore: World Scientific.

Pearl, R., and S. Parker. 1922. Experimental studies on the duration of life. II. Hereditary differences in duration of life in line-bred strains of drosophila. *American Naturalist* 56:174–187.

Pletcher, S. D., and J. W. Curtsinger. 1998. Mortality plateaus and the evolution of senescence: Why are old-age mortality rates so low? *Evolution* 52:454–464.

———. 2000. The influence of environmentally induced heterogeneity on age-specific genetic variance for mortality rates. *Genetical Research* 75:321–329.

Pletcher, S., D. Houle, and J. W. Curtsinger. 1998. Age-specific properties of spontaneous mutations affecting mortality in *Drosophila melanogaster. Genetics* 148:287–303.

———. 1999. The evolution of age-specific mortality rates in *Drosophila melanogaster*: Genetic divergence among unselected lines. *Genetics* 153:n813–n823.

Pletcher, S. D., S. J. Macdonald, R. Marguerie, U. Certa, S. C. Stearns, D. B. Goldstein, and L. Partridge. 2002. Genome-wide transcript profiles in aging and calorically restricted *Drosophila melanogaster. Current Biology* 12:712–723.

Promislow, D., M. Tatar, A. A. Khazaeli, and J. W. Curtsinger. 1996. Age specific patterns of genetic variance in *Drosophila melanogaster.* I. Mortality. *Genetics* 143:839–848.

Rauser, C. L., J. S. Hong, M. B. Cung, K. M. Pham, L. D. Mueller, and M. R. Rose. 2005. Testing whether male age or high nutrition causes the cessation of reproductive aging in female *Drosophila melanogaster* populations. *Rejuvenation Research* 8:86–95.

Rauser, C. L., L. D. Mueller, and M. R. Rose. 2003. Aging, fertility and immortality. *Experimental Gerontology* 38:27–33.

———. 2006a. The evolution of late life. *Ageing Research Reviews* 5:14–32.

Rauser, C. L., J. J. Tierney, S. M. Gunion, G. M. Covarrubias, L. D. Mueller, and M. R. Rose. 2006b. Evolution of late-life fecundity in *Drosophila melanogaster. Journal of Evolutionary Biology* 19:289–301.

Reed, D., and E. Bryant. 2000. The evolution of senescence under curtailed life span in laboratory populations of *Musca domestica* (the housefly). *Heredity* 85:115–121.

Reynolds, R. M., S. Temiyasathit, M. M. Reedy, E. A. Ruedi, J. M. Drnevich, J. Leips, and K. A. Hughes. 2007. Age specificity of inbreeding load in *Drosophila melanogaster* and implications for the evolution of late-life mortality plateaus. *Genetics* 173:587–595.

Reznick, D. N., H. Bryga, and J. A. Endler. 1990. Experimentally induced life-history evolution in a natural population. *Nature* 346:357–359.

Reznick, D. N., M. Bryant, and D. Holmes. 2006. The evolution of senescence and post-reproductive lifespan in guppies *(Poecilia reticulata)*. *PLoS Biology* 4(e7):1036–1043.

Reznick, D. N., M. J. Bryant, D. Roff, C. K. Ghalambor, and D. E. Ghalambor. 2004. Effect of extrinsic mortality on the evolution of senescence in guppies. *Nature* 431:1095–1099.

Reznick, D. N., F. H. Shaw, and R. G. Shaw. 1997. Evaluation of the rate of evolution in natural populations of guppies *(Poecilia reticulata)*. *Science* 275:1934–1936.

Rose, M. R. 1982. Antagonistic pleiotropy, dominance, and genetic variation. *Heredity* 48:63–78.

———. 1984. Laboratory evolution of postponed senescence in *Drosophila melanogaster*. *Evolution* 38:1004–1010.

———. 1985. Life-history evolution with antagonistic pleiotropy and overlapping generations. *Theoretical Population Biology* 28:342–358.

———. 1991. *Evolutionary Biology of Aging*. New York: Oxford University Press.

Rose, M. and B. Charlesworth. 1980. A test of evolutionary theories of senescence. *Nature* 287:141–142.

———. 1981a. Genetics of life history in *Drosophila melanogaster*. I. Sib analysis of adult females. *Genetics* 97:173–186.

———. 1981b. Genetics of life-history in *Drosophila melanogaster*. II. Exploratory selection experiments. *Genetics* 97:187–196.

Rose, M. R, M. D. Drapeau, P. G. Yazdi, K. H. Shah, D. B. Moise, R. R. Thakar, C. L. Rauser, and L. D. Mueller. 2002. Evolution of late-life mortality in *Drosophila melanogaster*. *Evolution* 56:1982–1991.

Rose, M. R., and A. D. Long. 2002. Ageing: The many-headed monster. *Current Biology* 12:R311–12.

Rose, M. R., H. B. Passananti, A. K. Chippindale, J. P. Phelan, M. Matos, H. Teotónio, and L. D. Mueller. 2005. The effects of evolution are local: Evidence from experimental evolution in *Drosophila*. *Integrative and Comparative Biology* 45:486–491.

Rose, M. R., H. B. Passananti, and M. Matos, eds. 2004. *Methuselah Flies: A Case Study in the Evolution of Aging*. Singapore: World Scientific.

Rose, M. R., C. L. Rauser, G. Benford, M. Matos, and L. D. Mueller. 2007. Hamilton's forces of natural selection after forty years. *Evolution* 61:1265–1276.

Service, P. M. 2000. Heterogeneity in individual morality risk and its importance for evolutionary studies of senescence. *American Naturalist* 156:1–13.

———. 2004. Demographic heterogeneity explains age-specific patterns of genetic variance in mortality rates. *Experimental Gerontology* 39:25–30.

Service, P. M., and M. R. Rose. 1985. Genetic covariation among life history components: The effect of novel environments. *Evolution* 39:943–945.

Shaw, F., D. E. Promislow, M. Tatar, K. A. Hughes, and C. J. Geyer. 1999. Toward reconciling inferences concerning genetic variation in senescence in *Drosophila melanogaster*. *Genetics* 152:553–566.

Simões, P., M. Pascual, J. Santos, M. R. Rose, and M. Matos. 2008. Evolutionary dynamics of molecular markers during local adaptation: a case study in *Drosophila subobscura*. *BMC Evolution and Biology* 8:e66.

Simões, P., M. R. Rose, A. Duarte, R. Gonçalves, and M. Matos. 2007. Evolutionary domestication in *Drosophila subobscura*. *Journal of Evolutionary Biology* 20:758–66.

Sokol, R. R. 1970. Senescence and genetic load: evidence from *Tribolium*. *Science* 167:1733–1734.

Stearns, S. C., M. Ackermann, M. Doebeli, and M. Kaiser. 2000. Experimental evolution of aging, growth, and reproduction in fruit flies. *Proceedings of the National Academy of Sciences of the USA* 97:3309–3313.

Steinsaltz, D. 2005. Re-evaluating a test of the heterogeneity explanation for mortality plateaus. *Experimental Gerontology* 40:101–113.

Stewart, E. J., R. Madden, G. Paul, and F. Taddei. 2005. Aging and death in an organism that reproduces by morphologically symmetric division. *PLoS Biology* 3:e45.

Tatar, M., J. R. Carey, and J. W. Vaupel. 1993. Long-term cost of reproduction with and without accelerated senescence in *Callosobruchus maculates*: Analysis of age-specific mortality. *Evolution* 47:1302–1312.

Toivonen, J., G. Walker, P. Martinez-Diaz, I. Bjedov, Y. Driege, H. Jacobs, D. Gems, and L. Partridge. 2007. No influence of *Indy* on lifespan in *Drosophila* after correction for genetic and cytoplasmic background effects. *PLoS Genetics* 3:e95.

Van Voorhies, W., J. W. Curtsinger, and M. R. Rose. 2006. Do longevity mutants always show trade-offs? *Experimental Gerontology* 41:1055–1058.

Vaupel, J. W., J. R. Carey, K. Christensen, T. E. Johnson, A. I. Yashin, N. V. Holm, I. A. Iachine, V. Kannisto, A. A. Khazaeli, P. Liedo, V. D. Longo, Y. Zeng, K. G. Manton, and J. W. Curtsinger. 1998. Biodemographic trajectories of longevity. *Science* 280:855–860.

Vaupel, J. W., T. E. Johnson, and G. J. Lithgow. 1994. Rates of mortality in populations of *Caenorhabditis elegans*. *Science* 266:826.

Vaupel, J. W., K. Manton, and E. Stallard. 1979. The impact of heterogeneity in individual frailty on the dynamics of mortality. *Demography* 16:439–454.

Wachter, K. W. 1999. Evolutionary demographic models for mortality plateaus. *Proceedings of the National Academy of Sciences of the USA* 96:10544–10547.

Wattiaux, J. M. 1968. Cumulative parental age effects in *Drosophila subobscura*. *Evolution* 22:406–421.

Weismann, A. 1891. *Essays upon Heredity and Kindred Biological Problems.* Vol. 1, 2nd ed. Oxford: Clarendon.

Williams, G. C. 1957. Pleiotropy, natural selection and the evolution of senescence. *Evolution* 11:398–411.

Yampolsky, L. Y., L. E. Pearse, and D. E. L. Promislow. 2002. Age-specific effects of novel mutations in *Drosophila melanogaster* I. Mortality. *Genetica* 110:11–29.

Zwaan, B. J., R. Bijlsma, and R. E. Hoekstra. 1995. Direct selection on lifespan in *Drosophila melanogaster*. *Evolution* 49:649–659.

# 19

# THEORETICAL AND EXPERIMENTAL APPROACHES TO THE EVOLUTION OF ALTRUISM AND THE LEVELS OF SELECTION

Benjamin Kerr

All mankind . . . is one volume; when one man dies, one chapter is not torn out of the book, but translated into a better language.

JOHN DONNE, MEDITATION XVII

One of the central themes in Donne's meditation is the interconnectivity between human beings. Our lives are not stand-alone chapters from an edited volume (like the one you are reading), but more like chapters from an elaborate novel, each setting the stage for chapters to come while simultaneously depending on chapters already read. Biological systems, from subcellular biochemical networks to multispecies food webs, display striking forms of interconnectivity in their parts. How does the theory of biological evolution by natural selection handle this interconnectivity? The simplest description of natural selection starts by *ignoring* interdependence. For simplicity, individuals in a population are assumed to affect neither each other's fitness nor the form of their environment. As an example of the logic, consider the giant anteater. A standard story of natural selection would maintain that anteaters with longer, stickier tongues have been selected because they are able to gather more ants and termites. After all, it is these food resources that improve survival and the production of (long-sticky-tongued) progeny. However, tongue length of one anteater is assumed *not* to affect the fitness of another, and tongue stickiness is assumed *not* to influence the behavior or morphology of the insect prey. In effect, each anteater is treated as "an island, entire of itself" (Donne 1624/1839).

Under the "organism-as-island" incarnation of natural selection, there is an extreme premium placed on personal fitness. An individual with a phenotype that best solves current environmental challenges "cashes in" by earning the highest personal fitness. Within this perspective, what are we to make of the existence of individuals that appear to sacrifice personal fitness to improve the fitness of others? Of course, the answer is that such behavior is fundamentally backward and should be swiftly eradicated by the action of natural selection. Individuals that exhibit restraint and self-sacrifice for the benefit of others are foolishly eroding the precious commodity of personal fitness and should be undone by selfish counterparts. Nonetheless, the biological world is filled with examples of altruistic leanings, from slime mold cells that sacrifice themselves to form the somatic stalks of fruiting bodies (on which reproductive spores sit) to the nonreproductive workers of eusocial insect colonies.

A second reaction to the existence of altruists is that the organism-as-island version of natural selection is misleading when considering such behaviors, as compared with simpler behaviors, such as locomotion in an activity wheel (Rhodes and Kawecki this volume; Swallow et al. this volume). By its very nature, altruism is a social activity. The fitness of an individual depends on the behavior of others. Social organisms do not simply solve the challenges of an external environment—they *are* the selective environment (or at least part of it). The interconnectivity ignored by the organism-as-island model is now front and center. Indeed, how organisms are connected turns out to be *the* critical issue. Most explanations for the evolution of altruism depend on altruists disproportionately finding themselves in the company of other altruists. For instance, organisms may interact with relatives (kin selection), organisms may condition their own altruistic behavior on the cooperative behavior of partners (reciprocal altruism), or organisms may exclude non-cooperating members from their pool of interactors (policing). If altruists are able to associate preferentially with other altruists, then the personal fitness cost can be viewed as the price of admission to a beneficial social milieu.

Sometimes cooperative interactions between the members of a group lead to a type of functionality and cohesiveness at the group level. For instance, eusocial insect colonies have been called "superorganisms" to underline the degree of interconnectivity between insects within the colony (e.g., Emerson 1939). However, does selection "act" at the level of groups in such cases? This question brings us to a subject that has occupied biologists and philosophers of biology for nearly half a century. While the existence of a biological hierarchy is uncontroversial (genes do line up on chromosomes, chromosomes are embedded in cells, cells do make up organisms, and organisms do interact in social groups), the level(s) targeted by natural selection is an extremely controversial subject. Indeed, a large literature has focused on this issue of levels of selection (e.g., Wynne-Edwards 1962; Maynard Smith 1964; Williams 1966; Price 1972; Uyenoyama and Feldman 1980; Sober and Wilson 1998; Michod 1999).

A particularly contentious part of this levels of selection debate concerns whether selection can operate on groups of organisms. Perhaps the best-known proponent of group-level selection was the Scottish ornithologist V. C. Wynne-Edwards. Wynne-Edwards interpreted specific social behaviors of individuals as group-level adaptations. For instance, he viewed animal territoriality as a mechanism of spacing out individuals such that the population would not overexploit critical resources (Wynne-Edwards 1962). The idea was that populations without mechanisms to curb overexploitation would run a higher risk of extinction, and these "short-sighted" populations would be supplanted by other populations that possessed mechanisms of control. Central to Wynne-Edwards's argument was that the process of natural selection could operate on a population of groups.

A significant blow was dealt to Wynne-Edwards's version of group selection with the publication of Williams's classic *Adaptation and Natural Selection* (1966). Williams admitted that group selection was a logical possibility; indeed, he even cited one study on house mice (Lewontin and Dunn 1960) that provided what he called "convincing evidence" of its operation. Furthermore, he claimed that group selection was *required* to explain group-level adaptation. He simply felt that group-level adaptations did not, as a rule, exist. First, Williams argued that the conditions necessary for the operation of group selection were rarely realized (e.g., turnover of groups relative to the turnover of individuals they contained was too slow, migration between groups was too high to maintain intergroup variation, numbers of groups within metapopulations was too low, etc.). Second, Williams argued that in many cases a simpler explanation of the evolutionary origin of a social behavior entailed adaptation at the individual level. For instance, territoriality in animals could evolve because the *individual* territory holder is able to secure more resources *for itself*. Invoking Occam's razor, Williams argued that the simpler explanation was preferable. It is difficult to overstate the impact of

Williams's book on the levels of selection controversy (Wilson 1983; Sober and Wilson 1998). Due in no small part to this publication, group selection became taboo in the mainstream of evolutionary biology, and indeed, this is still the state of affairs in many quarters.

Altruism is a recurring character in the group selection debate. Williams (1966) suggested that testing for group selection should involve "finding adaptations that promote group survival but are clearly neutral or detrimental to individual reproductive survival in within-group competition."[1] Wynne-Edwards (1962) specifically invoked group selection to explain how "short-term advantages of the individual [that] undermine the safety of the race" could be eliminated. The so-called "forces" of individual and group selection appear to be opposed in the evolution of prosocial, self-sacrificial behavior. Specifically, altruists are always at a relative disadvantage within groups; however, groups with more altruists are more productive or long-lived. Some authors maintain that both individual and group selection operate simultaneously in such cases (Wilson 1983; Sober and Wilson 1998). From this multilevel perspective, the evolution of altruism depends on the relative strengths of these opposing, concurrent forces: within-group selection for selfishness and between-group selection for altruism.

Part of Williams's argument against group selection for altruism was that he felt that the "within-group force" was relatively strong. He illustrated this point with an example of how a population of robins that exercised restraint in the use of common resources could be invaded by a "selfish" variant that was less prudent (Williams 1971). Williams concluded that an inexorable force for selfishness within groups would lead to the decrease of altruists globally. Figure 19.1 is a rejoinder to Williams's logic that shows that altruist frequency can globally increase even if the altruist frequency decreases within *every* group! This can occur if there is positive covariance between group output and altruist frequency within the group (i.e., more altruistic groups are more productive—see Sober and Wilson 1998 for a full discussion of this point). The example in figure 19.1 illustrates that the biological details of the system (e.g., how groups come together and how reproductive output depends on group composition) are crucial to determining how likely it will be that unselfish behavior evolves.

There was theoretical interest in exploring the conditions favoring the evolution of altruism before Williams's famous publication (e.g., Wright 1945; Hamilton 1964; Maynard Smith 1964). However, following Williams's critique, a large set of theoretical papers appeared exploring the evolution of altruism (Eshel 1972; Wilson 1975, 1977; Cohen and Eshel 1976; Matessi and Jayakar 1976; Uyenoyama and Feldman 1980; Karlin and Matessi 1983; Matessi and Karlin 1984). Most of these models identify conditions (e.g., the form of population structure) in which altruism can evolve. Because altruism is defined differently in different models, these conditions do not always coincide exactly. Furthermore, while some evolutionary explanations for altruism are pitched as individualistic alternatives to group selection explanations (e.g., kin selection: Hamilton 1963; reciprocal altruism: Axelrod and Hamilton 1981), other authors (and sometimes

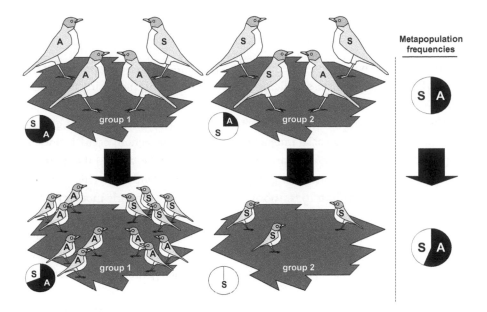

FIGURE 19.1

An illustration of the paradoxical result that altruists can decrease in frequency within groups but increase in frequency globally. Imagine two types of robins, self-restrained altruists and unrestrained selfish types. Suppose that the population of robins is split into two groups, where group 1 has three altruists and one selfish type and group 2 has three selfish types and a single altruist. Thus, the altruist frequency before selection occurs is 75 percent in group 1, 25 percent in group 2, and 50 percent globally (see pie charts). Over the selective episode, altruists have positive effects on their group mates and offspring are produced. We see that selfish types increase in frequency within both groups, but altruists increase in frequency globally (see pie charts). The reason behind this apparent paradox is that the group 1 (which started with more altruists) was more productive than group 2.

the same authors!) have found fundamental similarities between these "alternative" explanations (e.g., between kin selection and group selection: Hamilton 1975; Wade 1980c; Sober and Wilson 1998; Lehmann et al. 2007). Thus, for the casual observer, the precise connections between different theoretical ideas and the specific conditions favoring altruism may seem unclear.

In this chapter, I review some of the theoretical approaches to the study of altruism in an attempt to clarify some basic concepts. While I focus on the models and statistical tools that have appeared in the levels of selection literature, I attempt to make connections between different theoretical approaches in the process. With these theoretical ideas as a backdrop, I then proceed to describe laboratory and field experiments that have addressed the evolution of different forms of altruism, some directly and some indirectly. I end by discussing some philosophical issues in the debate over the levels of selection and the general impact of theoretical and empirical results for this debate.

## THEORETICAL BACKGROUND
### A SIMPLE TRAIT GROUP MODEL

Imagine an infinite population filled with two types of individuals, **A** types and **B** types. The **A** individuals are altruists, sacrificing personal fitness while increasing the fitness of others. The **B** individuals are selfish relative to the **A** types; they do not exhibit self-sacrifice. These individuals undergo the following life cycle: (1) individuals form groups of size $n$, (2) fitness-affecting social interactions (i.e., altruistic action) occur inside these groups and offspring are asexually produced, (3) the adults die, and (4) the groups dissolve. The offspring then form new groups and the cycle continues. This is a simple "trait group" scenario (Wilson 1975, 1980) used in several of the early models exploring the evolution of altruism (Cohen and Eshel 1976; Matessi and Jayakar 1976; Wilson 1977; Uyenoyama and Feldman 1980; Matessi and Karlin 1984). One noteworthy assumption of this model is that *discrete* groups are formed. Thus, the groups form well-defined entities and one question at hand is whether a (meta)population of these entities experiences a selective process.[1]

At generation $t$, the population-wide frequencies of **A** and **B** types (before groups are formed) are given by $\bar{p}(t)$ and $\bar{q}(t)$, respectively. The frequency of groups with $i$ **A** types at generation $t$ is given by $f_i(t)$. Note groups could form randomly, in which case the group frequency distribution would be binomial:

$$f_i(t) = \binom{n}{i} [\bar{p}(t)]^i [\bar{q}(t)]^{n-i} \tag{1}$$

for all $i \in \{0, 1, 2, \ldots n\}$. However, groups could also form nonrandomly (discussed later). Once groups have formed, the fitness of an **A** type in a group with $i$ **A** types is given by $\alpha_i$, while the fitness of a **B** type in a group with $i$ **A** types is given by $\beta_i$.

### DEFINING ALTRUISM

There are different definitions of altruism with interesting connections between them (see Nunney 1985; Wilson 1990; Kerr et al. 2004). In the context of our trait group model, we start with a popular definition of altruism, given by the following relations:

$$\alpha_{i+1} - \beta_i < 0 \tag{2}$$

$$(\alpha_{i+1} - \alpha_i)i + (\beta_{i+1} - \beta_i)(n - i - 1) > 0 \tag{3}$$

for all $i \in \{0, 1, 2, \ldots n-1\}$. We set $\alpha_0 = \beta_n = 0$ for condition (3). To see the origin of conditions (2) and (3), consider a selfish **B** type in a group with $i$ altruists. Suppose that this focal **B** individual switches types (a $B \rightarrow A$ conversion) so that there are $i + 1$ altruists in its group. The change in fitness of our focal individual is $\alpha_{i+1} - \beta_i$ and in order for altruism to be personally costly we require that this change is always negative (condition [2]). Now let's follow up on the change in fitness for the other $n - 1$ individuals ($i$ altruists and $n - i - 1$ selfish types) after the conversion of the focal. Their change in fitness is given by $(\alpha_{i+1} - \alpha_i)i + (\beta_{i+1} - \beta_i)(n - i - 1)$ . In order for altruism to benefit *others*, we require that this change is always positive (condition [3]). Conditions (2) and (3) define "focal-complement altruism" (as costs are measured on a focal converting individual and benefits are measured on the complement of the focal).

A second definition of altruism requires the following conditions:

$$\alpha_i < \beta_i \tag{4}$$

$$\alpha_i i + \beta_i(n - i) < \alpha_{i+1}(i + 1) + \beta_{i+1}(n - i - 1) \tag{5}$$

Condition (4) holds for all $i \in \{1, 2, \ldots n - 1\}$. Condition (5) holds for $i \in \{0, 1, 2, \ldots n - 1\}$, where we set $\alpha_0 = \beta_n = 0$. Condition (4) guarantees that in groups with both types, altruists have a lower fitness. Note that the cost of altruism is being gauged by *within-group* comparisons here. Condition (5) guarantees that the total reproductive output of the group increases with the fraction of altruists inside the group. Note that the benefit of altruism is measured by comparing the output of the *whole group* between groups that differ in altruist frequency. We term conditions (4) and (5) "multilevel altruism" (as costs are measured between individuals within groups and benefits are measured between groups). We note that neither focal-complement altruism nor multilevel altruism entails the other (Kerr et al. 2004).

## STATISTICAL ASSOCIATION AND THE EVOLUTION OF ALTRUISM

We defined focal-complement altruism by focusing on fitness costs to the focal altruist and fitness benefits to the focal individual's $n - 1$ "neighbors" (the complement). Let $X(t)$ and $Y(t)$ be random variables giving the number of $A$ types in the neighborhoods of a randomly chosen $A$ type and $B$ type, respectively, at generation $t$. If groups form randomly (i.e., equation [1] holds), then it can be shown:

$$\Pr\{X(t) = k\} = \Pr\{Y(t) = k\} = \binom{n - 1}{k} [\bar{p}(t)]^k [q(t)]^{n - 1 - k} \tag{6}$$

Thus, $A$ and $B$ individuals experience the same neighborhoods with the same probabilities if groups are randomly formed. This means that for any given generation, the selective environment experienced is, in a sense, constant across types. The altruist and selfish type are getting the same amount of average help from their neighbors; however, the altruist is giving up personal fitness, while the selfish type is not. Therefore, the logic of the organism-as-island version of natural selection would seem to apply. Specifically, it is as if different organisms are exposed to the same external selective environment and some "throw away" personal fitness. We expect those discarding fitness to be weeded out. Indeed, given random group formation, it can be proven (Cohen and Eshel 1976; Matessi and Jayakar 1976; Kerr and Godfrey-Smith 2002a) that condition (2) alone guarantees that altruists will be displaced by selfish types in our simple trait group framework.

For focal-complement altruism, we must have nonrandom formation of groups in order for altruists to stand a chance (this need not be the case for multilevel altruism). Specifically, altruists must tend to associate with other altruists (i.e., the group frequency distribution must be clumped). One way to measure clumping in the group frequency distribution is to ask how the actual variance in frequency of $A$ types deviates from what the variance would have been if groups formed randomly (see Kerr and Godfrey-Smith 2002b). Let the actual variance in the frequency of $A$ across groups at generation $t$ be given by $\sigma_p^2(t)$, and let the variance in frequency of $A$ if groups were to form randomly be $v^2(t) = \overline{p}(t)\overline{q}(t)/n$. Our "clumping index" is

$$\rho(t) = \frac{\sigma_p^2(t) - v^2(t)}{v^2(t)} \tag{7}$$

If groups actually form randomly, then $\sigma_p^2(t) = v^2(t)$, and the clumping index is zero. Large positive values of $\rho(t)$ correspond to frequency distributions with heavy tails, in which $A$ types often encounter other $A$ types and $B$ types encounter other $B$ types in their groups. Large negative values of $\rho(t)$ correspond to frequency distributions with heavy centers, in which many groups contain the same mixture of $A$ and $B$ types. Interestingly, the clumping index can also be written as

$$\rho(t) = E[X(t)] - E[Y(t)] \tag{8}$$

That is, this index also measures how many more altruistic neighbors an $A$ type can expect in its group when compared to a $B$ type.

Here we make use of our clumping index to illustrate the importance of association between altruists and to connect our framework to Hamilton's rule. We assume a simplified scenario in which each individual has a base fitness of $z$ and each altruist provides a fitness benefit $b$ to each of its $n - 1$ neighbors at a fitness cost $c$ to itself. These assumptions give the following linear fitness functions:

$$\alpha_i = z - c + b(i - 1) \tag{9}$$

$$\beta_i = z + bi \tag{10}$$

When $c > 0$ and $b > 0$ (as assumed), then equations (9) and (10) satisfy conditions (2) and (3), and we are thus dealing with focal-complement altruism (incidentally, if $b(n - 1) - c > 0$, then the linear fitness functions also qualify as multilevel altruism). It can be shown (see Wilson 1980; Kerr and Godfrey-Smith 2002b) that altruists will increase in frequency if

$$\frac{b}{c} > \frac{1}{\rho(t)}. \tag{11}$$

Equation (11) states that the ratio of benefits to costs of altruism must be greater than the reciprocal of the degree of clumping. Our simple model makes some of the same assumptions that Hamilton made (e.g., additivity of fitness costs/benefits of altruism), and equation (11) is structurally identical to Hamilton's famous rule for the increase of altruists (Hamilton 1963, 1964):

$$\frac{b}{c} > \frac{1}{r},\tag{12}$$

where $r$ is the coefficient of relatedness between actor and recipient. How is it that $r$ and $\rho(t)$ are playing similar roles? Both of these quantities are measuring statistical association: how likely altruists are to interact with other altruists. The larger these measures of association, the greater the chances for altruistic behavior. Hamilton (1975) himself emphasized that it was association, rather than relatedness per se, that was critical: "It makes no difference if altruists settle with altruists because they are related . . . or because they recognize fellow altruists as such, or settle together because of some pleiotropic effect of the [altruistic] gene on habitat preference . . . correlation between interactants is necessary if altruism is to receive positive selection." More recent incarnations of Hamilton's rule (e.g., Queller 1985; Fletcher and Zwick 2006) emphasize that it is actually the positive association between altruistic genotypes and the helping *phenotypes* of interactants that is of most general relevance (which extends the application of Hamilton's rule to reciprocal altruism and interspecific mutualism—see Fletcher and Zwick 2006; Fletcher et al. 2006). Of course, interactions between relatives (as occurs when individual offspring are deposited in a nest or colony) may be a particularly common way to achieve this association (Nunney 1985).

THE FITNESS STRUCTURE AND CONTEXT FORMATION

The population-wide frequency of the altruistic type $A$ at generation $t + 1$ is described by the following equation:

$$\bar{w}(t)\bar{p}(t + 1) = \sum_{i=1}^{n}\left(\frac{i}{n}\right)(\alpha_i)f_i(t),\tag{13}$$

where $\bar{w}(t)$ is average population-wide fitness of an individual at generation $t$:

$$\bar{w}(t) = \left\{\sum_{i=1}^{n}\left(\frac{i}{n}\right)(\alpha_i)f_i(t)\right\} + \left\{\sum_{i=0}^{n-1}\left(1 - \frac{i}{n}\right)(\beta_i)f_i(t)\right\}.\tag{14}$$

Thus, in order to predict change in the frequency of altruists, we must know at least two things: (1) the fitnesses of types in different social contexts (these are the $\alpha$'s and $\beta$'s) and (2) the way that social contexts come to be (given by the $f_i$'s). We label these two elements the *fitness structure* and *context formation*, respectively. The forms of both of these components will influence the prospects for altruism.

Equations (11) and (12) actually nicely separate terms representing the fitness structure ($b$ and $c$) from terms representing context formation ($\rho(t)$ or $r$). We will see later that some experiments exploring evolution in structured populations can be categorized by whether they manipulate factors affecting the fitness structure or context formation or both.

## THE PRICE EQUATION

For simplicity, let us stick with the trait group model, but we will introduce slightly different notation. For each group of size $n$, let us arbitrarily number the individuals from 1 to $n$. Let $w_{ij}$ be the fitness of the $j$th individual in a group with $i$ $A$ types (note that $w_{ij} = \alpha_i$ or $w_{ij} = \beta_i$). Let $w_{i\bullet}$ be the average fitness of an individual in a group with $i$ $A$ types (note that $w_{i\bullet} = (\alpha_i i + \beta_i(n - i))/n$). Let $\bar{w} = w_{\bullet\bullet}$ be the population-wide average individual fitness (given by equation [14]). Let $p_{ij}$ be "the frequency of $A$ types in the $j$th individual" in a group with $i$ $A$ types (note that $p_{ij} = 1$ if the $j$th individual is an $A$ type and $p_{ij} = 0$ if the $j$th individual is a $B$ type). Let $p_{i\bullet}$ be the frequency of $A$ types in a group with $i$ $A$ types (thus, $p_{i\bullet} = i/n$). Finally, let $\bar{p} = p_{\bullet\bullet}$ be the population-wide frequency of $A$ types, where we note

$$\bar{p}(t) = \sum_{i=1}^{n} \left(\frac{i}{n}\right) f_i(t). \tag{15}$$

Subtracting $\bar{w}(t)\,\bar{p}(t)$ from both sides of equation (13) and rearranging yields

$$\bar{w}\Delta\bar{p} = \text{cov}_{f_i}(p_{i\bullet}, w_{i\bullet}) + E_{f_i}[w_{i\bullet}\Delta p_{i\bullet}], \tag{16}$$

where we have dropped time arguments from the quantities (each quantity refers to generation $t$ except $\Delta\bar{p} = \bar{p}(t + 1) - \bar{p}(t)$ and $\Delta p_{i\bullet} = p_{i\bullet}(t + 1) - p_{i\bullet}(t)$), and the subscripts on cov and $E$ indicate that these are weighted by the group frequency distribution. Although not strictly appropriate, we leave the subscripts on our variables in equation (16) and hereafter for clarity. Equation (16) is a manifestation of the famous Price equation (Price 1970). The covariance term is often taken to represent the effect of "between-group" selection (Price 1972; Hamilton 1975; Sober and Wilson 1998) measuring how group output co-varies with group composition. The expectation term is often taken to represent "within-group" (or individual-level) selection, measuring the (weighted) expected change in altruist frequency within groups over the selective episode.

If there is simple asexual reproduction without mutation, equation (16) can be rewritten as

$$\bar{w}\Delta\bar{p} = \kappa_{w_{i\bullet}p_{i\bullet}}\text{var}_{f_i}(p_{i\bullet}) + E_{f_i}[\kappa_{w_{ij}p_{ij}}\text{var}_j(p_{ij})], \tag{17}$$

where $\kappa_{w_{i\bullet}p_{i\bullet}}$ is the regression coefficient of average fitness of a group on altruist frequency within a group, and $\kappa_{w_{ij}p_{ij}}$ is the regression coefficient of individual fitness on individual type. Thus, the "group-level" term is nonzero only if there is variance in group composition and if there is a nonzero relationship between group output and its composition. And the "individual-level" term is nonzero only if there are both types in some of the groups and if there is a nonzero relationship between individual fitness and type within these same groups.

For instance, using the linear fitness functions given by equations (9) and (10), we find

$$\kappa_{w_{i\bullet}p_{i\bullet}} = b(n - 1) - c \tag{18}$$

$$\kappa_{w_{ij}p_{ij}} = -(b + c) \tag{19}$$

Since $b > 0$ and $c > 0$, then within any mixed group altruists are selected against (i.e., "individual-level" selection is said to work against altruists). However, if the total benefit provided by a single altruist to all its neighbors, $b(n-1)$, outweighs the fitness cost of altruism, $c$, then more altruistic groups are more productive (i.e., "group-level" selection is said to work for altruists; see Wade 1980c). Whether altruists increase in frequency will be decided by the relative "strengths" of between-group selection for altruists (given by $\text{cov}_{f_i}(p_{i\bullet}, w_{i\bullet}) = \kappa_{w_i\bullet p_{i\bullet}}\text{var}_{f_i}(p_{i\bullet})$) versus within-group selection against altruists (given by $E_{f_i}[w_{i\bullet}\Delta p_{i\bullet}] = E_{f_i}[\kappa_{w_{ij}p_{ij}}\text{var}_j(p_{ij})]$). For multilevel altruism (but not focal-complement altruism), Price's first term is guaranteed to be nonnegative, while Price's second term is guaranteed to be nonpositive. By defining altruism with conditions (4) and (5), the "between-group" term will never work against altruists, and the "within-group" term will never work for altruists.

Price's equation also illustrates the role for association in the evolution of altruism. As $A$ types preferentially associate with other $A$ types, the variance in $A$ frequency across groups, $\text{var}_{f_i}(p_{i\bullet})$, increases. Under certain conditions (e.g., multilevel altruism), this works for the evolution of altruism through the first term in Price's equation—namely, $\kappa_{w_i\bullet p_{i\bullet}}\text{var}_{f_i}(p_{i\bullet})$. Note that $\sigma_p^2 = \text{var}_{f_i}(p_{i\bullet})$, and we have already seen that a larger variance in altruist frequency across groups can improve the chances of the evolution of altruism with linear fitness functions (see equations [7] and [11]).

## CONTEXTUAL ANALYSIS

Consider the following scenario (presented in alternative forms in Sober 1984; Nunney 1985; Heisler and Damuth 1987; Okasha 2006): groups of size $n$ form, but there are no meaningful interactions between individuals. Thus, there is no altruism present, and $A$ and $B$ are seen as simply two different types of asocial individuals. Suppose that $A$ always has two offspring and $B$ always has a single offspring (i.e., $\alpha_i = 2$ and $\beta_i = 1$ for all relevant $i$). Then if there is variation in the frequency of $A$ across groups, then Price's first term, $\text{cov}_{f_i}(p_{i\bullet}, w_{i\bullet})$, will be positive. This is because groups that happen to have more $A$ types are more productive. This is somewhat disheartening because Price's first term was supposed to capture group-level selection; yet here, where a description of pure individual selection seems apposite, the group-level term is nonzero. This appears to be a failure of the standard interpretation of Price's terms.

While Price's equation seems ill equipped to characterize this scenario, another statistical approach is ideal at handling it. Contextual analysis (Heisler and Damuth 1987; Goodnight et al. 1992; Okasha 2006) starts with the following linear regression model:

$$w_{ij} = \theta + \varphi p_{ij} + \Phi p_{i\bullet} + \varepsilon_{ij}. \tag{20}$$

Specifically, $\theta$ is the "base" fitness, $\varphi$ is the partial regression coefficient giving the effect of individual type, $p_{ij}$, on individual fitness, $w_{ij}$, (controlling for group composition), and

$\Phi$ is the partial regression coefficient giving the effect of group composition, $p_{i\bullet}$, on individual fitness (controlling for individual type). The term $\varepsilon_{ij}$ is the residual.

It can be shown that

$$\bar{w}\Delta\bar{p} = \Phi\,\mathrm{var}_{f_i}(p_{i\bullet}) + \varphi\,\mathrm{var}_{(f_{i,j})}(p_{ij}),\tag{21}$$

where the variance in individual types is given by

$$\mathrm{var}_{(f_{i,j})}(p_{ij}) = \sum_{i=0}^{n}\frac{f_i(t)}{n}\sum_{j=1}^{n}(p_{ij}-\bar{p})^2,\tag{22}$$

and the variance across groups in $A$'s frequency within groups, $\mathrm{var}_{f_i}(p_{i\bullet})$, is identical to its previous usage (see equation [17]). In equation (21), the first term is taken to give the effects of group composition on evolution (the "group-level" term), and the second term is taken to give the effects of individual type on evolution (the "individual-level" term). Again, for a nonzero effect in either term, both the partial regression coefficient and variance must be nonzero. For example, a group-level effect requires that group composition affects individual fitness (after controlling for individual type), $\Phi \neq 0$, and there must be variance in group composition, $\mathrm{var}_{f_i}(p_{i\bullet}) \neq 0$.

Revisiting our problematic case of constant $A$ and $B$ fitness, we see that the partial regression coefficient for group composition is zero. Given that $\Phi = 0$ there is no possibility for group-level effects. Thus, contextual analysis has yielded a result consistent with intuition about this case: evolution is wholly accounted for through individual-level effects.

Now let's return to the case of altruism and, in particular, to the linear fitness functions (9) and (10). Here, we can show that $\varphi = -(b + c)$ and $\Phi = bn$, and with $\theta = z$ and assuming $\varepsilon_{ij} = 0$, we have

$$w_{ij} = z - (b + c)p_{ij} + bnp_{i\bullet}.\tag{23}$$

As a quick check, by setting $p_{ij} = 1$ and $p_{i\bullet} = i/n$ in equation (23), we recover $\alpha_i$ from equation (9). Similarly, by setting $p_{ij} = 0$ and $p_{i\bullet} = i/n$ in equation (23), we recover $\beta_i$ from equation (10). Now, for the linear fitness functions, Price's terms have the following forms:

$$\mathrm{cov}_{f_i}(p_{i\bullet},w_{i\bullet}) = (\Phi + \varphi)\mathrm{var}_{f_i}(p_{i\bullet})\tag{24}$$

$$E_{f_i}[w_{i\bullet}\Delta p_{i\bullet}] = \varphi(\mathrm{var}_{(f_{i,j})}(p_{ij}) - \mathrm{var}_{f_i}(p_{i\bullet}))\tag{25}$$

Comparing equations (24) and (25) to equation (21) shows that both the group-level effects and individual-level effects are more extreme under contextual analysis than Price's equation for altruism with linear fitness functions. The main point is that contextual analysis and Price's equation give two different statistical perspectives on evolution.

We started this section with an example that was handled well by contextual analysis, but inappropriately by Price's equation. Another example illustrates the converse. Imagine a case of soft selection in which the average fitness of individuals within any group was constant ($w_{i\bullet} = w$). However, assume that individual fitness does vary across types within groups ($\alpha_i \neq \beta_i$). In this case, intuition suggests that terms measuring "between-group" selection (or group-level effects) should be zero (as groups do not vary in output). While Price's first term is indeed zero, the first term from the contextual analysis equation (21) may not be zero (because it is possible that $\Phi \neq 0$). Thus, each statistical equation seems to handle certain cases better. As we discuss experiments in the following section, we will refer back to these statistical approaches.

## EXPERIMENTAL RESEARCH

How do the above theoretical predictions fare for actual biological systems? There are several different empirical ways to explore the evolution of altruism (see also Huey and Rosenzweig this volume; Futuyma and Bennett this volume). One could approach the study of altruism from a phylogenetic perspective, looking whether certain traits predicted to favor altruism are likely to have been present before the radiation of a clade exhibiting altruism. For example, this approach pointed to the existence of high levels of inbreeding (promoting high coefficients of relatedness) *prior* to the origin of a soldier caste in eusocial gall thrip species (McLeish et al. 2006). Another approach to studying altruism is to check whether certain groups exhibiting altruism satisfy one of the theoretical criteria for the evolution of altruism. This has been an active approach within the kin selection literature, where much effort has been invested in computing coefficients of relatedness (all else being equal, higher relatedness between interactants is predicted to work for the evolution of altruism). As recent examples, high coefficients of relatedness have been found between wild turkey males that exhibit cooperative breeding (Krakauer 2005), within multiple-queen ant colonies (Bargum and Sundstrom 2007), and within groups of slime mold cells that form fruiting bodies with reproductive division of labor (Gilbert et al. 2007). A third approach involves experimental manipulation of factors predicted to affect the payoffs or evolutionary success of altruism (and the monitoring of different types). In some cases, this can be done over multiple generations, and the evolutionary loss or *de novo* gain of altruism can be monitored in real time. In this section, I focus on this third approach. A growing collection of artificial selection experiments, laboratory experiments, and field experiments have shed much light on the evolutionary circumstances favoring altruism and have provided food for thought within the levels of selection debate.

## ARTIFICIAL SELECTION EXPERIMENTS

It is no accident that the first chapter of *The Origin of Species* focuses on variation in plants and animals cultivated by humans (Darwin 1859). Darwin discussed morphologically divergent breeds of domesticated organisms coming from common ancestral stocks in order to convince the reader that *selection* (in this case, artificial) was a powerful agent of change. In the same vein, artificial selection for various properties of *groups* of organisms (rather than properties of individuals) has been experimentally explored as a way of illustrating the potential power of group-level selection.

The first group selection experiment was initiated by Wade (1977), using the red flour beetle, *Tribolium castaneum*. The group property selected in this experiment was population size. Obviously, in order to apply group selection, *multiple* groups (that vary in the selected property) must be maintained. For each of his experimental treatments, Wade propagated forty-eight populations of beetles over several population growth cycles. Every population was initiated with sixteen individual beetles. Different experimental treatments were defined by the nature of selection applied at the end of each population growth cycle (thirty-seven days). I will discuss three of his four treatments here.

In one treatment (Group Selection for Productivity), at the end of each cycle, the largest population was divided up into as many groups of sixteen individuals as possible, and these groups were used to initiate the next generation of populations. Thus, these "packets" of sixteen beetles served as "propagules" that seeded the next set of populations. If there were not enough propagules from the largest population to seed all forty-eight new populations (i.e., if $L/16 < 48$, where $L$ is the size of the largest population), then the second-most productive population was also used to supply propagules; and if there were still not enough, the third-most productive population was used; and so on. This same selective scheme was applied over several population growth cycles. A simplified version of this treatment is illustrated in figure 19.2a.

In a second treatment (Group Selection against Productivity), at the end of each cycle, the *smallest* population was divided up into as many propagules as possible. Then the second smallest population was divided into propagules, followed by the third smallest, and so on, until forty-eight propagules were obtained (enough to seed the entire set of future populations). Note that because smaller populations were selected in this treatment, the number of "parent" populations producing propagules was greater in this treatment than in the first treatment.

Finally, in a control treatment (Individual Selection), Wade allowed each population at the end of the population growth cycle to produce a single propagule. That is, each population in the next generation received all of its initial beetles from only a single "parent" population. This is a soft selection scheme, where different groups have the same *realized* productivity (despite their actual productivity at the end of the cycle). The reason this control treatment is termed "individual selection" can be illustrated with the Price equation described earlier. Given that Wade experimentally equalized group output across all populations, Price's first term ($\text{cov}_{f_i}(p_{i\bullet}, w_{i\bullet})$) is zero. If this first term is taken to represent group-level selection (notwithstanding the issues raised earlier), then any evolutionary change is attributed to Price's second term (representing selection within groups between individuals). A simplified version of this control treatment is shown in figure 19.2b.

Interestingly, over only a handful of population growth cycles (and thus selective episodes), Wade recorded large differences between his treatments (figure 19.2c). In the Group Selection for Productivity treatment, average population size (at the end of a growth cycle) remained high, while in the Group Selection against Productivity and Individual Selection treatments, average population size decreased. After nine rounds of selection, average population size differed by 158 beetles between the Group Selection for Productivity and Group Selection against Productivity treatments. Thus, selecting groups to provide propagules on the basis of their population size affects the population size in the future groups.

In the Individual Selection treatment, average population size did dramatically decrease (by over 150 beetles). One factor later determined to contribute to this decrease was an increase in the rate of cannibalism of pupae by adults in the Individual Selection treatment, whereas adult beetles in the Group Selection for Productivity treatment displayed a slightly lower rate of cannibalism (Wade 1979).[2] Thus, the direction of group

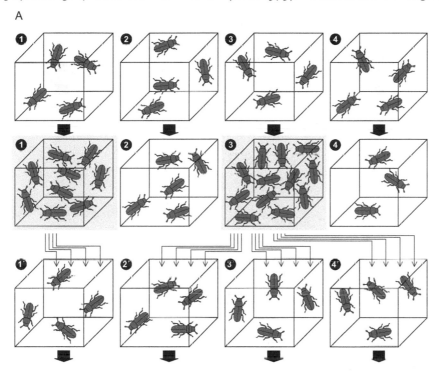

A

FIGURE 19.2

Wade's classic artificial group selection experiment. All subpopulations start with the same number of flour beetles and these subpopulations are incubated for a population growth cycle of thirty-seven days (the thick arrows in parts A and B). A, This schematic shows the Group Selection for Productivity treatment. The subpopulation producing the most individuals over a growth cycle is selected to "seed" the next set of subpopulations (in this case, it is subpopulation 3). Then the second-most productive subpopulation is used, and so on, until the next set of subpopulations have been filled and the next population growth cycle is initiated.

B

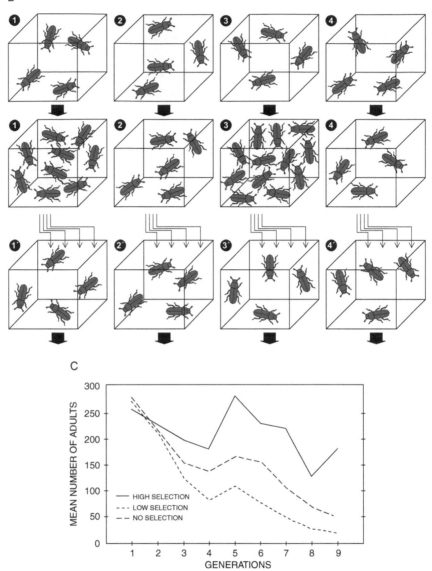

C

FIGURE 19.2 (*continued*)

B, This schematic shows the Individual Selection treatment. In this case, every group "seeds" one new group. Thus, there are no realized productivity differences between groups (despite differences that develop over the population growth cycle). C, The evolution of productivity in different selection treatments. High Selection is the Group Selection for Productivity treatment, Low Selection is the Group Selection against Productivity treatment; and No Selection is the Individual Selection treatment.

and individual selection in these cases appear to be in opposition. More specifically, when competition within groups occurs without realized productivity differences between groups (i.e., soft selection), some forms of cannibalism strengthen; however, when groups with higher actual productivity contribute more propagules to the next generation, some forms of cannibalism weaken.

An individual beetle that exercises restraint in its cannibalistic tendencies is an altruist (in this case, by the "multilevel" definition of conditions [4] and [5]). Here we analyze the evolution of this putative altruist under the Individual Selection and Group Selection for Productivity treatments using the Price equation. In each case, we measure evolutionary change starting from the set of populations before their growth cycle through their population growth, and ending with the next set of populations after a selective treatment is applied (the period illustrated from the top to the bottom of figures 19.2a and 19.2b). As mentioned, in the Individual Selection treatment, Price's first term, $\text{cov}_{f_i}(p_{i\bullet}, w_{i\bullet})$, measuring *realized* productivity differences between groups, will be zero due to the nature of selection. This means that the only term contributing to evolutionary change is Price's second term, which will be negative for multilevel altruism. Thus, the frequency of restrained cannibals (the altruists) is expected to decrease, leaving populations in the Individual Selection treatment more cannibalistic.

In the Group Selection for Productivity treatment, populations that are productive over the population growth cycle disproportionately contribute to the next set of populations. If cannibalistic restraint is multilevel altruism, then actual group productivity will increase with the fraction of restrained beetles. In this case, the regression of average group output on frequency, $\kappa_{w_{i\bullet}p_{i\bullet}}$, is positive and Price's first term can now work for the altruists. In this case, the group selection (which is artificially applied) is expected to lead to a relative increase in the frequency of restrained cannibals, leaving the population less cannibalistic.

Assume that Wade had used a selection scheme more closely modeled after the trait group life cycle described earlier. That is, all beetles from all populations would be mixed into a "migrant pool," and the next set of populations would be initiated with subsets from this pool. This is a "hard selection" scheme in which Price's first term is nonzero. That is, there are realized differences in group productivity (because more productive groups have a higher chance of contributing individuals from the migrant pool to the next generation of groups). Wade did not use such a migrant pool (but see note 4); rather, a set of propagules from the most productive groups were used in his Group Selection for Productivity treatment. This choice had two important consequences: (1) the magnitude of the regression coefficient $\kappa_{w_{i\bullet}p_{i\bullet}}$ will be larger if *only* the members of large groups (which tend to house more altruists) realize nonzero fitness; and (2) if there is variance (e.g., in frequency of altruists) between groups that are selected from the end of growth cycle $t$, then Wade's propagules will tend to maintain a higher variance between groups at the beginning of cycle $t + 1$ (because the mixing that occurs in a migrant pool will tend to equalize between-group variance in the next generation).[3]

Again, we see that Price's first term ($\mathrm{cov}_{f_i}(p_{i\bullet}, w_{i\bullet}) = \kappa_{w_{i\bullet}p_{i\bullet}}\mathrm{var}_{f_i}(p_{i\bullet})$) is larger under Wade's propagule selection scheme than under a migrant pool scheme. Thus, choosing circumstances in which groups send founder propagules to initiate new groups (over a trait group mixing/reformation phase) stacked the deck in favor of altruistic evolution in Wade's experiment. Indeed, in an experiment similar to Wade's, but with a livebearing fish, the failure of group selection to produce effects was attributed, in part, to the abandonment of propagule initiation of groups for a type of mixed migrant pool (Baer et al. 2000).[4]

## OTHER GROUP SELECTION EXPERIMENTS

Several other artificial group selection experiments on flour beetles followed Wade's pioneering work (Craig 1982; Wade 1979, 1980a, 1980b, 1982; Wade and McCauley 1980, 1984). Emigration rate and cannibalism rate were added as group properties under artificial selection. These studies also explored differences in population structure, random extinction, and propagule size on the effects of selection. As was the case for Wade's original experiments, all of these studies detected significant effects of artificial group selection.

Artificial group selection experiments were applied to the cress, *Arabidopsis thaliana*, by Goodnight (1985). Leaf area was the object of selection in this study. Goodnight mixed artificial individual selection and artificial group selection in a fully factorial manner. He selected groups of plants with the highest (or lowest) *mean* leaf area to serve as "propagule generators" for the next generation of groups. However, he also selected the individual plants (within the selected groups) that had the highest (or lowest) personal leaf area to serve as actual parents of plants in the next generation. "No selection" controls (i.e., picking random groups for propagules or random individual parents to fill those propagules) were included in the factorial design. Interaction between selection for individual and group properties could thus be gauged. While Goodnight detected strong positive responses to group selection, responses to individual selection were weak and in some cases negative. In particular, individual selection for high leaf area produced plants with *lower* leaf area when compared to the "no individual selection" control. Interactions between individual and group selection were also detected. Group selection was most effective in the absence of any individual selection. Specifically, group selection for increased mean leaf area within groups was muted when paired with individual selection for increased personal leaf area!

One of the explanations Goodnight offers for these unexpected results is that individual selection might select for plants that are able to interfere with the growth of their fellow group-mates. Picking the individuals with the highest leaf area (if this is associated with the most "interfering" type) might lead to a propagule filled with interfering types. This means that the group in the next generation coming from this propagule is filled with plants that are interfering with each other's growth. As a consequence, leaf area (along with general plant health) could decline. This could explain the result that individual selection for increased personal leaf area leads to decreased area when compared to the control of "no individual selection." The basic idea here is that no plant is an "island": by selecting specific plants, one is simultaneously selecting specific social environments for the next generation.

Selection for groups of plants with high mean area in Goodnight's experiment may tend to pick out plants that interfere less (or even facilitate) the growth of their neighbors (thus leading to green groups). If group selection for increased mean leaf area is not paired with individual selection for increased personal leaf area, then the most interfering members of the greenest groups are not preferentially selected. This could explain why the effect of group selection was stronger without the presence of individual selection. In general, competitive restraint in plants (like cannibalistic restraint in beetles) could be viewed as a form of altruism; if so, within and between group selective effects would be in opposition. How much competition, cannibalism, or interference evolves would then depend on the relative strengths of these effects and, thus, on the nature of selective episodes.

A result similar to Goodnight's was found in an interesting practical application of artificial group selection within the poultry industry (Craig and Muir 1996; Muir 1996). In order to improve harvesting efficiency, many egg-laying chickens are currently kept in cages with multiple hens. Aggression between chickens can be substantial and beaks are trimmed to prevent injuries inflicted by cage mates. If one always selects the most productive chickens from cages, then average egg productivity within cages can actually *decrease* (Craig et al. 1975). Like the aforementioned plants, chickens in these cages are not islands: by selecting the most productive hen, one may be favoring a more aggressive social environment in which future egg production suffers.[5] Muir (1996) instead selected for egg productivity at the level of the multihen cage (hens from the most productive cages were used as parents to "seed" the next set of cages). This artificial group selection scheme produced dramatic results after a small number of generations, with annual egg production increasing 160 percent. Part of this improvement is due to lower aggression and improved stress coping abilities within cages (Dennis et al. 2006). Indeed, aggression was low enough in these group-selected lines that beak trimming was no longer necessary. Thus, this group-selected line of chickens has led to the potential for substantial cost savings through decreased hen mortality, abandonment of beak trimming, and increased egg production. As Sober and Wilson (1998) put it: "If this strain becomes widely used in the poultry industry, the projected annual savings will far exceed the money spent by the U.S. government for basic research in evolutionary biology."

### ECOSYSTEM SELECTION EXPERIMENTS

The experiments described so far impose artificial selection on some property of groups of conspecific organisms. However, the same selection protocols can be applied to multispecies communities or ecosystems; and researchers have recently executed artificial selection experiments at these higher levels (Swenson et al. 2000a, 2000b). These authors placed soil and aquatic microbial communities into a set of microcosms and then selected on the basis of various ecosystem properties. In the case of the soil communities, the authors selected microbial communities on the basis of aboveground biomass of *Arabidopsis thaliana* growing in the soil (Swenson et al. 2000b). In the case of aquatic communities, the authors selected microbial communities on the basis of microcosm pH (Swenson et al. 2000b) or the degree to which 3-chloroaniline (an industrial waste product) was degraded (Swenson et al. 2000a). Note, that in these cases, the criterion for selection was *not* a direct property of the organisms being selected; rather, these microbes were being selected on the basis of their effects on other organisms (*A. thaliana*) or their effects on the physical environment (pH or levels of 3-chloroaniline).

The actual selective protocols were similar to Wade's classic experiment. For the pH selection experiment, from the twenty-four aquatic microcosms, the six with the highest (or lowest) pH were used to send propagules to four microcosms each in the next generation. In the other selective experiments (plant biomass and 3-chloroaniline breakdown), the authors mixed and redistributed the selected microcosms across generations (instead of transferring propagules). For each of these experiments, responses to the selection schemes were recorded (either between high and low selected lines or between selected and control lines). The authors also found that the chemical and organismal composition of their microcosms came to differ under different selective regimes. Putting aside whether and how such experiments inform us about evolution in natural systems, there are obvious practical implications of such research (e.g., in bioremediation and agriculture).

The central idea being exploited in these experiments is the premise behind the theories of niche construction and ecosystem engineering: organisms affect each other and their environments (Jones et al. 1994; Odling-Smee et al. 2003). Organisms are not islands with respect to social interactions with conspecifics, but they are also not islands with respect to interactions with heterospecifics or shared environments. In some cases, selecting community players from a series of multispecies configurations on the basis of some ecosystem-level property may produce a response that does not occur when an organism selected for the same property (e.g., in isolation) is returned to a multispecies community.

## QUASI-NATURAL SELECTION EXPERIMENTS

There have been several criticisms of the artificial selection experiments described in the previous section. First, some critics have argued that the selected properties seem arbitrary (e.g., community pH) and it is unclear how evolution would proceed with respect to such properties in natural systems. Second, critics have noted that a mechanistic understanding of responses to artificial selection is lacking in some cases. (Note, however, that this is a criticism that can be extended to many "individual-level" artificial selection experiments, as well; see, e.g., Swenson et al. 2000b.) Third, critics have maintained that the experimental conditions used are extremely unlikely to apply outside the lab (Harrison and Hastings 1996). For instance, if migration is simultaneously the way that new groups form and a force that homogenizes intergroup variance, then the very means of producing offspring groups leads to a collapse in their variation (and thus the strength of group selection—e.g., as represented by Price's first term). Thus, even if an experimentalist can decouple these effects, through a combination of group extinction and propagule transfer, the operation of this process in nature is proposed to be limited. Fourth, some critics maintain that none of this should be called "group" or "ecosystem" selection. After all, isn't it the individuals within the groups that are changing? Can't we simply talk about this as a more complicated form of individual-level selection? I will return to this last criticism in the discussion.

The process of selection is actually quite clear in the experiments described in the previous section—indeed, it is the hand of the experimenter that picks *groups* based on group properties. An alternative approach is to create structured worlds in which groups thrive, split, mix and die "by their own devices." Specifically, the experiment is set up so that there is not any one *single* group property (e.g., productivity) that is being used to sort groups. Rather, groups and the individuals they contain prosper or fail based on the eco-evolutionary conditions that apply to the created world. This second approach is what Scheiner (2002) calls "quasi-natural selection" experiments. These experiments often take place in the laboratory, so it is not the environmental conditions that are "natural." What is taken to be more natural is the idea that selection could be simultaneously "operating" on a *variety* of properties within the system. Some of the criticisms of the artificial selection scheme play out differently for quasi-natural selection experiments. For instance, ostensibly arbitrary properties are generally not the basis of selection.

### THE EVOLUTION OF COMPETITIVE RESTRAINT

A recent experiment illustrates the basic features of quasi-natural selection experiments. Kerr et al. (2006) explored evolution within a host-pathogen system consisting of the bacterium *Escherichia coli* (the host) and T4 bacteriophage (the viral pathogen). A meta-community was created by distributing the bacteria and phage into a large number of subpopulations (wells in multiwell plates). Using a high throughput liquid-handling robot, the entire metacommunity was serially propagated (transferring a small fraction of each well at the end of an incubation period to a corresponding well with fresh medium in a new plate to start the next incubation period). In addition, migration between wells occurred during these serial transfers. Within any subpopulation, phage T4 drives its bacterial host extinct over an incubation period; thus, the host and pathogen cannot coexist at a local scale. However, under moderate migration levels, the host and pathogen can coexist at the metapopulation scale due to dynamic asynchrony between wells. That is, at any point in time, some bacteria-filled wells are undergoing phage-driven extinction, while some phage-filled wells are becoming empty due to dilution in the absence of hosts, while some empty wells are being restocked with bacteria through recolonization.

Within these microbial metacommunities, Kerr et al. manipulated the pattern of migration between subpopulations. Thus, altered migration schemes distinguished the different created worlds in this quasi-natural selection experiment. One treatment constrained migration to take place between neighboring wells within the multiwell plates, termed the Restricted Migration treatment (figure 19.3a). A second treatment allowed migration to take place between any two wells within the metacommunity, termed the Unrestricted Migration treatment (figure 19.3b). The rate of migration was held constant between treatments (only the topology of migration differed). After several transfers, the authors measured phage productivity (number of progeny produced per parent phage particle in a well with bacteria over an incubation period) and phage competitive ability (relative fitness of the evolved T4 strain when in competition for shared host cells with a marked T4 mutant). They found that phage from the Restricted Migration treatment evolved high productivity, but low competitive ability. Meanwhile, phage from the Unrestricted Migration treatment evolved high competitive ability, but low productivity (figures 19.3c and 19.3d). Since the same phage strain was used to inoculate all treatments, these results suggested the selection of *de novo* phage mutations.

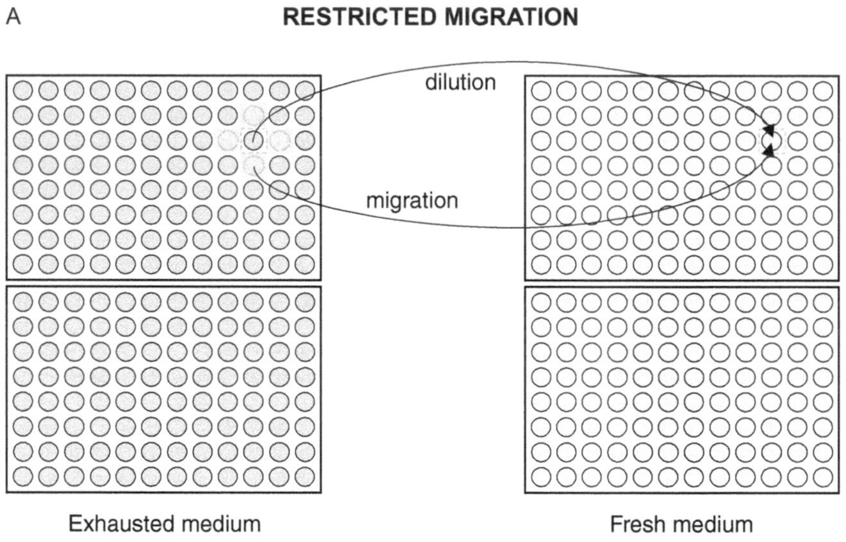

FIGURE 19.3

The metapopulation experiment of Kerr et al. (2006). Within two microtiter plates, 192 wells contain subpopulations of bacteria and phage. Every subpopulation is diluted into a well with fresh medium every twelve hours (a standard serial propagation at the metapopulation level). A, In the Restricted Migration treatment, immigration can occur into a focal well (boxed in a dashed line) from one of its nearest well neighbors (the wells that are highlighted to the north, south, east or west of the focal well).

# UNRESTRICTED MIGRATION

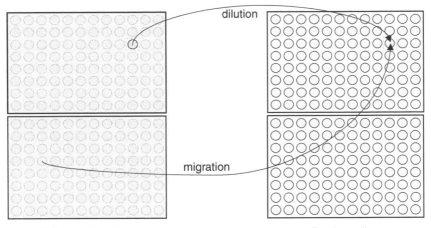

dilution

migration

Exhausted medium                    Fresh medium

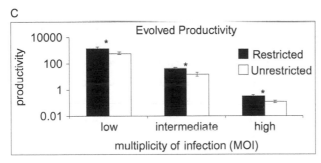

C

Evolved Productivity

■ Restricted
□ Unrestricted

multiplicity of infection (MOI)

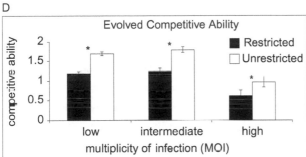

D

Evolved Competitive Ability

■ Restricted
□ Unrestricted

multiplicity of infection (MOI)

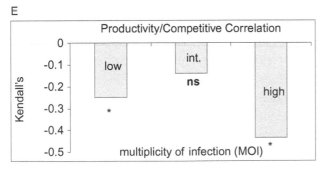

E

Productivity/Competitive Correlation

multiplicity of infection (MOI)

FIGURE 19.3 (continued)
B, In the Unrestricted Migration
treatment, immigration can
occur into a focal well from any
well in the entire metapopulation. C, Productivity of evolved
phage isolates was measured at
three different ratios of phage
to bacteria (the "multiplicity
of infection"). In all three cases,
phage from the Restricted
Migration treatment were
significantly more productive. D,
Phage from the Unrestricted
Migration treatment competed
significantly better for common
host resources when paired with
a marked phage strain. E, After
pooling the data, negative correlations between productivity and
competitive ability were discovered (significant in two out of
three cases).

In this experiment, bacteria are resources for the phage. Phage exhibiting unre-strained use of this resource (e.g., entering and killing their host quickly) may outcom-pete restrained phage for common host resources (Abedon et al. 2003). However, phage exercising restraint (e.g., spending a longer period inside their host) may extract more from their resources and consequently may be more productive when competitors are absent. In this microbial community, there appears to be a trade-off between productiv-ity and competitive ability (overall negative correlations were found between these prop-erties—figure 19.3e). Thus, this system has the ingredients for a "tragedy of the com-mons" (Hardin 1968), which occurs when unrestrained individuals displace restrained ones, leading to overexploitation of shared resources and a lower group productivity. A microbial version of this tragedy occurs when rapacious phage outcompete their pru-dent cousins and end up lowering group productivity in the process.

Why would the pattern of migration within a metapopulation influence the resolu-tion of this tragedy of the commons? As rapacious phage mutants are generated within a metacommunity, they will outcompete their prudent ancestors locally. However, these rapacious phage are less productive; therefore, given periodic dilution in the experiment, the rapacious phage are more extinction-prone. Prolonged survival of the rapacious phage depends on the virus continually finding fresh host bacteria to exploit. Such en-counters are more likely in the Unrestricted Migration scheme. Furthermore, mixing of different phage types into the same subpopulation is more likely within the Unrestricted Migration treatment. Both of these factors contribute to the success of the unrestrained strategy given unrestricted migration. In the Restricted Migration treatment, limitation to host access leaves the rapacious phage vulnerable to extinction and the phage popula-tion remains prudent by default (Kerr et al. 2006; Prado and Kerr 2008). In essence, the form of migration can determine whether local tragedies of the commons can become global tragedies of the commons.

Some of the first models exploring the evolution of cooperation (e.g., Maynard Smith 1964) assumed the potential for tragedies of the commons at the subpopulation scale. The altruist is the type displaying prudent use of resources. Indeed, relative to rapacious types, prudent phage from the experiment just described are "multilevel altruists" (i.e., they satisfy conditions [4] and [5]). Looking back to the Price equation, the first term, $\text{cov}_{f_i}(p_{i\bullet}, w_{i\bullet})$, will be positive. That is, as the frequency of prudent phage in a well with hosts increases, the absolute productivity of the well increases. However, Price's second term, $E_{f_i}[w_{i\bullet} \Delta p_{i\bullet}]$, will be negative. That is, the prudent phage always loses in competition to the rapacious type in a mixed subpopulation. If these statistical terms are interpreted as group-level and individual-level "selective forces," then the outcome of evolution depends on the strength of group selection for prudence versus individual selection for rapacity (see the discussion for an exploration of the causal adequacy of this account). What will influence the magnitude of these terms? One factor is migration. Specifically, restrictions to migration will tend to produce a clumped group frequency distribution (with rapacious phage isolated in some subpopulations and prudent phage isolated in others). This tends to tip the balance in favor of altruism (prudence). This is easiest to see when the fitnesses are linear (equations [9] and [10]), because the larger the "clumping index" (7), the more likely our generalized version of Hamilton's rule (11) is to hold. Thus, there are both theoretical and experimental reasons to think that the pattern of migration will influence the form of social evolution in metapopulations.

## THE EVOLUTION OF VIRULENCE VERSUS BENEVOLENCE

At a fundamental level, disease-causing organisms occupy metapopulations since their hosts constitute discrete sets of resources (i.e., potential subpopulations). The manner in which the pathogen uses the resources of its host will influence the progression and nature of the disease. One property influenced by pathogen traits is disease virulence. Here, I will define *virulence* to be the increase in the rate of death of a host when infected by its pathogen. The conventional wisdom is that pathogens should always evolve *reduced* virulence and that highly virulent pathogens result from evolutionarily *recent* introduction into a new host (see Bull 1994 for a full discussion). All else being equal, a pathogen that more rapidly kills its host is at a selective disadvantage relative to one that preserves its host (and future opportunities for transmission). A complication occurs if all else is *not* equal. Specifically, there may be functional relationships between virulence and other properties that will affect the success of a pathogen. For instance, when more virulent pathogens are transmitted at higher rates, selection may not favor the lowest level of virulence (Anderson and May 1982).

In some cases, virulence is a product of the schedule of pathogen reproduction within the host, where rapid reproduction yields higher virulence. In this context, avirulent pathogens reproduce slower. In pair-wise competition for common host resources, the virulent strain of the pathogen is expected to displace the avirulent strain. However, hosts with more virulent strains die sooner (by definition). This description of virulence and avirulence conforms nicely to multilevel altruism (equations [4] and [5]). The avirulent pathogen exercises relative restraint in the use of host resources, a strategy that is benevolent (toward both the host and co-occuring pathogens).

What are the conditions favoring virulence versus benevolence? There have now been several quasi-natural selection experiments on this subject. It turns out that the nature of transmission (its mode, spatial scale, and timing) can strongly influence evolutionary outcomes. Bull et al. (1991) and Messenger et al. (1999) used the filamentous phage f1 that infects *E. coli* to explore the role of the mode of transmission on the evolution of benevolence. This bacteriophage is unusual in that it establishes a permanent infection in which progeny phage are secreted through the bacterial envelope without killing the host. Infection is detrimental to the growth of the bacterial host however. Thus, it is not virulence by the above definition (effect on death rate), but rather the "degree of harm" to host growth rate that was monitored in these experiments. As the rate of production of secreted phage progeny increases, so, too, does the degree of harm to its host (Messenger et al. 1999).

These authors manipulated the degree of vertical versus horizontal transmission in this system. (Vertical transmission occurs when the viral genome is inherited from a parent host cell to its offspring, whereas horizontal transmission occurs when viral progeny from one cell infect a previously uninfected, unrelated cell.) They found that vertical transmission promotes the evolution of benevolence, whereas the phage was more harmful to its host (but more productive in terms of secreted progeny) under a selection scheme involving higher levels of horizontal transmission.

The standard explanation for these findings is that vertical transmission promotes partner fidelity, which tends to work for cooperation in the host-pathogen interaction. A complementary way to explain these results is to realize that there are two ways a phage strain can increase its numbers, through (1) passing its genome copies to the offspring of its current host (offspring cells of an infected parent cell are also infected) and (2) secreting packaged phage progeny that successfully infect formerly uninfected hosts (note there is no superinfection in this phage). The trade-off between production of secreted progeny and reproduction of the infected host translates to a trade-off between these two components of fitness. As vertical transmission becomes common, there may be a premium placed on promoting the welfare of the host. As horizontal transmission opportunities become abundant, selection may favor strains that invest in producing many packaged infective particles. These findings are not restricted to microbial host-pathogen systems. For instance, vertical transmission of algal symbionts in jellyfish promoted benevolence, whereas horizontal transmission favored the evolution of *parasitic* algal symbionts with negative effects on jellyfish reproduction and growth (Sachs and Wilcox 2006). As with the phage-bacterium system, a positive relationship between degree of harm to the host and the rate of expulsion of the symbiont seems to underlie these results.

One factor that is predicted to influence the mode of transmission is the migration rates or movement patterns of hosts and their pathogens. Generally, as migration or movement becomes less restricted the opportunities for horizontal transmission improve. This will tend to select for more virulent pathogens if virulence trade-offs exist (e.g., Kerr et al. 2006). A recent experiment probed the effects of host movement on pathogen evolution. Boots and Mealor (2007) altered the viscosity of microcosm environments containing larvae of a phycitiid moth and a species-specific granulosis virus. They found that the virus evolved higher infectivity when hosts could move easily within the microcosm (less viscous food medium) as compared to microcosms in which the host was more restricted in its movement (more viscous food medium). Infectivity is the proportion of hosts infected after exposure to the virus. While infectivity is a different property than virulence, there is an obvious connection between the two. In the same way that restricted migration between hosts can favor pathogens that are more prudent *within* their hosts (i.e., less virulent), restricted movement of hosts can favor pathogens that are more prudent with the *set* of hosts to which they are exposed (i.e., less infectious).

Another quasi-natural selection experiment explored the role of the timing of transmission on the evolution of virulence (Cooper et al. 2002). These authors used a viral pathogen of the gypsy moth. The experiment consisted of several cycles of transmission from infected larvae to uninfected larvae. Virus was obtained from infected larvae by homogenizing them at a specific time in their development (after which the virus was introduced to the next set of larvae to complete the transmission). In their Early treatment, virus was obtained from live infected larvae five days after infection. In their Late treatment, virus was obtained from live infected larvae nine days after infection. Virus evolved in the Early treatment was significantly more virulent than virus evolved in the Late treatment. An important component of their experimental protocol was that only *living* larvae could serve as sources of transmission at the time of transmission. While highly virulent viral strains outreproduce less virulent strains over short time periods (e.g., five days), these same virulent viral strains destroy their host over longer time periods (e.g., nine days). In the Late treatment, only those larvae containing less virulent viral pathogens would survive to the time of transmission.

The above study illustrates why virulence has popped up regularly in the discussions about levels of selection (Lewontin 1970; Sober and Wilson 1998; Wilson 2004). Of course, hosts often contain discrete *groups* of pathogens. When a pathogen's competitive ability within its host trades off with the longevity of the infected host, a multilevel perspective posits conflicting selective forces between and within groups. Specifically, the virulence of the pathogen determines the rate of group extinction (i.e., host death) such that groups of virulent pathogens are more extinction-prone. This is interpreted as a "between-group force" *against* virulence. On the flip side, when virulence is positively related to within-host competitive ability, more virulent pathogens should increase in frequency within hosts infected with different strains. This is interpreted as a "within-group force" *for* virulence. As before, the two terms from the Price equation nicely partition these components. Whether the pathogen becomes more virulent or more benevolent depends critically on how groups of pathogens form (e.g., do infections start with small numbers of pathogens that resist superinfection or does infection proceed with continual introduction of superinfecting pathogens?) and the exact nature of the relationships between virulence and other pathogen traits (i.e., trade-offs determining the fitness structure). As before, statistical association within hosts of benevolent pathogens will tend to work for lower virulence. It is interesting to note that Cooper et al. (2002) initialized each larval infection with a small number of viruses to "maintain within-host homogeneity." From a multilevel perspective, this has the effect of shrinking the within-group force, tipping the balance toward avirulence under conditions where there is a premium placed on group survival (in their Late treatment).

## THE EVOLUTION OF CONFLICT RESOLUTION

In 1966, microbiologist K. Jeon noticed that his cultures of amoeba were infected with an intracellular bacterial parasite (which he termed X-bacteria). The X-bacteria harmed its host (in terms of compromised growth, reduced reproduction, and increased mortality).

However, Jeon continued to propagate some of the infected amoeba. After five years, the amoeba populations fully rebounded. However, this recovery was not due to the exclusion of X-bacteria. Indeed, the bacterial symbiont was still very much present. Through surgical and chemical manipulations (Jeon 1972; Jeon and Hah 1977), Jeon and colleagues found that, in some cases, if X-bacteria and its amoeba host were separated, the *amoeba* would not be viable! A parasitic relationship had evolved into a mutualism.

In order for the change that Jeon witnessed to occur, conflict between interacting parties must be resolved. (See Sachs and Bull 2005 for another fascinating example involving viruses.) The issue of conflict resolution arises both in social interactions between members of different species (as in Jeon's system) and between individuals of the same species. Effective conflict resolution carries consequences for the origin and maintenance of altruism. Specifically, the fundamental barrier to the emergence and persistence of altruists is the intrinsic conflict with *cheaters*. Cheaters are taken to be individuals that derive benefits from interactions with altruists but do not contribute socially. In this section, I will focus on experiments with microbes that highlight the ways social conflict can be resolved (see Travisano and Velicer 2004 for a full discussion).

In microbial systems, the evidence for the existence of cheating and conflict is widespread, with examples in viruses, prokaryotes, and eukaryotes. For instance, in phage Φ6, Turner and Chao (1999) evolved a "cheating" strain that was able to outcompete its ancestor within a host cell (*Pseudomonas phaseolicola*) infected by both strains. However, this cheat had lower reproductive output in pure infections, in comparison to the output of its ancestor in pure infections. Another "tragedy of the commons" scenario was discovered in strains of *Saccharomyces cerevisiae* that excrete an enzyme that breaks down extracellular sucrose facilitating subsequent sugar uptake. Yeast strains defective in exoenzyme production were functionally "cheats," able to free-ride on the public goods provided by producers without incurring the cost of production (Greig and Travisano 2004). Similarly, cheats were detected in biofilms of *Pseudomonas flourescens* (Rainey and Rainey 2003). In this system, "cooperating" bacteria excrete a sticky polymer that allows a mat to form at the interface of liquid medium and the air. Cheating strains do not contribute to the public good (the sticky matrix). As a consequence, the cheating strains have a growth advantage within the mat but lead to premature collapse of the mat due to its weaker integrity. Perhaps the most famous microbial examples of cheating come from social microbes, such as the bacterium *Myxococcus xanthus* and the slime mold *Dictyostelium discoideum*. For both organisms, starvation causes aggregation of single cells into multicellular fruiting bodies. In these bodies, some cells sacrifice themselves as supportive "somatic tissue," while other cells form reproductive spores (the "germ line"). Cheating occurs when a strain is able to achieve disproportionate representation in the spore pool when mixed with other strains, but this strain is compromised in its ability to form functional fruiting bodies when it occurs alone. Such cheaters have been described in both species (Velicer et al. 1998; Dao et al. 2000; Strassmann et al. 2000).

Given the ubiquity of cheating types in these and other systems, how is it that cooperation does not succumb to these antisocial influences? Quasi-natural selection experiments have provided some answers. I will present two classes of explanations here: (1) passive cheater control due to favorable social context formation and (2) active cheater control due to exclusion. For the first class, the experimenter manipulates factors that influence how social context forms. For instance, Greig and Travisano (2004) found that cooperative yeast strains, which produce an exoenzyme that degrades sucrose, had a competitive edge over defectors, which don't produce the exoenzyme, when the mixed population was grown at low density in a spatially structured habitat. In this case, the benefits of cooperators are disproportionately experienced by cooperators, because clumped growth occurs in the structured habitat. On the other hand, defectors receive fewer such benefits due to the limited diffusion of the exoenzyme. Similarly, in an experiment on metapopulations of *Pseudomonas aeruginosa*, Griffin et al. (2004) reported that cooperating strains (producers of siderophores, extracellular iron-gathering compounds) were able to displace cheats (nonproducers) when each new subpopulation was established from a single clone under a hard selection scheme. Barring mutation, cooperating bacteria were guaranteed to be interacting with other cooperators (namely, relatives), whereas defectors were isolated in their own subpopulations. This form of social context formation is optimal for cooperation by maximizing Price's first term, while minimizing the second. Finally, in the study on phage T4 described earlier, Kerr et al. (2006) found that prudent phage evolved when migration was restricted. Under such restricted movement, cooperators and defectors are less likely to compete for the same resources. Again, this type of social context formation works for cooperation.

In these experiments, the experimenter controls social context formation, by controlling density, founder size, and migration. The second class of cheater control mechanisms involves active exclusion of would-be defectors by the cooperators themselves. In an experiment with marked strains of *Dictyostelium purpureum*, Mehdiabadi et al. (2006) monitored aggregations of different strains. They found that fruiting bodies consisted primarily of one strain, suggesting that a form of kin discrimination occurs during fruiting body formation. Working with *D. discoideum*, Queller et al. (2003) showed that the product of gene *csA*, a protein embedded in the membrane that is involved in cell-cell adhesion, enabled slime mold cells to exclude *csA*-knockout cells from fruiting bodies under natural conditions. *csA* thus qualifies as a "greenbeard" allele: (1) it displays a phenotypic trait (a membrane protein), (2) it recognizes the same phenotype in other individuals (through homophilic binding), and (3) it provides a disproportionate benefit to like types (under natural conditions, the adhesion engenders overrepresentation in aggregation streams). Interestingly, under laboratory conditions (e.g., on agar plates) where both *csA* and *csA*-knockouts can coaggregate, these authors showed that *csA*-knockouts are overrepresented in the spores! Thus, the *csA*-knockout can be thought of as a cheater. However, under natural conditions, this would-be cheater is kept in check by active exclusion.

Finally, in a tour de force experiment with *Myxococcus xanthus*, Fiegna et al. (2006) discovered cheater control originating from a most unexpected place. These authors followed the evolution of an obligate cheater—a strain that could not make functional fruiting bodies in isolation but preferentially moved into the spore pool when mixed with other strains. Starting with a mixture of a cooperating strain and this obligate cheater, the cheat increased in frequency; however, as it rose to prominence, its inability to form functional fruiting bodies caught up with it, because the authors required the population to go through a spore stage every propagation cycle. Once in the majority, the cheat's density dropped precipitously, a perfect example of a tragedy of the commons. However, a mutant then arose that the authors dubbed "Phoenix," in reference to the mythical bird. Phoenix is not an obligate cheater; in fact, it is a superior cooperator, in that it makes functional fruiting bodies and can actively exclude its cheating progenitor in mixed populations. Amazingly, full genome sequencing revealed that this social back-flip was mediated by a single mutation. Thus, in a somewhat ironic twist of events, cooperation and cheater control evolved directly from the cheat itself.

## NATURAL SELECTION EXPERIMENTS

The selection experiments of the previous section earned the adjective *quasi-natural*, because the arena of selection was often different from the natural conditions of the study organism. An argument could be made that, with respect to real-time evolution in the laboratory, the "artificial" conditions *were* the "natural" conditions. However, in understanding the traits of organisms in "the wild," it is sensible to study selection in the context under which those traits evolved. This means gathering data and performing experiments in the field.

In this section, I focus on a set of field experiments using plants. All of these field studies employed contextual analysis (discussed earlier; see also Heisler and Damith 1987; Goodnight et al. 1992; Goodnight and Stevens 1997). In contextual analysis, a relationship is sought between a proxy for the fitness of an individual (e.g., seed number or survival in a focal plant) and both "individual" and "contextual" traits. Individual traits belong to the focal individual (e.g., its size or height), whereas contextual traits belong to the group or neighborhood in which the focal individual is embedded (e.g., mean size or population density). A multiple-regression analysis teases apart the effects of individual properties and group properties on individual fitness. For instance, a statistically significant partial regression coefficient of individual fitness on a group trait is taken to indicate the operation of group selection.

In an observational study of the jewelweed, *Impatiens capensis*, Stevens et al. (1995) found that larger individuals had higher survival, more chasmogamous (open-pollinated) flowers, and more cleistogamous (self-pollinated) flowers. However, they also found that having smaller-sized neighbors led to higher survival and more cleistogamous flowers. Thus, according to the contextual analysis, there was individual selection for larger size working against group selection for smaller size. The authors discuss small plant size as a potential type of altruism: "The altruistic plant may forgo the individual advantage of large size 'for the sake of' the increased survival rate and reproduction of the group." Interestingly, in stands of *Impatiens*, relatedness is likely to be high due to high levels of selfing and low levels of seed dispersal. Thus, the type of statistical clumping (described earlier) needed to favor altruistic phenotypes may be present (see Stevens et al. 1995).

In the Stevens et al. study, the partial regression coefficients of cleistogamous flower production on individual plant size and the mean of plant size in the group are opposite in sign but equal in magnitude. This is consistent with soft selection on plant size in their populations (i.e., constant yield from groups despite composition or density). When group output does not vary with group composition, then Price's first term is zero (in the case of linear fitnesses, the relationship between equally opposing regression coefficients and Price's first term can be seen clearly in equation [24]). Thus, Price's analysis suggests that there is no group selection in the *Impatiens* system, whereas contextual analysis detects group selection. I claimed earlier that the presence of a group selection effect under soft selection schemes was a weakness of contextual analysis. However, this "weakness" is completely dependent on the way that "group selection" is defined. For instance, Stevens et al. (1995) define group selection as "variation in the fitness of an *individual* due to properties of the group or groups of which it is a member" (my emphasis). Given that definition, contextual analysis is the perfect tool for detecting group selection, because it isolates the effects of group context on individual fitness while controlling for individual effects. However, others have defined group selection as differential fitness of *groups* (e.g., Sober and Wilson 1998). That is, groups, rather than individuals, are the bearers of fitness. Given this second definition of group selection, contextual analysis may detect a group-level effect when there are no differences between groups in their outputs (i.e., there is a classic frequency-dependent soft selection scenario). Given that different statistical approaches package selection differently, claims about the presence of group selection may be wholly dependent on the definition one is employing.

Using contextual definitions of group selection, other researchers have experimentally explored multilevel selection in other plant systems. For instance, Donohue manipulated relatedness (Donohue 2003) and density (Donohue 2004) in experimental stands of the Great Lakes sea rocket, *Cakile edentula*. She found that both of these factors influenced the strength of group selection, where the strongest group-level effects occurred with high relatedness between a focal plant and its neighbors (Donohue 2003) and when density was at an intermediate level (Donohue 2004). Group selection on plant size was found to operate in concert with individual selection in sibling groups (Donohue 2003); however, group and individual selection were opposed at intermediate densities (Donohue 2004): shorter, heavier plants growing with taller, lighter neighbors had higher fitness. Weinig et al. (2007) also found an effect of stand density on group selection in *Arabidopsis thaliana* with a contextual analysis approach. These authors found that, at higher densities, the strength of group selection for size and branching patterns increased. In this study, group and individual selection effects were opposed, where individual selection favored an increase in size and elongation, while group selection favored a decrease. The idea of slower plant development as a form of altruistic restraint is particularly intriguing given the high levels of selfing and low levels of seed dispersal in natural populations of *A. thaliana*.

## DISCUSSION

Even though the simplest version of natural selection treats organisms as islands, entire of themselves, it is clear that virtually all organisms experience meaningful interactions with other organisms. In this context, altruism is of particular interest, as it most clearly highlights the shortcomings of the "organism-as-island" model. Common explanations for the evolution of altruism are based on the idea that an altruist has a different social experience than a nonaltruist. Specifically, the altruist somehow manages to enjoy a more altruistic social circle. This can occur if the altruist interacts with kin (Hamilton 1964), conditions its behavior on the previous behavior of its partner (Trivers 1971; Axelrod and Hamilton 1981), or punishes/excludes noncooperators in its social sphere (Frank 1995). Using these theoretical predictions as inspiration, a collection of elegant experiments in the laboratory and the field have provided a fuller picture of the evolution of altruism in biological systems.

The first empirical finding is that altruism exists. Of course, this statement needs to be qualified by the operational definition of altruism. However, by any one of a few definitions, altruism (or the potential for altruism) has been uncovered in many systems, including viruses (Turner and Chao 1999; Kerr et al. 2006), bacteria (Velicer et al. 1998; Rainey and Rainey 2003), protists (Dao et al. 2000; Strassmann et al. 2000), fungi (Greig and Travisano 2004), plants (Goodnight 1985; Stevens et al. 1995), and animals (Wade 1979; Muir 1996). The second result is that altruistic traits can increase in frequency in real-time experimental evolution (see Rainey and Rainey 2003; Fiegna et al. 2006; Kerr et al. 2006). As the theory predicts, factors influencing social context formation can be critical. Experimental manipulations of population density (Greig and Travisano 2004), interactor relatedness (Griffin et al. 2004), population viscosity (Boots and Mealor 2007), and migration pattern (Kerr et al. 2006) all have effects on the evolution of altruistic traits.

Given the existence of altruism and its potential for evolution by natural selection, what can we say about the level(s) of selection? Does selection act on individual organisms only? On groups of individuals? On groups and individuals simultaneously? Given our discussion of theoretical foundations and experimental findings, what can we add to the debate over the levels of selection?

In the process of answering these questions, I develop an analogy. In 1915, W. E. Hill published the picture shown in figure 19.4. This cartoon is meant to delight its viewer in

FIGURE 19.4
W. E. Hill's famous cartoon.

capturing both the image of an old woman and a young woman. A conversation between someone who can only see the young woman and someone who can only see the old woman would be a frustrating experience for both parties. "It is clearly a young woman that Hill has drawn!" protests the first. "How can you be so blind?!" clamors the second, "it is an *old* woman!" (If the reader is sympathetic with one of these individuals, the following hint may help: the chin and cheek of the young woman is the nose of the old woman.) Hill's cartoon is interesting because it is *simultaneously* a young woman and an old woman. A pluralist would maintain that it is a matter of the viewer's perspective as to which is seen.

For some, the idea of pluralism when it comes to the levels of selection debate is distasteful. How can it be that a case of selection in a group structured population can be seen as individual-level selection *or* group-level selection? Surely either group selection is occurring or it is not, right? A realist would claim that there is a single answer to the question: "At what level(s) is selection operating?" Note that this answer might be that selection is operating at the group and individual levels simultaneously, in that a realist can hold that multiple selective forces are in action. However, the realist would not claim that selection is operating at the group level *or* the individual level, depending on which way you look at it. This would be pluralist territory.

Why must we concern ourselves with the seemingly esoteric distinction between pluralism and realism? Part of the reason is that a good portion of the group selection debate has occurred between *realists*. Specifically, one set of realists maintains that group selection is not occurring in nearly all biological systems (e.g., Williams 1966), whereas another set of realists maintains that group selection occurs in plenty of biological systems (e.g., Sober and Wilson 1998). Is it possible that different realists have been arguing over a selective equivalent of Hill's cartoon? Is it simply a matter of perspective, such that selection in group-structured populations can be equivalently understood in two different ways? While I don't believe the situation is so simple, I do think that a thorough consideration of pluralism in the group selection debate leads to a fresh angle on different realist positions and a new take on specific empirical results.

The first question is whether pluralism is even possible in cases of group-structured selection. Let us revisit our simple trait group model. We define $\pi_i$ to be the fitness of a group with $i$ $A$ types:

$$\pi_i = \alpha_i i + \beta_i(n - i). \tag{26}$$

That is, $\pi_i$ is simply the total productivity of a group with $i$ altruists (note that $\pi_i = nw_{i\bullet}$). The altruist share of this productivity is simply

$$\phi_i = \frac{\alpha_i i}{\alpha_i i + \beta_i(n - i)}, \tag{27}$$

where $\phi_0 = 0$ and $\phi_n = 1$ always. Note that $\pi$'s and $\phi$'s constitute another way to parameterize the fitness structure (see earlier discussion). That is, if all the $\alpha$'s and $\beta$'s are known, then all the $\pi$'s and $\phi$'s can be derived through equations (26) and (27). Similarly, another set of equations can be derived to compute $\alpha$'s and $\beta$'s given $\pi$'s and $\phi$'s (see Kerr and Godfrey-Smith 2002a). Thus, there are two interchangeable perspectives (parametrically speaking) on selection in the trait group framework. Equations (26) and (27) are analogous to telling a perplexed viewer of figure 19.4 that the old woman's mouth is the young woman's necklace, the old woman's eye is the young woman's ear, and so on. Thus, pluralism is certainly *possible* for some cases of group-structured selection.

When working within the multilevel selection framework, the $\pi/\phi$ parameterization is a natural choice (see also Wilson 1990). One way to see this is to reconsider the multilevel definition of altruism (equations [4] and [5]) with this new parameterization:

$$\phi_i < i/n \tag{28}$$

$$\pi_i < \pi_{i+1} \tag{29}$$

The $\pi/\phi$ parameterization makes it crystal clear that altruist frequency drops within groups (equation [28]), but groups with more altruists are more productive (equation [29]). In the $\alpha/\beta$ parameterization, only individuals are the bearers of fitness. Groups *affect* individual fitness, but groups do not explicitly *have* fitness. Thus, the $\alpha/\beta$ perspective is a type of individualist parameterization (see Dugatkin and Reeve 1994; Sterelny 1996; Kerr and Godfrey-Smith 2002a). Statements about *group* productivity require parametric manipulation of individual productivities under the $\alpha/\beta$ perspective (namely, equation [26]), whereas such statements are made in terms of untouched parameters within the $\pi/\phi$ perspective. In this parameterization, groups are the immediate bearers of fitness, in the form of $\pi$ quantities.

The realist may respond that it is all fine and good that one can represent the selective process in different ways, but it is an empirical issue whether between-group differences contribute to the evolution of a trait (Wilson and Wilson 2007). A realist might further claim that only one perspective accurately represents the causal structure of the system, while the other distorts it. Indeed, causal language has frequently popped up in the defense of realist positions (e.g., Sober and Wilson 1998 vs. Maynard Smith 2002). The issue of causality is murky, but it is helpful for us to attempt one possible "causal test" here. Given multiple interchangeable parameterizations (e.g., our $\alpha/\beta$ and $\pi/\phi$ parameters), we will say that one parameterization is more "natural" if fewer parameters need to change to accommodate a slight change in the system being modeled. That is, the more causally appropriate parameterization represents small changes to the system in a more isolated way. Put differently, the more natural parameters readily "grab" the changes that can actually occur. Note that this definition of the more natural perspective is relative to the nature of the variant imagined, although there will often be solid empirical reasons for considering certain variants. A forthcoming manuscript describes this near-variant test in detail (P. Godfrey-Smith and B. Kerr, unpublished manuscript).

As an example of the near-variant test, let us revisit the scenario outlined earlier: a group-structured population in which no meaningful interactions occur between individuals; the *A* type has two offspring and the *B* type has a single offspring despite group context. We can describe this system using the $\alpha/\beta$ parameterization (i.e., $\alpha_i = 2$ and $\beta_i = 1$ for all $i$) or the $\pi/\phi$ parameterization (i.e., $\pi_i = n + i$ and $\phi_i = 2i/(n + i)$). Note that there are between-group differences in productivity ($\pi$ changes with $i$). However, even the most ardent defenders of group selection would not call this a case of group selection. Why? Because these group-level differences are nothing more than products of individual level differences. (The argument that Price's first term is spurious in this case has similar roots; see Okasha 2006.) This notion is captured cleanly by the near-variant test. Imagine a slight change to the system: say, the *A* type produces three offspring. In the individualist parameterization, only the $\alpha$'s change, whereas both $\pi$'s and $\phi$'s change in the multilevel parameterization. The $\alpha/\beta$ parameterization captures this variant in a more isolated way, because the parameters attach to the individuals and it is individual fitness that we envision as changeable. We *can* look at this selective episode from a multilevel perspective, but it seems more natural to do so from an individualistic one.

Similarly, there will be cases in which the multilevel perspective is more natural, even though the individualist perspective is available. It is interesting to note that game theory uses an individualist perspective that is identical to our $\alpha/\beta$ parameterization, while one-locus diploid population genetics uses a multilevel perspective, which is identical to our $\pi/\phi$ parameterization. This may reflect the tacit notion that context-dependent payoffs to individuals are changeable in many social games, while it is the fitness of groups of genes aggregated into genotypes that are changeable in many diploid population genetic cases (see Kerr and Godfrey-Smith 2002a for a discussion). The upshot of the near-variant test is to offer a rigorous way to back up a realist claim. That is, in cases where the pluralist has offered multiple perspectives, the near-variant test may adjudicate among them.

To continue our analogy with Hill's cartoon: Imagine that you find out that Hill was actually drawing a young woman without a necklace, and his daughter placed a dripping cup of coffee on the diagram in such a way as to produce a marking resembling a necklace for the young woman. Hill sees that this necklace can double as a mouth of a previously unseen old woman. Knowing this history, we might claim that Hill's picture more *naturally* depicts a young woman. This is because if his daughter had placed the coffee cup in another location (a near-variant), then the young woman might change slightly, but the old woman entirely disappears. The point is that pluralism need not be antithetical to realism. Indeed, a careful description of multiple perspectives combined with some knowledge about the system allows a judgment about whether one perspective has causal priority, at least by the near-variant test. How do these considerations play out for the experimental work described above?

In Wade's classic experiment on flour beetles, groups are literally being selected by Wade on the basis of their productivity ($\pi_i$). Thus, a multilevel perspective is extremely natural in this case, as the different selective treatments are explicitly defined on the basis of effective group productivity. In this case, the realist could legitimately claim that group selection for increased productivity through decreased cannibalism (as gauged by a difference in $\pi$'s) works against individual selection for decreased productivity through increased cannibalism (as gauged by a depression in $\phi$'s).

How about quasi-natural selection experiments, where groups are not chosen based on a specified group property? Wilson and Wilson (2007) discuss the results of the Kerr et al. (2006) study on the evolution of phage prudence as a clear case of group selection. Indeed, they call out the authors for not including the "g" word in their manuscript. That is, Wilson and Wilson see this case from a realist perspective, where group selection *is* occurring and a full understanding the evolutionary outcome depends on this recognition. In one way, I agree with Wilson and Wilson. It is not only possible to take a perspective that focuses on between-group differences ($\pi$'s) and within-group skewing ($\phi$'s), but I find this multilevel perspective particularly illuminating for the Kerr et al. system.[6] However, the heuristic value of a particular representation is different from its causal adequacy. *Must* we take a multilevel perspective in order to properly represent evolution within this experiment? Here, I must confess that I am not as certain as Wilson and Wilson. I freely admit that between-group differences exist in this system, but the question is whether prudence evolves *because* of these differences. To answer this question, causal adequacy must be rigorously defined. Recall that between-group differences can exist in the case of groups of non-interacting individuals; however, we would hesitate to chalk up individual change to these between-group differences.

I have provided one tentative approach to making causal statements more rigorous: the near-variant test. Interestingly, in many models of altruism (using a trait group framework), the near-variant test does *not* identify one parameterization as being more natural than the other (P. Godfrey-Smith and B. Kerr, unpublished manuscript). That is, multilevel and individualist parameterizations appear equally equipped to isolate changes. In one respect, this "gray zone" is reassuring. After all, it is these cases that have provoked much argument in the group selection debate, with some authors seeing these cases as clearly multilevel selection, while others see these cases as clearly individual-level selection. I would tentatively place the results of the Kerr et al. (2006) experiment and several other quasi-natural experiments in this gray zone. The multilevel perspective can be extremely helpful in understanding evolution, but there is another perspective available, one that does not focus on between-group differences, and it is unclear to me how to sort these different perspectives with regard to causal adequacy.

In the field experiments described earlier, group selection has a very specific meaning. These studies do *not* define differences in group productivity as group selection. Group selection is defined as the differences in an individual's fitness that are due to the collective traits of its group, after controlling for the effects of its own traits. These field studies are directly measuring the effects of social interaction on individual fitness. Thus, contextual analysis is actually pitched at the level of individuals. Altruism detected by contextual analysis could be called individual-centered altruism (see Kerr et al. 2004):

$$\alpha_i < \beta_i \tag{30}$$
$$\alpha_i < \alpha_{i+1} \tag{31}$$
$$\beta_i < \beta_{i+1} \tag{32}$$

Here, the cost of altruism is given by equation (30): as the focal individual switches to the altruistic type but maintains the composition of altruists in its group, its fitness drops. And the benefit is given by equations (31) and (32): as the frequency of altruists increase in the group, the unchanged focal individual improves in fitness. Note that this definition is most easily deployed in the $\alpha/\beta$ parameterization. Indeed, if groups are not discrete (e.g., as in Stevens et al. 1995) then it is hard to see how to define $\pi_i$ and $\phi_i$ (see Godfrey-Smith (2008) for a discussion of this point). Thus, it is not immediately clear that group selection detected through contextual analysis will readily jibe with other definitions of group selection, such as differences in group fitness.

To illustrate this issue, let's revisit the case of frequency-dependent soft selection in a group-structured population. Here, $\pi_i$ is constant for all values of $i$, whereas the $\phi_i$ is different from $i/n$. This means that all evolutionary change is captured by Price's second term, the term that is generally associated with individual-level selection within a multilevel perspective. However, a contextual analysis picks up a "group selection" effect in this case. Literally, the contextual analysis has picked up an effect of social context on individual fitness. Thus, it has correctly detected exactly what it was designed to detect. However, care should be exercised when communicating this form of "group selection" to an audience that may have something different in mind. It might be most straightforward to claim that individual fitness depends on group composition, which is consistent with the presence of meaningful social interactions, when a nonzero contextual partial regression coefficient is discovered. Thus, in the *I. capensis* and *A. thaliana* systems, it was found that group composition can affect the fitness of individual plants (Stevens et al. 1995; Weinig et al. 2007). This need not go hand-in-hand with fitness differences at the level of plant groups. Nevertheless, the statistical approach is telling us something important: the potential role of social interactions in determining fitness.

In both laboratory and natural systems, social interactions are ubiquitous. The individualist and multilevel perspectives discussed here highlight different aspects of these interactions. In the spirit of contextual analysis, the individualist perspective explicitly lays out how individuals are affected by their group mates. In the spirit of the Price equation, the multilevel perspective scales up to the group-level effects of these interactions. Kerr and Godfrey-Smith (2002a) suggested that it might be helpful to keep both perspectives on the table when dealing with social evolution in group-structured populations. Indeed, developing the ability to "gestalt switch" may lead to a richer understanding of the evolutionary process.

There is one area where gestalt switching may be particularly appealing. This is in discussions of the so-called major transitions in evolution (Maynard Smith and Szathmáry 1995; Michod 1999; Okasha 2006). When we ask any question about the level of selection in a biological hierarchy, we are taking the existence of the hierarchy for granted. However, an extremely interesting topic is how the hierarchy came into existence in the first place (Okasha 2005, 2006). A major transition can involve the creation of a new level (e.g., prokaryotes associating to produce eukaryotes, single-celled organisms giving rise to multicellular organisms, asocial individuals forming societies). When discussing a transition, we must consider the evolutionary process whereby interactions between previously autonomous lower-level entities generate a higher-level entity, which often can be characterized by lower-level altruism, cohesive integration, and division of labor.

Let us consider such a major transition. At the beginning, suppose that the lower-level entities are completely autonomous and do not interact in any meaningful way. Here an exclusively individualist perspective is natural, where parameters are context-independent fitnesses of lower-level entities. However, suppose that meaningful interactions start to occur within collections of the lower-level entities. We now enter the aforementioned gray zone where *both* an individualist perspective, with parameters that are context-*dependent* lower-level fitnesses, *and* a multilevel perspective, with parameters that are higher-level fitnesses and lower-level skew, are natural options. As the collection becomes more integrated, we enter a situation in which the multilevel perspective is the most natural. Indeed, one notion of "common fate" is that different lower-level entities simultaneously experience fitness changes, and this may be best captured by a parameterization explicitly representing "joint" fitness (e.g., the multilevel perspective's $\pi$).

In this sense, it seems likely that selection at higher levels, as captured by differences in higher-level fitness, was crucial to the completion of these major transitions. How precisely such transitions are accomplished is an active area of research, engaging both theoreticians and empirical biologists. Indeed, approaches have been proposed to empirically explore specific transitions (e.g., see Rainey 2007). Some of the same factors thought to influence the evolution of altruism also reappear in discussions of the major transitions. In particular, factors affecting the formation of social context are often seen as important. Thus, experiments manipulating such factors may contribute to a better understanding of the major transitions. Clearly organisms are not islands, but theoretical and experimental approaches to the evolution of social interactions may give us much more than this simple insight—such research may speak to the very origins of organisms themselves.

## SUMMARY

An orthodox perspective on natural selection maintains that any favored phenotype improves the fitness of the individual exhibiting it. From such a viewpoint, traits that involve personal sacrifice to increase the fitness of others should be swiftly eliminated. Despite this expectation, altruistic traits are found in many different natural systems, from the fruiting bodies of slime molds to the colonies of eusocial insects. This chapter explores the evolution of altruism using multilevel selection theory. Employing a simple trait group framework, the chapter reviews various definitions of *altruism* and some of the necessary conditions for the evolution of altruism under different definitions. In particular, it discusses the important role of association between altruists, which highlights certain connections between multilevel selection and kin selection. Different statistical approaches to partition selection within group-structured populations (e.g., the Price equation and contextual analysis) are also presented. The experimental evolution literature is explored next, with a survey of research ranging from artificial selection in the laboratory to natural selection in the field. Experiments with animals, plants, and microbes have provided much insight on the conditions promoting the evolution of altruism. The chapter ends with a reconsideration of the levels of selection controversy in light of the empirical and theoretical results.

## ACKNOWLEDGMENTS

I thank Ted Garland, Peter Godfrey-Smith, Michael Rose, Michael Wade, Karen Walag, and one anonymous reviewer for very useful feedback on this chapter. I am also grateful to Carl Bergstrom, Mark Borello, Brett Calcott, Michael Doebeli, Marc Ereshefsky, Marc Feldman, Jeremy Fox, Jeff Fletcher, Lisa Lloyd, John Maynard Smith, Len Nunney, Samir Okasha, Paul Rainey, Elliott Sober, Kim Sterelny, Ken Waters, David Sloan Wilson, and Rasmus Winther for past conversations about the topics of this chapter.

1. For some authors, the discreteness of groups is critical in discussions of group selection (e.g., Maynard Smith 1964, 1976). As Maynard Smith (1976) notes: "For kin selection . . . it is necessary that relatives live close to one another, but it is not necessary (although it may be favorable) that the population be divided into reproductively isolated groups. . . . For group selection, the division into groups which are partially isolated from one another is an essential feature." By Maynard Smith's account, when interactions between relatives are diffuse and overlapping (e.g., when individuals are distributed spatially and interact with neighboring relatives), kin selection is a possibility, but group selection is not (see Godfrey-Smith 2008 for a full discussion of this issue).

2. In another study, McCauley and Wade (1980) investigated factors contributing to high and low group productivity in *T. castaneum* and found that differences in egg fertility, time of development, and the sensitivity to crowding were important. In addition, they also found that differences in egg and pupal cannibalism by adults and larvae influenced group productivity.

3. How "mixing" affects between group variance has been explored elegantly in a separate experiment on the evolution of egg cannibalism (Wade 1980). Wade allowed *T. confusum* to evolve in metapopulations where two components of population structure were manipulated in a full factorial design: (1) mating (breeding took place exclusively within groups or at random) and (2) social interaction (larvae were offered eggs to potentially cannibalize of varying degrees of relatedness). When breeding was constrained to occur within groups, the beetles evolved lower cannibalism rates when encountering eggs with a higher degree of relatedness—consistent with kin selection theory. Under random mating, no significant differences in evolved cannibalism rates were discovered across social interaction treatments. Wade suggested that within-group breeding tended to promote a higher between-group variance in productivity, whereas the random breeding tended to homogenize groups (see Wade and Breden (1981) for a theoretical treatment of these issues).

4. Incidentally, Wade and colleagues (Wade 1982; Wade and McCauley 1984; Wade and Goodnight 1991) did perform a series of metapopulation experiments in which some migration between subpopulations accompanied the founding propagules. While the metapopulation treatments without migration generally promoted the largest variance in population size between demes (this variance is necessary for "interdemic" selection), significant variation was also discovered between demes in treatments with substantial migration (e.g., where 25 percent of each founding subpopulation were migrants). Furthermore, selection for population productivity produced a response in these higher migration treatments.

5. Griffing (1976a, 1976b) labels the influence of group mates on a focal individual as "associate effects." Using a theoretical approach, he discusses the benefits of different breeding strategies in the presence of such effects.

6. Of course, there are other valuable approaches available. West et al. (2008) argue that the Kerr et al. (2006) experiment can be viewed from a kin selection perspective. While noting connections between kin and group selection approaches, West et al. (2007, 2008) argue that there are general advantages to kin selection methodologies. In response, Wilson (2008) and Wilson and Wilson (2007) defend the use of multilevel selection approaches. It is telling how each set of authors describes the Kerr et al. (2006) experiment. West et al. (2007) state, "A more efficient use of host resources is favored when there is a higher relatedness between

the phage infecting a bacterium—local migration leads to a higher relatedness and hence selects for lower virulence," whereas Wilson (2007) states, "Biologically plausible migration rates enabled 'prudent' phage strains to outcompete more 'rapacious' strains in the metapopulation despite their selective disadvantage within each well, exactly as envisioned by Wynne-Edwards." I would argue that West et al. are focusing on social context formation (how different types find themselves in specific neighborhoods of interaction), while Wilson is focusing on the fitness structure (within-group disadvantage and between-group advantage of the prudent type). Of course, both social context formation and the fitness structure are critical to the prospects for the evolution of restraint (as I suspect all of these authors would readily admit). The original Kerr et al. (2006) manuscript mentioned neither group selection nor kin selection, but it does make statements that resonate with each approach, in part because each approach does focus on something important within this system.

## REFERENCES

Abedon, S. T., P. Hyman, and C. Thomas. 2003. Experimental examination of bacteriophage latent-period evolution as a response to bacterial availability. *Applied and Environmental Microbiology* 69:7499–7506.

Anderson, R. M., and R. M. May. 1982. Coevolution of hosts and parasites. *Parasitology* 85:411–426.

Axelrod, R., and W. D. Hamilton. 1981. The evolution of cooperation. *Science* 211:1390–1396.

Baer, C. F., J. Travis, and K. Higgins. 2000. Experimental evolution in *Heterandria formosa*, a livebearing fish: Group selection on population size. *Genetical Research* 76:169–178.

Bargum, K., and L. Sundstrom. 2007. Multiple breeders, breeder shifts and inclusive fitness returns in an ant. *Proceedings of the Royal Society B, Biological Sciences* 274:1547–1551.

Boots, M., and M. Mealor. 2007. Local interactions select for lower pathogen infectivity. *Science* 315:1284–1286.

Bull, J. J. 1994. Perspective: Virulence. *Evolution* 48:1423–1437.

Bull, J. J., I. J. Molineux, and W. R. Rice. 1991. Selection of benevolence in a host-parasite system. *Evolution* 45:875–882.

Cohen, D., and I. Eshel. 1976. On the founder effect and the evolution of altruistic traits. *Theoretical Population Biology* 10:276–302.

Cooper, V. S., M. H. Reiskind, J. A. Miller, K. A. Shelton, B. A. Walther, J. S. Elkinton, and P. W. Ewald. 2002. Timing of transmission and the evolution of virulence of an insect virus. *Proceedings of the Royal Society of London B, Biological Sciences* 269:1161–1165.

Craig, D. M. 1982. Group selection versus individual selection: An experimental analysis. *Evolution* 36:271–282.

Craig, J. V., M. L. Jan, C. R. Polley, A. L. Bhagwat, and A. D. Dayton. 1975. Changes in relative aggressiveness and social dominance associated with selection for early egg-production in chickens. *Poultry Science* 54:1647–1658.

Craig, J. V., and W. M. Muir. 1996. Group selection for adaptation to multiple-hen cages: Beak-related mortality, feathering, and body weight responses. *Poultry Science* 75:294–302.

Dao, D. N., R. H. Kessin, and H. L. Ennis. 2000. Developmental cheating and the evolutionary biology of *Dictyostelium* and *Myxococcus*. *Microbiology—UK* 146:1505–1512.

Darwin, C. 1859. On the Origin of Species. London: Murray.

Dennis, R. L., W. M. Muir, and H. W. Cheng. 2006. Effects of raclopride on aggression and stress in diversely selected chicken lines. *Behavioural Brain Research* 175:104–111.

Donne, J. 1624. Meditation XVII. Pages 574–575 *in* H. Alford, ed. *The Works of John Donne* III. London: Parker, 1839.

Donohue, K. 2003. The influence of neighbor relatedness on multilevel selection in the Great Lakes sea rocket. *American Naturalist* 162:77–92.

———. 2004. Density-dependent multilevel selection in the Great Lakes sea rocket. *Ecology* 85:180–191.

Dugatkin, L. A., and H. K. Reeve. 1994. Behavioral ecology and the levels of selection- dissolving the group selection controversy. Pages 101–133 *in Advances in the Study of Behavior, Vol. 23. Advances in the Study of Behavior.*

Emerson, A. E. 1939. Social coordination and the superorganism. *American Midland Naturalist* 21:182–209.

Eshel, I. 1972. On the neighbor effect and the evolution of altruistic traits. *Theoretical Population Biology* 3:258–277.

Fiegna, F., Y. T. N. Yu, S. V. Kadam, and G. J. Velicer. 2006. Evolution of an obligate social cheater to a superior cooperator. *Nature* 441:310–314.

Fletcher, J. A., and M. Zwick. 2006. Unifying the theories of inclusive fitness and reciprocal altruism. *American Naturalist* 168:252–262.

Fletcher, J. A., M. Zwick, M. Doebeli, and D. S. Wilson. 2006. What's wrong with inclusive fitness? *Trends in Ecology & Evolution* 21:597–598.

Frank, S. A. 1995. Mutual policing and repression of competition in the evolution of cooperative groups. *Nature* 377:520–522.

Gilbert, O. M., K. R. Foster, N. J. Mehdiabadi, J. E. Strassmann, and D. C. Queller. 2007. High relatedness maintains multicellular cooperation in a social amoeba by controlling cheater mutants. *Proceedings of the National Academy of Sciences of the USA* 104:8913–8917.

Godfrey-Smith, P. 2008. Varieties of population structure and the levels of selection. *British Journal for the Philosophy of Science* 59:25–50.

Goodnight, C. J. 1985. The influence of environmental variation on group and individual selection in a cress. *Evolution* 39:545–558.

Goodnight, C. J., J. M. Schwartz, and L. Stevens. 1992. Contextual analysis of models of group selection, soft selection, hard selection, and the evolution of altruism. *American Naturalist* 140:743–761.

Goodnight, C. J., and L. Stevens. 1997. Experimental studies of group selection: What do they tell us about group selection in nature? *American Naturalist* 150:S59–S79.

Greig, D., and M. Travisano. 2004. The Prisoner's Dilemma and polymorphism in yeast SUC genes. *Proceedings of the Royal Society of London B, Biological Sciences* 271:S25–S26.

Griffin, A. S., S. A. West, and A. Buckling. 2004. Cooperation and competition in pathogenic bacteria. *Nature* 430:1024–1027.

Griffing, B. 1976a. Selection in reference to biological groups. V. Analysis of full-sib groups. *Genetics* 82:703–722.

———. 1976b. Selection in reference to biological groups. VI. Use of extreme forms of nonrandom groups to increase selection efficiency. *Genetics* 82:723–731.

Hamilton, W. D. 1963. Evolution of altruistic behavior. *American Naturalist* 97:354–356.

———. 1964. Genetical evolution of social behavior I. *Journal of Theoretical Biology* 7:1–16.

———. 1975. Innate social aptitudes in man: An approach from evolutionary genetics. Pages 133–155 *in* R. Fox, ed. *Biosocial Anthropology.* New York: Wiley.

Hardin, G. 1968. Tragedy of the commons. *Science* 162:1243–1248.

Harrison, S., and A. Hastings. 1996. How effective is interdemic selection? Reply. *Trends in Ecology & Evolution* 11:299–299.

Heisler, I. L., and J. Damuth. 1987. A method for analyzing selection in hierarchically structured populations. *American Naturalist* 130:582–602.

Jeon, K. W. 1972. Development of cellular dependence on infective organisms: Microsurgical studies in amoebas. *Science* 176:1122.

Jeon, K. W., and J. C. Hah. 1977. Effect of chloramphenicol on bacterial endosymbiotes in a strain of *Amoeba proteus. Journal of Protozoology* 24:289–293.

Jones, C. G., J. H. Lawton, and M. Shachak. 1994. Organisms as ecosystem engineers. *Oikos* 69:373–386.

Karlin, S., and C. Matessi. 1983. Kin selection and altruism. *Proceedings of the Royal Society of London B, Biological Sciences* 219:327–353.

Kerr, B., and P. Godfrey-Smith. 2002a. Individualist and multilevel perspectives on selection in structured populations. *Biology & Philosophy* 17:477–517.

———. 2002b. On Price's equation and average fitness. *Biology & Philosophy* 17:551–565.

Kerr, B., P. Godfrey-Smith, and M. W. Feldman. 2004. What is altruism? *Trends in Ecology & Evolution* 19:135–140.

Kerr, B., C. Neuhauser, B. J. M. Bohannan, and A. M. Dean. 2006. Local migration promotes competitive restraint in a host-pathogen "tragedy of the commons." *Nature* 442:75–78.

Krakauer, A. H. 2005. Kin selection and cooperative courtship in wild turkeys. *Nature* 434:69–72.

Lehmann, L., L. Keller, S. West, and D. Roze. 2007. Group selection and kin selection: Two concepts but one process. *Proceedings of the National Academy of Sciences of the USA* 104:6736–6739.

Lewontin, R. C. 1970. The units of selection. *Annual Review of Ecology and Systematics* 1:1–18.

Lewontin, R. C., and L. C. Dunn. 1960. The evolutionary dynamics of a polymorphism in the house mouse. *Genetics* 45:705–722.

Matessi, C., and S. D. Jayakar. 1976. Conditions for the evolution of altruism under Darwinian selection. *Theoretical Population Biology* 9:360–387.

Matessi, C., and S. Karlin. 1984. On the evolution of altruism by kin selection. *Proceedings of the National Academy of Sciences of the USA* 81:1754–1758.

Maynard Smith, J. 1964. Group selection and kin selection. *Nature* 201:1145–1147.

———. 1976. Group selection. *Quarterly Review of Biology* 51:277–283.

———. 2002. Commentary on Kerr and Godfrey-Smith. *Biology & Philosophy* 17:523–527.

Maynard Smith, J., and E. Szathmáry. 1995. *The Major Transitions in Evolution.* Oxford: Oxford University Press.

McCauley, D. E., and M. J. Wade. 1980. Group selection: The genetic and demographic basis for the phenotypic differentiation of small populations of *Tribolium castaneum. Evolution* 34:813–821.

McLeish, M. J., T. W. Chapman, and B. J. Crespi. 2006. Inbreeding ancestors: The role of sibmating in the social evolution of gall thrips. *Journal of Heredity* 97:31–38.

Mehdiabadi, N. J., C. N. Jack, T. T. Farnham, T. G. Platt, S. E. Kalla, G. Shaulsky, D. C. Queller et al. 2006. Kin preference in a social microbe: Given the right circumstances, even an amoeba chooses to be altruistic towards its relatives. *Nature* 442:881–882.

Messenger, S. L., I. J. Molineux, and J. J. Bull. 1999. Virulence evolution in a virus obeys a trade-off. *Proceedings of the Royal Society of London B, Biological Sciences* 266:397–404.

Michod, R. E. 1999. *Darwinian Dynamics: Evolutionary Transitions in Fitness and Individuality.* Princeton, NJ: Princeton University Press.

Muir, W. M. 1996. Group selection for adaptation to multiple-hen cages: Selection program and direct responses. *Poultry Science* 75:447–458.

Nunney, L. 1985. Group selection, altruism, and structured-deme models. *American Naturalist* 126:212–230.

Odling-Smee, F. J., K. N. Laland, and M. W. Feldman. 2003. *Niche Construction: The Neglected Process in Evolution.* Princeton, NJ: Princeton University Press.

Okasha, S. 2005. Multilevel selection and the major transitions in evolution. *Philosophy of Science* 72:1013–1025.

———. 2006. *Evolution and the Levels of Selection.* Oxford: Oxford University Press.

Prado, F., and B. Kerr. 2008. The evolution of restraint in bacterial biofilms under nontransitive competition. *Evolution* 62: 538–548.

Price, G. R. 1970. Selection and covariance. *Nature* 227:520–521.

———. 1972. Extension of covariance selection mathematics. *Annals of Human Genetics* 35:485–490.

Queller, D. C. 1985. Kinship, reciprocity and synergism in the evolution of social behavior. *Nature* 318:366–367.

Queller, D. C., E. Ponte, S. Bozzaro, and J. E. Strassmann. 2003. Single-gene greenbeard effects in the social amoeba *Dictyostelium discoideum. Science* 299:105–106.

Rainey, P. B. 2007. Unity from conflict. *Nature* 446:616–616.

Rainey, P. B., and K. Rainey. 2003. Evolution of cooperation and conflict in experimental bacterial populations. *Nature* 425:72–74.

Sachs, J. L., and J. J. Bull. 2005. Experimental evolution of conflict mediation between genomes. *Proceedings of the National Academy of Sciences of the USA* 102:390–395.

Sachs, J. L., and T. P. Wilcox. 2006. A shift to parasitism in the jellyfish symbiont *Symbiodinium microadriaticum. Proceedings of the Royal Society of London B, Biological Sciences* 273:425–429.

Scheiner, S. M. 2002. Selection experiments and the study of phenotypic plasticity. *Journal of Evolutionary Biology* 15:889–898.

Sober, E. 1984. *The Nature of Selection.* Chicago: University of Chicago Press.

Sober, E., and D. S. Wilson. 1998. *Unto Others: The Evolution and Psychology of Unselfish Behavior.* Cambridge, MA: Harvard University Press.

Sterelny, K. 1996. The return of the group. *Philosophy of Science* 63:562–584.

Stevens, L., C. J. Goodnight, and S. Kalisz. 1995. Multilevel selection in natural populations of *Impatiens capensis. American Naturalist* 145:513–526.

Strassmann, J. E., Y. Zhu, and D. C. Queller. 2000. Altruism and social cheating in the social amoeba *Dictyostelium discoideum. Nature* 408:965–967.

Swenson, W., J. Arendt, and D. S. Wilson. 2000a. Artificial selection of microbial ecosystems for 3-chloroaniline biodegradation. *Environmental Microbiology* 2:564–571.

Swenson, W., D. S. Wilson, and R. Elias. 2000b. Artificial ecosystem selection. *Proceedings of the National Academy of Sciences of the USA* 97:9110–9114.

Travisano, M., and G. J. Velicer. 2004. Strategies of microbial cheater control. *Trends in Microbiology* 12:72–78.

Trivers, R. L. 1971. Evolution of reciprocal altruism. *Quarterly Review of Biology* 46:35–&.

Turner, P. E., and L. Chao. 1999. Prisoner's dilemma in an RNA virus. *Nature* 398:441–443.

Uyenoyama, M., and M. W. Feldman. 1980. Theories of kin and group selection: A population-genetics perspective. *Theoretical Population Biology* 17:380–414.

Velicer, G. J., L. Kroos, and R. E. Lenski. 1998. Loss of social behaviors by Myxococcus xanthus during evolution in an unstructured habitat. *Proceedings of the National Academy of Sciences of the USA* 95:12376–12380.

Wade, M. J. 1977. Experimental study of group selection. *Evolution* 31:134–153.

———. 1979. Primary characteristics of *Tribolium* populations group selected for increased and decreased population size. *Evolution* 33:749–764.

———. 1980a. An experimental study of kin selection. *Evolution* 34:844–855.

———. 1980b. Group selection, population growth rate, and competitive ability in the flour beetles, *Tribolium* species. *Ecology* 61:1056–1064.

———. 1980c. Kin selection: Its components. *Science* 210:665–667.

———. 1982. Group selection: migration and the differentiation of small populations. *Evolution* 36:949–961.

Wade, M. J., and F. Breden. 1981. Effect of inbreeding on the evolution of altruistic behavior by kin selection. *Evolution* 35:844–858.

Wade, M. J., and C. J. Goodnight. 1991. Wright's shifting balance theory: An experimental study. *Science* 253:1015–1018.

Wade, M. J., and D. E. McCauley. 1980. Group selection: The phenotypic and genotypic differentiation of small populations. *Evolution* 34:799–812.

———. 1984. Group selection: The interaction of local deme size and migration in the differentiation of small populations. *Evolution* 38:1047–1058.

Weinig, C., J. A. Johnston, C. G. Willis, and J. N. Maloof. 2007. Antagonistic multilevel selection on size and architecture in variable density settings. *Evolution* 61:58–67.

West, S. A., A. S. Griffin, and A. Gardner. 2007. Social semantics: Altruism, cooperation, mutualism, strong reciprocity and group selection. *Journal of Evolutionary Biology* 20:415–432.

———. 2008. Social semantics: How useful has group selection been? *Journal of Evolutionary Biology* 21:374–385.

Williams, G. C. 1966. *Adaptation and Natural Selection*. Princeton, NJ: Princeton University Press.

———. 1971. *Group Selection*. Chicago: Aldine Atherton.

Wilson, D. S. 1975. Theory of group selection. *Proceedings of the National Academy of Sciences of the USA* 72:143–146.

———. 1977. Structured demes and evolution of group-advantageous traits. *American Naturalist* 111:157–185.

———. 1980. *The Natural Selection of Populations and Communities*. Menlo Park, CA: Benjamin/Cummings.

———. 1983. The group selection controversy: History and current status. *Annual Review of Ecology and Systematics* 14:159–187.

———. 1990. Weak altruism, strong group selection. *Oikos* 59:135–140.

———. 2008. Social semantics: Toward a genuine pluralism in the study of social behaviour. *Journal of Evolutionary Biology* 21:368–373.

Wilson, D. S., and E. O. Wilson. 2007. Rethinking the theoretical foundation of sociobiology. *Quarterly Review of Biology* 82:327–348.

Wilson, R. A. 2004. Test cases, resolvability, and group selection: A critical examination of the myxoma case. *Philosophy of Science* 71:380–401.

Wright, S. 1945. Tempo and mode in evolution: A critical review. *Ecology* 26:415–419.

Wynne-Edwards, V. C. 1962. *Animal Dispersion in Relation to Social Behavior*. New York: Hafner.

# 20

# LABORATORY EXPERIMENTS ON SPECIATION

James D. Fry

KEY CONCEPTS

WHAT PAST EXPERIMENTS HAVE TAUGHT US

NEGLECTED QUESTIONS

GENERAL GUIDELINES FOR EXPERIMENTS ON SPECIATION

After neglecting the subject for nearly a century after the publication of *The Origin of Species*, evolutionary biologists have been intensively investigating mechanisms of speciation in the last few decades (reviewed in Barton 2001; Coyne and Orr 2004; Rundle and Nosil 2005; Noor and Feder 2006; Rieseberg and Willis 2007). Experimental evolution approaches have made an important contribution to this resurgence of interest in speciation, complementing theoretical, genetic, and comparative approaches (see also Futuyma and Bennett this volume). This chapter will review the literature on speciation experiments, identify neglected questions that could be addressed by new experiments, and suggest general guidelines for such experiments. Because several reviews of experiments on speciation have been published in recent years (Rice and Hostert 1993; Florin and Ödeen 2002; Kirkpatrick and Ravigné 2002; Coyne and Orr 2004), I will emphasize recent and overlooked experiments, and the prospects and challenges for new experiments.

Although not all definitions of the term *species* explicitly incorporate reproductive isolation (reviewed in Coyne and Orr 2004), whatever definition is adopted, some degree of reproductive isolation is necessary for sympatric species to coexist as distinguishable entities. Hence, understanding the origin of reproductive isolation is necessary for understanding the origin and maintenance of biological diversity. Laboratory experiments on speciation investigate the conditions under which reproductive isolation can evolve between members of what was initially a single, interbreeding population, as well as the conditions under which reproductive isolation between initially partly reproductively isolated populations can become intensified.

The organization of the chapter is as follows. The next section defines terms used in the speciation literature (mostly following Coyne and Orr 2004) and, in so doing, gives an overview of the questions that have been, or could be, addressed by experimental evolution approaches. The second section summarizes the main conclusions from past laboratory experiments on speciation. (I will not address the interesting recent experiments of Rieseberg et al. 1996 and Greig et al. 2002 on homoploid hybrid speciation, a relatively specialized mode of speciation.) The third section suggests questions that have been mostly neglected in past experiments, but which are ripe for further investigation. Finally, the fourth section gives some general guidelines for future laboratory experiments on speciation.

This chapter will focus exclusively on sexually reproducing eukaryotes—that is, those with meiosis and syngamy at some stage of the life cycle. Asexual lineages are automatically reproductively isolated except for occasional horizontal gene transfer (HGT) events. Although bacterial and archaeal lineages form clusters within which HGT occurs relatively easily, but between which HGT rarely occurs (Lawrence 2002), the evolution of such clusters has not, to my knowledge, been investigated experimentally. In experimental microcosms, single bacterial clones may differentiate into multiple, ecologically distinct forms that partition the available environment (Rainey and Travisano 1998; Travisano this volume), but this process does not involve the *de novo* evolution of reproductive isolation.

Reproductive isolating barriers can be classified as premating (e.g., lack of response of females to courtship signals of males of another species), postmating but prezygotic (e.g., inability of sperm to fertilize eggs of another species), or postzygotic (e.g., hybrid inviability or sterility, either in the $F_1$ or later generations). For each type of barrier, it is also useful to distinguish between those that depend on the abiotic or biotic environment ("extrinsic") and those that are relatively independent of the environment ("intrinsic"). For example, two species might form hybrids that are perfectly viable and fertile in the laboratory, but if the species are adapted to different niches, the hybrids would have low fitness in the wild. This is an example of what would be called "ecological isolation" (Coyne and Orr 2004). Ecological isolation can also apply to premating barriers; most notably, species that use different habitats or hosts may mate readily in the laboratory, but will rarely encounter each other in nature.

Reproductive barriers can also be classified by whether they arose in the presence or absence of gene flow between the diverging populations. Allopatric speciation, in which reproductive isolation evolves between geographically separated populations, is accepted as the predominant mode of speciation in most groups, but the possibility of speciation with either no geographic isolation (sympatric) or only partial geographic isolation (parapatric) has gained increasing, albeit far from universal, acceptance in recent years (e.g., Via 2001; but see Coyne and Orr 2004).

A final important set of distinctions involves the evolutionary factors responsible for the origin of reproductive isolation. Most broadly, reproductive isolation could evolve either due to selection or due to random genetic drift (and possibly their interaction); the relative importance of the two has long been controversial. In models that invoke selection, reproductive isolation can evolve either as a by-product of selection on other traits (due either to pleiotropy or linkage disequilibrium) or because of selection for reproductive isolation per se (most notably, in models of reinforcement, in which selection favors avoidance of mating with heterospecifics because of the low fitness of hybrids). Finally, drift can take several forms, such as single bottlenecks, repeated bottlenecks followed by population expansions ("founder-flush" cycles), and extended periods of low population size (see also Futuyma and Bennett this volume).

Thus, evolutionary experiments can be used to investigate such issues as the relative efficacy of natural selection and drift in generating reproductive isolation, the relative rates of evolution of the different types of reproductive barriers, the feasibility of sympatric and parapatric speciation, and the feasibility of reinforcement.

## WHAT PAST EXPERIMENTS HAVE TAUGHT US

*"Destroy all the hybrids" experiments show that the tendency to mate assortatively can be increased by selection*  For premating reproductive isolation to evolve between a pair of populations, genetic variation for the tendency to mate assortatively must be present in either or both.

This assumption has been tested in a series of experiments, often called "destroy all the hybrids" experiments, dating back to Koopman (1950). In these experiments, individuals of each sex from two different strains, subspecies, or incompletely reproductively isolated species are placed together and allowed to mate. Some mechanism is then used to ensure that only offspring resulting from "homogamic" (i.e., within-strain) matings are allowed to contribute to the next generation. (Most often, the two strains are homozygous for different recessive genetic markers, so that hybrids between them are immediately recognizable by their wild-type phenotype.) The frequency of homogamic and heterogamic matings is then monitored over successive generations.

I am aware of fourteen such experiments, twelve on flies and one each on maize and yeast, most of which were successful in selecting for increased premating reproductive isolation (table 20.1). Not surprisingly, some of the most rapid responses were in the experiments where the original populations were different species (Koopman 1950; Kessler 1966) or "semispecies" (Dobzhansky et al. 1976). In contrast, all three experiments that reported negative results used strains of *Drosophila melanogaster* as their starting material (of course, it is possible that there were other, unreported negative results). As noted by Coyne and Orr (2004), the lack of response in the experiment of Harper and Lambert (1983) could have been caused by lack of genetic variation in the strains used, which appear to have been "off-the-shelf" marker stocks. In the other two instances where no response was observed, the base populations were outbred, but closely related to one another (Robertson 1966b; Fukatami and Morikami 1970). In these cases, it is possible that there was no initial divergence between the strains in traits affecting mate choice for selection to augment. Nonetheless, the overall conclusion from the experiments summarized in table 20.1 is that there seems to be no shortage of genetic variation for the tendency to mate assortatively in most populations.

An important caveat about these experiments, however, is that they do not lend support to any particular model of speciation. In fact, because hybrids had zero fitness under the experimental regimes (or almost zero; Leu and Murray 2006), the populations were in effect completely reproductively isolated "species" from the outset. Formally, the experiments test for reproductive character displacement—the enhancement of premating isolation between already reproductively isolated species—rather than reinforcement, in which selection against the tendency to hybridize accelerates the speciation process (Butlin 1987). Nonetheless, the rapidity with which reproductive isolation evolved in many of the experiments suggests that experiments to test the reinforcement model might meet with success, though few such experiments have been attempted (discussed later).

*Divergent selection in allopatry often leads to the evolution of partial premating isolation*   The idea that reproductive isolation between allopatric populations often evolves as a by-product of adaptation to different environments has received strong support from several experiments on *Drosophila* and other flies. In these experiments, laboratory populations that had been

TABLE 20.1  Results of "Destroy All the Hybrids" Experiments

| Study | Species | Method[a] | Augmented Genetic Variation?[b] | Base Populations Closely Related?[c] | Response[d] |
|---|---|---|---|---|---|
| Koopman 1950 | *Drosophila persimilis* and *D. pseudoobscura* | VM | Yes | No | + |
| Kessler 1966 | *D. persimilis* and *D. pseudoobscura* | VM | Yes | No | + |
| Dobzhansky et al. 1976 | *D. paulistorum* (two "semispecies") | VM | No | No | + |
| Wallace 1953 | *D. melanogaster* | VM | No | No? | +/− |
| Knight et al. 1956 | *D. melanogaster* | VM | No | No? | + |
| Robertson 1966b | *D. melanogaster* | DO | Yes | Yes | − |
| Fukatami and Morikami 1970 | *D. melanogaster* | VM | Yes | Yes | − |
| Ehrman 1971, 1973, 1979 | *D. melanogaster* | HL | No | No? | +/− |
| Crossley 1974 | *D. melanogaster* | VM | Yes | Yes | + |
| Harper and Lambert 1983 | *D. melanogaster* | VM | No | No? | − |
| Hostert 1997 | *D. melanogaster* | VM | Yes | Yes | + |
| Regan et al. 2003 | *Musca domestica* | DO | Yes | Yes | + |
| Paterniani 1969 | *Zea mays* | VM | No | No? | + |
| Leu and Murray 2006 | *Saccharomyces cerivisiae* | HL | Yes | Yes | +[e] |

[a] VM = visible genetic markers; DO = direct observation of mating pairs; HL = hybrid lethal genetic system.

[b] Indicates whether special steps were taken prior to experiment to ensure that base populations were genetically variable (e.g., by backcrossing marker stocks to diverse wild-type lines).

[c] Indicates whether base populations were derived from the same wild source population.

[d] +, increase in premating isolation observed; −, no increase; +/−, ambiguous or inconsistent results.

[e] This study used a unique design. Evolving experimental populations (E) received immigrants from a reference, nonevolving strain (R) that had been genetically engineered to contain an inducible dominant suicide gene, in a ratio of 10 R cells to 1 E cell. After mating, the suicide gene was induced, killing most but not all R cells and R-E hybrids (approximately 2% of the survivors in the initial generations expressed markers from the R strain). This could be viewed as an experiment to test for reinforcement in a continent-island scenario (see main text's section on "How Readily Does Reinforcement Occur?"), except that the migrants greatly outnumbered residents, the opposite of what would be expected. For this reason, and because of the very strong selection against hybrids, the study is probably better viewed as a test for reproductive character displacement rather than as a test of a realistic reinforcement scenario.

divergently selected for traits as diverse as geotaxis (Soans et al. 1974; Hurd and Eisenberg 1975; Lofdahl et al. 1992), development time (Miyatake and Shimizu 1999), and tolerance of temperature and humidity extremes (Kilias et al. 1980) were found to show significant premating reproductive isolation in mating choice tests (see table 3.1 in Coyne and Orr, 2004, for a summary). In those experiments where two or more replicate lines were selected in the same direction, the replicates usually showed no reproductive isolation from each other, giving powerful evidence that the isolation that had evolved between the divergently selected lines was not the result of genetic drift.

Not surprisingly, not all divergent selection experiments have resulted in reproductive isolation. For example, *D. melanogaster* lines selected in opposite direction for abdominal (Koref-Santibañez and Waddington 1958) and sternopleural (Barker and Cummins 1969) bristle number showed no reproductive isolation, in spite of strong divergence for the selected traits. Thus, some traits apparently show stronger pleiotropic connections to mating behavior than others, although the reasons for the connections are seldom known (but see Miyatake and Shimizu 1999; Rundle et al. 2005). Negative results do not always indicate that the selected trait has no connection to mating behavior, however. In some cases (Mooers et al. 1999; Rundle 2003), the authors did not present evidence that the populations had adapted to their selection regimes. Without such evidence, it is hard to know whether the failure to observe isolation was caused by the choice of trait or by insufficient strength or duration of selection to produce a response. Similarly, in other studies where no reproductive isolation evolved (Robertson 1966b; Markow 1981; also see table 20.2), responses to selection clearly occurred, but selection was unidirectional rather than bidirectional (i.e., selected lines were tested for isolation

TABLE 20.2   Results of "Multiple-Choice" Tests for Premating Isolation among Ethanol-Adapted (HE, for "high ethanol") and Control (R, for "regular food") Populations of Fry et al. (2004)

| Population A | Population B | Observed Matings (Female × Male) | | | | $Y$ (S.E.)[a] | $X^2$ |
|---|---|---|---|---|---|---|---|
| | | $A \times A$ | $A \times B$ | $B \times A$ | $B \times B$ | | |
| HE1 | HE2 | 5 | 9 | 7 | 6 | −0.18 (0.19) | 0.90 |
| R1 | R2 | 10 | 6 | 7 | 7 | 0.13 (0.18) | 0.48 |
| HE1 | R1 | 10 | 8 | 5 | 8 | 0.17 (0.18) | 0.88 |
| HE1 | R2 | 9 | 4 | 7 | 5 | 0.12 (0.21) | 0.32 |
| HE2 | R1 | 6 | 9 | 7 | 12 | 0.03 (0.18) | 0.04 |
| HE2 | R2 | 12 | 7 | 10 | 12 | 0.18 (0.15) | 1.28 |

NOTE: For each replicate, a male and virgin female from each of two populations were placed together in an empty vial; flies from one of the populations were marked by prefeeding with medium containing red food coloring, a treatment shown to have no influence on mate choice in preliminary trials. Vials were placed on their side, and the first mating recorded. Although the sample sizes are low, the HE and R populations show no evidence of premating isolation in spite of their strong divergence in ethanol tolerance (Fry et al. 2004).

[a]Isolation index of Spieth and Ringo (1983); $Y$ significantly $>$ 0 implies assortative mating.

from an unselected base population). Bidirectional selection, by increasing the divergence between the lines, might be expected to be more likely to result in the evolution of reproductive isolation (cf. Florin and Ödeen 2002). (The negative results of Rundle et al. 2003 noted here might also have been due to the authors' use of unidirectional selection, rather than failure of the selected lines to respond to selection in these relatively long-term experiments.)

Of course, even if divergent selection on most traits does not cause reproductive isolation, allopatric populations in nature are likely to be divergently selected for many traits (selection is likely to be "multifarious"; Rice and Hostert 1993). Moreover, such selection takes place on far longer time scales than can be replicated in the laboratory. Thus, even if negative results are underreported, the readiness with which premating reproductive isolation evolves between divergently selected laboratory populations provides strong support for the by-product model of allopatric speciation.

*Disruptive selection on arbitrary traits usually does not cause the evolution of premating isolation in sympatry*   Motivated by the long-standing controversy over sympatric speciation, several sets of investigators have tested whether applying strong disruptive selection to a population can result in the evolution of premating reproductive isolation. In one famous case (Thoday and Gibson 1962), disruptive selection on sternopleural bristle number in *D. melanogaster* resulted in apparently complete assortative mating between the selected extremes within twelve generations. Several attempts to replicate this result with different stocks failed, however. Moreover, similar experiments with different traits and/or species have also usually been unsuccessful at producing reproductive isolation (for reviews, see Thoday and Gibson 1970; Scharloo 1970; Rice and Hostert 1993; Coyne and Orr 2004). The main exceptions to this pattern of negative results have been studies in which the selected trait is either known to be or is at least plausibly related to mate choice—in other words, where a tendency to mate assortatively on the basis of the trait was apparently already present in the base population. Most notably, disruptive selection on geotaxis in the house fly resulted in strong premating isolation (Hurd and Eisenberg 1975). Significantly, divergent selection on this trait in allopatry also caused the evolution of reproductive isolation, as noted above. In contrast, traits used for most of the "unsuccessful" experiments, such as bristle number, apparently had little connection to mate choice, as evidenced in some cases by the failure of divergent selection on these traits in allopatry to produce reproductive isolation (Robertson 1966b; Barker and Cummins 1969).

In general, disruptive selection should lead to selection for any mechanism to reduce mating between the selected extremes, because such matings lead to the production of hybrids with low fitness. In theory, therefore, even if the disruptively selected trait has no effect on mate choice, another trait that affects mate choice and that happens to be fortuitously associated with the selected trait could be recruited to serve as the basis for mating discrimination between the selected extremes. In formal terms, this requires that linkage disequilibrium be present between genes affecting the selected trait and those affecting the mating trait. This "two-trait" or "double-variation" model faces considerable theoretical difficulties, however, because gene flow between the selected extremes and the ensuing recombination continually erode the requisite linkage disequilibrium (Rice and Hostert 1993; Fry 2003). Thus, sympatric speciation is much more plausible when the disruptively selected trait(s) simultaneously serve as the basis of mate choice. This topic is considered in more detail later.

*Population bottlenecks, by themselves, seldom lead to the evolution of premating reproductive isolation*
Some theories of speciation posit a key role for genetic drift (e.g., Mayr 1963; Templeton 1980; Carson and Templeton 1984; for a critical review, see Coyne and Orr 2004). Motivated by these theories, several investigators have subjected populations of *Drosophila* or *Musca* to population bottlenecks and tested whether the bottlenecked lines developed premating isolation from each other and/or from nonbottlenecked control populations (Powell 1978; Dodd and Powell 1985; Ringo et al. 1985; Meffert and Bryant 1991; Galiana et al. 1993; Moya et al. 1995; Rundle et al. 1998; Mooers et al. 1999; Rundle 2003). In most cases, the experiments used multiple single-pair bottlenecks, after each of which the bottlenecked lines were allowed to expand to a large size. Although premating reproductive isolation was observed between some pairs of lines in these experiments, the vast majority of pairwise combinations of lines showed no isolation, and a few even showed negative assortative mating (reviewed in Coyne and Orr 2004). Given the readiness with which reproductive isolation evolves in destroy-all-the-hybrids experiments (table 20.1), the negative results of bottleneck experiments are not likely to have been caused by lack of genetic variation for mating-related traits. Although the relevance of some of these experiments to particular theories of drift-induced speciation can be debated (see especially Templeton 1999 and the reply, Rundle et al. 1999), overall the results indicate that genetic drift, by itself, is only rarely effective at generating premating isolation. Of course, rare events can be important in evolution, so the results do not rule out the possibility that drift-induced speciation sometimes occurs, but they do suggest that this mode of speciation is less common than speciation due to divergent natural selection. Another caveat is that all of the experiments were done on flies; it is possible that bottleneck experiments on other taxa would give different results.

## NEGLECTED QUESTIONS

*How readily does postmating isolation evolve?*   Most experimental studies of speciation have focused on premating isolation; we know remarkably little about how quickly postmating isolation can evolve in the laboratory, either due to selection or drift. For example, only five of the twenty or so studies which examined whether divergent selection can lead to premating isolation also reported tests for postmating isolation (see Coyne and Orr 2004, table 3.1). Similarly, most of the studies of bottlenecked lines also failed to report tests for postmating isolation. Of the handful of studies that have tested for postmating isolation, some are uninformative; for example, one simply reported the qualitative observation that F₁ hybrids between bottlenecked lines were fertile (Powell 1978), while another used ill-defined criteria for assessing reproductive isolation (de Oliveira and Cordeiro 1980).

This neglect of postmating isolation in experimental studies is surprising, because there is a widely accepted model by which selection, either alone or in combination with drift, could generate postmating incompatibilities between allopatric populations. The basic idea of the model, which is usually called the Dobzhansky-Muller model (reviewed in Coyne and Orr 2004), is that hybridization is expected to create genetic combinations which have never previously been "tested" by natural selection. Figure 20.1 gives simple examples potentially relevant to experiments in which populations are divergently selected or subjected to bottlenecks. In both cases, the base population is genotype *aabb*, and the mutations *A* and *B* arise and become fixed in one of the descendent populations. In figure 20.1A, *A* and *B* become fixed in different populations; when the populations are hybridized, the two alleles, which have never before been in the same individuals, do not "work" well together, reducing the fitness of hybrids. (The fitness consequences could be manifest in the F₁, as shown in the figure, or delayed until the F₂ or backcross generation, depending on whether the negative interaction between *A* and *B* involves dominant or recessive effects). Although the Dobzhansky-Muller model makes no assumptions about the causes of fixation of the alleles, it is easy to imagine fixation of different alleles in the different populations occurring in response to divergent selection. In figure 20.1B, both *A* and *B* become fixed in the same population; because *A* fixes before *B* arises, however, *B* never occurs together with *a* until the populations hybridize. Figure 20.1B could apply to the case where the first and second populations are selected and control populations, respectively. It might also apply to the situation where the first

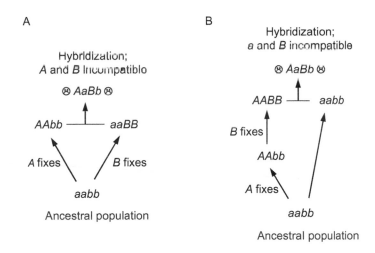

FIGURE 20.1
Dobzhansky-Muller incompatibilities. See text for explanation.

population is subjected to a population bottleneck and then allowed to expand. In this case, the bottleneck could cause fixation of a deleterious allele ($A$) by genetic drift; when the population expands, selection would then be expected to favor modifiers of the deleterious effects of $A$ ($B$ in the example).

I am aware of only eight studies that give useful data on the evolution of postmating reproductive isolation due to selection in the laboratory, and only a single informative study of the effects of population bottlenecks on postmating isolation. In the latter study (Ringo et al. 1985), eight replicate populations of *Drosophila simulans* were subjected to a series of single-pair bottlenecks, after each of which the populations were allowed to expand. Crosses between the bottlenecked populations and the base population showed a highly significant decline in productivity relative to within-population crosses with successive founder-flush cycles. Because the bottlenecked populations showed no evidence for premating isolation from the base population, the declining productivity of hybrid crosses gives evidence that partial postmating isolation evolved between the lines (either reduced $F_I$ viability or poor fertilization success in the hybrid matings). Because the base population was produced by crossing flies collected from widely separated localities, however, it is not clear that the incompatibilities arose *de novo*; the bottlenecked lines may have simply segregated out for variants that came from the different localities (cf. Rundle et al. 1998).

Ringo et al. (1985) also selected eight large populations for diverse traits and conducted similar tests of postmating isolation from the base population; none was observed. Kilias et al. (1980) similarly found little evidence for postmating isolation in crosses between a pair of *D. melanogaster* populations adapted to different regimes of temperature, light, and humidity, in spite of strong premating isolation between them. In contrast, Robertson (1966a) and Boake et al. (2003) created lines with different combinations of chromosomes from *D. melanogaster* populations selected for resistance to toxins (EDTA and DDT, respectively) and their respective controls, and they found that one of the "hybrid" chromosome combinations in each case had lower survival and/or fertility under control conditions than either parental population. Because there was only one selected and control population in each study, however, and the chromosome substitution experiments were themselves unreplicated, it is not clear that the selection treatments were responsible for the apparent genetic incompatibilities, if they were incompatibilities at all (e.g., it is possible that one of the parental lines in each study had a high frequency of a deleterious allele by chance, and this became fixed in the chromosome substitution process).

More convincing evidence that divergent selection can rapidly generate postmating isolation comes from an old but neglected study of mites (Overmeer 1966) and two recent studies of fungi (Dettman et al. 2007, 2008). Overmeer (1966), studying the haplodiploid plant pest *Tetranychus urticae*, crossed mites from two populations that had been independently selected for resistance to the pesticide Tedion back to the base population and found that the $F_I$ hybrids were partly sterile, laying many eggs that did not hatch (figure 20.2). Hybrids between the two selected populations had normal fertility, giving evidence that the isolation from the base population was the result of the resistance selection, not drift. A possible explanation for these results is that the resistance allele(s) initially had negative pleiotropic effects on embryo viability; this would have generated selection for modifiers of these effects in the selection lines. In eggs laid by $F_I$ females, however (particularly the haploid eggs laid by unfertilized females, which would normally develop into males), the modifiers would have become separated from the resistance alleles by recombination. The existence of negative fitness effects of pesticide resistance alleles, as well as modifiers of those effects, have been demonstrated in at least one other species (McKenzie and Game 1987). Interestingly, Overmeer's results suggest that some of the modifiers were cytoplasmic, because egg hatch was considerably lower when the cytoplasm of the $F_I$ females came from the base population than when it came from the resistant populations (figure 20.2). In contrast to Overmeer's result, Fry (1999) found no evidence for similar reproductive incompatibility between a *T. urticae* population selected for resistance to a toxic host plant and the control population.

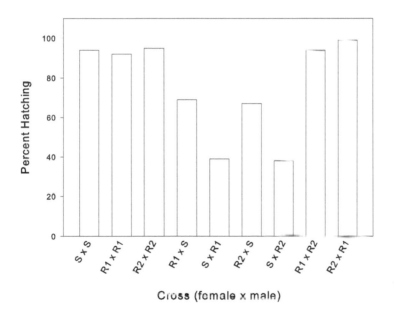

FIGURE 20.2

Hatching rates of haploid (unfertilized) eggs laid by $F_I$ females from crosses within and between populations of the haplodiploid mite *Tetranychus urticae* either selected for resistance to the acaracide Tedion (R1, R2) or not selected (S; from Overmeer 1966). Hatching rates are high in the five crosses between lines of the same type, lower when R line females are crossed to ancestral S line males, and lowest when R line males are crossed to S line females ($p < 0.001$, analysis of variance comparing the three types of crosses). Among crosses of a given type, the variation in hatching rates only slightly exceeds that expected due to binomial sampling ($N = 162 - 657$ eggs per cross).

Recent studies on yeast (Dettman et al. 2007) and the filamentous fungus *Neurospora* (Dettman et al. 2008) give additional evidence that adaptation to different environments can generate postmating isolation. In the yeast study, replicate populations derived from a single diploid progenitor were allowed to adapt to either of two stressful environments, high salinity and low glucose, for five hundred generations. Diploid hybrids between populations from different environments showed substantially lower meiotic efficiency (percentage of cells undergoing meiosis under conditions that normally elicit sexual reproduction) than hybrids between populations from the same environment. Mitotic (as opposed to meiotic) reproduction of the hybrids on permissive medium was normal, however. In a study on *Neurospora*, Dettman et al. (2008) allowed replicate populations derived from either an interspecific cross or an intraspecific cross to adapt to either high salinity or low temperature. Two measures of postmating isolation were obtained: perithecial production (essentially fertilization success) in crosses between populations, and the percentage viability of spores resulting from successful crosses. For the experiments involving both progenitors, average spore viability under permissive conditions was lower in crosses between lines from different regimes than in crosses between lines from the same regime. The authors' statistical analysis, however, failed to distinguish replicate lines within regimes; thus, the true statistical significance of this and other results they obtained is difficult to evaluate. A clear result, however, was that perithecial production in populations derived from the interspecific crosses was dramatically depressed when females came from the high-temperature treatment and males came from the high-salt treatment; genetic analysis showed that this was likely the result of a two-gene interaction. Because these lines were derived from an interspecific cross, with resulting high linkage disequilibrium, there is a strong possibility that fixation of the incompatible allele combinations in the different regimes was driven by hitchhiking, rather than direct selection. Moreover, the authors did not rule out the possibility that the incompatibility simply recapitulated a difference between the parental species.

Clearly, more experiments are needed to determine how rapidly postmating isolation can evolve due to either selection or drift. Quantifying postmating isolation in its diverse forms, however, is more difficult than simply testing whether a pair of lines mate assortatively; this probably explains why postmating isolation has been relatively neglected in laboratory experiments on speciation.

*How much gene flow is needed to prevent speciation by divergent selection?*   Although there is abundant evidence that divergent selection can lead to the evolution of reproductive isolation between allopatric populations, there is much less information on how readily reproductive isolation evolves between populations that are not completely separated by geography (Coyne and Orr 2004). An obvious experiment would be to subject laboratory populations to divergent selection with varying levels of gene flow between the selected extremes, choosing trait(s) for which divergent selection in allopatry is known to produce reproductive isolation. Surprisingly, this sort of experiment has rarely been done;

equally surprisingly, when it has been done, the results have suggested that substantial levels of gene flow need not greatly impede the evolution of reproductive isolation. Soans et al. (1974) and Hurd and Eisenberg (1975) subjected house fly populations to selection for positive and negative geotaxis with 0 percent (allopatric treatment; both studies), 30 percent (parapatric treatment; Soans et al.), and 50 percent (sympatric treatment; Hurd and Eisenberg) gene flow allowed between the selected extremes (note that the percentages refer to the potential gene flow that would occur in the absence of assortative mating and postmating isolation). In both studies, the treatments with gene flow were equally effective as the allopatric treatment in generating assortative mating. Grant and Mettler (1969) and Coyne and Grant (1972) selected *D. melanogaster* populations for high and low "escape" response (a measure that seems to combine negative geotaxis, positive phototaxis, and activity level) in allopatry and sympatry (Grant and Mettler) and later in parapatry (25 percent gene flow; Coyne and Grant). Highly significant premating isolation evolved in the allopatric treatment and in one of the two parapatric replicates, but not in the three sympatric replicates. While the results were not as striking as those of Hurd and Eisenberg (1975), the evolution of reproductive isolation in one of the two parapatric replicates in only ten generations suggests that lower but still substantial levels of gene flow (e.g., 5–10 percent) might not have impeded the evolution of reproductive isolation. More experiments of this type, with diverse organisms and traits, would help clarify the extent to which complete geographic separation is a prerequisite for the initial evolution of reproductive isolation.

*How feasible are models of sympatric speciation via divergence in host or habitat choice?* Specialization on different habitats or hosts can lead to premating reproductive isolation by causing populations to be physically separated at the time of mating. This has caused Bush (1975, 1994) and others (e.g., Via 2001) to champion the idea that host shifts can precipitate sympatric speciation in phytophagous insect species in which mating takes place on the host. Although there is evidence suggesting that host shifts have contributed to sympatric divergence in some groups (Via 2001) or at least help maintain divergence (Rundle and Nosil 2005), sympatric speciation is notoriously difficult to document in nature. Laboratory experiments can help clarify the feasibility of sympatric speciation via host or habitat shifts and identify the conditions under which it is most likely to occur.

Rice and Salt (1988, 1990) conducted an elegant test of one model of sympatric speciation via divergence in host or habitat preference, using *D. melanogaster*. The investigators built an elaborate maze that forced newly emerged flies to make three successive binary choices (light/dark, up/down, and odor 1/odor 2) before being able to find food and mate. This generated eight different artificial "hosts," each characterized by a unique set of stimuli. Only flies that chose two of the hosts, which required opposite sets of choices to locate (dark/up/odor 1 for host A, and light/down/odor 2 for host B), were allowed to contribute to the next generation, simulating the situation in which these were the only hosts suitable for development. Disruptive selection was simultaneously applied to development time, with only early-emerging flies that chose host A, and only late-emerging flies that chose host B, being allowed to breed. Within about thirty generations, gene flow between the two hosts ceased, because progeny of flies from host A were virtually never found in host B, and vice versa (Rice and Salt 1990).

Although Rice and Salt's experiment demonstrated that sufficiently strong disruptive selection on host or habitat choice can lead to sympatric divergence, a critical feature of the design was that only hosts requiring opposite sets of choices to locate were suitable for development. It is not clear how broadly applicable this scenario is to phytophagous insects and other host- and habitat-specific groups (to give a hypothetical example, if dark/wet and light/dry habitats are both suitable for development, why should dark/dry and light/wet habitats be lethal?). Moreover, the success of Rice and Salt's (1990) experiment seems to have been heavily dependent on the disruptive selection for development time, which can cause reproductive isolation only in species with nonoverlapping generations. From their figures 2–4, it does not appear that substantial divergence would have occurred based on selection for the opposite combinations of behavioral choices alone (with the possible exception of the treatment in which a somewhat artificial penalty of habitat switching was applied).

Interestingly, Bush's original verbal model of sympatric speciation in phytophagous insects (Bush 1975), which was partly inspired by the ideas of early entomologists, was based on very different assumptions from those simulated in Rice and Salt's experiment. Bush considered only two hosts, one ancestral and one novel. Mutations at separate loci arose which conferred acceptance of the new host, and ability to survive on it. A multilocus version of this model (Fry 2003) can indeed result in sympatric speciation, if alleles increasing viability on one host reduce viability on the other, and if there is sufficiently high genetic variance for viability and host preference, conferred by alleles with individually large effects.

A possible way to test the Bush model in the laboratory is diagrammed in figure 20.3. Emerging insects are allowed to choose between two hosts (real or artificial), after which mating takes place. Progeny of parents from the different hosts are kept separate and selected in the opposite direction for one or more quantitative traits, simulating the situation where the different hosts require opposite sets of traits for survival. After selection and emergence, insects from the two hosts are pooled and allowed to choose hosts again. This design differs from that of Rice and Salt (1990), because selection does not act directly on host choice (there are only two hosts, and insects are guaranteed to find one); instead, selection acts on the quantitative traits in a host-specific manner. This causes alleles affecting the selected traits to become associated with alleles affecting host preference (Diehl and Bush 1989; Fry 2003), resulting in indirect disruptive selection on host preference. In particular, even though genotypes with little or no host preference are no less successful at finding hosts than genotypes with strong host preference, they will tend to have intermediate values of the disruptively selected traits, and hence be maladapted to both hosts.

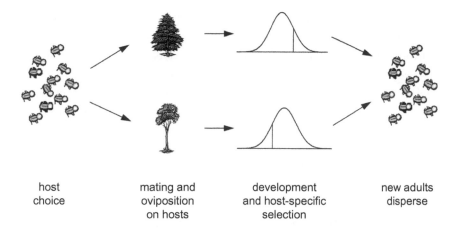

| host<br>choice | mating and<br>oviposition<br>on hosts | development<br>and host-specific<br>selection | new adults<br>disperse |

FIGURE 20.3

Hypothetical experiment for testing Bush's (1975) model of sympatric speciation. Nonmated insects are first allowed to choose between two real or artificial hosts; after host choice, the insects are allowed to mate and lay eggs. Depending on which host their parents chose, progeny are selected in opposite directions for one or more quantitative traits. The newly emerged adults are then allowed to choose hosts again, and the cycle repeated.

By using strong host-coupled disruptive selection on one or more highly heritable traits, the design in figure 20.3 mimics the situation where alleles increasing viability on one host reduce it on the other and where there is relatively high genetic variance for viability. Nonetheless, for population splitting ("speciation") to occur, there must be relatively high genetic variance for host preference as well. In a preliminary experiment on *D. melanogaster* in which flies were allowed to choose between two artificial hosts, however, genetic variance for host preference appeared to be too low to allow "speciation" (J.D.F., unpublished data). This does not necessarily sound the death knell for the Bush model of sympatric speciation, because *D. melanogaster* is not a host-specific phytophagous insect, and it may be a poor choice for such an experiment. A true phytophage population allowed to choose between two real hosts might have sufficiently high genetic variance for the scenario shown in figure 20.3 to result in population splitting. Indeed, crossing experiments on host races of *Rhagoletis* flies (Dambrowski et al. 2005) and *Nilaparvata* plant hoppers (Sezer and Butlin 1998) suggests that host preference in both groups is controlled by genes with large effects. Another possibility suggested by Fry's (2003) model, unfortunately not easily testable on the time scale of laboratory experiments, is that long-continued selection of the type shown in figure 20.3 would cause the genetic variance for preference to gradually increase, by favoring rare mutants with extremes of preference (see Fry 2003, figure 1) and/or suppressors of recombination between preference loci.

*How readily does reinforcement occur?*   In the traditional reinforcement scenario, two populations diverge in allopatry and develop partial, but not complete, postmating reproductive isolation. When the populations again come into contact, selection favors mechanisms that reduce the likelihood of mating between them, because such matings produce low fitness hybrids. In the mid-1900s, reinforcement was widely regarded as a common, and perhaps even necessary, final step in speciation (reviewed in Coyne and Orr 2004). Although serious objections to the feasibility of reinforcement were raised in the 1980s (e.g., Butlin 1987), more recent theoretical models have suggested that reinforcement is plausible, and empirical evidence indicates that a process resembling reinforcement has occurred in some groups, most notably *Drosophila*. Nonetheless, the evidence for reinforcement in natural populations is subject to alternative explanations (Coyne and Orr 2004), and the appropriateness of the assumptions of the various proposed models of reinforcement remains to be verified.

Surprisingly, there have been few if any attempts to test realistic reinforcement scenarios in the laboratory. As noted earlier, reinforcement needs to be distinguished from reproductive character displacement, in which selection favors the avoidance of mating between species which already show complete postmating isolation; thus, the "destroy all the hybrids" experiments reviewed here test for reproductive character displacement, not reinforcement. In three of the destroy all the hybrids studies (Robertson 1966b; Harper and Lambert 1983; Hostert 1997; see table 20.1), the investigators also created treatments in which some hybrids were retained, and hence some gene flow between the hybridizing populations could occur. In the first two of these studies, however, no increase in premating isolation was observed even in the treatments without gene flow, indicating that the base populations were untypical in lacking the requisite genetic variation for mating-related traits. In Hostert's (1997) study, a small amount of premating reproductive isolation evolved in the zero gene flow treatment in twenty-five generations, but none was observed in three treatments with nonzero gene flow (theoretical gene flow of 3.3 percent, 10 percent, and 50 percent). While this result gives evidence against reinforcement, it is questionable whether Hostert's design mimics a realistic reinforcement scenario in nature. The parental populations were formed by backcrossing two recessive markers into the same genetic background and therefore should have been genetically and phenotypically similar, except for the effects of the markers themselves. In contrast, reinforcement is usually thought to apply to populations that have diverged in allopatry for a long period; such populations, even if incompletely reproductively isolated, would be likely to have multiple differences in morphology, mating behavior, and so forth, which could serve as the basis for reinforcement. Moreover, the use of recessive markers, which are likely to be associated with reduced fitness, may have resulted in the actual level of gene flow being higher than the theoretical levels, because the wild-type hybrids would probably have been more vigorous than the pure population parents.

A more biologically relevant way to test the feasibility of reinforcement would be to start with populations from different parts of a species' range and that already show considerable phenotypic divergence, if not partial reproductive isolation. (An alternative would be to start with laboratory populations that had previously been divergently selected for one or more quantitative traits). The simplest design would be an island-continent scenario, in which an experimental "island" population receives a small number of immigrants each generation from a nonevolving "continental" population (cf. the model of Kirkpatrick and Servedio 1999). If the two populations differ initially in (say) body size, then selection on body size could be used to create selection against immigrants. By varying the rate of immigration and the strength of selection on body size, one could determine whether there are conditions that permit some initial gene flow (i.e., hybrid fitness greater than zero) but that nonetheless result in the evolution of enhanced mating discrimination of island females against continental males. One of the challenges in such an experiment would be to carefully control the intensity of selection, to make sure one was not inadvertently performing a "destroy all the hybrids" experiment.

*What is the role of sexual selection (including sexual conflict) in the evolution of reproductive isolation?* The evolution of behavioral premating isolation is generally thought to require some form of sexual selection (Coyne and Orr 2004). The precise way in which sexual selection generates reproductive isolation is unclear, however. Verbal and theoretical models have suggested ways in which "good-genes," runaway, and sexual conflict processes, among others, could all result in premating isolation between allopatric populations (reviewed in Coyne and Orr 2004). Laboratory experiments represent a promising way to test the predictions of these models.

Recent experiments on sexual conflict are a promising start in this direction. According to the theory of sexual conflict (Holland and Rice 1998), the optimal number of matings is lower for females than for males; as a result, females are selected to resist mating attempts by males, and males are selected for the ability to overcome female resistance. This can result in perpetual antagonistic coevolution between males and females, possibly resulting in reproductive isolation between allopatric populations (i.e., because the males of one population may not have the requisite traits for overcoming the mating resistance of females of another population). An elegant experiment by Martin and Hosken (2003) on the dung fly *Sepsis cynipsea* provided support for this scenario. Females from populations kept under enforced monogamy, a regime that eliminates the potential for sexual conflict, showed relatively little reluctance to mate with males from either their own population or other monogamous populations after thirty-five generations. In contrast, females from populations kept in containers with multiple flies of both sexes, and hence with the potential for sexual conflict over mating rates, showed greater reluctance to mate in general, but especially with males from other replicate populations. Reproductive isolation was greatest between experimental populations kept at high density, and hence with high potential for sexual conflict, than between populations kept at a lower density, contrary to the prediction of drift-based models.

Although two recent experiments (Wigby and Chapman 2006; Bacigalupe et al. 2007) found no evidence of reproductive isolation between *Drosophila* populations maintained with the potential for sexual conflict, this is entirely unsurprising: the multiple previous speciation experiments with *Drosophila* used conditions that allowed sexual conflict but (as we have seen) rarely resulted in reproductive isolation evolving between lines maintained under the same conditions. It is thus likely that the Martin and Hosken's (2003) different results with dung flies stem from biological differences between *Sepsis* and *Drosophila* (Wigby and Chapman 2006). Experiments testing for the relationship between sexual conflict and reproductive isolation in other groups would help elucidate the importance of sexual conflict in speciation.

Sexual selection is also likely to play a role in the evolution of premating isolation between populations subject to divergent selection. If a selection regime were to change the mating preference of females for a particular male trait (see Rundle et al. 2005 for an example), this would automatically create directional sexual selection on the trait. Similarly, if a selection regime were to substantially change the mean of a sexually selected male trait, this would create selection for females to be willing to mate with males with previously unpreferred trait values. In fact, in the absence of sexual selection, independent responses to selection of female preference and the male traits that are the subject of the preference would be just as likely to result in disassortative mating (e.g., a preference of females from a given selection regime for males from a different regime) as in assortative mating. Nonetheless, the way in which sexual selection interacts with nonsexual selection to produce premating isolation among populations subject to divergent selection needs to be clarified. The simplest prediction is that divergent selection with enforced monogamy, and hence no opportunity for sexual selection, should result in less sexual isolation than divergent selection with the opportunity for mate choice.

*What are the mechanisms by which reproductive isolation evolves?* Most studies on the experimental evolution of reproductive isolation have given no information on the traits responsible for the observed reproductive isolation. Two studies on the evolution of cuticular hydrocarbons (CHCs), a type of pheromone, in *Drosophila serrata* populations provide promising, but incomplete, exceptions in this regard (Higgie et al. 2000; Rundle et al. 2005). Higgie et al. (2000) maintained *D. serrata* populations in bottles with and without a closely related species, *D. birchii*, whose natural range overlaps that of *D. serrata*. When the *D. serrata* base populations came from regions lacking *D. birchii*, the presence of the latter species caused their CHC profiles to evolve to resemble that of *D. serrata* populations from regions of overlap with *D. birchii*, suggesting reproductive character displacement. The authors did not, however, investigate the mating behavior of the evolved lines. Rundle et al. (2005) showed that maintaining *D. serrata* populations on different diets caused changes in female CHCs and, more surprisingly, female preference for particular CHCs, but they did not test whether premating isolation between lines on different diets evolved as a result.

A basic mechanistic distinction that needs to be addressed in further work is that between intrinsic and extrinsic isolation. Most studies have tested for only one type of isolation and thus give no information on their relative rates of evolution. Rice and Salt's (1990) study on sympatric speciation provides an interesting exception; although the experimental populations evolved extrinsic premating isolation due to differences in habitat selection, they showed no evidence of intrinsic premating isolation in mating tests. In their study of *Neurospora*, Dettman et al. (2008) measured progeny viability of interline crosses under permissive conditions and under the stressful conditions of the selection lines (high salinity and low temperature). While the results under permissive conditions gave some evidence for intrinsic postzygotic isolation, as noted earlier, viability of hybrids was particularly low when measured in the low-temperature environment (but not in the high-salinity environment), indicating that some extrinsic postzygotic isolation had also evolved. More studies comparing the speed of evolution and magnitudes of intrinsic and extrinsic isolation are needed.

## GENERAL GUIDELINES FOR EXPERIMENTS ON SPECIATION

This review should make it clear that the literature on the experimental evolution of reproductive isolation has only scratched the surface of many important questions about speciation. Even those results that are relatively well replicated (e.g., that divergent selection often results in premating reproductive isolation, while population bottlenecks rarely do) come exclusively from experiments on flies. There is a clear need for innovative experiments on a broader range of species; recent studies on fungi (Greig et al. 2002; Leu and Murray 2006; Dettman et al. 2007, 2008) represent a promising start in this direction. Some questions (e.g., how often does postzygotic isolation evolve due to divergent selection?) could be addressed using selection lines created for other purposes, while others (e.g., how readily does reinforcement evolve?) will require experiments "from scratch."

To assist those designing new experiments, I offer a few general guidelines.

*Carefully consider the base population* The choice of base population is a crucial but often overlooked step in the design of any selection (or drift) experiment (see also Rhodes and Kawecki this volume; Rauser et al. this volume; Simões et al. this volume), experiments on speciation being no exception. Ideally, base populations should contain a broad sample of variation from a single, natural population. Starting a selection experiment from an inbred (whether deliberately or not) line will almost guarantee a weak response. At the other extreme, as pointed out by Rundle et al. (1998), the common practice of hybridizing lines from geographically diverse populations to establish the base population should also be avoided, because this may result in spurious outcomes with little relevance for natural populations. It may also be a good idea to allow the base population some time to adapt to laboratory conditions before the start of the experiment; otherwise, supposedly nonevolving "control" treatments may change rapidly due to laboratory adaptation (on the other hand, too long a period of laboratory adaptation may result in depletion of genetic variation; Templeton 1999). As noted earlier, experiments designed to test reinforcement-like scenarios arguably should start with populations that are differentiated to some extent.

*Maximize effective population sizes and duration of experiments*   The response to selection is an increasing function of both the effective population size of the selected lines and the number of generations of selection (Falconer and Mackay 1996). Moreover, these factors interact, because small populations typically reach a selection limit sooner, as variation is depleted, than large populations (Weber and Diggins 1990). Because speciation is typically thought to require thousands of generations, selection experiments designed to test hypotheses of speciation should strive to maximize both effective population size and duration. This requires careful choice of both organism and experimental methods. Yeast and other sexual microorganisms have obvious advantages in this regard. Moreover, whenever possible, selection schemes should be devised which minimize or eliminate the need for manually scoring or measuring individuals, either by automating these steps or by performing "quasi–natural selection" experiments (aka "laboratory natural selection"— Rose and Garland this volume; Futuyma and Bennett this volume; Gibbs and Gefen this volume; Huey and Rosenzweig this volume), in which populations are simply allowed to adapt to different environments (e.g., different media or temperatures), obviating the need for phenotypic scoring and sorting.

*Replication is crucial*   Two or more replicate lines per treatment should always be established (see also Rhodes and Kawecki this volume; Swallow et al. this volume). Different lines from the same treatment may behave very differently due to genetic drift. For example, Halliburton and Gall (1981) applied strong disruptive selection on pupal weight in the flour beetle, *Tribolium castaneum*; two replicate populations evolved strong reproductive isolation between the oppositely selected extremes, while the other two showed no evidence for reproductive isolation. Had the experiment not been replicated, the results might have tempted one to conclude that disruptive selection in this system is either likely or unlikely to result in reproductive isolation, depending on the result obtained. Moreover, in experiments to investigate whether divergent selection in allopatry produces reproductive isolation, a critical test is whether replicate lines from the same selection treatment show less reproductive isolation from each other than from lines from the opposite treatment. Without replicates, it would not be clear whether reproductive isolation observed between divergently selected lines resulted from the selection itself or from genetic drift (including "hitchhiking" of alleles causing reproductive isolation with those affecting the selected traits, which can be viewed as a form of drift in most instances).

In data presentation and statistical analysis, replicates must be clearly distinguished (see also Rhodes and Kawecki this volume). There is little point in creating replicate populations, only to later pool data from the different replicates for analysis (e.g., when conducting chi-square tests for nonrandom mating).

*Know the relevant literature*   Experiments should be designed in a way that takes advantage of previous methodological advances. For example, anyone conducting tests for premating isolation should consult the literature on the most effective way to conduct and analyze mating trials (e.g., Spieth and Ringo 1983; Coyne et al. 2005).

*Negative results should be reported* For the literature to give an accurate picture of the efficacy of a given treatment (e.g., divergent selection) in producing reproductive isolation, and to avoid needless duplication of effort, negative results should be reported, even if this requires publishing in relatively "minor" journals or other outlets.

## CONCLUSION

The last decade has seen a profusion of new approaches and ideas on speciation (reviewed in Barton 2001; Coyne and Orr 2004; Rundle and Nosil 2005; Noor and Feder 2006; Rieseberg and Willis 2007). A wealth of new theoretical models have been developed, genes contributing to reproductive isolation have been mapped and characterized, and new statistical methods for inferring the geographic mode of speciation have been applied. At the same time, experimental approaches have continued to give insights into mechanisms of speciation. While one should not lose sight of their limitations, particularly their short time scale and simplification of ecological conditions, laboratory experiments provide a powerful way to test the feasibility of theoretical models and to study the forces responsible for the initial evolution of reproductive isolation, something that is difficult to do with natural populations. They are therefore likely to continue to be an important part of the literature on speciation.

## SUMMARY

Laboratory experiments on speciation investigate the conditions under which reproductive isolation can evolve between members of what was initially a single population, as well as the conditions under which reproductive isolation between initially partly reproductively isolated populations can become intensified. This chapter reviews past speciation experiments, emphasizing recent and previously overlooked studies, identifies neglected questions that could be addressed by new experiments, and gives guidelines for such experiments. In past experiments, partial premating (i.e., behavioral) reproductive isolation has sometimes evolved as a by-product of divergent selection on allopatric populations, but it has rarely evolved due to genetic drift alone. In contrast to the results of divergent selection in allopatry, application of strong disruptive selection to an initially random-mating population, a situation promoting sympatric speciation according to some models, has only rarely resulted in the evolution of premating isolation between the selected extremes. Nonetheless, there is some evidence that the different results of the allopatric and sympatric studies may have been partly caused by the different traits used for selection, not just the homogenizing effect of gene flow in the sympatric studies. Experiments in which disruptive selection is applied to a single trait known to be related to mating behavior, with varying levels of gene flow between the selected extremes, could help clarify to what extent reproductive isolation can evolve in the face of ongoing gene flow. Other neglected subjects ripe for future experimental investigation include the roles of selection and drift in promoting postmating isolation, the feasibility of sympatric speciation via divergence in host or habitat preference, the conditions under which reinforcement can occur, and the role of sexual selection in the evolution of premating isolation. Speciation experiments have made important contributions to our understanding of mechanisms of speciation, and are likely to continue to do so, complementing comparative, genetic, and theoretical approaches.

## ACKNOWLEDGMENTS

I thank L. Meffert and an anonymous reviewer for helpful comments, and P. Corey for collecting the data in table 20.2. This work was supported by grant 0623268 from the National Science Foundation.

## REFERENCES

Bacigalupe, L. D., H. S. Crudgington, F. Hunter, A. J. Moore, and R. R. Snook. 2007. Sexual conflict does not drive reproductive isolation in experimental populations of *Drosophila pseudoobscura*. *Journal of Evolutionary Biology* 20:1763–1771.

Barker, J. S. F., and L. J. Cummins. 1969. The effect of selection for sternopleural bristle number on mating behaviour in *Drosophila melanogaster*. *Genetics* 61:713–719.

Barton, N. H. 2001. Speciation (introduction to special issue). *Trends in Ecology & Evolution* 16:325.

Boake, C. R. B., K. McDonald, S. Maitra, and R. Ganguly. 2003. Forty years of solitude: Life-history divergence and behavioural isolation between laboratory lines of *Drosophila melanogaster*. *Journal of Evolutionary Biology* 16:83–90.

Bush, G. L. 1975. Sympatric speciation in phytophagous parasitic insects. Pages 187–206 *in* P. W. Price, ed. *Evolutionary Strategies of Parasitic Insects and Mites*. New York: Plenum.

———. 1994. Sympatric speciation in animals: New wine in old bottles. *Trends in Ecology & Evolution* 8:285–288.

Butlin, R. 1987. Speciation by reinforcement. *Trends in Ecology & Evolution* 2:8–13.

Carson, H. L., and A. R. Templeton. 1984. Genetic revolutions in relation to speciation phenomena: The founding of new populations. *Annual Review of Ecology and Systematics* 15:97–131.

Coyne, J. A., and B. Grant. 1972. Disruptive selection on I-maze activity in *Drosophila melanogaster*. *Genetics* 71:185–188.

Coyne, J. A., and H. A. Orr. 2004. *Speciation*. Sunderland, MA: Sinauer.

Crossley, S. A. 1974. Changes in mating behavior produced by selection for ethological isolation between ebony and vestigial mutants of *Drosophila melanogaster*. *Evolution* 28:631–647.

Dambrowski, H. R., C. Linn, S. H. Berlocher, A. A. Forbes, W. Roelefs, and J. L. Feder. 2005. The genetic basis for fruit odor discrimination in *Rhagoletis* flies and its significance for sympatric host shifts. *Evolution* 59:1953–1964.

de Oliveira, A. K., and A. R. Cordeiro. 1980. Adaptation of *Drosophila willistoni* experimental populations to extreme pH medium. II. Development of incipient reproductive isolation. *Heredity* 44:123–130.

Dettman, J. R., J. B. Anderson, and L. M. Kohn. 2008. Divergent adaptation promotes reproductive isolation among experimental populations of the filamentous fungus *Neurospora*. *BMC Evolutionary Biology* 8:35.

Dettman, J. R., C. Sirjusingh, L. M. Kohn, and J. B. Anderson. 2007. Incipient speciation by divergent adaptation and antagonistic epistasis in yeast. *Nature* 447:585–589.

Diehl, S. R., and G. L. Bush. 1989. The role of habitat preference in adaptation and speciation. Pages 345–365 *in* D. Otte and J. Endler, eds. *Speciation and Its Consequences*. Sunderland, MA: Sinauer.

Dobzhansky, T., D. Pavlovsky, and S. R. Powell. 1976. Partially successful attempt to enhance reproductive isolation between semispecies of *Drosophila paulistorum*. Evolution 30:201–212.

Dodd, D. M. B., and J. R. Powell. 1985. Founder-flush speciation: an update of experimental results with *Drosophila*. *Evolution* 39:1388–1392.

Ehrman, L. 1971. Natural selection for the origin of reproductive isolation. *American Naturalist* 105:479–483.

———. 1973. More on natural selection for the origin of reproductive isolation. *American Naturalist* 107:318–319.

———. 1979. Still more on natural selection for the origin of reproductive isolation. *American Naturalist* 113:148–150.

Falconer, D. S., and T. F. C. Mackay. 1996. *Introduction to Quantitative Genetics*. Essex, U.K.: Longman.

Florin, A.-B., and A. Ödeen. 2002. Laboratory experiments are not conducive for allopatric speciation. *Journal of Evolutionary Biology* 15:10–19.

Fry, J. D. 1999. The role of adaptation to host plants in the evolution of reproductive isolation: Negative evidence from *Tetranychus urticae*. *Experimental and Applied Acarology.* 23:379.

———. 2003. Multilocus models of sympatric speciation: Bush vs. Rice vs. Felsenstein. *Evolution* 57:1735–1746.

Fry, J. D., C. M. Bahnck, M. Mikucki, N. Phadnis, and W. C. Slattery. 2004. Dietary ethanol mediates selection on aldehyde dehydrogenase activity in *Drosophila melanogaster*. *Integrative and Comparative Biology* 44:275–283.

Fukatami, A., and D. Morikami. 1970. Selection for sexual isolation in *Drosophila melanogaster* by a modification of Koopman's method. *Japanese Journal of Genetics* 45:193–204.

Galiana, A., A. Moya, and F. J. Ayala. 1993. Founder-flush speciation in *Drosophila pseudoobscura*: A large-scale experiment. *Evolution* 47:432–444.

Grant, B., and L. E. Mettler. 1969. Disruptive and stabilizing selection on the "escape" behavior of *Drosophila melanogaster*. *Genetics* 62:625–637.

Greig, D., E. J. Louis, R. H. Borts, and M. Travisano. 2002. Hybrid speciation in experimental populations of yeast. *Science* 298:1773–1775.

Halliburton, R., and G. A. E. Gall. 1981. Disruptive selection and assortative mating in *Tribolium castaneum*. *Evolution* 35:829–843.

Harper, A. A., and D. M. Lambert. 1983. The population genetics of reinforcing selection. *Genetica* 62:15–23.

Higgie, M., S. Chenoweth, and M. W. Blows. 2000. Natural selection and the reinforcement of mate recognition. *Science* 290:519–521.

Holland, B., and W. R. Rice. 1998. Chase-away sexual selection: Antagonistic seduction versus resistance. *Evolution* 52:1–7.

Hostert, E. E. 1997. Reinforcement: A new perspective on an old controversy. *Evolution* 51:697–702.

Hurd, L. E., and R. M. Eisenberg. 1975. Divergent selection for geotactic response and evolution of reproductive isolation in sympatric and allopatric populations of houseflies. *American Naturalist* 109:353–358.

Kessler, S. 1966. Selection for and against ethological isolation between *Drosophila pseudoobscura* and *Drosophila persimilis*. *Evolution* 20:634–645.

Kilias, G., S. N. Alahiotis, and M. Pelecanos. 1980. A multifactorial genetic investigation of speciation theory using *Drosophila melanogaster*. *Evolution* 34:730–737.

Kirkpatrick, M., and V. Ravigné. 2002. Speciation by natural and sexual selection: Models and experiments. *American Naturalist* 159:S22–S35.

Kirkpatrick, M., and M. R. Servedio. 1999. The reinforcement of mating preferences on an island. *Genetics* 151:865–884.

Knight, G. R., A. Robertson, and C. H. Waddington. 1956. Selection for sexual isolation within a species. *Evolution* 10:14–22.

Koopman, K. F. 1950. Natural selection for reproductive isolation between *Drosophila pseudoobscura* and *Drosophila persimilis*. *Evolution* 4:135–148.

Koref-Santibañez, S., and C. H. Waddington. 1958. The origin of sexual isolation between different lines within a species. *Evolution* 12:485–493.

Lawrence, J. G. 2002. Gene transfer in bacteria: speciation without species? *Theoretical Population Biology* 61:449–460.

Leu, J.-Y., and A. W. Murray. 2006. Experimental evolution of mating discrimination in budding yeast. *Current Biology* 16:280–286.

Lofdahl, K. L., D. Hu, L. Ehrman, J. Hirsch, and L. Skoog. 1992. Incipient reproductive isolation and evolution in laboratory *Drosophila melanogaster* selected for geotaxis. *Animal Behaviour* 44:783–786.

Markow, T. A. 1981. Mating preferences are not predictive of the direction of evolution in experimental populations of *Drosophila*. *Science* 213:1405–1407.

Martin, O. Y., and D. J. Hosken. 2003. The evolution of reproductive isolation through sexual conflict. *Nature* 423:979–982.

Mayr, E. 1963. *Animal Species and Evolution*. Cambridge, MA: Belknap.

McKenzie, J. A., and A. Y. Game. 1987. Diazinon resistance in *Lucilia cuprina*: Mapping of a fitness modifier. *Heredity* 59:371–381.

Meffert, L. M., and E. H. Bryant. 1991. Mating propensity and courtship behavior in serially bottlenecked lines of the housefly. *Evolution* 45:293–306.

Miyatake, T., and T. Shimizu. 1999. Genetic correlations between life-history and behavioral traits can cause reproductive isolation. *Evolution* 53:201–208.

Mooers, A. Ø., H. D. Rundle, and M. C. Whitlock. 1999. The effects of selection and bottlenecks on male mating success in peripheral isolates. *American Naturalist* 153:437–444.

Moya, A., A. Galiana, and F. J. Ayala. 1995. Founder-effect speciation theory: Failure of experimental corroboration. *Proceedings of the National Academy of Sciences of the USA* 92:3983–3986.

Noor, M. A. F., and J. L. Feder. 2006. Speciation genetics: evolving approaches. *Nature Review Genetics* 7:851–861.

Overmeer, W. P. J. 1966. Intersterility as a consequence of insecticide selections in *Tetranychus urticae* Koch (Acari: Tetranychidae). *Nature* 209:321.

Paterniani, E. 1969. Selection for reproductive isolation between two populations of maize, *Zea mays* L. *Evolution* 23:534–547.

Powell, J. R. 1978. The founder-flush speciation theory: An experimental approach. *Evolution* 32:465–474.

Rainey, P. B., and M. Travisano. 1998. Adaptive radiation in a heterogeneous environment. *Nature* 394:69–72.

Regan, J. L., L. M. Meffert, and E. H. Bryant. 2003. A direct experimental test of founder-flush effects on the evolutionary potential for assortative mating. *Journal of Evolutionary Biology* 16:302–312.

Rice, W. R., and E. E. Hostert. 1993. Laboratory experiments on speciation: What have we learned in 40 years? *Evolution* 47:1637–1653.

Rice, W. R., and G. W. Salt. 1988. Speciation via disruptive selection on habitat preference: Experimental evidence. *American Naturalist* 131:911–917.

———. 1990. The evolution of reproductive isolation as a correlated character under sympatric conditions: Experimental evidence. *Evolution* 44:1140–1152.

Rieseberg, L. H., B. Sinervo, C. R. Linder, M. C. Ungerer, and D. M. Arias. 1996. Role of gene interactions in hybrid speciation: Evidence from ancient and experimental hybrids. *Science* 272:741–745.

Rieseberg, L. H., and J. H. Willis. 2007. Plant speciation. *Science* 317:910–914.

Ringo, J., D. Wood, R. Rockwell, and H. Dowse. 1985. An experimental test of two hypotheses of speciation. *American Naturalist* 126:642–661.

Robertson, F. W. 1966a. The ecological genetics of growth in *Drosophila*. 8. Adaptation to a new diet. *Genetical Research, Cambridge* 8:165–179.

———. 1966b. A test of sexual isolation in *Drosophila*. *Genetical Research, Cambridge* 8:181–187.

Rundle, H. D. 2003. Divergent environments and population bottlenecks fail to generate premating isolation in *Drosophila pseudoobscura*. *Evolution* 57:2557–2565.

Rundle, H. D., S. F. Chenoweth, P. Doughty, and M. W. Blows. 2005. Divergent selection and the evolution of signal traits and mating preferences. *PLoS Biology* 3:e368.

Rundle, H. D., A. Ø. Mooers, and M. C. Whitlock. 1998. Single founder-flush events and the evolution of reproductive isolation. *Evolution* 52:1850–1855.

———. 1999. Experimental tests of founder-flush: A reply to Templeton. *Evolution* 53:1632–1633.

Rundle, H. D., and P. Nosil. 2005. Ecological speciation. *Ecology Letters* 8:336–352.

Scharloo, W. 1970. Reproductive isolation by disruptive selection: Did it occur? *American Naturalist* 105:83–88.

Sezer, M., and R. K. Butlin. 1998. The genetic basis of oviposition preference differences between sympatric host races of the brown planthopper *(Nilaparvata lugens)*. *Proceedings of the Royal Society of London B, Biological Sciences* 265:2399–2405.

Soans, A. B., D. Pimentel, and J. S. Soans. 1974. Evolution of reproductive isolation in allopatric and sympatric populations. *American Naturalist* 108:117–124.

Spieth, H. T., and J. M. Ringo. 1983. Mating behavior and sexual isolation in Drosophila. Pages 223–284 *in* M. Ashburner, H. L. Carson, and J. N. Thompson, eds. *The Genetics and Biology of* Drosophila. London: Academic Press.

Templeton, A. R. 1980. The theory of speciation via the founder principle. *Genetics* 94:1011–1038.

———. 1999. Experimental tests of genetic transilience. *Evolution* 53:1628–1632.

Thoday, J. M., and J. B. Gibson. 1962. Isolation by disruptive selection. *Nature* 193:1164–1166.

———. 1970. The probability of isolation by disruptive selection. *American Naturalist* 104:219–230.

Via, S. 2001. Sympatric speciation in animals: The ugly duckling grows up. *Trends in Ecology & Evolution* 16:381–390.

Wallace, B. 1953. Genetic divergence of isolated populations of *Drosophila melanogaster*. Page 761 *in* G. Montalenti and A. Chiarugi, eds. *Proceedings of the Ninth International Congress of Genetics, Bellagio, Florence, Italy.*

Weber, K. E., and L. R. Diggins. 1990. Increased selection response in larger populations: Selection for ethanol vapor resistance in *Drosophila melanogaster* at two population sizes. *Genetics* 125:585–597.

Wigby, S., and T. Chapman. 2006. No evidence that experimental manipulation of sexual conflict drives premating reproductive isolation in *Drosophila melanogaster*. *Journal of Evolutionary Biology* 19:1033–1039.

PART V

# CONCLUSION

21

# A CRITIQUE OF EXPERIMENTAL PHYLOGENETICS

### Todd H. Oakley

Already as an undergraduate, I had an inordinate fondness for phylogenetic trees, and few papers sparked my imagination more than one announcing the birth of experimental phylogenetics (Hillis et al. 1992). In that paper, Hillis and colleagues generated experimentally a phylogeny of viruses and used it to compare various phylogenetic methods. For the first time, researchers had at their disposal a phylogeny of "living" organisms generated in the lab for the express purpose of studying phylogenetic methods. This known phylogeny came at a time when the enterprise of testing phylogenetic methods was in its heyday. Even popular culture was enamored with the ability to simulate life, as the Maxis software company released their enormously popular video game *SimLife* in the same year. In 1992, I expected experimental studies to be a wave of the future in phylogenetics.

Sometimes crystal balls can be foggy. Despite the enthusiasm of a decade and a half ago, the field of experimental phylogenetics remains very small (see also Forde and Jessup this volume). Was my enthusiasm misplaced? Here, I will discuss what I believe to be the reasons why the field has barely grown since its inception fifteen years ago. Specifically, I will critique three primary arguments used to justify experimental phylogenetics. Most important, I conclude that experimental phylogenetics is an overly expensive simulation procedure. Even if experimental phylogenies have more biological realism than computer simulations, this realism comes at the considerable expense of decreased speed and potential for replication. This inherent trade-off between speed and biological realism is a recurring theme in experimental phylogenetics studies. Although an explicit understanding of the trade-off does not diminish the value of several previous studies, it may provide a guiding principle for those contemplating future contributions to experimental phylogenetics.

## THE PERCEIVED INFERIORITY OF HISTORICAL SCIENCE

One motivation in the literature for experimental phylogenetics has been a perceived inferiority of historical science, compared to experimental science. Here, I argue that there is no philosophical support for the claim that historical science is inferior to experimental science, thus negating one possible motivation for experimental evolution. Even though negating one motivation does not alone negate the entire rationale for experimental evolution, it is nevertheless important to promote a clearer understanding of historical science.

To some authors, experimental phylogenetics is motivated by the self-consciousness of historical scientists in the face of experimental science. We learn from an early age that "real" science relies on the possibility of unambiguously falsifying hypotheses. Yet specific events that happened in the past—such as the phylogenetic branching of mammals—can never be re-created. Like the legal system of the United States, historical science relies on demonstrating "beyond a reasonable doubt" that particular events did or did not occur. In science, reconstructing past events often takes the form of statistical/probability statements. Additionally, verifying specific historical occurrences may rely on various signatures left by historical events, such as the presence of a crater, high levels of iridium, and absence of previously prevalent fossils all dating to sixty-five million years ago, which congruently support the historical hypothesis of mass extinction by extraterrestrial impact. Although philosophers of science argue for the efficacy of such historical inference (Cleland 2001), there is still widespread perception of its inferiority.

This inferiority complex that burdens historical scientists is evident in the writing of Bull et al. (1993), illustrating it as a motivator for the field of experimental phylogenetics:

> From a cold and cruel perspective of the scientific method, the major weakness of this field is its difficulty in unambiguously falsifying hypotheses of phylogenetic relationships, and hence, of molecular evolution.

Here, the authors are stating that "the scientific method"—which I take to mean Popperian falsificationism—is the preferred way to perform science. A difficulty in falsifying historical hypotheses is seen by the authors as a major liability for phylogenetics and molecular evolution studies. If only we could actually test historical hypotheses through experimentation, the logic goes, this liability would be lessened. This attitude seems pervasive. For example, *Nature* editor Henry Gee (1999) wrote that historical hypotheses "can never be tested by experiment, and so they are unscientific. . . . No science can ever be historical." Yet another author, Skell (2005), wrote that "much of the evidence that might have established the theory [referring to "Darwin's theory of evolution"] on an unshakable empirical foundation, however, remains lost in the distant past." Skell's article makes many errors, especially the conflation of and unvalidated value judgments on historical and experimental scientific studies. Skell's article also naively equates all of evolution with a few "just so stories" about natural selection, and it ignores many practical applications of evolutionary theory, including gene function prediction and measures of biodiversity, to name just two of many. Unfortunately, that article was written by a member of the National Academy of Sciences, thereby suggesting scientific credibility on the issue, and it has been highlighted by the antievolution religious organization, the Discovery Institute.

Despite common perception, this inferiority complex for historical science is unwarranted for at least two reasons (Cleland 2001). First, despite what we learn in introductory science classes, there are problems with strict falsificationism. For example, probability statements are not falsifiable, yet they are still scientific because they are testable, indicating that a better theory of testability than falsificationism is required (Sober 2007). Furthermore, strict falsificationism is rarely followed, even by practicing experimental scientists. The reason is that, in many experiments, numerous variables are not controlled by the investigator. Even the seemingly simplest experiments do not control many potential variables (e.g., sun flares, humidity, season, etc.), because it is usually safe to assume that many variables do not affect the experiment at hand. As a result, the possibility always remains that an unsupported hypothesis is not supported because of one of these ancillary assumptions, even if the original hypothesis is true. Therefore, experimental scientists often examine these ancillary assumptions to show that they are responsible for the failure of the hypothesis at hand. For example, I remember many hypotheses about physical laws that were not supported by my experiments in Introductory Physics Lab. Rather than falsifying established laws of physics, I invoked the failure of ancillary assumptions, such as "this ancient and abused student balance produces reliable data."

A second reason to reject claims of inferiority for historical science, regardless of the status of falsificationism, is that historical hypotheses that explain observable phenomena provide predictions to be tested and are therefore scientific. In practice, these predictions often act as confirmatory hypotheses; historical scientists seek to demonstrate a "smoking gun"—strong evidence for a specific event (Cleland 2001). As an example, Darwin's historical hypothesis that all living organisms derive from a common ancestor has left numerous traces consistent with that hypothesis, including the use of RNA and DNA by all organisms, shared use of the same subset of all possible amino acids, and a nearly universal genetic code (for more detailed discussion of the hypothesis and difficulties in testing it, see Sober and Steel 2002). This "smoking gun" perspective is not necessarily falsificationist, yet it is clearly scientific by presenting testable hypotheses.

Another way that historical scientists work is to test ancillary assumptions of historical models. For example, Darwin hypothesized that natural selection gradually built complex eyes from simple precursors. This model assumes that functional intermediates exist at all stages between simple and complex eyes. Darwin (1859), and later Salvini-Plawen and Mayr (1977), provided support for this model by describing the functioning eyes of living animals at numerous stages of complexity. In addition, Nilsson and Pelger (1994) found strong support for another ancillary assumption of the natural selection hypothesis—that there has been sufficient time for gradual selection to build eyes of observed complexity. It is true that we cannot re-create the evolution of the human eye. Nevertheless, we can make models of how eye evolution proceeded and test the ancillary assumptions of that model. Clearly, historical inference is scientific and, while philosophically different than experimental science, should not be construed as inferior. Therefore, a perceived inferiority should not be used as a motivation for experimental phylogenetics.

Thus far, I have only negated one argument (the perceived inferiority of historical science) for experimental phylogenetics and as such have not yet provided any arguments against it, or for any alternative approach. The next two sections make explicit comparisons between experimental phylogenetics and the alternative approach of computer simulation. Before considering whether experimental phylogenetics allows for increased biological realism over computer simulation, I will consider the value of experimental phylogenetics for testing methods of phylogenetic inference.

## TESTING PHYLOGENETIC METHODS

Although claims for the inferiority of historical science do not have a sound philosophical basis, another motivation for experimental phylogenetics appears philosophically sound. Specifically, understanding the relative strengths and weaknesses of methods of inference is an important scientific endeavor, and experimental phylogenies can be used to attain these goals. However, simply realizing that experimental phylogenetics can be of use is not sufficient, because other approaches can be used to the same end. Therefore, a convincing argument for conducting experimental phylogenetics must provide justification over and above other possible approaches.

Computer simulation, statistical analysis, and congruence all can be used to assess the performance of phylogenetic methods (Hillis 1995). While a full review of methods and philosophies for testing phylogenetic methods is beyond the scope of this chapter, and they have been reviewed elsewhere (e.g., Hillis 1995; Grant 2002), I conclude here that generating biological phylogenies is an overly expensive enterprise, costing a prohibitively large amount of investigator time compared to computer simulation. Speed can be increased in specific situations, but perhaps at the expense of biological realism. The question then becomes whether increased biological realism overcomes the increased cost over computer simulation. I will argue that it does not, concurring with others who have pointed out that experimental phylogenetics is subject to the same constraints as simulations: in either situation, it is necessary to assume the evolutionary processes present in the tests apply universally (Sober 1993; Grant 2002). This assumption is especially true when trying to establish the efficacy of methods, as opposed to the shortcomings. Any one replicate history can call into question the reliability of a method, but because any single replicate could be nongeneral, establishing reliability of methods requires generating replicates under many different assumptions or parameter values.

## THE NEED FOR SPEED: COSTS AND CREATIVE SOLUTIONS

The goal of experimental phylogenetics is to generate clades of organisms (or genes or historical documents) with a known history and to examine the performance of methods for reconstructing that known history. Perhaps the most compelling advantage (discussed in detail later) of experimental phylogenetics over computer simulations comes down to the possibility for increased biological realism. As Hillis et al. (1993) wrote: "The point of the experimental approach is to avoid approximating biological evolution by examining actual cases of biological evolution." To be practical, experimental phylogenetics requires the ability to generate clades on a time scale of months or less, which in turn requires using systems with brief generation times and rapid rates of evolution. Obtaining such rapid rates of evolution restricts the set of organisms that can be utilized. This is the first cost of the need for speed: a reliance of the assumption that rapidly replicating biological systems faithfully model other systems, including those that evolve on long time scales. Even some of the most rapidly evolving systems have been further modified to increase their rate of evolution, leading to additional departures from natural biological systems. For example, the mutagen N-methyl-N'-nitrosoguanidine (NG) was added to increase the mutation rate of viruses in experimental phylogenetics (Hillis et al. 1992). The mutagen increases mutation rate, but also changes the mutational profile, causing

G → A or C → T changes to be most common (Bull et al. 1993). Here again, the altered mutational profile may be considered a deviation from biological realism that is a necessary by-product of increasing the speed of evolution.

As necessity is often the mother of invention, the demonstrated need for speed in experimental phylogenetics inspired some creative solutions. For example, Cunningham et al. (1997, 1998) produced a modular experimental phylogeny, which could be analyzed in multiple ways. Starting from a wild-type T7 bacteriophage, they evolved six separate lineages, each of which was bifurcated once. As a result, they were able to assemble multiple different four-taxon phylogenies with varying relative branch lengths, from a single original experiment (Cunningham et al. 1998). This highlights one major difference between testing methods of phylogenetic tree inference and methods of ancestral state reconstruction. Any phylogeny has multiple nodes, such that ancestral state reconstruction methods can be examined on each of them. For ancestral states, there is an automatic replication. For testing phylogenetic trees and for testing correlations between characters (correlative comparative methods; review in Garland et al. 2005), it may always be wise for the experiment to be modular, to allow for increased replication from the expensive experiment.

Another ingenious compromise between the need for speed in simulation studies and "biological realism" is hypermutagenic polymerase chain reaction (PCR). Instead of using living organisms or viruses, researchers have generated experimental phylogenies by utilizing the mutagenic properties inherent in copying DNA. By winnowing the evolving biological system to DNA and polymerase, the researchers have greatly increased the speed at which replicates can be generated. For example, Vartanian et al. (2001) copied a dihydrofolate reductase gene of *Escherichia coli* into a phylogeny of 124 "pseudogenes." Sanson et al. (2002) used similar methodology to generate sequence data (over 2,200 bp each) for an experimental phylogeny with fifteen ancestor and sixteen terminal sequences. However, just as in viral phylogenies, the increased speed in PCR-generated phylogenies comes at the expense of biological realism. In the PCR experiments, the biological system is reduced to an enzyme and DNA. The complexities of mutation and selection in the face of changing environments are greatly simplified in a PCR system compared to nature.

A third creative solution to the trade-off between speed and biological reality was parametric bootstrapping. Parametric bootstrapping involves estimating parameters of a model from real data, and using those parameter estimates and model to simulate multiple datasets (Efron 1985; Felsenstein 1988). Bull et al. (1993) estimated parameters for restriction site evolution from a bacteriophage experimental phylogeny. Using these parameters, they simulated by computer the evolution of multiple data sets to test methods of phylogeny reconstruction and molecular evolutionary inferences. Some may argue that this parametric bootstrap procedure provides a balance between biological realism and speed. Parameters are estimated from a biological system, and speed is gained by simulating multiple replicates by computer. However, the parameters of molecular evolution do not have to be estimated using experimental phylogenetic data; any comparative data set could be used to infer model parameters. Furthermore, if experiments on model selection are any guide, then model parameters might be well estimated even if the true phylogeny is not known precisely. That is, in simulation experiments, the specific starting tree had little effect on the models of molecular evolution chosen as statistically best fit (Posada and Crandall 2001; Posada and Buckley 2004), suggesting that the same might hold for parameter estimates of those models. In summary, parametric bootstrapping is a valuable tool that can extend the results gained from experimental phylogenetics (Bull et al. 1993). However, I remain unconvinced that experimental phylogenetic data are more valuable for parameter estimation than are comparative data from any naturally evolving system.

## INCREASED BIOLOGICAL REALISM

Perhaps the most plausible justification for the use of experimental phylogenetics relates to arguments that it provides increased biological realism. Unlike the previous arguments discussed here, this one is based on an explicit comparison between experimental phylogenetics and computer simulation. If experimental phylogenetics really does add increased biological realism over computer simulation, then this would be a powerful argument for the approach.

### WHAT IS BIOLOGICAL REALISM?

Experimental evolutionists take biological realism to mean elements that contribute to an evolving system that are not decided a priori by the investigator (see also Huey and Rosenzweig this volume). I will refer to this as the degree of specification. In a computer simulation, usually the only factor that is not specified by the investigator is one or more sequences of random numbers. Of course, these random numbers can be used to specify many elements of a simulation, such as timing of branching events, or rates of evolution. In experimental evolution, many elements are also specified—for example, the branching pattern of the phylogeny (Hillis et al. 1992). However, some aspects of the experiment are not specified by the investigator, such as the mutational process and the relationship between mutations and a phenotype like virus replication rate (Oakley and Cunningham 2000). The claim of proponents of the field is that these nonspecified elements increase biological realism over computer simulation.

### MY OWN BIOLOGICAL REALITY

These claims for increased realism may be difficult to assess with generality because they involve comparing a real-world system to a mathematical statistical model. We must decide, then, how well the models used in computer simulation account for real-world evolution. The models used in simulation, and the real-world trajectory of evolutionary history are so varied, it is difficult to know where to begin when attempting such a comparison. Nevertheless, this perspective suggests that the value of experimental phylogenies might be increased over computer simulations if experimental approaches are more likely to present the researcher with situations that are not explicitly modeled but that are produced by the nonspecified aspects evolutionary process itself.

Such a situation occurred in my only foray into experimental phylogenetics. I was using the bacteriophage phylogeny generated by Hillis et al. (1992) to study methods of ancestral state reconstruction for phenotypic traits (Oakley and Cunningham 2000). I found that virulence evolved in a way I didn't expect a priori—there were large amounts of homoplasy. Systematists often assume that characters should usually evolve phylogenetically, such that close relatives share traits that are more similar than distant relatives. This is the inherent assumption behind methods like independent contrasts (Felsenstein 1985; review in Garland et al. 2005), and it is an assumption that is often tested now (e.g., Abouheif 1999; Blomberg et al. 2003). However, simulated data are often neutral. In real-world systems, homoplasy may be very common, driven by structural and functional demands on organisms (reviewed in Conway Morris 2003).

In the case of the bacteriophage phylogeny, instead of close relatives being more similar in virulence characteristics than distant relatives, the character was highly convergent. I observed parallel decreases in virulence in all the experimental lineages, which was rapid enough to erase all phylogenetic signal of the character. For example, a nonphylogenetic model of character evolution (Lee et al. 2006; Mooers and Schluter 1998; Mooers et al. 1999; Oakley et al. 2005) is the best fit among nine Brownian motion–based models. Had I used neutral computer simulations exclusively in testing ancestral state reconstruction methods, I might not have modeled the evolutionary trajectory actually taken by the viruses. Here, the viruses might have provided more biological realism than computer simulation in that the biological system is arguably less specified than a computer simulation.

One counterargument to this discussion of the enhanced biological realism of experimental studies is that a wholly empirical system arrived at very similar conclusions to my study of ancestral virulence in bacteriophage: Webster and Purvis (2002) investigated extinct and living foraminifera and found that strong directional change in body size erased phylogenetic signal for this character. If an empirical system showed the same results, then perhaps an experimental system was not needed to find the results. Yet, appropriate fully empirical systems may be rare and may have higher costs than even experimental evolution in investigator time spent understanding the system.

## CONCLUSION

Despite enthusiasm in the early 1990s for a future of experimental phylogenetics, the field has stalled and produced very few papers and few novel insights. Part of this explanation is that phylogenetic methodologies have become rather standardized tools for evolutionary inference. However, as I have argued here, two other considerations point to fundamental flaws in the foundations of the field. First, historical science is not inferior to experimental science. Historical and experimental sciences are philosophically different, and historical science is not inferior or less scientific. Therefore, the perceived inferiority of historical science cannot be used to justify any experimental approach in science, including experimental phylogenetics. Second, I argued that experimental phylogenies are probably not inherently more valuable than any other "simulation," and they are vastly more expensive in terms of investigator time and resources. As such, experimental phylogenetic studies that are already conducted are no less valuable than any simulation study, but researchers contemplating new experimental phylogenetics should carefully weigh the costs. One possible saving grace for experimental phylogenetics is the possibility that computer simulations are highly specified, such that experimental approaches might be more likely to produce unanticipated but biologically realistic results (see also Swallow et al. this volume on one important value of replication in selection experiments—the possibility of finding "multiple solutions"). This is a difficult proposition to argue for or against quantitatively, but it certainly highlights the requirement that simulations must be based on as much biological knowledge as possible, which might limit generality and/or increase the cost of performing them. I hasten to point out that the critique presented here does not apply to experimental evolution in general, which can still serve as a valid demonstration of evolutionary processes. However, my own foray into experimental phylogenetics left me unsatisfied, and this chapter presents the reasons why.

## SUMMARY

The primary goal of the field of experimental phylogenetics is to generate branching histories of biological entities in the laboratory for use in testing methods of phylogenetic reconstruction. This chapter explores possible reasons why this field has remained small, despite hints of a bright future fifteen years ago. Specifically, it examines three primary arguments that researchers have used to motivate the field of experimental evolution. The first involves claims that hypotheses in phylogenetics and molecular evolution are difficult to unambiguously falsify, and therefore an experimental approach is required. These claims do not specifically motivate experimental phylogenetics because they are based on an incorrect interpretation of the philosophy of historical science, and they do not differentiate between experimental evolution and its competitor, computer simulation. A related argument is that experimental phylogenetics can be used to understand the strengths and limitations of various methods of historical inference. This is a valid argument but again does not distinguish between experimental evolution and computer simulation. In fact, this chapter argues that high replication under different conditions is most important for testing methods, putting a premium on speed and leading to a disadvantage of experimental phylogenetics compared to computer simulation. A third argument does compare experimental phylogenetics to computer simulation, claiming that experimental evolution has increased realism compared to computer simulation. For example, experimental phylogenies may present modes of evolution not often implemented by computer simulations, such as common parallel or generally convergent evolution. These arguments do not decrease the value of completed experimental phylogenetic studies, but they call for caution when weighing the costs of future studies that generate phylogenies in the lab.

## REFERENCES

Abouheif, E. 1999. A method for testing the assumption of phylogenetic independence in comparative data. *Evolutionary Ecology Research* 1:895–909.

Blomberg, S. P., T. Garland, Jr., and A. R. Ives. 2003. Testing for phylogenetic signal in comparative data: Behavioral traits are more labile. *Evolution: International Journal of Organic Evolution* 57:717–745.

Bull, J. J., C. W. Cunningham, I. J. Molineux, M. R. Badgett, and D. M. Hillis. 1993. Experimental molecular evolution of bacteriophage T7. *Evolution* 47:993–1007.

Cleland, C. 2001. Historical science, experimental science, and the scientific method. *Geology* 29:987–990.

Conway Morris, S. 2003. *Life's Solution: Inevitable Humans in a Lonely Universe*. Cambridge: Cambridge University Press.

Cunningham, C. W., K. Jeng, J. Husti, M. Badgett, I. J. Molineux, D. M. Hillis, and J. J. Bull. 1997. Parallel molecular evolution of deletions and nonsense mutations in bateriophage T7. *Molecular Biology and Evolution* 14:113–116.

Cunningham, C. W., H. Zhu, and D. M. Hillis. 1998. Best-fit maximum likelihood models for phylogenetic inference: Empirical tests with known phylogenies. *Evolution* 52:978–987.

Darwin, C. 1859. *On the Origin of the Species by Means of Natural Selection, or, The Preservation of Favoured Races in the Struggle for Life*. London: Murray.

Efron, B. 1985. Bootstrap confidence intervals for a class of parametric problems. *Biometrika* 72:45–58.

Felsenstein, J. 1985. Phylogenies and the comparative method. *American Naturalist* 125:1–15.

———. 1988. Phylogenies from molecular sequences: Inferences and reliability. *Annual Review of Genetics* 22:521–565.

Garland, T., Jr., A. F. Bennett, and E. L. Rezende. 2005. Phylogenetic approaches in comparative physiology. *Journal of Experimental Biology* 208:3015–3035.

Gee, H. 1999. *In Search of Deep Time: Beyond the Fossil Record to a New History of Life*. New York: Free Press.

Grant, T. 2002. Testing methods: The evaluation of discovery operations in evolutionary biology. *Cladistics* 18:94–111.

Hillis, D. M. 1995. Approaches for assessing phylogenetic accuracy. *Systematic Biology* 44:3–16.

Hillis, D. M., J. J. Bull, M. E. White, M. R. Badgett, and I. J. Molineux. 1993. Experimental approaches to phylogenetic analysis. *Evolution* 42:90–92.

————. 1992. Experimental phylogenetics: Generation of a known phylogeny. *Science* 255:589–592.

Lee, C., A. O. Mooers, S. Blay, A. Singh, and T. H. Oakley. 2006. CoMET: A mesquite package for comparison of continuous models of character evolution on phylogenies. *Evolutionary Bioinformatics* Online. 2:193–196.

Lee, Y., and J. Yin. 1996. Detection of evolving viruses. *Nature Biotechnology* 14:491–493.

Mooers, A. Ø., and D. Schluter. 1998. Fitting macroevolutionary models to phylogenies: An example using vertebrate body sizes. *Contributions to Zoology* 68:3–18.

Mooers, A. Ø., S. M. Vamosi, and D. Schluter. 1999. Using phylogenies to test macroevolutionary hypotheses of trait evolution in cranes (Gruinae). *American Naturalist* 154:249–259.

Nilsson, D. E., and S. Pelger. 1994. A pessimistic estimate of the time required for an eye to evolve. *Philosophical Transactions of the Royal Society of London B, Biological Sciences* 256:53–58.

Oakley, T. H., and C. W. Cunningham. 2000. Independent contrasts succeed where ancestor reconstruction fails in a known bacteriophage phylogeny. *Evolution* 54:397–405.

Oakley, T. H., Z. Gu, E. Abouheif, N. H. Patel, and W. H. Li. 2005. Comparative methods for the analysis of gene-expression evolution: An example using yeast functional genomic data. *Molecular Biology and Evolution* 22:40–50.

Posada, D., and T. Buckley. 2004. Model selection and model averaging in phylogenetics: Advantages of Akaike information criterion and Bayesian approaches over likelihood ratio tests. *Systematic Biology* 53:793–808.

Posada, D., and K. A. Crandall. 2001. Selecting the best-fit model of nucleotide substitution. *Systematic Biology* 50:580–601.

Salvini-Plawen, L. V., and E. Mayr. 1977. *On the Evolution of Photoreceptors and Eyes.* Evolutionary Biology, vol. 10. New York: Plenum.

Sanson, G. F., S. Y. Kawashita, A. Brunstein, and M. R. Briones. 2002. Experimental phylogeny of neutrally evolving DNA sequences generated by a bifurcate series of nested polymerase chain reactions. *Molecular Biology and Evolution* 19:170–178.

Skell, P. S. 2005. Evolutionary theory contributes little to experimental biology. *The Scientist* 19:10.

Sober, E. 1993. Experimental tests of phylogenetic inference methods. *Systematic Biology* 42:85–89.

————. 2007. What is wrong with intelligent design? *Quarterly Review of Biology* 82:3–8.

Sober, E., and M. Steel. 2002. Testing the hypothesis of common ancestry. *Journal of Theoretical Biology* 218:395–408.

Vartanian, J. P., M. Henry, and S. Wain-Hobson. 2001. Simulating pseudogene evolution in vitro: Determining the true number of mutations in a lineage. *Proceedings of the National Academy of Sciences of the USA* 98:13172–13176.

Webster, A. J., and A. Purvis. 2002. Testing the accuracy of methods for reconstructing ancestral states of continuous characters. *Proceedings of the Royal Society of London B, Biological Sciences* 269:143–149.

# 22

# LABORATORY EVOLUTION MEETS CATCH-22
## *Balancing Simplicity and Realism*

### Raymond B. Huey and Frank Rosenzweig

Everything should be made as simple as possible, but not simpler.

ATTRIBUTED TO A. EINSTEIN

This book lays out a clear and compelling message: selection experiments are remarkably powerful tools in the armamentarium of evolutionary biologists. We ourselves have often used selection experiments during our careers and certainly expect to use them in the future. In fact, the power and elegance of selection experiments applied to life-history evolution by Rose and colleagues (Rose and Charlesworth 1980; Service 1987) motivated one of us (R.B.H.) to switch from conducting descriptive evolutionary studies on lizards in the field to performing evolution experiments on *Drosophila* in the laboratory.

In this chapter, we look critically at a particular type of experimental evolution, often called laboratory natural selection. In this protocol, stocks of organisms are reared chronically under different conditions (e.g., different thermal or life-history regimes) and allowed to evolve by natural selection over many generations (Rose et al. 1987; Garland 2003). At intervals, phenotypes of population members can be compared in a "common garden" (i.e., reared under identical environmental conditions; Garland and Adolph 1991). Differences between selected and control lines—at least if observed consistently among replicates—represent either direct or indirect responses to the selective regime. This is an old and venerable type of experimental evolution (Dallinger 1887; see box).

Laboratory natural selection (LNS) is distinct from two other types of laboratory evolution (Rose et al. 1996, 1987; Garland 2003; Swallow and Garland 2005; Futuyma and Bennett this volume). In artificial selection, the experimenter actively measures and selects phenotypes to found the next generation. In laboratory culling, organisms are exposed to a lethal condition (e.g., no food, no water), and the longest-surviving individuals are used to found the next generation.

LNS experiments can provide insight into genetic architecture and correlations underlying traits of interest (Rose et al. 1990). They can also be used to evaluate the rate, tempo, and repeatability of evolutionary trajectories (Lenski and Travisano 1994; Ferea et al. 1999; Dunham et al. 2002; Cooper et al. 2003, 2001; Fong et al. 2005; Woods et al. 2006) and to assess how historical contingency (Travisano et al. 1995), sex (Grimberg and Zeyl 2005; Zeyl et al. 2005), sexual selection (Rundle et al. 2006), ploidy (Paquin and Adams 1983; Zeyl and Bell 1997; Zeyl et al. 2003), and life history (Zeyl et al. 2005) influence the outcome of those processes. LNS experiments are especially useful for testing functional hypotheses (Rose and Charlesworth 1980; Bennett and Lenski 1999; Gibbs 1999), as derived lines yield experimental subjects that have "exaggerated" or "novel" phenotypes (Gibbs 1999; Bennett 2003; Garland 2003; Futuyma and Bennett this volume).

LNS experiments are not only broadly applicable but also logistically advantageous (Rose et al. 1987; Futuyma and Bennett this volume). Effective population size can be manipulated over orders of magnitude, especially in microbes, largely eliminating genetic drift, if so desired. Experimental lines can readily be (should be!) replicated (Futuyma and Bennett this volume), and the intensity, frequency, uniformity, and duration of selection can carefully controlled. Selective agents of interest can be applied either singly or in concert. Moreover, selection is accomplished without direct intervention:

Probably the first experimental study of laboratory evolution was conducted in the 1880s by the Rev. W. H. Dallinger, who was President of the Royal Microscopical Society. Inspired by Darwin, Dallinger (1887) decided to determine experimentally "whether it was possible by change of environment . . . to superinduce changes of an adaptive character, if the observations extended over a sufficiently long period."

Dallinger studied "the lowest forms of the infusoria" (protists), because their life cycle was rapid. Apparently they thrived at 60°F. Using a very clever, temperature-controlled water bath, Dallinger gradually increased the temperature of the bath over seven years, backing off for a while when a new temperature seemed to induce stress (see figure 22.1).

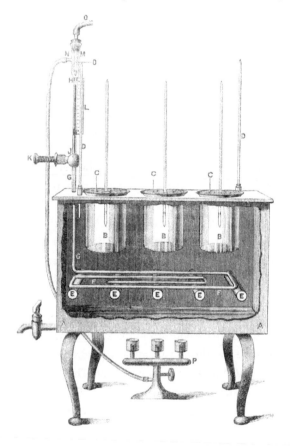

FIGURE 22.1

Temperature-controlled water bath that the Rev. W. H. Dallinger used to study adaptation of infusoria to increasingly high temperature. Temperature control was achieved via a valve to the gas supply. Heat from burners warmed the water bath, causing mercury to rise into the valve, eventually restricting the gas supply. This simple system enabled Dallinger to control temperature very precisely.

*(continued)*

Before the experiment was ended by an accident, the infusoria were able to tolerate 158°F! Dallinger noted (p. 199) that nonadapted individuals "are killed at 140°Fahr. But if the adapted organisms at 158°F were taken from that temperature and placed in . . . fluid at even 150°F they were finally destroyed."

Though a pedant could quibble that Dallinger failed to maintain a control line or to control for acclimation and cross-generation effects, all must recognize the pioneering brilliance of his work. Here for the first time was experimental evolution in action!

Interestingly, Dallinger corresponded with Darwin about the preliminary results from his selection results. Dallinger's article (1887, 191) contains an insightful evaluation from Darwin:

> I did not know that you were attending to the mutation of the lower organisms under changed conditions of life; and your results, I have no doubt, will be extremely curious and valuable. The fact which you mention about their being adapted to certain temperatures, but becoming gradually accustomed to much higher ones, is very remarkable. It explains the existence of algae in hot springs.

genotypes that do relatively well in a particular environment simply leave the most surviving offspring (Rose et al. 1987). In contrast, experimenters using artificial selection (AS), especially if family selection is involved (Scheiner and Lyman 1991), must tediously measure and select individuals each generation; the associated logistics can be daunting and may thus restrict sample sizes and increase the likelihood of drift.

Despite these conceptual and logistic advantages, LNS experiments have inherent problems and limitations. Many have been previously identified (Huey et al. 1991; Rose et al. 1996; Bennett and Lenski 1999; Gibbs 1999; Harshman and Hoffmann 2000; Prasad and Joshi 2003) but are nonetheless worth reiterating. Some are potentially so severe that they can compromise or confound evolutionary and functional interpretations. We describe these problems here and, where feasible, suggest ways to try to circumvent them.

Because LNS experiments have both strengths and weaknesses, researchers contemplating an LNS experiment face the classic catch-22 (or double-bind) situation immortalized in Joseph Heller's (1961) novel of the same name. They may well decide that LNS is the best way to test a given evolutionary hypothesis (Rose et al. 1996:236–238), but simultaneously they must accept the hard fact—and accept it in advance—that some of the inferences they draw from their LNS experiment may be of uncertain validity. Welcome to catch-22, where one is caught "between a rock and a hard place" (Harshman and Hoffmann 2000).

Researchers have several options when faced with such a bind. Yossarians of the world (Yossarian is the protagonist in *Catch-22*) would probably try to solve the problem by switching fields, hoping that they can publish many articles before they (or worse, before others) discover that their new field has its own catch-22s! Or, they can accept the situation but try to turn it to their advantage (Rose et al. 1996). With a bit of creativity, one can circumvent certain problems inherent in LNS or even use some of them as opportunities for interesting new studies. But despite one's best efforts, some issues are simply likely to remain bedeviling catch-22s. Nevertheless, as Rose et al. (1996:239) note, "from our encounter with this often confusing and unfair world, we can learn about our theories and improve them."

## USING LNS TO TEST HYPOTHESES DERIVED FROM COMPARATIVE STUDIES

LNS experiments are often used to test evolutionary hypotheses that have been derived from theory (Futuyma and Bennett this volume) or from observations in nature. We focus here on LNS experiments that are designed specifically to test hypotheses derived from comparative studies in nature. Of course, many issues discussed here will apply to LNS experiments testing theoretical hypotheses. We begin by describing how an LNS study might evolve from a comparative study in nature, and then illustrate some general difficulties.

Many traits show geographic clines. Once a cline is documented, two questions naturally arise (Endler 1977). Did this cline evolve by natural selection? And, if so, what selective factor(s) led to the cline?

Latitudinal clines in the frequency of various chromosomal inversions are well documented in the fly *Drosophila subobscura* (Krimbas 1993). Because these clines are generally in the same direction on three continents (discussed later), they likely evolved by natural selection (Prevosti et al. 1989). But what selective factor is responsible for the clines? Temperature is a reasonable guess: perhaps inversions that are relatively common at low latitudes contain alleles that are adapted to heat, whereas inversions common in high latitude have alleles adapted to cold. Of course, many other abiotic and biotic factors co-vary with latitude (photoperiod, intensity of competition and predation), but temperature is a reasonable first guess for an ectotherm.

To test the role of temperature as a selective agent, one might initially search for "natural experiments" occurring in the field (Endler 1986; Diamond 2001). Climate warming provides just such an opportunity. For decades, evolutionary geneticists have been scoring inversion frequencies of *D. subobscura* at many sites where climate is warming. If inversion clines are driven by temperature, then inversion frequencies at particular sites should shift as climate warms. Specifically, inversions common at low-latitude sites (i.e., presumably warm-adapted genotypes) should increase in frequency as climates warm. This is the case (Rodríguez-Trelles et al. 1998; Solé et al. 2002; Balanyá et al. 2006; see also Levitan and Etges 2005; Umina et al. 2005), consistent with our hypothesis. Nevertheless, because these patterns are still only correlational, we cannot be certain that temperature was indeed the selective factor responsible for the observed clines.

To challenge our hypothesis further, we might devise an experiment in which we manipulate temperature and then evaluate whether inversion frequencies change in a direction consistent with our comparative hypothesis. Inducing climate warming in the field might prove technically and politically challenging, so a more practical next step would be to induce it in the laboratory. A classical LNS approach would be to start with a large outbred population, set up replicated lines in population cages maintained across generations at different (fixed) temperatures (e.g., 12°, 18°, 22°C), and then at intervals monitor inversion frequencies in the different temperature treatments. Thus, we would use the 22°C constant-temperature treatment as a laboratory proxy of relatively low-latitude environments and 12°C one as a proxy of high-latitude environments. If our temperature hypothesis is correct, then inversions common at low latitudes should increase in frequency in the warm temperature treatment but decrease in frequency in the low temperature one. This experiment has been done (Santos et al. 2004, 2005, 2006), and we will return to it near the end of this chapter.

This LNS experiment seems like a logical and appealing test of our hypothesis, but its validity rests on at least two key assumptions. First, the direct agent(s) of selection in the laboratory must mimic—at least approximately—those in nature. For example, the experiment described here assumes that chronic exposure to high temperature in the laboratory approximates the selective impact of living in warm latitudes in nature. A priori this seems highly improbable. Low latitudes differ from high ones in many ways (Bradshaw and Holzapfel 2006), not just in temperature, and warm terrestrial environments (at least in temperate zones) are never chronically warm. Second, the genotypic variance of the large outbred (experimental) population should be roughly comparable to the combined variance of all natural populations along the cline. This ensures that our lab founder population has sufficient genetic potential for selection to realize phenotypic variance in the trait of interest.

Despite having concerns that these assumptions might be violated, our intrepid experimenters start the selection experiment. Let's imagine how they will interpret their eventual results. If they find that inversion frequencies shift as predicted, they will probably conclude (1) that these results support the hypothesis that latitudinal clines in nature are likely driven by temperature and (2) that their laboratory temperature regimes are a "good enough" approximation of latitudinal environments. No doubt they'll be able to publish their results in a fine journal—provided that they can convince reviewers that they have not committed a Type I error! But if they find patterns contrary to those expected, our investigators will find themselves on decidedly uneasy ground. They might conclude that the temperature hypothesis is false. Alternatively, they might conclude that their experimental design is fundamentally flawed: perhaps inversion frequencies in nature are driven by infrequent cold winter temperatures, not by average ones; or maybe they just chose the wrong temperature levels. Of course, such conclusions might be a Type II error.

LNS experiments have an even more serious problem. To simplify the experiment, our team manipulated only *one* environmental variable: thus, their experimental design

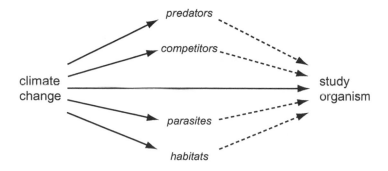

FIGURE 22.2
Effects of climate change on a study organism can be direct (solid arrow) or indirect (dashed arrows), via the impact of climate change on predators, competitors, parasites, and habitat.

tests only for direct effect of temperature on flies (figure 22.2), holding all else equal (in theory). In nature, however, latitudinal patterns of selection might reflect not only the direct effects (solid arrows) of temperature on our study organism, but also the direct and indirect (dashed arrows) effects of temperature on diverse abiotic and biotic factors, which indirectly impinge on our study organism (figure 22.2). Consequently, temperature might well be the selective agent responsible for inversion clines, but the mechanism could be direct or could involve complex interactions, or even all of the above. (And, of course, temperature might be a red herring, such that selection is really driven by other factors that co-vary with latitude [Bradshaw and Holzapfel 2006].) This complex pattern of selection in nature is fundamentally different from the simple and direct pattern of selection in an LNS experiment. Therefore, evolutionary trajectories driven by temperature in an LNS experiment might well differ from those in nature, even if temperature is driving both systems.

We see these problems as inevitable in an LNS experiment attempting to test comparative hypotheses. Although some have argued that "to seek patterns is to do science," documenting patterns is only the first step. We must go further and develop manipulative experiments that enable us to discover causal mechanisms (Paine 1994). But here again we meet a catch-22, for laboratory experiments are inherently artificial and simplistic; and any resulting conclusions must thus be accepted with caution (Rose et al. 1996). However, if we thoroughly understand the assumptions and limitations of our experiments, we should be able to design procedures that reveal rather than obscure mechanism.

A number of more specific issues bedevil LNS experiments. Some have to do with the adaptation of experimental stocks to the laboratory. Some have to do with selection protocols themselves. We itemize these in the following sections and try to suggest ways to circumvent them.

## SELECTING ON FIELD-FRESH LINES

Nature and the laboratory are different environments, so field-fresh organisms en-counter novel selective pressures when transferred to the laboratory (Service and Rose 1985; Matos et al. 2000b; Hoffmann et al. 2001; Simões et al. 2007, this volume). If an LNS experiment begins shortly thereafter, then the subjects will be adapting not only to the specific LNS regime but also to the general laboratory environment. The resulting conflation of selective factors can easily confound interpretations of responses to LNS, for example, yielding falsely positive genetic correlations between traits (see Service and Rose 1985; Clark 1987; Rose et al. 2005).

Service and Rose (1985) propose a solution: start selection only after the study organ-isms have adapted to the laboratory environment, indicated perhaps by a plateau of the selection response (Gilligan and Frankham 2003; Simões et al. 2007). Their proposition is reasonable in principle, but it leaves open the question of how much laboratory adap-tation is sufficient (Harshman and Hoffmann 2000). Unfortunately, we see no easy answer to this question, especially as different traits will adapt at different rates (Matos et al. 2000b). The pace of laboratory domestication is certain to be taxon-specific and to depend on the magnitude and nature of the deviation of the lab environment from that of the field, as well as the amount of standing genetic variation imported with the founder population. Moreover, adaptation to the laboratory may create special problems, as discussed in the next section.

## SELECTING ON LABORATORY-ADAPTED LINES

Service and Rose's (1985) proposal solves one problem but creates another. Here's the key issue: laboratory lineages may respond differently to experimental selection than do field-fresh stocks because genetic architecture will change as a result of laboratory adap-tation and random genetic drift (Clark 1987; Frankham et al. 1988; Harshman and Hoffmann 2000; Griffiths et al. 2005; but see Promislow and Tartar 1998; Krebs et al. 2001). Changes during laboratory adaptation can be profound. Consider the situation facing a research team contemplating an LNS experiment with the moth *Manduca sexta*. To sidestep the problems associated with field-fresh stocks (Service and Rose 1985), the team might choose experimental lines derived from a base laboratory stock that has been reared in the laboratory now for over thirty-three years (J. Kingsolver, personal commu-nication). This would be an appealing choice, as the lines should by now be well adapted to the laboratory. Moreover, a gold mine of physiological, endocrinological, neurological, and developmental information is available on these lines. So a National Science Foun-dation panel is likely to look favorably on this choice of stocks.

But if our goal is to learn how wild *Manduca* would respond to a specified type of selection, then we must question whether these laboratory stocks will be remotely representative of free-ranging *Manduca*. We suspect not. Laboratory *Manduca* have inadvertently been selected for fast growth to a large adult size (L. Riddiford, personal communication). They are now real "porkers," fly clumsily, have degraded vision (J. Sprayberry, personal communication), and have only five larval instars. In striking contrast, their free-ranging ancestors are sleek, agile, and visual, and they can even have six instars (Kingsolver 2007). Consequently, an LNS experiment starting with lab-adapted *Manduca* might yield very different results from those starting with more wild-type stocks. History can matter (Travisano et al. 1995; Moore and Woods 2006).

Again we encounter a catch-22: if we start selection with wild stocks, we conflate domestication with intentional selection; but if we start selection with lab stocks, we may observe unnatural evolutionary trajectories (Harshman and Hoffmann 2000). Experimental masochists might decide to conduct selection both on recently and on long-ago established lines (Harshman and Hoffmann 2000): parallel evolutionary trajectories would be reassuring. Nevertheless, Matos et al. (2000a) have criticized that approach—they see little to gain in studying recently sampled stocks.

## LABORATORY ENVIRONMENTS ARE TOO BENIGN

Genetic architecture can be modified by the benign nature of the laboratory itself. Consider our baseline fly experiment simulating climate change. Flies are raised for many generations on ample food and at constant (typically nonextreme) temperature, fixed photoperiod and humidity; they may encounter essentially no variation in the physical environment and have no interactions with predators or parasites or (interspecific) competitors, and food is near at hand (or wing). Life is good. Life is simple.

Such benign laboratory environments will likely weaken selection on many traits that in nature must deal with fluctuating physical environments, predators, parasites, or competitors (figure 22.2). As a result, the performance capacities for those traits might decay over time, as a result of either energy conservation (Regal 1977) or mutation accumulation (Mueller 1987; Promislow and Tartar 1998). Natural isolates of *C. elegans* show classical thermoregulatory behavior on a laboratory gradient, whereas the standard lab stock (N2) decidedly does not, suggesting that N2 has lost thermoregulatory abilities during laboratory adaptation at constant laboratory temperatures (Anderson et al. 2007). However, degradation in performance may not always be readily apparent. Kondrashov and Houle (1994) found that some fitness differences between control and mutation accumulation lines of *Drosophila* were apparent only in harsh environments. This problem may not be general, however, as Chang and Shaw (2003) observed no exaggerated decline in mean fitness of mutation accumulation lines of *Arabidopsis* when challenged in low-nutrient environments.

Moreover, benign environments permit evolutionary trajectories that would likely be maladaptive in nature (Gibbs 1999). For example, flies selected for starvation resistance (Chippindale et al. 1996; Hoffmann and Harshman 1999) quickly evolve enhanced levels of lipids and effectively become "butterballs": in one experiment (Harshman et al. 1999), starvation-selected lines were 21 percent heavier than control lines! Similarly, flies selected for desiccation resistance accumulate body water (table 22.1; Chippindale et al. 1998; Gibbs et al. 1997). Accumulating resources (lipids or water) during the larval period may be a viable evolutionary response for flies experiencing selection for starvation or desiccation resistance in the lab, but not for flies in nature: butterballs and water melons would be easy and tempting targets for predators, and furthermore, they probably have reduced ability to disperse (Gibbs 1999, 2714). Moreover, stress-selected lines had low preadult viability, suggesting that resource sequestration during the larval period has an associated cost (Chippindale et al. 1998). Not surprisingly, real desert flies do not accumulate water (table 22.1; Gibbs and Matzkin 2001). Therefore, mechanisms of adaptation to starvation or desiccation resistance in the laboratory may involve very different solutions than in nature (Gibbs 1999). Such "unnatural" LNS trajectories are still of academic interest and certainly may offer insight concerning genetic and physiological mechanisms, but they may not always be relevant to testing comparative hypotheses generated from field observations.

Laboratory adaptation can lead to other changes that might influence selective trajectories. When flies first are brought into the lab, they often pupate on the surface of the medium. But the remaining larvae continue to work the medium, such that pupae on the surface of the medium often die, presumably because they become buried. As a direct consequence, selection favors larvae that pupate on the walls of vials, especially in high-density regimes (Mueller and Sweet 1986). The shift is dramatic and rapid. But the shift means that LNS and wild pupae experience different environments, and this might (or might not) result in inadvertent selection on pupal traits (or on wandering larvae), altering diverse aspects of the genetic architecture and confounding overall evolutionary trajectories. A vial or bottle or even a population cage is not the field.

Among model microorganisms such as *Escherichia coli* and Baker's yeast, laboratory conditions appear to select against certain "wild-type" traits as well as select for others

TABLE 22.1    Increased Resistance for Desiccation Can Potentially
Be Achieved Several Ways

| Possible Response | Experimental Response of Lab-Selected Flies | Comparative Response of Flies from Nature |
|---|---|---|
| Store more water | Yes | No |
| Lose water more slowly | Yes | Yes |
| Tolerate greater water loss | No | No |
| Modify behavior | Not possible in the lab | Probably |

NOTE: Desert flies from nature rely primarily on losing water relatively slowly. However, flies selected for desiccation resistance in the laboratory rely mainly on storing more water, which would probably be disadvantageous in nature. Data summarized from Chippindale, Gibbs, et al. 1998; Gibbs et al. 1997; Gibbs and Matzkin 2001.

not commonly observed in the wild. Mikkola and Kurland (1992) imposed LNS on a set of natural *E. coli* isolates that were highly variable in their growth rate and translation efficiency. In fewer than three hundred generations, these diverse strains converged on growth and translation phenotypes that characterize laboratory "wild types."

Genome architecture can differ markedly between natural and laboratory *E. coli* strains. For example, the core genome of natural isolates is estimated to range from 2,800 (Fukiya et al. 2004) to 3,100 open reading frames (Dobrindt et al. 2003), but that of the nonpathogenic laboratory strain K12 contains 4,288 predicted open reading frames, many of which have unknown function (Kang et al. 2004). Even commensal natural isolates show enormous variation in the presence or absence of many virulence factors (Escobar-Paramo et al. 2006). Among natural isolates, genome size may vary by as much 20 percent between strains adapted to endocellular and extracellular lifestyles (Bergthorsson and Ochman 1999; Perna et al. 2001). These discrepancies point to the need for caution in generalizing results of LNS experiments using model lab strains.

By contrast, genome content appears remarkably conserved among *Saccharomyces* congeners (Kellis et al. 2003). And in *Saccharomyces cerevisiae*, systematic deletion of "nonessential" genes does not appear to confer competitive advantage (Sliwa and Korona 2005). Indeed, genomic studies have revealed widespread anueploidy among laboratory strains, including the widely used yeast "knock-out" collection (Hughes et al. 2000; Scherens and Goffeau 2004). Still, laboratory populations of *S. cerevisiae* differ from their wild conspecifics in many key respects, including pheromone response, as well as in the timing and location of daughter cell separation. Intriguingly, much of the variation in these particular features has been attributed to polymorphisms at the trans-acting regulatory loci GPA1 and AMN1 (Yvert et al. 2003; Ronald et al. 2006).

Environmental differences between nature and the lab can result in major differences in phenotypes. Free-living yeast and bacteria face the prospect of prolonged resource limitation and the threat of dehydration, and they mitigate these hazards by forming biofilms. However, microbes in the laboratory generally don't face these hazards and have evolved changes in both colony morphology and the associated transcriptional program that supports this quasi-multicellular habit (Kuthan et al. 2003; Palkova 2004).

Finally, and of perhaps greatest concern for LNS experiments, laboratory and wild microbes potentially differ in mutation rate. In *E. coli*, a gene's mutation rate differs according to chromosome location (Hudson et al. 2002). Because genome size and organization vary so widely among stocks, mutation rate in essential genes might well differ between lab and natural isolates, as well as among natural isolates. In *S. cerevisiae* strain S288c, recurrent bottlenecks and an overall relaxation in selection intensity are hypothesized to underlie its higher rates of nonsynonymous substitutions relative to its wild conspecific, YJM789 (Gu et al. 2005). Further analysis, using an additional wild isolate, has challenged the generality of this (Ronald et al. 2006).

As just noted, it is easy to conclude that LNS environments are too benign to be ecologically realistic—after all, predators or parasites are mercifully absent. But laboratory environments can be surprisingly stressful and potentially pathological in unexpected ways. It's a catch-22 all over again!

Unless replenished continuously (as in a chemostat), food quality will change over time. In a standard fly experiment, food is replaced at intervals; and food deteriorates as waste products accumulate and as food itself is depleted. In *D. melanogaster*, this can lead to a stable genetic polymorphism (Borash et al. 1998). One genotype evades these problems by evolving early emergence, which it achieves via elevated feeding rates. The other genotype feeds and grows more slowly, but it evolves greater tolerance of the waste product ammonia. This fascinating example shows that laboratory environment may not always be benign and can modify evolutionary trajectories in unanticipated—and unwanted—ways, including the evolution of enhanced (or possibly blunted) phenotypic plasticity (Garland and Kelly 2006).

When organisms are evolving in chemostats, their growth and reproduction are continuously substrate limited (Novick and Szilard 1951; Kubitschek 1970; Adams and Hansche 1974; Dykhuizen 1990). Under these conditions, one might expect populations to evolve by periodic selection of fittest clones, so that only one clone is likely common at any given time (Muller 1931; Williams 1975). However, chronic nutrient deprivation can promote stable genetic polymorphism (Helling et al. 1987). For example, when *E. coli* evolve on limited glucose, subdominant clones can quickly evolve the capacity to scavenge acetate, a fermentation by-product secreted by the dominant clone growing best on the limiting substrate (Treves et al. 1991; Rosenzweig et al. 1994). So both clones persist.

Laboratory environments are less than benign in other ways. Most LNS experiments (but see Bennett and Lenski 1993; Riehle et al. 2005) are conducted at constant temperatures. But to some organisms, constant temperatures appear to be physiologically pathological (Huey 1982) and can yield aberrant results (Brakefield and Mazzotta 1995). Similarly, light levels in incubators are low and often have nonnatural spectral qualities (G. Gilchrist, personal communication). Insects perceive the flicker of AC lights, and so their world resembles a "continuous disco" (J. W. Truman, personal communication). Dim light might suppress visual cues (important to behavioral interactions), modifying selection on behaviors; it also might reduce photochemical reactions (e.g., vitamin D synthesis), leading to physiological pathologies.

Finally, many *Drosophila* labs routinely maintain their stocks on constant twenty-four-hour light regimes (to eliminate time-of-day cues; M. R. Rose, personal communication); but a constant photoperiod will disrupt circadian patterns of behavior (Markow 1975, 1979; Paranjpe et al. 2004) and physiology (Pittendrigh 1960). In all these examples, LNS experiments will be selecting on lines that at least initially suffer laboratory induced (if inadvertently so) anomalies or even pathologies, such that the resulting evolutionary trajectories might differ from those of healthy lines. To be sure, lines evolving under constant photoperiod may adapt to such conditions (Sheeba et al. 1999), but whether those lines can serve as reliable models for "natural" organisms is uncertain.

One solution is to try to make laboratory environments more natural. Bradshaw and Holzapfel (2001) have done just that with pitcher plant mosquitoes. They use real pitcher plants as microhabitats, and they use natural photoperiods (including twilight) and thermoperiods. Unfortunately, simulating natural environments will not always be desirable in LNS experiments—consider the natural habitat of *E. coli*.

## SIMPLICITY CAN BE DECEIVING

LNS experiments are designed with a view toward simplicity: selection is reduced to one or few variables, selection is chronic, and selection is uniform across replicates (Cohan and Hoffmann 1986). Simplicity is desirable not just because it makes LNS logistically tractable or useful for model building and model testing, but also because simplicity has been a distinctive feature of the experimental method since Francis Bacon. Nonetheless, simplicity can spawn several problems that relate to the intensity of selection, as well as the temporal and spatial dimensions over which it is applied.

*Acute Shifts in Selection*    At the initiation of an LNS experiment, lines are usually transferred suddenly to different environments and maintained there for generations (figure 22.3a). The rationale for "steplike" shifts is compelling: the experiment is logistically simpler and is likely to foster a response to selection prior to the next grant cycle. Even so, steplike changes

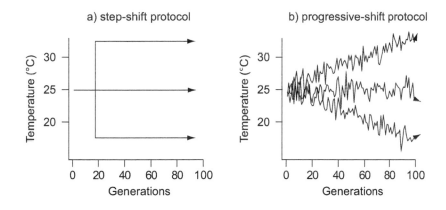

FIGURE 22.3

a, The standard protocol for an LNS experiment involves sudden and chronic shifts in the environmental conditions. b, An alternative protocol suggested by Brakefield (2003) would be to simulate a more gradual (and fluctuating) shift in conditions.

hardly mirror natural environmental changes, which are generally episodic (see also Garland and Kelly 2006 regarding the opportunity for plasticity to evolve). Brakefield (2003) argued that steplike protocols might yield misleading evolutionary trajectories and proposed LNS experiments in which environments are shifted "in a more gradual and realistic manner over generations" (figure 22.3b). Although obviously time-consuming for long-lived organisms, this approach is practical with microorganisms, as was demonstrated (see the box) over a century ago by Dallinger (1887). In any case, whether this issue is a real concern remains to be determined and is in fact an interesting opportunity for investigation.

*Chronic Selection*   A related simplicity concern is that environmental shifts in LNS experiments are typically chronic and sustained (figure 22.3a). And even if selective environments are eventually reversed (Estes and Teotónio this volume), they are still chronic. Do organisms in nature generally experience chronic and sustained environmental shifts? In some cases (e.g., if an organism emigrates to a cold environment), selection could indeed be somewhat chronic in nature. But at single localities, environmental factors such as temperature are highly variable, even in the face of a sustained environmental trend. After all, selection is anything but chronic in *Geospiza* finches on the Galapagos (Grant and Grant 2003).

Will chronic selection alter evolutionary trajectories relative to episodic selection? We suspect so. Although this question has not to our knowledge been systematically approached in any theoretical model, a useful comparison can be made between microbial populations undergoing clonal evolution in continuous serial dilution culture (SDC) versus those undergoing clonal evolution in continuous nutrient-limited chemostat culture (CC). These two selection regimes are commonly used, but few appreciate that selection in SDC is continuously varying, whereas selection in CC is constant. Nutrient limitation—and nutrient excess—are imposed regularly and episodically in the former, whereas nutrient limitation is imposed chronically in the latter (Kubitschek 1970). In the absence of antagonistic pleiotropy, fitness advantages can be secured in SDC culture by multiple adaptive mechanisms that include a decrease in lag time, an increase in maximum specific growth rate, and an increase in yield and/or survivorship in stationary phase. By contrast, fitness gains in CC are generally restricted to improvements in the capacity to scavenge limiting nutrient and/or to convert that limiting nutrient into cell mass (Brown et al. 1998). The different opportunities for evolutionary adaptation that exist under these different selection regimes may in part account for the differences observed between them in the pace and tempo of evolutionary change (Helling et al. 1987; Lenski and Travisano 1994), as well as the relative advantages that accrue to haploids versus diploids (Paquin and Adams 1983; Zeyl et al. 2003). Relative to episodic selection, chronic selection may increase the rate of response, and it may even alter the evolutionary trajectory.

A direct approach to the issue of evolutionary trajectories from chronic versus episodic selection could be achieved, at least in principle, by conducting a parallel selection experiment: in one treatment, selection would be applied chronically over hundreds of generations, while in another would be applied episodically every other generation, or even according to a randomized schedule. This would be practical only with a few organisms (e.g., bacteria and yeast), and the experimental design and analysis should take into account the fact that conditions defined as "relaxed selection" with respect to one agent, might constitute yet another type of selection.

*Duration of Selection*   The time scale of selection may also alter conclusions. Faced with pressures from funding agencies (or dissertation advisers), researchers will generally start to monitor responses immediately after beginning selection and will be tempted to publish once a trend becomes apparent. For a variety of reasons, however, evolutionary trajectories can shift and even reverse over time (Archer et al. 2003; Rose et al. 2005; Santos et al. 2006): (1) some adaptive responses may require novel genetic mutations, which take time to appear (Knies et al. 2006), and (2) trajectories may be modified by novel epistatic interactions.

*Behavioral Compensation Is Impossible*   In pursuit of simplicity, typical selection protocols prevent organisms from using behavior to help compensate for the imposed selection and thus may lead to aberrant evolutionary trajectories (see also Rhodes and Kawecki this volume). Consider the options available to ectotherms facing climate change (or other environmental challenges) in nature. Recall what might be called "Bartholomew's First Law of Physiological Ecology": namely, the first response of any animal will be to use behavioral adjustments to try to evade or at least ameliorate those changes (table 22.1; Bartholomew 1958, 1964; Slatkin and Kirkpatrick 1983). For example, many ectotherms shift habitat and time of activity along an altitudinal or latitudinal gradient (Hertz and Huey 1981; Clarke 1987; Pascual et al. 1993). As a result, their (activity) body temperatures are often remarkably similar across altitudes or latitudes (Jones et al. 1987; Huey et al. 2003). In fact, if behavioral thermoregulation is fully compensatory, then ectotherms facing climate change in nature might experience selection only on traits involved with the behavioral shifts, and not on thermal sensitivity per se (Bogert 1949; Huey et al. 2003).

Now consider an LNS experiment specifically designed to elucidate evolutionary responses to climate change. Typically, replicate lines would be maintained for many generations at different fixed temperatures in environmental chambers, where thermal heterogeneity is essentially nil. In such environments, the study subjects would have little or no opportunity to use behavioral thermoregulation to modify their temperatures from that of their resident thermal regimes (Gibbs 1999). As a result, LNS must act directly on their thermal sensitivity. In a very real sense then, LNS experiments transform mobile animals into "plants," organisms with relatively limited ability to use behavior to evade environmental challenges (Bartholomew 1958; Bradshaw 1972; Huey et al. 2002). Consequently, the evolutionary trajectories organisms follow in response to climate change in nature will likely differ from those evolving in response to fixed and forced temperatures in an LNS experiment.

Sometimes LNS experiments can be redesigned to solve (or at least help solve) this problem of behavioral imprisonment (Gibbs 1999, 2714). Davis et al. (1998) developed an ingenious experiment, which shows that ecological realism is indeed possible. They were interested in studying ecological responses of *Drosophila* species to climate change, but their methodology could easily be applied to evaluate evolutionary responses in an LNS experiment. One of their experiments involved sets of eight cages distributed among four incubators (thus two cages/incubator) that differed in temperature. To simulate climate warming, they set some incubators at 15°, 20°, 25°, or 30°C; and to simulate climate cooling, they set other incubators at 10°, 15°, 20°, and 25°C. In some experiments, the eight cages were connected in series via tubing; and flies could thus move among cages and chambers (e.g., between 20° and 25°C). In other sets, the tubing was blocked, so flies were held at fixed temperatures, as in a typical LNS experiment.

Davis et al. (1998) introduced three species of *Drosophila* either individually or simultaneously into the cages, and they even added parasitoids in some experiments. Consequently, these complex experiments enabled this team to monitor the ecological consequences of interactions involving behavior, temperature, interspecific competition, and parasitism.

To study the impact of climate warming, one could maintain this laboratory scenario (Davis et al. 1998) across many generations. We expect that flies would preferentially spend most of their time in the cages with the favored temperatures. However, competition for those favored thermal cages might force part of the population to occupy suboptimal thermal environments ("ideal free distribution" of Fretwell and Lucas 1970), perhaps thus modifying selection on thermal sensitivity itself (Levins 1968).

*Spatial Variation Is Eliminated*   Most LNS experiments use simple environments that attempt to eliminate any spatial variation in LNS. However, consideration of the Davis et al. (1998) study (discussed earlier) suggests that the presence or absence of spatial heterogeneity in LNS may sometimes influence results. A remarkable example is seen in a study of the bacterium *Pseudomonas fluorescens*. Rainey and Travisano (1998) studied how these bacteria evolved by LNS in unstirred (spatially heterogeneous) versus stirred (well-mixed, spatially homogeneous) microcosms. In the spatially heterogeneous microcosms, the bacteria underwent a rapid adaptive radiation, evolving visibly distinct morphs with marked niche preferences; but in the homogeneous microcosms, morphs stayed uniform. Obviously, anyone contemplating an LNS experiment on such microorganisms must decide in advance whether their stocks will be unshaken or stirred.

## A CASE STUDY: A SELECTION EXPERIMENT
## AT ODDS WITH FIELD STUDIES

So far our chapter has focused on problems that LNS studies face in testing hypotheses derived from comparative studies in nature. Although we have enumerated a variety of problems, we cannot be sure when these are trivial and when they are significant. This "academic" problem becomes very real when the results of an LNS experiment contradict a comparative hypothesis. Does such a lack of concordance mean that our hypothesis was flawed or that key aspects of our selection experiment were flawed?

Let's take a close-up look at an ambitious and excellent LNS experiment specifically designed to test a comparative hypothesis. The fly *D. subobscura* is native to a broad range of latitudes in the Old World from North Africa to Scandinavia, and so its populations experience a strong climatic gradient (Krimbas 1993). In the late 1970s, *D. subobscura* was accidentally introduced into South America (Brncic and Budnik 1980) and then into North America (Beckenbach and Prevosti 1986). It spread rapidly on both continents, where it now occurs over a broad latitudinal (climatic) range.

Evolutionary biologists soon recognized that the Old and New World flies provided an ideal opportunity for studying the evolution of geographic variation (Brncic et al. 1981; Prevosti et al. 1988; Ayala et al. 1989). The Old World flies provide a convenient evolutionary baseline, as these flies have had thousands of years to evolve clinal patterns. The New World flies serve as a "grand experiment in nature" (Ayala et al. 1989). Studies of the magnitude and patterns of geographic variation in these flies provides insight into the rates and predictability of evolution on a geographic scale (Prevosti et al. 1988; Ayala et al. 1989).

Two traits show pronounced latitudinal clines in the baseline Old World populations and have also been intensively studied in the New World. About eighty chromosomal inversions have been described in the Old World (Krimbas 1993), and the frequency of many show strong latitudinal patterns (Menozzi and Krimbas 1992). For example, the "standard" inversions of the various chromosomes are common in northern Europe, but rare to the south. Wing size also changes clinally (Prevosti 1955; Misra and Reeve 1964; Pegueroles et al. 1995; Huey et al. 2000; Gilchrist et al. 2004), as it does in many other *Drosophila* (Coyne and Beecham 1987; James et al. 1995; van't Land et al. 1999), and it is positively related to latitude.

The observed latitudinal patterns in the Old World suggest that inversions and wing size might be subject to selection from temperature or related climate factors. This hypothesis is reinforced by the discovery that similar latitudinal clines in inversion frequencies (Prevosti et al. 1988; Balanyà et al. 2003; Balanyà et al. 2004) and in wing size (Huey et al. 2000; Gilchrist et al. 2004) had evolved rapidly in both North and South America. More important, the frequency of "low-latitude" inversions at particular localities have increased over time, seemingly in accord with recent climate warming (Orengo and Prevosti 1996; Rodríguez-Trelles et al. 1996; Rodríguez-Trelles et al. 1998). For example, in twenty-two of twenty-six populations spread over three continents, climates have warmed over sample intervals; and low-latitude inversions have increased in frequency (Balanyà et al. 2006). All this comparative evidence strongly suggests that inversion and wing size clines are adaptive and that temperature (climate) is a key selective agent.

To test putative role of temperature in the evolution of these clines, evolutionary geneticists (Santos et al., 2004, 2005, 2006) in Barcelona developed an exemplary experiment in laboratory natural selection. They took a large and genetically heterogeneous stock collected from Puerto Montt, Chile, the likely site of the introduction into the New World. They then set up three replicate lines at each of three different (constant) temperatures (13°, 18°—the presumed optimum—and 22°C) and let the lines evolve by laboratory natural selection for several years. Importantly, they carefully controlled density, which might confound evolutionary trajectories (Santos et al. 2004). They scored inversion frequencies and wing size (and shape) when the lines were first established, and then at intervals following selection (experimental details in Santos et al. 2004).

Given the comparative evidence (as described earlier), one would expect to see the following patterns if temperature were the key selective agent: flies evolving at high temperature should evolve relatively small wings and should evolve inversion frequencies characteristic of low-latitude populations. In fact, wing size was independent of selected temperature (table 22.2; Santos et al. 2005). Chromosome inversion frequencies also shifted, but generally in ways inconsistent with expectations based on clinal patterns (table 22.2). Similarly, development rates did not match expectations (table 22.2; Santos et al. 2006). Santos et al. (2005) noted, "The most obvious feature was a general lack of correspondence between the outcomes from laboratory thermal selection and New World colonizations."

This conflict between nature and LNS can be interpreted in at least two ways. First, perhaps temperature is a red herring, such that observed latitudinal clines are really caused by some other environmental factor (Bradshaw and Holzapfel 2006; Santos et al. 2005, 269). If so, then expectations based on comparative patterns were simply wrong. This is possible but somewhat unlikely, given that inversion frequencies at single sites have shifted in the expected direction in response to recent climate warming (Balanyà et al. 2006). Second,

TABLE 22.2   Comparisons of Patterns of Average Trait Scores Predicted from Comparative Patterns (Latitudinal Clines) versus Results from Laboratory Natural Selection at 13°, 18°, or 22°C for *D. subobscura*

| Trait | Rank Order of Lines | Match Prediction | Reference |
|---|---|---|---|
| Predicted for all traits | 13° > 18° > 22° | — | |
| Wing size | 13° = 18° = 22° | No | Santos, Céspedes, et al. 2005 |
| Development rate at 13°C | 13° < 18° < 22° | No | Santos, Brites, et al. 2006 |
| Chromosome $A_{st}$ | 13° = 18° > 22° | Partially | Santos, Céspedes, et al. 2005 |
| Chromosome $E_{st}$ | 13° < 18° < 22° | No | Santos, Céspedes, et al. 2005 |
| Chromosome $J_{st}$ | 13° > 18° < 22° | No | Santos, Céspedes, et al. 2005 |
| Chromosome $O_{st}$ | 13° = 18° > 22° | Partially | Santos, Céspedes, et al. 2005 |
| Chromosome $U_{st}$ | 13° = 18° = 22° | No | Santos, Céspedes, et al. 2005 |

NOTE: For all traits, the prediction is that trait scores for the selection lines will be ranked 13° > 18° > 22°C (see text). In no case was the predicted pattern observed exactly.

perhaps the experimental conditions don't adequately mimic natural ones. As Santos et al. (2005) noted, their laboratory environments had fixed temperatures, whereas natural ones have daily and seasonal variation in temperature. Moreover, their flies had ample food resources, which sometimes will not be the case in nature; their flies were not exposed to interspecific interactions, which might co-vary with latitude; and their flies experienced constant densities, which might not reflect patterns in nature. Finally, flies in a population cage are probably not challenged physiologically in the same ways as flies in nature. For example, individual differences in flight performance may have little impact on fitness to fly in a small population cage, and thus laboratory selection on wing size might be weak or nonexistent.

At present, perhaps the safest conclusion is that evolutionary trajectories resulting from this experimental manipulation of temperature are inconsistent with the hypothesis that temperature drives clines in inversion frequency and in wing size. Whether this is the "fault" of temperature or of LNS is currently and frustratingly unclear.

## ON MODIFYING LNS EXPERIMENTS

Given that we need LNS to test comparative hypotheses, how can we improve LNS? Are there midflight corrections that solve (or at least reduce) some of the concerns? We think so, and we offer a few suggestions (see also Rose et al. 1996, 232–236).

At the risk of sounding professorial, we do advocate learning from others. There is a lot of accumulated wisdom in experimental evolution, and we can all learn from the mistakes of the past. Moreover, we would add that our ecological colleagues have gained extraordinary experience over the decades in experimental approaches, and we evolutionary biologists would do well to learn from their experiences.

Knowing the natural history of one's study organism is essential to the design of meaningful experiments, whether they be focused on ecology, or on evolution (Hairston 1989). Unfortunately, embarrassingly little is known about the natural history of the very organisms most suitable (Feder 1996) for experimental evolutionary studies (e.g., *Drosophila, E. coli, S. cerevisae, C. elegans, Mus*).

Earlier we addressed the catch-22 concerning selection on field-fresh versus laboratory-adapted stocks. Roff and Fairbairn (2006) have recently found a clever way to turn this disadvantage to an advantage. Their goal was to study evolutionary changes in the sand cricket (*Gryllus firmus*) during adaptation to the laboratory. This cricket has a striking wing dimorphism: long-winged morphs are migratory, but the short-winged ones are not. Given the known trade-off between migratory capability and fecundity in these morphs, Roff and Fairbairn predicted that domestication should result in a reduced frequency, an increased fecundity, and a decreased mass of flight muscles of long-winged females, but in little change in these traits of short-winged females. Importantly, they used quantitative genetic theory and measurements to predict evolutionary trajectory of these traits during domestication. Their predictions were verified.

An advantage of laboratory evolution is that one can control most variables and manipulate only one or a few. But probably most of the patterns that we seek to test are likely the result of selection involving many interacting processes (Quinn and Dunham 1983). Dunham and Beaupre (1998) argued that "the potential for multiple casual mechanisms must be incorporated into the construction of ecological theory and into the design of ecological experiments." The same should hold for the design of evolutionary experiments. Similarly, the possibility of responses with "multiple solutions" can be a key reason of including replicate lines in selection experiments (Garland 2003).

Consider the LNS experiment with *D. subobscura* described earlier. Perhaps humidity as well as temperature should have been manipulated, such that the conditions would range from cool and high relative humidity to warm and low relative humidity. In a humid environment, higher temperature will increase metabolism but won't increase evaporative water loss; but in a dry environment, higher temperatures will increase metabolism and water loss.

A call for greater ecological realism is not without precedent. Ecologists have developed sophisticated laboratory facilities that can mimic simple terrestrial ecosystems. At Silwood Park, for instance, the Ecotron consists of fifteen environmental chambers able to control and manipulate photoperiod, illumination (balanced spectrum, dawn/dusk simulation), temperature, humidity, rainfall, and even $CO_2$ (Lawton 1996). The chambers house multispecies ecosystems, allowing for complex ecological interactions of plants and animals. Using such a facility for experimental evolution would be expensive, but feasible. If ecologists can build and run an Ecotron, surely evolutionary biologists can build and run an Evotron!

Another option is to borrow from another experimental technique in ecology—the cattle tank. These are typically used for aquatic systems, but they are large enough to house salamanders, frogs, and fish. Potentially one could manipulate tanks and look at evolutionary shifts over time.

Of course, one might attempt LNS-type experiments in the field (Bennett and Lenski 1999). This has been done successfully in a few cases (Losos et al. 2004; Irschick and Reznick this volume). Such studies will be logistically challenging (and sometimes unfeasible). Moreover, their design may not enable genetic and environmental effects to be easily discriminated (Bennett and Lenski 1999).

An informative variant is to use field releases of LNS-engineered phenotypes. Kristensen et al. (2007) released *D. melanogaster* that had been selected for increased heat or cold resistance, and then they measured relative ability of the lines to reach baits under hot or cold field conditions. Such releases of experimental lines provide interesting tests of whether phenotypic shifts produced by laboratory selection result in enhanced fitness in nature.

Finally, experiments must be designed so that presumed causal mechanisms can in fact play a role (Dunham and Beaupre 1998). For example, if we wish to test experimentally the hypothesis that temperature is the selective agent, then we need to design an experiment that allows temperature to have a mechanistic impact. In the case of the clinal increase in wing size with latitude in *Drosophila*, we might assume that temperature exerts its force in nature via its mechanistic effect on flight dynamics. Therefore, to test whether temperature might drive the wing size cline, we need to design an experiment in which relative flight ability might influence fitness (Weber 1996; Marden et al. 1997). But would flight ability influence fitness in a small population cage? Probably not.

In some cases, mechanism may not be obvious a priori. For example, inversion frequencies change with latitude and temperature, but the mechanism (if in fact temperature is a causal agent) for a causal relationship is presently mysterious to us. For that reason, we cannot see how to design an experiment that realistically allows mechanism.

## CONCLUSION

Although we have focused on problems that can plague LNS experiments as emulations of natural selection in the wild, we do not mean to imply that LNS experiments are without utility. Quite the contrary. There are many ways to study evolution, some descriptive, some experimental. As has been noted repeatedly (Huey et al. 1991; Huey and Kingsolver 1993; Rose et al. 1996; Gibbs 1999; Garland 2003; Swallow and Garland 2005; Futuyma and Bennett this volume; Rose and Garland this volume), each method has its advantages, and each has its limitations. Moreover, an awareness of limitations can open opportunities for novel studies (e.g., chronic vs. nonchronic selection). In any case, a complete understanding of evolution will require the application of multiple integrated approaches. We see LNS as an essential tool for testing field-derived hypotheses, but one that must be handled thoughtfully, used along with other tools, and interpreted with care.

No matter how hard we work, no experiment or study will ever be perfect. We need to do away with the "Myth of Definitive Results" (Underwood 1998) and recognize that our view of evolution is deeper if we look at it through different and complementary glasses, not just though LNS ones. And we should try to improve the validity of each approach, learning as we go. As Underwood (1998, 345) noted, "The hallmark of progressive ideas is that they progress. Given that there is a good chance we are wrong quite often, we should be prepared to discover how wrong as fast as possible."

## SUMMARY

Experiments using laboratory natural selection (LNS) can illuminate the genetic architecture underlying complex traits, reveal evolutionary trajectories associated with different population structures and modes of selection, and provide derived lines with "exaggerated" or "novel" phenotypes. But LNS experiments have inherent problems and limitations, especially when used as to simulate natural selection in the wild. Certain problems can be so severe that they compromise or confound evolutionary and functional interpretations. Of these, some can be circumvented by modifying traditional experimental designs, but others cannot. Consequently, researchers contemplating LSN experiments face a classic catch-22 or double-bind. They know in advance that LNS may be an effective way to test a given evolutionary hypothesis, but they also recognize that the resulting conclusions may be of uncertain validity.

## ACKNOWLEDGMENTS

We thank T. Garland, Jr. and M. R. Rose for the invitation to participate in this project and for comments on the manuscript. Writing was facilitated by National Science Foundation grant IOB-0416843 to R.B.H. and by grants from the National Institutes of Health (NIH R01-HG003328-01 and R15 GM79762-01) and NASA (NNX07AJ28G) to F.R. We thank M. Dillon, M. Frazier, G. Gilchrist, M. Lakeman, M. Santos, P. Service, and G. Wang for discussions or comments.

## REFERENCES

Adams, J., and P. E. Hansche. 1974. Population studies in microorganisms I. Evolution of diploidy in *Saccharomyces cerevisae*. *Genetics* 76:327–338.

Anderson, J. L., L. Albergotti, S. Proulx, C. Peden, R. B. Huey, and P. C. Phillips. 2007. Thermal preference of *Caenorhabditis elegans*: A null model and empirical tests. *Journal of Experimental Biology* 210:3107–3116.

Archer, M. A., J. P. Phelan, K. A. Beckman, and M. R. Rose. 2003. Breakdown in correlations during laboratory evolution. II. Selection on stress resistance in *Drosophila* populations. *Evolution* 57:536–543.

Ayala, F. J., L. Serra, and A. Prevosti. 1989. A grand experiment in evolution: The *Drosophila subobscura* colonization of the Americas. *Genome* 31:246–255.

Balanyà, J., J. M. Oller, R. B. Huey, G. W. Gilchrist, and L. Serra. 2006. Global genetic change tracks global climate warming in *Drosophila subobscura*. *Science* 313:1773–1775.

Balanyà, J., L. Serra, G. W. Gilchrist, R. B. Huey, M. Pascual, F. Mestres, and E. Solé. 2003. Evolutionary pace of chromosomal polymorphism in colonizing populations of *Drosophila subobscura*: An evolutionary time series. *Evolution* 57:1837–1845.

Balanyà, J., E. Solé, J. M. Oller, D. Sperlich, and L. Serra. 2004. Long-term changes in the chromosomal inversion polymorphism of *Drosophila subobscura*. II. European populations. *Journal of Zoological Systematics and Evolutionary Research* 42:191–201.

Bartholomew, G. A. 1958. The role of physiology in the distribution of terrestrial vertebrates. Pages 81–95 *in* C. L. Hubbs, ed. *Zoogeography*. Publication No. 51. Washington, DC: American Association for the Advancement of Science.

———. 1964. The roles of physiology and behaviour in the maintenance of homeostasis in the desert environment. *Symposium of the Society for Experimental Biology* 18:7–29.

Beckenbach, A. T., and A. Prevosti. 1986. Colonization of North America by the European species *Drosophila subobscura* and *D. ambigua*. *American Midland Naturalist* 115:10–18.

Bennett, A. F. 2003. Experimental evolution and the Krogh Principle: Generating biological novelty for functional and genetic analyses. *Physiological and Biochemical Zoology* 76:1–11.

Bennett, A. F., and R. E. Lenski. 1993. Evolutionary adaptation to temperature. II. Thermal niches of experimental lines of *Escherichia coli*. *Evolution* 47:1–12.

———. 1999. Experimental evolution and its role in evolutionary physiology. *American Zoologist* 39:346–362.

Bergthorsson, U., and H. Ochman. 1999. Chromosomal changes during experimental evolution in laboratory populations of *Escherichia coli*. *Journal of Bacteriology* 181:1360–1363.

Bogert, C. M. 1949. Thermoregulation in reptiles, a factor in evolution. *Evolution* 3:195–211.

Borash, D. J., A. G. Gibbs, A. Joshi, and L. D. Mueller. 1998. A genetic polymorphism maintained by natural selection in a temporally varying environment. *American Naturalist* 151:148–156.

Bradshaw, A. D. 1972. Some of the evolutionary consequences of being a plant. *Evolutionary Biology* 5:25–47.

Bradshaw, W. E., and C. M. Holzapfel. 2001. Genetic shift in photoperiodic response correlated with global warming. *Proceedings of the National Academy of Sciences of the USA* 98:14509–14511.

Bradshaw, W. E., and C. M. Holzapfel. 2006. Evolutionary response to rapid climate change. *Science* 312:1477–1478.

Brakefield, P. M. 2003. Artificial selection and the development of ecologically relevant phenotypes. *Ecology* 84:1661–1671.

Brakefield, P. M., and V. Mazzotta. 1995. Matching field and laboratory environments: Effects of neglecting daily temperature variation on insect reaction norms. *Journal of Evolutionary Biology* 8:559–573.

Brncic, D., and M. Budnik. 1980. Colonization of *Drosophila subobscura* Collin in Chile. *Drosophila Information Service* 55:20.

Brncic, D., A. Prevosti, M. Budnik, M. Monclús, and J. Ocaña. 1981. Colonization of *Drosophila subobscura* in Chile I. First population and cytogenetic studies. *Genetica* 56:3–9.

Brown, C. J., K. M. Todd, and R. F. Rosenzweig. 1998. Multiple duplications of yeast hexose transport genes in response to selection in a glucose-limited environment. *Molecular Biology and Evolution* 15:931–942.

Chang, S.-M., and R. G. Shaw. 2003. The contributions of spontaneous mutation to variation in environmental response in *Arabidopsis thaliana*: Responses to nutrients. *Evolution* 57:984–994.

Chippendale, A. K., T. J. F. Chu, and M. R. Rose. 1996. Complex trade-offs and the evolution of starvation resistance in *Drosophila melanogaster*. *Evolution* 50:753–766.

Chippendale, A. K., A. G. Gibbs, M. Sheik, K. J. Yee, M. Djawdan, T. J. Bradley, and M. R. Rose. 1998. Resource acquisition and the evolution of stress resistance in *Drosophila melanogaster*. *Evolution* 52:1342–1352.

Clark, A. G. 1987. Senescence and the genetic-correlation hang-up. *American Naturalist* 129:932–940.

Clarke, A. 1987. Temperature, latitude, and reproductive effort. *Marine Ecology Progress Series* 38:89–99.

Cohan, F. M., and A. A. Hoffmann. 1986. Genetic divergence under uniform selection. II. Different responses to selection for knockdown resistance to ethanol among *Drosophila melanogaster* populations and their replicate lines. *Genetics* 114:145–163.

Cooper, T. F., D. E. Rozen, and R. E. Lenski. 2003. Parallel changes in gene expression after 20,000 generations of evolution in *E. coli*. *Proceedings of the National Academy of Sciences of the USA* 100:1072–1077.

Cooper, V. S., D. Schneider, M. Blot, and R. E. Lenski. 2001. Mechanisms causing rapid and parallel losses of ribose catabolism in evolving populations of *E. coli B*. *Journal of Bacteriology* 183:2834–3841.

Coyne, J. A., and E. Beecham. 1987. Heritability of two morphological characters within and among natural populations of *Drosophila*. *Genetics* 117:727–737.

Dallinger, W. H. 1887. Transactions of the Society. V. The president's address. *Journal of the Royal Microscopical Society* 1887:185–199.

Davis, A., J. Lawton, B. Shorrocks, and L. Jenkinson. 1998. Individualistic species responses invalidate simple physiological models of community dynamics under global environmental change. *Journal of Animal Ecology* 67:600–612.

Diamond, J. 2001. Dammed experiments! *Science* 294:1847.

Dobrindt, U., F. Agerer, K. Michaelis, A. Janka, X. Buchrieser, M. Samuelson, C. Svanborg, G. Gottschalk, H. Karch, and J. Hacker. 2003. Analysis of genome plasticity in pathogenic and commensal *Escherichia coli* isolates by use of DNA microarrays. *Journal of Bacteriology* 185:1831–1840.

Dunham, A. E., and S. J. Beaupre. 1998. Ecological experiments: Scale, phenomenology, mechanism and the illusion of generality. Pages 27–49 *in* W. J. Resetarits Jr. and J. Bernardo, eds. *Experimental Ecology: Issues and Perspectives*. New York: Oxford University Press.

Dunham, M. J., H. Badrane, T. Ferea, J. P. Adams, P. O. Brown, R. F. Rosenzweig, and D. Botstein. 2002. Characteristic genome rearrangements accompany experimental evolution of *S. cerevisiae*. *Proceedings of the National Academy of Sciences of the USA* 99:16144–16149.

Dykhuizen, D. E. 1990. Experimental studies of natural selection in bacteria. *Annual Review of Ecology and Systematics* 21:373–398.

Endler, J. A. 1977. *Geographic Variation, Speciation, and Clines*. Princeton, NJ: Princeton University Press.

———. 1986. *Natural Selection in the Wild*. Princeton, NJ: Princeton University Press.

Escobar-Paramo, P., A. LeMenach, T. LeGall, C. Amorin, S. Gouriou, B. Picard, D. Skurnik, and E. Denamur. 2006. Identification of forces shaping the commensal *Escherichia coli* genetic structure by comparing animal and human isolates. *Environmental Microbiology* 8:1975–1984.

Feder, M. F. Ecological and evolutionary physiology of stress proteins and the stress response: the *Drosophila melanogaster* model. Pages 79–102 *in* I. A. Johnston and A. F. Bennett, eds. *Animals and Temperature: Phenotypic and Evolutionary Adaptation*. Society of Experimental Biology Symposium Volume. Cambridge: Cambridge University Press.

Ferea, T. L., D. Botstein, P. O. Brown, and R. F. Rosenzweig. 1999. Systematic changes in gene expression patterns following adaptive evolution in yeast. *Proceedings of the National Academy of Sciences of the USA* 96:9721–9726.

Fong, S. S., A. R. Joyce, and B. O. Palsson. 2005. Parallel adaptive evolution cultures of Escherichia coli lead to convergent growth phenotypes with different expression states. *Genome Research* 1365–1372.

Frankham, R., B. H. Yoo, and B. L. Sheldon. 1988. Reproductive fitness and artificial selection in animal breeding: Culling on fitness prevents a decline in reproductive fitness in lines of *Drosophila melanogaster* selected for increased inebriation time. *Theoretical and Applied Genetics* 76:909–914.

Fretwell, D. S., and H. L. Lucas. 1970. On territorial behavior and other factors influencing habitat distribution in birds. *Acta Biotheoretica* 19:16–32.

Fukiya, S., H. Mizoguchi, T. Tobe, and H. Mori. 2004. Extensive genome diversity in pathogenic *Escherichia coli* and *Shigella* strains revealed by comparative genomic hybridization microarray. *Journal of Bacteriology* 186:3911–3921.

Garland, T., Jr. 2003. Selection experiments: An underutilized tool in biomechanics and organismal biology. *In* V. L. Bels, J.-P. Gasc and A. Casinos, eds. *Vertebrate Biomechanics and Evolution*. Oxford: BIOS Scientific.

Garland, T., Jr., and S. C. Adolph. 1991. Physiological differentiation of vertebrate populations. *Annual Review of Ecology and Systematics* 22:193–228.

Garland, T., Jr., and S. A. Kelly. 2006. Phenotypic plasticity and experimental evolution. *Journal of Experimental Biology* 209:2234–2261.

Gibbs, A. G. 1999. Laboratory selection for the comparative physiologist. *Journal of Experimental Biology* 220:2709–2718.

Gibbs, A. G., A. K. Chippindale, and M. R. Rose. 1997. Physiological mechanisms of evolved desiccation resistance in *Drosophila melanogaster*. *Journal of Experimental Biology* 200:1821–1832.

Gibbs, A. G., and L. M. Matzkin. 2001. Evolution of water balance in the genus *Drosophila*. *Journal of Experimental Biology* 204:2331–2338.

Gilchrist, G. W., R. B. Huey, J. Balanyà, M. Pascual, and L. Serra. 2004. A time series of evolution in action: Latitudinal cline in wing size in South American *Drosophila subobscura*. *Evolution* 58:768–780.

Gilligan, D. M., and R. Frankham. 2003. Dynamics of adaptation to captivity. *Conservation Genetics* 4:189–197.

Grant, R., and P. Grant. 2003. What Darwin's finches can teach us about the evolutionary origin and regulation of biodiversity. *BioScience* 53:965–975.

Griffiths, J. A., M. Schiffer, and A. A. Hoffmann. 2005. Clinal variation and laboratory adaptation in the rainforest species *Drosophila birchii* for stress resistance, wing size, wing shape and development time. *Journal of Evolutionary Biology* 18:213–222.

Grimberg, B., and C. Zeyl. 2005. The effects of sex and mutation rate on adaptation in test tubes and to mouse hosts by *Saccharomyces cerevisiae*. *Evolution* 59:431–438.

Gu, Z., L. David, D. Petrov, T. Jones, R. W. Davis, and L. M. Steinmetz. 2005. Elevated evolutionary rates in the laboratory of *Saccharomyces cerevisiae*. *Proceedings of the National Academy of Sciences of the USA* 102:1092–1097.

Hairston, N. G. 1989. *Ecological Experiments: Purpose, Design and Execution*. Cambridge: Cambridge University Press.

Harshman, L. G., and A. A. Hoffmann. 2000. Laboratory selection experiments using *Drosophila*: What do they really tell us? *Trends in Ecology & Evolution* 15:32–36.

Harshman, L. G., A. A. Hoffmann, and A. Clarke. 1999. Selection for starvation resistance in *Drosophila melanogaster*: Physiological correlates, enzyme activities and multiple stress responses. *Journal of Evolutionary Biology* 12:370–379.

Helling, R. B., C. N. Vargas, and J. Adams. 1987. Evolution of *Escherichia coli* during growth in a constant environment. *Genetics* 116:349–358.

Hertz, P. E., and R. B. Huey. 1981. Compensation for altitudinal changes in the thermal environment by some *Anolis* lizards on Hispaniola. *Ecology* 62:515–521.

Hoffmann, A. A., R. J. Hallas, C. Sinclair, and L. Partridge. 2001. Rapid loss of stress resistance in *Drosophila melanogaster* under adaptation to laboratory culture. *Evolution* 55:436–438.

Hoffmann, A. A., and L. G. Harshman. 1999. Desiccation and starvation resistance in *Drosophila*: Patterns of variation at the species, population and intrapopulation levels. *Heredity* 83:637–643.

Hudson, R. E., U. Bergthorsson, J. R. Roth, and H. Ochman. 2002. Effect of chromosome location on bacterial mutation rates. *Molecular Biology and Evolution* 19:85–92.

Huey, R. B. 1982. Temperature, physiology, and the ecology of reptiles. Pages 25–91 *in* C. Gans and F. H. Pough, eds. *Biology of the Reptilia. Vol. 12. Physiology.* London: Academic Press.

Huey, R. B., M. Carlson, L. Crozier, M. Frazier, H. Hamilton, H. Harley, A. Hoang, and J. G. Kingsolver. 2002. Plants versus animals: Do they deal with stress in different ways? *Integrative and Comparative Biology* 42:415–423.

Huey, R. B., G. W. Gilchrist, M. L. Carlson, D. Berrigan, and L. Serra. 2000. Rapid evolution of a geographic cline in an introduced species of fly. *Science* 287:308–309.

Huey, R. B., P. E. Hertz, and B. Sinervo. 2003. Behavioral drive versus behavioral inertia: A null model approach. *American Naturalist* 161:357–366.

Huey, R. B., and J. G. Kingsolver. 1993. Evolution of resistance to high temperature in ectotherms. *American Naturalist* 142:S21–S46.

Huey, R. B., L. Partridge, and K. Fowler. 1991. Thermal sensitivity of *Drosophila melanogaster* responds rapidly to laboratory natural selection. *Evolution* 45:751–756.

Hughes, T. R., C. J. Roberts, H. Dai, A. R. Jones, M. R. Meyer, D. Slade, J. Burchard, S. Dow, T. R. Ward, M. J. Kidd, S. H. Friend, and M. J. Marton. 2000. Widespread aneuploidy revealed by DNA microarray expression profiling. *Nature Genetics* 25:333–337.

James, A. C., R. B. R. Azevedo, and L. Partridge. 1995. Cellular basis and developmental timing in a size cline of *Drosophila melanogaster. Genetics* 140:659–666.

Jones, J. S., J. A. Coyne, and L. Partridge. 1987. Estimation of the thermal niche of *Drosophila melanogaster* using a temperature-sensitive mutant. *American Naturalist* 130:83–90.

Kang, Y., T. Durfee, J. D. Glasner, Y. Qiu, D. Frisch, and K. M. Winterberg. 2004. Systematic mutagenesis of the *Escherichia coli* genome. *Journal of Bacteriology* 186:4921–4930.

Kellis, M., N. Patterson, M. Endrizzi, B. Birren, and E. S. Lander. 2003. Sequencing and comparison of yeast species to identify genes and regulatory elements. *Nature* 423:241–254.

Kingsolver, J.G. 2007. Variation in growth and instar number in field and laboratory *Manduca sexta. Proceedings of the Royal Society of London B, Biological Sciences* 274:977–981.

Knies, J. L., R. Izem, K. L. Supler, J. G. Kingsolver, and C. L. Burch. 2006. The genetic basis of thermal reaction norm evolution in lab and natural phage populations. *PLoS Biology* 4:e201.

Kondrashov, A. S., and D. Houle. 1994. Genotype-environment interactions and the estimation of the genomic mutation rate in *Drosophila melanogaster. Proceedings of the Royal Society of London B, Biological Sciences* 258:221–227.

Krebs, R. A., S. P. Roberts, B. R. Bettencourt, and M. E. Feder. 2001. Changes in thermotolerance and Hsp70 expression with domestication in *Drosophila melanogaster*. *Journal of Evolutionary Biology* 14:75–82.

Krimbas, C. B. 1993. Drosophila subobscura: *Biology, Genetics and Inversion Polymorphism*. Hamburg: Kovac.

Kristensen, T. N., V. Loeschcke, and A. A. Hoffmann. 2007. Can artificially selected phenotypes influence a component of field fitness? Thermal selection and fly performance under thermal extremes. *Proceedings of the Royal Society of London B, Biological Sciences* 274:771–778.

Kubitschek, H. E. 1970. *Introduction to Research with Continuous Cultures*. Englewood Cliffs, NJ: Prentice Hall.

Kuthan, M., F. Devaux, B. Janderova, J. C. Slaninova, and Z. Palkova. 2003. Domestication of wild *Saccharomyces cerevisiae* is accompanied by changes in gene expression and colony morphology. *Molecular Microbiology* 47:745–754.

Lawton, J. H. 1996. The Ecotron facility at Silwood Park: The value of "big bottle" experiments. *Ecology* 77:665–669.

Lenski, R. E., and M. Travisano. 1994. Dynamics of adaptation and diversification: A 10,000-generation experiment with bacterial populations. *Proceedings of the National Academy of Sciences of the USA* 91:6808–6814.

Levins, R. 1968. *Evolution in Changing Environments*. Princeton, NJ: Princeton University Press.

Levitan, M., and W. J. Etges. 2005. Climate change and recent genetic flux in populations of *Drosophila robusta*. *BMC Evolutionary Biology* 5:4.

Losos, J. B., T. W. Schoener, and D. A. Spiller. 2004. Predator-induced behaviour shifts and natural selection in field-experimental lizard populations. *Nature* 432:505–508.

Marden, J. H., M. R. Wolf, and K. E. Weber. 1997. Aerial performance of *Drosophila melanogaster* from populations selected for upwind flight ability. *Journal of Experimental Biology* 200:2747–2755.

Markow, T. A. 1975. Effect of light on egg-laying rate and mating speed in phototactic strains of *Drosophila*. *Nature* 258:712–714.

———. 1979. A survey of intra- and interspecific variation for pupation height in *Drosophila*. *Behavior Genetics* 9:209–217.

Matos, M., C. Rego, A. Levy, H. Teotónio, and M. R. Rose. 2000a. An evolutionary no man's land. *Trends in Ecology and Evolution* 15:206.

Matos, M., M. R. Rose, M. T. Rocha Pite, C. Rego, and T. Avelar. 2000b. Adaptation to the laboratory environment in *Drosophila subobscura*. *Journal of Evolutionary Biology* 13:9–19.

Menozzi, P., and C. B. Krimbas. 1992. The inversion polymorphism of *D. subobscura* revisited: Synthetic maps of gene arrangement frequencies and their interpretation. *Journal of Evolutionary Biology* 5:625–641.

Mikkola, R., and C. G. Kurland. 1992. Selection of laboratory wild-type phenotype from natural isolates of *Escherichia coli* in chemostats. *Molecular Biology and Evolution* 9:394–402.

Misra, R. K., and E. C. R. Reeve. 1964. Clines in body dimensions in populations of *Drosophila subobscura*. *Genetical Research, Cambridge* 5:240–256.

Moore, F. B. G., and R. Woods. 2006. Tempo and constraint of adaptive evolution in *Escherichia coli* (Enterobacteriaceae, Enterobacteriales). *Proceedings of the Linnean Society* 88:403–411.

Mueller, L. D. 1987. Evolution of accelerated senescence in laboratory populations of *Drosophila*. *Proceedings of the National Academy of Sciences of the USA* 84:1974–1977.

Mueller, L. D., and V. G. Sweet. 1986. Density-dependent natural selection in *Drosophila*: Evolution of pupation height. *Evolution* 40:1354–1356.

Muller, H. J. 1931. Some genetic aspects of sex. *American Naturalist* 66:118–138.

Novick, A., and L. Szilard. 1951. Experiments on spontaneous and chemically induced mutations of bacteria growing in the chemostat. *Cold Spring Harbor Symposium on Quantitative Biology* 16:337–343.

Orengo, D. J., and A. Prevosti. 1996. Temporal changes in chromosomal polymorphism of *Drosophila subobscura* related to climatic changes. *Evolution* 50:1346–1350.

Paine, R. B. 1994. *Marine Rocky Shores and Community Ecology: An Experimentalist's Perspective*. Luhe, Germany: Oldendorf.

Palkova, Z. 2004. Multicellular microorganisms: Laboratory versus nature. *EMBO Reports* 5:470–476.

Paquin, C., and J. Adams. 1983. Frequency of fixation of adaptive mutations is higher in evolving diploid than haploid yeast populations. *Nature* 302:495–500.

Paranjpe, D. A., D. Anitha, V. K. Sharmar, and A. Joshi. 2004. Circadian clocks and life-history related traits: Is pupation height affected by circadian organization in *Drosophila melanogaster*? *Journal of Genetics* 83:73–77.

Pascual, M., F. J. Ayala, A. Prevosti, and L. Serra. 1993. Colonization of North America by *Drosophila subobscura*: Ecological analysis of three communities of drosophilids in California. *Zeitschrift für Zoologische Systematik und Evolutionsforschung* 31:216–226.

Pegueroles, G., M. Papaceit, A. Quintana, A. Guillén, A. Prevosti, and L. Serra. 1995. An experimental study of evolution in progress: Clines for quantitative traits in colonizing and Palearctic populations of *Drosophila*. *Evolutionary Ecology* 9:453–465.

Perna, N. T., G. Plunkett, III, V. Burland, B. Mau, J. D. Glasner, D. J. Rose, G. F. Mayhew, P. S. Evans, J. Gregor, H. A. Kirkpatrick, G. Posfai, J. Hackett, S. Klink, A. Boutin, Y. Shao, L. Miller, E. J. Grotbeck, N. W. Davis, A. Lim, E. T. Dimalanta, K. D. Potamousis, J. Apodaca, T. S. Anantharaman, J. Lin, G. Yen, D. C. Schwartz, R. A. Welch, and F. R. Blattner. 2001. Genome sequence of enterohaemorrhagic *Escherichia coli* O157:H7. *Nature* 409:529–533.

Pittendrigh, C. S. 1960. Circadian rhythms and the circadian organization of living systems. *Cold Spring Harbor Symposium on Quantitative Biology* 25:159–184.

Prasad, N. G., and A. Joshi. 2003. What have two decades of laboratory life-history evolution studies on *Drosophila melanogaster* taught us? *Journal of Genetics* 82:45–76.

Prevosti, A. 1955. Geographical variability in quantitative traits in populations of *Drosophila subobscura*. *Cold Spring Harbor Symposium on Quantitative Biology* 20:294–298.

Prevosti, A., G. Ribó, L. Serra, M. Aguadé, J. Balañà, M. Monclús, and F. Mestres. 1988. Colonization of America by *Drosophila subobscura*: Experiment in natural populations that supports the adaptive role of chromosomal-inversion polymorphism. *Proceedings of the National Academy of Sciences of the USA* 85:5597–5600.

Prevosti, A., L. Serra, M. Aguadé, G. Ribo, F. Mestres, J. Balañà, and M. Monclus. 1989. Colonization and establishment of the Palearctic species *Drosophila subobscura* in North and South America. Pages 114–129 *in* A. Fontdevila, ed. *Evolutionary Biology of Transient Unstable Populations*. Berlin: Springer.

Promislow, D. E. L., and M. Tartar. 1998. Mutation and senescence: Where genetics and demography meet. *Genetica* 102/103:299–314.

Quinn, J. F., and A. E. Dunham. 1983. On hypothesis testing in ecology and evolution. *American Naturalist* 122:602–617.

Rainey, P. B., and M. Travisano. 1998. Adaptive radiation in heterogeneous environments. *Nature* 394:69–72.

Regal, P. J. 1977. Evolutionary loss of useless features: Is it molecular noise suppression? *American Naturalist* 111:123–133.

Riehle, M. M., A. F. Bennett, and A. D. Long. 2005. Differential patterns of gene expression and gene complement in laboratory-evolved lines of *E. coli*. *Integrative and Comparative Biology* 45:532–538.

Rodríguez-Trelles, F., G. Alvarez, and C. Zapata. 1996. Time-series analysis of seasonal changes of the O inversion polymorphism of *Drosophila subobscura*. *Genetics* 142:179–187.

Rodríguez-Trelles, F., M. A. Rodríguez, and S. M. Scheiner. 1998. Tracking the genetic effects of global warming: *Drosophila* and other model systems. *Conservation Ecology* 2:2.

Roff, D. A., and D. J. Fairbairn. 2006. Laboratory evolution of the migratory polymorphism in the sand cricket: Combining physiology with quantitative genetics. *Physiological and Biochemical Zoology* 80:358–369.

Ronald, J., H. Tang, and R. B. Brem. 2006. Genome-wide evolutionary rates in laboratory and wild yeast. *Genetics* 174:541–544.

Rose, M., and B. Charlesworth. 1980. A test of evolutionary theories of senescence. *Nature* 287:141–142.

Rose, M. R., J. L. Graves, and E. W. Hutchinson. 1990. The use of selection to probe patterns of pleiotropy in fitness-characters. Pages 29–42 *in* F. Gilbert, ed. *Insect Life Cycles: Genetics, Evolution, and Co-ordination*. London: Springer.

Rose, M. R., T. J. Nusbaum, and A. K. Chippendale. 1996. Laboratory evolution: the experimental wonderland and the Cheshire cat syndrome. Pages 221–241 *in* M. R. Rose and G. V. Lauder, eds. *Adaptation*. San Diego, CA: Academic Press.

Rose, M. R., H. B. Passananti, A. K. Chippendale, J. P. Phelan, M. Matos, H. Teotónio, and L. D. Mueller. 2005. The effects of evolution are local: Evidence from experimental evolution in *Drosophila*. *Integrative and Comparative Biology* 45:486–491.

Rose, M. R., P. M. Service, and E. W. Hutchinson. 1987. Three approaches to trade-offs in life-history evolution. Pages 91–105 *in* V. Loeschcke, ed. *Genetic Constraints on Adaptive Evolution*. Berlin: Springer.

Rosenzweig, R. F., D. Treves, R. Sharp, and J. Adams. 1994. Microbial evolution in a simple unstructured environment: Genetic differentiation in *Escherichia coli*. *Genetics* 137:903–917.

Rundle, H. D., S. F. Chenoweth, and M. W. Blows. 2006. The roles of natural and sexual selection during adaptation to a novel environment. *Evolution* 60:2218–2225.

Santos, M., D. Brites, and H. Laayouni. 2006. Thermal evolution of pre-adult life history traits, geometric size and shape, and developmental stability in *Drosophila subobsura*. *Journal of Evolutionary Biology* 19:2006–2021.

Santos, M., W. Céspedes, J. Balanyà, V. Trotta, F. C. F. Calboli, A. Fontdevila, and L. Serra. 2005. Temperature-related genetic changes in laboratory populations of *Drosophila subobscura*: Evidence against simple climatic-based explanations for latitudinal clines. *American Naturalist* 165:258–273.

Santos, M., P. J. F. Iriarte, W. Céspedes, J. Balanyà, A. Fontdevila, and L. Serra. 2004. Swift laboratory thermal evolution of wing shape (but not size) in *Drosophila subobscura* and its relationship with chromosomal inversion polymorphism. *Journal of Evolutionary Biology* 17:841–855.

Scheiner, S. M., and R. F. Lyman. 1991. The genetics of phenotypic plasticity. II. Response to selection. *Journal of Evolutionary Biology* 4:23–50.

Scherens, B., and A. Goffeau. 2004. The uses of genome-wide yeast mutant collections. *Genome Biology* 5:229.

Service, P. M. 1987. Physiological mechanisms of increased stress resistance in *Drosophila melanogaster* selected for postponed senescence. *Physiological Zoology* 60:321–326.

Service, P. M., and M. R. Rose. 1985. Genetic covariations among life-history components: The effect of novel environments. *Evolution* 39:943–945.

Sheeba, V., V. K. Sharma, M. K. Chandrashekaran, and A. Joshi. 1999. Effect of different light regimes on pre-adult fitness in *Drosophila melanogaster* populations reared in constant light for over six hundred generations. *Biological Rhythm Research* 30:424–433.

Simões, P., M. R. Rose, D. Duarte, R. Gonçalves, and M. Matos. 2007. Evolutionary domestication in *Drosophila subobscura*. *Journal of Evolutionary Biology* 20:758–766.

Slatkin, M., and M. Kirkpatrick. 1983. Extrapolating quantitative genetic theory to evolutionary problems. Pages 283–293 *in* M. D. Huettel, ed. *Evolutionary Genetics of Invertebrate Behavior: Progress and Prospects*. New York: Plenum.

Sliwa, P., and R. Korona. 2005. Loss of dispensable genes is not adaptive in yeast. *Proceedings of the National Academy of Sciences of the USA* 102:17670–17674.

Solé, E., J. Balanyà, D. Sperlich, and L. Serra. 2002. Long-term changes of the chromosomal inversion polymorphism of *Drosophila subobscura*. I. Mediterranean populations from South-western Europe. *Evolution* 56:830–835.

Swallow, J. G., and T. Garland, Jr. 2005. Selection experiments as a tool in evolutionary and comparative physiology: Insights into complex traits: An introduction to the symposium. *Integrative and Comparative Biology* 45:387–390.

Travisano, M., J. A. Mongold, A. F. Bennett, and R. E. Lenski. 1995. Experimental tests of the roles of adaptation, chance, and history in evolution. *Science* 267:87–90.

Treves, D. S., S. Manning, and J. Adams. 1991. Repeated evolution of an acetate-crossfeeding polymorphism in long-term populations of *Escherichia coli*. *Molecular Biology and Evolution* 15:789–797.

Umina, P. A., A. R. Weeks, M. R. Kearney, S. W. McKechnie, and A. A. Hoffmann. 2005. A rapid shift in a classical clinal pattern in *Drosophila* reflecting climate change. *Science* 308:691–693.

Underwood, A. J. 1998. Design, implementation and analysis of ecological experiments: pitfalls in the maintenance of logic structures. Pages 325–349 *in* W. J. Resetarits Jr. and J. Bernardo, eds. *Experimental Ecology: Issues and Perspectives*. Oxford: Oxford University Press.

van't Land, J., P. van Putten, B. Zwaan, A. Kamping, and W. van Delden. 1999. Latitudinal variation in wild populations of *Drosophila melanogaster*: Heritabilities and reaction norms. *Journal of Evolutionary Biology* 12:222–232.

Weber, K.E. 1996. Large genetic change at small fitness cost in large populations of *Drosophila melanogaster* selected for wind tunnel flight: Rethinking fitness surfaces. *Genetics* 144:205–213.

Williams, G. C. 1975. *Sex and Evolution*. Princeton, NJ: Princeton University Press.

Woods, R., D. Schneider, C. L. Winkworth, M. A. Riley, and R. E. Lenski. 2006. Tests of parallel molecular evolution in a long-term experiment with *Escherichia coli*. *Proceedings of the National Academy of Sciences of the USA* 103:9107:9112.

Yvert, G., R. B. Brem, J. Whittle, J. M. Akey, E. Foss, E. N. Smith, R. Mackelprang, and L. Kruglyak. 2003. Trans-acting regulatory variation in *Saccharomyces cerevisiae* and the role of transcription factors. *Nature Genetics* 35:57–64.

Zeyl, C., and G. Bell. 1997. The advantage of sex in evolving yeast populations. *Nature* 388:465–468.

Zeyl, C., C. Curtin, K. Karnap, and E. Beauchamp. 2005. Tradeoffs between sexual and vegetative fitness in *Saccharomyces cerevisiae*. *Evolution* 59:2109–2115.

Zeyl, C., T. Vanderford, and M. Carter. 2003. An evolutionary advantage of haploidy in large yeast populations. *Science* 299:555–558.

# INDEX

Page numbers followed by *f* refer to figures, *n* refer to notes, and *t* refer to tables.

Christiansen, F. B., 202

chromosome extraction, 202–203, 573

chromosomes
and adaptive evolvability, 377–378
bacteria, 495
*Drosophila,* 510–514, 573, 640, 687, 688
*E. coli,* 681, 687
fitness, 202–203, 206, 207
insertion sequence elements, 373
and longevity, 229
regulatory sites, 292
segment reassortment, 491
viruses, 485
X, 227, 511*f,* 512–514
yeast, 377–378, 503, 505

chronic selection, 684–685

cichlids, 265–266

circadian rhythm, 232, 246–247, 682

*cis*-regulated genes, 282

Clark, A. G., 203

Clarke, B., 206

classical model of adaptation, 116

clay models, 176–179

cleistogamous (self-pollinated) flowers,
614–615

climate change (warming), 188, 189, 334,
675–677, 685–686, 687–689

clines, 529, 530, 675–677, 687, 691

clonal interference, 116, 487, 497–498,
505–506, 507–510, 514, 515

clumping index, 592, 609

co-adapted gene complexes, 482

coelacanths, 83

coevolution, 211, 232, 396, 406, 408–413,
605, 647

coexistence, 80–81, 84, 85, 408

cognitive ability, 269–270, 272–274

cohort, 271, 556, 567–568, 571, 576, 577

Colegrave, N., 509–510, 515

Collaborative Cross, 283

collared lizards, 178

colonization, 18, 185, 606

color preference, 272

common ancestry, 142, 146, 161, 662

common garden, 273, 357, 459, 672

comparative approach, 92, 100–101, 102, 104,
105, 112–113, 292, 381, 632

Comparative Genome Sequencing
(CGS), 366

comparative hypothesis, 675–677, 687

comparative method, 19–20, 99–102,
141, 664

compensation, 20, 154, 159, 160,
400–401, 685

compensatory mutations, 152–154, 158–159

competition theory, 79

competitive ability, 24, 100, 105, 118, 125, 208,
209, 606–608, 611

competitive fitness, 100, 308

competitive restraint, 605–609

complementation, 492–495

complementation test, 273, 281, 282

complex environments, 79–82

complexity of laboratory environments, 51, 174,
542, 543, 686

complex traits, 105, 302–303, 323, 435–440,
500, 691

computer simulations, 9, 160, 434, 663, 665,
667–668

conflict resolution, 611–614

congenic strains, 280

congruence, 663

conjugation, 482, 495

conservation genetics, 210

conservation programs, 104–105, 210

consistency of response, 220

constraint
developmental, 431–433, 435, 446, 451, 453,
457, 461
ecological, 333
functional, 21, 33, 35, 36, 51, 59
genetic correlations, 436
genomic approach, 370
on molecular evolution, 142
natural selection experiments, 276
on rate of fitness improvement, 119
and trade-offs, 174, 369

contextual analysis, 595–597, 614–615,
621, 622

contingency, 104, 146–150, 152, 161

control populations, 8, 16–19, 140–141

convergence, 92, 141–146, 149, 150, 156, 157,
158, 161, 365, 411

convergent evolution, 91–92, 668

epistatic variance ($V_I$), 25
eQTL, 281–282, 285
*Escherichia coli*
    antagonistic pleiotropic mutations, 151, 371
    candidate genes' fitness impact, 540
    competitive restraint, 605–609
    diversification, 117–123
    dominance, 75
    frequency-dependent selection, 155
    genome evolution, 359, 365, 367, 368,
        374, 379
    genotype-phenotype map, 82
    lactose metabolism, 69–82
    long-term evolutionary studies, 91
    natural vs. wild types, 681
    nutritional selection, 530–531
    phage experiments, 392, 403, 406,
        407, 408
    population growth rates, 210
    reverse evolution experiments, 143, 144–145,
        149, 153
    sex, 486, 488–489, 495–498
    temperature adaptation, 68
    temperature selection, 526–527
    trade-offs, 21–22
established populations, 94–98, 100
Estes, S., 151, 154
ethanol, 272, 304, 529
ethical issues, 189
eukaryotes, 158, 379–380, 499–514
*Eurytemora affinis*, 528
eusociality, 587
evo devo, Chapter 15
evolutionary allometry, 428
evolutionary biology
    behavior evolves first hypothesis, 265–266
    experimental study importance, 15–30, 175
    genomes, 354
    natural selection research patterns, 5–6
    phage experiments, 392, 413
    whole-organism performance, 302
evolutionary decline, 101
evolutionary domestication. *See* Domestication
evolutionary dynamics, 91, 92, 93, 99,
        100–101, 104, 149, 150, 188, 366, 392,
        396, 410
evolutionary ecology, 189

evolutionary improvement, 20–21, 92, 97, 98,
    102, 118, 119, 330, 407, 491
evolutionary pattern, 92, 93, 98, 99, 100, 102,
    113, 392
evolutionary physiology, 5, 218, 561
evolutionary process, 16, 32, 104, 126–127, 174,
    232, 264, 290, 392, 393, 398, 407, 411,
    498, 537, 622, 663, 667
evolutionary rate, 93, 98, 120, 141, 143
evolutionary theory
    aging, 499, 553, 555, 557, 561, 577
    late life, 569, 570–571, 576
    long-term evolution experiments, 129
    phage experiments for testing, 392–393
    reverse evolution, 136, 140, 147, 154, 159
    testing of, 18
evolutionary trajectory, 92–93, 101–104, 105,
    552, 672, 684–685
evolvability, 402–403
exaggerated (elaborated) trait, 441–442
exercise physiology, 313
experimental design
    *Drosophila subobscura* experiments, 93–95
    field experiments, 185
    importance of, 185
    late life, 572
    modeling approaches, 33
    phage experiments, 402
    population dynamics, 204
    reverse evolution, 140
experimental environment, 16, 367, 370
experimental evolution
    advantages and disadvantages of, 16–18,
        26–27
    in biological sciences, 10
    definition of, 6–8, 32
    evolutionary biology examples, 20–30
    faulty analysis, 126–127
    origins of, 355
    predictability, 58
    research goals, 27, 211
    *See also specific index headings*
experimental population biology, 198–200
expression plasticity, 539
expression profiling, 379–380
extreme age distribution, 575
extrinsic mortality rates, 562, 565

eyes, 328–329
eyespots, 250, 451
eye stalks, 440–444

Fairbairn, D. J., 34, 55, 236, 463, 689
Farmer, C. G., 319, 321
fat, 307, 312
fatty acids, 238, 240
feathers, 113
fecundity
    and adult size, 205
    and aging, 460, 569, 570, 571
    and body size, 447
    and bottle culture, 100
    density dependence of fitness, 206–208
    and dispersal capability, 235
    *Drosophila subobscura* real time studies,
        94, 95, 96–97, 98
    and energy storage, 226
    and feeding rate, 233
    and fitness, 104, 202
    *Gryllus,* 52, 53, 54, 55
    and hormone analogue, 233
    late-life patterns, 573, 576–577
    and lipid accumulation, 238
    long-established vs. recently introduced
        populations, 99
    and longevity, 218, 222, 233–234
    and population stability, 211
    starvation resistance, 102, 103*f*
    and stress resistance, 101
    and thermal environment, 526
    *Tribolium,* 204
    and virgin life span, 148
feeding, 209–210, 219, 231, 233–234, 238,
    265–266, 290, 528, 537, 541, 682
Fellowes, M. D. E., 231
Ferea, T. L., 377, 378
Ferenci, T., 367
fermentative metabolism, 377
Festing, Michael, 331
Fiegna, F., 614
field experiments, 173–193
    advantages of, 174
    alongside laboratory selection, 221
    altruism, 614–616
    behavior, 275–276

ethical issues, 189
    research opportunities, 189
field-fresh organisms, 678, 689
field introductions
    definition of, 6–7
    habitat alternations, 188–189
    of LNS-engineered phenotypes, 690
    methods, 175–185
    population establishment mechanisms,
        186–187
    rapid evolution, 187–188
    role of, 174, 186
field metabolic rate (FMR), 316
fish, 182, 185, 265–266, 273–276, 305, 421*f*,
    527–528, 556
Fisher-Muller Hypothesis, 483, 487, 488,
    497–498, 504*t*, 508, 514
Fisher, R. A., 83, 84, 199, 401, 482, 553
fishing, 274
fissile reproduction, 556, 565
fitness
    adaptive gain and correlated loss, 20–22
    in age-structured populations, 199
    and altruism, 591–597, 621–623
    antagonistic pleiotropy, 150–151
    biochemical basis of, 72–73, 84–85
    candidate genes' impact, 540
    competitive, 100, 308
    definition of, 161
    density dependence of, 206–208
    *Drosophila melanogaster,* 101
    *E. coli,* 117–119
    empirical measures of, 200–203
    higher-level, 622–623
    lactose system analysis, 71, 72–75
    life cycles and population regulation
        impact, 198
    phage, 395, 396–397, 399, 400–401
    vs. physical fitness, 333
    proxies for, 140
    recombination effects, 504*t*
    reverse evolution, 149–152, 153, 154,
        158–159
    sex effects on, 509–510
    sexuals vs. asexuals, 506–507
    single-trait model, 42
    and temperature selection, 526–527

top-down vs. bottom-up approach, 68–69
vegetative, 507
viruses, 488, 489, 490
fitness, higher-level, 622–623
fitness, physical, 312, 333
fitness plateau, 72–74, 76, 396, 397
fitness structure, 593, 611, 618, 625
fitness, vegetative, 507
Flatt, T., 233
flight muscles, 36, 37, 52, 53, 235–236, 241,
245f, 247–248, 459, 689
flight performance, 307–308, 328, 423, 444,
445, 447, 537, 689
flux, 71–75, 79, 82, 83, 238–241, 367, 371
flying speed, 275
FMR (field metabolic rate), 316
Foley, P. A., 233
Folk, D. G., 533–534
Fong, S. S., 365
food consumption patterns, 316
food quality, 441, 529, 541, 682
foot-and-mouth disease, 143
foraging behavior, 270, 272
"force" of selection, 553–555
Ford, E. B., 5
Forde, S. E., 409
form. *See* Shape variation
founder effects, 23, 25–26, 92–93, 104
founder-flush cycles, 633, 640
fox, 271
Frankham, R., 101
Frankino, W. A., 453, 455–456
free radical theory of aging, 228–229, 538
frequency-dependent selection, 83, 84–85,
155, 621–622
Froissart, R., 487, 492–495
fruit flies, 198, 202, 559, 562, 571, 576. *See also*
*Drosophila*
Fukatami, A., 635t
functional analyses, 220, 252, 253
*Functional Ecology*, 187
Fussman, G. F., 211
Futcher, B., 504t
Futuyma, D. J., 24

Galiana, Agustí, 24
Galton, F., 137

Gammie, S. C., 313
Garland, T., 284, 313, 381, 664
Gasser, M., 223–224
Gavrilets, S., 157
Gayon, J., 137
Gębczyński, A., 335
Gee, H., 661
Gefen, E., 535
gene expression
behavior, 267, 289
*Bicyclus*, 453
and DNA alteration, 265
*Drosophila*, 379, 536f
and enzyme activities, 241
genome structure changes, 372
heat loss, 322
*lac* operon, 69–70
mice, 379
microarrays, 377, 538, 566
parallel changes, 359, 365–366, 367
and physical activity, 313
profiling, 282–283
regulatory mutations, 368
and resource allocation, 218
reverse evolution, 158
starvation resistance, 539–540
gene flow, 409, 642–643
gene mapping, 281–283
gene number, and aging, 565–566
gene of major effect, 312, 538
generalists, 125, 369, 370, 405, 406, 407
generation time
adaptive evolution, 355
bacteria, 495
and choice of species in evolutionary
experiments, 276
guppies, 327
host-parasite interactions, 409
insects, 218
lactose system bottom-up analysis, 71, 81
live animal introductions, 185
and long-term evolution studies, 113, 114
microbes, 392
phages, 393
phylogenetics, 663
sexual vs. asexual populations, 483f
genetic adaptation, 99, 181, 265, 267

genetic architecture, 76, 83, 84, 140, 147, 186, 270, 449–450, 672, 679

genetic background, 23, 117, 189, 203, 482, 483, 492, 512–513, 532

genetic correlation or covariance
 aerobic capacity, 325
 animal shape, 436–438
 basal metabolic rate, 317, 318, 323
 behavioral traits, 313
 *Bicyclus* wing size and body size, 453–454
 definition of, 437
 *Drosophila* wing size and thorax size, 445
 field-fresh lines, 678
 genotype-by-environment interactions, 91, 104
 *Gryllus*, 56t
 JH regulation, 249
 and linkage disequilibrium, 562
 and long-term evolutionary studies, 113
 outbred population subject to laboratory environments, 152
 and pleiotropy, 34
 relatives, 562–563
 reverse evolution, 149
 shape evolution, 436–438
 stalk-eyed flies body size and eye span, 441–442
 starvation resistance, 98
 stress resistance and fecundity, 101
 variance-component models, 36
 virgin life span and early fecundity, 148

genetic drift, 19, 23–26, 104, 148, 267, 276, 277, 305, 316, 331, 433–435, 633, 678

genetic dynamics, 101

genetic effects, 77f, 78, 397, 502, 557, 559, 568

genetic engineering, 273, 283–284, 289, 292

genetic mechanisms, of reverse evolution, 157–159

genetic parasites, 373

genetic robustness, 402–403

genetic variance
 additive genetic variance ($V_A$), 25, 33, 34, 140, 280, 436
 age-specific, 562–563
 in base population, 221
 and directional selection, 36–37
 *Drosophila* wing shape, 450–451

fitness consequences, 208
in natural populations, 18
reverse evolution, 146, 147, 148, 150
and sympatric speciation, 644, 645
variance component models, 35

genome architecture, 354, 372–375, 381, 681

genome evolution, 353–388
 of bacteria, 355–376
 abbreviations glossary, 360–364t
 adaptive evolution trade-offs, 369–370
 large-scale changes in genome structure, 372–376
 loss of function, 371–372
 parallelism in adaptation, 357, 359, 365–367
 structural vs. regulatory mutation, 367–369
 future prospects, 380–381
 in higher eukaryotes, 379–380
 research considerations, 354–355
 of yeast, 376–379

genomic methods, 538, 566

genomic technologies, 354, 355, 358f, 372, 376, 380

genotype
 and adaptive radiation, 125–126
 definition of, 138
 and environment, 82, 405
 field introductions for establishing selection on, 189
 negative frequency-dependent selection, 155
 phage experiments, 393, 396, 397–399
 and phenotype, 397–399, 432
 population frequencies, 483
 scaling relationships, 439, 441
 *See also* Fitness; Reverse evolution

genotype-by-environment interaction, 76–79, 91, 98, 100, 104, 105, 140, 160, 268, 403–405, 508t, 558–559, 573

genotype-phenotype map, 68, 82, 84

genotyping, 280, 281, 283

geographic isolation, 633

geographic variation, 18, 19

geometric morphometrics, 426

geotaxis, 636, 637, 643

Gerrish, P. J., 487

Gibbs, A. G., 533, 535

and mass selection, 274
mortality, 574
natural populations, 18
in small populations, 37, 273
inbreeding depression, 58, 92, 96, 98, 99, 159, 273, 283
independent contrasts, 19, 412, 666
index selection, 269
indirect responses to selection, 91, 226, 233, 454, 461, 562, 563–564
individual-based models, 40, 59
infectivity, 218, 402, 409, 610
infinite population, 590
insertion sequence (IS) elements, 373–374
*Integrative and Comparative Biology*, 10
intelligence, 230
intensity of selection, 69, 79, 184, 210, 276, 647, 683
interdependence, 589
intermediary metabolism, 218, 229, 235, 236–242
intersexual selection, 326, 328
interspecific allometry, Chapter 15
interspecific comparisons, 9, 82–83, 463, 535, 543
interspecific competition, 24, 90
interspecific hybrids, 99
interspecific interactions, 689
intrasexual selection, 326
introductions of live animals, 179–188, 189, 275–276
invasive species, 6, 185, 189
inversion frequencies, 675–676, 687, 688, 689, 691
irreversibility, 156–157, 161
Irschick, Duncan, 178, 304
island population, 180, 181–182, 647
isofemale lines, 101, 573
isogenic, 118–119, 153, 229, 282–283
isolation, geographic, 633
isolation, reproductive. *See* Reproductive isolation
isometry, 424, 445, 453

*Jadera haematoloma*, 182–183
Jasnos, L., 504*t*
Jeon, K., 611–612

Johannsen, W., 138
Jones, A. G., 34
just-so story, 73–74, 661
juvenile hormone, 225, 233, 241, 243–244, 245, 459
juvenile hormone epoxide hydrolases (JHEHs), 225
juvenile hormone esterases (JHEs), 219, 225, 243–245, 247–249, 459–460
juvenile viability, 99, 105

Kacser, H., 71, 83
Karan, D., 529
Kawecki, T. J., 233, 281, 290
Keightley, P. D., 83
Kerr, B., 410–411, 605, 606, 620, 622, 624*n*6
Kessler, S., 635*t*
Kettlewell, H. B. D., 5
Kgwatalala, P. M., 320–321
Khazaeli, A. A., 227, 574
Kilias, G., 640
Kinnison, M. T., 187–188
kin selection, 586, 588–589, 597, 623, 624*n*1
Klingenberg, C. P., 450
Knies, J. L., 405, 413
Knight, G. R., 635*t*
knockdown temperature, 525, 538
Konarzewski, M., 314–315, 318–319, 335
Kondrashov, A. S., 495, 679
Koopman, K. F., 634, 635*t*
Korona, R., 504*t*
Koteja, P., 321
Kraaijveld, A. R., 231
Kristensen, T. N., 690
Kroll, E., 381
Kurland, C. G., 681
Kurlandzka, A., 375
Kyriacou, C. P., 273

laboratory adaptation
base population, 649
comparative studies, 99–104
*Drosophila*, 93, 561
problems with, 678–686
real-time studies, 92–98
laboratory-adapted organisms, 678–679, 689
Laboratory culling, 7, 16, 45, 219, 672

motivation, 268, 284–286, 291, 292, 310
mouse
  aerobic capacity, 332
  artificial selection experiments, 272
  basal metabolic rate, 314–319, 324–325
  behavior experimental methods, 268–270
  cross of inbred strains, 277, 279–280
  endurance and stress resistance, 307–308
  experimental evolution in different thermal
    environments, 306–307
  gene mapping, 282, 283
  group selection, 587
  heat loss, 320–321
  hypoxia, 331–332
  locomotor activity, 284
  macroevolution inferences from selection
    experiments, 8–9
  physical activity selection experiment,
    268–269
  selection experiments, 305
  strain differences, 331–332
  thermal environment, 307
  wheel running, 266, 282, 284–285, 292,
    308, 310–313, 379–380
Mouse Phenome Database, 283
Moya, A., 487
Mueller, L. D., 205, 206, 211, 560, 563–564,
  573, 575
Muir, W. M., 603
Muller, H. J., 482, 484
Muller's ratchet, 484, 487, 490–492, 514
Mulley, J. C., 99–100
multifarious selection, 637
multi-level selection, 615, 618, 621, 623,
  624–625n6
multiple solutions, 7, 220, 276, 306, 313,
  667, 690
multiple-trait models, 43–58
Murray, A. W., 504t, 635t
*Mus*, 266, 268–269, 279, 282, 283, 284, 307,
  314–319, 379, 587
muscle mass, 312
mutation
  clonal interference, 116
  compensatory, 152–154, 158
  context-dependence, 117
  developmental constraints, 432

direct mutation hypothesis, 126–127
P-element, 232
phage experiments, 396–403
regulatory, 354, 357, 367–369, 500
structural, 367–369
in yeast, 378–379
mutation accumulation
  aging, 558, 563–564, 569
  genetic mechanisms, 92
  late-life mortality patterns, 573
  vs. pleiotropy, 120–121, 369, 407
  reverse evolution, 158
  testing for, 106
  viruses, 491
mutation-accumulation hypothesis, 562–563
Mutational Deterministic Hypothesis, 483,
  487, 495–497, 504tt, 505, 508, 514
mutational robustness, 402–403
mutation rate
  *Bacillus*, 371, 532
  compensatory reverse evolution, 159
  *E. coli*, 145, 151, 497, 498, 681
  and genetic variability, 139
  and parallelism, 120
  phages, 393, 394
  viruses, 401, 402, 490, 663
  yeast, 505
mutualism, 612
*Myodes*, 315, 323, 324f
Myth of Definitive Results, 691
*Myxococcus*, 371–372, 612, 614

NASA Astrobiology Institute, 27
National Institutes of Health, 381
natural experiments, 179, 182–185, 188,
  189, 675
natural history, 689
natural populations
  *Arabidopsis* seed dispersal, 616
  *Chlamydomonas*, 514
  degenerative allele accumulation, 151
  *Drosophila*, 91, 231, 525, 534, 537
  egg laying behavior, 570
  genetic variation in, 203, 210, 529–530
  guppies, 327
  multiple trait evolution, 32
  pleiotropic effects of alleles, 559

natural populations *(continued)*
  polymorphisms, 538
  reinforcement, 646
  replacement methods, 175–176, 179
  reverse evolution, 156, 159
  studies on, 18–19
  thermal environment adaptation, 405
  viruses, 485
natural reward circuit, 279f, 285, 286,
     291, 292
natural selection
  altruism, 614–616
  Darwin's gradualism, 4–5, 662
  experimental research, 139
  "organism-as-island," 586
  quasi-natural selection experiments,
     604–614
  and reversions, 136–138
  *See also* Laboratory natural selection (LNS)
near-variant test, 619–621
negative epistasis, 160, 492, 493, 494–495,
     496, 497, 507, 509, 515
negative frequency-dependent selection, 155
nervous system, 266–267
nest building, 266, 284, 286
neurobiology, 291–292. *See also* Behavior
*Neurospora,* 642, 649
neutrality, 73, 76, 84, 116
neutral polymorphism, 76
Nevo, Evitar, 181
Newcombe, H. B., 355
niche expansion, 398
niche shifts, 21–22
niche specialization, 128
Nicholson, A. J., 204–205, 211
Nicholson, W. L., 532
Nielsen, M. K., 320, 321
Nijhout, H. F., 251, 439
Nilsson, D. E., 662
nonadditive effects, 34
noncompetitive fitness, 105
nonlinearity, 98, 104, 140, 458–460
novel environment, 16, 32, 98, 104, 143–144,
     182–184
novel phenotype, 434–435, 456, 462, 463,
     672, 691
null mutants, 75f, 283

nutrient acquisition, 234, 235–236, 251
nutrient limitation, 356f, 359, 369, 682, 684
nutrient transport, 121, 122
nutritional selection, 529, 530–532

Oakley, T. H., 411–412
Occam's razor, 587–588
odor preferences, 271, 274
olfactory learning ability, 281
*Oncorhynchus,* 182
*Onthophagus,* 425fig, 429fig, 458, 459
ontogenetic allometry, 428, 455
ontogeny, 431, 437, 455, 459
opsins, 83
organismal biology, Chapters 10, 11, 12, 17
organ size, 316
*The Origin of Species* (Darwin), 90, 597–598
ornaments, 326–329
Orr, H. A., 634
osmotic regulation, 304, 527–528
osmotic stress, 524, 527–528
Otto, S. P., 500
outcrossing, 481–482
*ovarioles,* 223–224
Overmeer, W. P. J., 640–641
oxidative damage, 218, 226, 228, 229, 538
oxygen availability, 540–541

pair bonding, 267, 284, 285, 287–290, 292
panmictic populations, 138
Pappas, C., 231
Paquin, C., 376, 377
parallel evolution, 119–123, 142–146, 162, 396,
     403, 538
parallelism, adaptive, 354, 357, 359, 365–367
parametric bootstrapping, 664–665
parapatric speciation, 633, 643
parasexual mechanisms, 482, 485
parasitism, 218, 231, 408–409, 611–612
parasitoids, 210, 231, 686
parental care, 273–274, 319, 481
Parsons, P. A., 528, 535
partial regression coefficient, 595–596,
     614, 615, 622
Partridge, L., 100, 202, 222
Passananti, H., 145, 560
*Passer domesticus,* 184

polyphenism, 243, 250, 458–460
Poon, A., 153, 400, 487
Popperian falsificationism, 661
population-based models, 33–34, 35
population biology, 198–200
population density, 185, 200, 205, 409,
576, 616
population dynamics, 174, 198, 199,
204–211, 410
population establishment mechanisms,
186–187, 189
population genetics, 150–155, 189, 198,
200–203, 355, 357
population genetics of aging and late life,
553–559, 562–566, 572–573
population growth, 23, 139, 185, 198–199,
206–210, 400, 402, 500, 553, 555,
598–601
population size
determination for experimental evolution
program, 32
effective, 650, 672
and epistasis, 509
and fitness, 398*f*, 510
mass selection, 274, 275
multiple-trait models, 46
natural populations, 18–19
single-trait models, 37
small, 273
variance-components models, 35
population stability, 205, 210–211
population structure, 90, 588, 602
postmating reproductive isolation, 633,
638–642, 645–646, 652
postponed aging, 563, 566
postzygotic isolating barriers, 23, 633, 649
Prasad, N. G., 211
predation, 175–179, 323, 327–328
predator-prey models, 187, 396
predictability, 58, 84
predictive science, 84
premating reproductive isolation, 633,
634, 636, 637, 638, 643, 646,
648, 649
Price Equation, 594–595, 598, 601–602, 608,
615, 622
principal component analysis, 426–427

proboscis extension reflex, 272, 281
productivity, 203, 598–602, 620
prokaryotes, 69–70, 153, 368–369, 373, 482,
612, 622
propagule selection, 602
protein expression experiments, 538
Prout, T., 198, 200–202, 203
*Pseudomonas*, 125–126, 155, 401, 408, 409,
490, 686
PTS (phosphotransferase system),
121–122, 123
public health, 291

QTL analysis, 227, 270, 273, 280, 281–282,
283, 285, 312, 321–322
quail, 272
quantitative-genetic parameters, 220, 563
quantitative genetics
correlation between relatives, 562–563
life-history traits, 218
linear additive model, 77–78, 83
modeling approaches, 33–36
shape evolution, 436–437
quantitative trait loci (QTL), 273, 280, 281–282,
321–322
quasi-natural selection experiments, 604–614,
620, 650
Queller, D. C., 613

rabbit, 462
radiation, adaptive, 112, 124–126
radioisotopic studies, 245
radiotracer studies, 238
Rainey, P. B., 155, 408, 409, 686
random genetic drift, 104, 267, 276, 277, 305,
316, 331, 433, 633, 678
rapid evolution, 186, 187–188, 189
rat, 5, 139, 269–270, 272, 305, 309–312, 334
rate of evolution, 187–188, 410, 663
rate of living theory, 226–228
Rauser, C. L., 569, 571, 572, 573, 577
reactive oxygen species (ROS), 229, 251
realism, 617–618, 619, 663–666
realized phenotypes, 35–36, 40, 43
real-time evolution approach, 92–98, 102
recessive allele, 312
reciprocal altruism, 586

thermal environment, 306–307, 316–317

thermal tolerance (thermoresistance), 304, 524–526, 636

thermoregulation, 317, 319, 679, 685

Thompson, D. W., 420

Thorpe, R. S., 181

threshold traits, 35–36

thyroid hormones, 321

Tolman, Edward, 269–270

Toma, D. P., 282

top-down approach, 68, 83–84, 220, 435, 436–438

toxin resistance, 640

trade-offs

    adaptive evolution requirements, 369–370

    antagonistic pleiotropy, 150–151, 557–558

    as correlated response to adaptation, 22–23

    definition of, 20–21

    *Drosophila* experiments, 91, 99

    *E. coli* experiment, 21–22

    experiment challenges, 21

    fecundity-mortality, 570

    field vs. laboratory experiments, 174

    growth rates, 208–209

    *Gryllus* experiments, 52, 53, 55, 234–247

    life-history traits, 218–219, 220, 233–234

    ornamental production, 326–327

    phage experiments, 406–407

    productivity-competitive ability, 608

    resource breadth and competitive ability, 125

    sexually selected traits and locomotor performance, 328–329

    somatic maintenance and reproduction, 557–558

    thermal, 22

tragedy of the commons, 608

trait exaggeration (elaboration), 326–329, 425

trait-group model, 590–591

transcriptome analysis, 232

transduction, 495

transformation, 495

transgenics, 221, 252, 281

transplant experiments, 179–188

*trans* regulated, 282

Tran, T. T., 487

Travisano, M., 155, 686

Treves, D. S., 144, 375

*Tribolium*, 25, 204, 205, 560, 598–602, 650

truncation selection, 37–40, 43, 46, 47f, 48–49

Tryon, Robert, 269–270

Turner, P. E., 407, 487

twofold cost of sex, 481, 499, 501

Type I error, 676

Type II error, 676

uncoupling protein, 322

Underwood, A. J., 691

unintentional selection (unintended selection), 540–542

unplanned experiments, 175, 182–185

up-regulation, 251, 332

use it or lose it principle, 500

vaccines, 413

van Tienderen, P. H., 34

variable environments, 526–527

variance-components models, 35–36, 59

Vartanian, J. P., 664

Vasi, F., 210

Vasi, F. K., 531

Vaupel, J. W., 570

vegetative fitness, 507

Velicer, G. J., 372

Vermuelen, C. J., 148

vertical transmission, 610

Via, S., 34

virulence, 354, 609–611

viruses, 143–144, 145, 158, 160, 485–495, 616. *See also* Bacteriophage

voles, 267, 285, 287–290, 315, 323

Vries, H. de, 5

*Vulpes vulpes*, 271

Wabrick-Smith, S., 529

Wade, M. J., 25, 598–602, 620, 624n2

Wagner, G. P., 83

Wallace, B., 635t

Wallace, Russel, 552

wasps, 268

water loss, 534, 535–536, 537

water storage, 533–534, 535

Wattiaux, J. M., 560

Weber, K. E., 307, 448–449, 453

WebQTL, 283
Weinberg, W., 138
Weiner, J., 335
Weinig, C., 615–616
Weinreich, D. M., 82
Weismann, August, 552
Weismann's principle, 505
Weissman, F. L. A., 137, 151
Weitz, J. S., 396
Weldon, W. F. R., 5
Wesolowski, S. R., 322
West, S. A., 624–625n6
whales, 265
Wheeler, D. E., 439
wheel running, 266, 268–269, 272, 277, 282,
    284–285, 308, 310–313, 379–380
Whitlock, M. C., 148, 159
whole-genome analyses, 157–158, 354, 357, 372,
    379, 538, 566
whole-organism physiology, 218, 302
Wichman, H. A., 144, 160, 379, 397, 399, 403
wild populations
    aging, 553, 556
    as base population, 277
    domestication, 90, 99
    vs. natural populations of mice, 279–280
    viruses, 492–495
    yeast, 503, 505
wild-type phenotypes, 455–458, 680–681
Wilkinson, G. S., 329, 442, 443
Williams, G. C., 553, 557, 565, 587–588
Wills, C., 504t
Wilson, D. S., 620, 624–625n6
Wilson, E. O., 620

wind tunnel flight, 307–308
wing dimorphism, 35, 51, 53, 689
wing loading, 444–448, 450, 453, 454, 456f
wing polymorphism, 52, 234–236, 243, 246, 270
wing shape, 444–445, 448–451, 454–455
wing size, 101, 183–184, 247, 444–448, 451–457
wing structure, 141, 148
wing vibration, 447–448
Wloch, D. M., 504t
Wolf, J., 463
Woods, R., 144
Woodworth, L. M., 99
workers, 586
Wynne-Edwards, V. C., 587, 588

X chromosome, 227, 511f, 512–514
Xu, J., 500

Yampolsky, L. Y., 564
yeast, 355, 376–379, 377, 499–507, 538, 642
Yin, J., 396
Yokoyama, Shozo, 82–83
You, L., 396

Zaklan, S. D., 450
zebra finch, 276
Zera, A. J., 236, 241
Zeyl, C., 500, 504t
Zhao, Z., 241
Zhong, S., 370, 374–375
Zhu, G., 82
Zilstra, W., 250
zone of coexistence, 80, 81f, 84, 85
zygote, 201, 501, 502f, 507, 511f

COMPOSITION: Michael Bass Associates
TEXT: 9.5/14 Scala
DISPLAY: Scala

Milton Keynes UK
Ingram Content Group UK Ltd.
UKHW010131080924
447992UK00007B/246